Bacterial Protein Secretion Systems

Bacterial Protein Secretion Systems

Editor: Abrielle Benton

AMERICAN
MEDICAL PUBLISHERS
www.americanmedicalpublishers.com

AMERICAN
MEDICAL PUBLISHERS
www.americanmedicalpublishers.com

Cataloging-in-Publication Data

Bacterial protein secretion systems / edited by Abrielle Benton.
p. cm.
Includes bibliographical references and index.
ISBN 978-1-63927-913-5
1. Bacterial proteins. 2. Secretion. 3. Microbial proteins. 4. Bacteria--Physiology.
5. Microbial metabolism. I. Benton, Abrielle.
QR92.P75 B33 2023
572.36--dc23

American Medical Publishers,
41 Flatbush Avenue,
1st Floor, New York,
NY 11217, USA

ISBN 978-1-63927-913-5 (Hardback)

Contents

Preface ... VII

Chapter 1 **Identification of a Large Family of Slam-Dependent Surface Lipoproteins in Gram-Negative Bacteria** ... 1
Yogesh Hooda, Christine C. L. Lai and Trevor F. Moraes

Chapter 2 **Confirmed and Potential Roles of Bacterial T6SSs in the Intestinal Ecosystem** ... 13
Can Chen, Xiaobing Yang and Xihui Shen

Chapter 3 **HpaR, the Repressor of Aromatic Compound Metabolism, Positively Regulates the Expression of T6SS4 to Resist Oxidative Stress in *Yersinia pseudotuberculosis*** ... 24
Zhuo Wang, Tietao Wang, Rui Cui, Zhenxing Zhang, Keqi Chen, Mengyun Li, Yueyue Hua, Huawei Gu, Lei Xu, Yao Wang, Yantao Yang and Xihui Shen

Chapter 4 **Tracking Proteins Secreted by Bacteria: What's in the Toolbox?** ... 39
Benoit Maffei, Olivera Francetic and Agathe Subtil

Chapter 5 **Type V Secretion Systems: An Overview of Passenger Domain Functions** ... 56
Ina Meuskens, Athanasios Saragliadis, Jack C. Leo and Dirk Linke

Chapter 6 **Enhancing Recombinant Protein Yields in the *E. coli* Periplasm by Combining Signal Peptide and Production Rate Screening** ... 75
Alexandros Karyolaimos, Henry Ampah-Korsah, Tamara Hillenaar, Anna Mestre Borras, Katarzyna Magdalena Dolata, Susanne Sievers, Katharina Riedel, Robert Daniels and Jan-Willem de Gier

Chapter 7 **Type VI Secretion Effectors: Methodologies and Biology** ... 86
Yun-Wei Lien and Erh-Min Lai

Chapter 8 **A Potential Late Stage Intermediate of Twin-Arginine Dependent Protein Translocation in *Escherichia coli*** ... 97
Hendrik Geise, Eyleen Sabine Heidrich, Christoph Stefan Nikolin, Denise Mehner-Breitfeld and Thomas Brüser

Chapter 9 **Regulation of Gram-Positive Conjugation** ... 107
Verena Kohler, Walter Keller and Elisabeth Grohmann

Chapter 10 **Biological Functions of the Secretome of *Neisseria meningitidis*** ... 120
Jan Tommassen and Jesús Arenas

Chapter 11 **A Type VI Secretion System Trans-Kingdom Effector is Required for the Delivery of a Novel Antibacterial Toxin in *Pseudomonas aeruginosa*** ... 142
Benjamin Berni, Chantal Soscia, Sarah Djermoun, Bérengère Ize and Sophie Bleves

Chapter 12 **Baseplate Component TssK and Spatio-Temporal Assembly of T6SS in *Pseudomonas aeruginosa*** 156
David Liebl, Mylène Robert-Genthon, Viviana Job, Valentina Cogoni and Ina Attrée

Chapter 13 **Type III Secretion in the Melioidosis Pathogen *Burkholderia pseudomallei*** 170
Charles W. Vander Broek and Joanne M. Stevens

Chapter 14 **The Dynamic ATP-Driven Mechanism of Bacterial Protein Translocation and the Critical Role of Phospholipids** 187
Ian Collinson

Chapter 15 **Distribution, Function and Regulation of Type 6 Secretion Systems of Xanthomonadales** 194
Ethel Bayer-Santos, Lucas de Moraes Ceseti, Chuck Shaker Farah and Cristina Elisa Alvarez-Martinez

Chapter 16 **Targeting the Type II Secretion System: Development, Optimization and Validation of a High-Throughput Screen for the Identification of Small Molecule Inhibitors** 203
Ursula Waack, Tanya L. Johnson, Khalil Chedid, Chuanwu Xi, Lyle A. Simmons, Harry L. T. Mobley and Maria Sandkvist

Chapter 17 **Prediction of T4SS Effector Proteins for *Anaplasma phagocytophilum* Using OPT4e, a New Software Tool** 215
Zhila Esna Ashari, Kelly A. Brayton and Shira L. Broschat

Chapter 18 **Enterococcal PcfF is a Ribbon-Helix-Helix Protein that Recruits the Relaxase PcfG Through Binding and Bending of the *oriT* Sequence** 224
Saima Rehman, Yang Grace Li, Andreas Schmitt, Lena Lassinantti, Peter J. Christie and Ronnie P.-A. Berntsson

Permissions

List of Contributors

Index

Preface

The main aim of this book is to educate learners and enhance their research focus by presenting diverse topics covering this vast field. This is an advanced book which compiles significant studies by distinguished experts in the area of analysis. This book addresses successive solutions to the challenges arising in the area of application, along with it; the book provides scope for future developments.

Secretion refers to the movement of materials from one point to another. In bacterial species, secretion is the transport of effector molecules like toxins, enzymes or proteins from the cytoplasm to exterior of the bacterial cell. It is a crucial mechanism in bacterial functioning and operation for their survival and adaptability in their natural surroundings. Bacterial secretion systems are the protein complexes on the cell membranes of bacteria for secretion of substances. These protein secretion systems play the role of cellular devices utilized by pathogenic bacteria to secrete their virulence factors to attack the host cells. They may be categorized on the basis of their composition, activity and distinct structure. Usually, proteins may be secreted via two distinct processes. In one process, cytoplasm of the bacteria is translocated and delivered directly in the host from cell membrane. Whereas, in the second process proteins are transported to the inner membrane, deposit on periplasm, and then transmitted to the host by the outer cell membrane. The topics included in this book on bacterial protein secretion systems are of utmost significance and bound to provide incredible insights to readers. With state-of-the-art inputs by acclaimed experts of this field, it targets medical students and professionals.

It was a great honour to edit this book, though there were challenges, as it involved a lot of communication and networking between me and the editorial team. However, the end result was this all-inclusive book covering diverse themes in the field.

Finally, it is important to acknowledge the efforts of the contributors for their excellent chapters, through which a wide variety of issues have been addressed. I would also like to thank my colleagues for their valuable feedback during the making of this book.

Editor

1

Identification of a Large Family of Slam-Dependent Surface Lipoproteins in Gram-Negative Bacteria

Yogesh Hooda, Christine C. L. Lai and Trevor F. Moraes *

Department of Biochemistry, University of Toronto, Toronto, ON, Canada

Correspondence:
Trevor F. Moraes
trevor.moraes@utoronto.ca

The surfaces of many Gram-negative bacteria are decorated with soluble proteins anchored to the outer membrane via an acylated N-terminus; these proteins are referred to as surface lipoproteins or SLPs. In *Neisseria meningitidis*, SLPs such as transferrin-binding protein B (TbpB) and factor-H binding protein (fHbp) are essential for host colonization and infection because of their essential roles in iron acquisition and immune evasion, respectively. Recently, we identified a family of outer membrane proteins called Slam (Surface lipoprotein assembly modulator) that are essential for surface display of neisserial SLPs. In the present study, we performed a bioinformatics analysis to identify 832 Slam related sequences in 638 Gram-negative bacterial species. The list included several known human pathogens, many of which were not previously reported to possess SLPs. Hypothesizing that genes encoding SLP substrates of Slams may be present in the same gene cluster as the Slam genes, we manually curated neighboring genes for 353 putative Slam homologs. From our analysis, we found that 185 (~52%) of the 353 putative Slam homologs are located adjacent to genes that encode a protein with an N-terminal lipobox motif. This list included genes encoding previously reported SLPs in *Haemophilus influenzae* and *Moraxella catarrhalis*, for which we were able to show that the neighboring Slams are necessary and sufficient to display these lipoproteins on the surface of *Escherichia coli*. To further verify the authenticity of the list of predicted SLPs, we tested the surface display of one such Slam-adjacent protein from *Pasteurella multocida*, a zoonotic pathogen. A robust Slam-dependent display of the *P. multocida* protein was observed in the *E. coli* translocation assay indicating that the protein is a Slam-dependent SLP. Based on multiple sequence alignments and domain annotations, we found that an eight-stranded beta-barrel domain is common to all the predicted Slam-dependent SLPs. These findings suggest that SLPs with a TbpB-like fold are found widely in *Proteobacteria* where they exist with their interaction partner Slam. In the future, SLPs found in pathogenic bacteria can be investigated for their role in virulence and may also serve as candidates for vaccine development.

Keywords: gram-negative bacteria, surface lipoproteins, outer membrane transporters, bacterial gene clusters, protein translocation pathways, flow cytometry

INTRODUCTION

Gram-negative bacteria contain an asymmetric outer membrane that protects the bacteria from environmental stress and acts as a shield from harmful chemicals (Silhavy et al., 2010). However, this creates a logistic challenge for the transport of various biomolecules to and from the extracellular milieu and across the outer membrane of the bacteria (Pagès et al., 2008; Delcour, 2009). Toward this end, Gram-negative bacteria have developed dedicated translocation systems that deliver various families of proteins and other biomolecules across the outer membrane (Nikaido, 2003; Costa et al., 2015; Geyter et al., 2016). The mechanism for this transport is distinct from translocation systems found in the inner membrane owing to the lack of ATP in the periplasm and is an active field of research (Karuppiah et al., 2011).

Surface lipoproteins or SLPs are a class of soluble proteins that are present on the surface of Gram-negative bacteria. They are anchored to the outer membrane via three fatty acyl chains that are post-translationally attached to their N-termini (Wilson and Bernstein, 2016). The first reported SLP was TraT, a protein of the F sex factor in *E. coli* (Manning et al., 1980). Several other SLPs were identified soon after within a few Gram-negative bacterial families including *Klebsiella* (Pugsley et al., 1986), *Neisseria* (Schryvers and Morris, 1988), and *Spirochetes* (Chamberlain et al., 1989; Brandt et al., 1990). Recently, there has been an increase in the number of reports of SLPs from different bacterial species with distinct structural folds and surface topologies (Konovalova and Silhavy, 2015). SLPs are involved in several important cellular pathways for nutrient acquisition, cellular adhesion and stress response (Zückert, 2014; Szewczyk and Collet, 2016; Wilson and Bernstein, 2016).

The discovery of SLPs in different bacteria has raised questions regarding the biosynthetic pathway used by these proteins for their synthesis and transport to the surface. SLPs are synthesized in the cytoplasm and transported to the periplasm by the Sec or Tat machinery based on the signal sequence present on the SLPs (Chatzi et al., 2013). Once in the periplasm, three enzymes in the inner membrane process the SLPs by cleaving the signal sequence and attaching three fatty acyl chains to the N-terminal cysteine residue (Szewczyk and Collet, 2016). Upon lipidation, most SLPs are transported across the periplasm to the inner leaflet of the outer membrane through the Lol system (Okuda and Tokuda, 2011). However, there are a few exceptions to this rule, including pullulanase that avoids the Lol system and moves to the surface through the Type-II secretion system (D'Enfert et al., 1987). Additionally, in *Borrelia* sp., SLPs are proposed to require a periplasmic "holding" chaperone that prevents premature folding of SLPs before reaching the outer membrane (Chen and Zückert, 2011; Zückert, 2014).

Upon insertion into the outer membrane, the translocation systems required for the movement of SLPs across the outer membrane remain poorly characterized. The first SLP for which the export pathway was characterized was pullulanase in *Klebsiella* sp. that utilizes the Type II secretion system (D'Enfert et al., 1987). More recent studies have shown that NalP (a neisserial SLP) functions as a Type Va "autotransporter" secretion system (Van Ulsen et al., 2003), while BamC (Webb et al., 2012) and RscF (Cho et al., 2014; Konovalova et al., 2014) in *E. coli* use the Bam complex to move across the outer membrane. Functional and mutagenesis studies in *Borrelia* sp. (Schulze et al., 2010; Chen and Zückert, 2011) and *Bacteroides* sp. (Lauber et al., 2016) have shown that the sorting rules used by these SLPs are distinct from other SLPs, indicating that different bacterial species may possess different translocation systems for the delivery of SLPs. Additionally, within *Neisseria* sp., *two* distinct SLP export pathways have been reported (Hooda et al., 2017), suggesting that multiple systems for the export of SLPs may exist in a single bacterial species.

The SLPs found in the genus *Neisseria* are amongst the most extensively studied SLPs. *N. meningitidis* and *N. gonorrhoeae* encode multiple SLPs that are involved in a variety of cellular pathways critical for survival of neisserial pathogens in humans (Hooda et al., 2017). In *N. meningitidis*, eight SLPs have been well-characterized, of which three [transferrin-binding protein B (TbpB) (Schryvers and Morris, 1988), lactoferrin-binding protein B (LbpB) (Pettersson et al., 1998) and hemoglobin-haptoglobin utilization protein (HpuA) (Lewis et al., 1997)] are involved in iron acquisition while two others: factor-H binding protein (fHbp) (Madico et al., 2006) and neisserial heparin binding antigen (NHBA) (Serruto et al., 2010) are involved in immune evasion. Other neisserial SLPs include *Neisseria* autotransporter protease (NalP) (Van Ulsen et al., 2003), anaerobically induced protein A (AniA) (Hoehn and Clark, 1992) and macrophage infectivity potentiator (MIP) (Leuzzi et al., 2005) which play roles in extracellular proteolysis, anaerobic growth and intracellular survival respectively. These SLPs have been shown to bind to different human factors and atomic resolution full-length or partial structures of these SLPs have aided in understanding their mechanism of action (Hooda et al., 2017). Recently, we described a family of outer membrane proteins called Slam or Surface lipoprotein assembly modulator that is essential for surface display of a subset of neisserial SLPs (Hooda et al., 2016). *N. meningitidis* contains two Slam proteins: Slam1 is necessary for the display of TbpB, LbpB, and fHbp, whereas Slam2 is specifically required for the SLP HpuA. Furthermore, Slam have been shown to potentiate the functional display of neisserial SLPs on the surface of laboratory strains of *E. coli* that do not possess any Slam or SLP homologs (Hooda et al., 2016). This work suggested that a subset of neisserial SLPs utilize a unique mechanism to get to the cell surface that is dependent on the Slam family of outer membrane proteins.

In our previous work, we discovered that genes encoding Slam2 and HpuA are adjacent to each other in multiple neisserial genomes (Hooda et al., 2016). Based on this observation, we searched and annotated genes upstream and downstream of 353 putative Slam homologs found in other proteobacterial species. This dataset showed that a large number of Slam related sequences are located adjacent to genes that encode putative lipoproteins with TbpB-like folds suggesting a genetic linkage between these two families of proteins. The bioinformatics analysis allowed us to identify TbpB-like SLPs in many bacterial species that were previously not known to possess SLPs, including the human pathogens

Acinetobacter baumannii and *Salmonella enterica* subsp. *arizonae*.

MATERIALS AND METHODS

Identification of Slam Homologs

To generate a database of Slam related sequences, iterative psi-blast searches were performed (March 4, 2016) against a non-redundant database containing all partial and complete bacterial genome sequences using the sequence of Slam1 protein (NMB0313) from *Neisseria meningitidis* strain MC58 as the query. Four independent psi-blast searches were performed for different clades of proteobacteria (alpha-, beta-, gamma-, and delta/epsilon/zeta-proteobacteria). The lists of putative Slam genes obtained from these four psi-blast searches were pooled and only unique representative Slam sequences were kept from a given bacterial species. The list was manually checked to remove the following: (i) partial sequences (containing premature stop codons or with partial gene sequence coverage), and (ii) sequences coding for only the N-terminal domain (Ntd) of Slam. This gave a final list (Supplementary Data 1) of 832 Slam sequences spanning 638 bacterial species.

To understand the distribution of Slam related sequences, a phylogenetic tree of different proteobacterial species was made using the 16S-RNA sequences obtained from the database Greengenes (DeSantis et al., 2006). One representative member was kept from each family of bacteria. In total 52 species (8: alpha-, 10: beta-, 23: gamma-, 5: delta-, 5: epsilon-, and 1: zeta-proteobacteria) were selected for the final tree. The tree was made using the PhyML plugin in the software Geneious (Kearse et al., 2012) with 100 bootstraps. The nodes were kept if they appeared in 60% of the bootstrap runs. The presence of Slam related sequences was mapped on the phylogenetic tree.

Analysis of Gene Neighborhoods Around Putative Slam Homologs

The list of 353 Slam related sequences generated in our previous study (Hooda et al., 2016) was used to further investigate the neighboring genes in the Slam gene clusters. This number is more than a third of the total Slam related sequences and covers all major bacterial phyla that possess Slam related sequences, except for epsilon- and zeta-proteobacteria (Supplementary Data 1). In selecting genomes for a given bacterial species, fully sequenced reference genomes were given preference. For each of the Slam related genes present in these species, the corresponding genomic record (NCBI genome) was used to identify genes upstream and downstream along with their corresponding functional annotations (NCBI protein database, Ensembl bacteria). In a few of the cases, no genes were predicted upstream or downstream as the Slam related genes were close to the beginning or the end of the contig respectively and these sequences were ignored.

Within the Slam related gene clusters, a number of the neighboring genes were predicted to encode lipoproteins (predicted by an N-terminal lipobox motif using LipoP and/or SignalP) and we also found many examples of genes encoding TonB dependent transporters (IPR000531). The putative lipoproteins were annotated as either GNA1870-related lipoproteins, TBP-like solute-binding proteins or pagP-beta barrel proteins (InterPro signature; IPR01490; IPR001677; IPR011250 respectively). All the genes with one of the above-mentioned annotations are included in Supplementary Data 2.

Bacterial Strains and Growth Conditions

Strains used in this study are summarized in Supplementary Table 1. *E. coli* were grown in LB media containing antibiotics when necessary (50 μg/mL kanamycin and 100 μg/mL ampicillin). Cloning procedures were carried out using *E. coli* MM294 competent cells. Protein expression was performed using *E. coli* C43 (DE3) cells for all the flow-cytometry and western blot analysis.

Generation of Plasmids for Expression of Slams and SLPs

For flow cytometry experiments, the three SLPs (*Haemophilus influenzae* TbpB, *Moraxella catarrhalis* TbpB and *Pasteurella multocida* PM1514) were cloned into pET52b (to make pET52b *Hinf* TbpB, *Mcat* TbpB or *Pmul* SLP) using the restriction-free (RF) cloning strategy (van den Ent and Löwe, 2006). The *tbpb* genes were amplified from the genomes of *H. influenzae* strain 86-028NP and *M. catarrhalis* strain O35E, and the *pm1514* gene was amplified from *P. multocida* strain h48. A FLAG tag was inserted on the C-terminus of *M. catarrhalis tbpb* using FastCloning (Li et al., 2011) to make pET52b Mcat TbpB-flag. pET52b PmSLP-flag and pET52b PmSLP-flag-Slam was cloned by replacing the *catarrhalis tbpb* gene with *pm1514* and *pm1515-pm1514* respectively in frame with the FLAG tag using RF cloning.

The corresponding Slams were inserted into pET26b (pET26 *Hinf* Slam1, *Mcat* Slam1, and *Pmul* Slam) using RF cloning (van den Ent and Löwe, 2006). A 6xHis-tag was inserted between the *pelB* and the mature Slam sequences.

Flow Cytometry

For the *E. coli* translocation assays, the display of an SLP was determined using flow cytometry. Pairs of SLP and Slam plasmids (shown in Supplementary Table 1) were transformed into C43 (DE3) cells and grown in 1 mL of auto-induction media (Studier, 2005) for 18 h at 37°C. *H. influenzae* Slam showed poor expression when grown overnight in autoinduction media. Hence, for *H. influenzae* TbpB flow cytometry assays were performed by growing cells at 37°C to an OD_{600} ~ 0.6 and then inducing protein expression by the addition of 1 mM isopropyl β-D-1-thiogalactopyranoside (IPTG). Upon induction, cells were grown at 18°C for 16–18 h. Cells were harvested, washed twice in PBS containing 1 mM $MgCl_2$, and incubated with α-Flag antibodies (1:200, Sigma), or biotinylated human transferrin (0.05 mg/ml, Sigma) for 1 h at 4°C. The cells were then washed twice with PBS containing 1 mM $MgCl_2$ and then labeled with R-phycoerythrin (R-PE) conjugated Streptavidin (0.5 mg/ml, Cedarlane) or R-PE conjugated α-mouse IgG (25 μg/mL, Thermo Fisher Scientific) for 1 h at 4°C. Following staining, cells were fixed in 2% formaldehyde for 20 min and further washed with PBS containing 1 mM

$MgCl_2$. Flow cytometry was performed with a Becton Dickinson FACSCalibur and the results were analyzed using FLOWJO software. Mean fluorescence intensity (MFI) from at least three biological replicates were used to compare surface exposure of a given SLP between different samples. Statistical significance was calculated by comparing MFI between different samples using the one-way ANOVA test available in the software Prism 6.

Western blots were used to test the expression levels of each of the constructs used for the flow cytometry experiments. α-Flag (1:5,000, Sigma) and α-His (1:5,000, Thermo Fisher Scientific) antibodies were used to test expression of the SLP and Slam constructs respectively. α-GroEL (1:10,000) antibodies were used as loading controls.

Sucrose Density Ultracentrifugation

E. coli C43(DE3) cells expressing pET52b PmSLP-flag and pET26b empty or pET26 PmSlam were grown overnight and then used to inoculate 50 ml LB with the appropriate antibiotic. The cells were grown to an $OD_{600} \sim 0.6$, induced with 1 mM IPTG and then grown for an additional 18 h. The cells were pelleted, resuspended in 20 mM Tris pH 8.0, 200 mM NaCl with fresh lysozyme (1 mg/ml), 2 mM PMSF and DNase I (0.05 mg/ml), lysed by sonication and then centrifuged at 10,000 r.c.f. to remove cell debris. The supernatant was centrifuged at 125,000 r.c.f. for 1 h to collect the cellular membranes. The membrane pellet was resuspended in 1 ml of 20 mM Tris pH 8.0, 200 mM NaCl using a micro-glass homogenizer.

The inner and the outer membrane of *E. coli* were separated using a modified sucrose density ultracentrifugation protocol that was previously described (Hooda et al., 2016). For this assay, 100 μl of the membrane pellet was applied on top of a 13.2 ml thin-wall polypropylene tube containing step gradients of 3 ml of 2.02 M, 6 ml of 1.44 M and 3 ml of 0.77 M sucrose. The tubes were centrifuged at 83,000 r.c.f. for 16 h. The outer membrane and inner membranes partitioned to the interface of the 2 M and 1.44 M sucrose cushions and 1.44 M and 0.77 M sucrose layers, respectively. Twelve 1 ml fractions were collected and subjected to SDS–PAGE followed by western blotting with α-Flag (1:10,000), α-LepB (1:10,000), and α-OmpA (1:40,000) antibodies.

RESULTS

Identification of Putative Slam Homologs in Gram-Negative Bacteria

In a previous study, we had used the *N. meningitidis* Slam1 sequence to perform psi-blast searches (Altschul et al., 1997) and identified 353 putative Slam homologs in 225 Gram-negative bacteria (Hooda et al., 2016). All identified Slam related sequences possessed an N-terminal domain (Ntd) predicted to contain tetratricopeptide repeats(TPR) and a C-terminal beta-barrel domain annotated as a DUF560 domain (**Figure 1A**). Since that study, a large number of bacterial genomes have been sequenced by next-generation sequencing techniques. Hence, we performed updated psi-blast searches and were able to identify 832 Slam related sequences in 638 Gram-negative bacteria (Supplementary Data 1). The Slam1 gene (*nmb0313*) was used as the search template and we manually analyzed the list to remove genes for which only partial sequences were available. We also removed several hits that contained a large single domain with TPR repeats that are similar to the TPR repeats found in Slam-Ntd. As was previously observed, all Slam sequences obtained in our dataset contained both the Ntd and DUF560 domains and no sequences containing only the DUF560 were obtained. With these additional genomic sequences, we were able to identify Slam related sequences in all clades of the phylum *Proteobacteria* (**Figure 1B**). Slam related sequences were identified in bacterial species living in diverse environments including free-living, commensal and/or pathogenic bacteria. Slam-like proteins are predicted to be found in many human pathogens such as *Vibrio cholerae, Salmonella enterica* subsp. *arizonae,* and *Acinetobacter baumannii.*

Slams Adjacent to TbpB in *M. catarrhalis* and *H. influenzae* Translocate Their Respective TbpBs to the Surface in *E. coli*

A number of potential Slam homologs identified in this study were found in bacterial species that colonize the upper respiratory tract of mammals. Human respiratory tract bacteria that contain Slam sequences include *N. meningitidis, M. catarrhalis* and the HACEK (*Haemophilus, Aggregatibacter, Cardiobacterium, Eikenella, Kingella*) group of bacteria (Nørskov-Lauritsen, 2014). Slam sequences were also identified in bacteria that colonize the upper respiratory tract of cattle (*Moraxella bovis, Mannheimia haemolytica,* and *Histophilus somni*) and pigs (*Actinobacillus pleuropneumoniae*). Many of these species have been reported to contain transferrin-binding surface lipoproteins that are homologs of TbpB in *N. meningitidis* (Gray-Owen and Schyvers, 1996). Not surprisingly, two TbpB homologs in *M. catarrhalis* (Yu and Schryvers, 1993) and *H. influenzae* (Gray-Owen et al., 1995) were found to be adjacent to putative Slam genes (**Figure 2A**). Both *M. catarrhalis* and *H. influenzae* are human pathogens and their TbpBs have been previously shown to bind to human transferrin. The presence of TbpB genes adjacent to a Slam gene in their genome strongly suggests that these bacteria use a Slam-dependent translocation system to deliver TbpBs to the bacterial cell surface.

To confirm that these TbpBs depend on their adjacent Slam for translocation to the cell surface, we introduced the gene encoding TbpB in *M. catarrhalis* in laboratory strains of *E. coli* with and without its neighboring Slam. To monitor the expression of *M. catarrhalis* TbpB, we introduced a FLAG-tag at the C-terminus of TbpB. We labeled the cells with biotinylated human transferrin or α-Flag antibodies (**Figure 2B**). Flow cytometry was used to quantify the amount of *M. catarrhalis* TbpB on the surface of the cell. An increase in *M. catarrhalis* TbpB was observed only in the presence of its neighboring Slam and was quantified using mean fluorescence intensity (MFI) (**Figure 2C**). Western blots with α-Flag and by α-His antibodies were used to test the expression of *M. catarrhalis* TbpB & Slam respectively (**Figure 2C**, lower panel). *M. catarrhalis* TbpB was robustly displayed on the surface of *E. coli* in the presence of Slam

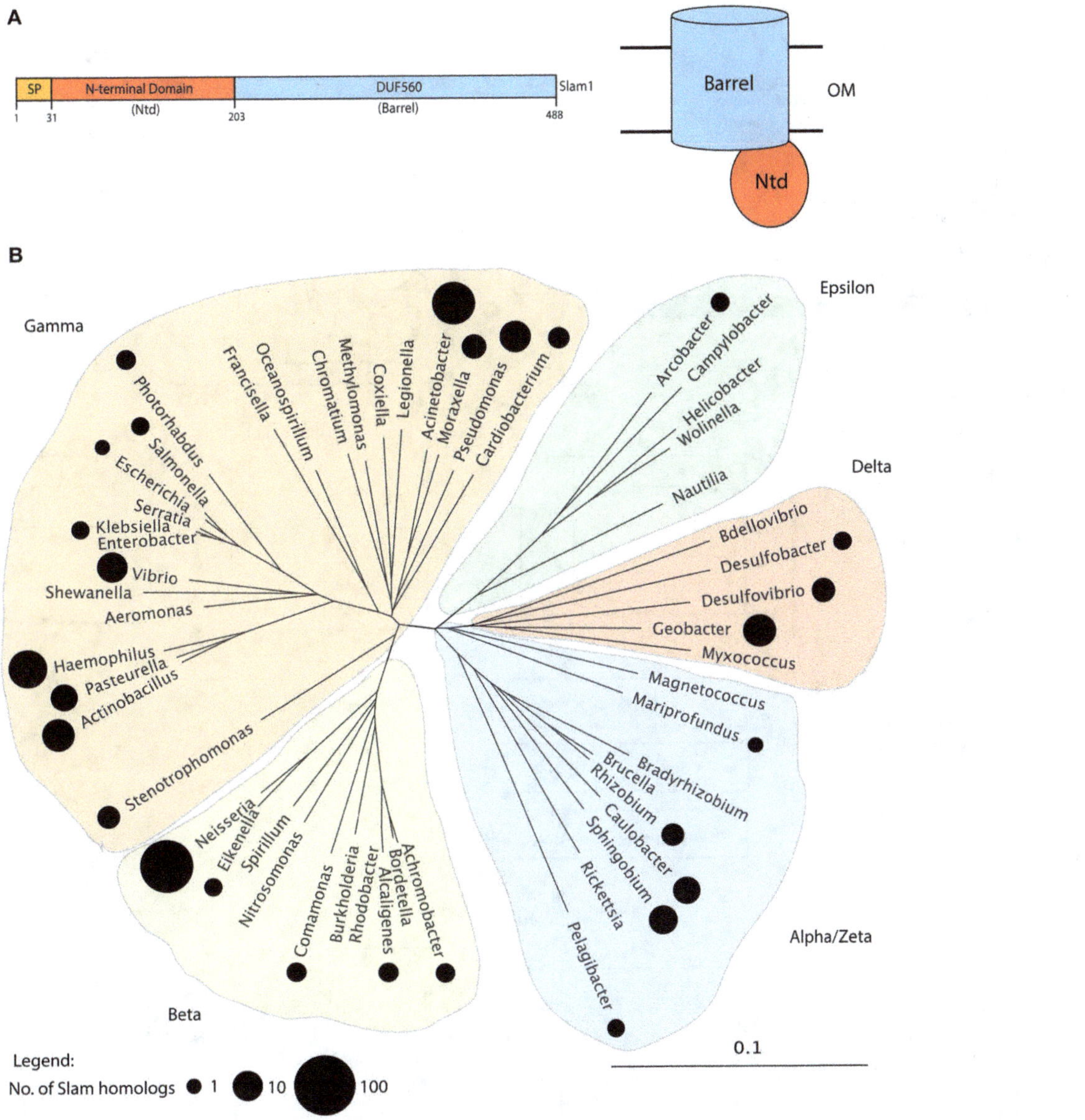

FIGURE 1 | Putative Slam family of proteins in Gram-negative bacteria. (A) Domain architecture of *N. meningitidis* Slam1. Slams possess 2 domains: a periplasmic N-terminal domain (Ntd) containing tetratricopeptide repeats and a predicted membrane bound 14-stranded barrel domain referred to as DUF560. **(B)** Distribution of Slam related sequences in *Proteobacteria*. A family tree of *Proteobacteria* was made using 16S-RNA sequences from 52 species representing the major bacterial families within *Proteobacteria*. The families containing at least 1 species with a Slam related sequence containing both the Ntd and DUF560 domains are highlighted by black circles. The size of the circle represents the number of Slams identified in a given bacterial family. Slams were found within all clades of *Proteobacteria*.

confirming that this TbpB homolog also uses a Slam-dependent system to reach the cell surface.

To verify that the FLAG-tag did not affect the Slam-dependent display of *M. catarrhalis* TbpB, we also tested the surface display of *M. catarrhalis* TbpB without the FLAG-tag in laboratory strains of *E. coli*. We labeled the cells using biotinylated human transferrin to test the functional display of the TbpB. Flow cytometry was used to quantify the amount of TbpB located on the surface of the cell. Similar to the results obtained for the FLAG-tagged construct, an increase in signal was obtained only in the presence of the neighboring Slam gene suggesting that the FLAG-tag does not affect translocation (**Figure 2D**). Additionally, we also tested the surface display of *H. influenzae* TbpB with or without its neighboring Slam using biotinylated human transferrin. In this case, a signal increase was obtained on the surface of *E. coli* in the presence of Slam, however the signal was weaker compared to *M. catarrhalis* TbpB. Westerns blot analysis using α-His antibody suggested that *M. catarrhalis* Slam (**Figure 2D**, lower panel) is expressed much more strongly in comparison to *H. influenzae* Slam (**Figure 2E**,

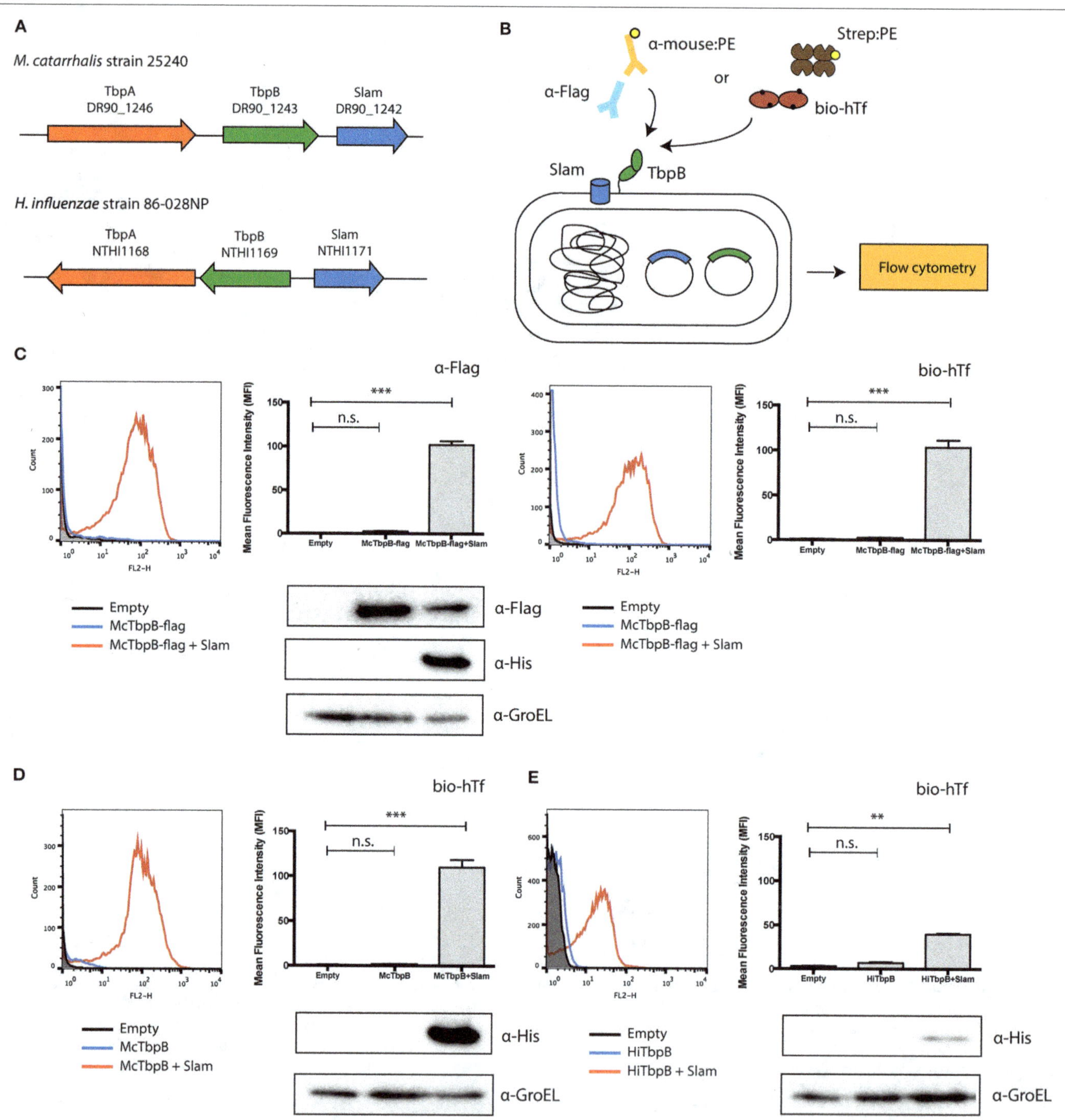

FIGURE 2 | Translocation assay with Slam and TbpB pairs from *Moraxella catarrhalis* and *Haemophilus influenzae*. (A) Slam and TbpB gene cluster in *M. catarrhalis* and *H. influenzae*. From the bioinformatics analysis, Slam genes were found adjacent to the known transferrin binding surface lipoprotein TbpB in the human pathogens *M. catarrhalis* and *H. influenzae*. **(B)** *E. coli* translocation assay used in this study. The Slam and TbpB genes were expressed in *E. coli* C43 (DE3) cells. The cells were labeled with either biotinylated human transferrin and streptavidin linked to R-PE, or mouse α-Flag antibody and α-mouse linked to R-PE. Surface display of TbpB was quantified using flow cytometry. **(C)** Flow cytometry profiles of C43 cells transformed with empty plasmid (black), *M. catarrhalis* TbpB-flag (blue), *M. catarrhalis* TbpB-flag + Slam (red) were obtained using α-Flag (left panel) and biotinylated human transferrin (right panel). An increase in fluorescence intensity was observed in the presence of Slam for both α-Flag and biotinylated human transferrin. Surface TbpB was quantified using mean fluorescence intensity. Western blots using α-Flag and α-His were used to quantify McTbpB and McSlam. α-GroEL is used as the loading control **(D)** Flow cytometry profile of TbpB homologs from *M. catarrhalis* (McTbpB) and **(E)** flow cytometry profile of *H. influenzae* (HiTbpB) using biotinylated human transferrin. Scatter plots of cell counts (Count) vs. R-PE fluorescence (FL2-H) in the presence (shown in red) or absence (shown in blue) of Slam are illustrated in the left panels. C43 cells transformed with empty plasmid (shown in black) were used as negative controls. Surface TbpB was quantified using mean fluorescence intensity (right panels). Western blots were done with α-His to examine Slam expression and α-GroEL was used as the loading control. Statistical significance was determined using one-way ANOVA. ***$p \leq 0.001$, **$p \leq 0.01$, n.s. $p > 0.05$.

lower panel), which may contribute to the lower signal obtained for *H. influenzae* TbpB.

Predicted SLP Genes Are Found Adjacent to Slam Genes in a Number of Gram-Negative Bacteria

For 353 of the 832 Slam related genes identified, a list of neighboring genes was analyzed using InterProScan (Jones et al., 2014) to identify Slam related gene clusters that also contain lipoprotein-encoding genes. From our analysis, 185 of the 353 (~52% of the clusters examined) Slam related genes contained lipoprotein-encoding genes in their gene clusters. This list of putative lipoproteins contained three experimentally confirmed SLPs (**Figure 3A**). These include two HpuA homologs in *Neisseria gonorrhoeae* and *Kingella denitrificans* that were functionally and structurally characterized by Wong et al. (2015). We also identified a putative human factor H binding protein from *H. influenzae,* referred to as protein H (PH) (Fleury et al., 2014) that has sequence homology to the Slam-dependent SLPs in *N. meningitidis*.

The 185 predicted lipoproteins were found in 129 different species distributed throughout *Proteobacteria*. One such lipoprotein was identified in the nematode pathogen *Xenorhabdus nematophila* that has not been previously shown to contain SLPs (**Figure 3B**). This lipoprotein, named NilC has been previously shown to be lipidated *in vivo*, present in the outer membrane and is important for host colonization (Cowles and Goodrich-Blair, 2004). The gene encoding NilB, a putative Slam homolog, is present next to the gene encoding NilC, and is shown to be required for host colonization (Bhasin et al., 2012). Our analysis suggests that NilC is a surface lipoprotein (SLP) that is dependent on the Slam homolog, NilB, for surface display.

Studying other genes in the Slam related gene clusters, we uncovered that 120 of the 185 clusters also possessed genes that encode proteins annotated as TonB-dependent receptors (TonBDR). This is not surprising as in *Neisseria* sp., three Slam-dependent SLPs (TbpB, LbpB, and HpuA) work in conjunction with a TonBDR to acquire iron (Hooda et al., 2017). The presence of TonBDRs gene in proximity to these putative SLPs supports their potential roles in nutrient acquisition. Furthermore, a small subset of Slam gene clusters (nine) contained multiple lipoprotein-encoding genes suggesting that they may be responsible for the display of multiple target SLPs.

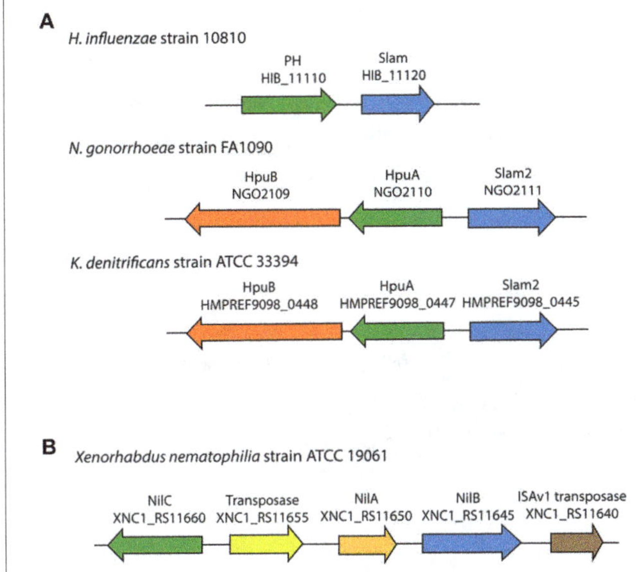

FIGURE 3 | Slam related gene clusters identified in this study with known lipoproteins. (A) Three SLP-containing Slam gene clusters were found in *Neisseria gonorrhoeae*, *Kingella denitrificans* and *Haemophilus influenzae*. **(B)** Lipoprotein-containing Slam gene cluster in *Xenorhabdus nematophilia*.

A Putative SLP Gene in *Pasteurella multocida* Is Displayed on the Surface of *E. coli* in a Slam-Dependent Manner

To further confirm the hypothesis that we have identified a large family of Slam-dependent SLPs, we sought to characterize the surface display of some of the remaining lipoproteins that were identified in this study for which no other functional data could be found. One such predicted lipoprotein was found in *P. multocida,* a zoonotic pathogen that resides in the normal respiratory microbiota of mammals. We identified the gene of the putative Slam (PM1515) adjacent to the predicted lipoprotein (PM1514) in all the sequenced strains of *P. multocida* (**Figure 4A**). The Slam displayed 32% identity to *N. meningitidis* Slam1 while the putative SLP showed no sequence similarity to any of the known Slam-dependent neisserial SLPs. Interestingly, the Slam gene cluster also included two other SLPs, PM1517 (PlpD), and PM1518 (PlpE) that have been investigated as potential vaccine antigens against *Pasteurella* infections (Nardini et al., 1998; Wu et al., 2007).

To test if this *pm1514* gene encoded a Slam-dependent SLP, we cloned the predicted Slam and lipoprotein genes into *E. coli* expression vectors (**Figure 4B**). We transformed *E. coli* cells with FLAG-tagged *P. multocida* lipoprotein and Slam, and used the FLAG-epitope to detect the SLP on the surface. As predicted, we saw an increase in FLAG signal upon expression of Slam as seen in flow cytometry profiles (**Figure 4C**) and quantified in mean fluorescence intensity (MFI) plots (**Figure 4D**). Furthermore, to confirm the translocation pathway used by the *P. multocida* SLP, we examined its outer membrane localization in the presence or absence of Slam by sucrose density ultracentrifugation (**Figure 4E**). Collectively, these findings suggest that PM1514 is a putative Slam-dependent SLP and that many other Slam-dependent SLPs on our list (Supplementary Data 2) likely use a similar translocation pathway as *N. meningitidis* TbpB to reach the surface (Hooda et al., 2016).

Comparison of Putative SLP Proteins Revealed a Conserved Structural Domain

With this dataset of putative Slam-dependent SLPs, we were interested in identifying structural features that are shared by this family of proteins. Structures for four neisserial Slam-dependent SLPs have been solved by X-ray crystallography

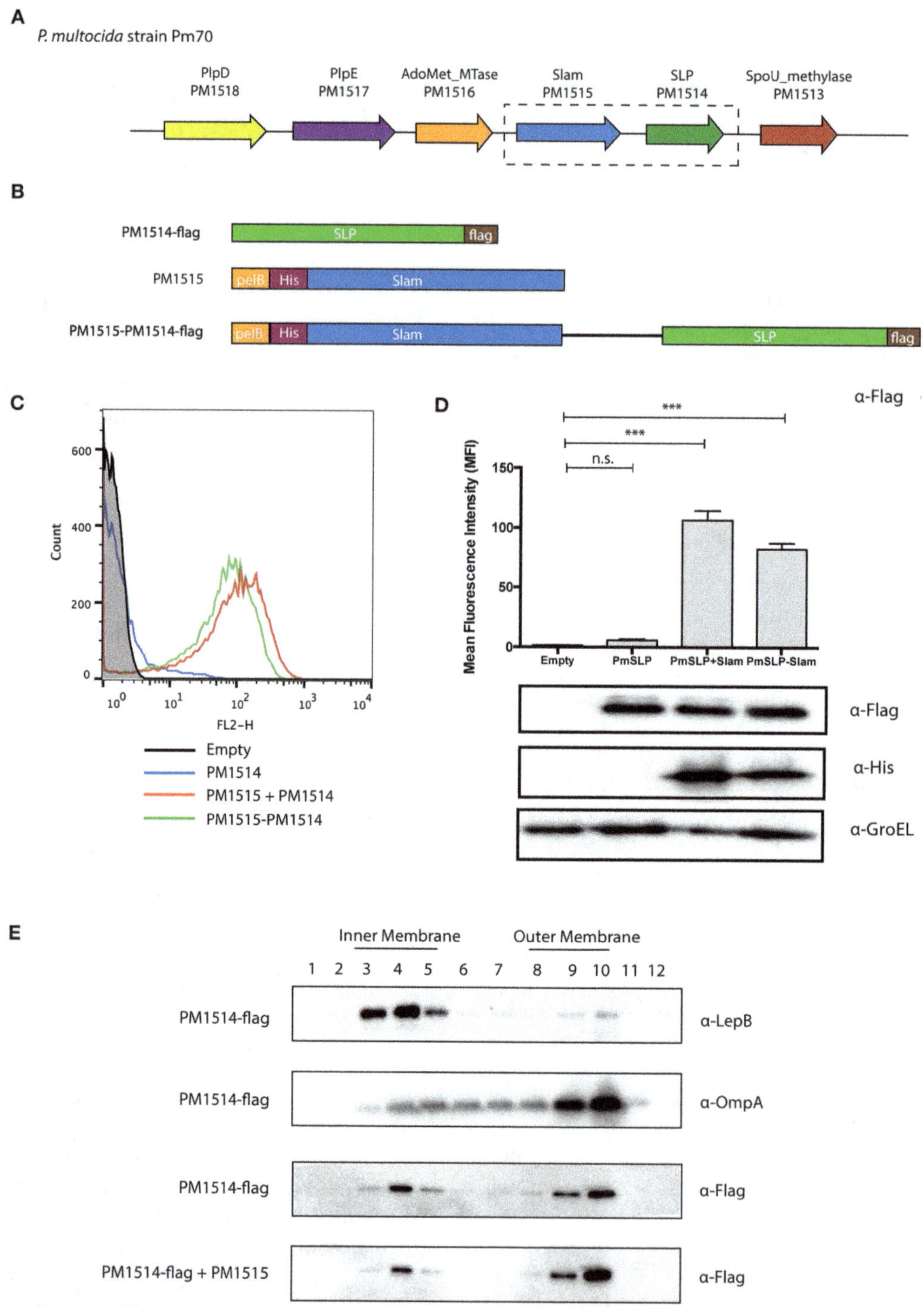

FIGURE 4 | Identification of a putative Slam-dependent surface lipoprotein in *Pasteurella multocida*. (A) Slam gene cluster in *P. multocida* strain Pm70. PM1515 (shown in blue) was identified as a putative Slam homolog in our bioinformatics search. PM1514 (shown in green) was annotated as a hypothetical protein with a predicted signal peptidase II cleavage site located within a putative lipobox motif. **(B)** *P. multocida* constructs utilized in an *E. coli* reconstituted translocation assay. To investigate if PM1514 is a Slam-dependent SLP, we cloned the following: PM1514 with a C-terminal FLAG-tag (PM1514-flag), PM1515 with an N-terminal His-tag and pelB signal sequence, and PM1515-PM1514-flag construct with both PM1515 and PM1514 region. **(C)** Flow cytometry profiles of *P. multocida* constructs. The 3 constructs were expressed in *E. coli* C43 (DE3) cells and labeled with α-Flag antibody and a secondary antibody directed against mouse IgG and linked to R-PE. Flow cytometry profiles of empty plasmid (black), PM1514-flag (blue), PM1515+PM1514-flag (red) and PM1515-PM1514-flag (green) are shown. **(D)** Mean fluorescence intensity plots for *P. multocida* constructs. Surface PM1514 was quantified using mean fluorescence intensity (MFI). Statistical significance was determined using one-way ANOVA. *** $p \leq 0.001$, n.s. $p > 0.05$. **(E)** Localization of *P. multocida* SLP using sucrose density ultracentrifugation. To test the localization of *P. multocida* SLP in *E. coli,* cells expressing PM1514-flag with or without PM1515 were harvested. Cell membranes were then isolated and layered on a sucrose gradient to separate the inner and outer membrane. Westerns blots were performed on different fractions with α-Flag, α-LepB, and α-OmpA antibodies to detect PmSLP, LepB (inner membrane control) and OmpA (outer membrane control) respectively.

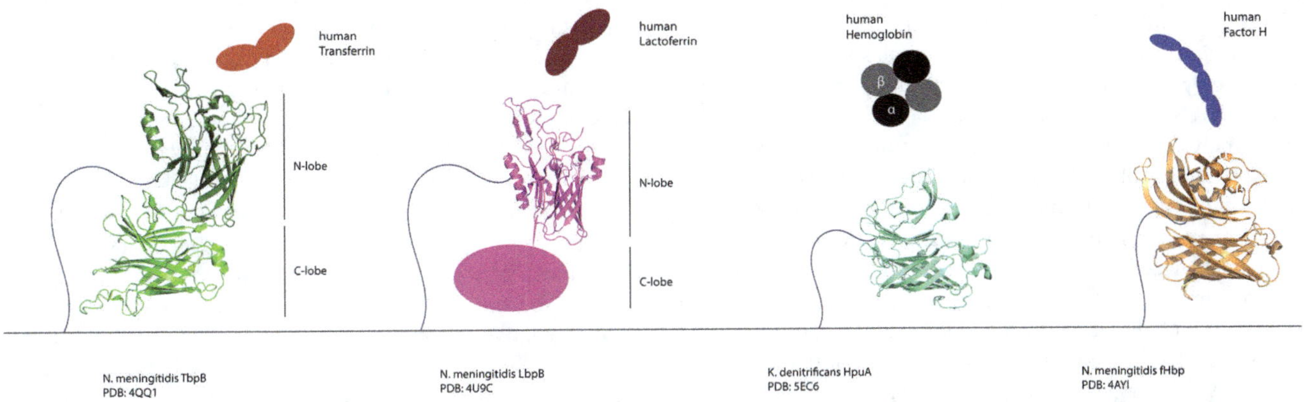

Species	PDB ID
Neisseria meningtidis	4QQ1, 3VE1, 3VE2, 3V8U
Actinobacillus suis	4O3W
Actinobacillus pleuropneumoniae	4O3X, 4O3Y, 4O3Z, 4O49, 3PQU, 3PQS, 3HOE, 3HOL
Hemophilus parasuis	4O4U, 4O4X

Species	PDB ID
Neisseria meningitidis	4U9C
Moraxella bovis	3UAQ

Species	PDB ID
Neisseria gonorrhoeae	5EE2
Kingella denitrificans	5EC6, 5EE4

Species	PDB ID
Neisseria meningitidis	4Z3T, 4AYD, 4AYE, 4AYI, 4AYM, 4AYN, 2Y7S, 3KVD, 2KC0, 2W80, 2W81

FIGURE 5 | Structures of known Slam-dependent SLPs. Structures of TbpB (green), LbpB N-lobe (magenta) HpuA (cyan) and fHbp (orange) are shown. The binding partner for each of the SLPs [human transferrin (red), lactoferrin (brown), hemoglobin (black), and Factor H (blue)] is also shown. Table below lists the accession numbers for all Slam-dependent SLP structures deposited to the Protein Data Bank (PDB). Primary references for TbpB (Moraes et al., 2009; Calmettes et al., 2011, 2012; Noinaj et al., 2012; Adamiak et al., 2015; Frandoloso et al., 2015), LbpB N-lobe (Arutyunova et al., 2012; Brooks et al., 2014), HpuA (Wong et al., 2015), and fHbp (Cantini et al., 2009; Schneider et al., 2009; Cendron et al., 2011; Scarselli et al., 2011; Johnson et al., 2012; Konar et al., 2015) structures are also included.

and NMR (**Figure 5**). While these proteins share no sequence similarity, they do share a protein domain composed of a flexible handle domain followed by an eight-stranded barrel domain. *N. meningitidis* TbpB and LbpB contain two lobes of the conserved domain, while fHbp and HpuA contain only one lobe.

To gain insight into the structure of Slam-dependent SLPs, we compared the predicted domains (InterProScan) on our list of putative Slam-dependent SLPs. We found that 58 contain either a lipoprotein GNA1870-related (InterPro signature: IPR014902) or a solute-binding protein, TBP-like (InterPro signature: IPR001677) domain. These domains are found on fHbp and TbpB respectively, both of which are Slam-dependent SLPs, and contain an eight-stranded beta barrel at their C-terminus. Further, 127 were predicted to encode a PagP-beta barrel (InterPro signature: IPR011250). PagP is an outer membrane enzyme that forms an eight-stranded barrel and is involved in catalyzing palmitate transfer from a phospholipid to a glucosamine unit of lipid A (Cuesta-Seijo et al., 2010). While PagP is an integral outer membrane protein, several soluble proteins are also predicted to contain an eight-stranded PagP-beta barrel. Most of the predicted Slam-adjacent lipoproteins also had a variable N-terminal region preceding the eight-stranded barrel that could form a handle-like domain seen in the representative three-dimensional structures of Slam-dependent SLPs.

DISCUSSION

With the increase in the number of bacterial genomic sequences, it has become evident that surface lipoproteins (or SLPs) are widespread in Gram-negative bacteria. One family of SLPs is characterized by a common structural architecture composed of an eight-stranded beta-barrel domain and a beta rich handle domain (TbpB, LbpB, HpuA, and fHbp). These proteins require a member of a unique family of proteins to traverse the outer membrane named Slam (Hooda et al., 2016). While we still do not know the exact role played by Slams, these proteins are specific for TbpB-like SLPs and may represent a novel class of outer membrane translocon or chaperone dedicated to the transport of SLPs from the inner leaflet of the outer membrane to the surface.

In this study, we performed a bioinformatic analysis of putative Slam homologs and identified a number of Slam-dependent SLPs in many different species of *Proteobacteria*. To our knowledge, this is the first systematic study to look at the distribution of SLPs in different Gram-negative bacteria. Previous attempts to look at TbpB-like lipoproteins have been stymied by the degree of variation that is found in these proteins. Since Slam sequences are more conserved, we were able to identify a large number of SLP homologs owing to the genetic linkage between Slams and TbpB-like SLPs. We have extended the number of bacteria that are now predicted to possess SLPs and also provide a framework to systematically search for these proteins. Based on this work, Slam-dependent SLPs represent the largest sub-family of SLPs reported thus far. However, our approach also has some limitations. To date, we have only identified genes for putative SLPs located in the immediate vicinity of genes for Slams. Using this method, we would have overlooked many known Slam-dependent SLPs in pathogenic *Neisseria* sp., which are transported in a Slam1-dependent manner but are encoded by genes not located in the vicinity of the gene for Slam1 (e.g., NmTbpB and NmLbpB). Therefore, our list of Slam-dependent SLPs is certainly incomplete; the bacterial species containing Slam homologs but no Slam-adjacent lipoproteins may also contain SLPs elsewhere in the genome.

Our analysis allowed the identification of a previously uncharacterized protein PM1514 in *Pasteurella multocida* as a putative SLP. Using localization assays we confirmed PM1514 is present in the outer membrane but is only detected on the surface of *E. coli* when co-expressed with the putative Slam PM1515. Taken together, these findings suggest that PM1515 is a Slam homolog and PM1514 is a Slam-dependent SLP. Additional work is required to further investigate this putative SLP, including a ^{3}H-palmitoyl labeling assay to confirm the lipidation of PM1514 in both *E. coli* and *P. multocida*. Flow cytometry and proteinase K shaving assay should be performed in *P. multocida* to test the surface display of PM1514 in its endogenous host. PM1514 is found in most sequenced strains of *P. multocida*, indicating that it may play an important role in the survival of *P. multocida* in its host organisms.

In our present search, we identified that Slam related sequences are found in *Proteobacteria* and not found in *Bacteroides*, *Borrelia,* and *Campylobacter*. Interestingly, we investigated the gene neighborhoods of four other known SLPs namely, JlpA in *Campylobacter* (Jin et al., 2001), HmuY (Wójtowicz et al., 2009) and SusD (Shipman et al., 2000) in *Bacteroides* and OspA/C in *Borrelia* (Schulze et al., 2010). Upon inspection none of these SLPs showed genetic linkage with any known/predicted outer membrane protein (data not shown) nor were any Slam related sequences found in *Bacteroides* or *Borrelia*. Coupled with the mechanistic studies completed on the SLP translocation pathways for *Bacteroides* (Lauber et al., 2016) and *Borrelia* (Chen and Zückert, 2011), it is likely that other translocation systems are used by *Bacteroides* and *Borrelia* to facilitate translocation of these SLPs to the cell surface.

Using the list of putative Slam-dependent SLPs, we have found that all these proteins contain a predicted eight-stranded barrel domain. All Slam-dependent SLPs contain either one or two copies of this barrel domain. Hence, we predict that the barrel domain may contain the translocation motif that is recognized by Slam. However, further experiments are required to tease apart the translocation motif present on these SLPs. Toward this end, the dataset generated by this study will be a valuable tool to identify secretion motif that is common amongst Slam-dependent SLPs.

The ability of Slam proteins to robustly potentiate the display of TbpB-like SLPs from a diverse set of bacteria in *E. coli* provides further evidence of their direct involvement in translocation. While more work is required to understand the mechanism of Slam function, the work done so far shows the efficacy of using Slam as a system to deliver proteins to the surface of Gram-negative bacteria and has potential applications in development of bacterial surface display technology (Ståhl and Uhlén, 1997). Taken together with our previous study (Hooda et al., 2016), our work suggests that TbpB-like SLPs and Slams form a “plug-and-play” cassette, reminiscent of the two-partner secretion systems (TPSS) or Type Vb secretion system (Jacob-Dubuisson et al., 2013). To date, we believe that Slams are specific for the delivery of lipidated proteins to the surface of Gram-negative bacteria. However, it will be interesting to see if they represent a more generalized secretion system. Finally, this study furthers the argument that SLPs are an important and yet under-appreciated family of proteins and their investigation may lead to identification of novel mechanisms utilized by many different bacteria to interact with their environmental surroundings and/or their hosts.

AUTHOR CONTRIBUTIONS

YH and TM designed the experiments and wrote the manuscript. YH did the bioinformatics analysis. YH and CL. performed the experiments. YH, CL, and TM. analyzed the data.

FUNDING

This work was supported by a Canadian Institutes of Health Research (CIHR) grants (PJT-148795 to TM), as well as with instrumentation and infrastructure support provided by Canadian Foundation for Innovation (CFI), and Ontario Ministry of Education and Innovation. TM is a Canada Research Chair in the Structural Biology of Membrane Proteins.

ACKNOWLEDGMENTS

We thank Drs. A.B. Schryvers (University of Calgary), Scott Gray-Owen (University of Toronto), JW de Gier (Stockholm University) and Walid Houry (University of Toronto) for plasmids, strains and antibodies. We also thank the members of the Moraes Lab for technical assistance and discussions.

REFERENCES

Adamiak, P., Calmettes, C., Moraes, T. F., and Schryvers, A. B. (2015). Patterns of structural and sequence variation within isotype lineages of the *Neisseria meningitidis* transferrin receptor system. *Microbiologyopen* 4, 491–504. doi: 10.1002/mbo3.254

Altschul, S. F., Madden, T. L., Schäffer, A. A., Zhang, J., Zhang, Z., Miller, W., et al. (1997). Gapped BLAST and PSI-BLAST: a new generation of protein database search programs. *Nucleic Acids Res.* 25, 3389–3402. doi: 10.1093/nar/25.17.3389

Arutyunova, E., Brooks, C. L., Beddek, A., Mak, M. W., Schryvers, A. B., and Lemieux, M. J. (2012). Crystal structure of the N-lobe of lactoferrin binding protein B from *Moraxella bovis*. *Biochem. Cell Biol. Biochim. Biol. Cell.* 90, 351–361. doi: 10.1139/o11-078

Bhasin, A., Chaston, J. M., and Goodrich-Blair, H. (2012). Mutational analyses reveal overall topology and functional regions of NilB, a bacterial outer membrane protein required for host association in a model of animal-microbe mutualism. *J. Bacteriol.* 194, 1763–1776. doi: 10.1128/JB.06711-11

Brandt, M. E., Riley, B. S., Radolf, J. D., and Norgard, M. V. (1990). Immunogenic integral membrane proteins of *Borrelia burgdorferi* are lipoproteins. *Infect. Immun.* 58, 983–991.

Brooks, C. L., Arutyunova, E., and Lemieux, M. J. (2014). The structure of lactoferrin-binding protein B from *Neisseria meningitidis* suggests roles in iron acquisition and neutralization of host defences. *Acta Crystallogr. Sect. F Struct. Biol. Commun.* 70, 1312–1317. doi: 10.1107/S2053230X14019372

Calmettes, C., Alcantara, J., Yu, R.-H., Schryvers, A. B., and Moraes, T. F. (2012). The structural basis of transferrin sequestration by transferrin-binding protein B. *Nat. Struct. Mol. Biol.* 19, 358–360. doi: 10.1038/nsmb.2251

Calmettes, C., Yu, R., Silva, L. P., Curran, D., Schriemer, D. C., Schryvers, A. B., et al. (2011). Structural variations within the transferrin binding site on transferrin-binding protein B, TbpB. *J. Biol. Chem.* 286, 12683–12692. doi: 10.1074/jbc.M110.206102

Cantini, F., Veggi, D., Dragonetti, S., Savino, S., Scarselli, M., Romagnoli, G., et al. (2009). Solution structure of the factor H-binding protein, a survival factor and protective antigen of *Neisseria meningitidis*. *J. Biol. Chem.* 284, 9022–9026. doi: 10.1074/jbc.C800214200

Cendron, L., Veggi, D., Girardi, E., and Zanotti, G. (2011). Structure of the uncomplexed *Neisseria meningitidis* factor H-binding protein fHbp (rLP2086). *Acta Crystallograph. Sect. F Struct. Biol. Cryst. Commun.* 67, 531–535. doi: 10.1107/S1744309111006154

Chamberlain, N. R., Brandt, M. E., Erwin, A. L., Radolf, J. D., and Norgard, M. V. (1989). Major integral membrane protein immunogens of *Treponema pallidum* are proteolipids. *Infect. Immun.* 57, 2872–2877.

Chatzi, K. E., Sardis, M. F., Karamanou, S., and Economou, A. (2013). Breaking on through to the other side: protein export through the bacterial sec system. *Biochem. J.* 449, 25–37. doi: 10.1042/BJ20121227

Chen, S., and Zückert, W. R. (2011). Probing the *Borrelia burgdorferi* surface lipoprotein secretion pathway using a conditionally folding protein domain. *J. Bacteriol.* 193, 6724–6732. doi: 10.1128/JB.06042-11

Cho, S.-H., Szewczyk, J., Pesavento, C., Zietek, M., Banzhaf, M., Roszczenko, P., et al. (2014). Detecting envelope stress by monitoring β-barrel assembly. *Cell* 159, 1652–1664. doi: 10.1016/j.cell.2014.11.045

Costa, T. R. D., Felisberto-Rodrigues, C., Meir, A., Prevost, M. S., Redzej, A., Trokter, M., et al. (2015). Secretion systems in Gram-negative bacteria: structural and mechanistic insights. *Nat. Rev. Microbiol.* 13, 343–359. doi: 10.1038/nrmicro3456

Cowles, C. E., and Goodrich-Blair, H. (2004). Characterization of a lipoprotein, NilC, required by *Xenorhabdus nematophila* for mutualism with its nematode host. *Mol. Microbiol.* 54, 464–477. doi: 10.1111/j.1365-2958.2004.04271.x

Cuesta-Seijo, J. A., Neale, C., Khan, M. A., Moktar, J., Tran, C. D., Bishop, R. E., et al. (2010). PagP crystallized from SDS/cosolvent reveals the route for phospholipid access to the hydrocarbon ruler. *Structure* 18, 1210–1219. doi: 10.1016/j.str.2010.06.014

D'Enfert, C., Chapon, C., and Pugsley, A. P. (1987). Export and secretion of the lipoprotein Pullulanase by *Klebsiella pneumoniae*. *Mol. Microbiol.* 1, 107–116.

Delcour, A. H. (2009). Outer membrane permeability and antibiotic resistance. *Biochim. Biophys. Acta* 1794, 808–816. doi: 10.1016/j.bbapap.2008.11.005

DeSantis, T. Z., Hugenholtz, P., Larsen, N., Rojas, M., Brodie, E. L., Keller, K., et al. (2006). Greengenes, a chimera-checked 16S rRNA gene database and workbench compatible with ARB. *Appl. Environ. Microbiol.* 72, 5069–5072. doi: 10.1128/AEM.03006-05

Fleury, C., Su, Y.-C., Hallström, T., Sandblad, L., Zipfel, P. F., and Riesbeck, K. (2014). Identification of a *Haemophilus influenzae* factor H-Binding lipoprotein involved in serum resistance. *J. Immunol.* 192, 5913–5923. doi: 10.4049/jimmunol.1303449

Frandoloso, R., Martínez-Martínez, S., Calmettes, C., Fegan, J., Costa, E., Curran, D., et al. (2015). Nonbinding site-directed mutants of transferrin binding protein B exhibit enhanced immunogenicity and protective capabilities. *Infect. Immun.* 83, 1030–1038. doi: 10.1128/IAI.02572-14

Geyter, J. D., Tsirigotaki, A., Orfanoudaki, G., Zorzini, V., Economou, A., and Karamanou, S. (2016). Protein folding in the cell envelope of *Escherichia coli*. *Nat. Microbiol.* 1:16107. doi: 10.1038/nmicrobiol.2016.107

Gray-Owen, S. D., Loosmore, S., and Schryvers, A. B. (1995). Identification and characterization of genes encoding the human transferrin-binding proteins from *Haemophilus influenzae*. *Infect. Immun.* 63, 1201–1210.

Gray-Owen, S. D., and Schyvers, A. B. (1996). Bacterial transferrin and lactoferrin receptors. *Trends Microbiol.* 4, 185–191. doi: 10.1016/0966-842X(96)10025-1

Hoehn, G. T., and Clark, V. L. (1992). The major anaerobically induced outer membrane protein of *Neisseria gonorrhoeae*, Pan 1, is a lipoprotein. *Infect. Immun.* 60, 4704–4708.

Hooda, Y., Lai, C. C.-L., Judd, A., Buckwalter, C. M., Shin, H. E., Gray-Owen, S. D., et al. (2016). Slam is an outer membrane protein that is required for the surface display of lipidated virulence factors in Neisseria. *Nat. Microbiol.* 1:16009. doi: 10.1038/nmicrobiol.2016.9

Hooda, Y., Shin, H. E., Bateman, T. J., and Moraes, T. F. (2017). Neisserial surface lipoproteins: structure, function and biogenesis. *Pathog. Dis.* 75:ftx010. doi: 10.1093/femspd/ftx010

Jacob-Dubuisson, F., Guérin, J., Baelen, S., and Clantin, B. (2013). Two-partner secretion: as simple as it sounds? *Res. Microbiol.* 164, 583–595. doi: 10.1016/j.resmic.2013.03.009

Jin, S., Joe, A., Lynett, J., Hani, E. K., Sherman, P., and Chan, V. L. (2001). JlpA, a novel surface-exposed lipoprotein specific to *Campylobacter jejuni*, mediates adherence to host epithelial cells. *Mol. Microbiol.* 39, 1225–1236. doi: 10.1111/j.1365-2958.2001.02294.x

Johnson, S., Tan, L., Veen, S., van der, Caesar, J., Jorge, E. G. D., Harding, R. J., et al. (2012). Design and evaluation of meningococcal vaccines through structure-based modification of host and pathogen molecules. *PLoS Pathog.* 8:e1002981. doi: 10.1371/journal.ppat.1002981

Jones, P., Binns, D., Chang, H.-Y., Fraser, M., Li, W., McAnulla, C., et al. (2014). InterProScan 5: genome-scale protein function classification. *Bioinformatics* 30, 1236–1240. doi: 10.1093/bioinformatics/btu031

Karuppiah, V., Berry, J.-L., and Derrick, J. P. (2011). Outer membrane translocons: structural insights into channel formation. *Trends Microbiol.* 19, 40–48. doi: 10.1016/j.tim.2010.10.006

Kearse, M., Moir, R., Wilson, A., Stones-Havas, S., Cheung, M., Sturrock, S., et al. (2012). Geneious basic: an integrated and extendable desktop software platform for the organization and analysis of sequence data. *Bioinformatics* 28, 1647–1649. doi: 10.1093/bioinformatics/bts199

Konar, M., Pajon, R., and Beernink, P. T. (2015). A meningococcal vaccine antigen engineered to increase thermal stability and stabilize protective epitopes. *Proc. Natl. Acad. Sci. U.S.A.* 112, 14823–14828. doi: 10.1073/pnas.1507829112

Konovalova, A., Perlman, D. H., Cowles, C. E., and Silhavy, T. J. (2014). Transmembrane domain of surface-exposed outer membrane lipoprotein RcsF is threaded through the lumen of β-barrel proteins. *Proc. Natl. Acad. Sci. U.S.A.* 111, E4350–E4358. doi: 10.1073/pnas.1417138111

Konovalova, A., and Silhavy, T. J. (2015). Outer membrane lipoprotein biogenesis: lol is not the end. *Phil. Trans. R. Soc. B* 370:20150030. doi: 10.1098/rstb.2015.0030

Lauber, F., Cornelis, G. R., and Renzi, F. (2016). Identification of a new lipoprotein export signal in gram-negative bacteria. *MBio* 7:01216. doi: 10.1128/mbio.01232-16

Leuzzi, R., Serino, L., Scarselli, M., Savino, S., Fontana, M. R., Monaci, E., et al. (2005). Ng-MIP, a surface-exposed lipoprotein of *Neisseria gonorrhoeae*, has a peptidyl-prolyl cis/trans isomerase (PPIase) activity and is involved in persistence in macrophages. *Mol. Microbiol.* 58, 669–681. doi: 10.1111/j.1365-2958.2005.04859.x

Lewis, L. A., Gray, E., Wang, Y. P., Roe, B. A., and Dyer, D. W. (1997). Molecular characterization of hpuAB, the haemoglobin-haptoglobin-utilization operon of *Neisseria meningitidis*. *Mol. Microbiol.* 23, 737–749. doi: 10.1046/j.1365-2958.1997.2501619.x

Li, C., Wen, A., Shen, B., Lu, J., Huang, Y., and Chang, Y. (2011). FastCloning: a highly simplified, purification-free, sequence- and ligation-independent PCR cloning method. *BMC Biotechnol.* 11:92. doi: 10.1186/1472-6750-11-92

Madico, G., Welsch, J. A., Lewis, L. A., McNaughton, A., Perlman, D. H., Costello, C. E., et al. (2006). The meningococcal vaccine candidate GNA1870 binds the complement regulatory protein factor H and enhances serum resistance. *J. Immunol. Baltim.* 177, 501–510. doi: 10.4049/jimmunol.177.1.501

Manning, P. A., Beutin, L., and Achtman, M. (1980). Outer membrane of *Escherichia coli*: properties of the F sex factor traT protein which is involved in surface exclusion. *J. Bacteriol.* 142, 285–294.

Moraes, T. F., Yu, R., Strynadka, N. C. J., and Schryvers, A. B. (2009). Insights into the bacterial transferrin receptor: the structure of transferrin-binding protein B from *Actinobacillus pleuropneumoniae*. *Mol. Cell* 35, 523–533. doi: 10.1016/j.molcel.2009.06.029

Nardini, P. M., Mellors, A., and Lo, R. Y. (1998). Characterization of a fourth lipoprotein from *Pasteurella haemolytica* A1 and its homology to the OmpA family of outer membrane proteins. *FEMS Microbiol. Lett.* 165, 71–77. doi: 10.1111/j.1574-6968.1998.tb13129.x

Nikaido, H. (2003). Molecular basis of bacterial outer membrane permeability revisited. *Microbiol. Mol. Biol. Rev.* 67, 593–656. doi: 10.1128/MMBR.67.4.593-656.2003

Noinaj, N., Easley, N. C., Oke, M., Mizuno, N., Gumbart, J., Boura, E., et al. (2012). Structural basis for iron piracy by pathogenic Neisseria. *Nature* 483, 53–58. doi: 10.1038/nature10823

Nørskov-Lauritsen, N. (2014). Classification, identification, and clinical significance of *Haemophilus* and *Aggregatibacter* species with host specificity for humans. *Clin. Microbiol. Rev.* 27, 214–240. doi: 10.1128/CMR.00103-13

Okuda, S., and Tokuda, H. (2011). Lipoprotein sorting in bacteria. *Annu. Rev. Microbiol.* 65, 239–259. doi: 10.1146/annurev-micro-090110-102859

Pagès, J.-M., James, C. E., and Winterhalter, M. (2008). The porin and the permeating antibiotic: a selective diffusion barrier in Gram-negative bacteria. *Nat. Rev. Microbiol.* 6, 893–903. doi: 10.1038/nrmicro1994

Pettersson, A., Prinz, T., Umar, A., Van Der Biezen, J., and Tommassen, J. (1998). Molecular characterization of LbpB, the second lactoferrin-binding protein of *Neisseria meningitidis*. *Mol. Microbiol.* 27, 599–610. doi: 10.1046/j.1365-2958.1998.00707.x

Pugsley, A. P., Chapon, C., and Schwartz, M. (1986). Extracellular pullulanase of Klebsiella pneumoniae is a lipoprotein. *J. Bacteriol.* 166, 1083–1088. doi: 10.1128/jb.166.3.1083-1088.1986

Scarselli, M., Aricò, B., Brunelli, B., Savino, S., Marcello, F. D., Palumbo, E., et al. (2011). Rational design of a meningococcal antigen inducing broad protective immunity. *Sci. Transl. Med.* 3, 91ra62–91ra62. doi: 10.1126/scitranslmed.3002234

Schneider, M. C., Prosser, B. E., Caesar, J. J. E., Kugelberg, E., Li, S., Zhang, Q., et al. (2009). *Neisseria meningitidis* recruits factor H using protein mimicry of host carbohydrates. *Nature* 458, 890–893. doi: 10.1038/nature07769

Schryvers, A. B., and Morris, L. J. (1988). Identification and characterization of the transferrin receptor from *Neisseria meningitidis*. *Mol. Microbiol.* 2, 281–288. doi: 10.1111/j.1365-2958.1988.tb00029.x

Schulze, R. J., Chen, S., Kumru, O. S., and Zückert, W. R. (2010). Translocation of *Borrelia burgdorferi* surface Lipoprotein OspA through the outer membrane requires an unfolded conformation and can initiate at the C-terminus. *Mol. Microbiol.* 76, 1266–1278. doi: 10.1111/j.1365-2958.2010.07172.x

Serruto, D., Spadafina, T., Ciucchi, L., Lewis, L. A., Ram, S., Tontini, M., et al. (2010). Neisseria meningitidis GNA2132, a heparin-binding protein that induces protective immunity in humans. *Proc. Natl. Acad. Sci. U.S.A.* 107, 3770–3775. doi: 10.1073/pnas.0915162107

Shipman, J. A., Berleman, J. E., and Salyers, A. A. (2000). Characterization of Four Outer membrane proteins involved in binding starch to the cell surface of *Bacteroides thetaiotaomicron*. *J. Bacteriol.* 182, 5365–5372. doi: 10.1128/JB.182.19.5365-5372.2000

Silhavy, T. J., Kahne, D., and Walker, S. (2010). The bacterial cell envelope. *Cold Spring Harb. Perspect. Biol.* 2:a000414. doi: 10.1101/cshperspect.a000414

Ståhl, S., and Uhlén, M. (1997). Bacterial surface display: trends and progress. *Trends Biotechnol.* 15, 185–192. doi: 10.1016/S0167-7799(97)01034-2

Studier, F. W. (2005). Protein production by auto-induction in high density shaking cultures. *Protein Expr. Purif.* 41, 207–234. doi: 10.1016/j.pep.2005.01.016

Szewczyk, J., and Collet, J.-F. (2016). "Chapter One - the journey of lipoproteins through the cell: one birthplace, multiple destinations," in *Advances in Microbial Physiology*, ed R. K. Poole (Academic Press), 1–50. Available online at: http://www.sciencedirect.com/science/article/pii/S0065291116300248 (Accessed October 24, 2016).

van den Ent, F., and Löwe, J. (2006). RF cloning: a restriction-free method for inserting target genes into plasmids. *J. Biochem. Biophys. Methods* 67, 67–74. doi: 10.1016/j.jbbm.2005.12.008

Van Ulsen, P., Van Alphen, L., Ten Hove, J., Fransen, F., Van Der Ley, P., and Tommassen, J. (2003). A *Neisserial autotransporter* NalP modulating the processing of other autotransporters. *Mol. Microbiol.* 50, 1017–1030. doi: 10.1046/j.1365-2958.2003.03773.x

Webb, C., Selkrig, J., Perry, A., Noinaj, N., Buchanan, S. K., and Lithgow, T. (2012). Dynamic association of BAM complex modules includes surface exposure of the lipoprotein BamC. *J. Mol. Biol.* 422, 545–555. doi: 10.1016/j.jmb.2012.05.035

Wilson, M. M., and Bernstein, H. D. (2016). Surface-Exposed Lipoproteins: an emerging secretion phenomenon in gram-negative bacteria. *Trends Microbiol.* 24, 198–208. doi: 10.1016/j.tim.2015.11.006

Wójtowicz, H., Guevara, T., Tallant, C., Olczak, M., Sroka, A., Potempa, J., et al. (2009). Unique structure and stability of HmuY, a novel heme-binding protein of *Porphyromonas gingivalis*. *PLoS Pathog.* 5:e1000419. doi: 10.1371/journal.ppat.1000419

Wong, C. T., Xu, Y., Gupta, A., Garnett, J. A., Matthews, S. J., and Hare, S. A. (2015). Structural analysis of haemoglobin binding by HpuA from the Neisseriaceae family. *Nat. Commun.* 6:10172. doi: 10.1038/ncomms10172

Wu, J.-R., Shien, J.-H., Shieh, H. K., Chen, C.-F., and Chang, P.-C. (2007). Protective immunity conferred by recombinant *Pasteurella multocida* lipoprotein E (PlpE). *Vaccine* 25, 4140–4148. doi: 10.1016/j.vaccine.2007.03.005

Yu, R., and Schryvers, A. B. (1993). The interaction between human transferrin and transferrin binding protein 2 from Moraxella (Branhamella) catarrhalis differs from that of other human pathogens. *Microb. Pathog.* 15, 433–445. doi: 10.1006/mpat.1993.1092

Zückert, W. R. (2014). Secretion of bacterial Lipoproteins: through the cytoplasmic membrane, the periplasm and beyond. *Biochim. Biophys. Acta* 1843, 1509–1516. doi: 10.1016/j.bbamcr.2014.04.022

2

Confirmed and Potential Roles of Bacterial T6SSs in the Intestinal Ecosystem

Can Chen[1†], Xiaobing Yang[2†] and Xihui Shen[2]*

[1] *Institute of Food and Drug Inspection, College of Life Science and Agronomy, Zhoukou Normal University, Zhoukou, China,*
[2] *State Key Laboratory of Crop Stress Biology for Arid Areas, Shaanxi Key Laboratory of Agricultural and Environmental Microbiology, College of Life Sciences, Northwest A&F University, Yangling, China*

***Correspondence:**
Xihui Shen
xihuishen@nwsuaf.edu.cn

The contact-dependent type VI secretion system (T6SS) in diverse microbes plays crucial roles in both inter-bacterial and bacteria-host interactions. As numerous microorganisms inhabit the intestinal ecosystem at a high density, it is necessary to consider the functions of T6SS in intestinal bacteria. In this mini-review, we discuss T6SS-dependent functions in intestinal microbes, including commensal microbes and enteric pathogens, and list experimentally verified species of intestinal bacteria containing T6SS clusters. Several seminal studies have shown that T6SS plays crucial antibacterial roles in colonization resistance, niche occupancy, activation of host innate immune responses, and modulation of host intestinal mechanics. Some potential roles of T6SS in the intestinal ecosystem, such as targeting of single cell eukaryotic competitors, competition for micronutrients, and stress resistance are also discussed. Considering the distinct activities of T6SS in diverse bacteria residing in the intestine, we suggest that T6SS research in intestinal microbes may be beneficial for the future development of new medicines and clinical treatments.

Keywords: type VI secretion system, intestinal microbiota, enteric pathogen, colonization resistance, competition

INTRODUCTION

The type VI secretion system (T6SS) is a contact-dependent transmembrane nanomachine that uses a contractile mechanism to inject effectors into adjacent prokaryotic or eukaryotic cells (Hood et al., 2010; Jani and Cotter, 2010; Cianfanelli et al., 2016). A typical T6SS apparatus is composed of 13 core subunits (TssA-TssM) and possesses a structure similar to that of a T4 bacteriophage tail (Boyer et al., 2009; Cascales and Cambillau, 2012; Zoued et al., 2013). In an integral T6SS apparatus, the baseplate complex (TssAEFGK) is anchored to the cell membrane by the membrane complex (TssJLM), providing structural support. The needle sheath (TssBC), tail tube (TssD/Hcp), and spike complex (TssI/VgrG) form the injection apparatus, which is responsible for effector secretion (Cascales and Cambillau, 2012; Silverman et al., 2012). A dynamic T6SS secretion process contains assembly/extension, contraction/puncture, and disassembly of the sheath. Briefly, the membrane complex is formed at the initial stage. Subsequently, baseplate proteins are recruited for sheath/tube anchoring and extension. When attacking, the membrane-puncturing spike pierces target cells through sheath contraction, accompanied by energy consumption and load transportation. Consequently, various effectors are delivered into target cells in a one-step manner (Basler et al., 2012; Brunet et al., 2015; Cianfanelli et al., 2016; Coulthurst, 2019). Although some pathogen-associated T6SSs have been reported to be involved in bacterial pathogenesis (Pukatzki et al., 2007; Aubert et al., 2016; Jiang et al., 2016), the primary function of T6SSs is to compete against rival bacteria in polymicrobial environments (Durand et al., 2014; Cianfanelli et al., 2016). However, other functions of T6SSs, such as resistance to amoeba predation (Zheng et al., 2011), biofilm formation (Gallique et al., 2017), and stress response (Weber et al., 2009; Zhang W. et al., 2013), have also been reported.

The intestinal ecosystem is composed of a densely populated community of multiple microorganisms that may be beneficial or harmful to host health (Walter and Ley, 2011). In intestinal environments where nutrients and space are limited, microbes must use various strategies to coexist or compete with the host and other bacteria. In recent years, the contact-dependent T6SS nanomachine has attracted much attention for its role in interbacterial competition in the mammalian gut (Sana et al., 2017; Verster et al., 2017). T6SS mediated antagonism in the mammalian gut not only benefits the microbiota-mediated colonization resistance by preventing invasion of pathogens, but also assists some pathogens to battle with the resident microbiota to invade an ecosystem and cause disease. Thus, it was stated that there is secret bacterial warfare in the gut, due to the T6SS roles both in the intestinal commensal microbes and enteric pathogens (Sana et al., 2017; Garcia-Bayona and Comstock, 2018). In this review, we have discussed recent studies on intestinal microorganisms with functional T6SS clusters and T6SS-dependent activities related to the intestinal microbial community and host health.

T6SS-DEPENDENT ACTIVITIES IN INTESTINAL ECOSYSTEM

Interference Competition Mediated by T6SS

Bacterial competition occurs as interference competition (killing of target cells with antibacterial weapons) or exploitation competition (consuming nutrients from the milieu) (Chassaing and Cascales, 2018). As an anti-bacterial weapon, T6SSs were primarily considered to mediate interference competition in the intestinal microbiota. The antibacterial function of T6SSs relies primarily on injection of "anti-bacterial" effectors that target conserved, essential features of the bacterial cell, such as peptidoglycan (Russell et al., 2011), membrane phospholipids (Russell et al., 2013), nucleic acids (Ma L.S. et al., 2014), NAD^+ (Whitney et al., 2015), and the critical cell division protein FtsZ (Ting et al., 2018).

Commensal bacteria form a protective barrier that protects the host from bacterial pathogens (Belkaid and Hand, 2014). In healthy adult individuals, Bacteroidales are the most abundant order of bacteria in the intestinal microbiota (Faith et al., 2013). Recently, by performing extensive bioinformatics analyses and creating hidden Markov models for Bacteroidales Tss proteins, Coyne and colleagues identified 130 T6SS loci within 205 human gut Bacteroidales genomes. As Bacteroidales comprise approximately 50% of all colonic bacteria in many people, this suggested that T6SSs are distributed in about 25% of bacteria in the human colon (Coyne et al., 2014; Coyne et al., 2016). Moreover, more than 10^9 T6SS-firing events were determined per minute per gram of colonic contents, further supporting the importance of this weapon in shaping gut microbiota composition (Wexler et al., 2016). T6SS loci of the human gut Bacteroidales species segregate into three distinct genetic architectures (GA), termed GA1-GA3. GA1 and GA2 loci are present on conserved integrative conjugative elements (ICE) and are transferred between diverse human gut Bacteroidales strains via ICE. GA3 loci are not contained on conserved ICE and are confined to *Bacteroides fragilis* (Coyne et al., 2014, 2016). The GA3 T6SSs of *B. fragilis* antagonizes human gut Bacteroidales species *in vivo* using previously undescribed effectors, likely to create a locally protected niche in the human gut (Wexler et al., 2016). The interbacterial interactions among symbiotic Bacteroidales species could be predicted according to the presence or absence of strain-specific effector/immunity (E/I) repertoires. Further, some of these strains may avoid contact-dependent killing by accumulating immunity genes to neutralize antibacterial effectors that they do not encode to persist in the gut (Wexler et al., 2016). Importantly, symbiotic non-toxigenic *B. fragilis* strains could restrict enteric colonization by an enterotoxigenic *B. fragilis* strain, dependent on a functional T6S, to protect the host against intestinal inflammatory disease, suggesting a novel role of T6SS in colonization resistance (Hecht et al., 2016; Casterline et al., 2017).

Type VI secretion systems are widely distributed and have been proposed to be present in about 25% of all sequenced Gram-negative bacteria (Boyer et al., 2009; Salomon and Orth, 2015), including enteric pathogenic microorganisms, e.g., *Vibrio cholerae*, *Salmonella enterica*, *Shigella sonnei*, and *Yersinia pseudotuberculosis*. Thus, T6SS mediated antagonism in the mammalian gut both benefits microbiota-mediated colonization resistance by preventing pathogen invasion and facilitates some pathogens to battle resident microbiota to invade the ecosystem and cause disease. Some enteric pathogens could utilize T6SS-mediated antibacterial weapons to kill symbionts and establish within the host gut. For example, *S. enterica* Serovar Typhimurium uses T6SS to kill commensal *Klebsiella oxytoca* and enhance colonization of the mouse gut (Sana et al., 2016). With a functional T6SS, *S. sonnei* showed an advantage in competing against *E. coli* and *Shigella flexneri* both *in vitro* and *in vivo*, which may explain the dominance and increasing global prevalence of *S. sonnei* in developed countries worldwide (Anderson et al., 2017). Constitutive T6SS expression provides *V. cholerae* with an advantage in intra-specific and inter-specific competition, such that T6SS-dependent toxicity toward other bacteria could enhance *V. cholerae* survival in the environment and/or during colonization of a host (MacIntyre et al., 2010; Unterweger et al., 2012; Fu et al., 2013). Through transcriptome sequencing (RNA-Seq), T6SSs and their associated toxins in the gut symbiont of 28 *Snodgrassella alvi* strains from diverse *Apis* and *Bombus* species were analyzed. T6SS-associated Rhs toxins with antibacterial activity could mediate both intraspecific and interspecific competition among *S. alvi* strains and other bee gut microbes. Furthermore, extensive recombination and horizontal transfer of toxin/immunity genes between strains in the gut microbiota have resulted in tremendous diversity in their toxin repertoires, which suggest that T6SS-mediated competition may be an important driver of coevolution (Steele et al., 2017). These studies showed that enteric pathogens could use their T6SSs for interbacterial competition *in vivo* and for niche occupancy.

Besides bacterial competitors, T6SS was also found to target some single cell eukaryotic competitors, including amoebae and fungi. In fact, T6SS was first identified by screening *V. cholerae* mutants that failed to kill the social amoeba *Dictyostelium discoideum*, and the lipid-binding effector VasX was found to be required for T6SS-mediated amoeba killing, possibly through plasma membrane perturbations (Pukatzki et al., 2006; Miyata et al., 2011; Zheng et al., 2011). T6SSs in *Pseudomonas syringae* and *Serratia marcescens* were reported to be required for competition against other single cell eukaryotic organisms, i.e., yeast and fungi (Haapalainen et al., 2012; Trunk et al., 2018). The first anti-fungal T6SS effectors, Tfe1 and Tfe2, were identified in *S. marcescens*. Tfe1 causes plasma membrane depolarization without formation of a specific pore, while Tfe2 disrupts nutrient uptake and amino acid metabolism, and induces autophagy (Trunk et al., 2018). Since single cell eukaryotes are important components of the intestinal microbiota (Coyne and Comstock, 2019), it is not surprising that intestinal bacteria might deploy anti-eukaryotic T6SSs to compete for nutrients and space against co-habiting microbial eukaryotes.

Exploitation Competition Meditated by T6SS

As essential micronutrients are involved in a wide range of cellular processes (e.g., DNA replication, respiration, and energy generation), transit of these metal ions is a crucial component of host-microbe interactions, including those in the gastrointestinal tract (Sorbara and Pamer, 2019). For example, *E. coli* mutants defective in catecholate siderophore production show impaired murine gut colonization (Pi et al., 2012). Without a high-affinity zinc transporter, *Campylobacter jejuni* is unable to replicate or colonize the gastrointestinal tract, as quantities of zinc in the gastrointestinal tract are reduced by the host (Gielda and DiRita, 2012). Thus, competition for micronutrients in the gut is intense.

Type VI secretion system has been reported to be involved in acquisition of essential micronutrients, such as zinc, manganese, and iron, indicating its novel function in increasing bacterial fitness through competition for essential nutrients. In *Pseudomonas aeruginosa*, the H3-T6SS secreted effector TseF facilitates iron acquisition by interacting with the iron-chelating molecule PQS (Pseudomonas quinolone signal) (Lin et al., 2017). In *Burkholderia thailandensis*, T6SS4 secretes zinc- and manganese-scavenging proteins to fulfill the increased cellular demand for these metal ions under oxidative stress (Si et al., 2017a,b). Similarly, in the enteric pathogen *Y. pseudotuberculosis*, T6SS4 functions to combat host nutritional immunity and multiple adverse stresses by translocating a zinc binding effector YezP (Wang et al., 2015). Distinct from the extensively studied contact-dependent "anti-bacterial" and "anti-eukaryotic" T6SSs, the "metal ion transporting" T6SS secretes metal-binding proteins into the extracellular milieu, independent of cell-cell contact (Si et al., 2017a). Thus, T6SS confers survival advantages to bacterium in niches with multiple bacterial species not only by delivering "anti-bacterial" toxins to kill competing cells for interference competition, but also by enhancing its ability to acquire essential micronutrients for exploitation competition. It will be interesting to determine the physiologic consequence of these metal acquisition T6SSs in the intestinal ecosystem in the future.

Interestingly, expression of *Y. pseudotuberculosis* T6SS4 was found to be nutrient status-dependently regulated by the stringent response factor RelA and the nutrient-dependent regulator RovM (Song et al., 2015; Yang et al., 2019). Further research is needed to reveal whether T6SS is involved in competition for other nutrients besides metal ions.

Activation of the Innate Immune Response and Virulence

Although it is well-known that enteric pathogens could utilize the antibacterial T6SS weapon to eradicate competing microbes *in vivo*, little is known about whether the T6SS-dependent interactions with commensal bacteria have additional effects on virulence. Recently, Zhao et al. (2018) discovered that the *V. cholerae* T6SS could compete against commensal *E. coli* strains in the mouse gut to facilitate colonization. Importantly, they found that this microbial antagonistic interaction could improve fitness of *V. cholera* by activating host innate immune responses and improving expression of bacterial virulence genes (i.e., cholera toxin) (Zhao et al., 2018). The authors suggested that the innate immune response was induced by bacterial debris derived from lysed commensals, which together with up-regulated cholera toxin (CT), resulted in elevated disease symptoms and increased fitness of *V. cholerae* as a pathogen. Fast et al. (2018) used the *Drosophila melanogaster* model of cholera to define the contribution of T6SS to *V. cholerae* pathogenesis. They found that interactions between T6SS and the gut commensal *Acetobacter pasteurianus* intensified disease symptoms and accelerated host death. The disease severity was attenuated by inactivation of T6SS, or removal of *A. pasteurianus*. Interestingly, mutation of the Immune Deficiency (IMD) pathway relieved T6SS dependent lethality, implicating innate defenses in T6SS-mediated host death (Fast et al., 2018). These studies established that interactions between T6SS and commensal bacteria activate innate immune responses and promote *V. cholera* pathogenesis.

In addition, a few T6SS toxic effectors result in metabolic disorders or diseases in eukaryotic host cells. VgrG-1 in *V. cholerae* was responsible for T6SS-dependent cytotoxicity of macrophages through covalent cross-linking of host cell actin (Pukatzki et al., 2007; Ma and Mekalanos, 2010). In the diarrheal isolate *Aeromonas hydrophila* SSU, T6SS-secreted VgrG1 exhibited actin ADP-ribosylating activity, which can induce a rounded phenotype in HeLa cells, resulting in apoptosis (Suarez et al., 2010). A non-VgrG T6SS effector in *Edwardsiella tarda*, EvpP, inhibited NLRP3 inflammasome activation through repression of the Ca^{2+}-dependent MAPK-Jnk pathway (Chen et al., 2017). T6SS effectors with phospholipase activity could cause damage to the cell membrane and accelerate the infection process (Dong et al., 2013; Russell et al., 2013). Notably, a few bacterial pathogens with the T6SS apparatus may spread through the digestive tract but cause diseases outside the gastrointestinal system. For example, in the newborn meningitis *E. coli* (NMEC) K1 strain, the Hcps of T6SS interacts with human brain microvascular endothelial cells (HBMEC) in a coordinated manner, first binding, then invading, and finally causing apoptosis of HBMEC (Zhou et al., 2012). The T6SS-secreted proteins VgrG1 and VgrG2 in *Helicobacter hepaticus* increases cellular innate pro-inflammatory responses, resulting in liver disease and intestinal inflammation (Bartonickova et al., 2013).

Modulation of Host Intestinal Mechanics

Recently, Logan and colleagues used a combination of microbial genetics, *in vitro* experiments, and quantitative *in vivo* imaging in zebrafish to determine the role of *V. cholerae* T6SS in gut colonization. They found that *V. cholerae* can expel resident microbiota of the genus *Aeromonas* in a T6SS-dependent manner. Unexpectedly, T6SS acted primarily to increase the strength of gut contractions, rather than killing the bacterial competitor. Coupling of T6SS activity to host contractions depended on an actin cross-linking domain (ACD) from the T6SS apparatus. When the ACD domain was deleted, *V. cholerae* could no longer induce enhanced host contractions, and dense *Aeromonas* communities remained in the gut. Deleting the ACD domain did not affect the ability of *V. cholerae* to kill *A. veronii in vitro* or enter and occupy the host intestine. These findings reveal a novel strategy by which enteric pathogens can modulate host intestinal mechanics to redefine gut communities and highlights the role of the host fluid-mechanical environment in shaping gut population dynamics (Logan et al., 2018). Whether this is a common tactic for gut-colonizing bacteria to invade the intestinal ecosystem especially in human needs to be verified in the future.

Stress Resistance

Within the intestinal lumen, enteric pathogens encounter severe stresses, such as bile salts, antimicrobial peptides, free fatty acids, enhanced osmolarity, and oxidative stress (Fang et al., 2016). To survive these adverse environments, bacteria have developed various delicate mechanisms. As a versatile molecular machine, T6SS has been found to play crucial roles in the bacterial stress response and contribute to cell survival under multiple environmental challenges. *Vibrio anguillarum* T6SS is regulated by the general stress response regulator RpoS and is involved in its resistance to hydrogen peroxide, ethanol, and low pH (Weber et al., 2009). Upon activation by RpoS and the global oxidative stress regulator OxyR, enterohaemorrhagic *Escherichia coli* (EHEC) uses its T6SS to secrete a Mn-containing catalase, KatN, thus providing a higher resistance to reactive oxygen species (ROS) produced by the host and enhancing survival levels in the host (Wan et al., 2017). *Y. pseudotuberculosis* T6SS4 was required for bacterial survival under osmotic, oxidative, and acidic stress and for resistance to bile salts, and its expression was regulated by various stress response regulators, such as the stationary growth phase stress σ factor RpoS, the global oxidative stress regulator OxyR, and the acid/osmotic regulator OmpR (Gueguen et al., 2013; Zhang W. et al., 2013; Guan et al., 2015; Wang et al., 2015). In *V. cholerae* O1, activation of T6SS has been shown to be dependent on the osmolarity of the milieu (Ishikawa et al., 2012). In addition, in the intestine, *V. cholerae* T6SS genes are overexpressed under mucin (intestinal host factors) and microbiota-modified bile salt conditions (Bachmann et al., 2015). Bile salts can also activate expression of T6SS genes in *Salmonella typhimurium* (Sana et al., 2016) and *C. jejuni* (Lertpiriyapong et al., 2012). These studies on gene expression regulation suggest that T6SS is expressed in the intestine and could provide resistance to multiple stresses, increasing bacterial adaptability or survival in the intestinal ecosystem.

Other Potential Functions

Recently, it was suggested that T6SS in *V. cholerae* serves as a predatory weapon and fosters horizontal gene transfer. For incorporation of DNA released by lysed cells into a competent predator cell, natural transformation and evolution can occur (Borgeaud et al., 2015). Thus, T6SS may promote co-evolution of bacterial microflora within their environments, including intestinal ecosystems.

Compelling data showed that *H. hepaticus* interacts with intestinal epithelial cells (IECs) and employs T6SS to limit within-host growth and virulence (Chow and Mazmanian, 2010). This feature was regarded as T6SS function of anti-virulence (Jani and Cotter, 2010). Nevertheless, this phenomenon is rarely seen.

To summarize, we constructed a schematic diagram showing the functions of T6SS in the intestinal ecosystem (**Figure 1**).

COMMENSAL INTESTINAL MICROBES WITH T6SS CLUSTERS

Bacteroidetes

Bacteroidetes, a phylum of Gram-negative bacteria, is highly abundant in the gastrointestinal tracts of humans and other mammals (Lozupone et al., 2012; Russell et al., 2014; Wexler and Goodman, 2017). Despite Bacteroidales being predominant in the mammalian gut, T6SSs in Bacteroidetes species were not identified until 2014, perhaps because the T6SS in Bacteroidetes does not have sufficient sequence similarity with core T6SS proteins of Proteobacteria for protein-protein comparisons (e.g., BLASTP) or protein-profile comparisons (e.g., Pfam, COG) (Russell et al., 2014; Coyne et al., 2016). Studies have revealed that the T6SS loci are segregated into three distinct genetic architectures (GA1-GA3), among which the T6SS loci GA1 and GA2 are contained on highly similar integrative conjugative element (ICE) (Coyne et al., 2016). Among the numerous mechanisms that compete in the extremely dense ecosystem of the gut, T6SSs are probably very prevalent antagonistic systems in gut Bacteroidales (Wilson et al., 2015; Chatzidaki-Livanis et al., 2016; Coyne et al., 2016; Coyne and Comstock, 2019), as the Bacteroides T6SS genes are widespread in human gut metagenomes (Verster et al., 2017). Given that T6SSs in Bacteroidetes mediate inter-bacterial antagonism against

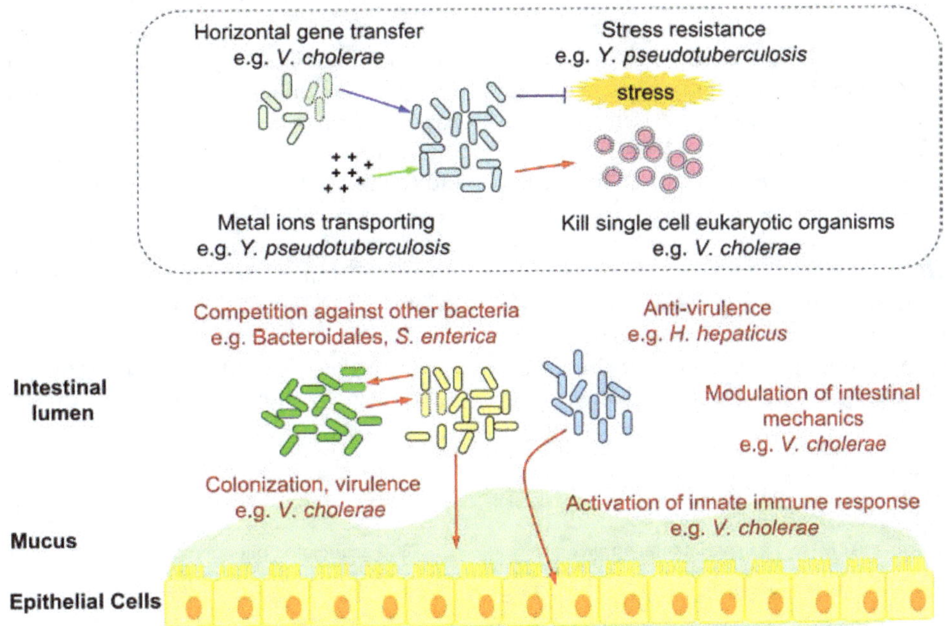

FIGURE 1 | Schematic diagram of bacterial T6SS functions in the intestinal ecosystem. T6SS functions verified in the gut (e.g., Competition against other bacteria, Colonization, Virulence, Anti-virulence, Modulation of intestinal mechanics, and Activation of innate immune response.) were marked in red. Deduced T6SS functions in the gut (e.g., Horizontal gene transfer, Metal ions transporting, Stress resistance, and Kill single cell eukaryotic organisms.) based on *in vitro* analysis were shown in the dotted box.

pathogens and in view of the observed stability of Bacteroidetes in the healthy human intestine, it predicts that T6SS in the stable Bacteroidetes community contributes to intestinal homeostasis. We believe that further research on T6SSs in Bacteroidales species will add to our understanding of microbial stability in the gut and to our ability to diagnose and treat gut microbial infections.

ENTERIC PATHOGENS WITH T6SS CLUSTERS

Vibrio cholerae

Vibrio cholerae is a Gram-negative pathogen consisting of over 200 serogroups that usually cause diarrhoeal diseases ranging from cholera to mild gastroenteritis after ingestion of contaminated food or water containing them (Kitaoka et al., 2011; Unterweger et al., 2014). All *V. cholerae* strains examined to date contain T6SS gene clusters (Unterweger et al., 2012). This microorganism provided several important and initial discoveries about T6SS, including the definition of the contact-dependent secretion model (Pukatzki et al., 2006; Basler et al., 2012), identification of T6SS-dependent effector-immunity pairs (Dong et al., 2013; Russell et al., 2013; Altindis et al., 2015), and induction of intestinal inflammation through actin cross-linking in host cells (Ma and Mekalanos, 2010). Recently, *in vivo* analysis presented new insight on the pathogenic mechanism of *V. cholerae* with T6SS. T6SS in *V. cholerae* acts on commensal bacteria *A. pasteurianus* to accelerate death of the *Drosophila* host (Fast et al., 2018). Another study revealed the roles of *V. cholerae* T6SS in antagonism against host commensal microbiota *in vivo* over their niche, which facilitated bacterial colonization of the mice gut (Zhao et al., 2018). Besides killing host gut bacterial symbionts, *V. cholerae* T6SS could modulate host intestinal mechanics, enhancing intestinal movements that led to expulsion of resident microbiota by the host (Logan et al., 2018).

Vibrio parahaemolyticus

Research indicated that vpT6SS2 contributed to adhesion of *V. parahaemolyticus* to host cells (Yu et al., 2012). In addition, the vpT6SS2 effector could induce autophagy in macrophage cells without causing apparent cytotoxicity (Yu et al., 2015).

Salmonella enterica

Experiments suggested that T6SS contributed to pathogenicity, at least in the *S. enterica* serotypes Gallinarum (Blondel et al., 2013), Enteritidis (Blondel et al., 2010; Troxell, 2018), Typhi (Wang et al., 2011), Dublin (Pezoa et al., 2014), and Typhimurium (Mulder et al., 2012). A recent study provided *in vivo* evidence that *S. typhimurium* used T6SS as a weapon to kill commensal bacteria (*K. oxytoca*) to successfully colonize the mouse gut (Sana et al., 2016).

Yersinia pseudotuberculosis

Studies of *Y. pseudotuberculosis* revealed roles for T6SS in responses to various stressors and identified corresponding regulators, including OmpR (Gueguen et al., 2013; Zhang W. et al., 2013), OxyR (Wang et al., 2015), ZntR (Wang et al., 2017),

RpoS (Guan et al., 2015), RovM (Song et al., 2015), and YpsI/YtbI (Zhang et al., 2011). Notably, T6SS4 in *Y. pseudotuberculosis* was found to be involved in zinc transportation by secreting a zinc-binding protein YezP to mitigate the impact of detrimental hydroxyl radicals under oxidative stress (Wang et al., 2015). The contribution of T6SS to ion transport was also confirmed in *B. thailandensis* and *P. aeruginosa* (Lin et al., 2017; Si et al., 2017a,b).

Escherichia coli

Escherichia coli are model bacteria that have been studied extensively. Although less abundant than Bacteroidales in the gastrointestinal tract, symbiotic gut *E. coli* play important roles in colonization resistance against enteric pathogens of the proteobacterial phylum (Coyne and Comstock, 2019). To date, more than 150 *E. coli* strains have been identified. Most are harmless, but several serotypes have been proposed to be pathogenic and may cause significant diarrheal and extraintestinal diseases (Croxen et al., 2013). The current knowledge regarding the prevalence, the assembly, the regulation, and the roles of the T6SS in *E. coli* has been reviewed (Journet and Cascales, 2016). It has been suggested that T6SS contributes to pathogenesis in *E. coli* strains. For example, T6SS-dependent Hcp1 in *E. coli* K1 induced actin cytoskeleton rearrangement, apoptosis, and release of interleukin-6 (IL-6) and IL-8 in human brain microvascular endothelial cells (HBMEC) (Zhou et al., 2012). Further, T6SS in avian pathogenic *E. coli* (APEC) strains contributes to APEC pathogenesis (Ma et al., 2018). Several

TABLE 1 | Intestinal bacteria with functional T6SSs that have been verified experimentally.

Bacteria species	No. of T6SS loci	Experimental strains	T6SS associated phenotypes	References
Aeromonas hydrophila	1	SSU	Induce host cell toxicity by ADP ribosylation of actin	Suarez et al., 2010
Bacteroides fragilis	1	TM4000, ATCC43858, ATCC43859	Virulence and strain competition	Casterline et al., 2017
		NCTC 9343, 638R	Competition	Chatzidaki-Livanis et al., 2016
		NCTC 9343	Interbacterial antagonism	Russell et al., 2014
Campylobacter jejuni	1	subsp. *jejuni* serotype O:3	Host cell adhesion, invasion, colonization; adaptation to DCA	Lertpiriyapong et al., 2012
		108	Cytotoxicity toward red blood cells	Bleumink-Pluym et al., 2013
Citrobacter freundii	1–2	CF74	Adhesion and cytotoxicity to host cells	Liu et al., 2015
Citrobacter rodentium	1–2	DBS100[ATCC51459]	Interbacterial killing	Gueguen et al., 2014
Edwardsiella tarda	1–2	EIB202	NLRP3 inflammasome inhibition, bacterial colonization *in vivo*	Chen et al., 2017
Enterobacter cloacae	1–2	subsp. *cloacae* ATCC13047	Interbacterial competition	Zhang H. et al., 2013
Escherichia coli	1–3	K12 W3110, EAEC 17-2	Bacterial growth competition	Brunet et al., 2013
		APEC SEPT362	Pathogenesis	de Pace et al., 2010
		APEC SEPT362	Enhance biofilm formation	de Pace et al., 2011
		RS218	Pathogenicity in HBMEC	Zhou et al., 2012
		TW-XM (O2:K1)	Pathogenic pathways	Ma J. et al., 2014
		EHEC EDL933	Virulence	Wan et al., 2017
		STEC	Virulence	Aas et al., 2018
Helicobacter hepaticus	1	ATCC51449	Modulate and exacerbate the innate pro-inflammatory effect	Bartonickova et al., 2013
		ATCC51449	Limiting colonization and intestinal inflammation	Chow and Mazmanian, 2010
Salmonella enterica	1–2	Typhimurium LT2	Pathogenesis	Mulder et al., 2012
		Gallinarum NCTC13346, Enteritidis NCTC13349	Chicken colonization	Blondel et al., 2010
		Typhi GIFU10007	Cytotoxicity of human epithelial cells	Wang et al., 2011
		Typhimurium 14028s	Chicken colonization	Blondel et al., 2013
		Dublin CT_02021853	Colonization of mice and chickens	Pezoa et al., 2014
		Enteritidis	Virulence	Troxell, 2018
		SL1344	Killing of commensal bacteria, Colonization of the host gut	Sana et al., 2016
Shigella sonnei	1	CIP106347	Interbacterial competition, niche occupancy	Anderson et al., 2017
Vibrio alginolyticus	1–2	EPGS, MVP01, ATCC33787	Regulated by QS and RpoN	Sheng et al., 2012
Vibrio anguillarum	1	NB10 serotype O1	Modulates quorum sensing and stress response	Weber et al., 2009

(Continued)

TABLE 1 | Continued

Bacteria species	No. of T6SS loci	Experimental strains	T6SS associated phenotypes	References
Vibrio cholerae	1	C6706	Microbial antagonistic interaction, intestinal colonization, virulence	Zhao et al., 2018
		O37 strain V52	Virulence, competition	Zheng et al., 2011
		O37 strain V52	Intra- and interspecific competition	Unterweger et al., 2012
		A1552	Foster horizontal gene transfer	Borgeaud et al., 2015
		O37 strain V52	Microbial pathogenesis	Pukatzki et al., 2006
		O37 strain V52	Induces an inflammatory diarrhea	Ma and Mekalanos, 2010
		O37 strain V52	Virulence	Miyata et al., 2011
		Multiple strains	Intraspecific competition	Unterweger et al., 2014
		O37 strain V52	Interspecies competition	MacIntyre et al., 2010
		C6706	Intermicrobial competition	Logan et al., 2018
Vibrio fluvialis	2	85003	Environmental survival	Pan et al., 2018
Vibrio parahaemolyticus	1–2	HZ	Cell adhesion and cytotoxicity	Yu et al., 2012
		RIMD 2210633	Enhance environmental fitness in marine environments, anti-bacterial activity	Salomon et al., 2013
		HZ	Induce autophagy in macrophages	Yu et al., 2015
Yersinia enterocolitica	1	subsp. *enterocolitica* 8081	Encumber bacteria invasion and survival in host cells	Jaakkola et al., 2015
Yersinia pseudotuberculosis	4–6	YpIII	Maintain intracellular pH homeostasis	Zhang W. et al., 2013
		IP31758	Survival in high osmolarity, resistance to deoxycholate	Gueguen et al., 2013
		YpIII	Transports Zn^{2+} to combat multiple stresses and host immunity	Wang et al., 2015

structural proteins of T6SS in *E. coli* have been studied to clarify the specific functions of these T6SS components (Felisberto-Rodrigues et al., 2011; Aschtgen et al., 2012; Durand et al., 2012). However, whether *E. coli* gut symbionts have T6SSs that function in colonization resistance against enteric pathogens, and how enteric pathogenic *E. coli* T6SSs function in the intestinal ecosystem still remain enigmatic.

Notably, the T6SS apparatus has a common evolutionary origin with phage tail-associated protein complexes. Yet, the involvement of phages in the evolution of bacterial T6SS remains unclear.

OTHER ENTERIC PATHOGENS

Type VI secretion system in some other enteric pathogens has also been studied. In *C. jejuni*, T6SS mediates host cell adhesion, invasion, colonization, and adaptation to deoxycholic acid (DCA) (Lertpiriyapong et al., 2012). In *Citrobacter freundii*, T6SS plays a wide-ranging role in the regulation of the flagellar system and in motility; it is also involved in the adherence and cytotoxicity to host cells (Liu et al., 2015). T6SS2 in *Vibrio fluvialis* is associated with anti-bacterial activity and contributes to bacteria survival in highly competitive environments (Pan et al., 2018). The T6SS effector VgrG1 from *A. hydrophila* induces host cell toxicity by ADP ribosylation of actin (Suarez et al., 2010). In *E. tarda*, the T6SS effector EvpP significantly inhibits NLRP3 inflammasome activation by inhibiting the Ca^{2+}-dependent MAPK-Jnk pathway (Chen et al., 2017). Additionally, functional T6SSs have been verified in the enteric pathogens *Enterobacter cloacae* (Whitney et al., 2014), *Edwardsiella ictaluri* (Rogge and Thune, 2011), *Yersinia enterocolitica* (Jaakkola et al., 2015), and *V. anguillarum* (Weber et al., 2009). To facilitate retrieval for future research, we have listed the experimentally verified intestinal bacteria that contain functional T6SS loci in **Table 1**.

CONCLUDING REMARKS

Although the composition and function of the intestinal microbiota has been well documented, the underlying mechanisms of its assembly remain poorly understood. T6SS is a contact-dependent molecular weapon primarily used for interbacterial competition. Considering that T6SS loci exist in 50% of intestinal Bacteroidetes, which are the dominant microflora in the gastrointestinal system, and in some other commensal intestinal microbes and enteric pathogens, T6SS might be widely used in the intestinal ecosystem to shape microbiota composition. Actually, several seminal studies have greatly advanced our understanding of the importance of T6SS in the intestine. These studies showed that the T6SS antibacterial weapons are not only used by pathogens to colonize their hosts, but also by gut commensals to prevent pathogen colonization. Interestingly, two studies in *V. cholera* further indicated that microbial antagonistic interactions mediated by T6SS could elevate disease symptoms by activating host innate immune responses (Fast et al., 2018; Zhao et al., 2018). Moreover, using a zebrafish model, Logan et al. (2018) revealed that the T6SS of *V. cholera* could modulate host intestinal mechanics to redefine the gut microbial composition. These findings provide new insights into the mechanisms used by enteric pathogens for gut colonization. As for the gut commensals, Verster et al. (2017) investigate the prevalence and roles of the T6SS in the human gut microbiome by using metagenomic analyses. In addition to the compatibility with species GA1 and GA2, GA3 is associated with increased Bacteroides abundance in infant microbiomes, and confers an advantage in Bacteroides-rich ecosystems. Thus, it revealed the prevalence and potential role of T6SS-dependent competition in shaping human gut microbial composition (Verster et al., 2017).

It is noteworthy that sometimes a few invertebrate models and small animal models (e.g., Drosophila, zebrafish) were used to analyze T6SS functions in the intestinal ecosystem (Fast et al., 2018; Logan et al., 2018). These results should be further verified in mammalian models. In addition, further *in vivo* studies are urgently needed to confirm whether anti-fungal, stress-resistant and metal ion-acquiring T6SSs are functional in the intestinal ecosystem. Studies of T6SS in the intestinal ecosystem are just beginning and have a long way to go. As T6SS in several bacterial pathogens contributes to virulence against the host, the T6SS apparatus could be exploited as a therapeutic drug target. Based on this possibility, a novel natural antimicrobial was identified that can reduce the pathogenicity of *Campylobacter* T6SS *in vitro* and also decrease its colonization *in vivo* (Sima et al., 2018). These findings suggest that studies on T6SS functions in the intestinal ecosystem may provide a theoretical basis for the development of new medicines for various pathogenic infections in the future.

AUTHOR CONTRIBUTIONS

CC and XY collected and assessed the references. CC and XS contributed in the proposal and guidelines of the review. CC and XY wrote this manuscript.

FUNDING

This work was funded by the National Natural Science Foundation of China (Nos. 31725003 and 31670053) and the Open Project Program of the State Key Laboratory of Pathogen and Biosecurity (No. SKLPBS1825).

REFERENCES

Aas, C. G., Drablos, F., Haugum, K., and Afset, J. E. (2018). Comparative transcriptome profiling reveals a potential role of type VI secretion system and fimbriae in virulence of Non-O157 shiga toxin-producing *Escherichia coli*. *Front. Microbiol.* 9:1416. doi: 10.3389/fmicb.2018.01416

Altindis, E., Dong, T., Catalano, C., and Mekalanos, J. (2015). Secretome analysis of *Vibrio cholerae* type VI secretion system reveals a new effector-immunity pair. *mBio* 6:e00075. doi: 10.1128/mBio.00075-15

Anderson, M. C., Vonaesch, P., Saffarian, A., Marteyn, B. S., and Sansonetti, P. J. (2017). *Shigella sonnei* encodes a functional T6SS used for interbacterial competition and niche occupancy. *Cell Host Microbe* 21, 769.e3–776.e3. doi: 10.1016/j.chom.2017.05.004

Aschtgen, M. S., Zoued, A., Lloubes, R., Journet, L., and Cascales, E. (2012). The C-tail anchored TssL subunit, an essential protein of the enteroaggregative *Escherichia coli* Sci-1 Type VI secretion system, is inserted by YidC. *Microbiologyopen* 1, 71–82. doi: 10.1002/mbo3.9

Aubert, D. F., Xu, H., Yang, J., Shi, X., Gao, W., Li, L., et al. (2016). A *Burkholderia* type VI effector deamidates Rho GTPases to activate the pyrin inflammasome and trigger inflammation. *Cell Host Microbe* 19, 664–674. doi: 10.1016/j.chom.2016.04.004

Bachmann, V., Kostiuk, B., Unterweger, D., Diaz-Satizabal, L., Ogg, S., and Pukatzki, S. (2015). Bile salts modulate the mucin-activated type VI secretion system of pandemic *Vibrio cholerae*. *PLoS Negl. Trop. Dis.* 9:e0004031. doi: 10.1371/journal.pntd.0004031

Bartonickova, L., Sterzenbach, T., Nell, S., Kops, F., Schulze, J., Venzke, A., et al. (2013). Hcp and VgrG1 are secreted components of the *Helicobacter hepaticus* type VI secretion system and VgrG1 increases the bacterial colitogenic potential. *Cell Microbiol.* 15, 992–1011. doi: 10.1111/cmi.12094

Basler, M., Pilhofer, M., Henderson, G. P., Jensen, G. J., and Mekalanos, J. J. (2012). Type VI secretion requires a dynamic contractile phage tail-like structure. *Nature* 483, 182–186. doi: 10.1038/nature10846

Belkaid, Y., and Hand, T. W. (2014). Role of the microbiota in immunity and inflammation. *Cell* 157, 121–141. doi: 10.1016/j.cell.2014.03.011

Bleumink-Pluym, N. M., van Alphen, L. B., Bouwman, L. I., Wosten, M. M., and van Putten, J. P. (2013). Identification of a functional type VI secretion system in *Campylobacter jejuni* conferring capsule polysaccharide sensitive cytotoxicity. *PLoS Pathog.* 9:e1003393. doi: 10.1371/journal.ppat.1003393

Blondel, C. J., Jimenez, J. C., Leiva, L. E., Alvarez, S. A., Pinto, B. I., Contreras, F., et al. (2013). The type VI secretion system encoded in *Salmonella* pathogenicity island 19 is required for *Salmonella enterica* serotype Gallinarum survival within infected macrophages. *Infect. Immun.* 81, 1207–1220. doi: 10.1128/IAI.01165-12

Blondel, C. J., Yang, H. J., Castro, B., Chiang, S., Toro, C. S., Zaldivar, M., et al. (2010). Contribution of the type VI secretion system encoded in SPI-19 to chicken colonization by *Salmonella enterica* serotypes Gallinarum and Enteritidis. *PLoS One* 5:e11724. doi: 10.1371/journal.pone.0011724

Borgeaud, S., Metzger, L. C., Scrignari, T., and Blokesch, M. (2015). The type VI secretion system of *Vibrio cholerae* fosters horizontal gene transfer. *Science* 347, 63–67. doi: 10.1126/science.1260064

Boyer, F., Fichant, G., Berthod, J., Vandenbrouck, Y., and Attree, I. (2009). Dissecting the bacterial type VI secretion system by a genome wide in silico analysis: what can be learned from available microbial genomic resources? *BMC Genomics* 10:104. doi: 10.1186/1471-2164-10-104

Brunet, Y. R., Espinosa, L., Harchouni, S., Mignot, T., and Cascales, E. (2013). Imaging type VI secretion-mediated bacterial killing. *Cell Rep.* 3, 36–41. doi: 10.1016/j.celrep.2012.11.027

Brunet, Y. R., Zoued, A., Boyer, F., Douzi, B., and Cascales, E. (2015). The Type VI secretion TssEFGK-VgrG phage-like baseplate is recruited to the TssJLM membrane complex via multiple contacts and serves as assembly platform for tail tube/sheath polymerization. *PLoS Genet.* 11:e1005545. doi: 10.1371/journal.pgen.1005545

Cascales, E., and Cambillau, C. (2012). Structural biology of type VI secretion systems. *Philos. Trans. R. Soc. Lond. B Biol. Sci.* 367, 1102–1111. doi: 10.1098/rstb.2011.0209

Casterline, B. W., Hecht, A. L., Choi, V. M., and Bubeck Wardenburg, J. (2017). The *Bacteroides fragilis* pathogenicity island links virulence and strain competition. *Gut Microbes* 8, 374–383. doi: 10.1080/19490976.2017.1290758

Chassaing, B., and Cascales, E. (2018). antibacterial weapons: targeted destruction in the microbiota. *Trends Microbiol.* 26, 329–338. doi: 10.1016/j.tim.2018.01.006

Chatzidaki-Livanis, M., Geva-Zatorsky, N., and Comstock, L. E. (2016). *Bacteroides fragilis* type VI secretion systems use novel effector and immunity proteins to antagonize human gut Bacteroidales species. *Proc. Natl. Acad. Sci. U.S.A.* 113, 3627–3632. doi: 10.1073/pnas.1522510113

Chen, H., Yang, D., Han, F., Tan, J., Zhang, L., Xiao, J., et al. (2017). The bacterial T6SS effector EvpP prevents NLRP3 inflammasome activation by inhibiting the Ca2+-dependent MAPK-Jnk pathway. *Cell Host Microbe* 21, 47–58. doi: 10.1016/j.chom.2016.12.004

Chow, J., and Mazmanian, S. K. (2010). A pathobiont of the microbiota balances host colonization and intestinal inflammation. *Cell Host Microbe* 7, 265–276. doi: 10.1016/j.chom.2010.03.004

Cianfanelli, F. R., Monlezun, L., and Coulthurst, S. J. (2016). Aim, load, fire: the type VI secretion system, a bacterial nanoweapon. *Trends Microbiol.* 24, 51–62. doi: 10.1016/j.tim.2015.10.005

Coulthurst, S. (2019). The Type VI secretion system: a versatile bacterial weapon. *Microbiology* 165, 503–515. doi: 10.1099/mic.0.000789

Coyne, M. J., and Comstock, L. E. (2019). Type VI secretion systems and the gut microbiota. *Microbiol Spectr* 7:PSIB-0009-2018. doi: 10.1128/microbiolspec.PSIB-0009-2018

Coyne, M. J., Roelofs, K. G., and Comstock, L. E. (2016). Type VI secretion systems of human gut Bacteroidales segregate into three genetic architectures, two of which are contained on mobile genetic elements. *BMC Genomics* 17:58. doi: 10.1186/s12864-016-2377-z

Coyne, M. J., Zitomersky, N. L., McGuire, A. M., Earl, A. M., and Comstock, L. E. (2014). Evidence of extensive DNA transfer between bacteroidales species within the human gut. *mBio* 5, e01305–e01314. doi: 10.1128/mBio.01305-14

Croxen, M. A., Law, R. J., Scholz, R., Keeney, K. M., Wlodarska, M., and Finlay, B. B. (2013). Recent advances in understanding enteric pathogenic *Escherichia coli*. *Clin. Microbiol. Rev.* 26, 822–880. doi: 10.1128/CMR.00022-13

de Pace, F., Boldrin de Paiva, J., Nakazato, G., Lancellotti, M., Sircili, M. P., Guedes Stehling, E., et al. (2011). Characterization of IcmF of the type VI secretion system in an avian pathogenic *Escherichia coli* (APEC) strain. *Microbiology* 157(Pt 10), 2954–2962. doi: 10.1099/mic.0.050005-0

de Pace, F., Nakazato, G., Pacheco, A., de Paiva, J. B., Sperandio, V., and da Silveira, W. D. (2010). The type VI secretion system plays a role in type 1 fimbria expression and pathogenesis of an avian pathogenic *Escherichia coli* strain. *Infect. Immun.* 78, 4990–4998. doi: 10.1128/IAI.00531-10

Dong, T. G., Ho, B. T., Yoder-Himes, D. R., and Mekalanos, J. J. (2013). Identification of T6SS-dependent effector and immunity proteins by Tn-seq in *Vibrio cholerae*. *Proc. Natl. Acad. Sci. U.S.A.* 110, 2623–2628. doi: 10.1073/pnas.1222783110

Durand, E., Cambillau, C., Cascales, E., and Journet, L. (2014). VgrG, Tae, Tle, and beyond: the versatile arsenal of Type VI secretion effectors. *Trends Microbiol.* 22, 498–507. doi: 10.1016/j.tim.2014.06.004

Durand, E., Zoued, A., Spinelli, S., Watson, P. J., Aschtgen, M. S., Journet, L., et al. (2012). Structural characterization and oligomerization of the TssL protein, a component shared by bacterial type VI and type IVb secretion systems. *J. Biol. Chem.* 287, 14157–14168. doi: 10.1074/jbc.M111.338731

Faith, J. J., Guruge, J. L., Charbonneau, M., Subramanian, S., Seedorf, H., Goodman, A. L., et al. (2013). The long-term stability of the human gut microbiota. *Science* 341:1237439. doi: 10.1126/science.1237439

Fang, F. C., Frawley, E. R., Tapscott, T., and Vazquez-Torres, A. (2016). Bacterial stress responses during host infection. *Cell Host Microbe* 20, 133–143. doi: 10.1016/j.chom.2016.07.009

Fast, D., Kostiuk, B., Foley, E., and Pukatzki, S. (2018). Commensal pathogen competition impacts host viability. *Proc. Natl. Acad. Sci. U.S.A.* 115, 7099–7104. doi: 10.1073/pnas.1802165115

Felisberto-Rodrigues, C., Durand, E., Aschtgen, M. S., Blangy, S., Ortiz-Lombardia, M., Douzi, B., et al. (2011). Towards a structural comprehension of bacterial type VI secretion systems: characterization of the TssJ-TssM complex of an *Escherichia coli* pathovar. *PLoS Pathog.* 7:e1002386. doi: 10.1371/journal.ppat.1002386

Fu, Y., Waldor, M. K., and Mekalanos, J. J. (2013). Tn-Seq analysis of *Vibrio cholerae* intestinal colonization reveals a role for T6SS-mediated antibacterial activity in the host. *Cell Host Microbe* 14, 652–663. doi: 10.1016/j.chom.2013.11.001

Gallique, M., Decoin, V., Barbey, C., Rosay, T., Feuilloley, M. G., Orange, N., et al. (2017). Contribution of the *Pseudomonas* fluorescens MFE01 Type VI secretion system to biofilm formation. *PLoS One* 12:e0170770. doi: 10.1371/journal.pone.0170770

Garcia-Bayona, L., and Comstock, L. E. (2018). Bacterial antagonism in host-associated microbial communities. *Science* 361:eaat2456. doi: 10.1126/science.aat2456

Gielda, L. M., and DiRita, V. J. (2012). Zinc competition among the intestinal microbiota. *mBio* 3:e00171-12. doi: 10.1128/mBio.00171-12

Guan, J., Xiao, X., Xu, S., Gao, F., Wang, J., Wang, T., et al. (2015). Roles of RpoS in *Yersinia pseudotuberculosis* stress survival, motility, biofilm formation and type VI secretion system expression. *J. Microbiol.* 53, 633–642. doi: 10.1007/s12275-015-0099-6

Gueguen, E., Durand, E., Zhang, X. Y., d'Amalric, Q., Journet, L., and Cascales, E. (2013). Expression of a *Yersinia pseudotuberculosis* type VI secretion system is responsive to envelope stresses through the OmpR transcriptional activator. *PLoS One* 8:e66615. doi: 10.1371/journal.pone.0066615

Gueguen, E., Wills, N. M., Atkins, J. F., and Cascales, E. (2014). Transcriptional frameshifting rescues *Citrobacter rodentium* type VI secretion by the production of two length variants from the prematurely interrupted tssM gene. *PLoS Genet.* 10:e1004869. doi: 10.1371/journal.pgen.1004869

Haapalainen, M., Mosorin, H., Dorati, F., Wu, R. F., Roine, E., Taira, S., et al. (2012). Hcp2, a secreted protein of the phytopathogen *Pseudomonas syringae* pv. tomato DC3000, is required for fitness for competition against bacteria and yeasts. *J. Bacteriol.* 194, 4810–4822. doi: 10.1128/JB.00611-12

Hecht, A. L., Casterline, B. W., Earley, Z. M., Goo, Y. A., Goodlett, D. R., and Bubeck Wardenburg, J. (2016). Strain competition restricts colonization of an enteric pathogen and prevents colitis. *EMBO Rep.* 17, 1281–1291. doi: 10.15252/embr.201642282

Hood, R. D., Singh, P., Hsu, F., Guvener, T., Carl, M. A., Trinidad, R. R., et al. (2010). A type VI secretion system of *Pseudomonas aeruginosa* targets a toxin to bacteria. *Cell Host Microbe* 7, 25–37. doi: 10.1016/j.chom.2009.12.007

Ishikawa, T., Sabharwal, D., Broms, J., Milton, D. L., Sjostedt, A., Uhlin, B. E., et al. (2012). Pathoadaptive conditional regulation of the type VI secretion system in *Vibrio cholerae* O1 strains. *Infect. Immun.* 80, 575–584. doi: 10.1128/IAI.05510-11

Jaakkola, K., Somervuo, P., and Korkeala, H. (2015). Comparative genomic hybridization analysis of *Yersinia enterocolitica* and *Yersinia pseudotuberculosis* identifies genetic traits to elucidate their different ecologies. *Biomed. Res. Int.* 2015:760494. doi: 10.1155/2015/760494

Jani, A. J., and Cotter, P. A. (2010). Type VI secretion: not just for pathogenesis anymore. *Cell Host Microbe* 8, 2–6. doi: 10.1016/j.chom.2010.06.012

Jiang, F., Wang, X., Wang, B., Chen, L., Zhao, Z., Waterfield, N. R., et al. (2016). The *Pseudomonas aeruginosa* type VI secretion PGAP1-like effector induces host autophagy by activating endoplasmic reticulum stress. *Cell Rep.* 16, 1502–1509. doi: 10.1016/j.celrep.2016.07.012

Journet, L., and Cascales, E. (2016). The type VI secretion system in *Escherichia coli* and related species. *EcoSal Plus* 7, 1–20. doi: 10.1128/ecosalplus.ESP-0009-2015

Kitaoka, M., Miyata, S. T., Unterweger, D., and Pukatzki, S. (2011). Antibiotic resistance mechanisms of *Vibrio cholerae*. *J. Med. Microbiol.* 60(Pt 4), 397–407. doi: 10.1099/jmm.0.023051-0

Lertpiriyapong, K., Gamazon, E. R., Feng, Y., Park, D. S., Pang, J., Botka, G., et al. (2012). *Campylobacter jejuni* type VI secretion system: roles in adaptation to deoxycholic acid, host cell adherence, invasion, and in vivo colonization. *PLoS One* 7:e42842. doi: 10.1371/journal.pone.0042842

Lin, J., Zhang, W., Cheng, J., Yang, X., Zhu, K., Wang, Y., et al. (2017). A *Pseudomonas* T6SS effector recruits PQS-containing outer membrane vesicles for iron acquisition. *Nat. Commun.* 8:14888. doi: 10.1038/ncomms14888

Liu, L., Hao, S., Lan, R., Wang, G., Xiao, D., Sun, H., et al. (2015). The type VI secretion system modulates flagellar gene expression and secretion in *Citrobacter freundii* and contributes to adhesion and cytotoxicity to host cells. *Infect. Immun.* 83, 2596–2604. doi: 10.1128/IAI.03071-14

Logan, S. L., Thomas, J., Yan, J., Baker, R. P., Shields, D. S., Xavier, J. B., et al. (2018). The *Vibrio cholerae* type VI secretion system can modulate host intestinal mechanics to displace gut bacterial symbionts. *Proc. Natl. Acad. Sci. U.S.A.* 115, E3779–E3787. doi: 10.1073/pnas.1720133115

Lozupone, C. A., Stombaugh, J. I., Gordon, J. I., Jansson, J. K., and Knight, R. (2012). Diversity, stability and resilience of the human gut microbiota. *Nature* 489, 220–230. doi: 10.1038/nature11550

Ma, A. T., and Mekalanos, J. J. (2010). *In vivo* actin cross-linking induced by *Vibrio cholerae* type VI secretion system is associated with intestinal inflammation. *Proc. Natl. Acad. Sci. U.S.A.* 107, 4365–4370. doi: 10.1073/pnas.0915156107

Ma, J., Bao, Y., Sun, M., Dong, W., Pan, Z., Zhang, W., et al. (2014). Two functional type VI secretion systems in avian pathogenic *Escherichia coli* are involved in different pathogenic pathways. *Infect. Immun.* 82, 3867–3879. doi: 10.1128/IAI.01769-14

Ma, L. S., Hachani, A., Lin, J. S., Filloux, A., and Lai, E. M. (2014). *Agrobacterium tumefaciens* deploys a superfamily of type VI secretion DNase effectors as weapons for interbacterial competition in planta. *Cell Host Microbe* 16, 94–104. doi: 10.1016/j.chom.2014.06.002

Ma, J., Sun, M., Pan, Z., Song, W., Lu, C., and Yao, H. (2018). Three Hcp homologs with divergent extended loop regions exhibit different functions in avian pathogenic *Escherichia coli*. *Emerg. Microbes Infect.* 7:49. doi: 10.1038/s41426-018-0042-0

MacIntyre, D. L., Miyata, S. T., Kitaoka, M., and Pukatzki, S. (2010). The *Vibrio cholerae* type VI secretion system displays antimicrobial properties. *Proc. Natl. Acad. Sci. U.S.A.* 107, 19520–19524. doi: 10.1073/pnas.1012931107

Miyata, S. T., Kitaoka, M., Brooks, T. M., McAuley, S. B., and Pukatzki, S. (2011). Vibrio cholerae requires the type VI secretion system virulence factor VasX to kill *Dictyostelium discoideum*. *Infect. Immun.* 79, 2941–2949. doi: 10.1128/IAI.01266-10

Mulder, D. T., Cooper, C. A., and Coombes, B. K. (2012). Type VI secretion system-associated gene clusters contribute to pathogenesis of *Salmonella enterica serovar Typhimurium*. *Infect. Immun.* 80, 1996–2007. doi: 10.1128/IAI.06205-11

Pan, J., Zhao, M., Huang, Y., Li, J., Liu, X., Ren, Z., et al. (2018). Integration host factor modulates the expression and function of T6SS2 in *Vibrio fluvialis*. *Front. Microbiol.* 9:962. doi: 10.3389/fmicb.2018.00962

Pezoa, D., Blondel, C. J., Silva, C. A., Yang, H. J., Andrews-Polymenis, H., Santiviago, C. A., et al. (2014). Only one of the two type VI secretion systems encoded in the *Salmonella enterica* serotype Dublin genome is involved in colonization of the avian and murine hosts. *Vet. Res.* 45:2. doi: 10.1186/1297-9716-45-2

Pi, H., Jones, S. A., Mercer, L. E., Meador, J. P., Caughron, J. E., Jordan, L., et al. (2012). Role of catecholate siderophores in gram-negative bacterial colonization of the mouse gut. *PLoS One* 7:e50020. doi: 10.1371/journal.pone.0050020

Pukatzki, S., Ma, A. T., Revel, A. T., Sturtevant, D., and Mekalanos, J. J. (2007). Type VI secretion system translocates a phage tail spike-like protein into target cells where it cross-links actin. *Proc. Natl. Acad. Sci. U.S.A.* 104, 15508–15513. doi: 10.1073/pnas.0706532104

Pukatzki, S., Ma, A. T., Sturtevant, D., Krastins, B., Sarracino, D., Nelson, W. C., et al. (2006). Identification of a conserved bacterial protein secretion system in *Vibrio cholerae* using the *Dictyostelium* host model system. *Proc. Natl. Acad. Sci. U.S.A.* 103, 1528–1533. doi: 10.1073/pnas.0510322103

Rogge, M. L., and Thune, R. L. (2011). Regulation of the *Edwardsiella ictaluri* type III secretion system by pH and phosphate concentration through EsrA, EsrB, and EsrC. *Appl. Environ. Microbiol.* 77, 4293–4302. doi: 10.1128/AEM.00195-11

Russell, A. B., Hood, R. D., Bui, N. K., LeRoux, M., Vollmer, W., and Mougous, J. D. (2011). Type VI secretion delivers bacteriolytic effectors to target cells. *Nature* 475, 343–347. doi: 10.1038/nature10244

Russell, A. B., LeRoux, M., Hathazi, K., Agnello, D. M., Ishikawa, T., Wiggins, P. A., et al. (2013). Diverse type VI secretion phospholipases are functionally plastic antibacterial effectors. *Nature* 496, 508–512. doi: 10.1038/nature12074

Russell, A. B., Wexler, A. G., Harding, B. N., Whitney, J. C., Bohn, A. J., Goo, Y. A., et al. (2014). A type VI secretion-related pathway in *Bacteroidetes* mediates interbacterial antagonism. *Cell Host Microbe* 16, 227–236. doi: 10.1016/j.chom.2014.07.007

Salomon, D., Gonzalez, H., Updegraff, B. L., and Orth, K. (2013). *Vibrio parahaemolyticus* type VI secretion system 1 is activated in marine conditions to target bacteria, and is differentially regulated from system 2. *PLoS One* 8:e61086. doi: 10.1371/journal.pone.0061086

Salomon, D., and Orth, K. (2015). Type VI secretion system. *Curr. Biol.* 25, R265–R266. doi: 10.1016/j.cub.2015.02.031

Sana, T. G., Flaugnatti, N., Lugo, K. A., Lam, L. H., Jacobson, A., Baylot, V., et al. (2016). *Salmonella Typhimurium* utilizes a T6SS-mediated antibacterial weapon to establish in the host gut. *Proc. Natl. Acad. Sci. U.S.A.* 113, E5044–E5051. doi: 10.1073/pnas.1608858113

Sana, T. G., Lugo, K. A., and Monack, D. M. (2017). T6SS: the bacterial "fight club" in the host gut. *PLoS Pathog.* 13:e1006325. doi: 10.1371/journal.ppat.1006325

Sheng, L., Gu, D., Wang, Q., Liu, Q., and Zhang, Y. (2012). Quorum sensing and alternative sigma factor RpoN regulate type VI secretion system I (T6SSVA1) in fish pathogen *Vibrio alginolyticus*. *Arch. Microbiol.* 194, 379–390. doi: 10.1007/s00203-011-0780-z

Si, M., Wang, Y., Zhang, B., Zhao, C., Kang, Y., Bai, H., et al. (2017a). The type VI secretion system engages a redox-regulated dual-functional heme transporter for zinc acquisition. *Cell Rep.* 20, 949–959. doi: 10.1016/j.celrep.2017.06.081

Si, M., Zhao, C., Burkinshaw, B., Zhang, B., Wei, D., Wang, Y., et al. (2017b). Manganese scavenging and oxidative stress response mediated by type VI secretion system in *Burkholderia thailandensis*. *Proc. Natl. Acad. Sci. U.S.A.* 114, E2233–E2242. doi: 10.1073/pnas.1614902114

Silverman, J. M., Brunet, Y. R., Cascales, E., and Mougous, J. D. (2012). Structure and regulation of the type VI secretion system. *Annu. Rev. Microbiol.* 66, 453–472. doi: 10.1146/annurev-micro-121809-151619

Sima, F., Stratakos, A. C., Ward, P., Linton, M., Kelly, C., Pinkerton, L., et al. (2018). A novel natural antimicrobial can reduce the *in vitro* and *in vivo* pathogenicity of T6SS positive *Campylobacter jejuni* and *Campylobacter coli* chicken isolates. *Front. Microbiol.* 9:2139. doi: 10.3389/fmicb.2018.02139

Song, Y., Xiao, X., Li, C., Wang, T., Zhao, R., Zhang, W., et al. (2015). The dual transcriptional regulator RovM regulates the expression of AR3- and T6SS4-dependent acid survival systems in response to nutritional status in *Yersinia pseudotuberculosis*. *Environ. Microbiol.* 17, 4631–4645. doi: 10.1111/1462-2920.12996

Sorbara, M. T., and Pamer, E. G. (2019). Interbacterial mechanisms of colonization resistance and the strategies pathogens use to overcome them. *Mucosal Immunol.* 12, 1–9. doi: 10.1038/s41385-018-0053-0

Steele, M. I., Kwong, W. K., Whiteley, M., and Moran, N. A. (2017). Diversification of type VI secretion system toxins reveals ancient antagonism among bee gut microbes. *MBio* 8:e01630-17. doi: 10.1128/mBio.01630-17

Suarez, G., Sierra, J. C., Erova, T. E., Sha, J., Horneman, A. J., and Chopra, A. K. (2010). A type VI secretion system effector protein, VgrG1, from *Aeromonas hydrophila* that induces host cell toxicity by ADP ribosylation of actin. *J. Bacteriol.* 192, 155–168. doi: 10.1128/JB.01260-09

Ting, S. Y., Bosch, D. E., Mangiameli, S. M., Radey, M. C., Huang, S., Park, Y. J., et al. (2018). Bifunctional immunity proteins protect bacteria against FtsZ-targeting ADP-ribosylating toxins. *Cell* 175, 1380.e14–1392.e14. doi: 10.1016/j.cell.2018.09.037

Troxell, B. (2018). A type 6 secretion system (T6SS) encoded gene within *Salmonella enterica* serovar Enteritidis contributes to virulence. *Virulence* 9, 585–587. doi: 10.1080/21505594.2017.1421829

Trunk, K., Peltier, J., Liu, Y. C., Dill, B. D., Walker, L., Gow, N. A. R., et al. (2018). The type VI secretion system deploys antifungal effectors against microbial competitors. *Nat. Microbiol.* 3, 920–931. doi: 10.1038/s41564-018-0191-x

Unterweger, D., Kitaoka, M., Miyata, S. T., Bachmann, V., Brooks, T. M., Moloney, J., et al. (2012). Constitutive type VI secretion system expression gives *Vibrio cholerae* intra- and interspecific competitive advantages. *PLoS One* 7:e48320. doi: 10.1371/journal.pone.0048320

Unterweger, D., Miyata, S. T., Bachmann, V., Brooks, T. M., Mullins, T., Kostiuk, B., et al. (2014). The *Vibrio cholerae* type VI secretion system employs diverse effector modules for intraspecific competition. *Nat. Commun.* 5:3549. doi: 10.1038/ncomms4549

Verster, A. J., Ross, B. D., Radey, M. C., Bao, Y., Goodman, A. L., Mougous, J. D., et al. (2017). The landscape of type VI secretion across human gut microbiomes reveals its role in community composition. *Cell Host Microbe* 22, 411.e14–419.e14. doi: 10.1016/j.chom.2017.08.010

Walter, J., and Ley, R. (2011). The human gut microbiome: ecology and recent evolutionary changes. *Annu. Rev. Microbiol.* 65, 411–429. doi: 10.1146/annurev-micro-090110-102830

Wan, B., Zhang, Q., Ni, J., Li, S., Wen, D., Li, J., et al. (2017). Type VI secretion system contributes to Enterohemorrhagic *Escherichia coli* virulence by secreting catalase against host reactive oxygen species (ROS). *PLoS Pathog.* 13:e1006246. doi: 10.1371/journal.ppat.1006246

Wang, M., Luo, Z., Du, H., Xu, S., Ni, B., Zhang, H., et al. (2011). Molecular characterization of a functional type VI secretion system in *Salmonella enterica* serovar Typhi. *Curr. Microbiol.* 63, 22–31. doi: 10.1007/s00284-011-9935-z

Wang, T., Chen, K., Gao, F., Kang, Y., Chaudhry, M. T., Wang, Z., et al. (2017). ZntR positively regulates T6SS4 expression in *Yersinia pseudotuberculosis*. *J. Microbiol.* 55, 448–456. doi: 10.1007/s12275-017-6540-2

Wang, T., Si, M., Song, Y., Zhu, W., Gao, F., Wang, Y., et al. (2015). Type VI secretion system transports Zn2+ to combat multiple stresses and host immunity. *PLoS Pathog.* 11:e1005020. doi: 10.1371/journal.ppat.1005020

Weber, B., Hasic, M., Chen, C., Wai, S. N., and Milton, D. L. (2009). Type VI secretion modulates quorum sensing and stress response in *Vibrio anguillarum. Environ. Microbiol.* 11, 3018–3028. doi: 10.1111/j.1462-2920.2009.02005.x

Wexler, A. G., Bao, Y., Whitney, J. C., Bobay, L. M., Xavier, J. B., Schofield, W. B., et al. (2016). Human symbionts inject and neutralize antibacterial toxins to persist in the gut. *Proc. Natl. Acad. Sci. U.S.A.* 113, 3639–3644. doi: 10.1073/pnas.1525637113

Wexler, A. G., and Goodman, A. L. (2017). An insider's perspective: *Bacteroides* as a window into the microbiome. *Nat. Microbiol.* 2:17026. doi: 10.1038/nmicrobiol.2017.26

Whitney, J. C., Beck, C. M., Goo, Y. A., Russell, A. B., Harding, B. N., De Leon, J. A., et al. (2014). Genetically distinct pathways guide effector export through the type VI secretion system. *Mol. Microbiol.* 92, 529–542. doi: 10.1111/mmi.12571

Whitney, J. C., Quentin, D., Sawai, S., LeRoux, M., Harding, B. N., Ledvina, H. E., et al. (2015). An interbacterial NAD(P)(+) glycohydrolase toxin requires elongation factor Tu for delivery to target cells. *Cell* 163, 607–619. doi: 10.1016/j.cell.2015.09.027

Wilson, M. M., Anderson, D. E., and Bernstein, H. D. (2015). Analysis of the outer membrane proteome and secretome of *Bacteroides fragilis* reveals a multiplicity of secretion mechanisms. *PLoS One* 10:e0117732. doi: 10.1371/journal.pone.0117732

Yang, X., Song, Y., Dai, Q., Zhang, H., Song, L., Wang, Z., et al. (2019). The stringent response factor, RelA, positively regulates T6SS4 expression through the RovM/RovA pathway in *Yersinia pseudotuberculosis. Microbiol. Res.* 220, 32–41. doi: 10.1016/j.micres.2018.12.002

Yu, Y., Fang, L., Zhang, Y., Sheng, H., and Fang, W. (2015). VgrG2 of type VI secretion system 2 of *Vibrio parahaemolyticus* induces autophagy in macrophages. *Front. Microbiol.* 6:168. doi: 10.3389/fmicb.2015.00168

Yu, Y., Yang, H., Li, J., Zhang, P., Wu, B., Zhu, B., et al. (2012). Putative type VI secretion systems of *Vibrio parahaemolyticus* contribute to adhesion to cultured cell monolayers. *Arch. Microbiol.* 194, 827–835. doi: 10.1007/s00203-012-0816-z

Zhang, H., Gao, Z. Q., Wei, Y., Xu, J. H., and Dong, Y. H. (2013). Insights into the cross-immunity mechanism within effector families of bacteria type VI secretion system from the structure of StTae4-EcTai4 complex. *PLoS One* 8:e73782. doi: 10.1371/journal.pone.0073782

Zhang, W., Wang, Y., Song, Y., Wang, T., Xu, S., Peng, Z., et al. (2013). A type VI secretion system regulated by OmpR in *Yersinia pseudotuberculosis* functions to maintain intracellular pH homeostasis. *Environ. Microbiol.* 15, 557–569. doi: 10.1111/1462-2920.12005

Zhang, W., Xu, S., Li, J., Shen, X., Wang, Y., and Yuan, Z. (2011). Modulation of a thermoregulated type VI secretion system by AHL-dependent quorum sensing in *Yersinia pseudotuberculosis. Arch. Microbiol.* 193, 351–363. doi: 10.1007/s00203-011-0680-2

Zhao, W., Caro, F., Robins, W., and Mekalanos, J. J. (2018). Antagonism toward the intestinal microbiota and its effect on *Vibrio cholerae* virulence. *Science* 359, 210–213. doi: 10.1126/science.aap8775

Zheng, J., Ho, B., and Mekalanos, J. J. (2011). Genetic analysis of anti-amoebae and anti-bacterial activities of the type VI secretion system in *Vibrio cholerae. PLoS One* 6:e23876. doi: 10.1371/journal.pone.0023876

Zhou, Y., Tao, J., Yu, H., Ni, J., Zeng, L., Teng, Q., et al. (2012). Hcp family proteins secreted via the type VI secretion system coordinately regulate *Escherichia coli* K1 interaction with human brain microvascular endothelial cells. *Infect. Immun.* 80, 1243–1251. doi: 10.1128/IAI.05994-11

Zoued, A., Durand, E., Bebeacua, C., Brunet, Y. R., Douzi, B., Cambillau, C., et al. (2013). TssK is a trimeric cytoplasmic protein interacting with components of both phage-like and membrane anchoring complexes of the type VI secretion system. *J. Biol. Chem.* 288, 27031–27041. doi: 10.1074/jbc.M113.499772

3

HpaR, the Repressor of Aromatic Compound Metabolism, Positively Regulates the Expression of T6SS4 to Resist Oxidative Stress in *Yersinia pseudotuberculosis*

Zhuo Wang†, Tietao Wang†, Rui Cui, Zhenxing Zhang, Keqi Chen, Mengyun Li, Yueyue Hua, Huawei Gu, Lei Xu, Yao Wang, Yantao Yang and Xihui Shen**

State Key Laboratory of Crop Stress Biology for Arid Areas, Shaanxi Key Laboratory of Agricultural and Environmental Microbiology, College of Life Sciences, Northwest A&F University, Yangling, China

***Correspondence:**
Yantao Yang
yyt811113@nwsuaf.edu.cn
Xihui Shen
xihuishen@nwsuaf.edu.cn*

HpaR, a MarR family transcriptional regulator, was first identified in *Escherichia coli* W for its regulation of the *hpa-meta* operon. Little else is known regarding its functionality. Here, we report that in *Yersinia pseudotuberculosis*, HpaR negatively regulates the *hpa-meta* operon similar to in *E. coli* W. To investigate additional functions of HpaR, RNA sequencing was performed for both the wild-type and the Δ*hpaR* mutant, which revealed that the type VI secretion system (T6SS) was positively regulated by HpaR. T6SS4 is important for bacteria resisting environmental stress, especially oxidative stress. We demonstrate that HpaR facilitates bacteria resist oxidative stress by upregulating the expression of T6SS4 in *Y. pseudotuberculosis*. HpaR is also involved in biofilm formation, antibiotic resistance, adhesion to eukaryotic cells, and virulence in mice. These results greatly expand our knowledge of the functionality of HpaR and reveal a new pathway that regulates T6SS4.

Keywords: HpaR, type VI secretion system, *Yersinia pseudotuberculosis*, oxidative stress, biofilm, aromatic compounds degradation

INTRODUCTION

HpaR has been characterized as a repressor of the *hpa-meta* operon (Galan et al., 2003). It has been shown that *Escherichia coli* strains B, C, and W can use 4-hydroxyphenylacetic acid (4-HPA) as a carbon source via the *hpa-meta* pathway, while *E. coli* K-12 does not have this ability (Cooper and Skinner, 1980). Studies of the *hpa-meta* cluster have focused mainly on *E. coli* W (Galan et al., 2001). The *hpa-meta* cluster of *E. coli* W is composed of 11 genes in two putative operons (Prieto et al., 1996; Galan et al., 2003). The *upper* operon consists of *hpaBC*, which encodes the two-component 4-HPA monooxygenase, which converts 4-HPA to 3,4-dihydroxyphenylacetic acid (3,4-HPA) (Prieto and Garcia, 1994; Galan et al., 2000). Transcription of the *hpaBC* operon is controlled by HpaA, which belongs to the AraC/XylS family and functions as an activator to activate the expression of *hpaBC*, except for the *meta* operon in the presence of 4-HPA or 3-HPA (Prieto and Garcia, 1997).

The *meta* operon is composed of *hpaGEDFHI*, which encodes enzymes that catalyze 3,4-HPA in the Krebs cycle (Roper et al., 1993; Prieto et al., 1996). The expression of both the *meta* operon and *hpaR* are regulated by HpaR (Galan et al., 2003). HpaR was first identified as a repressor in *E. coli* W and is identical to HpcR in *E. coli* C (Roper et al., 1993; Galan et al., 2003). The *hpa-meta* cluster has also been found in other bacteria. In *Pseudomonas putida* U, the *hpa* pathway is coupled with the *tyn* pathway to degrade tyramine and dopamine (Barbour and Bayly, 1981; Arcos et al., 2010). Together these form the 3,4-HPA catabolon, in which 3,4-HPA is the central intermediate. The *hpa-meta* pathway in *P. putida* U is composed of the genes *hpaRBCIHXFDEG2G1AY*. HpaR of *P. putida* U is similar to that in *E. coli* in that it represses the *meta* operon (Arcos et al., 2010). Interestingly, a second *hpaR* (*hpaY*) was found in *P. putida* U with similar functionality, ensuring stronger control of the *meta* operon (Arcos et al., 2010). In *Burkholderia xenovorans* LB400, the *hpa-meta* pathway is made up of *hpaG1G2EDFHI* and *hpaBC*, while *hpaA* and *hpaX* are absent and *hpaBC* is not adjacent to the *meta* operon (Mendez et al., 2011). HpaR of *B. xenovorans* LB400 also regulates the *meta* operon as a repressor (Mendez et al., 2011). However, previous research has defined HpaR as a repressor that negatively controls expression of the *meta* operon, but not *hpaBC*. According to sequence comparison and a three-dimensional model, HpaR is classified into the MarR transcriptional regulator family, which is involved in various physiological processes (Alekshun et al., 2001; Galan et al., 2003; Grove, 2013). However, whether HpaR might exert other functions in addition to acting as a repressor of the *hpa-meta* operon remains unknown.

Yersinia pseudotuberculosis is a Gram-negative enteric pathogen of animals and humans that causes a variety of diseases such as acute ileitis, mesenteric lymphadenitis, and septicemia (Brubaker, 1991; Smego et al., 1999; Fallman and Gustavsson, 2005). During infection, environmental stress and host immunity reactions can cause an increase in reactive oxygen species (ROS) levels of *Y. pseudotuberculosis* (Green et al., 2016). Elevated cellular ROS levels lead to oxidative stress, which induces oxidative damage to macromolecules such as proteins, lipids, and DNA (D'Autreaux and Toledano, 2007). Protection against the adverse effects of ROS is vital—bacteria have developed a wide range of systems including antioxidant enzymes, i.e., peroxidase, superoxide dismutase, glutaredoxin, and thioredoxin; low molecular weight antioxidants, i.e., the tripeptide glutathione (GSH) and β-carotene; and vitamins, i.e., vitamins C and E (D'Autreaux and Toledano, 2007; Si et al., 2015, 2017a; Staerck et al., 2017). Recently, we found that the type VI secretion system (T6SS) in *Y. pseudotuberculosis* was also involved in resistance to oxidative stress; it secretes a zinc-binding protein that imports zinc to mitigate ROS (Wang et al., 2015).

T6SS is a versatile transmembrane machine used by many Gram-negative bacteria to inject effector proteins into cells, either prokaryotic or eukaryotic, or the extracellular milieu (Durand et al., 2014; Russell et al., 2014; Basler, 2015). Although traditionally T6SS is recognized as a contact-dependent bacterial weapon for interspecies competition, some T6SSs from diverse species are also found to play roles in bacterial pathogenesis, biofilm formation, and stress response (Durand et al., 2014; Russell et al., 2014; Yang et al., 2018). For example, a *Vibrio anguillarum* T6SS regulated by the general stress response regulator RpoS is involved in resistance to hydrogen peroxide, ethanol, and low pH (Weber et al., 2009). Enterohemorrhagic *E. coli* (EHEC) uses its T6SS to deliver KatN, an Mn-containing catalase, into the host cytosol, resulting in reduced levels of intracellular ROS and greater survival of the pathogen (Wan et al., 2017). In *B. thailandensis*, T6SS4 exports TseZ and TseM into the extracellular medium to acquire the antioxidant metal ions Zn^{2+} and Mn^{2+}, respectively, to combat oxidative stress by reducing intracellular ROS levels (Si et al., 2017b,c). The avian pathogenic *E. coli* (APEC) strain TW-XM harbors two functional T6SSs. The first, T6SS1, plays versatile roles in adherence to host cells, biofilm formation, and bacterial competition; the second, T6SS2, is responsible only for cerebral infection (Ma et al., 2014; Navarro-Garcia et al., 2019).

Despite being reported to control expression of the *hpa-meta* operon, knowledge of HpaR in *Y. pseudotuberculosis* is lacking. Unexpectedly, in this study, we found that HpaR not only acts as a repressor of the *hpa-meta* operon in a similar manner to that of *E. coli* but also acts as an activator to upregulate the expression of T6SS4 to acquire Zn^{2+} and resist oxidative stress. HpaR is also involved in biofilm formation, antibiotic resistance, adhesion to eukaryotic cells, and virulence in mice. These results greatly expand our current knowledge of the functions of HpaR.

MATERIALS AND METHODS

Bacterial Strains and Growth Conditions

Bacterial strains and plasmids used in this study are listed in **Supplementary Table S1**. *Escherichia coli* strains were cultured at 37°C in Lysogeny Broth (LB) or LB plates. *Y. pseudotuberculosis* strains were grown in Yersinia-Lysogeny-Broth (YLB) broth (tryptone 1%, yeast extract 0.5%, NaCl 0.5%) or YLB plates at 30 or 26°C. The *Y. pseudotuberculosis strain* YPIII was the parent of all derivatives used in this study. To construct the *hpaR* in-frame deletion mutant, the wild-type *Y. pseudotuberculosis* was mated with *E. coli* S17-1λpir carrying pDM4-*hpaR* and chromosomal integration was selected by plating on YLB agar plates supplemented with nalidixic acid and chloramphenicol. The *hpaR* deletion mutant was subsequently screened on YLB agar plates with 20% sucrose and confirmed by polymerase chain reaction (PCR) and DNA sequencing. Appropriate antibiotics were included in growth medium at the following concentrations: ampicillin, 100 μg/ml; kanamycin, 50 μg/ml; nalidixic acid, 20 μg/ml; chloramphenicol, 20 μg/ml.

Plasmid Construction

Primers used in this study are listed in **Supplementary Table S2**. The *lacZ* fusion reporter vector pDM4-T6SS4*p:lacZ* was made in previous study (Zhang et al., 2013). To obtain the expression plasmid, *hpaR*-F-*Bam*HI/*hpaR*-R-*Sal*I primer pair was used to amplify *hpaR* gene fragment from *Y. pseudotuberculosis* genomic DNA by PCR. The *hpaR* gene fragment were digested with *Bam*HI/*Sal*I and then inserted into similar digested pET28a yielding the pET28a-*hpaR*. The suicide plasmid pDM4-*hpaR* used to construct the *hpaR* mutant was prepared by overlap PCR. Briefly, *hpaR*-UF-*Xba*I/*hpaR*-UR and *hpaR*-DF/*hpaR*-DR-*Spe*I primers were used to amplify the 1000 bp upstream fragment and 1000 bp downstream fragment of *hpaR*, respectively. Then, the amplified fragments were fused together by overlap PCR with *hpaR*-UF-*Xba*I/*hpaR*-DR-*Spe*I. The fused PCR products were digested with *Xba*I/*Spe*I and cloned into similar digested pDM4 resulting pDM4-*hpaR*. In order to construct the complementary plasmid pKT100-*hpaR*, *hpaR*-F-*Bam*HI/*hpaR*-R-*Sal*I primers were used to amplify *hpaR* gene from template. The PCR product was digested with *Bam*HI/*Sal*I and inserted into similarly digested pKT100 to produce pKT100-*hpaR*, which is subsequently electroporated into Δ*hpaR* to construct the complementation strains. To construct the *hpaG1* promoter reporter vector, *hpaG1p*-F-*Sal*I/*hpaG1p*-R-*Xba*I primer pair was used to amplify the 387 bp promoter fragment from *Y. pseudotuberculosis*. The product of PCR was digested with *Sal*I/*Xba*I and cloned into similar digested pDM4-T6SS4*p:lacZ* to produce pDM4-*hpaG1p:lacZ*. To construct the T6SS4 promoter fragments with HpaR binding site mutations, overlap PCR was performed to replace the HpaR binding site with identical amount of irrelevant base pairs. T6SS4pM_{HpaR}-R and T6SS4pM_{HpaR}-F were designed to contain 22 bp overlapping DNA fragment (ATTTGTTAGATTCCGAACCGTC) used to replace the HpaR binding site. T6SS4*p*-F-*Sal*I/T6SS4pM_{HpaR}-R and T6SS4pM_{HpaR}-F/T6SS4*p*-R-*Xba*I primers used to amplify the up-fragment and down-fragment of T6SS4*p* promoter, respectively. The PCR products were ligated with T6SS4*p*-F-*Sal*I/T6SS4*p*-R-*Xba*I by overlap PCR to produce the mutant promoter fragment T6SS4pM_{HpaR}. The T6SS4pM_{HpaR} fragment were digested with *Sal*I/*Xba*I and ligated into similarly digested pDM4-T6SS4*p:lacZ* to produce pDM4-T6SS4pM_{HpaR}*:lacZ*. The validity of all the plasmids constructed above was confirmed by DNA sequencing.

Sequence Data Analysis

Sequence alignment was performed as previously described (Shen et al., 2005; Huang et al., 2008). The sequence of *hpaRGEDFHIXABC* (Accession No. Z37980.2) of *E. coli* W were retrieved from GenBank. Sequence comparisons were carried out using BLAST program at National Center for Biotechnology Information website[1]. Multiple protein sequence alignment was made by CLUSTAL W. The sequences used in alignment have been deposited in the GenBank database [Accession No. EcC *E. coli* C (S56952.1), EcW *E. coli* W (Z37980.2), Yptb *Y. pseudotuberculosis* YPIII (ACA68739.1), PpU *P. putida* U (FJ904934.1), BxLB400 *B. xenovorans* LB400 (ABE33958.1)]. The result was exported by ESPript[2].

Construction of Chromosomal Fusion Reporter Strains and β-Galactosidase Activity Assay

The *lacZ* fusion reporter vector pDM4-T6SS4*p:lacZ*, pDM4-T6SS4pM_{HpaR}*:lacZ* and pDM4-*hpaG1p:lacZ* was transformed into *E. coli* S17-1λpir and then introduced into *Y. pseudotuberculosis* by conjugation as described (Xu et al., 2014). All constructed *lacZ* fusion reporter strains were grown in YLB broth at 26°C and *o*-nitrophenyl-β-galactoside (ONPG) as substrate for measuring β-galactosidase activities as described by Miller (1992). The β-galactosidase results shown represent the mean of one representative assay performed in triplicate, and error bars represent standard deviation. Statistical analysis was carried out with Student's *t*-test.

Protein Expression and Purification

To express and purify His_6- tagged HpaR, the constructed expression vector pET28a-*hpaR* were transformed into BL21(DE3). Single colony was cultured in 5 ml LB broth at 37°C overnight and diluted 100-fold into 500 ml LB. Until the OD_{600} = 0.5, the culture was shifted to 22°C, induced with 0.3 mM IPTG and further cultivated for 12 h to express the recombinant proteins. Cell pellet was collected by centrifugation, washed and resuspended in His binding buffer, and lysed by sonication. The recombinant proteins were purified with the His·Bind Ni-NTA resin (Novagen) according to the manufacturer's instructions. Eluted recombinant proteins were dialyzed against the appropriate buffer at 4°C for 4 h and then stored at −80°C until used.

Electrophoretic Mobility Shift Assay (EMSA)

EMSA was performed as previously described (Zhang et al., 2013; Wang et al., 2015). Bio-T6SS4p_{HpaR}-F/Bio-T6SS4p_{HpaR}-R primers were used to amplify the biotin 5′-end-labeled promoter probes (Bio-T6SS4p_{HpaR}) from *Y. pseudotuberculosis* genomic DNA. The unlabeled T6SS4p_{HpaR} was amplified with T6SS4p_{HpaR}-F/T6SS4p_{HpaR}-R from template, which used as a competitor. In addition, the unrelated protein BSA was used as negative control. All promoter probes were purified by EasyPure Quick Gel Extraction Kit (TransGen Biotech). According to the manufacturer's protocol (LightShift Chemiluminescent EMSA Kit; Thermo Fisher Scientific), each 20 μl EMSA reaction solution was prepared as follows: 1 × binding buffer, 50 ng poly (dI-dC), 2.5% glycerol, 0.05% Nonidet P-40, 5 mM $MgCl_2$, 3 ng Bio-T6SS4p_{HpaR} DNA, 1 ng T6SS4p_{HpaR} DNA as competitor, and different concentration of protein (0, 0.6, 0.8, 0.8 μg). Reaction solutions were incubated for 20 min at room temperature. The samples were loaded onto a 6% polyacrylamide native gel and transferred to a Biodyne B nylon membrane (Thermo Fisher Scientific). The biotin-labeled DNA bands were visualized by chemiluminescent substrate according to the manufacturer's protocol. *hpaG1p*-F/*hpaG1p*-R primers were used to amplify the *hpaG1p* probes from template. An unrelated DNA (URD) in the similar length or bovine serum albumin (BSA) was included in the binding assay system to serve as negative controls. Increasing

[1]https://www.ncbi.nlm.nih.gov/

[2]http://espript.ibcp.fr/ESPript/cgi-bin/ESPript.cgi

concentrations of purified His6-HpaR (0, 0.4, 0.8, 1.2, 1.2 μg) were incubated with 20 ng *hpaG1p* probes in EMSA buffer (20 mM Tris-HCl [pH 7.4], 4 mM $MgCl_2$, 100 mM NaCl, 1 mM dithiothreitol, 10% glycerol). After incubation for 20 min at 26°C, the sample was subjected to electrophoresis on a 6% polyacrylamide native gel. Then the DNA probe was detected using SYBR green.

Stress Survival Assay

Yersinia pseudotuberculosis strains grown in YLB broth until mid-exponential are diluted 50-fold into M9 medium (Na_2HPO_4, 6 g/L; KH_2PO_4, 3 g/L; NaCl, 0.5 g/L; NH_4Cl, 1 g/L; $MgSO_4$, 1 mM; $CaCl_2$, 0.1 mM; glucose 0.2%) with or without H_2O_2 (1.5 mM), cumene hydroperoxide (CHP, 0.5 mM), ampicillin (0.5 μg/ml), gentamicin (0.2 μg/ml) at 26°C for 1 h at 100 rpm. After treatment, the cultures were diluted 1000-fold and plated onto YLB agar plates with nalidixic acid. Colonies were counted after 24 h at 26°C. The survival percentage of *Y. pseudotuberculosis* is calculated by dividing the number of CFU of stressed cells by the number of CFU of cells without stress. All these assays were performed in triplicate at least three times.

Fluorescence Dye-Based Intracellular ROS Detection

The intracellular ROS levels were detected by the fluorescent reporter dye 5-(and-6)-chloromethyl-2′,7′-dichlorodihydro fluorescein diacetate, acetylester (CM-H_2DCFDA, Invitrogen) as previously reported (Si et al., 2017c). Briefly, 1 ml culture were collected after treatment, washed with M9 medium and resuspended in 1 ml of M9 medium containing 10 μM CM-H_2DCFDA. Samples were incubated at 26°C in the dark for 20 min. Then the cells were pelleted, washed with M9 medium and resuspended in 1 ml of PBS. Then 200 μl samples were transferred to a dark 96-well plate. Fluorescence signals were measured using a SpectraMax M2 Plate Reader (Molecular Devices) with excitation/emission wavelengths of 495/520 nm. The results shown represented the mean of one representative assay performed in triplicate, and error bars represent the standard deviation (SD).

Protein Secretion Assay

Secretion of YezP (encoded by *ypk_3459*) was detected as previously described (Wang et al., 2015). Briefly, all strains were grown in 180 ml YLB broth at 26°C on a rotary shaker (220 rpm) until OD_{600} = 1.6. A 2 ml culture was centrifuged and the cell pellet was resuspended in 100 μl SDS-loading buffer to serve as the total cell pellet sample. A total of 150 ml culture was centrifuged at low speed (5000 rpm) for 10 min to remove most cell pellets, and the supernatant was further centrifuged at high speed (8000 rpm) for 15 min to eliminate bacteria. To remove the remaining bacteria, the supernatant was filtered through a 0.22-μm filter (Millipore, Billerica, MA, United States). The supernatant was filtered three times through a nitrocellulose filter (BA85) (Whatman) to collect all secretion proteins. The filter was cut into pieces in 1.5 ml tubes and resuspended in 100 μl SDS-loading buffer for 20 min at 65°C to recover the proteins. All samples were normalized to the OD_{600} of the culture and volume used in preparation.

Western Blot Analysis

Western blot analysis was performed as described previously (Shen et al., 2009; Lin et al., 2017). Samples were resuspended in SDS-Loading buffer and separated in 15% polyacrylamide gel, then transferred onto PVDF membranes (Millipore). The membrane was blocked in 5% (w/v) BSA for 4 h at room temperature, and incubated with primary antibody at 4°C overnight. The primary antibodies were anti-VSVG (sc-365019, Santa Cruz biotechnology, United States), 1:1000 and anti-ICDH, 1:6000. The ICDH antisera were made in our previous study (Xu et al., 2010). The membrane was washed five times with TBST buffer (50 mM Tris, 150 mM NaCl, 0.05% Tween 20, pH 7.4), and incubated with 1:5,000 dilution of horseradish peroxidase conjugated secondary antibody (Shanghai Genomics) for 4 h at 4°C. Signals were detected using the ECL plus kit (GE Healthcare, Piscataway, NJ, United States) according to the manufacturer's specified protocol.

DNase I Footprinting Assay

DNase I footprinting assay was performed as described preciously (Zhao et al., 2017). Briefly, *hpaG1p* and T6SS4*p* were amplified by PCR with *hpaG1p*-FP-F/*hpaG1p*-FP-R and T6SS4*p*-FP-F/T6SS4*p*-FP-R primers, respectively. The PCR products were cloned into the pMD-18T vector (TaKaRa) as template to further amplify the fluorescent FAM labeled probes with primers M13R(FAM-labeled) and M13F(-47). The FAM-labeled probes were purified by the Wizard SV Gel and PCRClean-Up System (Promega) and quantified with NanoDrop 2000C (Thermo). For DNase I footprinting assay, 400 ng probes were incubated with various amount of His_6-HpaR in 40 μl of each sample. After binding at 30°C for 30 min, 10 μl solution containing about 0.010 unit DNase I (Promega) and 100 nmol $CaCl_2$ were added and further incubated for 1 min at 25°C. 140 μl DNase I stop solution (200 mM unbuffered sodium acetate, 30 mM EDTA and 0.15% SDS) was added into samples to stop the reaction. Samples were extracted by phenol/chloroform to remove proteins, precipitated with ethanol, washed and dissolved in 35 μl MiniQ water. The preparation of the DNA ladder, electrophoresis, and data analysis were similar to described previously (Wang et al., 2012).

RNA-seq Experiments

Yersinia pseudotuberculosis and the Δ*hpaR* mutant were used for RNA-Seq transcriptomics analysis (Wang et al., 2017). Two strains (three biological replicates) were grown in YLB medium at 26°C to OD_{600} = 1.6. Total RNA was extracted from each sample for cDNA library construction by using bacteria total RNA isolation kit (Tiangen) and analyzed with the Bioanalyzer 2100 system (Agilent Technologies). Library sequencing was completed at the Beijing Genomic Institute (BGI-Shenzhen). The result of sequencing was aligned with the reference genome of *Y. pseudotuberculosis* and RPKM (Reads per kilobase transcriptome per million mapped reads) was used to normalize the expression level of genes. The differential expressed genes were shown as fold change calculated by $\log_2$ (RPKM of Δ*hpaR*/WT).

Quantitative Real-Time PCR

All strains were cultured at mid-exponential, harvested, washed with PBS. Total RNA was isolated from each sample using the RNAprep Pure Cell/Bacteria Kit, treated with RNase-free DNase (Tiangen) and measured the purity and concentration by gel electrophoresis and spectrophotometer (NanoDrop, Thermo Scientific). 1.5 μg of total RNA was used to synthesis the first-strand cDNA by reverse transcriptase (TransGen Biotech). Quantitative real-time PCR (qRT-PCR) was performed in CFX96 Real-Time PCR Detection System (Bio-Rad) with TransStart Green qPCR SuperMix (TransGen Biotech). The qRT-PCR primers were listed in **Supplementary Table S2** and qRT-PCR parameter was set as follows: 95°C for 30 s followed by 40 cycles of 94°C for 15 s, 52°C for 30 s. The relative abundance of 16S rRNA was used as the internal standard for standardization of results.

Determination of Intracellular Ion Content

Intracellular ion content was determined as described previously (Wang et al., 2015). Briefly, *Y. pseudotuberculosis* wild-type, Δ*hpaR* and Δ*hpaR* (*hpaR*) strains were grown in YLB medium to mid-exponential. 20 ml culture were collected, washed with PBS, resuspended in 20 ml PBS buffer containing 0.4% glucose, 1.5 mM H_2O_2 and 1 μM Zn^{2+}, and then incubated further for 20 min at 26°C. After centrifugation at low speed (4000 rpm) for 10 min, the supernatant was removed and the cell pellet weight was measured. Then the pellet was resuspended and chemically lysed on a rotating mixer at a slow setting for 20 min by using Bugbuster (Novagen, Madison, WI, United States) according to the manufacturer's instructions. The protein concentration was analyzed with NanoDrop ND-1000 spectrophotometer (NanoDrop Technologies) according to the manufacturer's instructions. The lysis solution was diluted 100-fold in 2% molecular grade nitric acid to 10 ml. These samples were analyzed by Inductively Coupled Plasma Mass Spectrometry (ICP-MS) (Varian 802-MS), and the results were corrected using the appropriate buffers for reference and dilution factors. Triplicate cultures of each strain were analyzed during a single experiment and the experiment was repeated at least three times.

Biofilm Formation Assay

Biofilm formation was measured using the test tube, performed as previously described (Guan et al., 2015). *Y. pseudotuberculosis* strains were grown in YLB broth, and then transferred into 3 ml of M9 medium containing 0.4% glucose on a rotary shaker (220 rpm) at 26°C. After 24 h, the test tube was washed gently three times with PBS and stained with 1% crystal violet for 15 min, then washed three times with PBS again to remove redundant crystal violet. The remaining crystal violet was resolved with 95% ethanol and 200 μl of samples were measured the absorbance at 595 nm by a microplate reader (BioTek Instruments, Inc.).

Cell Adhesion Assay

The adhesion ability of bacteria was measured as previously reported (Tan et al., 2017). HeLa cells were seeded into 12-well plate at a concentration of 1×10^5 cells/ml before infection. *Y. pseudotuberculosis* wild-type, Δ*hpaR* and Δ*hpaR* (pKT100-*hpaR*) strains were grown at 26°C till OD_{600} = 1.0 and then collected, washed and resuspended in DMEM. The bacterial suspensions were serial diluted and plated onto YLB agar plates to count the actual number of the bacteria. HeLa cells were infected with bacterial strains at a multiplicity of infection (MOI) of 100 and then centrifuged for 5 min at 1500 rpm to promote contact between bacteria and HeLa cells. After 2 h, HeLa cells were washed for three times using prewarmed PBS. Cell-associated bacteria were liberated by treatment with 0.1% Triton X-100 in sterile PBS for 5 min. The number of cell-associated bacteria was divided by the total number of bacteria in a well to calculate the adhesion percentage.

Mouse Infection

All mice were maintained and handled in accordance with the Regulations for the Administration of Affairs Concerning Experimental Animals approved by the State Council of People's Republic of China. The protocol was approved by the Animal Welfare and Research Ethics Committee of Northwest A&F University (protocol number: NWAFUSM2018001). All strains growing in YLB broth at 26°C until mid-exponential were harvested, washed, resuspended in sterilized PBS to a final concentration of 3×10^9 bacteria for survival assays. The 6–8 weeks old female C57BL/6 mice were orogastric infected with *Y. pseudotuberculosis* using a ball-tipped feeding needle. The survival rate of the mice was determined by monitoring the survival everyday for 24 days.

Statistical Analysis

Statistical analyses of gene expression data, LacZ activity, survival assay, ROS determination, intracellular ion content determination, and cell adhesion assay were performed using paired two-tailed Student's *t*-test. Statistical analyses were performed using GraphPad Prism software (GraphPad Software).

RESULTS

Regulation of the *hpa* Gene Cluster in *Y. pseudotuberculosis*

The 4-hydroxyphenylacetic acid (4-HPA) catabolic pathway was determined by alignment with *E. coli* W (ATCC 11105) while searching for the complete genome sequence of *Y. pseudotuberculosis* (**Figure 1A**). Analyses of the open reading frames of the gene cluster suggested that the 4-HPA metabolic cluster is composed of 11 genes in *Y. pseudotuberculosis*, similar to that of *E. coli* W (Roper et al., 1993; Prieto and Garcia, 1994). These genes include the genes of 4-HPA monooxygenase (*hpaBC*), the genes of homoprotocatechuate (HPC) *meta* cleavage enzymes (*hpaG1G2EDFHI*), the HPA transport protein gene (*hpaX*), and one regulatory gene (*hpaR*).

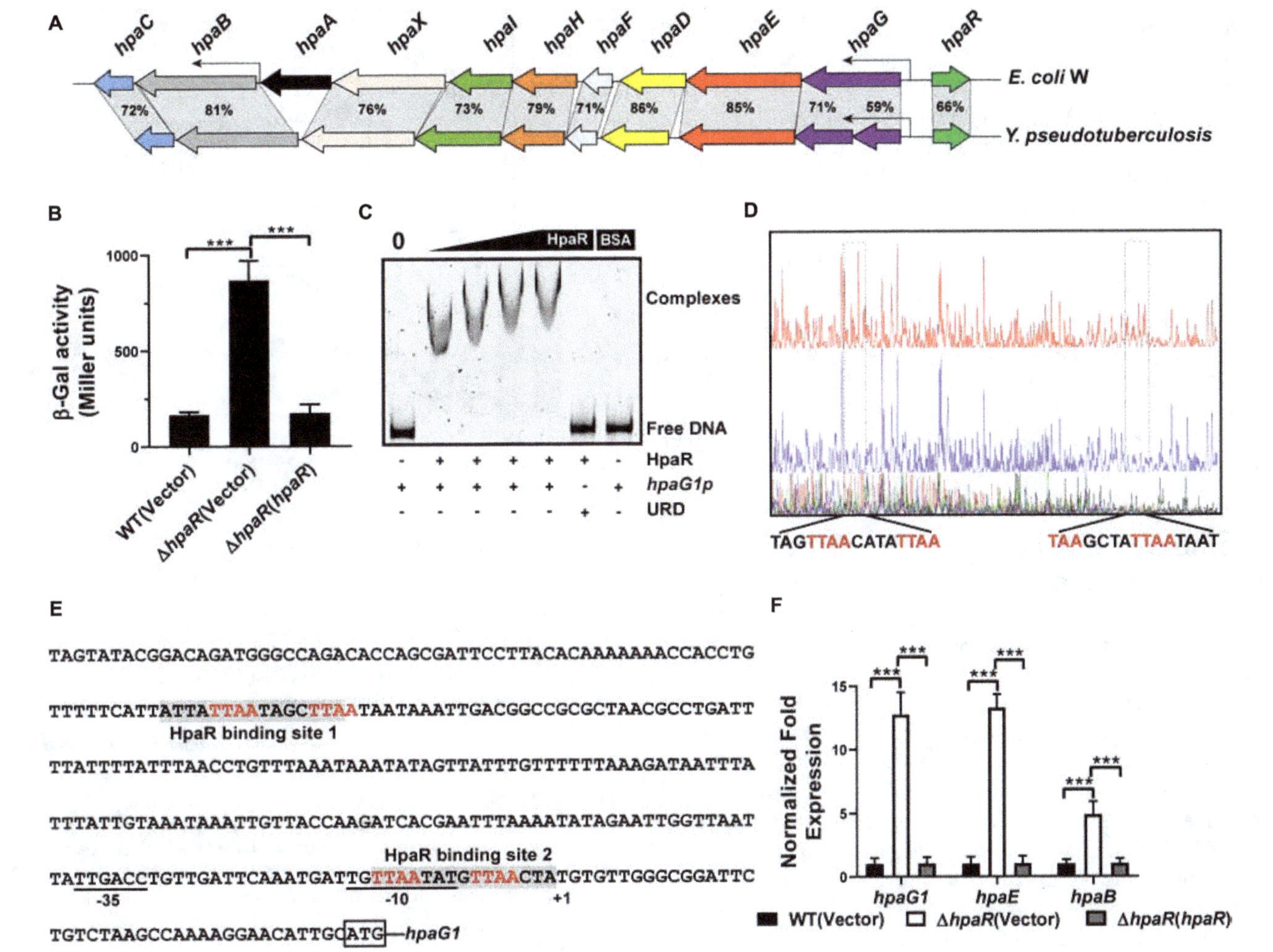

FIGURE 1 | HpaR negatively regulates *hpa-meta* operon in *Yersinia pseudotuberculosis*. **(A)** Gene organization of *hpa-meta* cluster and identities of protein sequence in *Escherichia coli* W and *Y. pseudotuberculosis*. **(B)** β-Galactosidase activity of the *hpaG1* promoter in *Y. pseudotuberculosis* Wild-type, Δ*hpaR* mutant, and its complementary strain Δ*hpaR*(*hpaR*). Data shown are the average of three independent experiments. The error bars indicate standard deviations of the average. $^{***}P < 0.001$. **(C)** EMSA was performed to analyze the interactions between His_6-HpaR and the *hpaG1* promoter. URD, unrelated DNA fragment. **(D)** Identification of the HpaR-binding site within the *hpaG1* promoter using the DNase I footprinting assay. Letters in red denote inverted repeat sequences. **(E)** The HpaR binding sites detected by DNase I footprinting assay were shown in the *hpaG1* promoter region by shading. Letters in red denote inverted repeat sequences. **(F)** Relative expression levels of *hpaG1*, *hpaE*, and *hpaB* were measured by quantitative RT-PCR in wild-type, Δ*hpaR* mutant, and complementary strain Δ*hpaR*(*hpaR*).

All genes (*hpaG1G2EDFHIXBC*) lay in a continuous row in the same direction as in *E. coli*. The translation products of these genes were identified as the homologs of the 4-HPA hydroxylases in *E. coli* W with high identities (**Figure 1A**). HpaBC, the HPA monooxygenase, oxidizes 4-HPA or 3-HPA to HPC. HpaD (HPC 2,3-dioxygenase), HpaE (CHMS dehydrogenase), HpaF (CHM isomerase), HpaG (OPET decarboxylase), HpaH (hydratase), and *Hpa*I (HHED aldolase) form the HPA catabolic pathway, which degrades HPC in the pyruvic acid or succinic semialdehyde pathways. The HpaR protein shares 66, 66, 48, and 34% amino acid sequence identities with identified HpaR in *E. coli* C, *E. coli* W, *P. putida* U, and *B. xenovorans* LB400, respectively (**Supplementary Figure S1**). Thus, we propose the gene cluster from *ypk_2452* to *ypk_2462* as an *hpa-meta* gene cluster in *Y. pseudotuberculosis.*

In *E. coli* W, the *meta* cleavage enzymes (HpaGEDFHI) that catabolize 4-HPA are repressed by HpaR, and the HPA monooxygenase (HpaBC) is positively regulated by HpaA (Prieto and Garcia, 1997). To verify the role of HpaR in expression of the *hpa* gene cluster in *Y. pseudotuberculosis*, a single-copy *hpaG1p:lacZ* fusion reporter was introduced into the chromosomes of wild-type, Δ*hpaR* mutant, and complemented Δ*hpaR*(*hpaR*) strains. The β-galactosidase activity of the three strains was quantitatively measured (**Figure 1B**). The deletion of *hpaR* significantly increased the activity of the *hpa* promoter, which could be fully restored by introducing a complementary plasmid expressing HpaR (**Figure 1B**). Next, we analyzed the interaction between HpaR and the putative operator with an electrophoresis mobility shift assay (EMSA). Incubation of HpaR with a probe containing the promoter of the *hpa-meta* operon (*hpaG1p*) led to the formation of DNA–protein complexes (**Figure 1C**), which suggests direct interaction of HpaR and the *hpaG1p* promoter. DNase I footprinting analysis further revealed that the putative HpaR binding site was protected from digestion in DNA–HpaR complexes. Two HpaR-protected sites upstream of the transcriptional start site in the *hpaG1p* promoter were identified (**Figures 1D,E**). Thus, HpaR specifically recognizes an operator within the *hpa-meta* promoter, likely influencing its expression. Indeed, negative regulation of the *hpa-meta* operon by HpaR was confirmed by qRT-PCR analyses. The data indicated that the expression of *hpaG1*, *hpaE*, and *hpaB* was upregulated in Δ*hpaR*, and such an increase could be restored to wild-type levels by complementation with the expression plasmid pKT100-*hpaR* (**Figure 1F**). Overall, in *Y. pseudotuberculosis*, HpaR functions as a repressor and negatively regulates the *hpa-meta* gene cluster in a similar manner to *E. coli* W.

Genome-Wide Analysis of the Genes Regulated by HpaR

The multiple antibiotic resistance regulator (MarR) family transcriptional regulator has a variety of biological functions, including resistance to multiple antibiotics and other toxic chemicals such as organic solvents, household disinfectants, and oxidative stress agents (Alekshun and Levy, 1999; Wei et al., 2007). As a MarR family regulator, we speculate that HpaR may play other regulatory roles besides aromatic degradation. Thus, RNA sequencing (RNA-seq) was employed to detect the genes regulated by HpaR, except the *hpa-meta* gene cluster.

To identify the HpaR-dependent expression genes in *Y. pseudotuberculosis*, RNA was extracted for RNA-seq from both wild-type and Δ*hpaR* mutant strains in mid-exponential phase (SRA accession: PRJNA612340). After RNA-seq, quality control and gene expression level analysis were performed by fragments per kilobase of transcript per million mapped reads (FPKM). A total of 106 gene candidates that exhibited differential expression were found between wild-type and Δ*hpaR* mutant strains, including 57 upregulated and 49 downregulated genes (**Supplementary Table S3**). To verify the RNA-seq data, qRT-PCR analysis of 11 representative genes was conducted. The log_2-transformed values of each gene were in good agreement with the RNA-seq data (**Figure 2A**), which confirmed the credibility of the RNA-seq data. Subsequently, the functions of differentially expressed genes were identified by KEGG pathway analysis. As shown in **Figure 2B**, 17 different pathways were found in the DGE (Different Gene Expression), in which eight pathways were up-regulated and nine were down-regulated. These included degradation of aromatic compounds, metabolism, microbial metabolism in a diverse environment, purine metabolism, bacterial secretion system, and tyrosine metabolism. In the RNA-seq data, two gene clusters were found to be fully regulated by HpaR. One cluster was the *hpa-meta* operon, which was 2.3–4.56-fold upregulated in the Δ*hpaR* mutant (**Table 1**). The other was the T6SS4 gene cluster in *Y. pseudotuberculosis.* T6SS is a newly described secretion system in many Gram-negative bacteria (Bingle et al., 2008; Jani and Cotter, 2010). The RNA-seq data revealed that the genes in the T6SS4 operon were 0.54–0.96-fold downregulated in the Δ*hpaR* mutant (**Table 2**). Hemolysin co-regulated protein (Hcp), which forms the hexameric rings for the T6SS syringe needle, was 0.68-fold downregulated. Valine-glycine repeat protein G (VgrG) was 0.54-fold downregulated. The periplasmic domain protein IcmF was 0.62-fold downregulated. ImpA/B (VipA/VipB), which serve as tail sheath proteins in T6SS, were 0.73/0.94-fold downregulated. The same was true with other structural genes. These data suggest that the *hpa-meta* operon was negatively regulated by HpaR, consistent with the transcription of chromosomal *hpaG1p:lacZ* fusions. The data also suggest that T6SS4 is positively regulated by the transcriptional regulator HpaR.

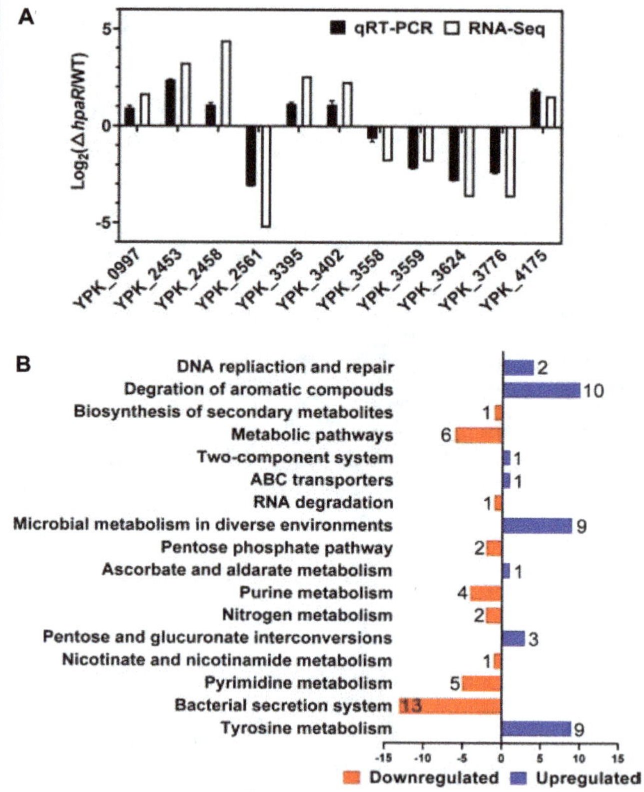

FIGURE 2 | Transcriptomics analysis of HpaR regulated genes in *Y. pseudotuberculosis*. **(A)** Validation of RNA-Seq data using qRT-PCR. Eleven representative genes were evaluated for validation of the RNA-seq data using qRT-PCR. **(B)** KEGG pathway analysis of differentially expressed genes (*hpaR* mutant VS wild-type). The red and blue bars represent up- and down-regulated genes, respectively, and the numeric labels represent the number of genes with corresponding function.

HpaR Positively Regulates T6SS4 Expression by Directly Binding to Its Promoter

Based on the RNA-seq data, we found HpaR could positively regulate the expression of T6SS4. To further determine the role of HpaR in the expression of T6SS4, a single copy T6SS4*p:lacZ* fusion was introduced into the chromosomes of wild-type, *hpaR*

TABLE 1 | The *hpa-meta* operon genes regulated by HpaR in *Y. pseudotuberculosis*.

Locus tag	Product	Gene description	Fold change*
YPK_2452	HpaC	4-Hydroxyphenylacetate-3-monooxygenase small chain	2.31
YPK_2453	HpaB	4-Hydroxyphenylacetate-3-monooxygenase large chain	3.24
YPK_2454	HpaX	4-Hydroxyphenylacetate permease	3.37
YPK_2455	HpaI	2,4-Dihydroxyhept-2-ene-1,7-dioic acid aldolase	3.83
YPK_2456	HpaH	2-Oxo-hept-3-ene-1,7-dioate hydratase	3.91
YPK_2457	HpaF	5-Carboxymethyl-2-hydroxymuconate isomerase	4.25
YPK_2458	HpaD	3,4-Dihydroxyphenylacetate 2,3-dioxygenase	4.42
YPK_2459	HpaE	5-Carboxymethyl-2-hydroxymuconic-semialdehyde dehydrogenase	4.45
YPK_2460	HpaG2	2-Hydroxyhepta-2,4-diene-1,7-dioate isomerase	4.56
YPK_2461	HpaG1	2-Hydroxyhepta-2,4-diene-1,7-dioate isomerase	4.50

*Fold change: $log_2(\Delta hpaR/WT)$.

TABLE 2 | T6SS4 genes regulated by HpaR in *Y. pseudotuberculosis*.

Locus tag	Product	Gene description	Fold change*
YPK_3550	ImpL	Type VI secretion system protein ImpL	−1.52
YPK_3551	ImpK	Type VI secretion system protein ImpK	−1.53
YPK_3552	ImpJ	Type VI secretion system protein ImpJ	−1.63
YPK_3556		Pentapeptide repeat protein	−1.74
YPK_3557		Pentapeptide repeat protein	−1.77
YPK_3558	VgrG	Type VI secretion system secreted protein VgrG	−1.78
YPK_3559	VasG	Type VI secretion system protein VasG	−1.70
YPK_3560	ImpH	Type VI secretion system protein ImpH	−1.52
YPK_3561	ImpG	Type VI secretion system protein ImpG	−1.79
YPK_3564	ImpC	Type VI secretion system protein ImpC	−1.77
YPK_3565	ImpB	Type VI secretion system protein ImpB	−1.69
YPK_3566	ImpA	Type VI secretion system protein ImpA	−1.90

*Fold change: $log_2(\Delta hpaR/WT)$.

mutant, and complemented Δ*hpaR*(*hpaR*) strains and LacZ activity of the resulting strains was quantitatively measured (**Figure 3A**). Compared to the wild-type strain, the T6SS4*p:lacZ* fusion was decreased significantly in Δ*hpaR*; this decrease could be restored by the complementary plasmid (pKT100-*hpaR*). This suggests that HpaR positively regulates T6SS4 expression. To test whether the T6SS4 gene clusters were regulated by HpaR directly, the interaction between HpaR and the T6SS4 promoter was examined by EMSA. Incubation of a probe containing the T6SS4 promoter with HpaR led to the formation of DNA–protein complexes (**Figure 3C**). DNase I footprinting analysis revealed that the binding site was protected from digestion in DNA–HpaR complexes, further indicating the recognition of this DNA element by HpaR (**Figure 3D**). An HpaR-protected site (CCTCTTATTTTGGCTATTCATCCACGTCATCGTGCTA) upstream of the -35 and -10 elements was identified (**Figure 3E**).

The interactions between HpaR and the T6SS4 promoter appeared to be site-specific, as mutation of the HpaR binding site in the T6SS4 promoter prevented formation of the protein–DNA complex (**Figure 3C**) and abolished HpaR dependent T6SS4 regulation (**Figure 3B**). The positive regulation of T6SS4 by HpaR was further confirmed by qRT-PCR analysis (**Figure 3F**). The data indicated that the expression of T6SS4 components *clpV4*, *hcp4*, and *vgrG4* were reduced in Δ*hpaR*, which could be restored by plasmid (pKT100-*hpaR*). Thus, HpaR recognizes and binds to the promoter of T6SS4 specifically to activate expression.

HpaR Affects the Antioxidant Activity of *Y. pseudotuberculosis* by Regulating T6SS4

In our previous study, T6SS4 of *Y. pseudotuberculosis* was discovered to play important roles in multiple stress defenses, especially oxidative stress (Wang et al., 2015). The regulation of T6SS4 by HpaR prompted us to examine whether HpaR plays a role in oxidative stress resistance via T6SS4. To test the hypothesis that HpaR exerts antioxidant activity by regulating T6SS4, the survival rate and ROS level in cells treated with cumene hydroperoxide (CHP) or hydrogen peroxide (H_2O_2) were measured. After incubation with CHP, approximately 60% of wild-type cells survived. The survival rate of the mutant lacking *hpaR* decreased to 16%, and a 13% survival rate was observed for the double mutant Δ*hpaR*Δ*clpV4*. Expressing HpaR could fully rescue such a decrease in the Δ*hpaR*, but only increased the survival rate slightly in the double mutant Δ*hpaR*Δ*clpV4* (**Figure 4A**). The ROS level under oxidative stress was also examined using CM-H_2DCFDA fluorescent dye. The data showed that the ROS level in Δ*hpaR* and Δ*hpaR*Δ*clpV4*

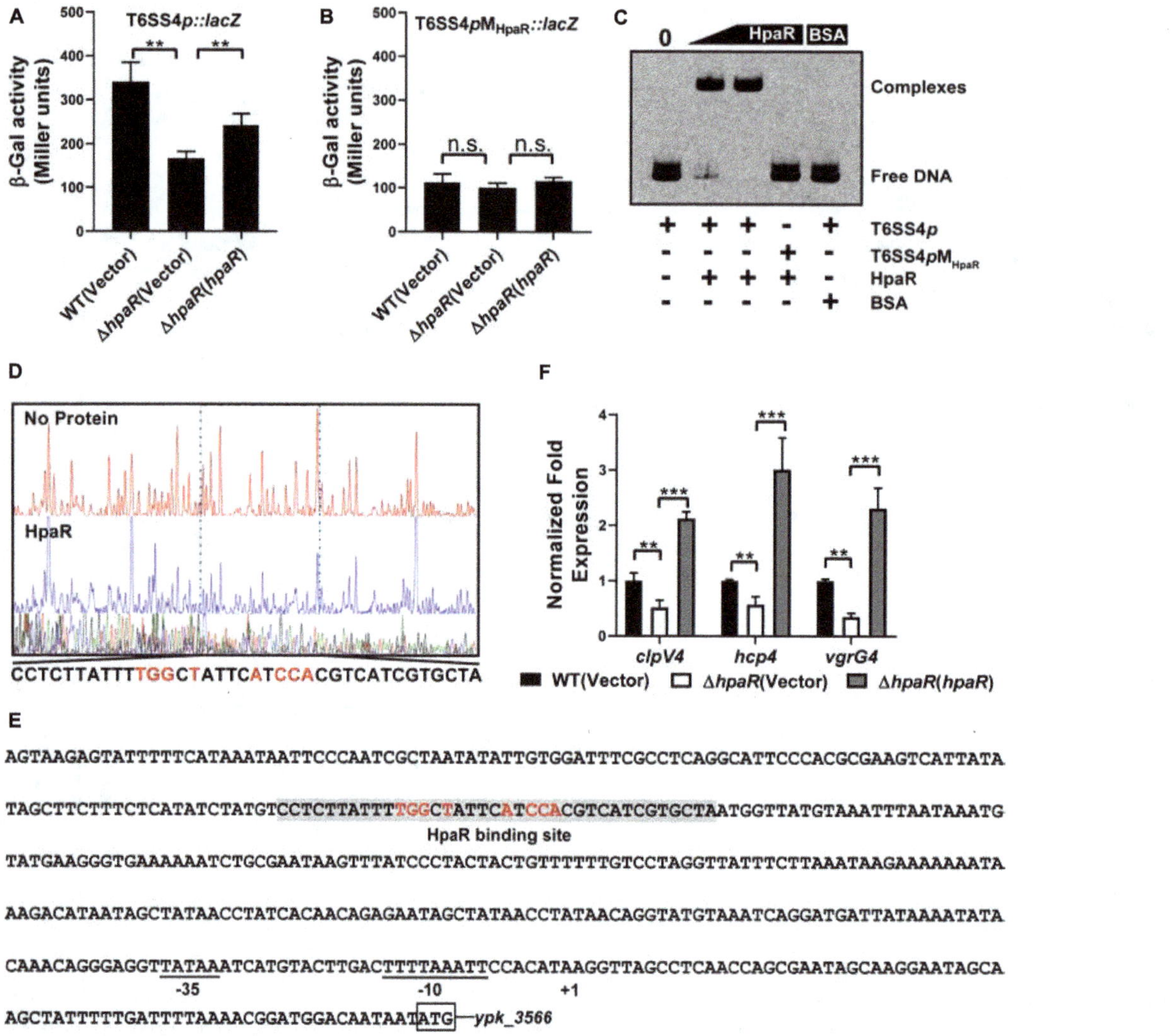

FIGURE 3 | HpaR positively regulates the T6SS4 expression in *Y. pseudotuberculosis*. β-galactosidase activity of the T6SS4 promoter **(A)** and T6SS4 promoter with mutations in the HpaR binding site **(B)** in *Y. pseudotuberculosis* Wild-type, Δ*hpaR* mutant, and its complementary strain Δ*hpaR*(*hpaR*). Data shown are the average of three independent experiments. The error bars indicate standard deviations of the average. $^{**}P < 0.01$; n.s., not significant. **(C)** EMSA was performed to analyze the interactions between His_6-HpaR and the T6SS4 promoter (T6SS4*p*) or the promoter mutant identified His_6-HpaR binding region (T6SS4*p*M_{HpaR}).
(D) Identification of the HpaR-binding site within the T6SS4 promoter using the DNase I footprinting assay. Letters in red denote inverted repeat sequences. **(E)** The HpaR binding site detected by DNase I footprinting assay were shown in the T6SS4 promoter region by shading. Letters in red denote inverted repeat sequences. **(F)** Relative expression levels of *clpV4*, *hcp4,* and *vgrG4* were measured by qRT-PCR. The error bars indicate the standard deviation from three independent experiments. $^{***}P < 0.001$; $^{**}P < 0.01$.

was higher than the wild-type strain. Complementation with the plasmid pKT100-*hpaR* could reduce the ROS level in Δ*hpaR*, but not in the Δ*hpaR*Δ*clpv4* double mutant (**Figure 4B**). Similar results were obtained with H_2O_2 treatment (**Figures 4C,D**). T6SS4 could secrete the zinc-binding protein YezP for zinc import in *Y. pseudotuberculosis* (Wang et al., 2015). Thus, YezP-VSVG was expressed in relevant *Y. pseudotuberculosis* strains to test the secretion of YezP. Significant amounts of YezP-VSVG could be readily detected in culture supernatant from wild-type bacteria. The secretion of YezP decreased by 32% in the Δ*hpaR* mutant and was completely restored in a mutant complemented with HpaR (**Figure 4E**). The zinc contents in bacteria challenged with H_2O_2 were measured using inductively coupled plasma mass spectrometry (ICP-MS). The results revealed that the deletion of *hpaR* significantly lowered the intracellular Zn^{2+} level and expression of HpaR restored the defects (**Figure 4F**). These results established that HpaR benefited *Y. pseudotuberculosis* resisting oxidative stress by upregulating T6SS4.

HpaR Contributes to Biofilm Formation and Adhesion to Epithelial Cells

Similar to other Gram-negative bacteria, *Y. pseudotuberculosis* is capable of forming biofilms on abiotic surfaces. The biofilm formation ability of wild-type, Δ*hpaR* mutant, and complemented Δ*hpaR*(*hpaR*) strains were observed in a test tube

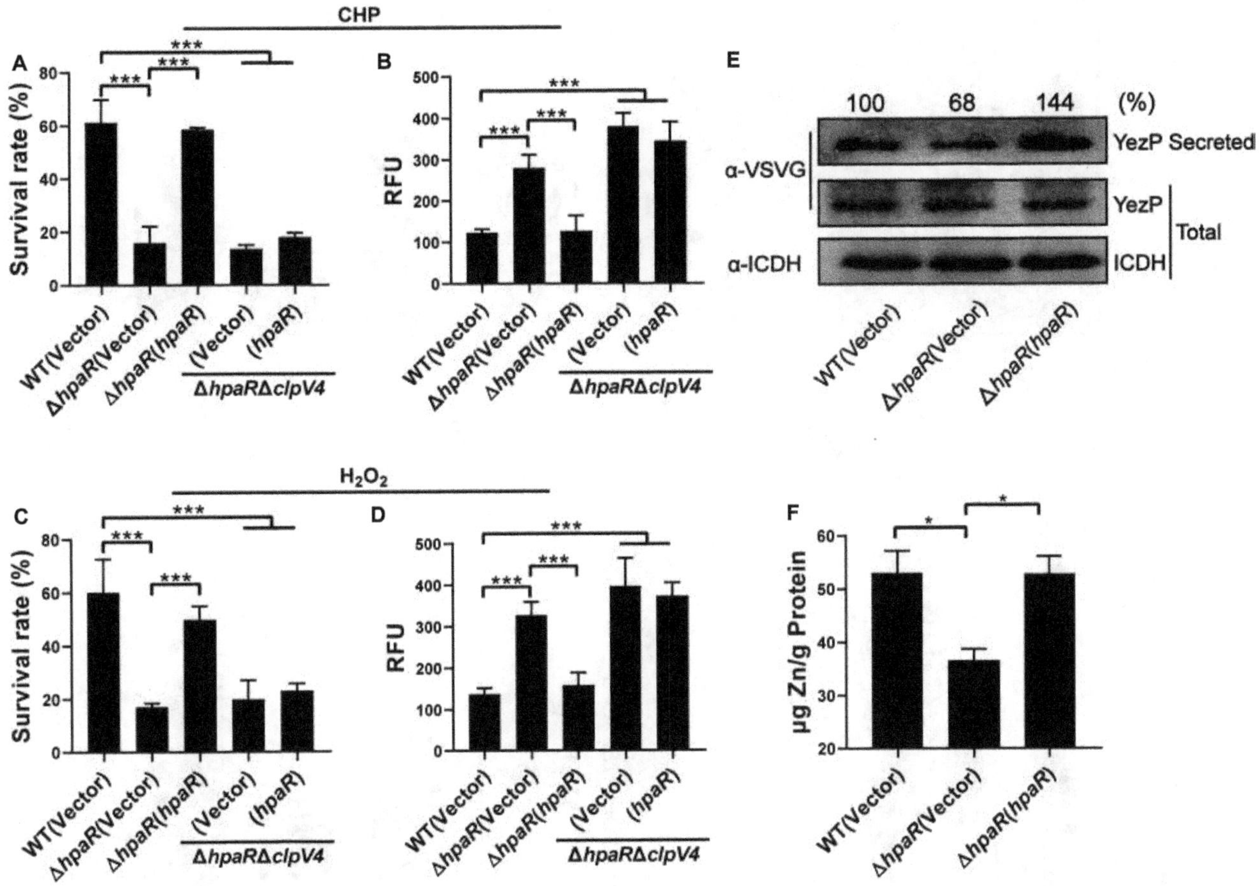

FIGURE 4 | HpaR plays a role in resisting oxidative stress via T6SS4. Survival rate of indicated *Y. pseudotuberculosis* strains were determined after treatment with CHP **(A)** or H_2O_2 **(C)** for 1 h. **(B,D)** Intracellular ROS of indicated strains were stained with H_2DCFDA dye (Invitrogen) and fluorescence signals were measured using a SpectraMax M2 Plate Reader (Molecular Devices) with excitation/emission wavelengths of 495/520. **(E)** Western blot detection of YezP secretion in *Y. pseudotuberculosis* wild-type, Δ*hpaR* mutant, and its complementary strain Δ*hpaR*(*hpaR*). Proteins in culture supernatant of relevant *Y. pseudotuberculosis* strains expressing YezP-VSVG were probed for VSVG (upper) by immunoblotting. **(F)** Mid-exponential phase of *Y. pseudotuberculosis* strains were exposed to 1.5 mM H_2O_2 for 30 min in PBS containing 1 μM $ZnCl_2$. Zn^{2+} associated with bacterial cells was measured by inductively coupled plasma mass spectrometry (ICP-MS). Data shown are the average of three independent experiments. The error bars indicate the standard deviation from three independent experiments. ***$P < 0.001$; **$P < 0.01$; *$P < 0.05$.

by staining with crystal violet (**Figure 5A**) and quantified by resolving with 95% ethanol and measuring the absorbance at 595 nm (**Figure 5B**). The Δ*hpaR* mutant showed attenuated biofilm formation when compared to wild-type and the decrease could be restored by expressing *hpaR*. Note that there's no significant difference between the growth activities of these strains (**Supplementary Figure S2**). Biofilm formation plays an important role in the pathogen lifecycle. It is imperative in chronic infection, antibiotic resistance, and adhesion to host cells. Thus, ampicillin and gentamicin, two commonly used antibiotics in clinics, were used to verify viability of the *hpaR* mutant in antibiotic resistance. The survival rates of the Δ*hpaR* strain were only 34 and 18% following gentamicin and ampicillin treatment, respectively, while approximately 72 and 51% of wild-type cells survived under the same conditions (**Figure 5C**). This suggests that the Δ*hpaR* mutant was more sensitive to antibiotics than the wild-type. Adhesion to host cells is a crucial step in the infection process and host colonization. Thus, the adherence of wild-type, *hpaR* mutant, and complemented strains to HeLa cells was evaluated. As shown in **Figure 5D**, the Δ*hpaR* strain had fewer adherences to HeLa cells than wild-type. Overall, *hpaR* may assist in the maintenance of biofilm formation activity, antibiotic resistance, and adhesion to eukaryotic cells.

Y. pseudotuberculosis Mutant Lacking *hpaR* Is Defective in Virulence in Mice

Biofilm formation is helpful for the survival and colonization of *Y. pseudotuberculosis* in hosts. In addition, oxidative burst is an important microbial-killing mechanism in phagocytes. The findings that HpaR affects the antioxidant activity, biofilm formation and the ability of adhesion to eukaryotic cells point to its role in the virulence of *Y. pseudotuberculosis.* To examine this, C57BL/6 mice were inoculated orogastrically with relevant bacterial strains. Wild-type bacteria caused more than 90% lethality within 3 weeks of inoculation. By contrast,

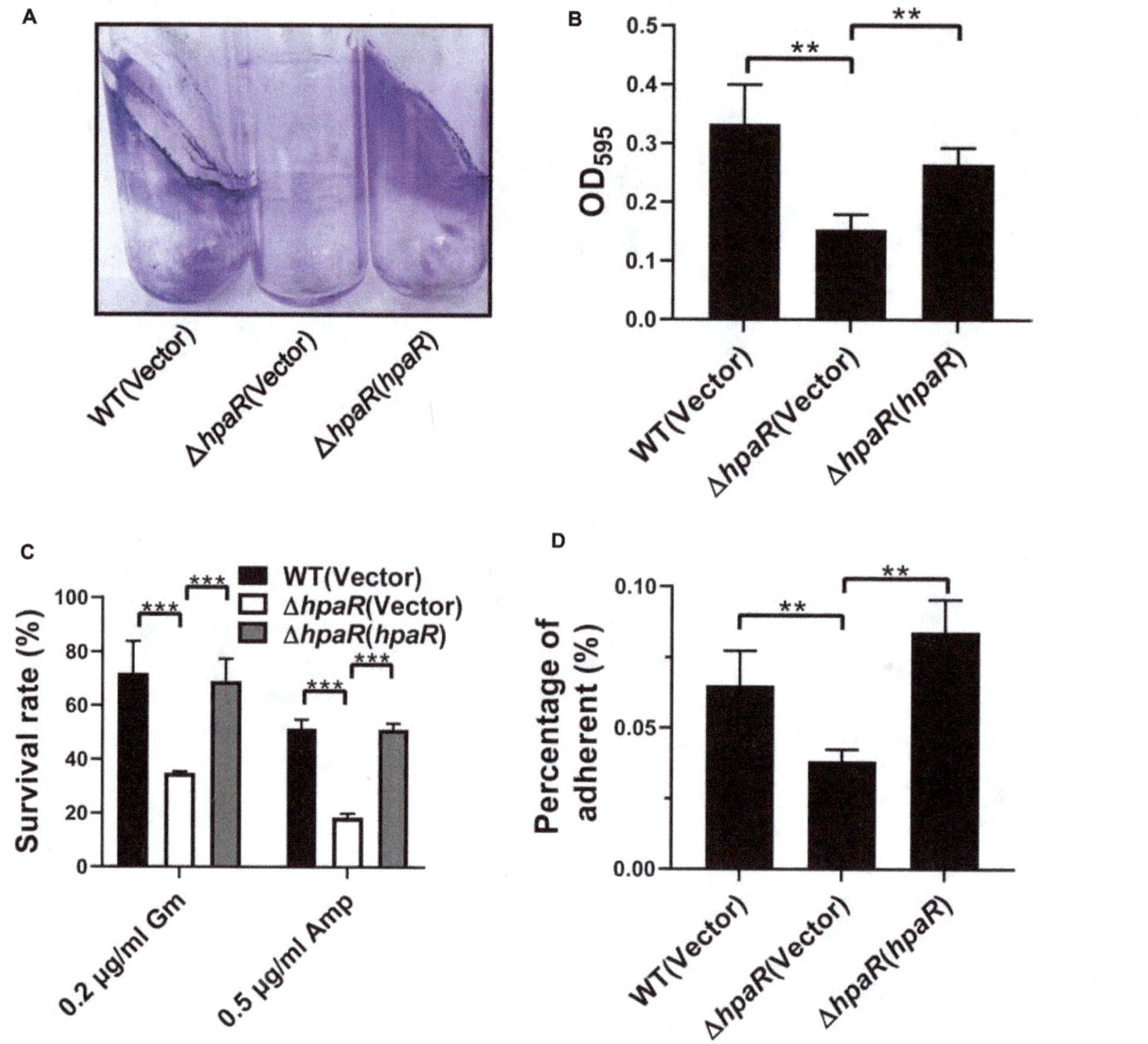

FIGURE 5 | HpaR is involved in biofilm formation, antibiotic resistance and adhesion to eukaryotic cells. **(A)** Direct observation of *Y. pseudotuberculosis* biofilm formation by Wild-type, Δ*hpaR* mutant, and its complementary strain Δ*hpaR* (*hpaR*). The results shown the biofilm biomass stained with crystal violet in glass test tubes. **(B)** Quantitative measurement of biofilm formation by resolving in 95% ethanol and measuring the absorbance at OD_{595}. **(C)** The survival rate of Wild-type, Δ*hpaR* mutant, and its complementary strain Δ*hpaR* (*hpaR*) under 0.2 μg/ml gentamicin (Gm) or 0.5 μg/ml ampicillin (Amp) treatment. **(D)** Adhesion of the strains to HeLa cells. Adherence level is expressed as the percentage of adherent bacteria relative to the inoculum count. Data shown are the average of three independent experiments; the error bars indicate the standard deviation from three independent experiments. ***$P < 0.001$; **$P < 0.01$; *$P < 0.05$; n.s., not significant.

consistent with the hypothesis, mice infected with mutants lacking *hpaR* or T6SS4 had better survival rates than the wild-type (**Figure 6**). The data indicate that *hpaR* affects the virulence of *Y. pseudotuberculosis* in the infection of mammalian hosts.

DISCUSSION

MarR family transcriptional regulators are widely distributed in bacteria and archaea (Perez-Rueda and Collado-Vides, 2001; Perez-Rueda et al., 2004). They play crucial roles in the bacterial lifecycle as they control various genes involved in antibiotic resistance, aromatic compound catabolism, oxidative stress responses, and produce of virulence factors (Alekshun and Levy, 1999; Fiorentino et al., 2007; Deochand and Grove, 2017). Some of these regulators are involved in the degradation of specific aromatic compounds (Diaz and Prieto, 2000). For example, BadR in *Rhodopseudomonas palustris* regulates anaerobic benzoate degradation in cooperation with AadR (Egland and Harwood, 1999). In *Comamonas testosteroni* BR60, CbaR controls the *cbaABC* operon for converting 3-chlorobenzoate to protocatechuate (PCA) and 5-Cl-PCA (Providenti and Wyndham, 2001). HpaR was first identified in *E. coli* W (Roper et al., 1993). As a repressor, HpaR regulates

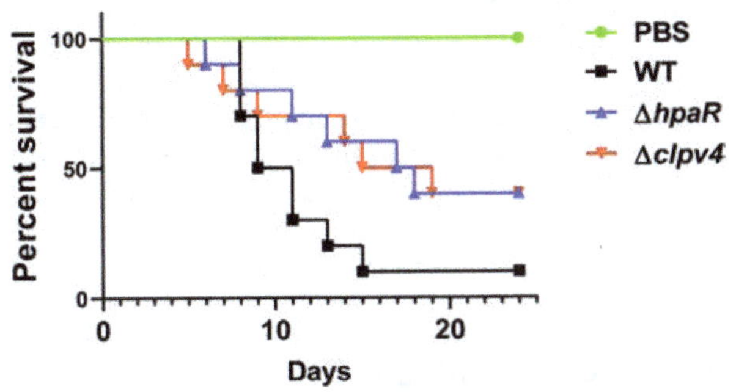

FIGURE 6 | Yptb mutants lacking *hpaR* or T6SS4 are defective in virulence against mice. Bacterial strains grown in YLB were washed twice in sterilized PBS and used for orogastric infection of 6–8 weeks old female C57BL/6 mice using a ball-tipped feeding needle. For survival assays 3×10^9 bacteria of each strain were applied to different groups of mice (n = 10/strain), and the survival rate of the mice was determined by monitoring the survival daily for 24 days.

the *hpa-meta* operon in 4-HPA catabolism (Prieto et al., 1996; Galan et al., 2003). However, whether HpaR has other functions remains unclear. Here, we report that in *Y. pseudotuberculosis*, HpaR acts as a dual-functional regulator. It not only negatively regulates the *hpa-meta* cluster but also positively controls T6SS4 expression to resist oxidative stress, which maintains cellular ROS levels.

The *hpa-meta* cluster in *E. coli* W is composed of 11 genes within two operons, *hpaBC* and *hpaGEDFHI*, which are controlled by HpaA and HpaR, respectively (Galan et al., 2001). Based on sequence alignment with *E. coli* W, we identified the *hpa-meta* cluster in *Y. pseudotuberculosis*. An HpaA homolog was not identified in *Y. pseudotuberculosis*. The *hpa-meta* cluster of one operon consisted of 11 genes, *hpaG1G2EDFHIXBC*, and one regulator, HpaR (**Figure 1A**). Using a β-galactosidase activity assay and qRT-PCR, we revealed that HpaR represses expression of the *hpaG1G2EDFHIXBC* operon in *Y. pseudotuberculosis*, similar to *E. coli* W. HpaR binding sites 1 and 2, both composed of a 15 bp region, were identified on the *hpaG1* promoter containing a palindromic sequence of 4 bp (TTAA-XXXX-TTAA) on each side separated by 4 bp similar to *E. coli* W (**Figure 1E**). Site 1 was located -189 bp relative to the transcription start site + 1 of the *hpaG1* promote. Site 2 was centered at position -8 of the *hpaG1* promoter, overlapping the -10 element, which implies that HpaR inhibits the binding of RNA polymerase on the promoter to repress transcription of the *hpa-meta* cluster.

So far HpaR was specifically recognized as a repressor of the 4-HPA catabolic pathway (Galan et al., 2003; Arcos et al., 2010; Mendez et al., 2011). Unexpectedly, in the RNA-seq based transcriptomics analysis we found that in addition to the *hpa-meta* cluster genes, genes of the T6SS4 cluster were also found to be regulated by HpaR. Based on the β-galactosidase activity assay and qRT-PCR results, we confirmed that HpaR acts as an activator rather than a repressor to regulate T6SS4 (**Figure 3A**). We also identified the HpaR binding site on the T6SS4 promoter located -241 bp relative to the transcription start site + 1 of the T6SS4 promote, containing an imperfect invert sequence TGGXT-XXXX-AXCCA on each side separated by 4 bp (**Figure 3E**). The HpaR binding site of *hpaG1* and T6SS4 promoter both contain a 4–5 bp palindromic sequence on each side separated by 4 bp but the site on T6SS4 promoter is imperfect. The binding site of T6SS4 promoter is far from the

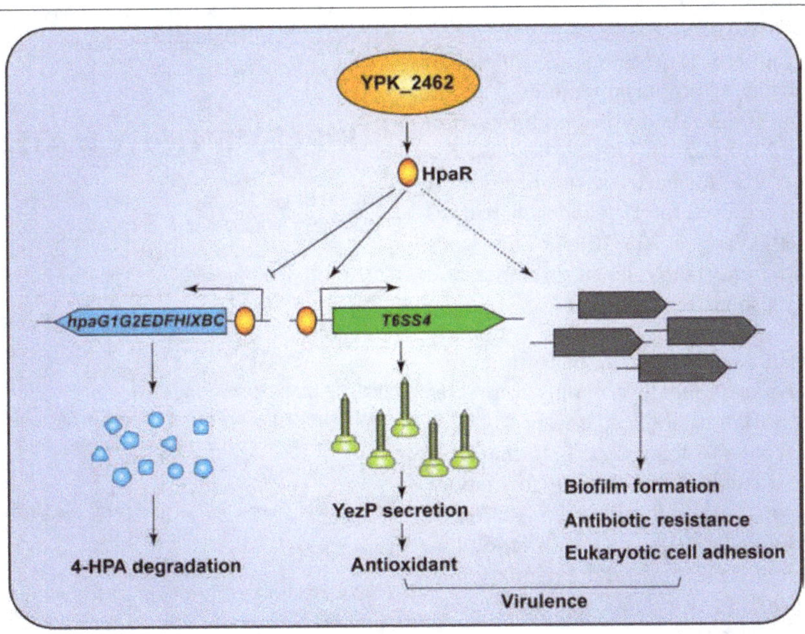

FIGURE 7 | Hypothetical model for the function of HpaR in *Y. pseudotuberculosis*. HpaR represses the 4-HPA catabolic pathway but activates the transcription of T6SS4 to maintain ROS levels in response to oxidative stress. HpaR also affects the biofilm formation, eukaryotic cell adhension, antibiotic resistance, and virulence of *Y. pseudotuberculosis*.

transcription start site. Besides, the GC content of binding site on T6SS4 promoter is higher than that on *hpaG1* promoter. We presume that HpaR may bind to the stable site upstream of the -35 and -10 elements recruiting RNA polymerase to activate the transcription of T6SS4.

T6SS has been recognized as an anti-bacterial weapon to attack target cells in a contact-dependent manner (Durand et al., 2014; Ho et al., 2014; Russell et al., 2014). However, our previous report showed that the T6SS4 in *Y. pseudotuberculosis* is related to resistance to multiple environmental stresses, including high osmolarity, low pH, and oxidative stress (Song et al., 2015; Wang et al., 2015). We therefore speculated that HpaR might protect against oxidative stress via T6SS4. And we confirmed this by observing the survival rate and determining the intracellular ROS levels of wild-type and mutant strains under oxidative stress (**Figures 4A–D**). *Y. pseudotuberculosis* could import Zn^{2+} via secretion of YezP through T6SS4 to maintain intracellular ROS levels (Wang et al., 2015). Our results showed that the levels of YezP secreted by T6SS4 and the intracellular Zn^{2+} levels of Δ*hpaR* were also decreased under H_2O_2 stress (**Figures 4E,F**), indicating that HpaR regulates the expression of T6SS4 to resist oxidative stress.

Biofilm formation was also affected by HpaR (**Figures 5A,B**). However, no other biofilm-related gene was found to be differentially expressed in the RNA-seq results. Biofilm formation is complex—it may be indirectly regulated by other mechanisms that will require further investigation. Biofilm formation is beneficial to bacterial survival in a harmful environment and is associated with adhesion to eukaryotic cells (Flemming and Wingender, 2010; Hall and Mah, 2017). Our results also showed that HpaR was involved in antibiotic resistance and adhered to HeLa cells (**Figures 5C,D**). *Y. pseudotuberculosis* is an enteric pathogen of animals and humans (Smego et al., 1999). In host cells, the oxidative burst is an important method to kill bacteria and biofilm formation is essential for bacterial infection (Flemming and Wingender, 2010; Slauch, 2011). Besides, the adhesion to host cell is also crucial for bacterial colonization. It has been shown that T6SS4 is required for *Y. pseudotuberculosis* virulence in a mouse model (Wang et al., 2015). This work revealed that HpaR also plays important roles in virulence and biofilm formation, potentially also via regulation of T6SS4.

In summary, we revealed that HpaR is a dual-functional regulator in *Y. pseudotuberculosis*. HpaR not only represses expression of genes of the 4-HPA catabolic pathway but also activates the transcription of T6SS4 to acquire Zn^{2+} to maintain ROS levels in response to oxidative stress. Biofilm formation, antibiotic resistance, eukaryotic cell adhesion, and bacterial virulence were also affected by *hpaR* (**Figure 7**). This study has greatly expanded the function of HpaR beyond the regulation of aromatic compound catabolism and revealed a new pathway to regulate T6SS4 in response to environmental stress.

ETHICS STATEMENT

All mice were maintained and handled in accordance with the Regulations for the Administration of Affairs Concerning Experimental Animals approved by the State Council of People's Republic of China. The protocol was approved by the Animal Welfare and Research Ethics Committee of Northwest A&F University (protocol number: NWAFUSM2018001).

AUTHOR CONTRIBUTIONS

ZW, TW, YY, and XS conceived the study, designed the experiments, analyzed the data, and wrote the manuscript. ZW, TW, RC, ZZ, KC, ML, YH, HG, LX, and YW performed the experimental work.

FUNDING

This work was supported by the grant of National Key R&D Program of China (2018YFA0901200 to XS), the National Natural Science Foundation of China (31670053 and 31725003 to XS and, 31970114 and 31671292 to YW), the Open Project Program of the State Key Laboratory of Pathogen and Biosecurity (SKLPBS1825 to XS). LX was supported by China Postdoctoral Science Foundation (2018M631201) and Shaanxi Postdoctoral Science Foundation (2018BSHTDZZ20).

ACKNOWLEDGMENTS

We thank the Teaching and Research Core Facility at College of Life Sciences, NWAFU for their support in this work.

SUPPLEMENTARY MATERIAL

FIGURE S1 | Protein sequence alignment of HpaR homologs by CLUSTAL W. The sequences used in alignment have been deposited in the GenBank database [Accession No. EcC *Escherichia coli* C (S56952.1), EcW *Escherichia coli* W (Z37980.2), Yptb *Yersinia pseudotuberculosis* YPIII (ACA68739.1), PpU *Pseudomonas putida* U (FJ904934.1), BxLB400 *Burkholderia xenovorans* LB400 (ABE33958.1)]. The result was exported by ESPript (http://espript.ibcp.fr/ESPript/cgi-bin/ESPript.cgi).

FIGURE S2 | Growth curves of the wild-type (WT), Δ*hpaR* mutant, and the complemented strain Δ*hpaR* (*hpaR*) under normal condition. The growth of the indicated strains in YLB medium was monitored by measuring OD_{600} at indicated time points under 26°C.

TABLE S1 | The strains and plasmids used in this study.

TABLE S2 | Primers used in this study.

TABLE S3 | Genes differentially transcribed in Δ*hpaR* mutant compared to the *Y. pseudotuberculosis* wild-type detected by RNA-seq.

REFERENCES

Alekshun, M. N., and Levy, S. B. (1999). The mar regulon: multiple resistance to antibiotics and other toxic chemicals. *Trends Microbiol.* 7, 410–413. doi: 10.1016/s0966-842x(99)01589-1589

Alekshun, M. N., Levy, S. B., Mealy, T. R., Seaton, B. A., and Head, J. F. (2001). The crystal structure of MarR, a regulator of multiple antibiotic resistance, at 2.3 A resolution. *Nat. Struct. Biol.* 8, 710–714. doi: 10.1038/90429

Arcos, M., Olivera, E. R., Arias, S., Naharro, G., and Luengo, J. M. (2010). The 3,4-dihydroxyphenylacetic acid catabolon, a catabolic unit for degradation of biogenic amines tyramine and dopamine in *Pseudomonas putida* U. *Environ. Microbiol.* 12, 1684–1704. doi: 10.1111/j.1462-2920.2010.02233.x

Barbour, M. G., and Bayly, R. C. (1981). Control of meta-cleavage degradation of 4-hydroxyphenylacetate in *Pseudomonas putida*. *J. Bacteriol.* 147, 844–850. doi: 10.1128/jb.147.3.844-850.1981

Basler, M. (2015). Type VI secretion system: secretion by a contractile nanomachine. *Philos. Trans. R. Soc. Lond. B Biol. Sci.* 370, 1679. doi: 10.1098/rstb.2015.0021

Bingle, L. E., Bailey, C. M., and Pallen, M. J. (2008). Type VI secretion: a beginner's guide. *Curr. Opin. Microbiol.* 11, 3–8. doi: 10.1016/j.mib.2008.01.006

Brubaker, R. R. (1991). Factors promoting acute and chronic diseases caused by yersiniae. *Clin. Microbiol. Rev.* 4, 309–324. doi: 10.1128/cmr.4.3.309

Cooper, R. A., and Skinner, M. A. (1980). Catabolism of 3- and 4-hydroxyphenylacetate by the 3,4-dihydroxyphenylacetate pathway in *Escherichia coli*. *J. Bacteriol.* 143, 302–306. doi: 10.1128/jb.143.1.302-306.1980

D'Autreaux, B., and Toledano, M. B. (2007). ROS as signalling molecules: mechanisms that generate specificity in ROS homeostasis. *Nat. Rev. Mol. Cell Biol.* 8, 813–824. doi: 10.1038/nrm2256

Deochand, D. K., and Grove, A. (2017). MarR family transcription factors: dynamic variations on a common scaffold. *Crit. Rev. Biochem. Mol. Biol* 52, 595–613. doi: 10.1080/10409238.2017.1344612

Diaz, E., and Prieto, M. A. (2000). Bacterial promoters triggering biodegradation of aromatic pollutants. *Curr. Opin. Biotechnol.* 11, 467–475. doi: 10.1016/s0958-1669(00)00126-129

Durand, E., Cambillau, C., Cascales, E., and Journet, L. (2014). VgrG, Tae, Tle, and beyond: the versatile arsenal of Type VI secretion effectors. *Trends Microbiol.* 22, 498–507. doi: 10.1016/j.tim.2014.06.004

Egland, P. G., and Harwood, C. S. (1999). BadR, a new MarR family member, regulates anaerobic benzoate degradation by Rhodo *Pseudomonas palustris* in concert with AadR, an Fnr family member. *J. Bacteriol.* 181, 2102–2109. doi: 10.1128/jb.181.7.2102-2109.1999

Fallman, M., and Gustavsson, A. (2005). Cellular mechanisms of bacterial internalization counteracted by Yersinia. *Int. Rev. Cytol.* 246, 135–188. doi: 10.1016/S0074-7696(05)46004-46000

Fiorentino, G., Ronca, R., Cannio, R., Rossi, M., and Bartolucci, S. (2007). MarR-like transcriptional regulator involved in detoxification of aromatic compounds in *Sulfolobus solfataricus*. *J. Bacteriol.* 189, 7351–7360. doi: 10.1128/JB.00885-887

Flemming, H. C., and Wingender, J. (2010). The biofilm matrix. *Nat. Rev. Microbiol.* 8, 623–633. doi: 10.1038/nrmicro2415

Galan, B., Diaz, E., Prieto, M. A., and Garcia, J. L. (2000). Functional analysis of the small component of the 4-hydroxyphenylacetate 3-monooxygenase of *Escherichia coli* W: a prototype of a new Flavin:NAD(P)H reductase subfamily. *J. Bacteriol.* 182, 627–636. doi: 10.1128/jb.182.3.627-636.2000

Galan, B., Kolb, A., Garcia, J. L., and Prieto, M. A. (2001). Superimposed levels of regulation of the 4-hydroxyphenylacetate catabolic pathway in *Escherichia coli*. *J. Biol. Chem.* 276, 37060–37068. doi: 10.1074/jbc.M103033200

Galan, B., Kolb, A., Sanz, J. M., Garcia, J. L., and Prieto, M. A. (2003). Molecular determinants of the hpa regulatory system of *Escherichia coli*: the HpaR repressor. *Nucleic Acids Res.* 31, 6598–6609. doi: 10.1093/nar/gkg851

Green, E. R., Clark, S., Crimmins, G. T., Mack, M., Kumamoto, C. A., and Mecsas, J. (2016). Fis is essential for yersinia pseudotuberculosis virulence and protects against reactive oxygen species produced by phagocytic cells during infection. *PLoS Pathog.* 12:e1005898. doi: 10.1371/journal.ppat.1005898

Grove, A. (2013). MarR family transcription factors. *Curr. Biol.* 23, R142–R143. doi: 10.1016/j.cub.2013.01.013

Guan, J., Xiao, X., Xu, S., Gao, F., Wang, J., Wang, T., et al. (2015). Roles of RpoS in *Yersinia pseudotuberculosis* stress survival, motility, biofilm formation and type VI secretion system expression. *J. Microbiol.* 53, 633–642. doi: 10.1007/s12275-015-0099-96

Hall, C. W., and Mah, T. F. (2017). Molecular mechanisms of biofilm-based antibiotic resistance and tolerance in pathogenic bacteria. *FEMS Microbiol. Rev.* 41, 276–301. doi: 10.1093/femsre/fux010

Ho, B. T., Dong, T. G., and Mekalanos, J. J. (2014). A view to a kill: the bacterial type VI secretion system. *Cell Host Microb.* 15, 9–21. doi: 10.1016/j.chom.2013.11.008

Huang, Y., Zhao, K. X., Shen, X. H., Jiang, C. Y., and Liu, S. J. (2008). Genetic and biochemical characterization of a 4-hydroxybenzoate hydroxylase from *Corynebacterium glutamicum*. *Appl. Microbiol. Biotechnol.* 78, 75–83. doi: 10.1007/s00253-007-1286-1280

Jani, A. J., and Cotter, P. A. (2010). Type VI secretion: not just for pathogenesis anymore. *Cell Host Microb.* 8, 2–6. doi: 10.1016/j.chom.2010.06.012

Lin, J., Zhang, W., Cheng, J., Yang, X., Zhu, K., Wang, Y., et al. (2017). A *Pseudomonas* T6SS effector recruits PQS-containing outer membrane vesicles for iron acquisition. *Nat. Commun.* 8:14888. doi: 10.1038/ncomms14888

Ma, J., Bao, Y., Sun, M., Dong, W., Pan, Z., Zhang, W., et al. (2014). Two functional type VI secretion systems in avian pathogenic *Escherichia coli* are involved in different pathogenic pathways. *Infect. Immun.* 82, 3867–3879. doi: 10.1128/IAI.01769-1714

Mendez, V., Agullo, L., Gonzalez, M., and Seeger, M. (2011). The homogentisate and homoprotocatechuate central pathways are involved in 3- and 4-hydroxyphenylacetate degradation by *Burkholderia xenovorans* LB400. *PLoS One* 6:e17583. doi: 10.1371/journal.pone.0017583

Miller, J. H. (1992). *A Short Course in Bacterial Genetics - A Laboratory Manual and Handbook for Escherichia coli and Related Bacteria*. Cold Spring Harbor, NY: Cold Spring Harbor Laboratory Press.

Navarro-Garcia, F., Ruiz-Perez, F., Cataldi, A., and Larzabal, M. (2019). Type VI secretion system in pathogenic *Escherichia coli*: structure, role in virulence, and acquisition. *Front. Microbiol.* 10:1965. doi: 10.3389/fmicb.2019.01965

Perez-Rueda, E., and Collado-Vides, J. (2001). Common history at the origin of the position-function correlation in transcriptional regulators in archaea and bacteria. *J. Mol. Evol.* 53, 172–179. doi: 10.1007/s002390010207

Perez-Rueda, E., Collado-Vides, J., and Segovia, L. (2004). Phylogenetic distribution of DNA-binding transcription factors in bacteria and archaea. *Comput. Biol. Chem.* 28, 341–350. doi: 10.1016/j.compbiolchem.2004.09.004

Prieto, M. A., Diaz, E., and Garcia, J. L. (1996). Molecular characterization of the 4-hydroxyphenylacetate catabolic pathway of *Escherichia coli* W: engineering a mobile aromatic degradative cluster. *J. Bacteriol.* 178, 111–120. doi: 10.1128/jb.178.1.111-120.1996

Prieto, M. A., and Garcia, J. L. (1994). Molecular characterization of 4-hydroxyphenylacetate 3-hydroxylase of *Escherichia coli*. A two-protein component enzyme. *J. Biol. Chem.* 269, 22823–22829.

Prieto, M. A., and Garcia, J. L. (1997). Identification of a novel positive regulator of the 4-hydroxyphenylacetate catabolic pathway of *Escherichia coli*. *Biochem. Biophys. Res. Commun.* 232, 759–765. doi: 10.1006/bbrc.1997.6368

Providenti, M. A., and Wyndham, R. C. (2001). Identification and functional characterization of CbaR, a MarR-like modulator of the cbaABC-encoded chlorobenzoate catabolism pathway. *Appl. Environ. Microbiol.* 67, 3530–3541. doi: 10.1128/AEM.67.8.3530-3541.2001

Roper, D. I., Fawcett, T., and Cooper, R. A. (1993). The *Escherichia coli* C homoprotocatechuate degradative operon: hpc gene order, direction of transcription and control of expression. *Mol. Gen. Genet.* 237, 241–250. doi: 10.1007/bf00282806

Russell, A. B., Peterson, S. B., and Mougous, J. D. (2014). Type VI secretion system effectors: poisons with a purpose. *Nat. Rev. Microbiol.* 12, 137–148. doi: 10.1038/nrmicro3185

Shen, X., Banga, S., Liu, Y., Xu, L., Gao, P., Shamovsky, I., et al. (2009). Targeting eEF1A by a *Legionella pneumophila* effector leads to inhibition of protein synthesis and induction of host stress response. *Cell Microbiol.* 11, 911–926. doi: 10.1111/j.1462-5822.2009.01301.x

Shen, X. H., Jiang, C. Y., Huang, Y., Liu, Z. P., and Liu, S. J. (2005). Functional identification of novel genes involved in the glutathione-independent gentisate pathway in *Corynebacterium glutamicum*. *Appl. Environ. Microbiol.* 71, 3442–3452. doi: 10.1128/AEM.71.7.3442-3452.2005

Si, M., Wang, T., Pan, J., Lin, J., Chen, C., Wei, Y., et al. (2017a). Graded response of the multifunctional 2-cysteine peroxiredoxin, CgPrx, to increasing levels of hydrogen peroxide in *Corynebacterium glutamicum*. *Antioxid. Redox. Signal.* 26, 1–14. doi: 10.1089/ars.2016.6650

Si, M., Wang, Y., Zhang, B., Zhao, C., Kang, Y., Bai, H., et al. (2017b). The type VI secretion system engages a redox-regulated dual-functional heme transporter for zinc acquisition. *Cell Rep.* 20, 949–959. doi: 10.1016/j.celrep.2017.06.081

Si, M., Zhao, C., Burkinshaw, B., Zhang, B., Wei, D., Wang, Y., et al. (2017c). Manganese scavenging and oxidative stress response mediated by type VI secretion system in *Burkholderia thailandensis*. *Proc. Natl. Acad. Sci. U.S.A.* 114, E2233–E2242. doi: 10.1073/pnas.1614902114

Si, M., Zhang, L., Chaudhry, M. T., Ding, W., Xu, Y., Chen, C., et al. (2015). *Corynebacterium glutamicum* methionine sulfoxide reductase A uses both mycoredoxin and thioredoxin for regeneration and oxidative stress resistance. *Appl. Environ. Microbiol.* 81, 2781–2796. doi: 10.1128/AEM.04221-4214

Slauch, J. M. (2011). How does the oxidative burst of macrophages kill bacteria? Still an open question. *Mol. Microbiol.* 80, 580–583. doi: 10.1111/j.1365-2958.2011.07612.x

Smego, R. A., Frean, J., and Koornhof, H. J. (1999). Yersiniosis I: microbiological and clinicoepidemiological aspects of plague and non-plague Yersinia infections. *Eur. J. Clin. Microbiol. Infect. Dis.* 18, 1–15. doi: 10.1007/s100960050219

Song, Y., Xiao, X., Li, C., Wang, T., Zhao, R., Zhang, W., et al. (2015). The dual transcriptional regulator RovM regulates the expression of AR3- and T6SS4-dependent acid survival systems in response to nutritional status in *Yersinia pseudotuberculosis*. *Environ. Microbiol.* 17, 4631–4645. doi: 10.1111/1462-2920.12996

Staerck, C., Gastebois, A., Vandeputte, P., Calenda, A., Larcher, G., Gillmann, L., et al. (2017). Microbial antioxidant defense enzymes. *Microb. Pathog.* 110, 56–65. doi: 10.1016/j.micpath.2017.06.015

Tan, Y., Liu, W., Zhang, Q., Cao, S., Zhao, H., Wang, T., et al. (2017). *Yersinia pestis* YopK inhibits bacterial adhesion to host cells by binding to the extracellular matrix adaptor protein matrilin-2. *Infect. Immun.* 85:e01069-16. doi: 10.1128/IAI.01069-1016

Wan, B., Zhang, Q., Ni, J., Li, S., Wen, D., Li, J., et al. (2017). Type VI secretion system contributes to enterohemorrhagic *Escherichia coli* virulence by secreting catalase against host reactive oxygen species (ROS). *PLoS Pathog* 13:e1006246. doi: 10.1371/journal.ppat.1006246

Wang, T., Chen, K., Gao, F., Kang, Y., Chaudhry, M. T., Wang, Z., et al. (2017). ZntR positively regulates T6SS4 expression in *Yersinia pseudotuberculosis*. *J. Microbiol.* 55, 448–456. doi: 10.1007/s12275-017-6540-6542

Wang, T., Si, M., Song, Y., Zhu, W., Gao, F., Wang, Y., et al. (2015). Type VI secretion system transports Zn2+ to combat multiple stresses and host immunity. *PLoS Pathog.* 11:e1005020. doi: 10.1371/journal.ppat.1005020

Wang, Y., Cen, X. F., Zhao, G. P., and Wang, J. (2012). Characterization of a new GlnR binding box in the promoter of amtB in *Streptomyces coelicolor* inferred a PhoP/GlnR competitive binding mechanism for transcriptional regulation of amtB. *J. Bacteriol.* 194, 5237–5244. doi: 10.1128/JB.00989-912

Weber, B., Hasic, M., Chen, C., Wai, S. N., and Milton, D. L. (2009). Type VI secretion modulates quorum sensing and stress response in *Vibrio anguillarum*. *Environ. Microbiol.* 11, 3018–3028. doi: 10.1111/j.1462-2920.2009.02005.x

Wei, K., Tang, D. J., He, Y. Q., Feng, J. X., Jiang, B. L., Lu, G. T., et al. (2007). hpaR, a putative marR family transcriptional regulator, is positively controlled by HrpG and HrpX and involved in the pathogenesis, hypersensitive response, and extracellular protease production of *Xanthomonas campestris* pathovar campestris. *J. Bacteriol.* 189, 2055–2062. doi: 10.1128/JB.01331-1336

Xu, L., Shen, X., Bryan, A., Banga, S., Swanson, M. S., and Luo, Z. Q. (2010). Inhibition of host vacuolar H+-ATPase activity by a *Legionella pneumophila* effector. *PLoS Pathog.* 6:e1000822. doi: 10.1371/journal.ppat.1000822

Xu, S., Peng, Z., Cui, B., Wang, T., Song, Y., Zhang, L., et al. (2014). FliS modulates FlgM activity by acting as a non-canonical chaperone to control late flagellar gene expression, motility and biofilm formation in *Yersinia pseudotuberculosis*. *Environ. Microbiol.* 16, 1090–1104. doi: 10.1111/1462-2920.12222

Yang, X., Pan, J., Wang, Y., and Shen, X. (2018). Type VI secretion systems present new insights on pathogenic yersinia. *Front. Cell Infect. Microbiol.* 8:260. doi: 10.3389/fcimb.2018.00260

Zhang, W., Wang, Y., Song, Y., Wang, T., Xu, S., Peng, Z., et al. (2013). A type VI secretion system regulated by OmpR in *Yersinia pseudotuberculosis* functions to maintain intracellular pH homeostasis. *Environ. Microbiol.* 15, 557–569. doi: 10.1111/1462-2920.12005

Zhao, R., Song, Y., Dai, Q., Kang, Y., Pan, J., Zhu, L., et al. (2017). A starvation-induced regulator, RovM, acts as a switch for planktonic/biofilm state transition in *Yersinia pseudotuberculosis*. *Sci. Rep.* 7:639. doi: 10.1038/s41598-017-00534-539

4

Tracking Proteins Secreted by Bacteria: What's in the Toolbox?

Benoit Maffei[1,2], *Olivera Francetic*[3,4*] *and Agathe Subtil*[1,2*]

[1] *Unité de Biologie Cellulaire de l'Infection Microbienne, Institut Pasteur, Paris, France,* [2] *Centre National de la Recherche Scientifique UMR3691, Paris, France,* [3] *Unité de Biochimie des Interactions Macromoléculaires, Institut Pasteur, Paris, France,* [4] *Centre National de la Recherche Scientifique ERL6002, Paris, France*

Correspondence:
Olivera Francetic
ofrancet@pasteur.fr
Agathe Subtil
asubtil@pasteur.fr

Bacteria have acquired multiple systems to expose proteins on their surface, release them in the extracellular environment or even inject them into a neighboring cell. Protein secretion has a high adaptive value and secreted proteins are implicated in many functions, which are often essential for bacterial fitness. Several secreted proteins or secretion machineries have been extensively studied as potential drug targets. It is therefore important to identify the secretion substrates, to understand how they are specifically recognized by the secretion machineries, and how transport through these machineries occurs. The purpose of this review is to provide an overview of the biochemical, genetic and imaging tools that have been developed to evaluate protein secretion in a qualitative or quantitative manner. After a brief overview of the different tools available, we will illustrate their advantages and limitations through a discussion of some of the current open questions related to protein secretion. We will start with the question of the identification of secreted proteins, which for many bacteria remains a critical initial step toward a better understanding of their interactions with the environment. We will then illustrate our toolbox by reporting how these tools have been applied to better understand how substrates are recognized by their cognate machinery, and how secretion proceeds. Finally, we will highlight recent approaches that aim at investigating secretion in real time, and in complex environments such as a tissue or an organism.

Keywords: reporter, secretion signal, exoproteome, secretion machinery, live imaging

Secretion refers to the capacity, shared by all cells, to release a selected subset of the proteins they produce beyond the membrane that defines them as individual entities. In bacteria, secreted proteins are implicated in many essential functions such as nutrient uptake and catabolism, biodegradation of polymers, respiration, motility, cell attachment to the substratum or to other cells to allow beneficial or detrimental contacts, and biofilm formation. In pathogenic bacteria, the major virulence factors are typically secreted into the milieu or injected into neighboring target or host cells to change their integrity or function. These multiple roles are often essential for bacterial fitness, and several secreted proteins have been studied as potential drug targets. Optimization of the secretion process is also key for the production of many bioengineered products. Protein secretion, which allows communication with the external world, is also a fascinating biological problem for which numerous mechanisms have evolved, each one adapted to different, and often multiple, biological functions.

The purpose of this review is to provide an overview of the available biochemical, genetic and imaging tools that have been developed to evaluate protein secretion in a qualitative or quantitative manner. The different secretion machineries in bacteria are briefly presented in **Figure 1** and we

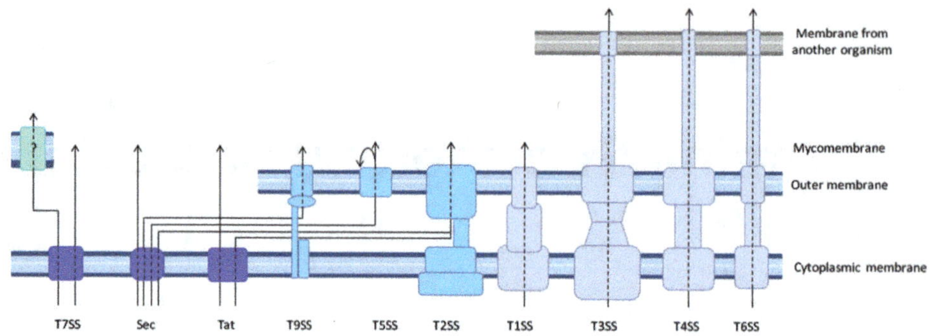

FIGURE 1 | Diversity of the bacterial secretion systems. Schematic representation of the secretion systems identified in bacteria, partially based on structural data. In monoderm bacteria **(left)**, protein export (synonymous for secretion in that case) follows the Sec or Tat pathway, or the signal peptide independent T7SS. In diderm-non-LPS bacteria such as *Mycobacteria* or *Corynebacteria* **(far-left)**, it is unknown whether the T7SS system results in protein secretion in one or two steps (question mark). In diderm-LPS bacteria **(center and right)**, secreted proteins can reach the external environment through a one-step process via T1SS, T3SS, T4SS, or T6SS. Other secreted proteins are first exported to the periplasm via the Sec system (T2SS, T5SS, or T9SS) or the Tat system (for T2SS only).

refer the reader to several excellent reviews for mechanical insight on these processes (Desvaux et al., 2009; Korotkov et al., 2012; Leyton et al., 2012; Lycklama A Nijeholt and Driessen, 2012; Palmer and Berks, 2012; Kanonenberg et al., 2013; Christie et al., 2014; Ho et al., 2014; Basler, 2015; Costa et al., 2015; Ates et al., 2016; Green and Mecsas, 2016). Importantly, different tools are applicable to different secretion systems, and the choice of tools is oriented by the question asked. Therefore, after a brief overview of the different tools available, we will illustrate their advantages and limitations through a discussion of some of the current open questions related to protein secretion. We will start with the question of the identification of secreted proteins, which for many bacteria remains a critical initial step toward a better understanding of their interactions with the environment. We will then illustrate our toolbox by describing how these tools have been applied to better understand how substrates are recognized by their cognate machinery, and how secretion proceeds. Finally, we will highlight recent technical developments that aim at investigating secretion in real time and in complex environments such as a tissue or an organism. For space limitations, we will not discuss here the tools to study pilus assembly systems, wherein secretion is coupled to polymerization of protein subunits into fibers, and which include type I pili (Remaut et al., 2008), type IV pili (Berry and Pelicic, 2015), or curli (Van Gerven et al., 2015).

To discuss protein transport across bacterial membranes, we need to clarify the terminology, as the term "protein secretion" is commonly used to describe three distinct processes. We will use the term "export" to describe the translocation of proteins across the cytoplasmic membrane (also called inner membrane (IM) in diderm bacteria). In monoderm bacteria, exported proteins are surface exposed, thus export is equivalent to "secretion." However, in diderm bacteria, exported proteins typically remain intracellular, and are therefore not "secreted," unless another machinery takes them across the outer membrane (OM). Several trans-envelope machineries (classified by numbers, often reflecting the order of their discovery) can perform this second translocation step, including the Type 2, 5, 7, and 9 secretion systems (T2SS, T5SS, T7SS, and T9SS). Other secretion machineries that span the entire envelope of diderm bacteria promote protein secretion in a single step directly from the cytoplasm (T1SS, T3SS, T4SS, and T6SS). Note that some secreted proteins may remain associated with cell surface, and are therefore not synonymous with the "exoproteome" that describes the subset of proteins present in the extracellular medium (Desvaux et al., 2009). Finally, the third process commonly covered by the term "secretion" leads to the injection of the protein beyond a third membrane, that of a neighboring cell, so that the translocated protein becomes inserted in this membrane or is released in the cytoplasm of the neighboring cell. While this process is more accurately described by the word "injection," the term secretion is also suitable.

OVERVIEW OF THE TOOLBOX

The different methods to investigate protein secretion follow two main strategies. In the first, secreted proteins are identified after a fractionation step that isolates the compartment into which they are targeted. The second strategy is to keep cells intact, and use assays based on accessibility to probes, or on the activity of the secreted proteins, to monitor secretion. This second category includes several microscopy-based assays, and is amenable to the study of secretion in living cells. It also includes genetic or chemical screens aimed at the identification of components of secretion machineries and of secretion signals (see Identification of components of secretion machineries and secretion signals). Complementary to these approaches, bioinformatics tools are being used to identify putative secretion substrates based on their sequence (see Bioinformatics tools).

Fractionation-Based Assays

Fractionation-based assays are most appropriate to study proteins that are released from the bacteria, either free in the extracellular medium, or injected into a host cell. However, coupled to strategies to purify the OM, such assays can also be used to study proteins that remain associated with the bacteria after secretion.

The fractionation step can be very straightforward, like separating bacteria from the culture medium by centrifugation, or more complicated, for instance in the case of a protein translocated into a given eukaryotic compartment, that will require isolation of that compartment. The readout for secretion after the fractionation step depends on the protein of interest (detection with antibodies, detection of localized enzymatic activity, use of an engineered chimera with a readily-detectable reporter protein etc.), if the strategy is used to follow a given protein. Progress in protein identification by mass spectrometry now enables the use of fractionation-based strategies to characterize bacterial exoproteomes and to detect protein modifications after secretion, without prior information on the secretion mechanism involved. One limitation of this strategy is that setting up the appropriate protocol for the isolation of the compartment of interest to limit contamination by proteins from other compartments can be time-consuming. The high sensitivity of the technique raises the issue of false-positives and requires further experimental validation. For example, a bacterial protein can only be considered as secreted if most of it is found in an extra-cytoplasmic compartment or environment and if such a location is compatible with its function. Furthermore, identification of membrane proteins by mass spectrometry remains technically challenging, and the approach is therefore best suited for soluble secreted proteins.

Whole-Cell Based Assays

Identification and characterization of secretion system components usually starts with genetic analyses such as deletion or insertion mutagenesis that can define the roles of individual components in more or less complex secretion machineries. Ideally, secretion should be specifically linked to a phenotype to allow for screening or selection; for example, secretion of an amylase is required for growth on starch, while secretion of a hemolysin produces a halo on blood agar plates (see Analysis of protein secretion signals and machineries for examples of genetic and chemical screens). Other, more complex phenotypes can be studied *in vitro*, such as killing of target bacteria *via* the T6SS effectors in mixed bacterial cultures (Brunet et al., 2013) (see When is protein injection activated?). Appearance of a given protein on the bacterial surface can sometimes be assessed directly using antibodies or reporter systems, as illustrated below. For proteins secreted into the extracellular medium, or translocated into a neighboring cell, use of a reporter system is usually the method of choice, in particular when secretion is measured using microscopy to achieve spatial and temporal resolution.

IDENTIFICATION OF SECRETED PROTEINS

Secreted proteins are ambassadors, mediating most of the interactions of a bacterium with its surrounding environment. Cataloguing the secreted proteins is often an obligatory step toward a comprehensive understanding of how a given bacterium deals with its environment. Some of the tools that can be used to identify secreted proteins, like the bioinformatics approaches described below, are specific to a given secretion machinery. Others, like proteomics-based approaches, or phage display, do not require information on the secretion mechanism. The tools illustrated below are complementary. Typically, global approaches generate lists of secreted proteins candidates, which are later validated using targeted secretion assays, often based on reporter fusion systems.

Bioinformatics Tools

Type 1 to type 6 secretion systems (**Figure 1**) are sufficiently well documented and conserved to predict the secretion machinery repertoire in newly sequenced bacterial genomes. One recent study built online and standalone computational tools to predict protein secretion systems and related appendages accurately in bacteria with an OM containing lipopolysaccharide, retrieving ~10,000 candidate systems amongst which T1SS and T5SS were by far the most abundant and widespread (Abby et al., 2016). The identification of the substrates of these secretion machineries is more difficult, and novel secretion substrates generally cannot be identified unambiguously from genomic sequence alone. However, in many cases, sequence similarity with a known secretion substrate, and/or the presence of a "signal peptide" (see below), and/or genomic localization in proximity to genes coding for a secretion machinery, provide strong indications of novel secretion substrates. This is often not sufficient, especially for secretion substrates of pathogenic bacteria that are tailored for a very specific target, and are therefore often specific to a single bacterial species. To identify these elusive secretion substrates, machine-learning approaches have been implemented for use with T3SS and T4SS, for which the data base is sufficiently large. Globally, secretion substrates fall into two categories, depending on the presence or absence of a so-called signal peptide.

First Scenario: Presence of a Signal Peptide

Two machineries export proteins across the IM: the Sec translocon and the twin-arginine translocation (Tat) machinery. Proteins that are targeted to these export machineries have N-terminal extensions called signal peptides. Canonical signal peptides have a tripartite structure with a basic region at the N-terminus, a central hydrophobic region and a polar carboxyl terminus with a consensus cleavage site (AXA) (von Heijne, 1990). The Tat signal peptides differ somewhat from the Sec-targeting signals in that they possess an extended N-terminal region with a conserved twin-arginine motif TRRxFLK that is crucial for targeting to Tat export pathway (Palmer and Berks, 2012). Importantly, the Tat pathway is capable of transporting folded proteins and protein complexes; therefore, proteins that lack a signal peptide but form complexes with partner subunits that have twin-arginine signal peptides can also be exported in a "piggy-back fashion" through this pathway. Furthermore, it is important to note that some bacterial genomes have a strong base compositional bias and, consequently, encode Sec-dependent proteins with non-canonical signal peptides (Payne et al., 2012).

Several bioinformatics programs can be used to predict the presence of cleavable Sec or Tat signal peptides, such as SignalP (http://www.cbs.dtu.dk/services/SignalP) (Petersen et al., 2011),

PSort (http://www.psort.org), which conveniently also provides a list of links toward other subcellular prediction programs, Pred-Tat, TatP or TatFind (see Berks, 2015, for comparison). Lipoprotein signal peptides are a distinct class of Sec dependent signal peptides characterized by a C-terminal consensus sequence, the lipobox, which ends with an absolutely conserved cysteine residue that, after fatty acylation, becomes the first residue of the mature protein (www.cbs.dtu.dk/services/LipoP/) (Juncker et al., 2003). Experimental validation is necessary, however, to demonstrate a tentatively-identified secreted protein uses a given transport pathway to exit the producing cell, as illustrated in parts Analysis of protein secretion signals and machineries and Resolution of secretion in time and space.

As explained earlier, protein transport across the IM by the Sec or Tat pathways in monoderm bacteria results in protein secretion. However, in diderm bacteria, proteins face an additional barrier, the OM that, together with the IM, defines the periplasmic compartment with distinct properties, composition and content. While the periplasm is the final destination for many proteins, others are inserted into the OM or cross it entirely to reach the bacterial surface using specialized transport systems. Protein secretion pathways that include a periplasmic intermediate are called two-step pathways.

The T2SS is a typical two-step pathway that takes up specific periplasmic substrates in a folded state. T2SS substrates can therefore be recognized by the presence of an N-terminal signal sequence in their precursors. However, despite many extensive analyses, the recognition events and signals that mediate the second transport step have not been elucidated and appear to vary between different bacterial species and substrates (Korotkov et al., 2012).

T5SS substrates, formerly called auto-transporters, are made of a translocator and passenger domain; these domains are usually encoded by a single gene, but can also be separate polypeptides in the so-called two-partner secretion systems (Leyton et al., 2012). The conserved and mandatory C-proximal translocator domain with characteristics typical of most outer beta-barrel membrane proteins, following a large N-proximal domain with a signal sequence is usually sufficient to predict the latter as the passenger (secreted) domain (Abby et al., 2016).

Another two-step secretion pathway is the recently discovered T9SS, found exclusively in the Bacteroidetes phylum. Substrates of this pathway are secreted in a folded state, and, in addition to the signal sequence, share a conserved C-terminal domain harboring the secretion signal (Sato et al., 2010; de Diego et al., 2016).

Second Scenario: Absence of a Signal Peptide

Proteins that are not made as precursors with an amino terminal signal peptide can still be secreted, by T1SS, T3SS, T4SS, T6SS, or T7SS. It proved difficult or impossible to identify sequence features in secretion substrates that indicate that they will use one or other of these pathways. Several machine-learning techniques have been developed recently for this purpose based on datasets of known T3SS and T4SS secretion substrates. They differ in terms of the machine learning methods, the curated data sets and the features used, and reach different levels of prediction performance. Several, but not all (Meyer et al., 2013), of these tools were recently compared (An et al., 2016). All predictors are flawed, to various degrees, with false positives (secretion signal identified in non-secreted proteins) and false negatives (documented secretion substrates not predicted as such), so while they can be precious in orienting research, experimental validation is required.

T1SS substrates also contain a Sec-independent secretion sequence, which is either located at the N-terminus (certain bacteriocins or colicins) or at the C-terminus (all other systems) of the substrate. Like in T3SS and T4SS substrates, these sequences lack recognizable features, so T1SS substrates cannot be identified based on the recognition of a characteristic secretion signal (Kanonenberg et al., 2013). However, prediction tools cannot yet be developed, because the data base is too small to build training sets. The same is true for T6SS substrates for which some properties (size, isoelectric point, operon structure) have been used to orient genome searches, but robust bioinformatics based methods do not exist (Ho et al., 2014).

Finally, the T7SS, initially identified in mycobacteria, is still poorly understood. Mycobacteria and related genera have an external membrane composed of unique and complex lipids and are therefore diderm, despite the fact that, like monoderms, they stain Gram-positive. They secrete several proteins by a Sec-independent mechanism, and it is not known whether protein secretion occurs in one or two steps. Substrates of this secretion machinery share a loosely defined C-terminal secretion signal, which includes the consensus motif YxxxD/E (Ates et al., 2016). The T7SS is also present in a few monoderm bacteria such as *Staphylococcus aureus*, where it promotes secretion of proteins with antibacterial activities (Cao et al., 2016).

Identification of Secreted Proteins through Proteomics

One method of choice to identify new secreted proteins in the extracellular medium, irrespective of the secretion mechanism, is mass spectrometry. Combined with the use of mutant strains, or specific culture conditions, it can also identify the substrates of a given secretion machinery. For instance, comparison of the exoproteome of *Pseudomonas aeruginosa* in conditions where one T6SS machinery was on or off allowed the identification of three novel T6SS secretion substrates (Hood et al., 2010).

Quantitative proteomics not only can lead to identification of novel effectors, but also provide information on the regulation of secretion, as was recently shown with a focused exoproteome analysis of the T3SS in *Ralstonia*. The use of secretion mutants revealed that secretion is finely tuned and identified specific subsets of effectors with different secretion patterns (Lonjon et al., 2016).

Combined with a fractionation method to isolate a specific compartment of a target cell, proteomics can also identify novel translocated proteins in their target location. For example, this approach was used to draw up a list of putative nuclear effectors of *Anaplasma phagocytophilum* (Sinclair et al., 2015), and to identify *Chlamydia trachomatis* proteins associated to lipid droplets (Saka et al., 2015). Obviously, the method is not sufficiently sensitive if the effector is present in minute amounts and is overwhelmed by eukaryotic proteins. One way to circumvent this limitation is to enrich for bacterial proteins using non-canonical amino acid tagging (Mahdavi et al., 2014). Selective labeling of bacterial proteins is accomplished via translational incorporation of a methionine surrogate. The technique requires the introduction of a gene coding for a mutant form of the methionyl-tRNA synthetase into the bacterium. During mixed culture with the bacterium, the host cells do not produce the altered methionyl-tRNA synthetase and do not incorporate the methionine surrogate, whereas the bacteria do. Bacterial proteins with the methionine surrogate are then enriched from the eukaryotic cells and identified by mass-spectrometry. This approach was used to identify *Yersinia* proteins that were secreted into the medium and translocated into cells (Mahdavi et al., 2014). In addition, pulse labeling with the methionine surrogate can be used to achieve temporal resolution (see Resolution of secretion in time and space).

Integral membrane proteins are notoriously difficult to detect by mass spectrometry because detergents interfere with the analyses. In a pioneering work in which alkaline sodium carbonate was used instead of detergent, Molloy et al identified the majority of integral *Escherichia coli* OM proteins and several diacyl-glyceride attached lipoproteins (Molloy et al., 2000). Proteomics is currently the method of choice to obtain an overall view of the secretion capacity of a given bacterium. For instance, it was recently applied to characterize the OM proteome and the exoproteome of *Bacteroides fragilis*, highlighting striking differences with *Proteobacteria*, from which most of the current information on bacterial secretion was derived so far (Wilson et al., 2015).

Reporter-Based Assays

A large part of the "secretion toolbox" relies on reporter-based assays, which use genetic tools to tag a given secretion substrate with a readily detectable but otherwise neutral reporter and follow its secretion through a dedicated assay. In most cases, these assays are applied with a specific secretion machinery in mind, which orients the choice of the tag and the tagging strategy. A few rules are listed below, and most tags currently in use are briefly described in **Table 1**, with examples of their application.

Tag Flexibility

For each new tag, one prerequisite is to ensure that the tag itself is neutral with regard to the secretion capacity of the machinery under inspection. Some secretion processes require the secretion substrate to unfold, and proteins that fold very rapidly upon synthesis are inappropriate as tags. For instance, green fluorescence protein (GFP) (Jaumouille et al., 2008) and glutathione S-transferase (Riordan et al., 2008), which fold rapidly block T3SS. Although GFP-tagged proteins are successfully exported through the Sec pathway, GFP fluorophore

TABLE 1 | Reporter systems to track protein secretion.

Tag	Read-out	Tested for	Examples of application
Calmodulin-dependent adenylate cyclase (Cya)	cAMP production, Western-blot	T3S, T4S	Defining components and signals required for secretion (Sory and Cornelis, 1994) Genome wide screens for T3S and T4S effectors candidates (Subtil et al., 2005; Carey et al., 2011)
Alkaline phosphatase (PhoA)	Enzymatic assay		Protein export reporter (Manoil and Beckwith, 1985).
Amino-peptidase (AP)	Enzymatic assay	Sec dependent export (Gram+ bacteria)	Optimization of secretion (Guan et al., 2016)
β-1,4-mannanase (ManB)	Enzymatic assay	Sec dependent export (Gram+ bacteria)	Optimization of secretion (Lin et al., 2015)
staphylococcal nuclease (NucA)	Enzymatic assay	Sec dependent export (Gram+ bacteria)	Validation of predicted signal peptides and optimization of secretion (Mathiesen et al., 2009)
Green fluorescent protein (GFP)	fluorescence	T2S	Assembly pathway of the T2S complex (Lybarger et al., 2009)
Split GFP	fluorescence	T3S	Localization of secreted effectors in eukaryotic host (Van Engelenburg and Palmer, 2010)
Tetracysteine motif	fluorescence	T3S	Kinetics of effector translocation (Enninga et al., 2005)
LOV	fluorescence	T3S	Detection of effector translocation (Gawthorne et al., 2016)
Glutamyl carboxypeptidase	fluorescence	T3S	Detection of protein secretion (Yount et al., 2010)
TEM-1 β-lactamase	fluorescence (FRET)	T3S, T4S	Genome wide screen for T4S effector candidates (Zhu et al., 2011), study of translocation dynamics (Mills et al., 2013)
Gaussia princeps luciferase (Gluc)	luminescence	T1S	Detection of secreted fusion protein in culture supernatant (Wille et al., 2012)
Bacteriophage P1 Cre recombinase	luminescence/ fluorescence	T3S, T4S	Detection of effector translocation (Luo and Isberg, 2004; Briones et al., 2006)
Phosphorylation target	Western blot: detection of phosphorylated tag	T3S, T4S	Detection of effector translocation (Day et al., 2003; Garcia et al., 2006)
Nucleoskeletal-like protein (Nsp)	Western-blot	Flagellar secretion apparatus	Identification of export signal (Wang et al., 2016)

is inefficiently folded in the periplasm (Feilmeier et al., 2000). On the other hand, GFP is an ideal reporter for the Tat system, since it can be exported in a folded form (Thomas et al., 2001). The GFP re-folding problems have been solved using superfolder GFP variant (Choi et al., 2017) or monomeric red fluorescent protein (mRFP) and its derivatives (Shaner et al., 2004). Several strategies were developed to find alternatives to GFP to facilitate the tracking of effector injection into a host cell with fluorescence (**Figure 2**, and see part Resolution of secretion in time and space for live imaging with temporal resolution).

Tag Position

The position at which the tag is fused to the secreted protein is important because the tag can compromise the recognition of the secretion signal. For instance, T3SS substrates have an amino-terminal signal that must remain unaltered in the tagged construct. Typically, C-terminal tags are used for proteins with an N-terminal signal sequence.

Secretion Readout

Many tags are associated with an enzymatic activity, which provides quantitative data on secretion. Tags that are enzymatically active only in a given environment facilitate quantification. For example, alkaline phosphatase and beta-lactamase are active only in the periplasm, and have been used extensively to probe IM protein topology and to identify periplasmic proteins by gene fusion approaches. The membrane impermeable beta-lactamase substrate nitrocephin is a useful probe for surface exposed beta-lactamase fusions that remain cell associated upon secretion (Sauvonnet and Pugsley, 1996). One widely used reporter of protein injection (via T3SS or T4SS) into a host cell is the calmodulin-dependent adenylyl cyclase of *Bordetella pertussis*. Since bacteria do not produce calmodulin, whereas eukaryotic cells do, accumulation of cyclic AMP marks the injection of the reporter in the cytoplasmic compartment of the target cell (Sory and Cornelis, 1994). Reporter tags that become phosphorylated in the eukaryotic cytosol have also been used (Day et al., 2003; Garcia et al., 2006). In the last ten years, the development of fluorogenic beta lactamase substrates allowed to follow injection of effectors fused with beta lactamase (TEM1), and this enzymatic-based assay has also largely been used, including very recently, *in vivo* (see Resolution of secretion in time and space).

In contrast to global approaches using bioinformatics or proteomics described above, reporter-based assays are only amenable in genetically tractable microorganisms. However, secretion machineries and secretion signals are well conserved, allowing the use of heterologous secretion systems to identify secretion substrates in non-genetically tractable bacteria (Subtil et al., 2001).

Reporter-based assays are mostly used for candidate-based approaches, because they require the generation of genetically modified organisms. They are typically used to validate candidates indicated by secretion signal predictors or by proteomic approaches. However, high-throughput cloning strategies and simplification of the read-outs have allowed the application of these assays to screen for novel secretion substrates in genome wide approaches (Subtil et al., 2005; Carey et al., 2011; Zhu et al., 2011).

Functional Screens in Yeast

Many proteins injected into a eukaryotic host cell target proteins and pathways that are highly conserved in all eukaryotic cells. Based on this observation, expression libraries of bacterial genes under inducible promoters have been screened in yeast. Most screens selected bacterial genes that inhibited yeast growth (Campodonico et al., 2005; Slagowski et al., 2008), but atypical localization of bacterial proteins (for instance in the nucleus)

FIGURE 2 | Comparison of the different tools to image effector secretion into living cells. Illustrations reflect different experimental set-ups, please refer to the indicated reference for description of the bacterium, effector and time scale. Bacteria are represented by an oval, cells by a rectangle. Colors represent the fluorescent signal recorded before **(left)** and after **(right)** the effector translocation has started. Discussion on the pros and cons of some of these assays can be found in Ehsani et al. (2009) and Zuverink and Barbieri (2015).

(Sisko et al., 2006) or other particular phenotypes such as interference with the secretory pathway can also be screened for (Heidtman et al., 2009). The ease of manipulating yeast genome, and conservation of the molecular pathways with higher eukaryotes, can then facilitate the identification of specific targets of the bacterial effectors identified by this approach.

Phage Display Technology

Jacobsson and Frykberg were the first to take advantage of the power of shot-gun filamentous bacteriophage display (display of random fragments of bacterial genomic DNA) to identify *S. aureus* proteins that interact with components of the extracellular matrix and immunoglobulins (Jacobsson and Frykberg, 1996). Since then, many genes encoding proteins involved in host-microbial interactions have been identified using this technology. Coupled to next generation sequencing, it can also be applied to the identification of secretomes, in particular in mixed microbial populations. This powerful technology was recently reviewed in detail and will not be discuss further (Gagic et al., 2016).

ANALYSIS OF PROTEIN SECRETION SIGNALS AND MACHINERIES

Identification of Components of Secretion Machineries and Secretion Signals

Genetic and Chemical Screens

The first protein export signals identified were the cleavable N-terminal signal peptides that target proteins to the endoplasmic reticulum in eukaryotic cells (Jackson and Blobel, 1977). In *E. coli*, studies of genes involved in maltose and lactose transport or catabolism have provided a wealth of genetic tools to study protein export (**Figure 3**). Powerful genetic approaches have been designed based on special properties of beta-galactosidase (LacZ) fusions (Shuman and Silhavy, 2003). Selections devised to identify mutations that affect protein export were based on the Lac- phenotype of strains producing fusions of the periplasmic maltose-binding protein MalE with the cytoplasmic LacZ. These fusions, which target beta-galactosidase to the export pathway, are enzymatically inactive and confer a Lac- phenotype to bacteria, allowing for selection of Lac+ export-defective mutants. The mutations frequently mapped to the proximal region of gene fusions coding for the signal peptide (Bedouelle et al., 1980), introducing charged residues in their hydrophobic segment. Importantly, Lac+ selection also yielded several classes of mutants with pleiotropic export defects. Many of those were conditional lethal mutations in genes encoding novel protein export factors, including the preprotein translocase SecA, or the translocation channel SecY (Ito et al., 1983). Suppressor (*prl*) mutations that restored export of proteins with defective signal sequences also mapped in the *secA* and *secY* genes, strongly suggesting that signal sequences interacted with their gene products. Indeed, this was confirmed by many studies and by structural data that revealed how signal sequences bind to SecA (Gelis et al., 2007) or SecY and the lipid bilayer (Li et al., 2016). Although they do not share sequence homology and appear to be interchangeable between proteins, signal sequences differ in their ability to promote efficient export, as illustrated in a genome-wide

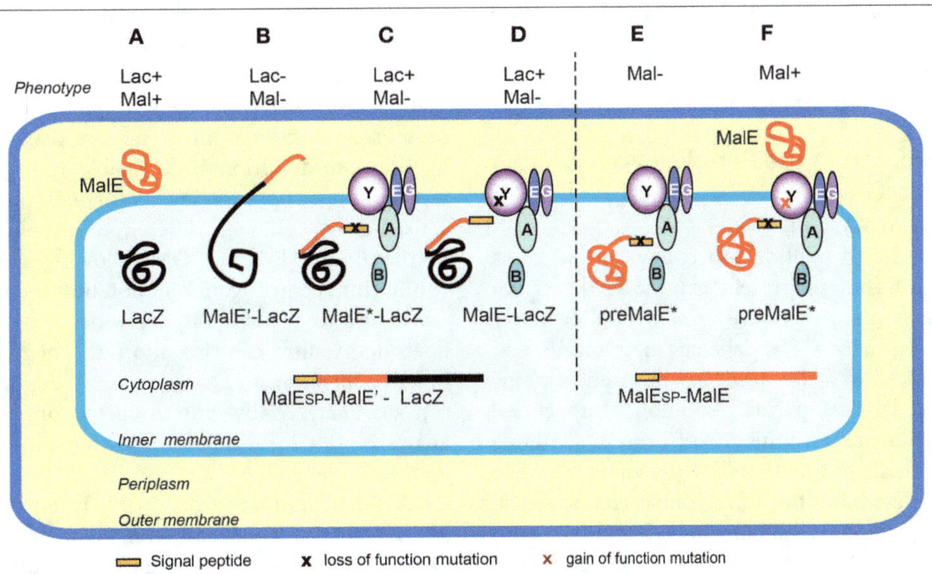

FIGURE 3 | Genetic selections of export-defective mutants based of MalE-LacZ fusions. In wild type *E. coli* the correct localization of LacZ (cytoplasmic) and MalE (periplasmic) allows lactose catabolism and maltose uptake, respectively, conferring the ability to ferment these sugars, and the red colony phenotype on MacConkey indicator plates **(A)**. The MalE-LacZ fusion proteins directed to the periplasm confer a Lac- phenotype **(B)**, which served as a basis for selection of spontaneous Lac+ mutants on lactose tetrazolium plates. These strains either contained mutations in MalE signal sequence (MalE*) **(C)**, or the loss-of- function mutations in the *sec* genes encoding export factors, five of which, (SecA, B, Y, E and G) are depicted **(D)**. Bacteria producing the full-length MalE precursor with a signal sequence mutation are export defective and Mal- **(E)**, allowing for selection of Mal+ suppressor mutations (gain-of-function *prl* alleles) mapping in several *sec* genes, (e.g., *secY*) that promote export of MalE* variants with signal sequence mutations **(F)**. Note that the MalE signal sequence (yellow rectangle) is absent from periplasmic MalE-LacZ or MalE, as it is cleaved and degraded upon export across the IM.

study of *Lactobacillus* signal peptides that assessed their ability to promote export of nuclease reporter (Mathiesen et al., 2009). Successful approaches to improve signal sequence efficiency have been reported. For example, combinatorial mutagenesis of the signal sequence-coding region resulted in variants with increased export and production of beta-lactamase, probably due to an overlap of export signals with elements of translational or post-translational regulation (Heggeset et al., 2013).

The Tat signal was also largely investigated through genetic screens. Although a native Sec substrate, MalE was successfully used to monitor the activity of the Tat-dependent signal peptide of TorA, by following maltose utilization on pH indicator media (indicating maltose fermentation) or by growth on minimal maltose plates (Kreutzenbeck et al., 2007). This allowed the use of powerful genetic approaches to identify suppressors of Tat signal sequence changes that restored MalE export, affecting genes encoding the TatB and TatC export machinery components (Lausberg et al., 2012). In a genome-wide screen for Tat-dependent exo-proteins of *P. aeruginosa*, the authors used the *E. coli* amidase AmiA as a reporter to validate the functionality of a newly identified Tat signal peptide (Ball et al., 2016). Tat-mediated AmiA export is required for the correct separation of daughter cells during cell division, and defects in this process render *E. coli* hypersensitive to detergents, providing a simple plate test.

In monoderm bacteria, surface proteins of the "LPxTG" family are anchored to the cell wall in a process mediated by the sortase enzymes (Schneewind and Missiakas, 2014). Since sortase substrates include major virulence factors, small molecule screening has been used to identify sortase inhibitors (Maresso et al., 2007). In *S. aureus*, transposon mutagenesis was used to look for mutants with defective surface anchoring of the protein A (SpA), using detection of fluorescently labeled Anti-SpA antibodies and flow cytometry (Frankel et al., 2010).

Assays of Protein Export to the Periplasm or Secretion across the OM

Signal pepides and the Sec system

Measuring the rate of signal peptide processing can serve as a good quantitative indicator of protein export, as the signal peptidase cleavage takes place in the periplasm. Radioactively labeled amino acids (usually ^{35}S - labeled methionine and cysteine) are incorporated into the newly synthesized proteins during a short "pulse" labeling period. Addition of an excess of unlabeled amino acids and an inhibitor of protein synthesis during a "chase" period allows one to follow precursor cleavage with precision and to establish the time course characteristic for a given strain (**Figure 4**; Kumamoto and Gannon, 1988). In addition, strong export defects, such as those caused by signal sequence mutations, can be observed in steady state by the presence of precursor forms of a protein by denaturing SDS-PAGE and immuno-detection. Fractionation in steady state can reveal partitioning of the precursors with the cell fraction and mature forms to the periplasm (Francetic et al., 2007). This approach has been used to characterize the strongly hydrophobic signal sequences that direct proteins like thioredoxin or DsbA to the co-translational export *via* the signal recognition particle (SRP) (Huber et al., 2005). All of these assays are target-based and rely on the availability of specific antibodies or antigen tags for immuno-precipitation of radioactively labeled protein or for immuno-detection.

Lipoproteins

Lipoproteins are exported proteins that undergo fatty acylation during biogenesis and remain anchored in membranes (Zückert, 2014). Lipoprotein signal peptides can be predicted by bioinformatics approaches, mostly thanks to the presence of a conserved lipobox motif with a consensus LAGC. This motif probably interacts with Lgt (Pallier et al., 2012), the first component of the biogenesis pathway that adds the diacyl-glycerol moiety to the invariant Cys residue. In the next step, the signal peptide is removed by a dedicated lipoprotein signal peptidase, Lsp. Accumulation of unprocessed precursors in the presence of globomycin, an Lsp inhibitor, can serve as a tool for experimental validation of predicted lipoproteins, at least in *E. coli*. Another method frequently used to demonstrate protein fatty-acylation *in vivo* is metabolic labeling with radioactive fatty acids, usually palmitate (Jackowski and Rock, 1986).

In *E. coli* and related bacteria, lipoproteins either remain anchored in the IM or are taken to the OM *via* the Lol sorting machinery (Zückert, 2014). The sorting signals that determine retention in the IM or extraction and transport to the OM typically reside within the four N-terminal residues of the mature lipoprotein. Systematic analysis of model lipoproteins in an *in vitro* membrane release assay allowed Hara and co-workers to characterize these signals in *E. coli* providing the basis for bioinformatics predictions (Hara et al., 2003). However, the sorting rules are not valid for all bacteria and there are many exceptions, even in well-studied enterobacterial species. Membrane fractionation in sucrose density gradients followed by immuno-detection is a reliable tool to determine *in vivo* localization of lipoproteins, as for integral membrane proteins. This method relies on differences in composition and density of the *E. coli* IM and OM, allowing good separation using equilibrium centrifugation in flotation sucrose density gradients, where the total membranes are deposited on the bottom and float upon ultra-centrifugation to their equilibrium density. Relative positions of specific proteins in gradient fractions can be analyzed by immunodetection, as in the systematic analysis of +2 residue substitutions in a model lipoprotein lipoMalE (Seydel et al., 1999). In addition, lipoMalE conferred a Mal+ phenotype to *E. coli* when localized in the IM but not in the OM, providing a plate assay for lipoprotein localization based on utilization of maltose as a carbon source. As a different approach, *in vivo* analysis of lipoproteins fused to the monomeric fluorescent protein mCherry can be performed following plasmolysis, which swells the periplasm and facilitates direct visualization of IM or OM lipoprotein association: while OM lipoproteins follow the contours of the bacterial cell, the irregularly shaped plasmolysis bays are decorated with IM associated lipoproteins (Lewenza et al., 2006).

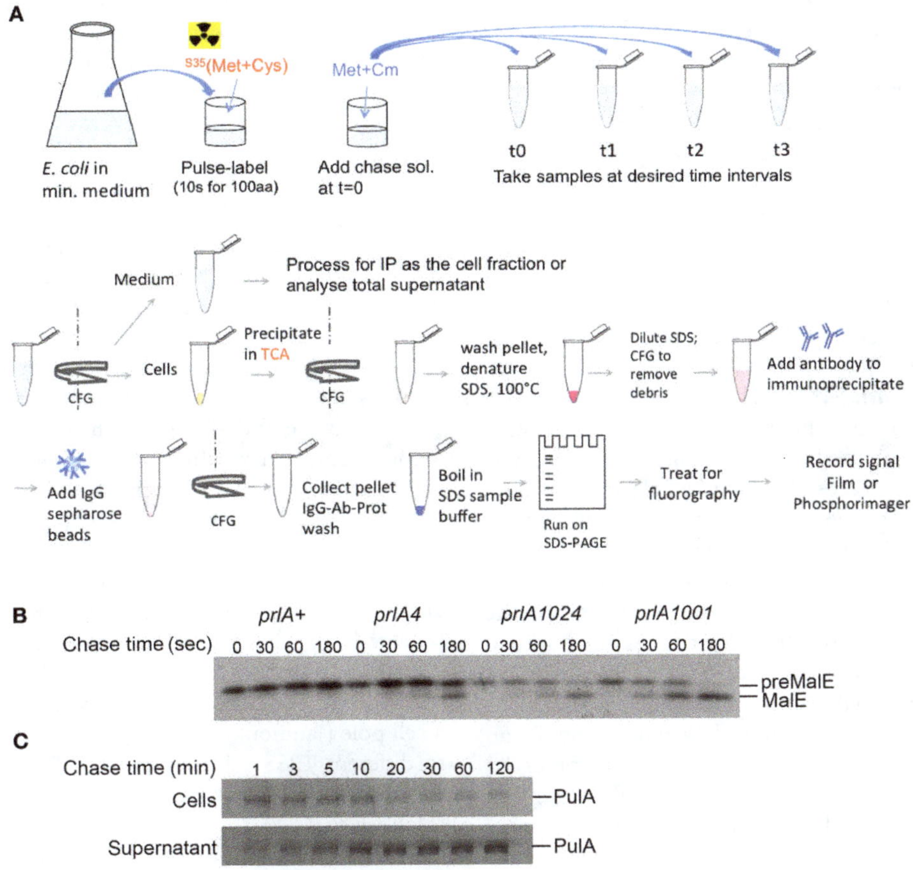

FIGURE 4 | Pulse-chase assay to analyze protein export or secretion rates. **(A)** Bacteria are grown in minimal medium in conditions inducing the expression of the gene of interest and pulse labeled with ^{35}S-methionine and cysteine for a short period (30 s–2 min, depending on the size of the protein under study), followed by addition of cold methionine and chloramphenicol (Cm) to stop protein synthesis. Samples are collected at indicated times and bacterial cultures are either precipitated with TCA or separated from the medium prior to precipitation of each fraction. The collected precipitates are washed with acetone, dissolved in SDS buffer and boiled to denature proteins. Upon the removal of cell debris by centrifugation and dilution of SDS, antibodies are added for immuno-precipitation. Antigen-antibody complexes are adsorbed on protein A-sepharose beads, washed and eluted in SDS sample buffer for analysis by SDS-gel electrophoresis and fluorography. **(B)** Kinetics of signal sequence processing in preMalE variant carrying a signal sequence mutation, reflecting the kinetics of MalE export. While the preMalE export and processing are blocked in wild type *E. coli* (*prlA+*), they are partially restored in strains carrying different suppressor *prlA* alleles of the *secY* gene encoding the translocation channel. Bacteria were labeled for 20 s with radioactive methionine and a chase with excess cold methionine was performed for the indicated times. After immuno-precipitation of total cell extracts with anti-MalE antibodies, proteins were separated on SDS-PAGE, and analyzed by fluorography (modified from Francetic et al., 1993). **(C)** Pulse-chase and fractionation were used to follow the kinetics of pullulanase (PulA) secretion *via* the T2SS. The bacteria were cultured as in **(A)** and pulse-labeled for 3 min. Samples were collected after the indicated times of chase with cold methionine, and cell and supernatant fractions were separated by centrifugation, prior to immunoprecipitation and SDS-PAGE analysis as in **(A)** (modified from Francetic and Pugsley, 2005).

While lipoproteins represent an important group of surface proteins in monoderm bacteria, in diderms they generally reside in the periplasmic membrane leaflets. However, recent investigations have revealed lipoproteins that are fully or partially surface-exposed, either through known secretion systems (T2SS, T5SS) (Leyton et al., 2012; Rondelet and Condemine, 2013), or by novel mechanisms (Wilson and Bernstein, 2016). In the Lyme disease agent *Borrelia burgdorferi* many lipoproteins that play a role in virulence are exposed on the cell surface. Their localization has been studied using mRFP as a fluorescent reporter. Using the model lipoprotein OspA-mRFP, a cell-sorting based mutant screen led to identification of residues required for the surface exposure presumably participating in a flipping step whose cellular determinants remain to be identified (Kumru et al., 2010). Some *E. coli* lipoproteins, including Lpp (Cowles et al., 2011) and RcsF, are partially exposed on the bacterial surface. This unusual feature has been tested using the so-called epitope walking approach of the OM lipoprotein RcsF. The FLAG tag was inserted at multiple positions in the RcsF sequence, and a dot blot analysis of whole and lysed cells using monoclonal anti-FLAG antibody allowed for identification of protein regions that are accessible in intact cells. The FLAG appears to be a useful and neutral tag that does not seem to interfere with transport across membranes (Konovalova et al., 2014). This approach revealed how RcsF uses OM proteins as portals for surface exposure, and how this mechanism allows the bacteria to monitor the functional state of their Beta-barrel Assembly Machinery (BAM) (Konovalova et al., 2014).

Surface exposed and associated proteins

A subset of exported proteins is inserted in the outer membrane and partially exposed on the cell surface to perform diverse functions, including solute uptake, macromolecule transport or proteolysis. The OM insertion generally relies on the signals encoded in the mature protein, notably the propensity to form beta-barrels, and is generally mediated by the BAM complex (Voulhoux et al., 2003; Wu et al., 2005). The BAM machinery is probably responsible for OM insertion of the beta-barrel domain of T5SSs, which provide the portal for secretion of so-called "passenger domains" (Leyton et al., 2012; Bernstein, 2015). Although the passenger domains can be cleaved and released into the medium to perform diverse extracellular functions, many of them remain surface bound. The surface exposure of specific proteins or protein domains can be assessed through analysis of non-permeabilized cells by immunofluorescence or by analyzing protease accessibility in whole cells, with a comparative assessment of general protease susceptibility of the same substrate in lysed bacteria (Besingi and Clark, 2015). For enzymes that degrade biopolymers, plate assays for protease, lipase, cellulase or chitinase activities, to name a few, rely on the visualization of a halo zone of substrate degradation surrounding the secreting colonies. Secretion of one of the first identified T2SS substrates, the lipoprotein pullulanase (d'Enfert et al., 1987), has been assessed by growth on minimal media containing pullulan as the sole carbon source, through degradation of chromogen-tagged pullulan, or by semi-quantitative enzymatic assays that determine the fraction of hydrolytic activity present in intact compared to the lysed bacteria by measuring the reducing sugar as the reaction product. All these methods are semi-quantitative, end-point assays, which do not provide kinetic information on protein secretion.

Similar approaches have been used to study surface protein secretion via T5SS or the recently discovered T9SS. In *Porphyromonas gingivalis* the T9SS is required for the black pigment of colonies on blood agar, linked to heme acquisition mediated by secreted proteases gingipains. Substrates of this pathway are exported to the periplasm via the Sec pathway and their N-terminal signal peptide. Their secretion in a folded state requires a C-terminal signal within a specific beta-sandwich domain, which is able to promote secretion of folded GFP (de Diego et al., 2016).

Mechanistic Approaches

Understanding a secretion mechanism requires detailed structural knowledge of the secretion machinery, its composition, biogenesis and dynamics during interactions with the secreted substrate. A few assays have been designed to understand how secretion proceeds. They are complementary to microscopic observations of secretion machineries, itself an expanding field of research (Costa et al., 2015; Li et al., 2016). The question concerning energy requirements for the process has been addressed by depleting ATP or by dissipating the proton motive force, two main energy sources required for membrane transport (e.g., Possot et al., 1997). Many secretion systems use at least one ATPase as an essential component, and ATP hydrolysis as a mechanical force generator (Costa et al., 2015).

Capture of Protein Interactions and of Intermediate States in the Secretion Process

In many systems, a bacterial two-hybrid approach is an excellent tool to map protein-protein interactions between dynamic components in secretion systems, in particular membrane embedded elements in trans-envelope complexes (Karimova et al., 1998). Fusions to cytoplasmic fragments of the CyaA reporter can effectively block some transport intermediates in the membrane and allow assessment of interactions that are transient in the native system. Examples include studies of complexes in T2SS (Nivaskumar et al., 2016) and T6SS (Logger et al., 2016).

Specific inhibitors of secretion can also be used to block secretion at a specific step (Moir et al., 2011). Chemical libraries have been used to identify compounds that specifically inhibit elastase and Plc secretion in *Pseudomonas* T2SS using a colorimetric enzymatic screen of culture supernatants (Moir et al., 2011). Once such inhibitors are identified, the major challenge is to identify their specific protein targets and modes of inhibition.

In T1SS and T3SS, substrates are secreted in an unfolded state and bulky domains fused to the substrate might block secretion, potentially providing important clues about intermediates in the secretion process. In T3SS, blocking the secretion using a GFP fusion with a secretion substrate revealed that secretion occurs at a cell pole (Jaumouille et al., 2008). A bulky "knot" region fused to different T3SS substrate allowed Dohlich and coworkers to demonstrate that the substrate passes through the T3SS channel (Dohlich et al., 2014).

In many cases substrates of blocked or incomplete secretion systems are degraded *in vivo* due to the absence of specific partners, chaperones or cellular structures. In T6SS, for example, the component of an inner tube HCP binds specific folded substrates in the bacterial cytoplasm and is required for their stability (Silverman et al., 2013). Folded substrate PulA of the T2SS (East et al., 2016) or TcpF secreted by the assembly system of TCP pili (Kirn et al., 2003) are also prone to degradation if their secretion is compromised. Since protein secretion efficiency is typically assessed in steady state by combining fractionation and substrate detection using antibodies or activity assays, it is important to keep in mind that proteolysis of non-secreted substrate may skew quantification of secretion efficiency. While the bulk of secretion-defective variants will be degraded, a small amount of extracellular protein that escapes degradation might give an impression of full secretion efficiency. The use of radiolabeling in pulse-chase assays coupled to fractionation might help overcome this problem (**Figure 4C**). Radioactive labeling is a powerful tool in secretion analysis due to its unsurpassed sensitivity. Selective labeling of bacterial proteins in cell culture has helped to identify proteins from enteropathogenic *E. coli* injected into eukaryotic cells *via* T3SS (Kenny and Finlay, 1995).

A number of gene fusion and mutagenesis approaches have been employed using different T2SS substrates to elucidate the molecular nature of the secretion signal and the component of the secretion machinery with which they interact. As already discussed, the sequence diversity of exoproteins, including those using the same secretion system, and systems makes this task difficult, however. In view of their heterogeneity, it is reasonable to assume that T2SS components that are interchangeable between systems do not make specific contacts with secretion substrates. A site-directed *in vivo* cross-linking approach has been used to identify secretion motifs in PelI and their interactions with the T2SS of the plant pathogen *Dickeya dadantii*. The unnatural photo-cross-linkable amino acid pBPA incorporated at specific sites in exoproteins was used as a tool to capture transient complexes *in vivo*, providing evidence for PelI binding to OutD forming the OM channel of the T2SS, and to OutC, which interacts with OutD (Pineau et al., 2014). A similar approach has been used to track secretion intermediates in T5SS (Ieva et al., 2011).

In vitro Reconstitution

In an advanced stage of analysis, secretion systems could be reconstituted *in vitro* to gain insights into the transport process. With the notable exception of the Sec system that has been functionally reconstituted *in vitro* (Duong and Wickner, 1997), few other systems have been studied at this level, due to their complexity and difficulties to extract them from the bacterial envelope in a functional state. Nevertheless, *in vitro* transcription-translation systems have been used successfully to study biogenesis of specific transport components including the OM channel called secretin that self-assembles and insert into liposomes (Guilvout et al., 2008) or the IM prepilin peptidase, both components of the T2SS (Aly et al., 2013).

RESOLUTION OF SECRETION IN TIME AND SPACE

Several tools have been developed in the last two decades to improve the spatio-temporal resolution of techniques aimed at tracking secretion. They have mostly been applied to effector injection into a neighboring cell through the T3SS and T4SS, and to some extent to the T6SS.

When Is Protein Injection Activated?

Protein injection in a neighboring cell is a highly regulated process that is typically constitutively turned off and only activated by specific signals. In many cases, activation occurs at least in part at the transcriptional level (Mavris et al., 2002; Urbanowski et al., 2005), so the activity of promoters has been used as a read-out for secretion activity. Monteiro et al. (2012) used the luciferase reporter under the control of the promoter of *hrpB*, the transcriptional regulator that controls the expression of *Ralstonia solanacearum* T3SS genes, in order to track the activity of the T3SS *in planta*. This approach revealed that T3SS activation was important not only for the first stages of infection, to manipulate host plant defenses, but also during late stages of infection (Monteiro et al., 2012). In a somewhat different set-up using a fast-maturing GFP under the control of the transcription activator MxiE, Campbell-Valois and collaborators provided evidence that *Shigella flexneri* T3SS goes through two waves of activation: one upon cell contact, during the invasion process, and a second, concomitant with the motile stage of the infection cycle, when bacteria move throughout the cytoplasm through actin cytoskeleton remodeling. The T3SS was switched off between these two phases (Campbell-Valois et al., 2014).

Another readout of active protein injection can be found in the effects on the target, when this effect is rapid and easy to detect. The T6SS functions as a contractile nanomachine, called the molecular crossbow, that punctures target cells to deliver lethal effectors. Time-lapse fluorescence microscopy of cocultures demonstrated that prey cells were killed upon contact with predator cells, and that prey lysis occurred within minutes after sheath contraction (Brunet et al., 2013). A killing assay was designed to understand how *Proteus mirabilis* coordinates multicellular swarming behavior and discriminates itself from another *Proteus* species during swarming. This assay, together with live-cell microscopy, demonstrated that T6SS-mediated lethality is unique to morphologically distinct swarmer cells, and that it requires direct cell-cell contact (Alteri et al., 2013).

The strategies illustrated above aim at measuring the activation of a given secretion machinery. They do not provide information on the nature of translocated effectors, or on the kinetics of secretion of a given effector, questions that are addressed below.

How Fast Does Protein injection Take Place and What Is the Hierarchy of Substrate Secretion?

Highly sensitive and time-resolutive tools were needed to answer these questions in order to focus on the very early secreted effectors, typically during the first 15 min following the attachment of a bacterium to its target cell. The current tools are mostly microscopy-based, allowing for single-cell measurements.

A pioneering work in the field focused on the secretion of SipA, an early effector of *Salmonella* Typhimurium (Schlumberger et al., 2005). The kinetics of secretion of this effector were measured by two complementary approaches. First, using live microscopy and a GFP-tagged version of the SipA chaperone, InvB, produced by a genetically engineered eukaryotic host cell, the secretion kinetics were determined by monitoring and quantifying InvB-GFP recruitment to contact sites between *Salmonella* and the cell. This elegant approach can not be applied to all effectors, since it requires the identification of a high-affinity partner (here the chaperone). As the second readout, the redistribution of SipA from the bacterial cytoplasm to its periphery was measured, on samples fixed at various times after live microscopy, from the moment of contact between the bacteria and the cell. Although this could in theory be automated to some extent, it requires very clean antibodies and extensive image acquisition. Another disadvantage was that detection of SipA with antibodies required sample fixation, preventing live imaging and analysis of later events.

In a different approach, two early effectors of *S. flexneri*, IpaB, and IpaC, were tagged with a tetracysteine tag. By loading the bacteria with fluorescent FlAsH probes, it was demonstrated that both effectors were secreted instantly after contact with host cells, with a half maximal rate of 4 min in both cases (Enninga et al., 2005). FlAsH labeling yields a somewhat poor signal-to-noise ratio, and is probably not appropriate for effectors of moderate or low abundance. The signal was only poorly detectable in the host cell, and loss of effector from the bacteria was recorded.

One current limitation of all the live imaging techniques is that they allow tracking of only one effector at a time. In order to gain insight into the hierarchy of secretion of different effectors, it is therefore necessary to compare the kinetics of secretion measured separately. For instance, the rate of secretion of tetracysteine-tagged SopE2 was found to be about 2-fold faster than that of SptP, another *Salmonella* effector, explaining how two effectors with antagonistic effects on the host cell could cooperate during the infectious process (Van Engelenburg and Palmer, 2008).

Currently, the most widely used assay is based on the development of fluorogenic substrates of beta-lactamase, and has been extensively applied to monitor secretion in T3SS and T4SS. Effector proteins are fused to beta-lactamase, while host cells are pre-loaded with the membrane-permeable substrate coumarin cephalosporin fluorescein (CCF2 or CCF4). The injection of the effector/beta-lactamase fusion protein into the host cytosol is detected by the loss of FRET upon cleavage of the fluorogenic substrate, inducing a switch in the fluorescence from green to blue (Charpentier and Oswald, 2004; Zuverink and Barbieri, 2015; **Figure 5**). This system has for instance been used with the beta-lactamase reporter encoded chromosomally in fusion to twenty different enteropathogenic *E. coli* (EPEC) effectors (Mills et al., 2013). Changes in the fluorescence of the CCF2 were monitored at 90 s intervals, and the secretion kinetics of ten different effectors was determined through this approach.

Although less resolutive in time, measurements on whole populations can provide information as to the order of effector secretion under some circumstances. *Chlamydia trachomatis* secretes several effectors, e.g., TarP and TepP, upon contact with the host cell. Once in the cytosol, TarP and TepP are tyrosine phosphorylated by host kinases. TarP phosphorylation was detectable as early as 5 min post infection whereas TepP phosphorylation occurred only between 5 and 15 min post infection, indicating that TarP injection occurs first (Chen et al., 2014). A similar strategy, combined with the addition of tags that become phosphorylated in the eukaryotic environment (Day et al., 2003; Garcia et al., 2006), could, in theory, be applied to several effectors, provided that they differ in size. Using an identical tag for each effector would limit the possibility that a difference in phosphorylation rates only reflects substrate preference by the host kinases involved.

On a larger scale, bio-orthogonal non-canonical amino acid tagging (BONCAT), combined with pulse labeling of the methionine surrogate, introduced the possibility to track the injection of endogenous proteins. Using this elegant strategy, Mahdavi and collaborators were able to characterize the order and pace of secretion, over 3 h of infection, of 11 Yop effectors from *Yersinia enterocolitica* (Mahdavi et al., 2014).

When Does Secretion Occur in a Complex Environment?

Although the complexity of secretion has been explored experimentally *in vitro* at a single cell level, fewer studies have tried to unravel the secretion complexity in more complex systems such as a population of cells and bacteria or even during infection in animal models.

The heterogeneity in secretion rates between different effectors led Mills and collaborators to pay attention not only to effectors but also to the target cells. By single cell tracking in cultures loaded with CCF2 and infected with either *Salmonella*

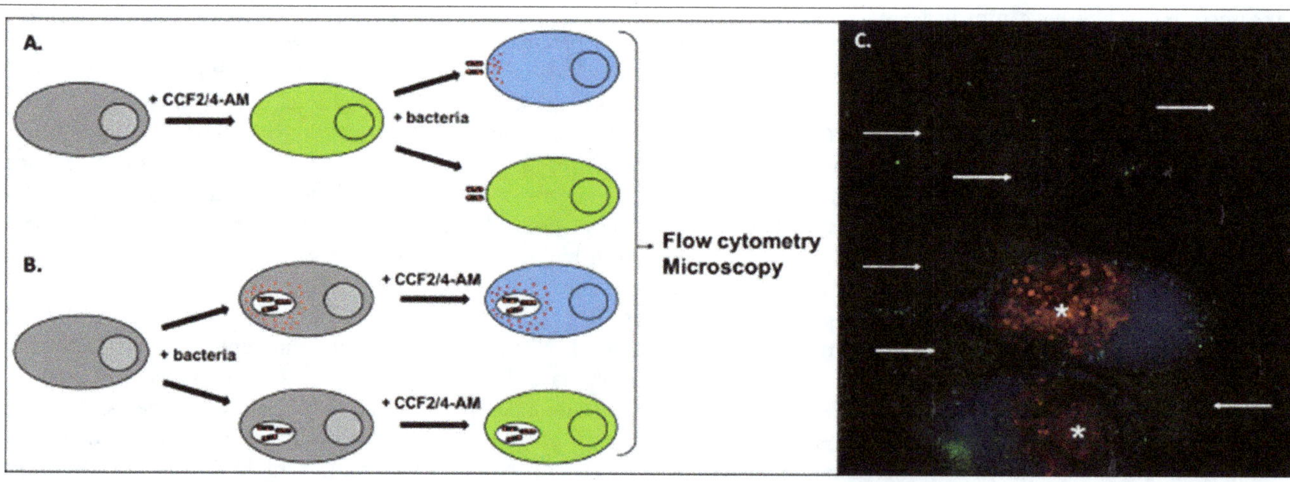

FIGURE 5 | Detection of effector translocation using the beta-lactamase/CCF2/4-AM system. **(A,B)** Schematic view of the experimental set-up to test the translocation of a bacterial protein fused to beta-lactamase (red dots). To monitor secretion from extracellular bacteria, or from intracellular bacteria at an early stage of infection, cells are usually pre-loaded with the CCF2/4-AM **(A)**. Alternatively, the probe can be added to cells after infection, to monitor secretion events that occur later in infection **(B)**. Conversion of the fluorescent probe can be measured by microscopy or by flow cytometry. **(C)** Illustration of the analysis using microscopy. HeLa cells were infected for 40 h with *Chlamydia trachomatis* stably transformed with a plasmid expressing the translocated protein TarP in fusion with beta-lactamase and mCherry (Mueller and Fields, 2015). In the last 2 h of incubation the cells were loaded with CCF4-AM probe, before fixation. In uninfected cells the probe emits green fluorescence (arrows). In infected cells (bacteria in red, asterisks) translocation of the beta-lactamase activity into the cytosol is revealed by the appearance of a blue fluorescent signal, corresponding to the cleaved probe.

Typhimurium or EPEC producing effectors fused to beta-lactamase, they identified a population of cells that were resistant to secretion, for which, despite the contact with bacteria, the fluorescence of the CCF2 remained unchanged even a couple of hours after infection. Whether this resistance to secretion is conferred by the host cell or is due to heterogeneities in the bacterial population remains unclear (Mills et al., 2013).

Only few studies have analyzed bacterial protein secretion in whole organisms. One example is the study of the activation of T3SS in whole plants, as was mentioned above (Monteiro et al., 2012). The early production of the T3S effector ExoU of *P. aeruginosa* and its secretion were shown to be critical for the development of the pathology in the lung of infected mice, using inducible production of the protein and immunolabelling on histological sections with a specific antibody to ExoU (Howell et al., 2013).

In a more complex system, Rolán and collaborators tackled the challenge of identifying the targets of the *Yersinia* effector YopH *in vivo* in mice, the main difficulty being the isolation of the few neutrophils targeted by the bacteria. In order to achieve this, they fused the beta-lactamase reporter to the first 100 amino acids of YopE. The splenocytes were then purified and loaded with CCF4 *ex vivo*, and the neutrophils injected with YopE were sorted from non-targeted cells on the basis of their fluorescence profile. This strategy allowed the authors to identify YopH-targeted signal transduction pathways that impair neutrophil responses *in vivo* (Hortensia Rolán et al., 2013).

Very recently, a genome-wide method, named EXIT (exported *in vivo* technology), was developed to identify proteins that are exported by bacteria during infection (Perkowski et al., 2017). EXIT utilizes the TEM beta-lactamase reporter lacking its native signal peptide, which, when fused in-frame to an export signal (i.e., signal peptide or transmembrane domain) confers beta-lactam resistance. By combining a comprehensive library of in-frame TEM fusions with the ability to select bacteria exporting fusion proteins *in vivo* and next-generation sequencing en masse of the recovered fusions, EXIT identified 593 proteins exported by *Mycobacterium tuberculosis* during infection in mice (Perkowski et al., 2017). Fifty-seven percent of these hits were predicted integral membrane proteins and 38% contained a predicted signal peptide. EXIT also identified 32 proteins (5%) lacking *in silico* predicted export signals.

CONCLUDING REMARKS

In the last decade, the repertoires of bacteria known to secrete proteins and of secreted proteins have expanded, and several novel secretion machineries have been discovered. These findings were mostly driven by technological advances, such as the discovery of secretion substrates through phage display coupled to next generation sequencing or through proteomics, independently of the secretion mechanism used, or by the development of probes that allowed sensitive detection of effector injection into a neighboring cell. Machine learning approaches have also pointed to new potential secretion substrates, and, with the expansion of the training sets (i.e., validated substrates and characterized secretion machineries) these approaches will likely be applicable to more secretion systems. In parallel, remarkable progress has been made in the exploration of the structure of the multicomponent secretion machineries using cryo-electron microscopy and tomography that should allow us in the future to understand their molecular mechanisms and their dynamic behavior. Still, very fundamental questions regarding the specificity of these machines remain unanswered. In many cases, the nature of the signal(s) that designate a protein as a secretion substrate and its recognition by the given machinery are still poorly understood. Insights into substrate recognition require detailed structural knowledge of a dynamic process that likely involves a series of intermediates. Capturing these intermediates has been extremely challenging, and their structure was resolved in part only in the Sec system. Cross-linking or mutagenesis approaches might enable one to stabilize and study these intermediate states. Computational approaches, including structural modeling and molecular dynamics have the potential to predict the details of these transient interactions and, combined with other validation tools, improve our understanding of transport processes. Finally, recent developments in high-resolution microscopy and tomography have provided important information on the architecture of secretion systems *in situ*, as a valuable basis for future studies (Chang et al., 2016; Hu et al., 2017). More studies are needed to understand how the secretion of different substrates by the same machinery is regulated. The design of probes that would allow one to track the secretion of two proteins simultaneously in living cells would certainly also help in addressing this question. Some of the tools described in this review are amenable to single-cell studies, and might reveal heterogeneity in the secretory behavior of bacteria, which has not been addressed so far. However, the next main challenge seems to be to probe bacterial secretion in complex environments such as biofilms, mixed microbial populations or within living hosts. Only a few of the approaches described here can be used in complex environments: our toolbox needs yet more new tools.

AUTHOR CONTRIBUTIONS

BM, OF, and AS conceived and wrote the manuscript.

ACKNOWLEDGMENTS

We thank Tony Pugsley for critical reading of the manuscript. This work was supported by the European Research Council Starting Grant NUChLEAR N°282046 to AS, the Institut Pasteur and the Centre National de la Recherche Scientifique. The OF group is funded by the French ANR grant FiberSpace N°ANR-14-CE09-0004.

REFERENCES

Abby, S. S., Cury, J., Guglielmini, J., Neron, B., Touchon, M., and Rocha, E. P. (2016). Identification of protein secretion systems in bacterial genomes. *Sci. Rep.* 6:23080. doi: 10.1038/srep23080

Alteri, C. J., Himpsl, S. D., Pickens, S. R., Lindner, J. R., Zora, J. S., Miller, J. E., et al. (2013). Multicellular bacteria deploy the Type VI secretion system to preemptively strike neighboring cells. *PLoS Path.* 9:e1003608. doi: 10.1371/journal.ppat.1003608

Aly, K. A., Beebe, E. T., Chan, C. H., Goren, M. A., Sepulveda, C., Makino, S., et al. (2013). Cell-free production of integral membrane aspartic acid proteases reveals zinc-dependent methyltransferase activity of the *Pseudomonas aeruginosa* prepilin peptidase PilD. *Microbiologyopen* 2, 94–104. doi: 10.1002/mbo3.51

An, Y., Wang, J., Li, C., Leier, A., Marquez-Lago, T., Wilksch, J., et al. (2016). Comprehensive assessment and performance improvement of effector protein predictors for bacterial secretion systems III, IV and VI. *Brief. Bioinformat.* doi: 10.1093/bib/bbw100. [Epub ahead of print].

Ates, L. S., Houben, E. N., and Bitter, W. (2016). Type VII secretion: a highly versatile secretion system. *Microbiol. Spectr.* 4. doi: 10.1128/microbiolspec.VMBF-0011-2015

Ball, G., Antelmann, H., Imbert, P. R., Gimenez, M. R., Voulhoux, R., and Ize, B. (2016). Contribution of the Twin Arginine Translocation system to the exoproteome of *Pseudomonas aeruginosa*. *Sci. Rep.* 6:27675. doi: 10.1038/srep27675

Basler, M. (2015). Type VI secretion system: secretion by a contractile nanomachine. *Philos. Trans. R. Soc. Lond. B. Biol. Sci.* 370: 20150021. doi: 10.1098/rstb.2015.0021

Bedouelle, H., Bassford P. J. Jr., Fowler, A. V., Zabin, I., Beckwith, J., and Hofnung, M. (1980). Mutations which alter the function of the signal sequence of the maltose binding protein of *Escherichia coli*. *Nature* 285, 78–81. doi: 10.1038/285078a0

Berks, B. C. (2015). The twin-arginine protein translocation pathway. *Annu. Rev. Biochem.* 84, 843–864. doi: 10.1146/annurev-biochem-060614-034251

Bernstein, H. D. (2015). Looks can be deceiving: recent insights into the mechanism of protein secretion by the autotransporter pathway. *Mol. Microbiol.* 97, 205–215. doi: 10.1111/mmi.13031

Berry, J. L., and Pelicic, V. (2015). Exceptionally widespread nanomachines composed of type IV pilins: the prokaryotic Swiss Army knives. *FEMS Microbiol. Rev.* 39, 134–154. doi: 10.1093/femsre/fuu001

Besingi, R. N., and Clark, P. L. (2015). Extracellular protease digestion to evaluate membrane protein cell surface localization. *Nat. Protoc.* 10, 2074–2080. doi: 10.1038/nprot.2015.131

Briones, G., Hofreuter, D., and Galan, J. E. (2006). Cre reporter system to monitor the translocation of type III secreted proteins into host cells. *Infect. Immun.* 74, 1084–1090. doi: 10.1128/IAI.74.2.1084-1090.2006

Brunet, Y. R., Espinosa, L., Harchouni, S., Mignot, T., and Cascales, E. (2013). Imaging type VI secretion-mediated bacterial killing. *Cell Rep.* 3, 36–41. doi: 10.1016/j.celrep.2012.11.027

Campbell-Valois, F. X., Schnupf, P., Nigro, G., Sachse, M., Sansonetti, P. J., and Parsot, C. (2014). A Fluorescent Reporter Reveals On/Off Regulation of the Shigella Type III Secretion Apparatus during Entry and Cell-to-Cell Spread. *Cell Host Microbe* 15, 177–189. doi: 10.1016/j.chom.2014.01.005

Campodonico, E. M., Chesnel, L., and Roy, C. R. (2005). A yeast genetic system for the identification and characterization of substrate proteins transferred into host cells by the *Legionella pneumophila* Dot/Icm system. *Mol. Microbiol.* 56, 918–933. doi: 10.1111/j.1365-2958.2005.04595.x

Cao, Z., Casabona, M. G., Kneuper, H., Chalmers, J. D., and Palmer, T. (2016). The type VII secretion system of *Staphylococcus aureus* secretes a nuclease toxin that targets competitor bacteria. *Nat. Microbiol.* 2, 1–11. doi: 10.1038/nmicrobiol.2016.183

Carey, K. L., Newton, H. J., Lührmann, A., and Roy, C. R. (2011). The Coxiella burnetii Dot/Icm System Delivers a Unique Repertoire of Type IV Effectors into Host Cells and Is Required for Intracellular Replication. *PLoS Path.* 7:e1002056. doi: 10.1371/journal.ppat.1002056

Chang, Y. W., Rettberg, L. A., Treuner-Lange, A., Iwasa, J., Sogaard-Andersen, L., and Jensen, G. J. (2016). Architecture of the type IVa pilus machine. *Science* 351:aad2001. doi: 10.1016/j.bpj.2015.11.2509

Charpentier, X., and Oswald, E. (2004). Identification of the secretion and translocation domain of the enteropathogenic and enterohemorrhagic *Escherichia coli* effector Cif, using TEM-1 beta-lactamase as a new fluorescence-based reporter. *J. Bacteriol.* 186, 5486–5495. doi: 10.1128/JB.186.16.5486-5495.2004

Chen, Y. S., Bastidas, R. J., Saka, H. A., Carpenter, V. K., Richards, K. L., Plano, G. V., et al. (2014). The *Chlamydia trachomatis* type III secretion chaperone Slc1 engages multiple early effectors, including TepP, a tyrosine-phosphorylated protein required for the recruitment of CrkI-II to nascent inclusions and innate immune signaling. *PLoS Pathog.* 10:e1003954. doi: 10.1371/journal.ppat.1003954

Choi, J. Y., Jang, T. H., and Park, H. H. (2017). The mechanism of folding robustness revealed by the crystal structure of extra-superfolder GFP. *FEBS Lett.* 591, 442–447. doi: 10.1002/1873-3468.12534

Christie, P. J., Whitaker, N., and Gonzalez-Rivera, C. (2014). Mechanism and structure of the bacterial type IV secretion systems. *Biochim. Biophys. Acta* 1843, 1578–1591. doi: 10.1016/j.bbamcr.2013.12.019

Costa, T. R., Felisberto-Rodrigues, C., Meir, A., Prevost, M. S., Redzej, A., Trokter, M., et al. (2015). Secretion systems in Gram-negative bacteria: structural and mechanistic insights. *Nat. Rev. Microbiol.* 13, 343–359. doi: 10.1038/nrmicro3456

Cowles, C. E., Li, Y., Semmelhack, M. F., Cristea, I. M., and Silhavy, T. J. (2011). The free and bound forms of Lpp occupy distinct subcellular locations in *Escherichia coli*. *Mol. Microbiol.* 79, 1168–1181. doi: 10.1111/j.1365-2958.2011.07539.x

Day, J. B., Ferracci, F., and Plano, G. V. (2003). Translocation of YopE and YopN into eukaryotic cells by *Yersinia pestis* yopN, tyeA, sycN, yscB and lcrG deletion mutants measured using a phosphorylatable peptide tag and phosphospecific antibodies. *Mol. Microbiol.* 47, 807–823. doi: 10.1046/j.1365-2958.2003.03343.x

de Diego, I., Ksiazek, M., Mizgalska, D., Koneru, L., Golik, P., Szmigielski, B., et al. (2016). The outer-membrane export signal of *Porphyromonas gingivalis* type IX secretion system (T9SS) is a conserved C-terminal beta-sandwich domain. *Sci. Rep.* 6:23123. doi: 10.1038/srep23123

d'Enfert, C., Chapon, C., and Pugsley, A. P. (1987). Export and secretion of the lipoprotein pullulanase by *Klebsiella pneumoniae*. *Mol. Microbiol.* 1, 107–116. doi: 10.1111/j.1365-2958.1987.tb00534.x

Desvaux, M., Hebraud, M., Talon, R., and Henderson, I. R. (2009). Secretion and subcellular localizations of bacterial proteins: a semantic awareness issue. *Trends Microbiol.* 17, 139–145. doi: 10.1016/j.tim.2009.01.004

Dohlich, K., Zumsteg, A. B., Goosmann, C., and Kolbe, M. (2014). A substrate-fusion protein is trapped inside the Type III secretion system channel in *Shigella flexneri*. *PLoS Path.* 10:e1003881. doi: 10.1371/journal.ppat.1003881

Duong, F., and Wickner, W. (1997). Distinct catalytic roles of the SecYE, SecG and SecDFyajC subunits of preprotein translocase holoenzyme. *EMBO J.* 16, 2756–2768. doi: 10.1093/emboj/16.10.2756

East, A., Mechaly, A. E., Huysmans, G. H., Bernarde, C., Tello-Manigne, D., Nadeau, N., et al. (2016). Structural basis of pullulanase membrane binding and secretion revealed by X-Ray crystallography, molecular dynamics and biochemical analysis. *Structure* 24, 92–104. doi: 10.1016/j.str.2015.10.023

Ehsani, S., Rodrigues, C. D., and Enninga, J. (2009). Turning on the spotlight - using light to monitor and characterize bacterial effector secretion and translocation. *Curr. Opin. Microbiol.* 12, 24–30. doi: 10.1016/j.mib.2008.11.007

Enninga, J., Mounier, J., Sansonetti, P., and and, G., Tran Van Nhieu (2005). Secretion of type III effectors into host cells in real time. *Nat. Methods* 2, 959–965. doi: 10.1038/nmeth804

Feilmeier, B. J., Iseminger, G., Schroeder, D., Webber, H., and Phillips, G. J. (2000). Green fluorescent protein functions as a reporter for protein localization in *Escherichia coli*. *J. Bacteriol.* 182, 4068–4076. doi: 10.1128/JB.182.14.4068-4076.2000

Francetic, O., Buddelmeijer, N., Lewenza, S., Kumamoto, C. A., and Pugsley, A. P. (2007). Signal recognition particle-dependent inner membrane targeting of the PulG Pseudopilin component of a type II secretion system. *J. Bacteriol.* 189, 1783–1793. doi: 10.1128/JB.01230-06

Francetic, O., Hanson, M. P., and Kumamoto, C. A. (1993). prlA suppression of defective export of maltose-binding protein in secB mutants of *Escherichia coli*. *J. Bacteriol.* 175, 4036–4044. doi: 10.1128/jb.175.13.4036-4044.1993

Francetic, O., and Pugsley, A. P. (2005). Towards the identification of type II secretion signals in a nonacylated variant of pullulanase from *Klebsiella oxytoca*. *J. Bacteriol.* 187, 7045–7055. doi: 10.1128/JB.187.20.7045-7055.2005

Frankel, M. B., Wojcik, B. M., DeDent, A. C., Missiakas, D. M., and Schneewind, O. (2010). ABI domain-containing proteins contribute to surface protein display and cell division in *Staphylococcus aureus*. *Mol. Microbiol.* 78, 238–252. doi: 10.1111/j.1365-2958.2010.07334.x

Gagic, D., Ciric, M., Wen, W. X., Ng, F., and Rakonjac, J. (2016). Exploring the secretomes of microbes and microbial communities using filamentous phage display. *Front. Microbiol.* 7:429. doi: 10.3389/fmicb.2016.00429

Garcia, J. T., Ferracci, F., Jackson, M. W., Joseph, S. S., Pattis, I., Plano, L. R., et al. (2006). Measurement of effector protein injection by type III and type IV secretion systems by using a 13-residue phosphorylatable glycogen synthase kinase tag. *Infect. Immun.* 74, 5645–5657. doi: 10.1128/IAI.00690-06

Gawthorne, J. A., Audry, L., McQuitty, C., Dean, P., Christie, J. M., Enninga, J., et al. (2016). Visualizing the translocation and localization of bacterial Type III effector proteins by using a genetically encoded reporter system. *Appl. Environ. Microbiol.* 82, 2700–2708. doi: 10.1128/AEM.03418-15

Gelis, I., Bonvin, A. M., Keramisanou, D., Koukaki, M., Gouridis, G., Karamanou, S., et al. (2007). Structural basis for signal-sequence recognition by the translocase motor SecA as determined by NMR. *Cell* 131, 756–769. doi: 10.1016/j.cell.2007.09.039

Green, E. R., and Mecsas, J. (2016). Bacterial secretion systems: an overview. *Microbiol. Spectr.* 4. doi: 10.1128/microbiolspec.VMBF-0012-2015

Guan, C. R., Cui, W. J., Cheng, J. T., Liu, R., Liu, Z. M., Zhou, L., et al. (2016). Construction of a highly active secretory expression system via an engineered dual promoter and a highly efficient signal peptide in *Bacillus subtilis*. *N. Biotechnol.* 33, 372–379. doi: 10.1016/j.nbt.2016.01.005

Guilvout, I., Chami, M., Berrier, C., Ghazi, A., Engel, A., Pugsley, A. P., et al. (2008). *in vitro* multimerization and membrane insertion of bacterial outer membrane secretin PulD. *J. Mol. Biol.* 382, 13–23. doi: 10.1016/j.jmb.2008.06.055

Hara, T., Matsuyama, S., and Tokuda, H. (2003). Mechanism underlying the inner membrane retention of *Escherichia coli* lipoproteins caused by Lol avoidance signals. *J. Biol. Chem.* 278, 40408–40414. doi: 10.1074/jbc.M307836200

Heggeset, T. M., Kucharova, V., Naerdal, I., Valla, S., Sletta, H., Ellingsen, T. E., et al. (2013). Combinatorial mutagenesis and selection of improved signal sequences and their application for high-level production of translocated heterologous proteins in *Escherichia coli*. *Appl. Environ. Microbiol.* 79, 559–568. doi: 10.1128/AEM.02407-12

Heidtman, M., Chen, E. J., Moy, M. Y., and Isberg, R. R. (2009). Large-scale identification of *Legionella pneumophila* Dot/Icm substrates that modulate host cell vesicle trafficking pathways. *Cell. Microbiol.* 11, 230–248. doi: 10.1111/j.1462-5822.2008.01249.x

Ho, B. T., Dong, T. G., and Mekalanos, J. J. (2014). A view to a kill: the bacterial type VI secretion system. *Cell Host Microbe* 15, 9–21. doi: 10.1016/j.chom.2013.11.008

Hood, R. D., Singh, P., Hsu, F., Guvener, T., Carl, M. A., Trinidad, R. R., et al. (2010). A type VI secretion system of *Pseudomonas aeruginosa* targets a toxin to bacteria. *Cell Host Microbe* 7, 25–37. doi: 10.1016/j.chom.2009.12.007

Hortensia Rolán, G., Enrique Durand, A., and Mecsas, J. (2013). Identifying Yersinia YopH-Targeted signal transduction pathways that impair neutrophil responses during *in vivo* murine infection. *Cell Host Microbe* 14, 306–317. doi: 10.1016/j.chom.2013.08.013

Howell, H. A., Logan, L. K., and Hauser, A. R. (2013). Type III secretion of ExoU is critical during early *Pseudomonas aeruginosa* pneumonia. *mBio* 4:e00032-13. doi: 10.1128/mBio.00032-13

Hu, B., Lara-Tejero, M., Kong, Q., Galán, J. E., and Liu, J. (2017). *In situ* molecular architecture of the salmonella Type III secretion machine. *Cell* 168, 1065–1074.e10. doi: 10.1016/j.cell.2017.02.022

Huber, D., Boyd, D., Xia, Y., Olma, M. H., Gerstein, M., and Beckwith, J. (2005). Use of thioredoxin as a reporter to identify a subset of *Escherichia coli* signal sequences that promote signal recognition particle-dependent translocation. *J. Bacteriol.* 187, 2983–2991. doi: 10.1128/JB.187.9.2983-2991.2005

Ieva, R., Tian, P., Peterson, J. H., and Bernstein, H. D. (2011). Sequential and spatially restricted interactions of assembly factors with an autotransporter beta domain. *Proc. Natl. Acad. Sci. U.S.A.* 108, E383–E391. doi: 10.1073/pnas.1103827108

Ito, K., Wittekind, M., Nomura, M., Shiba, K., Yura, T., Miura, A., et al. (1983). A temperature-sensitive mutant of *E. coli* exhibiting slow processing of exported proteins. *Cell* 32, 789–797. doi: 10.1016/0092-8674(83)90065-X

Jackowski, S., and Rock, C. O. (1986). Transfer of fatty acids from the 1-position of phosphatidylethanolamine to the major outer membrane lipoprotein of *Escherichia coli*. *J. Biol. Chem.* 261, 11328–11333.

Jackson, R. C., and Blobel, G. (1977). Post-translational cleavage of presecretory proteins with an extract of rough microsomes from dog pancreas containing signal peptidase activity. *Proc. Natl. Acad. Sci. U.S.A.* 74, 5598–5602. doi: 10.1073/pnas.74.12.5598

Jacobsson, K., and Frykberg, L. (1996). Phage display shot-gun cloning of ligand-binding domains of prokaryotic receptors approaches 100% correct clones. *Biotechniques* 20, 1070-6, 1078, 1080–1081.

Jaumouille, V., Francetic, O., Sansonetti, P. J., and Tran Van Nhieu, G. (2008). Cytoplasmic targeting of IpaC to the bacterial pole directs polar type III secretion in Shigella. *EMBO J.* 27, 447–457. doi: 10.1038/sj.emboj.7601976

Juncker, A. S., Willenbrock, H., Von Heijne, G., Brunak, S., Nielsen, H., and Krogh, A. (2003). Prediction of lipoprotein signal peptides in Gram-negative bacteria. *Protein Sci.* 12, 1652–1662. doi: 10.1110/ps.0303703

Kanonenberg, K., Schwarz, C. K., and Schmitt, L. (2013). Type I secretion systems - a story of appendices. *Res. Microbiol.* 164, 596–604. doi: 10.1016/j.resmic.2013.03.011

Karimova, G., Pidoux, J., Ullmann, A., and Ladant, D. (1998). A bacterial two-hybrid system based on a reconstituted signal transduction pathway. *Proc. Natl. Acad. Sci. U.S.A.* 95, 5752–5756. doi: 10.1073/pnas.95.10.5752

Kenny, B., and Finlay, B. B. (1995). Protein secretion by enteropathogenic *Escherichia coli* is essential for transducing signals to epithelial cells. *Proc. Natl. Acad. Sci. U.S.A.* 92, 7991–7995. doi: 10.1073/pnas.92.17.7991

Kirn, T. J., Bose, N., and Taylor, R. K. (2003). Secretion of a soluble colonization factor by the TCP type 4 pilus biogenesis pathway in *Vibrio cholerae*. *Mol. Microbiol.* 49, 81–92. doi: 10.1046/j.1365-2958.2003.03546.x

Konovalova, A., Perlman, D. H., Cowles, C. E., and Silhavy, T. J. (2014). Transmembrane domain of surface-exposed outer membrane lipoprotein RcsF is threaded through the lumen of beta-barrel proteins. *Proc. Natl. Acad. Sci. U.S.A.* 111, E4350–E4358. doi: 10.1073/pnas.1417138111

Korotkov, K. V., Sandkvist, M., and Hol, W. G. (2012). The type II secretion system: biogenesis, molecular architecture and mechanism. *Nat. Rev. Microbiol.* 10, 336–351. doi: 10.1038/nrmicro2762

Kreutzenbeck, P., Kroger, C., Lausberg, F., Blaudeck, N., Sprenger, G. A., and Freudl, R. (2007). *Escherichia coli* twin arginine (Tat) mutant translocases possessing relaxed signal peptide recognition specificities. *J. Biol. Chem.* 282, 7903–7911. doi: 10.1074/jbc.M610126200

Kumamoto, C. A., and Gannon, P. M. (1988). Effects of *Escherichia coli* secB mutations on pre-maltose binding protein conformation and export kinetics. *J. Biol. Chem.* 263, 11554–11558.

Kumru, O. S., Schulze, R. J., Slusser, J. G., and Zuckert, W. R. (2010). Development and validation of a FACS-based lipoprotein localization screen in the Lyme disease spirochete *Borrelia burgdorferi*. *BMC Microbiol.* 10:277. doi: 10.1186/1471-2180-10-277

Lausberg, F., Fleckenstein, S., Kreutzenbeck, P., Frobel, J., Rose, P., Muller, M., et al. (2012). Genetic evidence for a tight cooperation of TatB and TatC during productive recognition of twin-arginine (Tat) signal peptides in *Escherichia coli*. *PLoS ONE* 7:e39867. doi: 10.1371/journal.pone.0039867

Lewenza, S., Vidal-Ingigliardi, D., and Pugsley, A. P. (2006). Direct visualization of red fluorescent lipoproteins indicates conservation of the membrane sorting rules in the family Enterobacteriaceae. *J. Bacteriol.* 188, 3516–3524. doi: 10.1128/JB.188.10.3516-3524.2006

Leyton, D. L., Rossiter, A. E., and Henderson, I. R. (2012). From self sufficiency to dependence: mechanisms and factors important for autotransporter biogenesis. *Nat. Rev. Microbiol.* 10, 213–225. doi: 10.1038/nrmicro2733

Li, L., Park, E., Ling, J., Ingram, J., Ploegh, H., and Rapoport, T. A. (2016). Crystal structure of a substrate-engaged SecY protein-translocation channel. *Nature* 531, 395–399. doi: 10.1038/nature17163

Lin, J., Zou, Y., Ma, C., She, Q., Liang, Y., Chen, Z., et al. (2015). Heterologous expression of mannanase and developing a new reporter gene system in *Lactobacillus casei* and *Escherichia coli*. *PLoS ONE* 10:e0142886. doi: 10.1371/journal.pone.0142886

Logger, L., Aschtgen, M. S., Guerin, M., Cascales, E., and Durand, E. (2016). Molecular dissection of the interface between the Type VI secretion TssM cytoplasmic domain and the TssG baseplate component. *J. Mol. Biol.* 428, 4424–4437. doi: 10.1016/j.jmb.2016.08.032

Lonjon, F., Turner, M., Henry, C., Rengel, D., Lohou, D., van de Kerkhove, Q., et al. (2016). Comparative secretome analysis of *Ralstonia solanacearum* Type 3 secretion-associated mutants reveals a fine control of effector delivery, essential for bacterial pathogenicity. *Mol. Cell. Proteomics* 15, 598–613. doi: 10.1074/mcp.M115.051078

Luo, Z. Q., and Isberg, R. R. (2004). Multiple substrates of the *Legionella pneumophila* Dot/Icm system identified by interbacterial protein transfer. *Proc. Natl. Acad. Sci. U.S.A.* 101, 841–846. doi: 10.1073/pnas.0304916101

Lybarger, S. R., Johnson, T. L., Gray, M. D., Sikora, A. E., and Sandkvist, M. (2009). Docking and assembly of the type II secretion complex of *Vibrio cholerae*. *J. Bacteriol.* 191, 3149–3161. doi: 10.1128/JB.01701-08

Lycklama A Nijeholt, J. A., and Driessen, A. J. (2012). The bacterial Sec-translocase: structure and mechanism. *Philos. Trans. R. Soc. Lond. B. Biol. Sci.* 367, 1016–1028. doi: 10.1098/rstb.2011.0201

Mahdavi, A., Szychowski, J., Ngo, J. T., Sweredoski, M. J., Graham, R. L., Hess, S., et al. (2014). Identification of secreted bacterial proteins by noncanonical amino acid tagging. *Proc. Natl. Acad. Sci. U.S.A.* 111, 433–438. doi: 10.1073/pnas.1301740111

Manoil, C., and Beckwith, J. (1985). TnphoA: a transposon probe for protein export signals. *Proc. Natl. Acad. Sci. U.S.A.* 82, 8129–8133. doi: 10.1073/pnas.82.23.8129

Maresso, A. W., Wu, R., Kern, J. W., Zhang, R., Janik, D., Missiakas, D. M., et al. (2007). Activation of inhibitors by sortase triggers irreversible modification of the active site. *J. Biol. Chem.* 282, 23129–23139. doi: 10.1074/jbc.M701857200

Mathiesen, G., Sveen, A., Brurberg, M. B., Fredriksen, L., Axelsson, L., and Eijsink, V. G. (2009). Genome-wide analysis of signal peptide functionality in Lactobacillus plantarum WCFS1. *BMC Genomics* 10:425. doi: 10.1186/1471-2164-10-425

Mavris, M., Page, A. L., Tournebize, R., Demers, B., Sansonetti, P., and Parsot, C. (2002). Regulation of transcription by the activity of the *Shigella flexneri* type III secretion apparatus. *Mol. Microbiol.* 43, 1543–1553. doi: 10.1046/j.1365-2958.2002.02836.x

Meyer, D. F., Noroy, C., Moumene, A., Raffaele, S., Albina, E., and Vachiery, N. (2013). Searching algorithm for type IV secretion system effectors 1.0: a tool for predicting type IV effectors and exploring their genomic context. *Nucleic Acids Res.* 41, 9218–9229. doi: 10.1093/nar/gkt718

Mills, E., Baruch, K., Aviv, G., Nitzan, M., and Rosenshine, I. (2013). Dynamics of the Type III secretion system activity of enteropathogenic *Escherichia coli*. *mBio* 4:e00303-13. doi: 10.1128/mBio.00303-13

Moir, D. T., Di, M., Wong, E., Moore, R. A., Schweizer, H. P., Woods, D. E., et al. (2011). Development and application of a cellular, gain-of-signal, bioluminescent reporter screen for inhibitors of Type II secretion in *Pseudomonas aeruginosa* and *Burkholderia pseudomallei*. *J. Biomol. Screen.* 16, 694–705. doi: 10.1177/1087057111408605

Molloy, M. P., Herbert, B. R., Slade, M. B., Rabilloud, T., Nouwens, A. S., Williams, K. L., et al. (2000). Proteomic analysis of the *Escherichia coli* outer membrane. *Eur. J. Biochem.* 267, 2871–2881. doi: 10.1046/j.1432-1327.2000.01296.x

Monteiro, F., Genin, S., van Dijk, I., and Valls, M. (2012). A luminescent reporter evidences active expression of Ralstonia solanacearum type III secretion system genes throughout plant infection. *Microbiol. Sgm* 158, 2107–2116. doi: 10.1099/mic.0.058610-0

Mueller, K. E., and Fields, K. A. (2015). Application of β-lactamase reporter fusions as an indicator of effector protein secretion during infections with the obligate intracellular pathogen *Chlamydia trachomatis*. *PLoS ONE* 10:e0135295. doi: 10.1371/journal.pone.0135295

Nivaskumar, M., Santos-Moreno, J., Malosse, C., Nadeau, N., Chamot-Rooke, J., Tran Van Nhieu, G., et al. (2016). Pseudopilin residue E5 is essential for recruitment by the type 2 secretion system assembly platform. *Mol. Microbiol.* 101, 924–941. doi: 10.1111/mmi.13432

Pallier, J., Aucher, W., Pires, M., and Buddelmeijer, N. (2012). Phosphatidyllycerol::prolipoprotein diacylglyceryl transferase (Lgt) of *Escherichia coli* has seven transmembrane segments and its essential residues are embedded in the membrane. *J. Bacteriol.* 194, 2142–2151. doi: 10.1128/JB.06641-11

Palmer, T., and Berks, B. C. (2012). The twin-arginine translocation (Tat) protein export pathway. *Nat. Rev. Microbiol.* 10, 483–496. doi: 10.1038/nrmicro2814

Payne, S. H., Bonissone, S., Wu, S., Brown, R. N., Ivankov, D. N., Frishman, D., et al. (2012). Unexpected diversity of signal peptides in prokaryotes. *mBio* 3: e00339-12. doi: 10.1128/mBio.00339-12

Perkowski, E. F., Zulauf, K. E., Weerakoon, D., Hayden, J. D., Ioerger, T. R., Oreper, D., et al. (2017). The EXIT strategy: an approach for identifying bacterial proteins exported during host infection. *mBio* 8:e00333-17. doi: 10.1128/mBio.00333-17

Petersen, T. N., Brunak, S., von Heijne, G., and Nielsen, H. (2011). SignalP 4.0: discriminating signal peptides from transmembrane regions. *Nat. Methods* 8, 785–786. doi: 10.1038/nmeth.1701

Pineau, C., Guschinskaya, N., Robert, X., Gouet, P., Ballut, L., and Shevchik, V. E. (2014). Substrate recognition by the bacterial type II secretion system: more than a simple interaction. *Mol. Microbiol.* 94, 126–140. doi: 10.1111/mmi.12744

Possot, O. M., Letellier, L., and Pugsley, A. P. (1997). Energy requirement for pullulanase secretion by the main terminal branch of the general secretory pathway. *Mol. Microbiol.* 24, 457–464. doi: 10.1046/j.1365-2958.1997.3451726.x

Remaut, H., Tang, C., Henderson, N. S., Pinkner, J. S., Wang, T., Hultgren, S. J., et al. (2008). Fiber formation across the bacterial outer membrane by the chaperone/usher pathway. *Cell* 133, 640–652. doi: 10.1016/j.cell.2008.03.033

Riordan, K. E., Sorg, J. A., Berube, B. J., and Schneewind, O. (2008). Impassable YscP substrates and their impact on the *Yersinia enterocolitica* type III secretion pathway. *J. Bacteriol.* 190, 6204–6216. doi: 10.1128/JB.00467-08

Rondelet, A., and Condemine, G. (2013). Type II secretion: the substrates that won't go away. *Res. Microbiol.* 164, 556–561. doi: 10.1016/j.resmic.2013.03.005

Saka, H. A., Thompson, J. W., Chen, Y. S., Dubois, L. G., Haas, J. T., Moseley, A., et al. (2015). *Chlamydia trachomatis* infection leads to defined alterations to the lipid droplet proteome in epithelial cells. *PLoS ONE* 10:e0124630. doi: 10.1371/journal.pone.0124630

Sato, K., Naito, M., Yukitake, H., Hirakawa, H., Shoji, M., McBride, M. J., et al. (2010). A protein secretion system linked to bacteroidete gliding motility and pathogenesis. *Proc. Natl. Acad. Sci. U.S.A.* 107, 276–281. doi: 10.1073/pnas.0912010107

Sauvonnet, N., and Pugsley, A. P. (1996). Identification of two regions of *Klebsiella oxytoca* pullulanase that together are capable of promoting beta-lactamase secretion by the general secretory pathway. *Mol. Microbiol.* 22, 1–7. doi: 10.1111/j.1365-2958.1996.tb02650.x

Schlumberger, M. C., Muller, A. J., Ehrbar, K., Winnen, B., Duss, I., Stecher, B., et al. (2005). Real-time imaging of type III secretion: Salmonella SipA injection into host cells. *Proc. Natl. Acad. Sci. U.S.A.* 102, 12548–12553. doi: 10.1073/pnas.0503407102

Schneewind, O., and Missiakas, D. (2014). Sec-secretion and sortase-mediated anchoring of proteins in Gram-positive bacteria. *Biochim. Biophys. Acta* 1843, 1687–1697. doi: 10.1016/j.bbamcr.2013.11.009

Seydel, A., Gounon, P., and Pugsley, A. P. (1999). Testing the '+2 rule' for lipoprotein sorting in the *Escherichia coli* cell envelope with a new genetic selection. *Mol. Microbiol.* 34, 810–821. doi: 10.1046/j.1365-2958.1999.01647.x

Shaner, N. C., Campbell, R. E., Steinbach, P. A., Giepmans, B. N., Palmer, A. E., and Tsien, R. Y. (2004). Improved monomeric red, orange and yellow fluorescent proteins derived from Discosoma sp. red fluorescent protein. *Nat. Biotechnol.* 22, 1567–1572. doi: 10.1038/nbt1037

Shuman, H. A., and Silhavy, T. J. (2003). The art and design of genetic screens: *Escherichia coli*. *Nat. Rev. Genet.* 4, 419–431. doi: 10.1038/nrg1087

Silverman, J. M., Agnello, D. M., Zheng, H., Andrews, B. T., Li, M., Catalano, C. E., et al. (2013). Haemolysin coregulated protein is an exported receptor and chaperone of type VI secretion substrates. *Mol. Cell* 51, 584–593. doi: 10.1016/j.molcel.2013.07.025

Sinclair, S. H., Garcia-Garcia, J. C., and Dumler, J. S. (2015). Bioinformatic and mass spectrometry identification of *Anaplasma phagocytophilum* proteins translocated into host cell nuclei. *Front. Microbiol.* 6:55. doi: 10.3389/fmicb.2015.00055

Sisko, J. L., Spaeth, K., Kumar, Y., and Valdivia, R. H. (2006). Multifunctional analysis of *Chlamydia*-specific genes in a yeast expression system. *Mol. Microbiol.* 60, 51–66. doi: 10.1111/j.1365-2958.2006.05074.x

Slagowski, N. L., Kramer, R. W., Morrison, M. F., LaBaer, J., and Lesser, C. F. (2008). A functional genomic yeast screen to identify pathogenic bacterial proteins. *PLoS Pathog.* 4:e9. doi: 10.1371/journal.ppat.0040009

Sory, M. P., and Cornelis, G. R. (1994). Translocation of a hybrid YopE-adenylate cyclase from *Yersinia enterocolitica* into HeLa cells. *Mol. Microbiol.* 14, 583–594. doi: 10.1111/j.1365-2958.1994.tb02191.x

Subtil, A., Delevoye, C., Bala,-á, M. E., Tastevin, L., Perrinet, S., and Dautry-Varsat, A. (2005). A directed screen for chlamydial proteins secreted by a type III mechanism identifies a translocated protein and numerous other new candidates. *Mol. Microbiol.* 56, 1636–1647. doi: 10.1111/j.1365-2958.2005.04647.x

Subtil, A., Parsot, C., and Dautry-Varsat, A. (2001). Secretion of predicted Inc proteins of *Chlamydia pneumoniae* by a heterologous type III machinery. *Mol. Microbiol.* 39, 792–800. doi: 10.1046/j.1365-2958.2001.02272.x

Thomas, J. D., Daniel, R. A., Errington, J., and Robinson, C. (2001). Export of active green fluorescent protein to the periplasm by the twin-arginine translocase (Tat) pathway in *Escherichia coli*. *Mol. Microbiol.* 39, 47–53. doi: 10.1046/j.1365-2958.2001.02253.x

Urbanowski, M. L., Lykken, G. L., and Yahr, T. L. (2005). A secreted regulatory protein couples transcription to the secretory activity of the *Pseudomonas aeruginosa* type III secretion system. *Proc. Natl. Acad. Sci. U.S.A.* 102, 9930–9935. doi: 10.1073/pnas.0504405102

Van Engelenburg, S. B., and Palmer, A. E. (2008). Quantification of real-time Salmonella effector type III secretion kinetics reveals differential secretion rates for SopE2 and SptP. *Chem. Biol.* 15, 619–628. doi: 10.1016/j.chembiol.2008.04.014

Van Engelenburg, S. B., and Palmer, A. E. (2010). Imaging type-III secretion reveals dynamics and spatial segregation of Salmonella effectors. *Nat. Methods* 7, 325–330. doi: 10.1038/nmeth.1437

Van Gerven, N., Klein, R. D., Hultgren, S. J., and Remaut, H. (2015). Bacterial amyloid formation: structural insights into curli biogensis. *Trends Microbiol.* 23, 693–706. doi: 10.1016/j.tim.2015.07.010

von Heijne, G. (1990). The signal peptide. *J. Membr. Biol.* 115, 195–201. doi: 10.1007/BF01868635

Voulhoux, R., Bos, M. P., Geurtsen, J., Mols, M., and Tommassen, J. (2003). Role of a highly conserved bacterial protein in outer membrane protein assembly. *Science* 299, 262–265. doi: 10.1126/science.1078973

Wang, G. Q., Xia, Y. J., Xiong, Z. Q., Zhang, H., and Ai, L. Z. (2016). Use of a novel report protein to study the secretion signal of Flagellin in *Bacillus subtilis*. *Curr. Microbiol.* 73, 242–247. doi: 10.1007/s00284-016-1054-4

Wille, T., Blank, K., Schmidt, C., Vogt, V., and Gerlach, R. G. (2012). Gaussia princeps luciferase as a reporter for transcriptional activity, protein secretion, and protein-protein interactions in *Salmonella enterica* Serovar Typhimurium. *Appl. Environ. Microbiol.* 78, 250–257. doi: 10.1128/AEM.06670-11

Wilson, M. M., Anderson, D. E., and Bernstein, H. D. (2015). Analysis of the outer membrane proteome and secretome of *Bacteroides fragilis* reveals a multiplicity of secretion mechanisms. *PLoS ONE* 10:e0117732. doi: 10.1371/journal.pone.0117732

Wilson, M. M., and Bernstein, H. D. (2016). Surface-Exposed Lipoproteins: an emerging secretion phenomenon in gram-negative bacteria. *Trends Microbiol.* 24, 198–208. doi: 10.1016/j.tim.2015.11.006

Wu, T., Malinverni, J., Ruiz, N., Kim, S., Silhavy, T. J., and Kahne, D. (2005). Identification of a multicomponent complex required for outer membrane biogenesis in *Escherichia coli*. *Cell* 121, 235–245. doi: 10.1016/j.cell.2005.02.015

Yount, J. S., Tsou, L. K., Dossa, P. D., Kullas, A. L., van der Velden A. W., and Hang, H. C. (2010). Visible fluorescence detection of type III protein secretion from bacterial pathogens. *J. Am. Chem. Soc.* 132, 8244–8245. doi: 10.1021/ja102257v

Zhu, W., Banga, S., Tan, Y., Zheng, C., Stephenson, R., Gately, J., et al. (2011). Comprehensive identification of protein substrates of the Dot/Icm Type IV Transporter of *Legionella pneumophila*. *PLoS ONE* 6:e17638. doi: 10.1371/journal.pone.0017638

Zückert, W. (2014). Secretion of bacterial lipoproteins: through the cytoplasmic membrane, the periplasm and beyond. *Biochem. Biophys. Acta Mol. Cell Res.* 1843, 1509–1516. doi: 10.1016/j.bbamcr.2014.04.022

Zuverink, M., and Barbieri, J. T. (2015). From GFP to β-lactamase: advancing intact cell imaging for toxins and effectors. *Pathog. Dis.* 73:ftv097. doi: 10.1093/femspd/ftv097

Type V Secretion Systems: An Overview of Passenger Domain Functions

*Ina Meuskens, Athanasios Saragliadis, Jack C. Leo and Dirk Linke**

Department of Biosciences, Section for Genetics and Evolutionary Biology, University of Oslo, Oslo, Norway

Correspondence:
Dirk Linke
dirk.linke@ibv.uio.no

Bacteria secrete proteins for different purposes such as communication, virulence functions, adhesion to surfaces, nutrient acquisition, or growth inhibition of competing bacteria. For secretion of proteins, Gram-negative bacteria have evolved different secretion systems, classified as secretion systems I through IX to date. While some of these systems consist of multiple proteins building a complex spanning the cell envelope, the type V secretion system, the subject of this review, is rather minimal. Proteins of the Type V secretion system are often called autotransporters (ATs). In the simplest case, a type V secretion system consists of only one polypeptide chain with a β-barrel translocator domain in the membrane, and an extracellular passenger or effector region. Depending on the exact domain architecture of the protein, type V secretion systems can be further separated into sub-groups termed type Va through e, and possibly another recently identified subtype termed Vf. While this classification works well when it comes to the architecture of the proteins, this is not the case for the function(s) of the secreted passenger. In this review, we will give an overview of the functions of the passengers of the different AT classes, shedding more light on the variety of functions carried out by type V secretion systems.

Keywords: secretion systems, AT, virulence, bacterial outer membrane, Gram-negative microorganisms

INTRODUCTION

Bacteria in general display a great variety of proteins on their cell surface, serving functions in nutrient transport, signaling, adhesion, or virulence. The proteins and protein complexes responsible for secretion in Gram-negative bacteria can be divided into categories, termed type I through type IX secretion systems (Costa et al., 2015; Green and Mecsas, 2016). These secretion systems differ in their complexity, with some of them consisting of only one polypeptide chain, like some of the type V secretion systems, to very intricate machineries consisting of multiple proteins building a complex, sometimes spanning several membranes. Compared with multi-subunit secretion systems, type V secretion systems seem somewhat peculiar. In comparison to most of the other secretion systems, they are much smaller, and only span the Gram-negative outer membrane (OM) (Leo et al., 2012). Type V secretion systems have no obvious energy source for transport, as there is no chemical energy such as ATP available in the periplasm, and no stable proton or other ion gradients exist across the OM. This has led to the name "autotransporter" (AT) that suggests a completely self-sufficient system for secretion (Klauser et al., 1993). Today we know of multiple factors that are involved in the secretion of ATs, but the source of energy for the secretion is still a matter of debate (Thanassi et al., 2005; Kang'ethe and Bernstein, 2013; Drobnak et al., 2015; Oberhettinger et al., 2015).

Type V secretion systems come in different forms depending on their structural features and domain organization (**Figure 1**). Type V ATs are therefore divided into sub-classes, type Va through type Ve, and possibly the very recently suggested type Vf (Grijpstra et al., 2013). While sub-classification according to the domain structure of ATs is useful to show differences in general principles of their organization and biogenesis, this does not usually reflect the secreted passengers' function(s). AT passengers function in very diverse ways, ranging from adhesins or enzymes to toxic proteins. **Table 1** gives a general overview of functionalities of passengers from the different subclasses. While many reviews concentrate on the topology and biogenesis of ATs, we deliberately focus on the functions of the passenger domains of ATs. We also give a short overview of the different topologies of AT sub-classes and their biogenesis.

TOPOLOGY OF ATs

Autotransporters consist of two distinct regions, a secreted passenger and a β-barrel domain that resides in the bacterial OM. The transmembrane domain typically is C-terminal to the passenger, but in type Ve ATs this domain order is inverted (**Figure 1**). Both regions are found in a single polypeptide chain with the exception of type Vb secretion systems, where the moieties are separate polypeptide chains (Guérin et al., 2017). While this broad separation into two functional regions is conserved among all type V systems, additional functional features have been identified. Examples include the PL-region (pertactin-like region), stable core or autochaperone region, all describing the same features of the membrane-proximal part of AT passengers that have special functions in folding and transport of the rest of the passenger (Drobnak et al., 2015). To further complicate the issue, the passenger itself has been referred to as the α-domain and the transmembrane β-barrel as the translocator or β-domain (Pohlner et al., 1995; Henderson et al., 2004; Drobnak et al., 2015). To avoid confusion, we will only refer to the β-barrel and the passenger in this review according to Drobnak et al. (2015). In the following section, we will give a short overview over the different structural features of the different sub-classes of ATs.

Type Va (Classical Autotransporters)

Type Va ATs are commonly known as classical ATs. They have been studied extensively, both functionally and structurally. Well studied members are the IgA protease from *Neisseria meningitidis* and EstA, a lipase from *Pseudomonas aeruginosa* (Henderson et al., 2004). Type Va ATs consist of a 12-stranded β-barrel domain, which functions as a C-terminal anchor in the OM and which is required for the transport of the N-terminal passenger to the extracellular environment. The passenger usually adopts a repetitive β-helix fold extending away from the bacterial cell surface, as demonstrated by the crystal structure of the Pertactin passenger (Emsley et al., 1996). Other forms of passengers are possible as well, as exemplified by EstA folding into a predominantly α-helical passenger (Brzuszkiewicz et al., 2009). The passenger harbors the specific function of the protein, and most model systems that have been studied in different species are important virulence factors. The diversity of passenger functions and specifically of protease functions among type Va passengers has given rise to classifications into SPATE (serine protease autotransporters of Enterobacteriaceae) proteases, SPATE-like and non-SPATE proteases (Yen et al., 2008; Ruiz-Perez and Nataro, 2014). In some cases the passenger domain of type Va ATs can be cleaved off after secretion. Passengers with enzymatic activity, like SPATE proteases, more often belong to the group of cleaved passengers than adhesin passengers, though cleavage has been observed also in adhesins such as AIDA-I (Charbonneau et al., 2006; Barnard et al., 2007; Dautin et al., 2007). Other examples are the SAATs (self-aggregating ATs) such as Ag43 from *E. coli* (Klemm et al., 2004).

Type Vb (Two-Partner Secretion)

Type Vb secretion systems consist of two distinct polypeptide chains encoded in one operon, e.g., the *Bordetella* filamentous hemagglutinin FHA (Chevalier et al., 2004; Jacob-Dubuisson et al., 2013). Due to this, they are also called two-partner secretion systems (TPSSs). TPSSs are composed of two proteins, one functioning as the translocator (TpsB) and the other as the secreted passenger (TpsA). TpsB is a 16-stranded, OM integral β-barrel protein with two periplasmic POTRA (polypeptide transport-associated) domains (Kajava et al., 2001). Due to the separation of the β-barrel and the passenger into two separate polypeptide chains, the passenger is released into the cell's environment after transport without any need for release by proteolytic cleavage. The fate of the passenger after secretion can differ. Some TpsB proteins stay attached to the OM in a non-covalently bound manner as exemplified by the *Haemophilus influenzae* proteins HMW1 and HMW2, while others are secreted, such as the *Serratia marcescens* protein ShlA (Braun et al., 1993; St. Geme, 1994).

Type Vc (Trimeric Autotransporter Adhesins)

Type Vc secretion systems have the same topology as type Va systems, but form highly intertwined trimeric structures. For this reason, and because all examples studied so far are adhesins, they are often referred to as trimeric autotransporter adhesins (TAAs). YadA, the *Yersinia* adhesin A from *Yersinia enterocolitica* and *Yersinia pseudotuberculosis* is the best studied member of this class of ATs (Mühlenkamp et al., 2015). These proteins consist of three identical polypeptide chains, and in their final folded form are composed of a C-terminal 12 stranded β-barrel (4 β-strands per monomer), and a passenger, which is also trimeric and typically folds into a lollipop-like structure with a coiled coil stalk and a globular head domain at the N-terminus of the protein (Hoiczyk, 2000; Linke et al., 2006; Wollmann et al., 2006).

Type Vd

The type Vd ATs are a fairly recently discovered class of ATs that resembles a hybrid of type Va and type Vb systems with PlpD

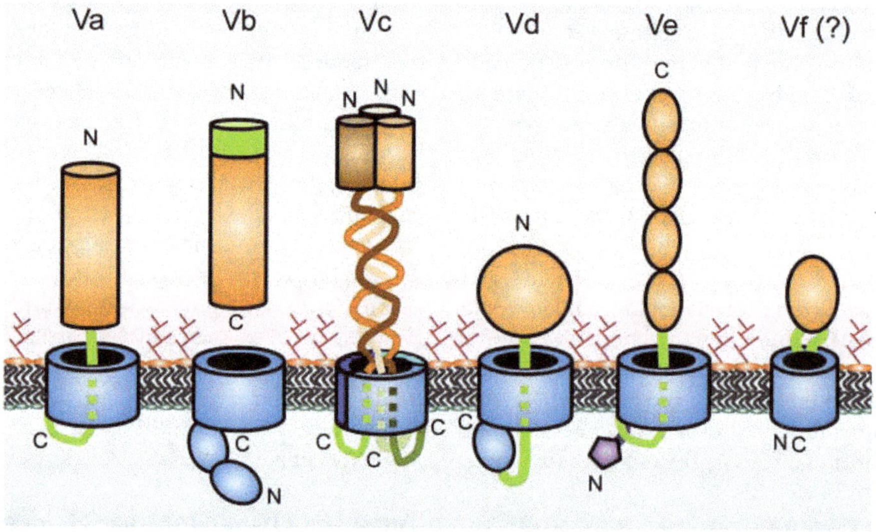

FIGURE 1 | Schematic of type V secretion system subclasses. β-barrels and POTRA domains are shown in blue, linkers and TPS domains in green, and passengers in orange. The periplasmic extension of type Ve proteins is in purple. The positions of the N- and C-termini are indicated. Type Vf is not fully established as part of the type V secretion scheme; this is denoted by the question mark.

from *P. aeruginosa* and FplA from *Fusobacterium nucleatum* as prototypical members (Salacha et al., 2010; Casasanta et al., 2017). The C-terminal β-barrel domain of type Vd ATs consist of 16 β-strands which is similar to the β-barrel of TpsB proteins. However, type Vd ATs have only one POTRA domain, whereas TpsB proteins have two POTRA domains for binding their TpsA substrate for secretion (Leo et al., 2012). Whereas type Va ATs have a passenger that typically folds into a β-helical structure, the passenger domains of type Vd ATs, which have been found to harbor lipase activity, adopt an α/β-hydrolase fold (Emsley et al., 1996; Da Mata Madeira et al., 2016). Note though that there are also some type Va ATs with passengers that have an α/β hydrolase fold, e.g., EstA (Brzuszkiewicz et al., 2009). A key difference between Va and type Vd passengers seems to be that type Va passengers have a multitude of different folds and functionalities, while type Vd passengers characterized so far only function as lipases/esterases (Da Mata Madeira et al., 2016; Casasanta et al., 2017).

Type Ve (Inverse Autotransporters)

Type Ve ATs share obvious similarities to type Va ATs, with a modular architecture including a 12-stranded β-barrel domain and a secreted, monomeric passenger that remains attached after translocation. The major difference to type Va ATs is that the type Ve ATs have an inverted domain order, with the β-barrel at the N-terminal end and the passenger at the C-terminus as shown for intimin and invasin from *Escherichia coli* and *Y. enterocolitica* (Leo et al., 2012, 2015b). This has led to the name "inverse autotransporters" (Oberhettinger et al., 2015). The passenger of inverse ATs typically contains domains with Ig-like or lectin-like folds, and some exemplars have long, repetitive stretches of Ig-like domains that are capped with a lectin-like domain (Leo et al., 2015b). Some type Ve ATs have an additional periplasmic domain which is not found in other types of ATs. This periplasmic domain aids in dimerization as well as in interactions with peptidoglycan, possibly anchoring it and helping in receptor interactions during host invasion (Leo et al., 2015a).

Type Vf?

The type Vf secretion systems were very recently described as a new class of ATs, with BapA as the prototypical member, and appear to be unique to *Helicobacter pylori*. This proposed class of ATs has a surface-exposed domain inserted into the N-terminal region between the first and second β-strand of an 8-stranded β-barrel domain, and contains no additional passenger at either terminus of the protein. Thus, the proposed passenger actually is an extended loop of the β-barrel domain and the β-barrel is smaller than that of any other AT (Coppens et al., 2018). Though BapA and related proteins have been proposed to be part of the AT family, their topology is very different from the other types of ATs. It is therefore questionable whether these proteins should be considered ATs, and further investigation of the secretion mechanism is required to find out whether these proteins actually self-export in a similar fashion to other ATs.

BIOGENESIS OF ATs

Transport Across Membranes and β-Barrel Insertion

Like most OM proteins, ATs follow a conserved pathway in their biogenesis.

Autotransporters are translated in the cytosol where the polypeptide chain is kept in an unfolded state by the help of chaperones and translocated across the inner membrane (IM) into the periplasm by the SecYEG translocon (Sijbrandi et al., 2003; Tsirigotaki et al., 2017). An N-terminal signal sequence ensures proper recognition of the AT as a Sec target, and targeting

TABLE 1 | Functions of type V secretion system passenger domains.

Function	Mode of action	Class	Group	Example	AT-Type	References
Enzyme	Protease	SPATE	Group I	EspC (*Escherichia coli*)	Va	Stein et al., 1996
				EspP (*Escherichia coli*)		Roman-Hernandez et al., 2014
			Group II	Hbp (*Escherichia coli*)	Va	Soprova et al., 2010
				Pic (*Escherichia coli*)		Abreu et al., 2016
		SPATE-like		IgA protease (*Neisseria gonorrhoeae*)	Va	Hulks and Plaut, 1978; Diebel et al., 2004
		Non-SPATE		NalP (*Neisseria meningitidis*)	Va	Arenas et al., 2013
				Ssph1/2 (*Serratia marcescens*)		Ohnishi et al., 1997; Kida et al., 2008
				LepA (*Pseudomonas aeruginosa*)	Vb	Ohnishi and Horinouchi, 2009
	Lipase/esterase			EstA (*Pseudomonas aeruginosa*)	Va	Wilhelm et al., 2007
				McaP (*Moraxella catarrhalis*)		Lipski et al., 2007
				EstA (*Serratia liquefaciens*)		Riedel et al., 2003
				ApeE (*Salmonella* Typhimurium)		Carinato et al., 1998
				PlpD (*Pseudomonas aeruginosa*)	Vd	Salacha et al., 2010; Da Mata Madeira et al., 2016
				FplA (*Fusobacteriumnucleatum*)		Casasanta et al., 2017
CDI	Growth-inhibition of competing bacteria			CdiA/B (Enterobacteriaceae)	Vb	Diner et al., 2012
				BcpA/B (*Burkholderia pseudomallei*)		Garcia et al., 2013
Alteration of Host Cell Processes				BadA (*Bartonella henselae*)	Vb	Riess et al., 2004
				InvD (*Yersinia pseudotuberculosis*)	Ve	Sadana et al., 2018
Immune evasion	Circumventing host immune response			IgA protease (*Escherichia coli*)	Va	Zinchenko et al., 2018
				EtpA (*Escherichia coli*)	Vb	Roy et al., 2009
				YadA (*Yersinia* spp.)	Vc	Schindler et al., 2012
				Eib (*Escherichia coli*)		Leo and Goldman, 2009; Leo et al., 2011
Cyto-/Hemolysis				VacA (*Helicobacterpylori*)	Va	Cover and Blanke, 2005
				ShlA (*Serratia marcescens*)	Vb	Reboud et al., 2017
				ExlA (*Pseudomonas aeruginosa*)		Yang and Braun, 2000
Adhesin	Adhesion to surfaces/receptors			AIDA-I (*Escherichia coli*)	Va	Laarmann and Schmidt, 2003;
				EhaA (*Escherichia coli*)		Charbonneau et al., 2006
				Pertactin (*Bordetella pertussis*)		Aricò et al., 1993
				FHA (*Bordetella pertussis*)	Vb	Serra et al., 2011
				HMW1/2 (*Haemophilus influenzae*)		Buscher et al., 2006
				YadA (*Yersinia* spp.)	Vc	Tertti et al., 1992; Mühlenkamp et al., 2017
				Eib (*Escherichia coli*)		Leo and Goldman, 2009
				Intimin (*Escherichia coli*)	Ve	Kenny et al., 1997
				InvA (*Yersinia* spp.)		Isberg et al., 2000
	Auto-Agglutination and biofilm formation			Ag43 (*Escherichia coli*)	Va	Sherlock et al., 2006
				FHA (*Bordetella pertussis*)	Vb	Serra et al., 2011
				EtpA (*Escherichia coli*)		Roy et al., 2009
				YadA (*Yersinia enterocolitica*)	Vc	Trunk et al., 2018
				Eib (*Escherichia coli*)		Leo and Goldman, 2009
Intracellular motility	Activation of actin polymerases			IcsA (*Shigella flexneri*)	Va	Goldberg and Theriot, 1995
				YapV (*Yersinia* spp.)		Besingi et al., 2013
				BimA (*Burkholderia* spp.)	Vc	Benanti et al., 2015

and secretion through the IM and signal peptide cleavage after transport works in the same way as for other Sec-secreted proteins (Papanikou et al., 2007). Some ATs, like Hbp and AIDA-I, show an extended Sec signal sequence which might aid in slowing down IM translocation and thus in prevention of premature folding and aggregation of the AT within the periplasm (Henderson et al., 1998; Szabady et al., 2005; Jacob-Dubuisson et al., 2013). For type Vb systems, it has been shown that some TpsA passengers aggregate much faster than others and therefore retaining the AT bound to the Sec is beneficial; Otp is a protein which is not prone to aggregation and therefore does not require fast transport to the OM (Choi and Bernstein, 2010). In other systems like FHA, quick secretion is of importance as degradation of unfolded FHA by DegP is more likely due to the length of the FHA precursor (Baud et al., 2009).

In the periplasm, ATs are kept unfolded but in a folding-competent state, shielded from aggregation by periplasmic chaperones like SurA, Skp and DegP (Baud et al., 2009; Ieva and Bernstein, 2009; Oberhettinger et al., 2012; Pavlova et al., 2013; Weirich et al., 2017). Insertion of the β-barrel domain of ATs is then facilitated by the β-barrel assembly machinery (BAM) complex (Jain and Goldberg, 2007; Leo et al., 2012). In *E. coli*, it is composed of five subunits, BamA through BamE. This complex interacts with most if not all OM integral β-barrel proteins (Lee et al., 2018). The 16-stranded β-barrel integral membrane protein BamA helps in insertion of the substrate barrel into the OM by a not yet entirely understood mechanism (Schiffrin et al., 2017). For type Va ATs, it has been clearly shown by crosslinking experiments that the 12-stranded β-barrel membrane anchor folds and inserts into the OM aided directly by the BAM complex. The passenger of EspP, an *E. coli* AT, for example, can be crosslinked to periplasmic chaperones, as well as to its β-barrel domain and to BamA (Ieva and Bernstein, 2009; Pavlova et al., 2013). Similarly, type Vc and Ve ATs interact with the Bam complex, as shown for YadA and Invasin (Roggenkamp et al., 2003; Oberhettinger et al., 2015).

Passenger Secretion

While most other bacterial secretion systems have access to energy sources like proton gradients across the IM or are directly energized by cytoplasmic ATP, ATs only span the OM, which is too leaky for ion gradients, and the periplasm is devoid of ATP (Nikaido and Vaara, 1985; Silhavy et al., 2006). Various models for how the secretion and folding process of passengers is energized have been proposed. One plausible explanation is that the energy for transport comes from the intrinsic folding capacity of the AT itself, either directly driving export or leading to a Brownian ratchet model where, once secreted, the passenger cannot slide back into the periplasm and is therefore driven to move outside the cell and fold (Henderson et al., 2004; Choi and Bernstein, 2010). Furthermore, asymmetric charge distribution within the passenger has been put forward as a possible driving factor for passenger secretion (Kang'ethe and Bernstein, 2013).

Passenger transport and secretion differ slightly between the various AT subclasses due to differences in domain organization. In type Va ATs, the passenger is transported via a C-terminus-first mechanism. According to the widely accepted hairpin-loop model of secretion, a hairpin-loop is formed at the C-terminus of the passenger in the interior of the β-barrel, followed by sequential folding of the passenger on the cell surface starting from the C-terminus (Junker et al., 2006). This was shown for multiple members of the type Va AT subclass, including Pertactin, Hbp and EspP (Junker et al., 2009; Peterson et al., 2010; Soprova et al., 2010).

For type Vb secretion, models are somewhat different since in the TPSSs the β-barrel domain is separated from the passenger domain. After the TpsB transporter is properly inserted into the OM by the BAM complex, recognition of TpsA by TpsB is provided by interaction of the TpsB POTRAs and the N-terminal TPS signal of TpsA (Baud et al., 2009). The TPS signal is a conserved stretch with an amphipathic character that remains unfolded in the periplasm. Association and dissociation rates of the TPS signal with the TpsB POTRA domains are high based on surface plasmon resonance experiments, making the interaction transient, and helping in later release of the TpsA substrate from its transporter (Delattre et al., 2010; Guérin et al., 2017). NMR experiments have shown similar highly dynamic interactions (Garnett et al., 2015). Crosslinking experiments have further shown that the TPS signal interacts with the TpsB POTRA domains, as well as some central amino acids within the barrel lumen (Baud et al., 2014). Similarly to all other Type V secretion systems, it is assumed that during transport, TpsA is unfolded as it passes through the central pore of the TpsB barrel and that folding of the substrate occurs during exit from the transporter barrel.

There are two different models for how the export of the TpsA is initiated: one is that, like in other ATs, a hairpin is formed within the barrel pore driving folding of the secreted substrate in a C-to-N direction. Release of the TpsB-bound TPS domain would then occur at the end of secretion, after major parts of TpsA have already folded (Pavlova et al., 2013; Norell et al., 2014). In this case, the high on/off rate between the PORTA domains and the TPS signal domain would facilitate the release that is based on the pulling forces generated by the folding process itself (Guérin et al., 2017). According to the second model, the N-terminal TPS domain nucleates folding, i.e., the TPS domain is exported first and the rest of the protein folds N-to-C (Hodak and Jacob-Dubuisson, 2007). The fact that the TpsA proteins' N-terminal domain can also fold independently bolsters this argument (Clantin et al., 2004, 2007).

In type Vc ATs, passenger secretion is more intricate due to the trimeric nature of the proteins. Three passenger polypeptide chains have to be orchestrated through a comparatively narrow β-barrel domain. After formation of the 12-stranded β-barrel, the passenger is transported to the exterior of the cell starting with the formation of a hairpin loop of each of the three passenger domains followed by folding of the coiled coil stalk (Linke et al., 2006; Szczesny and Lupas, 2008; Mikula et al., 2012; Chauhan et al., 2019). Transport of three distinct polypeptide chains in a hairpin loop conformation across a comparably small barrel might be sterically challenging. The interior of type Vc β-barrels contains many glycine and alanine residues which have small side chains, and it has been suggested that this facilitates passage of multiple chains though the barrel interior (Mikula et al., 2012). Additionally, β-barrel proteins are not necessarily fully rigid pores. The capacity of "breathing" movement without breakage of the hydrogen bonding has already been shown for the usher protein FimD, which in its apo-structure is more narrow than when bound to a transport substrate (Phan et al., 2011). Similar breathing behavior would be necessary in type Vc autotransport to accommodate all chains simultaneously. An additional problem comes with the highly intertwined passenger structure in type Vc systems. Sequential folding after initial hairpin formation would build up mechanical strain. It has been shown for some examples that an YxD/RxD motif toward the C-terminus of the passenger helps in initiation of passenger folding and folding outside the membrane anchor, potentially by releasing mechanical strain. YxD motifs furthermore stabilize right-handed coiled-coils whereas RxD motifs support left-handed coiled-coils (Alvarez et al., 2010). In addition, while the core residues of coiled-coil proteins are generally hydrophobic, some trimeric AT passengers contain hydrophilic residues in these positions. These residues can coordinate anions, which might allow sequences that are otherwise not easily folded to interact and stabilize (Hartmann et al., 2009; Leo et al., 2011).

It is not yet entirely clear how passenger secretion works in type Vd systems, and what role the POTRA domain plays in this (Salacha et al., 2010). It might function either as a chaperone for the passenger, aiding in secretion, or aiding in the recruitment of proteases for passenger cleavage. In some strains of *F. nucleatum* the passenger domain of FplA seems to be cleaved off while in other strains this could not be shown (Casasanta et al., 2017). It is unclear whether proteolytic cleavage of the passenger of type Vd ATs is achieved via autoproteolysis, like in some type Va ATs, or via an independent protease, like in the example of the NalP cleaving the type Va AT IgA protease for release from its β-barrel domain (Salacha et al., 2010; Casasanta et al., 2017). However, the fact that type Vd passengers remain uncleaved in some strains and when heterologously expressed in *E. coli* supports the latter interpretation (Salacha et al., 2010).

The biogenesis of type Ve ATs is similar to the one of type Va ATs. Although the topology of type Ve ATs is inverted, the β-barrel functions as a transport pore in an analogous way via formation of a hairpin-loop, and the passenger is secreted in a very similar fashion to the passenger secretion of classical ATs, but in the opposite direction (N-to-C rather than C-to-N) (Oberhettinger et al., 2012, 2015). Folding is energized by sequential folding of the extracellular Ig-like domains, as shown for the example of Intimin (Leo et al., 2016).

FUNCTIONS OF AT PASSENGERS

Research on ATs has traditionally focused on a single protein and its function (often in pathogenesis) (**Figure 2**), or on individual subclasses based on topology and biogenesis (**Figure 1**). The latter has led to the systematic sub-classification into type Va to Ve secretion systems, but this classification does not reflect the function of the passenger. Passengers from different subclasses can have similar functions despite structural differences, and some individual passengers mediate multiple functions. Therefore, a systematic differentiation between passenger functions is harder to achieve (Dautin and Bernstein, 2007; Drobnak et al., 2015). Furthermore, while some functions can be described as general virulence traits and can be found in a wide variety of Gram-negative bacteria, such as protease activity (Yen et al., 2008; Dautin, 2010), other functions are rather unique and are involved in tasks specific to the bacterium and its lifestyle, such as intracellular mobility and nutrient acquisition in limited environments (Luckett et al., 2012; Benanti et al., 2015).

Tables 1, 2 show an overview of the functions and exemplary members of the functional groups.

Enzymatic Activities

Passengers with enzymatic functions can directly alter host cell processes, be involved in immune evasion or help in the establishment and colonization of niches. The large diversity of possible enzymatic functions includes protease activity, lipase activity, but also contact-dependent growth inhibition (CDI) mediated by ADP ribosyl cyclases, adenosine deaminases, or nicking endonucleases.

Lipases and Esterases

Lipase and esterase activity can be found in type Va as well as in type Vd passenger domains. It has been proposed that these enzymes aid in niche establishment especially for intracellular bacteria, but also in alteration of host cell signaling by phosphoinositide (PI) cleavage (Casasanta et al., 2017). Though there are some ideas and models for the role that ATs with lipase and esterase function play in virulence, their exact function is not clear, and it is quite possible that AT lipases of different bacteria act on different targets in the host (Da Mata Madeira et al., 2016; Casasanta et al., 2017).

A prototypical example for a passenger with a lipase domain is EstA, a type Va AT of *P. aeruginosa.* In this protein, the passenger is not cleaved from its β-barrel domain (Wilhelm et al., 1999). EstA is implicated in the cleavage of rhamnolipids, which by themselves are important for biofilm formation and have toxic properties (Davey et al., 2003; Klausen et al., 2003). When deleting the *estA* gene, rhamnolipids in biofilms shifted from di-rhamnolipids to mono-rhamnolipids, indicating that rhamnolipids might be a target of EstA (Wilhelm et al., 2007; Tielen et al., 2010). Knockouts of *estA* also showed alterations in motility. While swimming and swarming are absent in *estA* deletion strains, twitching motility is enhanced, which seems somewhat puzzling taking into account that only swarming, not swimming or twitching motility is influenced by the rhamnolipid content of a biofilm (Kohler et al., 2000; Déziel et al., 2003; Wilhelm et al., 2007). Other differences in motility can thus not be explained by rhamnolipids as targets of EstA alone (Wilhelm et al., 2007). Other type Va AT with comparable lipolytic functions can be found in *Moraxella catarrhalis* (Mcap), *Serratia liquefaciens* (EstA) and also *Salmonella typhimurium*

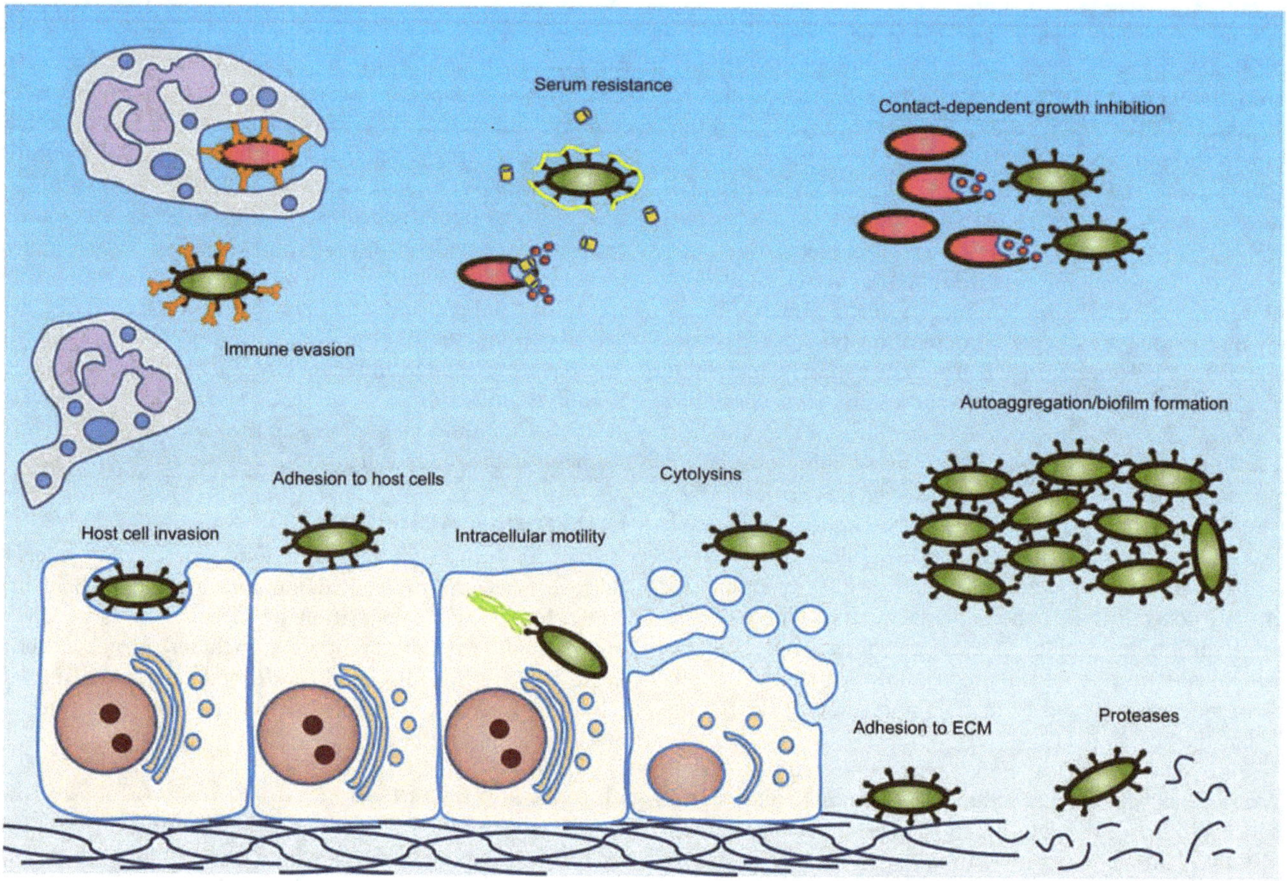

FIGURE 2 | Typical activities of autotransporter proteins. The major activities found in autotransporters are shown schematically. Green bacteria express autotransporters, but red bacteria do not. Autotransporters mediate activities such as adhesion to host cells or the extracellular matrix (ECM), invasion of host cells, immune evasion and serum resistance, contact-dependent growth inhibition, toxicity toward host cells, intracellular mobility, protease activity, and autoaggregation and biofilm formation. Cytolysin activity can be mediated by lipases or pore-forming toxins. Immune evasion can be accomplished by a variety of mechanisms, such as by binding to the constant regions of antibodies to prevent opsonisation (depicted here), cleavage of antibodies, mediating phagocytosis resistance or modulating immune cell signaling and gene regulation. In serum resistance, bacteria become resistant to the bactericidal effects of complement (shown as orange cylinders) by binding to complement-regulatory factors such as Factor H (shown as yellow ribbons). For full descriptions, see main text.

(ApeE) (Carinato et al., 1998). EstA of *S. liquefaciens*, for example, is involved in cellular signaling by providing the cells with enough lipids for synthesis of second messenger molecules (Riedel et al., 2003). This function could not be verified for EstA in *P. aeruginosa*, however (Wilhelm et al., 2007).

In type Vd ATs, all passengers described so far have lipase activity. Though lipase activity can also be found in type Va ATs, the structure and domain organization of type Vd ATs is very distinct with a single POTRA domain fusing the β-barrel and the passenger even though both passengers have a α/β hydrolase fold (Salacha et al., 2010; Casasanta et al., 2017). Furthermore, the catalytically active site in type Va lipases like the GDSL lipase EstA usually consists of a catalytic triad while type Vd ATs show a catalytic dyad (Desvaux et al., 2005; Brzuszkiewicz et al., 2009; van den Berg, 2010). Prototypic for type Vd ATs are PlpD from *P. aeruginosa* and FplA from *F. nucleatum* (Salacha et al., 2010; Casasanta et al., 2017). The name of the *Pseudomonas* proteins comes from structural similarity of the passenger to patatin from potatoes; hence the name patatin-like protein D. So far more than 200 Gram-negative bacterial species have been found to encode PlpD orthologues, including a wide variety of pathogenic as well as non-pathogenic, environmental bacteria (Salacha et al., 2010). The patatin-like domain of PlpD harbors A1 phospholipase activity (Da Mata Madeira et al., 2016). Similarly, the PlpD homologue from *F. nucleatum*, FplA, has phospholipase A1 activity and displays very high hydrolytic activity toward artificial substrates (Casasanta et al., 2017). The structure of the PlpD passenger has been solved, and the active site offers enough space to accommodate C16 to C20 acyl chains (Da Mata Madeira et al., 2016). The catalytic dyad in PlpD consists of Ser60 and Asp207, which are in close proximity to a hydrophobic helix within the cleft containing the active site, which probably stabilizes the lipid interaction (Casasanta et al., 2017). Similarly, in FplA, the catalytic residues are Ser98 and Asp243 (Casasanta et al., 2017). Biochemical experiments with FplA showed no binding affinity for phospholipids like phosphatidycholine (PC), phosphatidyethanolamine (PE), and phospatidic acid (PA) but

TABLE 2 | Adhesins and adhesion targets of autotransporters.

Type	Host bacterium	Binding partners	References
Va			
AIDA-I	*Escherichia coli*	glycoprotein	Laarmann and Schmidt, 2003
EhaB		collagen, laminin	Chagnot et al., 2012
Tsh		collagen, fibronectin	Kostakioti and Stathopoulos, 2004
Hap	*Haemophilus influenzae*	fibronectin, laminin, collagen	Fink et al., 2003
MisL	*Salmonella enterica*	collagen, fibronectin	Dorsey et al., 2005
ShdA			Kingsley et al., 2002, 2004
Vb			
FHA	*Bordetella* spp.	heparin, dextran	Menozzi et al., 1994
EtpA	*Escherichia coli*	flagellin	Roy et al., 2009
HMW1/2	*Haemophilus influenzae*	sialylated glycoprotein/ unknown	St. Geme, 1994; St. Geme and Yeo, 2009
MhaB1	*Moraxella catarrhalis*	mammalian cells	Balder et al., 2007
HrpA	*Neisseria meningitidis*	mammalian cells	Schmitt et al., 2007
Vc			
BadA	*Bartonella henselae*	collagen, laminin, fibronectin	Riess et al., 2004
EhaG	*Escherichia coli*	laminin, fibronectin,	Totsika et al., 2012
UpaG		fibrinogen, collagen	Valle et al., 2008; Totsika et al., 2012
DsrA	*Haemophilus ducreyi*	fibronectin	Leduc et al., 2008
NcaA		Collagen	Fulcher et al., 2006
Hia	*Haemophilus influenzae*	cells	Spahich, 2011
UspA	*Moraxella catharralis*	fibronectin, laminin	Tan et al., 2005, 2006
NadA	*Neisseria meningiditis*	cells	Tavano et al., 2011
NhhA		laminin, heparan sulfate, cells	Scarselli et al., 2006
SadA	*Salmonella typhimurium*	cells (no ECM)	Raghunathan et al., 2011
YadA	*Yersinia enterocolitica*	collagen, fibronectin, laminin	El Tahir and Skurnik, 2001
Ve			
FdeC	*Escherichia coli*	cells, collagen	Nesta et al., 2012
Intimin		Tir protein	Batchelor et al., 2000
Invasin	*Yersinia* spp.	β1-integrins	Isberg et al., 2000
InvD		IgG/IgA	Sadana et al., 2018

it seems to bind to phosphatidyinositol (PI) phosphates which acts in signaling of host cells (Casasanta et al., 2017). It has been proposed that lipase activity is important for establishing a niche especially for intracellularly living bacteria by recognizing phosphorylated PIs and by their cleavage altering cell signaling (Casasanta et al., 2017). Furthermore, this lipase activity might also help in cytosolic release of the bacteria from the vacuole and phagosomal survival within the host cell, which are important for niche establishment and an intracellular lifestyle of this bacterium (Casasanta et al., 2017).

Proteolytic Activity

Proteolytic cleavage of host target proteins can be beneficial for the bacterium in terms of virulence. A number of proteolytically active type Va AT passengers have been discovered. The first AT discovered is Immunoglobulin A (IgA) protease from *Neisseria gonorrhoeae* and *N. meningitidis*. The passenger of IgA protease can be divided into a 106 kDa protease domain, a more C-terminal 3 kDa γ-peptide, a 12–44 kDa α-peptide, a linker region and the C-terminal β-barrel membrane anchor (Halter et al., 1984). Following export, the passenger is cleaved off from the β-barrel transporter to function as a protease in virulence. Cleavage of the passenger can happen in different ways: either the passenger is autoproteolyzed or cleaved by another AT called NalP in a phase-dependent fashion (Roussel-Jazédé et al., 2010). NalP-dependent cleavage results in the release of the full passenger domain, including the protease domain, the α-peptide, the γ-peptide and the linker region, while autoproteolysis results in the release of fragments of different sizes: either the protease, the protease and the γ-peptide or the complete passenger. The autoproteolytic cleavage seems to be strain-dependent and can have different effects on the role the released passenger has in virulence.

More recently, it has been shown that IgA protease can also modulate host gene expression (Hulks and Plaut, 1978). If the passenger is cleaved off together with the α-peptide, the α-peptide via its nuclear localization-like sequence can guide the IgA protease domain to the host cell nucleus. The IgA protease domain can then cleave NF-κB and p65/RelA in the nucleus and thereby modulate the entire host cell response to pathogenic stressors (Pohlner et al., 1995; Besbes et al., 2015).

Furthermore, IgA proteases have been shown to cleave LAMP1, a highly glycosylated endosomal and lysosomal membrane protein which normally protects the membrane from degradation (Hauck and Meyer, 1997). Degradation of LAMP1 in turn seems beneficial for the growth of intracellular bacteria, as *Neisseria* lacking IgA protease grow slower intracellularly than wild-type bacteria (Lin et al., 1997). *Neisseria* pilus proteins and porins influence the calcium intake of epithelial cells, leading to the exocytosis of endosomes (Ayala et al., 2002). This shows that IgA proteases secreted by *Neisseria* have multiple targets and play an important role in survival and growth of *Neisseria* in infection contexts.

Autotransporter proteases do not only play a role in host immune response circumvention and establishing a niche in the host, but also in nutrient acquisition. As an example, AaaA from *P. aeruginosa* is involved in nitrogen acquisition from peptides in chronic infections. It acts as an aminopeptidase aiding in long-term infections in mice (Luckett et al., 2012). Another group of type Va ATs with protease activity are the SPATEs (Serine Protease Autotransporters of *Enterobacteriaceae*) proteins. Their targets can be very diverse; one prominent example is Hbp (hemoglobin protease or hemoglobin binding protein) from *E. coli.* Hbp employs an active Ser residue for cleavage of hemoglobin, but shows no specificity for other proteins such as albumin (Otto et al., 2005). Hbp, like all SPATEs, is cleaved off after passenger transport (Dautin, 2010). Functionally, Hbp is used by enterohemorrhagic *E. coli* strains for heme acquisition (Dautin, 2010).

Contact-Dependent Growth Inhibition (CDI)

Enzymatically active passenger domains can also be found in type Vb ATs. Like in the type Va ATs, there can be dramatic differences in enzymatic functions. One major function of the passengers of type Vb ATs is CDI. Here, the passenger domains can function as nucleases, deaminases, and also as metallopeptidases. The name of this functional class of type Vb ATs is potentially misleading, since it is still unclear whether growth inhibition of competing bacterial strains is the main evolutionary purpose of this system (Guérin et al., 2017).

Contact-dependent growth inhibitions are transporter-effector pairs called CdiB/A pairs, where CdiA corresponds to TpsA and CdiB to TpsB (see section on type Vb transport, above) (Ruhe et al., 2013a). Upon secretion of CdiA by CdiB, CdiA acts as a toxin that can inhibit the growth of other bacteria, typically of the same species. In order to render themselves immune against their own CdiA toxin, the *cdi* gene cluster encodes an additional immunity protein named CdiI (Aoki et al., 2005). The receptor for *E. coli* CdiA is BamA, the β-barrel protein of the BAM (Aoki et al., 2008). BamA is an essential protein in all bacterial species, but shows considerable sequence variability in its extracellular loops, allowing discrimination between closely and more distantly or unrelated species for CDI (Ruhe et al., 2013b).

Growth inhibition by the cytotoxic C-terminus of CdiA (Cdi-CT) is then facilitated by diverse mechanisms such as nuclease activity, adenosine deaminase activity, metallopeptidase activity, or ADP ribosyl cyclase activity (Ruhe et al., 2013a). One well-studied example of a CDI system is the system of *E. coli* UPEC 536, which encodes a CdiA-CT that harbors tRNase activity (Aoki et al., 2010). Interestingly, this protein is not active until in its host, where it binds to CysK, a protein involved in the catalysis of L-serine to L-cysteine (Diner et al., 2012). Only upon binding to CysK is CdiA able to cleave tRNA. Normally, CysK interacts with CysE, but CdiA has a common amino acid motif with CysE that allows it to bind CysK and become active (Diner et al., 2012).

Another example of a CDI system is the *cdiAIB* system of *Burkholderia pseudomallei* coding for BcpAIB. Here, BcpA is the secreted protein that plays a role in cooperative bacterial communication during biofilm formation. Architecturally, BcpA also harbors a toxic C-terminus, but secretion of BcpA enhances the formation of stable biofilms, possibly by influencing the amount of extracellular DNA as part of the stable biofilm. It has also been proposed that BcpA has nickase activity which might consolidate the microbial community in a biofilm by crosslinking eDNA or attaching the bacteria to eDNA within a biofilm (Garcia et al., 2013; Ruhe et al., 2013a). This example suggests that CDI systems not only provide a growth advantage over other bacterial strains in a competitive environment, but can also play more complex roles.

Immune Evasion

Immune evasion is a collective term for a number of highly specialized mechanisms employed by some bacteria to escape from host immune responses. ATs can participate in immune evasion by interfering with different components of the host immune system.

The first level of immune evasion is serum resistance, which means the ability to survive the action of the complement system that is part of the innate immune response. Serum resistance is achieved by binding to and/or inactivating different components of the complement cascade, such as by binding to Factor H and C3 through C9 products. Binding complement proteins via adhesins like the YadA from *Y. enterocolitica* or the classical AT Vag8 from *Bordetella pertussis* inhibit the full cascade and thus formation of the terminal complement complex which would lead to lysis of the pathogen (Marr et al., 2011; Schindler et al., 2012). YadA is able to sequester different factors, like Factor H and C4b-binding protein, involved in complement regulation (Grosskinsky et al., 2007; Biedzka-Sarek et al., 2008; Kirjavainen et al., 2008). In addition, YadA can bind directly to C3b and iC3b, which in turn promotes Factor H binding and allows *Y. enterocolitica* to escape killing by complement (Schindler et al., 2012). YadA-expressing *Yersinia* can also bind to a variety of host surfaces, initiating virulence processes like secretion of *Yersinia* outer proteins (Yops) via the type III secretion system and thereby killing host immune cells like neutrophils (Rosqvist et al., 1991; Bliska et al., 1993).

Another, more indirect means of host immune evasion is the disguise of immunogenic structures on the bacterial surface. EtpA from *E. coli* for example binds to flagella, which may serve in protection against the host immune system by covering FliC, the main flagellar protein, as an antigen; this may help the bacterium to colonize the host (Roy et al., 2009). EtpA serves a double purpose here, as it also mediates binding to host surfaces and thus bacterial adhesion and biofilm formation (Roy et al., 2009). ATs can also bind factors of the adaptive immune response. The Eib proteins from *E. coli* evade the host immune system by binding to Fc antibody fragments of IgG, possibly in order to avoid opsonization and subsequent phagocytosis (Leo and Goldman, 2009). A more prominent and well-studied example for interactions of ATs with the adaptive immune system is IgA protease of *Neisseria* spp. Here, the serine protease domain exerts endopeptidase activity, recognizing and cleaving the TPPTPSPS motif in the hinge region of human IgA1 and IgA2 and releasing the antigen-binding Fab region from the Fc region. This allows the bacteria to evade opsonization (Plaut et al., 1975, 1977).

Alteration of Other Host Cell Processes

Autotransporters can also influence other host processes beyond the host immune response, typically to promote pathogenic processes. A well-studied example is the TAA BadA, which activates hypoxia-inducible factor 1 (HIF-1), which in turn leads to the release of vaso-endothelial growth factor (VEGF) inducing endothelial proliferation (Kempf et al., 2001). Interaction of BadA-expressing *Bartonella henselae* with endothelial cells has been shown to inhibit apoptosis, supporting the effect of VEGF induced vaso-endothelial proliferation (Kempf et al., 2005). Interestingly, a comparative study investigating the interplay of BadA with the type IV secretion system VirB/D4 in different clinical isolates showed that the enormous length of the BadA stalk (240 nm) might interfere with injection of toxic proteins into the host cell by the type IV secretion system in *B. henselae* (Riess et al., 2004; Müller et al., 2011). This at first seems puzzling, but in light of the intracellular lifestyle of these bacteria, encouraging endothelial growth and inhibiting apoptosis as a natural host cell reaction to infection would benefit the bacteria in a way that the bacteria promote host cell growth for intracellular replication (Schülein et al., 2001).

Alterations in host immune responses not only benefit bacteria with an intracellular lifestyle but also bacteria invading deeper tissues, such as some *Yersiniae*. InvD from *Y. pseudotuberculosis* has recently been shown to utilize host antibody binding during acute infections of the intestinal tract for virulence (Sadana et al., 2018). InvD is a type Ve AT which specifically binds to the Fab region of IgA antibodies. This binding might enable InvD expressing *Yersinia* to alter the host immune response. This way immune exclusion, normally preventing bacteria from crossing the mucosal barrier, can be circumvented and thus allows the bacteria to invade deeper tissues (Sadana et al., 2018). Similarly, the *Shigella flexneri* AT SIgA is involved in invasion of the Peyer's patches, though *Shigella* is not usually found in underlying tissues (Mantis et al., 2002; Favre et al., 2005).

Cyto- and Hemolysins

Passenger domains functioning as cyto- and hemolysins are so far mostly known in type Vb ATs. Here, toxins like ShlA from *S. marcescens* and ExlA from *P. aeruginosa* are secreted and lead to leakage from or lysis of host cells. The cyto- and hemolytic activity of the TpsA C-termini is most often conferred by pore formation within the target cell membrane, triggering a cascade of downstream events that cause the cells to lose their intercellular integrity and to die of ATP depletion. Both ShlA and ExlA form pores in epithelial and endothelial cells causing a massive influx of cations that reduces or abolishes the membrane potential. This leads to ATP depletion and to the activation of eukaryotic proteases cleaving cadherin, leading to breakage of the epithelial barrier (Hertle et al., 1999; Reboud et al., 2017). A classical AT with a cytotoxic passenger can be found in *H. pylori* and is named VacA (vacuolating cytotoxin A). The passenger of VacA is cleaved off and acts in induction of host cell vacuolation, which does not seem to be lethal for the host cell but is important for efficient colonization of the bacteria within the host (Cover and Blanke, 2005). VacA can also act on mitochondria and thus induce apoptosis as part of VacA cell toxicity (Cover and Blanke, 2005).

Adhesion

Adhesion to host cells is a major attribute of many different ATs. Interactions with host cells are usually conferred by adhesins belonging to ATs of types Va, b, c, and e. Importantly, "adhesin" serves as a collective term for different types of interactions, with host cells, tissues, or non-living surfaces. In pathogenesis, depending on the individual adhesin and host organism, this can range from diffuse adhesion to a variety of surface molecules to very specific, high-affinity interactions with a given receptor as shown in **Table 2**.

Adhesion to Surfaces

In order for bacteria to thrive in a competitive environment or during an ongoing infection, the ability to adhere to biotic and abiotic surfaces is pivotal to the bacteria, both in direct contact with host cells and as a first step toward biofilm formation. Being able to adhere to a surface ensures not only interaction with a target cell or surface but also helps the bacteria to regulate gene expression optimal for biofilm conditions. AIDA-I is a major adhesin in *E. coli* that belongs to the class of classical ATs (Benz and Schmidt, 1989). It has been shown to confer diffuse adhesion to human cells and also to other cell types by interaction with surface glycoproteins of the host cells (Laarmann and Schmidt, 2003; Sherlock et al., 2004). Both the receptors and AIDA-I itself are glycosylated. The protein responsible for glycosylation of AIDA-I is encoded directly upstream of the *aidA* gene and is named *aah* (Benz and Schmidt, 2001). Without glycosylation, AIDA-I does not function as an adhesin anymore, though glycosylation appears to be dispensable for autoaggregation, another virulence trait conferred by AIDA-I (Benz and Schmidt, 2001; Sherlock et al., 2004). The passenger of AIDA-I can be cleaved *in vitro*, and it has been hypothesized that cleavage of the passenger might contribute to persistence of the bacterial infection as released AIDA-I passengers facilitate host cell entry (Charbonneau et al., 2006; Pizarro-Cerdá and Cossart, 2006).

In contrast to the diversity of target structures of AIDA-I, the type Vb adhesins FHA from *B. pertussis* and EtpA from *E. coli* target a smaller set of host cell structures in order to confer adhesion. FHA binds specifically to certain surface glycans in human lung epithelial cells (Tuomanen et al., 1988; Relman et al., 1990). FHA favors some carbohydrates over others on the surface of ciliated cells, leading to adhesion to the target and later infection of the lung epithelial cells. Also, EtpA has a more specific mode of action by binding to flagellin, as described in Section "Immune Evasion" (Roy et al., 2009).

Adhesin functions have also been found in essentially all type Vc ATs studied so far, leading to the term trimeric autotransporter adhesin ("TAA") (Linke et al., 2006). The best-studied adhesin belonging to this group is YadA, the *Yersinia* Adhesin A. This protein is encoded on the 70 kb pYV virulence plasmid in all human pathogenic *Yersinia* species including *Y. enterocolitica*, *Y. pseudotuberculosis*, and *Y. pestis*. However, in *Y. pestis* YadA is not expressed at all due to a frameshift in its gene (Rosqvist et al., 1988; Skurnik and Wolf-Watz, 1989), and the role of *Y. pseudotuberculosis* YadA has been elusive since it seems to be dispensable for full virulence (Bolin and Wolf-Watz, 1984). In *Y. enterocolitica*, YadA is important for a multitude of virulence-associated traits. The YadA β-roll head domain at the very N-terminus of the passenger plays an important role in binding to different ECM proteins in the host (El Tahir and Skurnik, 2001). YadA associates with general binding to ECM molecules, including collagen, laminin, fibronectin, and vitronectin (Emody et al., 1989; Flügel et al., 1994; El Tahir et al., 2000; Heise and Dersch, 2006; Mühlenkamp et al., 2017). The interaction of YadA with collagen relies on binding to a triple-helical structure of collagen rich in iminoacids and poor in charged residues (Leo et al., 2008). Because of the high density of YadA on the cell surface during infections it is possible for YadA to bind to collagen strongly despite the low affinity of the interaction (Leo et al., 2010).This contributes to host cell attachment and tissue invasion. For the latter, expression of both YadA and Invasin seems important (Bliska et al., 1993; Eitel et al., 2002). Entry into epithelial cells is needed to further disseminate to underlying tissues and to Peyer's patches and subsequent infection of the liver and spleen (Isberg et al., 1987). BadA from *B. henselae* is among the largest characterized trimeric AT adhesins, with an overall length of 240 nm and a molecular weight of 984 kDa of the trimer (Riess et al., 2004). Similarly, to YadA, BadA binds to ECM proteins including collagen, fibronectin, and laminin (Goldman and Linke, 2011; Müller et al., 2011). The stalk domain of BadA is important for binding to fibronectin, as BadA mutants lacking most of the stalk cannot bind fibronectin anymore (Kaiser et al., 2012). Apart from the stalk binding to fibronectin, most of the adhesive properties lie within the BadA head domain (Müller et al., 2011; Kaiser et al., 2012).

Adhesion to Receptors

Not all ATs bind to multiple surfaces in a promiscuous fashion. Some ATs also bind to specific host receptors, or even to bacterial receptors injected into the host. Intimate contact initiated by Intimin, an *E. coli* type Ve AT, employs such a bacterial receptor inserted into the host membrane named Translocated intimin receptor (Tir). During infection of the small intestine of the host, enteropathogenic *E. coli* and enterohemorrhagic *E. coli* (EHEC and EPEC) induce attaching and effacing (A/E) lesions of the mucosal membrane. These A/E lesions are characterized by polymerization of actin and other cytoskeletal proteins leading to the demise of the cells and at the same time to building an actin pedestal for adhesion of the pathogen. In order for the bacterium to achieve this infection platform, EHEC and EPEC genomes include a pathogenicity island termed LEE (locus of enterocyte effacement), which among other factors encodes for the Tir receptor and intimin (Kenny et al., 1997; Deibel et al., 1998; Gauthier et al., 2000). Though the exact mechanism by which the Tir receptor is inserted into the host cell plasma membrane is still unclear, host cell delivery is facilitated by the bacterial type III secretion system, which is also encoded within the LEE island (Gauthier et al., 2000; Frankel et al., 2001). While Tir is important as a receptor for the Intimin passenger, the Intimin-Tir complex also has implications in cytoskeletal dynamics in the host cell and is thus a multifactorial virulence factor of *E. coli* (Kenny, 1999; Goosney et al., 2001).

Another important member of the type Ve ATs is Invasin of *Yersinia* spp. In contrast to Intimin, it does not bind via a bacterial receptor in the host cell, but instead binds directly to β_1-integrins expressed on the apical side of gut epithelial cells (Isberg et al., 1987, 2000; Schulte et al., 2000). As a consequence of this binding, *Yersinia* spp. are internalized into the cell via endocytosis and can infect the underlying tissue (Isberg and Leong, 1990; Pepe, 1993; Grassl et al., 2003). Subsequently, Invasin-expressing bacteria can infect the lymph nodes and disseminate into other tissue types (Isberg et al., 2000).

Other subclasses of ATs can also use specific receptors expressed on host cells for adhesion. UspA1, a *M. catarrhalis* TAA, binds to carcinoembryonic antigen-related cell adhesion molecule 1 (CAECAM-1), a cell surface protein displayed by epithelial cells (Hill and Virji, 2003). Interestingly, UspA1 is an example of an AT where the binding function lies within the coiled-coil stalk of the protein (Conners et al., 2008). Due to the length of UspA1 and the density with which UspA1 covers the surface of *M. catarrhalis*, this interaction requires bending of the stalk in order to interact with CAECAM-1 as a receptor (Conners et al., 2008). In addition to the interaction with CAECAM-1, UspA1 also binds to laminin and fibronectin via the head domain in a similar fashion as other type Vc ATs (Tan et al., 2005).

Autoaggregation and Biofilm Formation

Biofilms act as a protective mesh that protects bacterial communities from outside influences. Biofilms are a lifestyle that is distinct from that of planktonic bacteria; this is reflected in very different gene expression patterns and by signaling processes inside the biofilm that lead to this adaption. Once established, biofilms can be very hard to remove and are a special challenge in infection and hygiene contexts (Monds and O'Toole, 2009; Trunk et al., 2018).

Both initial adhesion to surfaces and autoaggregation are necessary for establishing biofilms on surfaces. A model system for self-recognition leading to autoaggregation of the bacteria is Ag43 (Antigen 43), a type Va AT from *E. coli*. Ag43 confers intercellular binding by a self-recognizing handshake interaction. By this mechanism the bacteria flocculate, which is beneficial in colonization, immune evasion and persistence in the host (Klemm et al., 2004; Sherlock et al., 2006).

Also FHA aids in the formation of biofilms and thus directly contributes to host colonization and persistence during pathogenesis in *B. pertussis*, together with other proteinaceous factors regulated by the BvgAS system, like fimbriae and ACY, which negatively regulates biofilm formation by interaction with FHA (Irie et al., 2004). In this context, FHA seems to be important in initiation of the attachment of bacteria to the surface, thereby aiding in building micro-colonies that later become part of a bacterial biofilm. FHA also seems to play a major role in maintaining the integrity of the biofilm (Serra et al., 2011). Surprisingly, free FHA actually inhibits the formation of biofilms. This has been speculated to play a regulatory role in biofilm formation during pathogenesis of *B. pertussis* (Serra et al., 2011). In *E. coli*, the type Vb protein EtpA plays a major role in adhesion and biofilm formation. In contrast to free FHA that regulates biofilm formation, secreted EtpA plays a bridging role, aiding in the interaction of flagella with intestinal cells during enterobacterial infection. EtpA binds conserved domains of the flagellar protein FliC at the tip of the flagellum. EtpA then guides the bacterium to gut epithelial cells, where it interacts with mucin producing cells and aids in biofilm formation (Roy et al., 2009).

YadA form entero-pathogenic *Yersiniae* contributes to the induction of microabscesses by autoaggregation. Bacteria expressing YadA tend to aggregate following a zipper-like interaction of YadA proteins displayed on the bacterial surface. This autoaggregation seems beneficial during infection of the Peyer's patches since induction of microabscesses aids in bacterial persistence (El Tahir and Skurnik, 2001). Autoaggregation by YadA also seems to play a role in general biofilm formation, which in turn contributes to persistence of the infection and immune evasion (Trunk et al., 2018). Thus, YadA is a premier example of how trimeric AT adhesins can take part in multiple virulence-associated tasks including adhesion, immune evasion, autoaggregation, and biofilm formation.

Adhesion to Abiotic Surfaces

Adhesion to abiotic surfaces such as rubber, glass, or plastic is a major problem when it comes to working with primary, sterile material in hospital settings or the food industry as adhesion to surfaces is the first step in formation of stable biofilms (Lindsay and von Holy, 2006). Some ATs are universally sticky and interact with a variety of surfaces. Especially TAAs, for example YadA and BadA, interact with various surfaces such as plastic and glass (Schulze-Koops et al., 1993; Müller et al., 2011; Berne et al., 2015). This depends on various factors including the nature of the surface as well as the growth conditions (static vs. flow). There are also specific differences: BadA, for example, interacts more strongly with plastic than YadA (Müller et al., 2011). Together with other modes of surface adhesion and auto-aggregation, YadA, and BadA can thus induce formation of biofilms on a variety of materials. Another well-studied TAA, AtaA from *Acinetobacter* sp., does not appear to have specific binding partners. Like YadA, AtaA has been shown to function in adhesion to collagen and laminin but in general displays low specificity for any particular substrate – instead, it seems to be "universally sticky" adhering strongly to a variety of biotic and abiotic surfaces, including, e.g., polyurethane, steel, and glass (Ishikawa et al., 2012; Hori et al., 2015; Koiwai et al., 2016). Similar features are known for YeeJ, a type Ve inverse AT from *E. coli* involved in adhesion to numerous abiotic surfaces and biofilm formation (Moriel et al., 2017). Unfortunately, though many AT adhesins exhibit the ability to interact with abiotic surfaces, information regarding affinities and systematic studies on which surface materials are preferred are sparse.

INTRACELLULAR MOTILITY

Some bacteria have the ability to invade cells and live within the host cells. This is beneficial for virulence because the bacterium is protected from host immune responses, has access to nutrients, and is influenced less by the harsh environment within the host. Examples of ATs involved in this process have already been mentioned above, including YadA and IgA protease that are implicated in tissue invasion and in release into the host cytosol after uptake into the host cells, respectively. Once inside the cell, bacteria can use ATs in different ways to confer intracellular motility employing the host cell actin cytoskeleton. This motility can be conveyed through the interaction with actin polymerases of the host or by mimicking polymerizing factors themselves (Sitthidet et al., 2011). Both mechanisms can be found in BimA, a type Vc AT of *B. pseudomallei*, *Burkholderia mallei*, and *Burkholderia thailandensis* (Stevens et al., 2005). The mechanism employed for actin polymerization differs slightly between the different species. *B. mallei* and *B. pseudomallei* BimA have a WH2 domain (Wiskott-Aldrich syndrome protein homology domain) that directly mimics WASP (Wiskott-Aldrich syndrome protein) as an actin polymerase. The *B. thailandensis* BimA has a CA domain (central and acidic) that can bind and activate Arp2/3, a complex that activates actin polymerization, in addition to a WH2 domain (Sitthidet et al., 2011; Benanti et al., 2015).

A slightly different mechanism for actin polymerization is used by IcsA, also called VirG, a type Va AT of *S. flexneri* and YapV, a type Va AT of *Y. pseudotuberculosis* and *Y. pestis*. While BimA either binds Arp2/3 by mimicking WASP through a WH2 domain or activate actin polymerization directly by a CA domain, IcsA and YapV seem to interact with N-WASP (neural WASP) (Besingi et al., 2013; Chauhan et al., 2016). IcsA is mostly located at the bacterial pole (Goldberg et al., 1993). Its passenger binds to N-WASP which then can activate Arp2/3 which subsequently functions as an actin polymerase (Bernardini et al., 1989; Goldberg and Theriot, 1995). YapV from *Y. pestis* functions in a similar fashion. Like IcsA, YapV also binds to N-WASP, an actin polymerization factor in order to influence actin polymerization and utilize it for intracellular movement (Besingi et al., 2013; Chauhan et al., 2016; Leupold et al., 2017).

CONCLUSION AND OUTLOOK

The different subclasses of type V secretion systems, or ATs, display similarities in their biogenesis and mode of passenger secretion, but the functional implications of their passengers in virulence as well as in symbiosis are very diverse. These functions can extend from surface adhesion via enzymatic activity to complex interactions with cellular factors directly influencing host cell behavior. These diverse functions do not cluster with the secretion system (sub)classification; similar functions can be found across some or all subclasses. Detailed mechanistic knowledge about functions of AT passengers is available only for a few well-studied examples, and while the biogenesis pathway(s) are conserved across all species that harbor AT genes, the specific AT functions are often not.

Though well studied, open questions remain in the biogenesis of ATs. It is not entirely clear how the β-barrel domain is inserted into the membrane, and specifically, what role the BAM complex plays in this. BAM is involved in the insertion of all β-barrel membrane proteins in the OM of Gram-negative bacteria, but several AT biogenesis models suggest additional functions for BAM also in passenger secretion (Sauri et al., 2009; Leo et al., 2012; Leo and Linke, 2018). Partly as a consequence of this debate, the mechanism of passenger secretion by the β-barrel is not fully understood. While type Va and Ve ATs secrete their passengers via a hairpin-loop intermediate, the question of how this works, e.g., in trimeric ATs is still under debate (Leo and Linke, 2018). As to whether all three passenger polypeptides are transported at the same time or sequentially has still to be shown. The presence of trimeric helper proteins that may act as chaperones to coordinate export of type Vc ATs suggest simultaneous export and recent study points toward contemporaneous transport of all passenger polypeptides (Grin et al., 2014; Chauhan et al., 2019).

Likewise, the molecular mechanisms of many AT functions have still not been entirely elucidated.

In many cases of adhesins bind to a variety of surfaces promiscuously, but it is still not known as to whether different affinities toward different ECM molecules have a biological consequence. Furthermore, it is still not known how exactly differential binding works and which residues plays a role in differentiation of the binding targets. Another example for how ill-defined the functions of some passenger domains are protease and lipase targets. Although ATs with protease and lipase function have been studied for decades - and in fact belong to the best studied ATs - most of the host cellular targets have not been found yet. Even if targets have been defined in many cases the implications of targeting these structures is mostly unclear.

A lot of virulence potential lies within the secreted passengers of ATs, which would make many ATs potential targets for, e.g., drug and vaccine development (Xin et al., 2010; Olvera et al., 2011; Bentancor et al., 2012). Currently, there are already two major recombinant vaccines using ATs on the market. These contain FHA and Pertactin of *B. pertussis*, as well as NadA from *N. meningitidis* (Hellwig et al., 2003; Malito et al., 2014). Since ATs oftentimes belong to the virulence factors initiating an infection, for example Intimin and Invasin, one could also think of them as targets for anti-infective drugs (Durand et al., 2009; Heras et al., 2015).

In this review, we picked some prominent examples to illustrate the variety of passenger functions in ATs. Future research will undoubtedly lead to a more detailed picture of the variety of passenger functions and their involvement in infections as well as in symbiotic or environmental lifestyles.

AUTHOR CONTRIBUTIONS

IM wrote the first draft. All the authors contributed to complete the manuscript.

FUNDING

This work was funded by the Horizon 2020 Innovative Training Network "ViBrANT" (to DL) and the Norwegian Research Council (to JL). Contributions by the University of Oslo are gratefully acknowledged. AS is partly funded from Novo Nordisk Fonden grant NNF18OC0032818.

REFERENCES

Abreu, A. G., Abe, C. M., Nunes, K. O., Moraes, C. T. P., Chavez-Dueñas, L., Navarro-Garcia, F., et al. (2016). The serine protease Pic as a virulence factor of atypical enteropathogenic *Escherichia coli*. *Gut Microbes* 7, 115–125. doi: 10.1080/19490976.2015.1136775

Alvarez, B. H., Gruber, M., Ursinus, A., Dunin-Horkawicz, S., Lupas, A. N., and Zeth, K. (2010). A transition from strong right-handed to canonical left-handed supercoiling in a conserved coiled-coil segment of trimeric autotransporter adhesins. *J. Struct. Biol.* 170, 236–245. doi: 10.1016/j.jsb.2010.02.009

Aoki, S. K., Diner, E. J., De Roodenbeke, C. T. K., Burgess, B. R., Poole, S. J., Braaten, B. A., et al. (2010). A widespread family of polymorphic contact-dependent toxin delivery systems in bacteria. *Nature* 468, 439–442. doi: 10.1038/nature09490

Aoki, S. K., Malinverni, J. C., Jacoby, K., Thomas, B., Pamma, R., Trinh, B. N., et al. (2008). Contact-dependent growth inhibition requires the essential outer membrane protein BamA (YaeT) as the receptor and the inner membrane transport protein AcrB. *Mol. Microbiol.* 70, 323–340. doi: 10.1111/j.1365-2958.2008.06404.x

Aoki, S. K., Pamma, R., and Hernday, A. D. (2005). Contact-dependent inhibition of growth in *Escherichia coli*. *Science* 309, 1245–1249.

Arenas, J., Nijland, R., Rodriguez, F. J., Bosma, T. N. P., and Tommassen, J. (2013). Involvement of three meningococcal surface-exposed proteins, the heparin-binding protein NhbA, the α-peptide of IgA protease and the autotransporter protease NalP, in initiation of biofilm formation. *Mol. Microbiol.* 87, 254–268. doi: 10.1111/mmi.12097

Aricò, B., Nuti, S., Scarlato, V., and Rappuoli, R. (1993). Adhesion of *Bordetella pertussis* to eukaryotic cells requires a time-dependent export and maturation of filamentous hemagglutinin. *Proc. Natl. Acad. Sci. U.S.A.* 90, 9204–9208. doi: 10.1073/pnas.90.19.9204

Ayala, P., Vasquez, B., Wetzler, L., and So, M. (2002). *Neisseria gonorrhoeae* Porin P1.B induces endosome exocytosis and a redistribution of lamp1 to the plasma membrane. *Infect. Immun.* 70, 5965–5971. doi: 10.1128/iai.70.11.5965-5971.2002

Balder, R., Hassel, J., Lipski, S., and Lafontaine, E. R. (2007). *Moraxella* catarrhalis strain O35E expresses two filamentous hemagglutinin-like proteins that mediate adherence to human epithelial cells. *Infect. Immun.* 75, 2765–2775. doi: 10.1128/iai.00079-07

Barnard, T. J., Dautin, N., Lukacik, P., Bernstein, H. D., and Buchanan, S. K. (2007). Autotransporter structure reveals intra-barrel cleavage followed by conformational changes. *Nat. Struct. Mol. Biol.* 14, 1214–1220. doi: 10.1038/nsmb1322

Batchelor, M., Prasannan, S., Daniell, S., Reece, S., Connerton, I., Bloomberg, G., et al. (2000). Structural basis for recognition of the translocated intimin receptor (Tir) by intimin from enteropathogenic *Escherichia coli*. *EMBO J.* 19, 2452–2464. doi: 10.1093/emboj/19.11.2452

Baud, C., Guérin, J., Petit, E., Lesne, E., Dupré, E., Locht, C., et al. (2014). Translocation path of a substrate protein through its Omp85 transporter. *Nat. Commun.* 5:5271. doi: 10.1038/ncomms6271

Baud, C., Hodak, H., Willery, E., Drobecq, H., Locht, C., Jamin, M., et al. (2009). Role of DegP for two-partner secretion in Bordetella. *Mol. Microbiol.* 74, 315–329. doi: 10.1111/j.1365-2958.2009.06860.x

Benanti, E. L., Nguyen, C. M., and Welch, M. D. (2015). Virulent burkholderia species mimic host actin polymerases to drive actin-based motility. *Cell* 161, 348–360. doi: 10.1016/j.cell.2015.02.044

Bentancor, L. V., Routray, A., Bozkurt-Guzel, C., Camacho-Peiro, A., Pier, G. B., and Maira-Litrán, T. (2012). Evaluation of the trimeric autotransporter ata as a vaccine candidate against *Acinetobacter baumannii* infections. *Infect. Immun.* 80, 3381–3388. doi: 10.1128/IAI.06096-11

Benz, I., and Schmidt, M. A. (1989). Cloning and expression of an adhesin (AIDA-I) involved in diffuse adherence of enteropathogenic *Escherichia coli. Infect. Immun.* 57, 1506–1511.

Benz, I., and Schmidt, M. A. (2001). Glycosylation with heptose residues mediated by the aah gene product is essential for adherence of the AIDA-I adhesin. *Mol. Microbiol.* 40, 1403–1413. doi: 10.1046/j.1365-2958.2001.02487.x

Bernardini, M. L., Mounier, J., d'Hauteville, H., Coquis-Rondon, M., and Sansonetti, P. J. (1989). Identification of icsA, a plasmid locus of *Shigella* flexneri that governs bacterial intra- and intercellular spread through interaction with F-actin. *Proc. Natl. Acad. Sci. U.S.A.* 86, 3867–3871. doi: 10.1073/pnas.86.10.3867

Berne, C., Ducret, A., Hardy, G. G., and Brun, Y. V. (2015). Adhesins involved in attachment to abiotic surfaces by Gram-negative bacteria. *Microbiol. Spectr.* 3, 1–45. doi: 10.1128/microbiolspec.MB-0018-2015

Besbes, A., Le Goff, S., Antunes, A., Terrade, A., Hong, E., Giorgini, D., et al. (2015). Hyperinvasive Meningococci Induce Intra-nuclear Cleavage of the NF-κB Protein p65/RelA by Meningococcal IgA Protease. *PLoS Pathog.* 11:e1005078. doi: 10.1371/journal.ppat.1005078

Besingi, R. N., Chaney, J. L., and Clark, P. L. (2013). An alternative outer membrane secretion mechanism for an autotransporter protein lacking a C-terminal stable core. *Mol. Microbiol.* 90, 1028–1045. doi: 10.1111/mmi.12414

Biedzka-Sarek, M., Jarva, H., Hyytiäinen, H., Meri, S., and Skurnik, M. (2008). Characterization of complement factor H binding to *Yersinia enterocolitica* serotype O:3. *Infect. Immun.* 76, 4100–4109. doi: 10.1128/IAI.00313-08

Bliska, J. B., Copass, M. C., and Falkow, S. (1993). The *Yersinia pseudotuberculosis* adhesin YadA mediates intimate bacterial attachment to and entry into HEp-2 cells. *Infect. Immun.* 61, 3914–3921.

Bolin, I., and Wolf-Watz, H. (1984). Molecular cloning of the temperature-inducible outer membrane protein 1 of *Yersinia pseudotuberculosis. Infect. Immun.* 43, 72–78.

Braun, V., Ondraczek, R., and Hobbie, S. (1993). Activation and secretion of Serratia hemolysin. *Zentralbl. Bakteriol.* 278, 306–315. doi: 10.1016/s0934-8840(11)80847-9

Brzuszkiewicz, A., Nowak, E., Dauter, Z., Dauter, M., Cieśliski, H., Dlługolł?cka, A., et al. (2009). Structure of esta esterase from psychrotrophic pseudoalteromonas sp. 643A covalently inhibited by monoethylphosphonate. *Acta Crystallogr. Sect. F Struct. Biol. Cryst. Commun.* 65, 862–865. doi: 10.1107/S1744309109030826

Buscher, A. Z., Grass, S., Heuser, J., Roth, R., and St. Geme, J. W. (2006). Surface anchoring of a bacterial adhesin secreted by the two-partner secretion pathway. *Mol. Microbiol.* 61, 470–483. doi: 10.1111/j.1365-2958.2006.05236.x

Carinato, M. E., Collin-Osdoby, P., Yang, X., Knox, T. M., Conlin, C. A., and Miller, C. G. (1998). The apeE gene of *Salmonella* typhimurium encodes an outer membrane esterase not present in *Escherichia coli. J. Bacteriol.* 180, 3517–3521.

Casasanta, M. A., Yoo, C. C., Smith, H. B., Duncan, A. J., Cochrane, K., Varano, A. C., et al. (2017). A chemical and biological toolbox for Type Vd secretion: characterization of the phospholipase A1 autotransporter FplA from *Fusobacterium nucleatum. J. Biol. Chem.* 292, 20240–20254. doi: 10.1074/jbc.M117.819144

Chagnot, C., Listrat, A., Astruc, T., and Desvaux, M. (2012). Bacterial adhesion to animal tissues: protein determinants for recognition of extracellular matrix components. *Cell. Microbiol.* 14, 1687–1696. doi: 10.1111/cmi.12002

Charbonneau, M. È, Berthiaume, F., and Mourez, M. (2006). Proteolytic processing is not essential for multiple functions of the *Escherichia coli* autotransporter adhesin involved in diffuse adherence (AIDA-I). *J. Bacteriol.* 188, 8504–8512. doi: 10.1128/jb.00864-06

Chauhan, N., Hatlem, D., Orwick-Rydmark, M., Schneider, K., Floetenmeyer, M., van Rossum, B., et al. (2019). Insights into the autotransport process of a trimeric autotransporter, Yersinia Adhesin A (YadA). *Mol. Microbiol.* 111, 844–862. doi: 10.1111/mmi.14195

Chauhan, N., Wrobel, A., Skurnik, M., and Leo, J. C. (2016). *Yersinia* adhesins: an arsenal for infection. *Proteomics Clin. Appl.* 10, 949–963. doi: 10.1002/prca.201600012

Chevalier, N., Moser, M., Koch, H. G., Schimz, K. L., Willery, E., Locht, C., et al. (2004). Membrane targeting of a bacterial virulence factor harbouring an extended signal peptide. *J. Mol. Microbiol. Biotechnol.* 8, 7–18. doi: 10.1159/000082076

Choi, P. S., and Bernstein, H. D. (2010). Sequential translocation of an Escherchia coli two-partner secretion pathway exoprotein across the inner and outer membranes. *Mol. Microbiol.* 75, 440–451. doi: 10.1111/j.1365-2958.2009.06993.x

Clantin, B., Delattre, A., Rucktooa, P., Saint, N., Méli, A. C., Locht, C., et al. (2007). Structure of the membrane protein transporter superfamily. *Science* 416, 3–7.

Clantin, B., Hodak, H., Willery, E., Locht, C., Jacob-Dubuisson, F., and Villeret, V. (2004). The crystal structure of filamentous hemagglutinin secretion domain and its implications for the two-partner secretion pathway. *Proc. Natl. Acad. Sci. U.S.A.* 101, 6194–6199. doi: 10.1073/pnas.0400291101

Conners, R., Hill, D. J., Borodina, E., Agnew, C., Daniell, S. J., Burton, N. M., et al. (2008). The *Moraxella* adhesin UspA1 binds to its human CEACAM1 receptor by a deformable trimeric coiled-coil. *EMBO J.* 27, 1779–1789. doi: 10.1038/emboj.2008.101

Coppens, F., Castaldo, G., Debraekeleer, A., Subedi, S., Moonens, K., Lo, A., et al. (2018). Hop-family *Helicobacter* outer membrane adhesins form a novel class of Type 5-like secretion proteins with an interrupted β-barrel domain. *Mol. Microbiol.* 110, 33–46. doi: 10.1111/mmi.14075

Costa, T. R. D., Felisberto-Rodrigues, C., Meir, A., Prevost, M. S., Redzej, A., Trokter, M., et al. (2015). Secretion systems in Gram-negative bacteria: Structural and mechanistic insights. *Nat. Rev. Microbiol.* 13, 343–359. doi: 10.1038/nrmicro3456

Cover, T. L., and Blanke, S. R. (2005). *Helicobacter* pylori VacA, a paradigm for toxin multifunctionality. *Nat. Rev. Microbiol.* 3, 320–332. doi: 10.1038/nrmicro1095

Da Mata Madeira, P. V., Zouhir, S., Basso, P., Neves, D., Laubier, A., Salacha, R., et al. (2016). Structural basis of lipid targeting and destruction by the type v secretion system of *Pseudomonas aeruginosa. J. Mol. Biol.* 428, 1790–1803. doi: 10.1016/j.jmb.2016.03.012

Dautin, N. (2010). Serine protease autotransporters of *Enterobacteriaceae* (SPATEs): Biogenesis and function. *Toxins* 2, 1179–1206. doi: 10.3390/toxins2061179

Dautin, N., Barnard, T. J., Anderson, D. E., and Bernstein, H. D. (2007). Cleavage of a bacterial autotransporter by an evolutionarily convergent autocatalytic mechanism. *EMBO J.* 26, 1942–1952. doi: 10.1038/sj.emboj.7601638

Dautin, N., and Bernstein, H. D. (2007). Protein secretion in gram-negative bacteria via the autotransporter pathway. *Annu. Rev. Microbiol.* 61, 89–112. doi: 10.1146/annurev.micro.61.080706.093233

Davey, M. E., Caiazza, N. C., and O'Toole, G. A. (2003). Rhamnolipid surfactant production affects biofilm architecture in *Pseudomonas aeruginosa* PAO1. *J. Bacteriol.* 185, 1027–1036. doi: 10.1128/jb.185.3.1027-1036.2003

Deibel, C., Krämer, S., Chakraborty, T., and Ebel, F. (1998). EspE, a novel secreted protein of attaching and effacing bacteria, is directly translocated into infected host cells, where it appears as a tyrosine-phosphorylated 90 kDa protein. *Mol. Microbiol.* 28, 463–474. doi: 10.1046/j.1365-2958.1998.00798.x

Delattre, A. S., Clantin, B., Saint, N., Locht, C., Villeret, V., and Jacob-Dubuisson, F. (2010). Functional importance of a conserved sequence motif in FhaC, a prototypic member of the TpsB/Omp85 superfamily. *FEBS J.* 277, 4755–4765. doi: 10.1111/j.1742-4658.2010.07881.x

Desvaux, M., Khan, A., Beatson, S. A., Scott-Tucker, A., and Henderson, I. R. (2005). Protein secretion systems in *Fusobacterium nucleatum*: Genomic identification of Type 4 piliation and complete Type V pathways brings new insight into mechanisms of pathogenesis. *Biochim. Biophys. Acta* 1713, 92–112. doi: 10.1016/j.bbamem.2005.05.002

Déziel, E., Lépine, F., Milot, S., and Villemur, R. (2003). rhlA is required for the production of a novel biosurfactant promoting swarming motility in *Pseudomonas aeruginosa*: 3-(3-hydroxyalkanoyloxy)alkanoic acids (HAAs), the precursors of rhamnolipids. *Microbiology* 149, 2005–2013. doi: 10.1099/mic.0.26154-0

Diebel, L. N., Liberati, D. M., Baylor, A. E., Brown, W. J., and Devlin, J. (2004). Immunoglobulin A protease is a virulence factor for gram-negative pneumonia. *Surgery* 136, 937–943. doi: 10.1016/j.surg.2004.06.029

Diner, E. J., Beck, C. M., Webb, J. S., Low, D. A., and Hayes, C. S. (2012). Identification of a target cell permissive factor required for contact-dependent growth inhibition (CDI). *Genes Dev.* 26, 515–525. doi: 10.1101/gad.182345.111

Dorsey, C. W., Laarakker, M. C., Humphries, A. D., Weening, E. H., and Bäumler, A. J. (2005). *Salmonella enterica* serotype Typhimurium MisL is an intestinal colonization factor that binds fibronectin. *Mol. Microbiol.* 57, 196–211. doi: 10.1111/j.1365-2958.2005.04666.x

Drobnak, I., Braselmann, E., Chaney, J. L., Leyton, D. L., Bernstein, H. D., Lithgow, T., et al. (2015). Of linkers and autochaperones: an unambiguous nomenclature to identify common and uncommon themes for autotransporter secretion. *Mol. Microbiol.* 95, 1–16. doi: 10.1111/mmi.12838

Durand, E., Verger, D., Rêgo, A. T., Chandran, V., Meng, G., Fronzes, R., et al. (2009). Structural biology of bacterial secretion systems in gram-negative pathogens - Potential for new drug targets. *Infect. Disord. Drug Targets* 9, 518–547. doi: 10.2174/187152609789105722

Eitel, J., Eitel, J., Dersch, P., and Dersch, P. (2002). The YadA protein of *Yersinia pseudotuberculosis* mediates high-efficiency uptake into human cells under environmental conditions in which Invasin is repressed. *Society* 70, 4880–4891. doi: 10.1128/iai.70.9.4880-4891.2002

El Tahir, Y., Kuusela, P., and Skurnik, M. (2000). Functional mapping of the *Yersinia enterocolitica* adhesin YadA. Identification of eight NSVAIG - S motifs in the amino-terminal half of the protein involved in collagen binding. *Mol. Microbiol.* 37, 192–206. doi: 10.1046/j.1365-2958.2000.01992.x

El Tahir, Y., and Skurnik, M. (2001). YadA, the multifaceted *Yersinia* adhesin. *Int. J. Med. Microbiol.* 291, 209–218. doi: 10.1078/1438-4221-00119

Emody, L., Heesemann, J., Wolf-Watz, H., Skurnik, M., Kapperud, G., O'Toole, P., et al. (1989). Binding to collagen by *Yersinia enterocolitica* and *Yersinia pseudotuberculosis*: evidence for yopA-mediated and chromosomally encoded mechanisms. *J. Bacteriol.* 171, 6674–6679. doi: 10.1128/jb.171.12.6674-6679.1989

Emsley, P., Charles, I. G., Fairweather, N. F., and Isaacs neil, W. (1996). Structure of *Bordatella pertussis* virulence factor P.69 pertactin. *Nature* 381, 90–92. doi: 10.1038/381090a0

Favre, L., Spertini, F., and Corthesy, B. (2005). Secretory IgA possesses intrinsic modulatory properties stimulating mucosal and systemic immune responses. *J. Immunol.* 175, 2793–2800. doi: 10.4049/jimmunol.175.5.2793

Fink, D. L., Buscher, A. Z., Green, B., Fernsten, P., and St. Geme, J. W. (2003). The *Haemophilus influenzae* Hap autotransporter mediates microcolony formation and adherence to epithelial cells and extracellular matrix via binding regions in the C-terminal end of the passenger domain. *Cell. Microbiol.* 5, 175–186. doi: 10.1046/j.1462-5822.2003.00266.x

Flügel, A., Schulze-Koops, H., Heesemann, J., Kühn, K., Sorokin, L., Burkhardt, H., et al. (1994). Interaction of enteropathogenic *Yersinia enterocolitica* with complex basement membranes and the extracellular matrix proteins collagen type IV, laminin-1 and -2, and nidogen/entactin. *J. Biol. Chem.* 269, 29732–29738.

Frankel, G., Phillips, A. D., Trabulsi, L. R., Knutton, S., Dougan, G., and Matthews, S. (2001). Intimin and the host cell - Is it bound to end in Tir(s)? *Trends Microbiol.* 9, 214–218. doi: 10.1016/s0966-842x(01)02016-9

Fulcher, R. A., Cole, L. E., Janowicz, D. M., Toffer, K. L., Fortney, K. R., Katz, B. P., et al. (2006). Expression of *Haemophilus ducreyi* collagen binding outer membrane protein NcaA is required for virulence in swine and human challenge models of chancroid. *Infect. Immun.* 74, 2651–2658. doi: 10.1128/iai.74.5.2651-2658.2006

Garcia, E. C., Anderson, M. S., Hagar, J. A., and Cotter, P. A. (2013). BurkholderiaBcpA mediates biofilm formation independently of interbacterial contact-dependent growth inhibition. *Mol. Microbiol.* 89, 1213–1225. doi: 10.1111/mmi.12339

Garnett, J. A., Muhl, D., Douse, C. H., Hui, K., Busch, A., Omisore, A., et al. (2015). Structure-function analysis reveals that the *Pseudomonas aeruginosa* Tps4 two-partner secretion system is involved in CupB5 translocation. *Protein Sci.* 24, 670–687. doi: 10.1002/pro.2640

Gauthier, A., De Grado, M., and Finlay, B. B. (2000). Mechanical fractionation reveals structural requirements for enteropathogenic *Escherichia coli* Tir insertion into host membranes. *Infect. Immun.* 68, 4344–4348. doi: 10.1128/iai.68.7.4344-4348.2000

Goldberg, M. B., Barzu, O., Parsot, C., and Sansonetti, P. J. (1993). Unipolar localization and ATPase activity of IcsA, a *Shigella* flexneri protein involved in intracellular movement. *J. Bacteriol.* 175, 2189–2196. doi: 10.1128/jb.175.8.2189-2196.1993

Goldberg, M. B., and Theriot, J. A. (1995). *Shigella* flexneri surface protein IcsA is sufficient to direct actin-based motility. *Proc. Natl. Acad. Sci. U.S.A.* 92, 6572–6576. doi: 10.1073/pnas.92.14.6572

Goldman, A., and Linke, D. (2011). Bacterial Adhesion. *Adva. Exp. Med. Biol.* 715, 271–284. doi: 10.1007/978-94-007-0940-9_17

Goosney, D. L., DeVinney, R., and Finlay, B. B. (2001). Recruitment of cytoskeletal and signaling proteins to enteropathogenic and enterohemorrhagic *Escherichia coli* pedestals. *Infect. Immun.* 69, 3315–3322. doi: 10.1128/iai.69.5.3315-3322.2001

Grassl, G. A., Bohn, E., Müller, Y., Bühler, O. T., and Autenrieth, I. B. (2003). Interaction of *Yersinia enterocolitica* with epithelial cells: Invasin beyond invasion. *Int. J. Med. Microbiol.* 293, 41–54. doi: 10.1078/1438-4221-00243

Green, E. R., and Mecsas, J. (2016). Bacterial Secretion Systems – An overview CHAPTER SUMMARY. *Microbiol. Spectr.* 4, 1–32.

Grijpstra, J., Arenas, J., Rutten, L., and Tommassen, J. (2013). Autotransporter secretion: varying on a theme. *Res. Microbiol.* 164, 562–582. doi: 10.1016/j.resmic.2013.03.010

Grin, I., Hartmann, M. D., Sauer, G., Alvarez, B. H., Schütz, M., Wagner, S., et al. (2014). A trimeric lipoprotein assists in trimeric autotransporter biogenesis in enterobacteria. *J. Biol. Chem.* 289, 7388–7398. doi: 10.1074/jbc.M113.513275

Grosskinsky, U., Schütz, M., Fritz, M., Schmid, Y., Lamparter, M. C., Szczesny, P., et al. (2007). A conserved glycine residue of trimeric autotransporter domains plays a key role in *Yersinia* adhesin A autotransport. *J. Bacteriol.* 189, 9011–9019. doi: 10.1128/jb.00985-07

Guérin, J., Bigot, S., Schneider, R., Buchanan, S. K., and Jacob-Dubuisson, F. (2017). Two-partner secretion: combining efficiency and simplicity in the secretion of large proteins for bacteria-host and bacteria-bacteria interactions. *Front. Cell. Infect. Microbiol.* 7:148. doi: 10.3389/fcimb.2017.00148

Halter, R., Pohlner, J., and Meyer, T. F. (1984). IgA protease of *Neisseria gonorrhoeae*: isolation and characterization of the gene and its extracellular product. *EMBO J.* 3, 1595–1601. doi: 10.1002/j.1460-2075.1984.tb02016.x

Hartmann, M. D., Ridderbusch, O., Zeth, K., Albrecht, R., Testa, O., Woolfson, D. N., et al. (2009). A coiled-coil motif that sequesters ions to the hydrophobic core. *Proc. Natl. Acad. Sci. U.S.A.* 106, 16950–16955. doi: 10.1073/pnas.0907256106

Hauck, C. R., and Meyer, T. F. (1997). The lysosomal/phagosomal membrane protein h-lamp-1 is a target of the IgA1 protease of *Neisseria gonorrhoeae*. *FEBS Lett.* 405, 86–90. doi: 10.1016/S0014-5793(97)00163-4

Heise, T., and Dersch, P. (2006). Identification of a domain in Yersinia virulence factor YadA that is crucial for extracellular matrix-specific cell adhesion and uptake. *Proc. Natl. Acad. Sci. U.S.A.* 103, 3375–3380. doi: 10.1073/pnas.0507749103

Hellwig, S. M. M., Rodriguez, M. E., Berbers, G. A. M., van de Winkel, J. G. J., and Mooi, F. R. (2003). Crucial role of antibodies to Pertactin in *Bordetella pertussis* immunity. *J. Infect. Dis.* 188, 738–742.

Henderson, I. R., Navarro-Garcia, F., Desvaux, M., Fernandez, R. C., and Ala'Aldeen, D. (2004). Type V protein secretion pathway: the autotransporter story. *Microbiol. Mol. Biol. Rev.* 68, 692–744. doi: 10.1128/mmbr.68.4.692-744.2004

Henderson, I. R., Navarro-García, F., and Nataro, J. P. (1998). The great escape: structure and function of the autotransporter proteins. *Trends Microbiol.* 6, 370–378. doi: 10.1016/s0966-842x(98)01318-3

Heras, B., Scanlon, M. J., and Martin, J. L. (2015). Targeting virulence not viability in the search for future antibacterials. *Br. J. Clin. Pharmacol.* 79, 208–215. doi: 10.1111/bcp.12356

Hertle, R., Hilger, M., Weingardt-Kocher, S., and Walev, I. (1999). Cytotoxic action of *Serratia marcescens* hemolysin on human epithelial cells. *Infect. Immun.* 67, 817–825.

Hill, D. J., and Virji, M. (2003). A novel cell-binding mechanism of *Moraxella* catarrhalis ubiquitous surface protein UspA: specific targeting of the N-domain of carcinoembryonic antigen-related cell adhesion molecules by UspA1. *Mol. Microbiol.* 48, 117–129. doi: 10.1046/j.1365-2958.2003.03433.x

Hodak, H., and Jacob-Dubuisson, F. (2007). Current challenges in autotransport and two-partner protein secretion pathways. *Res. Microbiol.* 158, 631–637. doi: 10.1016/j.resmic.2007.08.001

Hoiczyk, E. (2000). Structure and sequence analysis of *Yersinia* YadA and *Moraxella* UspAs reveal a novel class of adhesins. *EMBO J.* 19, 5989–5999. doi: 10.1093/emboj/19.22.5989

Hori, K., Ohara, Y., Ishikawa, M., and Nakatani, H. (2015). Effectiveness of direct immobilization of bacterial cells onto material surfaces using the bacterionanofiber protein AtaA. *Appl. Microbiol. Biotechnol.* 99, 5025–5032. doi: 10.1007/s00253-015-6554-9

Hulks, M. H., and Plaut, A. G. (1978). IgA protease production as a characteristic distinguishing pathogenic from harmless Neisseriaceae. *N. Engl. J. Med.* 299, 973–976. doi: 10.1056/nejm197811022991802

Ieva, R., and Bernstein, H. D. (2009). Interaction of an autotransporter passenger domain with BamA during its translocation across the bacterial outer membrane. *Proc. Natl. Acad. Sci. U.S.A.* 106, 19120–19125. doi: 10.1073/pnas.0907912106

Irie, Y., Mattoo, S., and Yuk, M. H. (2004). The Bvg virulence control system regulates biofilm formation in Bordetella bronchiseptica. *J. Bacteriol.* 186, 5692–5698. doi: 10.1128/jb.186.17.5692-5698.2004

Isberg, R. R., Hamburger, Z., and Dersch, P. (2000). Signaling and invasin-promoted uptake via integrin receptors. *Microbes Infect.* 2, 793–801. doi: 10.1016/s1286-4579(00)90364-2

Isberg, R. R., and Leong, J. M. (1990). Multiple β1 chain integrins are receptors for invasin, a protein that promotes bacterial penetration into mammalian cells. *Cell* 60, 861–871. doi: 10.1016/0092-8674(90)90099-z

Isberg, R. R., Voorhis, D. L., and Falkow, S. (1987). Identification of invasin: a protein that allows enteric bacteria to penetrate cultured mammalian cells. *Cell* 50, 769–778. doi: 10.1016/0092-8674(87)90335-7

Ishikawa, M., Nakatani, H., Hori, K., Hori, K., Matsumoto, S., Soto, G., et al. (2012). AtaA, a new member of the Trimeric Autotransporter Adhesins from *Acinetobacter* sp. Tol 5 mediating high adhesiveness to various abiotic surfaces. *PLoS One* 7:e48830. doi: 10.1371/journal.pone.0048830

Jacob-Dubuisson, F., Guérin, J., Baelen, S., and Clantin, B. (2013). Two-partner secretion: As simple as it sounds? *Res. Microbiol.* 164, 583–595. doi: 10.1016/j.resmic.2013.03.009

Jain, S., and Goldberg, M. B. (2007). Requirement for YaeT in the outer membrane assembly of autotransporter proteins. *J. Bacteriol.* 189, 5393–5398. doi: 10.1128/jb.00228-07

Junker, M., Besingi, R. N., and Clark, P. L. (2009). Vectorial transport and folding of an autotransporter virulence protein during outer membrane secretion. *Mol. Microbiol.* 71, 1323–1332. doi: 10.1111/j.1365-2958.2009.06607.x

Junker, M., Schuster, C. C., McDonnell, A. V., Sorg, K. A., Finn, M. C., Berger, B., et al. (2006). Pertactin beta-helix folding mechanism suggests common themes for the secretion and folding of autotransporter proteins. *Proc. Natl. Acad. Sci. U.S.A.* 103, 4918–4923. doi: 10.1073/pnas.0507923103

Kaiser, P. O., Linke, D., Schwarz, H., Leo, J. C., and Kempf, V. A. J. (2012). Analysis of the BadA stalk from *Bartonella henselae* reveals domain-specific and domain-overlapping functions in the host cell infection process. *Cell. Microbiol.* 14, 198–209. doi: 10.1111/j.1462-5822.2011.01711.x

Kajava, A. V., Cheng, N., Cleaver, R., Kessel, M., Simon, M. N., Willery, E., et al. (2001). Beta-helix model for the filamentous haemagglutinin adhesin of *Bordetella pertussis* and related bacterial secretory proteins. *Mol. Microbiol.* 42, 279–292. doi: 10.1046/j.1365-2958.2001.02598.x

Kang'ethe, W., and Bernstein, H. D. (2013). Charge-dependent secretion of an intrinsically disordered protein via the autotransporter pathway. *Proc. Natl. Acad. Sci. U.S.A.* 110, E4246–E4255. doi: 10.1073/pnas.1310345110

Kempf, V. A. J., Schairer, A., Neumann, D., Grassl, G. A., Lauber, K., Lebiedziejewski, M., et al. (2005). *Bartonella henselae* inhibits apoptosis in Mono Mac 6 cells. *Cell. Microbiol.* 7, 91–104. doi: 10.1111/j.1462-5822.2004.00440.x

Kempf, V. A. J., Volkmann, B., Schaller, M., Sander, C. A., Alitalo, K., Rieß, T., et al. (2001). Evidence of a leading role for VEGF in *Bartonella henselae*-induced endothelial cell proliferations. *Cell. Microbiol.* 3, 623–632. doi: 10.1046/j.1462-5822.2001.00144.x

Kenny, B. (1999). Phosphorylation of tyrosine 474 of the enteropathogenic *Escherichia coli* (EPEC) Tir receptor molecule is essential for actin nucleating activity and is preceded by additional host modifications. *Mol. Microbiol.* 31, 1229–1241. doi: 10.1046/j.1365-2958.1999.01265.x

Kenny, B., DeVinney, R., Stein, M., Reinscheid, D. J., Frey, E. A., and Finlay, B. B. (1997). Enteropathogenic *E. coli* (EPEC) transfers its receptor for intimate adherence into mammalian cells. *Cell* 91, 511–520. doi: 10.1016/s0092-8674(00)80437-7

Kida, Y., Higashimoto, Y., Inoue, H., Shimizu, T., and Kuwano, K. (2008). A novel secreted protease from *Pseudomonas aeruginosa* activates NF-KB through protease-activated receptors. *Cell. Microbiol.* 10, 1491–1504. doi: 10.1111/j.1462-5822.2008.01142.x

Kingsley, R. A., Ghanem, D. A., Puebla-Osorio, N., Keestra, A. M., Berghman, L., and Bäumler, A. J. (2004). Fibronectin binding to the *Salmonella enterica* serotype typhimurium ShdA autotransporter protein is inhibited by a monoclonal antibody recognizing the A3 repeat. *J. Bacteriol.* 186, 4931–4939. doi: 10.1128/jb.186.15.4931-4939.2004

Kingsley, R. A., Santos, R. L., Keestra, A. M., Adams, L. G., and Bäumler, A. J. (2002). *Salmonella enterica* serotype Typhimurium ShdA is an outer membrane fibronectin-binding protein that is expressed in the intestine. *Mol. Microbiol.* 43, 895–905. doi: 10.1046/j.1365-2958.2002.02805.x

Kirjavainen, V., Jarva, H., Biedzka-Sarek, M., Blom, A. M., Skurnik, M., and Meri, S. (2008). *Yersinia enterocolitica* serum resistance proteins YadA and ail bind the complement regulator C4b-binding protein. *PLoS Pathog.* 4:e1000140. doi: 10.1371/journal.ppat.1000140

Klausen, M., Aaes-Jørgensen, A., Molin, S., and Tolker-Nielsen, T. (2003). Involvement of bacterial migration in the development of complex multicellular structures in *Pseudomonas aeruginosa* biofilms. *Mol. Microbiol.* 50, 61–68. doi: 10.1046/j.1365-2958.2003.03677.x

Klauser, T., Pohlner, J., and Meyer, T. F. (1993). The secretion pathway of IgA protease-type proteins in gram-negative bacteria. *Bioessays* 15, 799–805. doi: 10.1002/bies.950151205

Klemm, P., Hjerrild, L., Gjermansen, M., and Schembri, M. A. (2004). Structure-function analysis of the self-recognizing Antigen 43 autotransporter protein from *Escherichia coli*. *Mol. Microbiol.* 51, 283–296. doi: 10.1046/j.1365-2958.2003.03833.x

Kohler, T., Curty, L. K., Barja, F., Van Delden, C., and Pechere, J. C. (2000). Swarming of *Pseudomonas aeruginosa* is dependent on cell-to-cell signaling and requires flagella and pili. *J. Bacteriol.* 182, 5990–5996. doi: 10.1128/jb.182.21.5990-5996.2000

Koiwai, K., Hartmann, M. D., Linke, D., Lupas, A. N., and Hori, K. (2016). Structural basis for toughness and flexibility in the C-terminal passenger domain of an *Acinetobacter* trimeric autotransporter adhesin. *J. Biol. Chem.* 291, 3705–3724. doi: 10.1074/jbc.M115.701698

Kostakioti, M., and Stathopoulos, C. (2004). Functional analysis of the Tsh autotransporter from an avian pathogenic *Escherichia coli* strain. *Infect. Immun.* 72, 5548–5554. doi: 10.1128/iai.72.10.5548-5554.2004

Laarmann, S., and Schmidt, M. A. (2003). The *Escherichia coli* AIDA autotransporter adhesin recognizes an integral membrane glycoprotein as receptor. *Microbiology* 149, 1871–1882. doi: 10.1099/mic.0.26264-0

Leduc, I., White, C. D., Nepluev, I., Throm, R. E., Spinola, S. M., and Elkins, C. (2008). Outer membrane protein DsrA is the major fibronectin-binding determinant of *Haemophilus ducreyi*. *Infect. Immun.* 76, 1608–1616. doi: 10.1128/IAI.00994-07

Lee, J., Sutterlin, H. A., Wzorek, J. S., Mandler, M. D., Hagan, C. L., Grabowicz, M., et al. (2018). Substrate binding to BamD triggers a conformational change in BamA to control membrane insertion. *Proc. Natl. Acad. Sci. U.S.A.* 115, 201711727. doi: 10.1073/pnas.1711727115

Leo, C. J., and Linke, D. (2018). A unified model for BAM function that takes into account type Vc secretion and species differences in BAM composition. *AIMS Microbiol.* 4, 455–468. doi: 10.3934/microbiol.2018.3.455

Leo, J. C., Elovaara, H., Bihan, D., Pugh, N., Kilpinen, S. K., Raynal, N., et al. (2010). First analysis of a bacterial collagen-binding protein with collagen toolkits: promiscuous binding of YadA to collagens may explain how YadA interferes with host processes. *Infect. Immun.* 78, 3226–3236. doi: 10.1128/IAI.01057-09

Leo, J. C., Elovaara, H., Brodsky, B., Skurnik, M., and Goldman, A. (2008). The *Yersinia* adhesin YadA binds to a collagenous triple-helical conformation but without sequence specificity. *Protein Eng. Des. Sel.* 21, 475–484. doi: 10.1093/protein/gzn025

Leo, J. C., and Goldman, A. (2009). The immunoglobulin-binding Eib proteins from *Escherichia coli* are receptors for IgG Fc. *Mol. Immunol.* 46, 1860–1866. doi: 10.1016/j.molimm.2009.02.024

Leo, J. C., Grin, I., and Linke, D. (2012). Type V secretion: mechanism(S) of autotransport through the bacterial outer membrane. *Philos. Trans. R. Soc. B Biol. Sci.* 367, 1088–1101. doi: 10.1098/rstb.2011.0208

Leo, J. C., Lyskowski, A., Hattula, K., Hartmann, M. D., Schwarz, H., Butcher, S. J., et al. (2011). The structure of *E. coli* IgG-binding protein D suggests a general model for bending and binding in trimeric autotransporter adhesins. *Structure* 19, 1021–1030. doi: 10.1016/j.str.2011.03.021

Leo, J. C., Oberhettinger, P., Chaubey, M., Schütz, M., Kühner, D., Bertsche, U., et al. (2015a). The Intimin periplasmic domain mediates dimerisation and binding to peptidoglycan. *Mol. Microbiol.* 95, 80–100. doi: 10.1111/mmi.12840

Leo, J. C., Oberhettinger, P., Schütz, M., and Linke, D. (2015b). The inverse autotransporter family: intimin, invasin and related proteins. *Int. J. Med. Microbiol.* 305, 276–282. doi: 10.1016/j.ijmm.2014.12.011

Leo, J. C., Oberhettinger, P., Yoshimoto, S., Gupta Udatha, D. B. R. K., Morth, J. P., Schu, M., et al. (2016). Secretion of the intimin passenger domain is driven by protein folding. *J. Biol. Chem.* 291, 20096–20112. doi: 10.1074/jbc.M116.731497

Leupold, S., Büsing, P., Mas, P. J., Hart, D. J., and Scrima, A. (2017). Structural insights into the architecture of the *Shigella flexneri* virulence factor IcsA/VirG and motifs involved in polar distribution and secretion. *J. Struct. Biol.* 198, 19–27. doi: 10.1016/j.jsb.2017.03.003

Lin, L., Ayala, P., Larson, J., Mulks, M., Fukuda, M., Carlsson, S. R., et al. (1997). The Neisseria type 2 IgA1 protease cleaves LAMP1 and promotes survival of bacteria within epithelial cells. *Mol. Microbiol.* 24, 1083–1094. doi: 10.1046/j.1365-2958.1997.4191776.x

Lindsay, D., and von Holy, A. (2006). Bacterial biofilms within the clinical setting: what healthcare professionals should know. *J. Hosp. Infect.* 64, 313–325. doi: 10.1016/j.jhin.2006.06.028

Linke, D., Riess, T., Autenrieth, I. B., Lupas, A., and Kempf, V. A. J. (2006). Trimeric autotransporter adhesins: variable structure, common function. *Trends Microbiol.* 14, 264–270. doi: 10.1016/j.tim.2006.04.005

Lipski, S. L., Akimana, C., Timpe, J. M., Wooten, R. M., and Lafontaine, E. R. (2007). The *Moraxella* catarrhalis autotransporter McaP is a conserved surface protein that mediates adherence to human epithelial cells through its N-terminal passenger domain. *Infect. Immun.* 75, 314–324. doi: 10.1128/iai.01330-06

Luckett, J. C. A., Darch, O., Watters, C., AbuOun, M., Wright, V., Paredes-Osses, E., et al. (2012). A novel virulence strategy for *Pseudomonas aeruginosa* mediated by an autotransporter with arginine-specific aminopeptidase activity. *PLoS Pathog.* 8:e1002854. doi: 10.1371/journal.ppat.1002854

Malito, E., Biancucci, M., Faleri, A., Ferlenghi, I., Scarselli, M., Maruggi, G., et al. (2014). Structure of the meningococcal vaccine antigen NadA and epitope mapping of a bactericidal antibody. *Proc. Natl. Acad. Sci. U.S.A.* 111, 17128–17133. doi: 10.1073/pnas.1419686111

Mantis, N. J., Cheung, M. C., Koteswara, R., Rey, J., Corthésy, B., and Neutra, R. (2002). Selective adherence of IgA to Murine Peyer's Patch M cells: evidence for a novel IgA receptor. *J. Immunol.* 169, 1844–1851. doi: 10.4049/jimmunol.169.4.1844

Marr, N., Shah, N. R., Lee, R., Kim, E. J., and Fernandez, R. C. (2011). *Bordetella pertussis* autotransporter Vag8 binds human C1 esterase inhibitor and confers serum resistance. *PLoS One* 6:e20585. doi: 10.1371/journal.pone.0020585

Menozzi, F. D., Mutombo, R., Renauld, G., Gantiez, C., Hannah, J. H., Leininger, E., et al. (1994). Heparin-inhibitable lectin activity of the filamentous hemagglutinin adhesin of *Bordetella pertussis*. *Infect. Immun.* 62, 769–778.

Mikula, K. M., Leo, J. C., Łyskowski, A., Kedracka-Krok, S., Pirog, A., and Goldman, A. (2012). The translocation domain in *trimeric autotransporter adhesins* is necessary and sufficient for trimerization and autotransportation. *J. Bacteriol.* 194, 827–838. doi: 10.1128/JB.05322-11

Monds, R. D., and O'Toole, G. A. (2009). The developmental model of microbial biofilms: ten years of a paradigm up for review. *Trends Microbiol.* 17, 73–87. doi: 10.1016/j.tim.2008.11.001

Moriel, D. G., Rackaityte, E., Goh, K. G. K., Martinez-Gil, M., Audrain, B., Sakamoto, C., et al. (2017). YeeJ is an inverse autotransporter from *Escherichia coli* that binds to peptidoglycan and promotes biofilm formation. *Sci. Rep.* 7:11326. doi: 10.1038/s41598-017-10902-0

Mühlenkamp, M., Oberhettinger, P., Leo, J. C., Linke, D., and Schütz, M. S. (2015). Yersinia adhesin A (YadA) - Beauty & beast. *Int. J. Med. Microbiol.* 305, 252–258. doi: 10.1016/j.ijmm.2014.12.008

Mühlenkamp, M. C., Hallström, T., Autenrieth, I. B., Bohn, E., Linke, D., Rinker, J., et al. (2017). Vitronectin binds to a specific stretch within the head region of *Yersinia* adhesin a and thereby modulates *Yersinia enterocolitica* host interaction. *J. Innate Immun.* 9, 33–51. doi: 10.1159/000449200

Müller, N. F., Kaiser, P. O., Linke, D., Schwarz, H., Riess, T., Schäfer, A., et al. (2011). Trimeric autotransporter adhesin-dependent adherence of *Bartonella henselae*, *Bartonella quintana*, and *Yersinia enterocolitica* to matrix components and endothelial cells under static and dynamic flow conditions. *Infect. Immun.* 79, 2544–2553. doi: 10.1128/IAI.01309-10

Nesta, B., Spraggon, G., Alteri, C., Moriel, D. G., Rosini, R., Veggi, D., et al. (2012). FdeC, a novel broadly conserved *Escherichia coli* adhesin eliciting protection against urinary tract infections. *mBio* 3:e00010-12.

Nikaido, H., and Vaara, M. (1985). Molecular basis of bacterial outer membrane permeability. *Microbiol. Rev.* 49, 1–32.

Norell, D., Heuck, A., Tran-Thi, T.-A., Götzke, H., Jacob-Dubuisson, F., Clausen, T., et al. (2014). Versatile in vitro system to study translocation and functional integration of bacterial outer membrane proteins. *Nat. Commun.* 5:5396. doi: 10.1038/ncomms6396

Oberhettinger, P., Leo, J. C., Linke, D., Autenrieth, I. B., and Schütz, M. S. (2015). The inverse autotransporter intimin exports its passenger domain via a hairpin intermediate. *J. Biol. Chem.* 290, 1837–1849. doi: 10.1074/jbc.M114.604769

Oberhettinger, P., Schütz, M., Leo, J. C., Heinz, N., Berger, J., Autenrieth, I. B., et al. (2012). Intimin and invasin export their C-terminus to the bacterial cell surface using an inverse mechanism compared to classical autotransport. *PLoS One* 7:e47069. doi: 10.1371/journal.pone.0047069

Ohnishi, Y., Beppu, T., and Horinouchi, S. (1997). Two genes encoding serine protease homologues in *Serratia marcescens* and characterization of their products in *Escherichia coli*. *J. Biochem.* 121, 902–913. doi: 10.1093/oxfordjournals.jbchem.a021672

Ohnishi, Y., and Horinouchi, S. (2009). Extracellular production of a *Serratia marcescens* serine protease in *Escherichia coli*. *Biosci. Biotechnol. Biochem.* 60, 1551–1558.

Olvera, A., Pina, S., Pérez-Simó, M., Aragón, V., Segalés, J., and Bensaid, A. (2011). Immunogenicity and protection against Haemophilus parasuis infection after vaccination with recombinant virulence associated trimeric autotransporters (VtaA). *Vaccine* 29, 2797–2802. doi: 10.1016/j.vaccine.2011.01.105

Otto, B. R., Sijbrandi, R., Luirink, J., Oudega, B., Heddle, J. G., Mizutani, K., et al. (2005). Crystal structure of hemoglobin protease, a heme binding autotransporter protein from pathogenic *Escherichia coli*. *J. Biol. Chem.* 280, 17339–17345. doi: 10.1074/jbc.m412885200

Papanikou, E., Karamanou, S., and Economou, A. (2007). Bacterial protein secretion through the translocase nanomachine. *Nat. Rev. Microbiol.* 5, 839–851. doi: 10.1038/nrmicro1771

Pavlova, O., Peterson, J. H., Ieva, R., and Bernstein, H. D. (2013). Mechanistic link between barrel assembly and the initiation of autotransporter secretion. *Proc. Natl. Acad. Sci. U.S.A.* 110, E938–E947. doi: 10.1073/pnas.1219076110

Pepe, J. C. (1993). *Yersinia enterocolitica* invasin: a primary role in the initiation of infection. *Proc. Natl. Acad. Sci. U.S.A.* 90, 6473–6477. doi: 10.1073/pnas.90.14.6473

Peterson, J. H., Tian, P., Ieva, R., Dautin, N., and Bernstein, H. D. (2010). Secretion of a bacterial virulence factor is driven by the folding of a C-terminal segment. *Proc. Natl. Acad. Sci. U.S.A.* 107, 17739–17744. doi: 10.1073/pnas.1009491107

Phan, G., Remaut, H., Wang, T., Allen, W. J., Pirker, K. F., Lebedev, A., et al. (2011). Crystal structure of the FimD usher bound to its cognate FimC-FimH substrate. *Nature* 474, 49–53. doi: 10.1038/nature10109

Pizarro-Cerdá, J., and Cossart, P. (2006). Bacterial adhesion and entry into host cells. *Cell* 124, 715–727. doi: 10.1016/j.cell.2006.02.012

Plaut, A. G., Gilbert, J. V., Artenstein, M. S., and Capra, J. D. (1975). *Neisseria gonorrhoeae* and *Neisseria meningitidis*: extracellular enzyme cleaves human immunoglobulin A. *Science* 190, 1103–1105. doi: 10.1126/science.810892

Plaut, A. G., Gilbert, J. V., and Wistar, R. (1977). Loss of antibody activity in human immunoglobulin A exposed to extracellular immunoglobulin A proteases of *Neisseria gonorrhoeae* and *Streptococcus sanguis*. *Infect. Immun.* 17, 130–135.

Pohlner, J., Langenberg, U., Wölk, U., Beck, S. C., and Meyer, T. F. (1995). Uptake and nuclear transport of Neisseria IgA1 protease-associated α-proteins in human cells. *Mol. Microbiol.* 17, 1073–1083. doi: 10.1111/j.1365-2958.1995.mmi_17061073.x

Raghunathan, D., Wells, T. J., Morris, F. C., Shaw, R. K., Bobat, S., Peters, S. E., et al. (2011). SadA, a trimeric autotransporter from *Salmonella enterica* serovar Typhimurium, can promote biofilm formation and provides limited protection against infection. *Infect. Immun.* 79, 4342–4352. doi: 10.1128/IAI.05 592-11

Reboud, E., Bouillot, S., Patot, S., Béganton, B., Attrée, I., and Huber, P. (2017). *Pseudomonas aeruginosa* ExlA and *Serratia marcescens* ShlA trigger cadherin cleavage by promoting calcium influx and ADAM10 activation. *PLoS Pathog.* 13:e1006579. doi: 10.1371/journal.ppat.1006579

Relman, D., Tuomanen, E., Falkow, S., Golenbock, D. T., Saukkonen, K., and Wright, S. D. (1990). Recognition of a bacterial adhesin by an integrin: Macrophage CR3 (αMβ2, CD11b CD18) binds filamentous hemagglutinin of *Bordetella pertussis*. *Cell* 61, 1375–1382. doi: 10.1016/0092-8674(90)90701-f

Riedel, K., Talker-Huiber, D., Givskov, M., Schwab, H., and Eberl, L. (2003). Identification and characterization of a GDSL esterase gene located proximal to the swr quorum-sensing system of *Serratia liquefaciens* MG1. *Appl. Environ. Microbiol.* 69, 3901–3910. doi: 10.1128/aem.69.7.3901-3910.2003

Riess, T., Andersson, S. G. E., Lupas, A., Schaller, M., Schäfer, A., Kyme, P., et al. (2004). *Bartonella* adhesin a mediates a proangiogenic host cell response. *J. Exp. Med.* 200, 1267–1278. doi: 10.1084/jem.20040500

Roggenkamp, A., Ackermann, N., Jacobi, C. A., Truelzsch, K., Hoffmann, H., and Heesemann, J. (2003). Molecular analysis of transport and oligomerization of the *Yersinia enterocolitica* adhesin YadA. *J. Bacteriol.* 185, 3735–3744. doi: 10.1128/jb.185.13.3735-3744.2003

Roman-Hernandez, G., Peterson, J. H., and Bernstein, H. D. (2014). Reconstitution of bacterial autotransporter assembly using purified components. *eLife* 3:e04234. doi: 10.7554/eLife.04234

Rosqvist, R., Forsberg, A., and Wolf-Watz, H. (1991). Intracellular targeting of the *Yersinia* YopE cytotoxin in mammalian cells induces actin microfilament disruption. *Infect. Immun.* 59, 4562–4569.

Rosqvist, R., Skurnik, M., and Wolf-Watz, H. (1988). Increased virulence of *Yersinia pseudotuberculosis* by two independent mutations. *Nature* 334, 522–525. doi: 10.1038/334522a0

Roussel-Jazédé, V., Jongerius, I., Bos, M. P., Tommassen, J., and Van Ulsen, P. (2010). NalP-mediated proteolytic release of lactoferrin-binding protein B from the meningococcal cell surface. *Infect. Immun.* 78, 3083–3089. doi: 10.1128/IAI. 01193-09

Roy, K., Hilliard, G. M., Hamilton, D. J., Luo, J., Ostmann, M. M., and Fleckenstein, J. M. (2009). Enterotoxigenic *Escherichia coli* EtpA mediates adhesion between flagella and host cells. *Nature* 457, 594–598. doi: 10.1038/nature07568

Ruhe, Z. C., Low, D. A., and Hayes, C. S. (2013a). Bacterial contact-dependent growth inhibition. *Trends Microbiol.* 21, 230–237. doi: 10.1016/j.tim.2013.02. 003

Ruhe, Z. C., Wallace, A. B., Low, D. A., and Hayes, C. S. (2013b). Receptor polymorphism restricts contact-dependent growth inhibition to members of the same species. *mBio* 4:e00480-13. doi: 10.1128/mBio.00480-13

Ruiz-Perez, F., and Nataro, J. P. (2014). Bacterial serine proteases secreted by the autotransporter pathway: classification, specificity, and role in virulence. *Cell. Mol. Life Sci.* 71, 745–770. doi: 10.1007/s00018-013-1355-8

Sadana, P., Geyer, R., Pezoldt, J., Helmsing, S., Huehn, J., Hust, M., et al. (2018). The invasin D protein from *Yersinia pseudotuberculosis* selectively binds the Fab region of host antibodies and affects colonization of the intestine. *J. Biol. Chem.* 293, 8672–8690. doi: 10.1074/jbc.RA117.001068

Salacha, R., Kovačić, F., Brochier-Armanet, C., Wilhelm, S., Tommassen, J., Filloux, A., et al. (2010). The *Pseudomonas aeruginosa* patatin-like protein PlpD is the archetype of a novel Type V secretion system. *Environ. Microbiol.* 12, 1498–1512. doi: 10.1111/j.1462-2920.2010.02174.x

Sauri, A., Soprova, Z., Wickström, D., de Gier, J. W., Van der Schors, R. C., Smit, A. B., et al. (2009). The Bam (Omp85) complex is involved in secretion of the autotransporter haemoglobin protease. *Microbiology* 155, 3982–3991. doi: 10.1099/mic.0.034991-0

Scarselli, M., Serruto, D., Montanari, P., Capecchi, B., Adu-Bobie, J., Veggi, D., et al. (2006). Neisseria meningitidis NhhA is a multifunctional trimeric autotransporter adhesin. *Mol. Microbiol.* 61, 631–644. doi: 10.1111/j.1365-2958.2006.05261.x

Schiffrin, B., Brockwell, D. J., and Radford, S. E. (2017). Outer membrane protein folding from an energy landscape perspective. *BMC Biol.* 15:123. doi: 10.1186/ s12915-017-0464-5

Schindler, M. K. H., Schutz, M. S., Muhlenkamp, M. C., Rooijakkers, S. H. M., Hallstrom, T., Zipfel, P. F., et al. (2012). *Yersinia enterocolitica* YadA mediates complement evasion by recruitment and inactivation of C3 products. *J. Immunol.* 189, 4900–4908. doi: 10.4049/jimmunol.1201383

Schmitt, C., Turner, D., Boesl, M., Abele, M., Frosch, M., and Kurzai, O. (2007). A functional two-partner secretion system contributes to adhesion of *Neisseria meningitidis* to epithelial cells. *J. Bacteriol.* 189, 7968–7976. doi: 10.1128/jb. 00851-07

Schülein, R., Seubert, A., Gille, C., Lanz, C., Hansmann, Y., Piémont, Y., et al. (2001). Invasion and persistent intracellular colonization of erythrocytes. *J. Exp. Med.* 193, 1077–1086. doi: 10.1084/jem.193.9.1077

Schulte, R., Kerneis, S., Klinke, S., Bartels, H., Preger, S., Kraehenbuhl, J. P., et al. (2000). Translocation of *Yersinia enterocolitica* across reconstituted intestinal epithelial monolayers is triggered by *Yersinia* invasin binding to? beta1 integrins apically expressed on M-like cells. *Cell. Microbiol.* 2, 173–185. doi: 10.1046/j. 1462-5822.2000.00047.x

Schulze-Koops, H., Burkhardt, H., Heesemann, J., Kirsch, T., Swoboda, B., Bull, C., et al. (1993). Outer membrane protein YadA of enteropathogenic yersiniae mediates specific binding to cellular but not plasma fibronectin. *Infect. Immun.* 61, 2513–2519.

Serra, D. O., Conover, M. S., Arnal, L., Sloan, G. P., Rodriguez, M. E., Yantorno, O. M., et al. (2011). FHA-mediated cell-substrate and cell-cell adhesions are critical for *Bordetella pertussis* biofilm formation on abiotic surfaces and in the mouse nose and the trachea. *PLoS One* 6:e28811. doi: 10.1371/journal.pone. 0028811

Sherlock, O., Dobrindt, U., Jensen, J. B., Vejborg, R. M., and Klemm, P. (2006). Glycosylation of the self-recognizing *Escherichia coli* Ag43 autotransporter protein. *J. Bacteriol.* 188, 1798–1807. doi: 10.1128/jb.188.5.1798-1807.2006

Sherlock, O., Schembri, M. A., Reisner, A., and Klemm, P. (2004). Novel roles for the AIDA adhesin from diarrheagenic *Escherichia coli*: cell aggregation and biofilm formation. *J. Bacteriol.* 186, 8058–8065. doi: 10.1128/jb.186.23.8058-8065.2004

Sijbrandi, R., Urbanus, M. L., Ten Hagen-Jongman, C. M., Bernstein, H. D., Oudega, B., Otto, B. R., et al. (2003). Signal recognition particle (SRP)-mediated targeting and Sec-dependent translocation of an extracellular *Escherichia coli* protein. *J. Biol. Chem.* 278, 4654–4659. doi: 10.1074/jbc.m2116 30200

Silhavy, T. J., Ruiz, N., and Kahne, D. (2006). Advances in understanding bacterial outer-membrane biogenesis. *Nat. Rev. Microbiol.* 4, 57–66. doi: 10.1038/ nrmicro1322

Sitthidet, C., Korbsrisate, S., Layton, A. N., Field, T. R., Stevens, M. P., and Stevens, J. M. (2011). Identification of motifs of *Burkholderia pseudomallei* BimA required for intracellular motility, actin binding, and actin polymerization. *J. Bacteriol.* 193, 1901–1910. doi: 10.1128/JB.01455-10

Skurnik, M., and Wolf-Watz, H. (1989). Analysis of the yopA gene encoding the Yop1 virulence determinants of *Yersinia* spp. *Mol. Microbiol.* 3, 517–529. doi: 10.1111/j.1365-2958.1989.tb00198.x

Soprova, Z., Sauri, A., Van Ulsen, P., Tame, J. R. H., Den Blaauwen, T., Jong, W. S. P., et al. (2010). A conserved aromatic residue in the autochaperone domain of the autotransporter Hbp is critical for initiation of outer membrane translocation. *J. Biol. Chem.* 285, 38224–38233. doi: 10.1074/jbc.M110. 180505

Spahich, N. (2011). Structure and function of the *Haemophilus influenzae* autotransporters. *Front. Cell. Infect. Microbiol.* 1:5. doi: 10.3389/fcimb.2011. 00005

St. Geme, J. W. (1994). The HMW1 adhesin of nontypeable *Haemophilus influenzae* recognizes sialylated glycoprotein receptors on cultured human epithelial cells. *Infect. Immun.* 62, 3881–3889.

St. Geme, J. W., and Yeo, H. J. (2009). A prototype two-partner secretion pathway: the *Haemophilus influenzae* HMW1 and HMW2 adhesin systems. *Trends Microbiol.* 17, 355–360. doi: 10.1016/j.tim.2009.06.002

Stein, M., Kenny, B., Stein, M. A., and Finlay, B. B. (1996). Characterization of EspC, a 110-kilodalton protein secreted by enteropathogenic *Escherichia coli* which is homologous to members of the immunoglobulin A protease-like family of secreted proteins. *J. Bacteriol.* 178, 6546–6554. doi: 10.1128/jb.178. 22.6546-6554.1996

Stevens, M. P., Stevens, J. M., Jeng, R. L., Taylor, L. A., Wood, M. W., Hawes, P., et al. (2005). Identification of a bacterial factor required for actin-based motility of *Burkholderia pseudomallei*. *Mol. Microbiol.* 56, 40–53. doi: 10.1111/j.1365-2958.2004.04528.x

Szabady, R. L., Peterson, J. H., Skillman, K. M., and Bernstein, H. D. (2005). From The Cover: an unusual signal peptide facilitates late steps in the biogenesis of a bacterial autotransporter. *Proc. Natl. Acad. Sci. U.S.A.* 102, 221–226. doi: 10.1073/pnas.0406055102

Szczesny, P., and Lupas, A. (2008). Domain annotation of trimeric autotransporter adhesins - daTAA. *Bioinformatics* 24, 1251–1256. doi: 10.1093/bioinformatics/btn118

Tan, T. T., Forsgren, A., and Riesbeck, K. (2006). The respiratory pathogen *moraxella* catarrhalis binds to laminin via ubiquitous surface proteins A1 and A2. *J. Infect. Dis.* 194, 493–497. doi: 10.1086/505581

Tan, T. T., Nordström, T., Forsgren, A., and Riesbeck, K. (2005). The respiratory pathogen *Moraxella catarrhalis* adheres to epithelial cells by interacting with fibronectin through ubiquitous surface proteins A1 and A2. *J. Infect. Dis.* 1029–1038. doi: 10.1086/432759

Tavano, R., Capecchi, B., Montanari, P., Franzoso, S., Marin, O., Sztukowska, M., et al. (2011). Mapping of the *Neisseria meningitidis* NadA cell-binding site: Relevance of predicted?α-helices in the NH2-terminal and dimeric coiled-coil regions. *J. Bacteriol.* 193, 107–115. doi: 10.1128/jb.00430-10

Tertti, R., Skurnik, M., Vartio, T., and Kuusela, P. (1992). Adhesion protein YadA of *Yersinia* species mediates binding of bacteria to fibronectin. *Infect. Immun.* 60, 3021–3024.

Thanassi, D. G., Stathopoulos, C., Karkal, A., and Li, H. (2005). Protein secretion in the absence of ATP: the autotransporter, two-partner secretion and chaperone/usher pathways of Gram-negative bacteria. *Mol. Membr. Biol.* 22, 63–72. doi: 10.1080/09687860500063290

Tielen, P., Rosenau, F., Wilhelm, S., Jaeger, K. E., Flemming, H. C., and Wingender, J. (2010). Extracellular enzymes affect biofilm formation of mucoid *Pseudomonas aeruginosa*. *Microbiology* 156, 2239–2252. doi: 10.1099/mic.0.037036-0

Totsika, M., Wells, T. J., Beloin, C., Valle, J., Allsopp, L. P., King, N. P., et al. (2012). Molecular characterization of the EhaG and UpaG trimeric autotransporter proteins from pathogenic *Escherichia coli*. *Appl. Environ. Microbiol.* 78, 2179–2189. doi: 10.1128/AEM.06680-11

Trunk, T., Khalil, H. S., and Leo, J. C. (2018). Bacterial autoaggregation. *AIMS Microbiol.* 4, 140–164. doi: 10.3934/microbiol.2018.1.140

Tsirigotaki, A., De Geyter, J., Šoštarić, N., Economou, A., and Karamanou, S. (2017). Protein export through the bacterial Sec pathway. *Nat. Rev. Microbiol.* 15, 21–36. doi: 10.1038/nrmicro.2016.161

Tuomanen, E., Towbin, H., Rosenfelder, G., Braun, D., Larson, G., Hansson, G. C., et al. (1988). Receptor analogs and monoclonal antibodies that inhibit adherence of *Bordetella pertussis* to human ciliated respiratory epithelial cells. *J. Exp. Med.* 168, 267–277. doi: 10.1084/jem.168.1.267

Valle, J., Mabbett, A. N., Ulett, G. C., Toledo-Arana, A., Wecker, K., Totsika, M., et al. (2008). UpaG, a new member of the trimeric autotransporter family of adhesins in uropathogenic *Escherichia coli*. *J. Bacteriol.* 190, 4147–4161. doi: 10.1128/JB.00122-08

van den Berg, B. (2010). Crystal structure of a full-length autotransporter. *J. Mol. Biol.* 396, 627–633. doi: 10.1016/j.jmb.2009.12.061

Weirich, J., Bräutigam, C., Mühlenkamp, M., Franz-Wachtel, M., Macek, B., Meuskens, I., et al. (2017). Identifying components required for OMP biogenesis as novel targets for antiinfective drugs. *Virulence* 8, 1170–1188. doi: 10.1080/21505594.2016.1278333

Wilhelm, S., Gdynia, A., Tielen, P., Rosenau, F., and Jaeger, K. E. (2007). The autotransporter esterase EstA of *Pseudomonas aeruginosa* is required for rhamnolipid production, cell motility, and biofilm formation. *J. Bacteriol.* 189, 6695–6703. doi: 10.1128/jb.00023-07

Wilhelm, S., Tommassen, J., and Jaeger, K. E. (1999). A novel lipolytic enzyme located in the outer membrane of *Pseudomonas aeruginosa*. *J. Bacteriol.* 181, 6977–6986.

Wollmann, P., Zeth, K., Lupas, A. N., and Linke, D. (2006). Purification of the YadA membrane anchor for secondary structure analysis and crystallization. *Int. J. Biol. Macromol.* 39, 3–9. doi: 10.1016/j.ijbiomac.2005.11.009

Xin, K. W., Chow, V. T. K., and Alonso, S. (2010). Exploring the versatility of the autotransporter BrkA for the presentation of enterovirus 71 vaccine candidates at the surface of attenuated *Bordetella pertussis*. *Procedia Vaccinol.* 2, 64–70.

Yang, F. L., and Braun, V. (2000). ShlB mutants of *Serratia marcescens* allow uncoupling of activation and secretion of the ShlA hemolysin. *Int. J. Med. Microbiol.* 290, 529–538. doi: 10.1016/S1438-4221(00)80018-1

Yen, Y. T., Kostakioti, M., Henderson, I. R., and Stathopoulos, C. (2008). Common themes and variations in serine protease autotransporters. *Trends Microbiol.* 16, 370–379. doi: 10.1016/j.tim.2008.05.003

Zinchenko, A. A., Kotelnikova, O. V., Gordeeva, E. A., Prokopenko, Y. A., Razgulyaeva, O. A., Serova, O. V., et al. (2018). Immunogenic and protective properties of *Neisseria meningitidis* IgA1 protease and of its truncated fragments. *Russ. J. Bioorg. Chem.* 44, 64–72. doi: 10.1134/s1068162018010193

6

Enhancing Recombinant Protein Yields in the *E. coli* Periplasm by Combining Signal Peptide and Production Rate Screening

Alexandros Karyolaimos[1†], Henry Ampah-Korsah[1†], Tamara Hillenaar[1], Anna Mestre Borras[1], Katarzyna Magdalena Dolata[2], Susanne Sievers[2], Katharina Riedel[2], Robert Daniels[1] and Jan-Willem de Gier[1]*

[1] Department of Biochemistry and Biophysics, Center for Biomembrane Research, Stockholm University, Stockholm, Sweden, [2] Institute of Microbiology, University of Greifswald, Greifswald, Germany

***Correspondence:**
Jan-Willem de Gier
degier@dbb.su.se

Proteins that contain disulfide bonds mainly mature in the oxidative environment of the eukaryotic endoplasmic reticulum or the periplasm of Gram-negative bacteria. In *E. coli*, disulfide bond containing recombinant proteins are often targeted to the periplasm by an N-terminal signal peptide that is removed once it passes through the Sec-translocon in the cytoplasmic membrane. Despite their conserved targeting function, signal peptides can impact recombinant protein production yields in the periplasm, as can the production rate. Here, we present a combined screen involving different signal peptides and varying production rates that enabled the identification of more optimal conditions for periplasmic production of recombinant proteins with disulfide bonds. The data was generated from two targets, a single chain antibody fragment (BL1) and human growth hormone (hGH), with four different signal peptides and a titratable rhamnose promoter-based system that enables the tuning of protein production rates. Across the screen conditions, the yields for both targets significantly varied, and the optimal signal peptide and rhamnose concentration differed for each protein. Under the optimal conditions, the periplasmic BL1 and hGH were properly folded and active. Our study underpins the importance of combinatorial screening approaches for addressing the requirements associated with the production of a recombinant protein in the periplasm.

Keywords: *Escherichia coli*, recombinant protein, periplasm, signal peptide, protein production rate, protein production screen

INTRODUCTION

The bacterium *Escherichia coli* is one of the most widely used hosts to produce recombinant proteins (Rosano and Ceccarelli, 2014). While the majority of recombinant proteins are produced in the cytoplasm of *E. coli*, it is common to produce recombinant proteins that contain disulfide bonds in the periplasm. By targeting these proteins to the periplasm, it is possible to take advantage of the oxidizing environment and the resident disulfide bond formation (Dsb)-system to facilitate the proper disulfide bonds (Bardwell et al., 1991; Baneyx and Mujacic, 2004; Landeta et al., 2018). Although using the periplasm is more ideal for disulfide bond containing recombinant proteins, the bottlenecks associated with the targeting across the cytoplasmic membrane can substantially limit periplasmic yields (Baneyx and Mujacic, 2004; De Geyter et al., 2016).

To reach the periplasm, recombinant proteins are most often equipped with an N-terminal signal peptide that guides it to the Sec-translocon, which is a protein-conducting channel in the cytoplasmic membrane (Denks et al., 2014; Crane and Randall, 2017; Tsirigotaki et al., 2017). The targeting to the Sec-translocon can occur either co-translationally *via* the SRP-pathway or post-translationally in a chaperone-dependent or -independent manner (Kim et al., 2000; Tsirigotaki et al., 2017). Currently, it is not clear how the particular signal peptide, the mature part of the secretory protein, or even the mRNA affect the mode of targeting to the Sec-translocon (Gouridis et al., 2009; Chatzi et al., 2017; Tsirigotaki et al., 2018). Independent of the targeting pathway, proteins directed to the Sec-translocon are ultimately translocated across the cytoplasmic membrane in an unfolded conformation (Arkowitz et al., 1993), and the signal peptide is removed by leader peptidase (Paetzel, 2014; Tsirigotaki et al., 2017). Due to the requirements for translocation, the folding reaction is limited until these proteins reach the periplasm where the Dsb-system can mediate disulfide bond formation and various catalysts can guide the folding process (Inaba, 2009; Landeta et al., 2018). Different signal peptides from both targeting pathways as well as engineered signal peptides have been used for the production of recombinant proteins in the periplasm (Low et al., 2013; Zhou et al., 2016; Freudl, 2018; Selas Castineiras et al., 2018; Zhang et al., 2018). Thus far, it has not been possible to predict which signal peptide is optimal for the production of a particular recombinant protein in the periplasm.

Usually, the gene encoding a recombinant protein is expressed at the highest level possible (Wagner et al., 2008). For recombinant proteins that carry a signal peptide this can lead to saturation of the Sec-translocon capacity which can negatively affect biomass formation and protein production yields (Schlegel et al., 2013; Hjelm et al., 2017; Baumgarten et al., 2018). To overcome this bottleneck, our laboratory developed a rhamnose promoter-based system, which enables the precise regulation of protein production rates (Hjelm et al., 2017). Recently, we have shown that when a rhamnose promoter is used to govern the expression of the gene encoding a recombinant protein in a RhaT-mediated rhamnose transport and rhamnose catabolism deficient double mutant background, rhamnose promoter-based protein production rates can be regulated in a rhamnose concentration-dependent manner. This setup has successfully been used to avoid saturation of the Sec-translocon capacity during the production of a secretory recombinant protein, which leads to enhanced periplasmic protein production levels.

Numerous studies have shown that signal peptides and secretory protein production rates can independently influence the yields of periplasmic proteins, but these two aspects have not been examined in combination. The aim of this study was to examine the effects on periplasmic protein production when combining these two aspects. Hence, we produced two recombinant proteins containing disulfide bonds, the single chain variable fragment (scFv) BL1 and human growth hormone (hGH), using four signal peptides at different protein production rates. To vary the protein production rates aforementioned rhamnose promoter-based setup was used. For both target proteins a setup for enhanced production was identified using the signal peptide and production rate-based combinatorial screening approach.

MATERIALS AND METHODS

Construction of *E. coli* W3110Δ*rha*Δ*lac*

To delete the *rha* operon and the *lac* operon in W3110 (obtained from the American Type Culture Collection) the Red-swap-method was used (Datsenko and Wanner, 2000). In short, kanamycin cassettes with regions homologous to the 5′ and 3′ flanking regions of the *rha* operon and the *lac* operon were generated by PCR using the pKD13 plasmid as a template and the primer pairs listed in **Supplementary Table S1**. The template was digested with *Dpn*I (NEB cutsmart) and the molecular weight of the PCR products were verified using an agarose gel and isolated using the Thermo Fisher Scientific gel extraction kit. To generate the W3110*rha*::Km^R and W3110*lac*::Km^R strains, the purified PCR products were electroporated into W3110 cells harboring pKD46 that had been cultured at 30°C in standard Lysogeny broth (LB) medium (Difco) containing 0.2% arabinose. Kanamycin-resistant clones (kan: 50 μg/ml final concentration) were then screened for the proper kanamycin cassette insertion by PCR using the primer pairs listed in **Supplementary Table S1**. Using P1-mediated generalized transduction, the region of interest of the strains exposed to the lambda Red system were transferred to cells that had not been exposed to the lambda Red system (Thomason et al., 2007). Upon successful transduction of the genetic region of interest, cells were transformed with pCP20 to remove the kanamycin cassette from the genome using FLP-recombinase (Datsenko and Wanner, 2000) and removal of the cassette was verified by PCR/sequencing. Finally, the cells were cured from pCP20 by a prolonged cultivation at 37°C. To generate the W3110Δ*rha*Δ*lac* strain, which is referred to in the text as *E. coli*Δ*rha*, the *rha* operon in W3110 was deleted and then the *lac* operon was deleted from the resulting strain. It is of note that the removal of the *lac* operon prevents any secondary effects on the model recombinant protein scFv BL1 that could occur from binding to its substrate *E. coli* β-galactosidase (Schlegel et al., 2013).

Construction of Expression Vectors

To create the expression vectors for the signal peptide-BL1-His_6 constructs, i.e., $DsbA_{sp}BL1His_6$, $Hbp_{sp}BL1His_6$, $OmpA_{sp}BL1His_6$, and $PhoA_{sp}BL1His_6$ the gene encoding BL1 was amplified with forward primers containing the signal peptide coding sequence with an *Eco*RI site as an overhang (**Supplementary Table S1**) and a reverse primer that contained the His_6-tag coding sequence with a *Hin*dIII site overhang (Schlegel et al., 2013). The PCR products encoding BL1 with a C-terminal His_6-tag and the four different N-terminal signal peptides were then cloned into a pRha67Km^R-derived vector (Giacalone et al., 2006; Hjelm et al., 2017). To create the expression vectors for the signal peptide-hGH-His$_6$ fusion constructs, i.e., DsbA$_{sp}$hGHHis$_6$, Hbp$_{sp}$hGHHis$_6$, OmpA$_{sp}$hGHHis$_6$, and PhoA$_{sp}$hGHHis$_6$ and, the BL1 plasmids described above were used as templates for Gibson assembly where the BL1 coding region was replaced by a gene encoding hGH that is optimized for expression in *E. coli* (Browning et al., 2017). This resulted in four plasmids where hGH was N-terminally fused to a different signal peptide and C-terminally fused to a His$_6$-tag. The primers for the Gibson assembly are listed in **Supplementary Table S1**. All plasmids were verified by sequencing. Plasmid sequences will be provided upon request.

Protein Production Experiments

For all protein production experiments the *E. coli* strain W3110Δ*rha*Δ*lac* (referred to as *E. coli*Δ*rha* in the main text) was used. Cells were transformed with the expression vectors described in the previous section or an empty expression vector that served as a control. Protein production screening was done in a standard 24-well plate format (culture volume 4 ml). Cells were grown aerobically at 30°C and 200 rpm (New Brunswick Innova 42R shaker with an orbit diameter of 1.9 cm), in LB medium supplemented with 50 μg/ml kanamycin. To precultures, 0.2% glucose was added to prevent background expression of the genes encoding the target proteins. Growth was monitored by measuring the A_{600} with a UV-1300 spectrophotometer (Shimadzu). At an A_{600} of ~0.5 target gene expression was induced by the addition of rhamnose. For scFv BL1 production screening the following rhamnose concentrations were used: 0, 50, 100, 250, 500, and 5000 μM. For hGH production screening the following rhamnose concentrations were used: 0, 10, 50, 100, 250, and 5000 μM. Cells were harvested at the indicated time points for further analysis. Standard deviations shown in figures of culturing experiments are based on at least three independent biological replicates. Cultures for the isolation of hGH and BL1 were done in 2.5 L shake flasks (Tunair™) containing 1 L of LB medium.

SDS-PAGE, Coomassie Staining and Immuno-Blotting

To monitor the production of the scFv BL1 and hGH, proteins in whole-cell lysates were separated by sodium dodecyl sulfate-polyacrylamide gel electrophoresis (SDS-PAGE) followed by either Coomassie blue staining or immuno-blotting. scFv BL1 was analyzed on Tris Glycine 8–16% (Invitrogen) gels and hGH was analyzed using Tricine 16% (Invitrogen) gels. Lysates equivalent to 0.04 A_{600} units were loaded onto gels for Coomassie staining and lysates equivalent to 0.02 A_{600} units were loaded onto gels that were analyzed by immuno-blotting. For volumetric comparisons, cell pellets derived from 1 ml of culture were solubilized in 300 μl of sample buffer and 8 μl was loaded for Coomassie blue staining and 4 μl for immuno-blotting. In the blotting experiments, secretory/secreted BL1 and hGH were detected based on their C-terminal His$_6$-tag using HisProbe-HRP Conjugate (Pierce) or the AlexaFluor 647 conjugated anti-HIS antibody (Invitrogen). The HisProbe-HRP Conjugate was used for mere detection and the AlexaFluor 647 conjugated anti-HIS antibody for quantification. The Western Bright Sirius kit (Advansta) was used to visualize immuno-blots stained with the HisProbe-HRP Conjugate. Chemiluminescence and fluorescence signals were detected using an Azure c600 imaging system (Azure Biosystems).

Protein Isolation

Cells producing the scFv BL1 and hGH from 1 L cultures were sedimented at 6000 × *g* for 20 min, 4°C. The supernatant was discarded and the cell pellets were stored at −80°C. The cell pellets were thawed on ice and subsequently lysed in 25 ml buffer A [50 mM Tris–HCl pH 7.4 (for BL1) or pH 8 (for hGH), 0.1% Triton X-100, 1 mg/ml lysozyme, 1× Roche protease inhibitor cocktail, 1 mM phenylmethylsulfonyl fluoride (PMSF)] using the EmusiFlex C3 homogenizer (Avestin) in a cold room. The cell lysate was sedimented (10.000 × *g* for 20 min, 4°C) and the supernatant was transferred into a 70 Ti ultracentrifugation tube and sedimented again (100.000 × *g* for 1 h, 4°C). The high-spin supernatant was transferred to a 50 ml falcon tube, adjusted to a final concentration of 10 mM imidazole and 1 ml of a Ni-NTA agarose beads slurry was added that was pre-equilibrated with buffer B [1× phosphate buffered saline (PBS) pH 7.4, 150 mM NaCl, 0.1% Triton X-100, 10% Glycero]) containing 10 mM imidazole. The mixture was incubated in a cold room overnight with gentle tumbling and then transferred into a 10 ml polyprep chromatographic column. The column was washed with 20 column volumes (CV) of buffer B supplemented with 20 mM imidazole and the BL1 or hGH was eluted with buffer B supplemented with 300 mM imidazole. The eluted protein samples were pooled and loaded onto a HiLoad 16/60 Superdex 200 size exclusion chromatography (SEC) column mounted on an ÄKTAprime plus (GE Healthcare) that was pre-equilibrated with buffer C (50 mM Tris–HCl pH 7.4 (for BL1) or pH 8 (for hGH), 150 mM NaCl, 10% Glycerol). The column was run at a flow-rate of 1 ml/min, the protein was monitored at a wavelength of 280 nm, and the fractions corresponding to the peak were collected and stored at −20°C. All purification steps were done in the cold room. Protein concentration and purity were determined with the BCA method (Pierce) and SDS-PAGE followed by Coomassie staining. Protein folding was monitored using reducing and non-reducing SDS-PAGE followed by Coomassie staining and immuno-blotting. For BL1 analysis 2 μg of protein was loaded for Coomassie staining and 0.5 μg of protein for immuno-blotting and for hGH analysis 2 μg of protein was loaded for Coomassie staining and 1 μg of protein for immuno-blotting.

BL1 Activity Assay

The proper folding of the isolated BL1 was assayed by the recognition of its substrate, *E. coli* β-galactosidase, using a dot-blot assay (Schlegel et al., 2013). For the activity assay, 100 μl of a 1:2 serial dilution of β-galactosidase (100, 50, 25, 12.5, 6.25, 3.125, 1.563, and 0 μg/ml) was spotted directly onto a

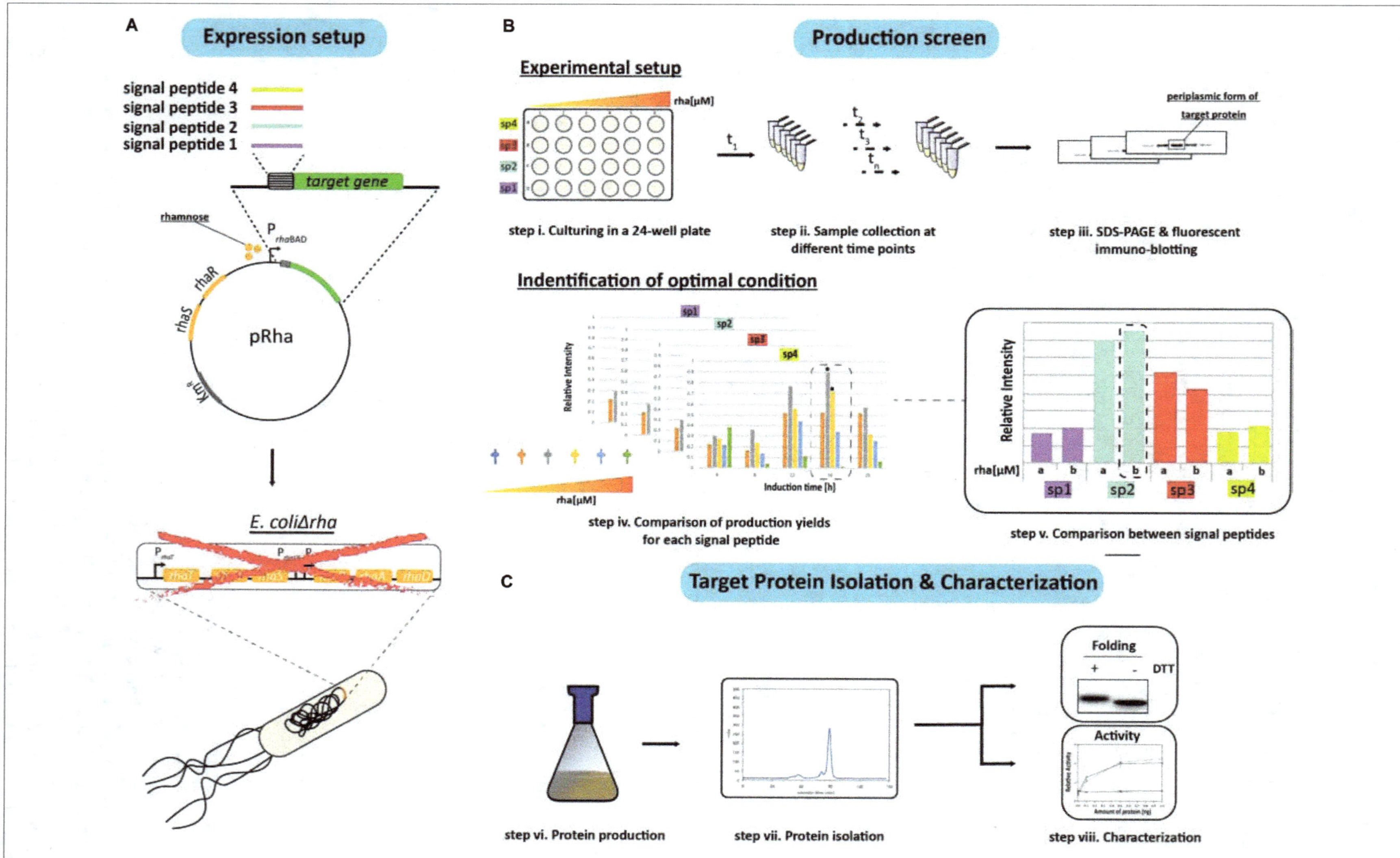

FIGURE 1 | Setup of the signal peptide and production rate-based combinatorial screening approach to enhance periplasmic protein production yields in *E. coli*. **(A)** The gene encoding the protein to be produced in the periplasm is fused to the genetic information encoding different signal peptides. The genetic fusions are subsequently cloned in a rhamnose promoter-based expression vector (pRha). The expression vectors harboring the genes encoding the different signal peptide target protein fusions are subsequently transformed into an *E. coli* strain lacking the rhamnose operon (*E. coli*Δ*rha*). **(B)** Protein production screening is done in standard 24-well plates (step i). Expression of the genes encoding the different signal peptide target protein fusions is induced using varying amounts of rhamnose. Protein production in the periplasm is monitored over time (t_1, t_n) (step ii) by separating proteins present in equal volumes of culture by SDS-PAGE followed by immuno-blotting using a fluorescently labeled antibody recognizing the target protein (step iii). This enables to make for each of the used signal peptides a relative comparison of the protein production yields in the periplasm (step iv). The two highest periplasmic protein production yields are indicated with a black hexagon. For each signal peptide the two setups leading to the highest production yields are compared directly (step v). **(C)** This exercise is used to pick the condition used for scaling up target protein production (step vi) and the subsequent isolation of the target protein (step vii). The isolated protein is analyzed using SDS-PAGE in the presence and absence of the reductant DTT followed by Coomassie staining and immuno-blotting as well as a protein specific activity assay (step viii).

nitrocellulose membrane (Millipore) using a BIO-DOT device (Bio-Rad). As a negative control, equal amounts of bovine serum albumin (BSA) were spotted on the same membrane. The membranes were blocked with a solution of tris buffered saline containing 0.05% Tween 20 (TBS-T) and 5% milk for 1 h at room temperature (RT), washed three times for 15 min with TBS-T and incubated overnight at 4°C with 17 μg/ml of the isolated BL1 in buffer C. The bound BL1 was labeled with the HisProbe-HRP Conjugate and the membrane was developed using the Western Bright Sirius kit and an Azure c600 imaging system.

Human Growth Hormone Activity Assay

The activity of the purified hGH protein was ascertained by its ability to promote the growth of prolactin-dependent Nb2 lymphoma cell line (Nb2-11 cells) as previously described with slight modifications (Sonoda and Sugimura, 2008; Kim et al., 2013). Nb2-11 cells (Sigma) were cultured in RPMI 1640 medium (Gibco/Thermo Fisher Scientific) containing 10% fetal bovine serum (FBS), 10% horse serum (Gibco/Thermo Fisher Scientific) and 1% penicillin-streptomycin (Gibco/Thermo Fisher Scientific) at 37°C in a humidified atmosphere containing 5% CO_2. To begin the assay, the cells were sedimented (500 × *g* for 3 min), washed once in medium without FBS and re-suspended in medium without FBS. The cell density was then determined with a Countess II automated cell counter (Life Technologies), diluted to approximately 2×10^5 cells/ml and 100 μl of the cell suspension was added to each well of a 96 well plate. Each well then received the following amounts of the purified hGH protein: 0, 0.1, 0.5, and 1 ng. After mixing the solution on a shaker at 100 rpm for 1 min, the cells were placed in the 37°C humidified incubator and the cell proliferation 48 h after the hGH addition was used to determine its activity. To measure the cell growth, 20 μl of CellTiter 96®aqueous one solution reagent (Promega) was added to each well, the plate was placed back in the 37°C humidified incubator for 2.5 h and then the absorbance was measured at 490 nm using a SpectraMax M2e microplate reader (Molecular Devices).

Quantification of the scFv BL1 and hGH Amounts Produced in the Periplasm

Cells from 1 ml of either a 4 ml or 1 L culture were sedimented and lysed using 0.3 ml of sample buffer. The amounts of periplasmic BL1 and hGH in the samples were determined by measuring the mature levels of the proteins by immuno-blotting as follows. For BL1 samples a Tris Glycine 8–16% gel was loaded with 3 μl and a two-fold dilution of the whole cell lysate followed by a two-fold dilution series of the isolated BL1 containing 2, 1, 0.5, 0.25, 0.1, and 0.05 μg of protein. The samples were then resolved, transferred to a PVDF membrane, and the BL1 was detected with a fluorescently labeled antibody that recognizes the His_6-tag. Images were acquired using an Azure c600 system. The band intensities were quantified with ImageJ and the values from the standard curve were used to determine the amount of BL1 in the two samples. The amount of hGH was determined similarly with the exception that a Bis Tris 4–12% gel was loaded with 10 μl and a two-fold dilution of the whole cell lysate and the dilution series of isolated hGH had the following amounts of protein: 3, 2, 1, 0.5, 0.25, 0.1, and 0.05 μg.

RESULTS AND DISCUSSION

Combinatorial Signal Peptide and Production Rate Screening Approach

In an effort to identify the ideal condition for producing an oxidized recombinant protein in the *E. coli* periplasm, a combinatorial screen was set up using a panel of different signal peptides and a titratable rhamnose promoter system for controlling protein production rates. To initiate the screen, the genes encoding the target proteins with a C-terminal His_6-tag were fused to sequences encoding for the signal peptides from the *E. coli* proteins DsbA, OmpA, PhoA, and the hemoglobin protease (Hbp) autotransporter, and inserted into the rhamnose promoter-based expression vector pRha (**Figure 1A**). These signal peptides were chosen because they are (i) from *E. coli* proteins, (ii) are commonly used for periplasmic targeting of recombinant proteins, and/or (iii) the corresponding mature proteins are reported to be rather abundant in the *E. coli* cell (Kendall et al., 1986; Schierle et al., 2003; Sijbrandi et al., 2003; Schlegel et al., 2013; Schmidt et al., 2016; Hjelm et al., 2017; Baumgarten et al., 2018) (see for more information **Supplementary Table S2**). The plasmids were then transformed into an *E. coli* strain (W3110Δ*rha*Δ*lac*) that enables tuning of the target production rates by varying the rhamnose concentration. The rhamnose tunability of the strain, herein referred to as *E.coli*Δ*rha*, was achieved by deleting the whole *rha* operon, which includes the removal of the genes involved in rhamnose transport (*rhaT*) and catabolism (*rhaB*) that were previously shown to create a rhamnose titratable protein production strain (Hjelm et al., 2017).

Due to sample number limitations, the initial screen for each target protein with a particular signal peptide was analyzed independently using 24-well plates and a range of rhamnose concentrations to induce the target gene expression (**Figure 1B**). Following the induction, the cells from equal culture volumes were harvested at different time intervals and the periplasmic protein production was monitored by measuring the mature protein amounts using immuno-blots stained with fluorescently labeled antibodies (**Figure 1B**). The resulting data was then used to identify the best conditions for producing the target protein with each of the four signal peptides. To validate that the observed improvements in periplasmic production were not associated with the culture volume analysis, the production levels in equal cell amounts was also examined using Coomassie staining and immuno-blotting.

Based on the results from the initial screen, the two best conditions for each target protein and signal peptide combination were compared directly to determine the rhamnose concentration and signal peptide that provided the highest periplasmic yields (**Figure 1B**). The most effective signal peptide and rhamnose concentration was then tested using large scale cultures where the target proteins were isolated, quantified and characterized for disulfide bond formation and activity (Francis et al., 2002; **Figure 1C**). Two targets were used throughout this study, the single chain antibody fragment BL1 (scFv BL1), which recognizes the *E. coli* β-galactosidase (Schlegel et al., 2013), and human growth hormone (hGH), which stimulates cell proliferation (Kim et al., 2013). Both targets were chosen because they contain two disulfide bonds in their native state and a C-terminal His_6-tag was incorporated for isolation and detection.

ScFv BL1 Production Screen

During the initial screen the temporal production of the scFv BL1 in *E. coliΔrha* was examined using the four different signal peptides ($DsbA_{SP}$, Hbp_{SP}, $OmpA_{SP}$, and $PhoA_{SP}$) and a range of rhamnose concentrations (**Figure 2**). The BL1 amounts, which were determined for equal culture volumes by SDS-PAGE immune-blots (**Figure 1B**), showed that for all signal peptides the highest BL1 yields were obtained 16 h post-induction with rhamnose. The production of BL1 with each signal peptide was then examined at 16 h post-induction using equal amounts of biomass (**Supplementary Figures S1**, **S2**). While the optimal rhamnose concentration slightly changed for some signal peptides in the equal biomass analysis based on Coomassie staining and immuno-blotting against the C-terminal His_6, the preference for lower rhamnose concentrations was confirmed.

Next, the BL1 production yields for each signal peptide at the two best rhamnose conditions were directly compared using equal culture volumes (**Figure 3A**). The results from this volumetric analysis revealed that for BL1 the highest periplasmic production yields were achieved with the OmpA signal peptide and 100 μM rhamnose. To monitor if BL1 produced with the OmpA signal peptide had formed disulfide bonds in the periplasm, the whole cell lysates were separated by SDS-PAGE in the presence and the absence of the reductant DTT followed by Coomassie staining and immuno-blotting (**Figure 3B**). Indicative that BL1 is properly folded with two intramolecular disulfide bonds, BL1 migrated more slowly when it was reduced by DTT.

To more thoroughly examine the periplasmic BL1 that was produced with the most optimal signal peptide and rhamnose concentration, BL1 was produced in a larger scale culture and subsequently isolated using immobilized-metal affinity chromatography (IMAC) followed by size exclusion chromatography (SEC). The isolated BL1 showed the expected migration shift in the presence of DTT (**Figure 4A**). To confirm that the isolated BL1 is indeed properly folded, we examined its ability to bind

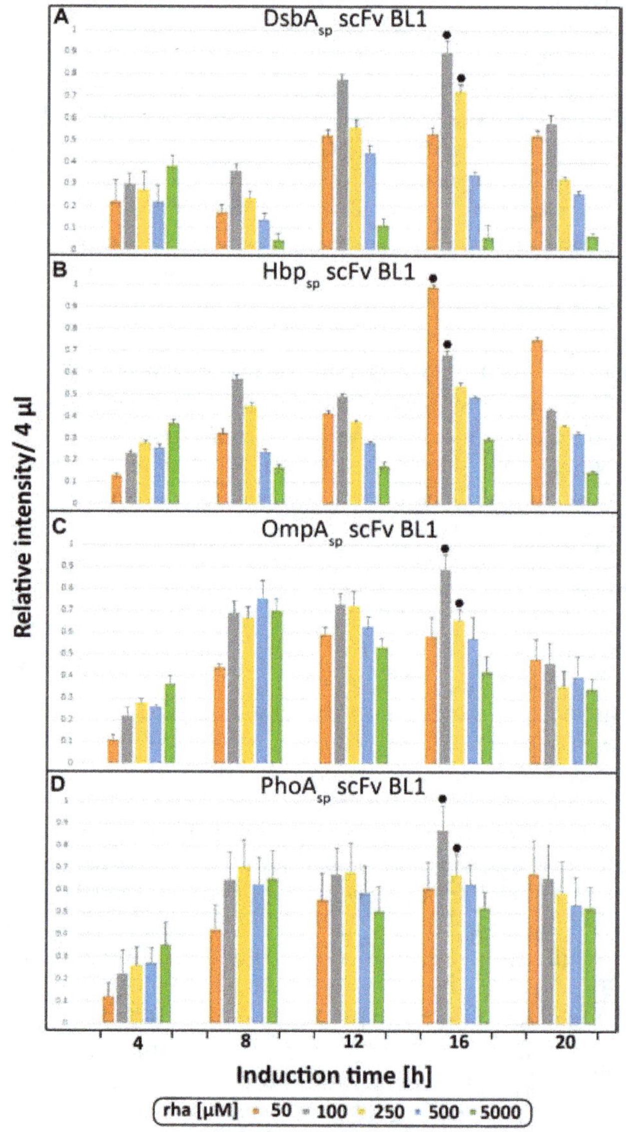

FIGURE 2 | Periplasmic production of the scFv BL1 using different signal peptides and inducer concentrations monitored volumetrically over time. Production of the scFv BL1 N-terminally fused to the DsbA **(A)**, the Hbp **(B)**, the OmpA **(C)**, and the PhoA **(D)** signal peptide, in *E. coliΔrha* using varying amounts of rhamnose, was monitored over time (4, 8, 12, 16, and 20 h after induction of target gene expression). Proteins present in equal amounts of culture volume were separated by SDS-PAGE followed by immuno-blotting using a fluorescently labeled antibody recognizing the His_6-tag fused to the C-terminus of BL1 (see section "Materials and Methods"). Fluorescent signals representing BL1 (i.e., the processed/periplasmic form of the protein) were quantified using ImageJ and peak intensities were used for comparison. To enable the comparison of BL1 production between different time-points for each signal peptide, standard curves were made using peak intensities from serial dilutions of the optimal condition of each signal peptide. For each signal peptide used the two highest periplasmic BL1 production yields are indicated with a black hexagon.

β-galactosidase (**Figure 4B**; Schlegel et al., 2013). For this analysis, β-galactosidase and BSA were spotted in decreasing concentrations on a nitrocellulose membrane.

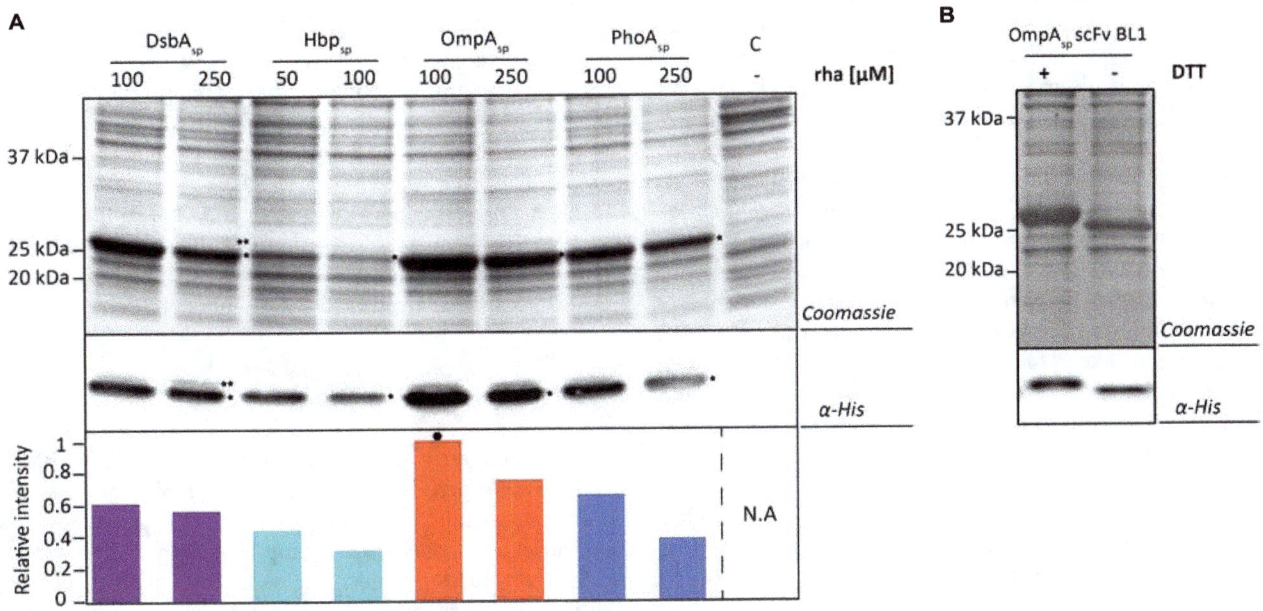

FIGURE 3 | Identifying the optimal setup for the periplasmic production of the scFv BL1. **(A)** For each signal peptide used the two setups leading to the highest periplasmic production yields of BL1 were compared using SDS-PAGE followed by Coomassie staining and immuno-blotting using a fluorescently labeled antibody recognizing the His_6-tag at the C-terminus of BL1. Importantly, equal amounts of culture volume were loaded per lane since this enables to directly compare the amount of BL1 produced in the periplasm volumetrically. Relative intensities of the fluorescent signals are shown below the blot. The highest periplasmic BL1 production yield is indicated with a black hexagon (see also **Supplementary Figure S3**). The precursor form, i.e., BL1 with the signal peptide still attached, is indicated with ** and the mature form, i.e., BL1 without signal peptide, is indicated with *. **(B)** The use of the OmpA-BL1 fusion resulted in the highest production of BL1 in the periplasm. To assess the folding of the BL1 produced in the periplasm cells were solubilized in sample buffer with and without the reductant DTT prior to SDS-PAGE followed by Coomassie staining and immuno-blotting using HisProbe.

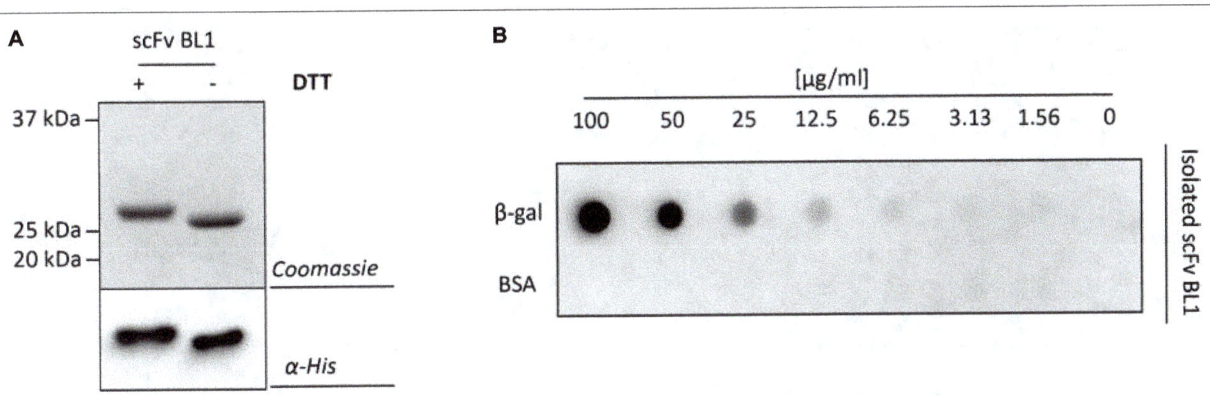

FIGURE 4 | Characterization of the isolated scFv BL1. For the isolation of BL1, the OmpA signal peptide and 100 μM of rhamnose for the induction of target gene expression were used (see **Figure 3A**). BL1 was isolated using a combination of IMAC and SEC. **(A)** The isolated BL1 was incubated in sample buffer with and without the reductant DTT prior to SDS-PAGE followed by Coomassie staining and immuno-blotting using HisProbe. **(B)** A nitrocellulose membrane containing increasing amounts of β-galactosidase as well as increasing amounts of BSA (control) was incubated with the isolated BL1. Binding of BL1 to the β-galactosidase spotted on the nitrocellulose membrane was detected using HisProbe recognizing the His_6-tag at the C-terminus of BL1.

The membrane was then incubated with the isolated BL1 and the bound BL1 was detected using HisProbe (see section "Materials and Methods"). As expected, the BL1 produced under the optimal conditions was capable of binding to its substrate and showed no affinity for the BSA negative control.

Together, these results show that by screening different signal peptides and protein production rates for the scFv BL1 we were able to identify conditions capable of producing functional BL1 in the periplasm at around 200 μg per ml of culture (**Supplementary Figure S4**).

hGH Production Screen

To test if the target can have an influence on the yields obtained with each signal peptide and the different production rates, the screen was repeated using hGH and the two best production conditions were identified for each signal peptide (**Figure 5A** and **Supplementary Figures S5, S6**). Similar to the results using BL1, the highest hGH yields were observed at 16 h post-induction with rhamnose for all of the signal peptides. For hGH the use of the Hbp signal peptide and 50 μM rhamnose resulted in the highest production of hGH in the periplasm. To monitor if hGH produced with the Hbp signal peptide had formed disulfide bonds in the periplasm whole cell lysates were analyzed using SDS-PAGE in the presence and absence of DTT (**Figure 5B**). As expected, the mature hGH showed slower mobility in the presence of DTT, indicating that it likely folds correctly and obtains its two intramolecular disulfide bonds. Thus, both signal peptide and rhamnose concentration leading to the highest production of hGH in the periplasm are different from the ones observed for BL1.

The optimal Hbp-signal peptide-based hGH production setup was scaled up and a combination of IMAC and SEC was used to isolate hGH. Isolated hGH was analyzed by means of SDS-PAGE in the presence and the absence of the reductant DTT followed by Coomassie staining and immuno-blotting (**Figure 6A**). The shift of isolated hGH toward a lower molecular weight upon omitting DTT is an indication that the hGH produced is properly folded. Next, we examined the ability of the isolated hGH to promote the growth of prolactin-dependent Nb2 lymphoma cell line (Nb2-11 cells) using commercially available hGH as a reference and BSA as a negative control (**Figure 6B**). Using this set-up, we showed that hGH produced under the optimal conditions was capable of promoting the growth of prolactin-dependent Nb2 lymphoma cell line equally well as commercially purchased hGH.

Taken together, here we show for hGH that screening different signal peptides and protein production rates leads to the identification of a setup that enables to produce functional hGH in the periplasm at around 30 μg per ml of culture.

CONCLUDING REMARKS

Here, to enhance the production of the scFv BL1 and hGH in the periplasm we used a combinatorial screening approach using four different signal peptides and a rhamnose promoter-based setup enabling to precisely set protein production rates. Main finding is that the signal peptide and rhamnose concentration that provide the highest production yields in the periplasm are different for the two target proteins that were tested. Although the induction times for the optimal conditions are the same for both targets, this may be a mere coincidence and

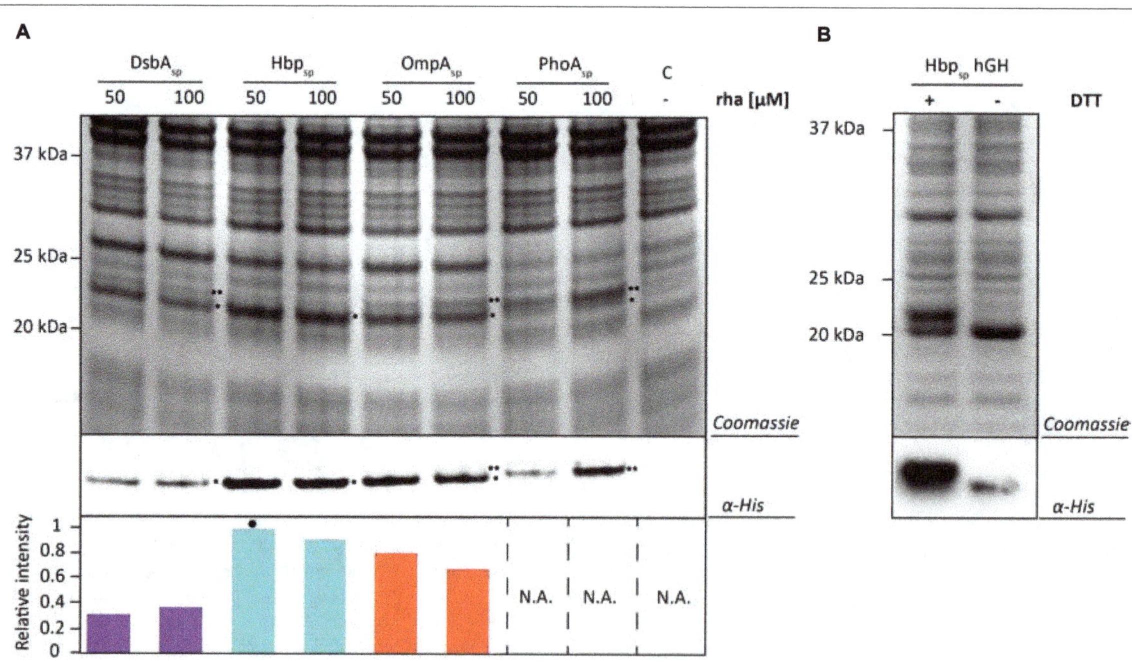

FIGURE 5 | Identifying the optimal setup for the periplasmic production of hGH. **(A)** For each signal peptide used, the two setups leading to the highest periplasmic production yields of hGH were compared by SDS-PAGE followed by Coomassie staining and immuno-blotting using a fluorescently labeled antibody recognizing the His_6-tag at the C-terminus of hGH Importantly, equal amounts of culture volume were loaded per lane since this enables to compare the amount of hGH produced in the periplasm volumetrically. Relative intensities of the fluorescent signals are shown below the blot. The highest periplasmic hGH production yield is indicated with a black hexagon (see also **Supplementary Figure S7**). The precursor form, i.e., hGH with the signal peptide still attached, is indicated with ** and the mature form, i.e., hGH without signal peptide, is indicated with *. **(B)** Whole cell lysates derived from cells producing hGH fused to the Hbp signal peptide at 50 μM of rhamnose isolated 16 h after the addition of rhamnose, were analyzed by means of SDS-PAGE followed by Coomassie staining and immuno-blotting (see section "Materials and Methods"). To assess the folding of the hGH produced in the periplasm cells were dissolved in sample buffer with and without the reductant DTT prior to SDS-PAGE (see also **Supplementary Figure S8**).

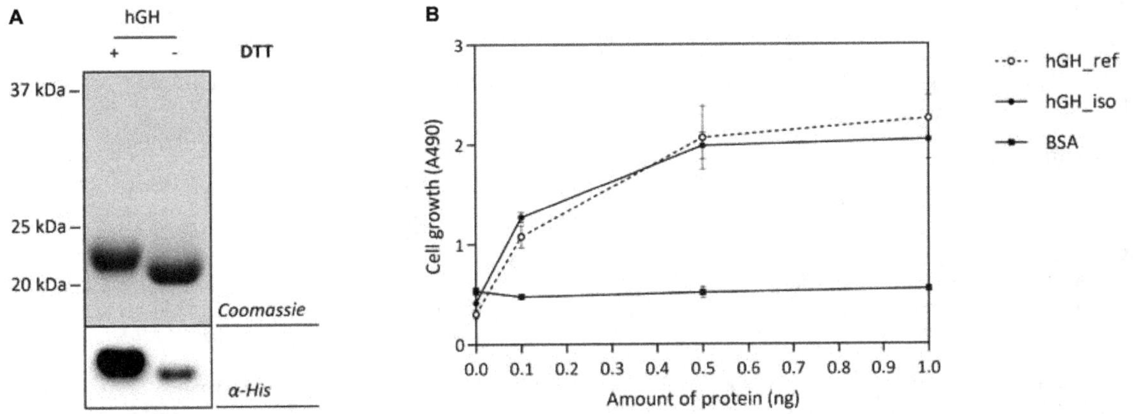

FIGURE 6 | Characterization of isolated hGH. For the isolation of hGH, the Hbp signal peptide and 50 μM of rhamnose for the induction of target gene expression were used (see **Figure 5**). hGH was isolated using a combination of IMAC and SEC. **(A)** To assess the folding of the isolated hGH, it was solubilized in sample buffer with and without the reductant DTT prior to SDS-PAGE followed by Coomassie staining and immuno-blotting using HisProbe (see also **Supplementary Figure S8**). **(B)** The activity of the isolated hGH was assessed by monitoring its ability to promote growth of an Nb2-11 cell line. Resuspended Nb2-11 cells in medium without FBS were incubated with various amounts of isolated hGH (filled circle), hGH reference (empty circle), and BSA (filled square) in a 96-well plate and cultured at 37°C in a humidified, 5% CO_2 atmosphere for 48 h. Cell proliferation was determined colorimetrically with MTT reagent. hGH reference and BSA were used as positive and negative control, respectively. All experiments were done in triplicate and standard deviations are depicted by error bars.

needs to be confirmed using additional targets before this parameter can be fixed.

Previously, it has been shown that the mature part of *E. coli* secretory proteins can contain information that is important for their secretion and that a signal peptide can affect the folding/targeting of a protein destined for the periplasm (Liu et al., 1989; Rusch et al., 2002; Gouridis et al., 2009; Singh et al., 2013; Chatzi et al., 2017; Tsirigotaki et al., 2018). How these observations translate to the production of heterologous proteins in the periplasm of *E. coli* is still *terra incognita*. However, the effects observed for BL1 and hGH production using different signal peptides strongly suggest that signal peptide - target protein combinations can have a profound effect on the production of a protein in the periplasm. Striking example is that the use of the PhoA signal peptide leads to good - although not the best - periplasmic production of BL1, whereas hardly any hGH is produced in the periplasm when the PhoA signal peptide is used. Recently, for PhoA it has been shown that its efficient translocation depends on so-called mature domain targeting signals (Chatzi et al., 2017). Interestingly, using the MatureP predictor, which was trained using the *E. coli* proteome, such a mature domain targeting signal in BL1 could be identified, whereas in hGH such a signal could not be identified (Orfanoudaki et al., 2017). This may explain why the PhoA signal peptide only promotes the efficient export of BL1. These observations raise the question if it would be possible to develop predictors which can streamline screening for the enhanced production of recombinant proteins in the periplasm.

In conclusion, our study shows that combinatorial screening of different signal peptides and protein production rates opens up a new avenue to enhance protein production yields in the periplasm of *E. coli*.

AUTHOR CONTRIBUTIONS

AK, HA-K, RD, and J-WG conceived and designed the experiments. AK, HA-K, TH, AM, and KD conducted the experiments. AK, HA-K, TH, AM, KD, RD, and J-WG analyzed the data. AK, HA-K, KMD, SS, KR, RD, and J-WG wrote the manuscript.

FUNDING

This work was supported by grants from the Swedish Research Council (2015-05288) and the Carl Tryggers Foundation (CTS17:114) to J-WG, a Marie Curie Initial Training Network Grants (Horizon 2020, ProteinFactory, 642863) to the Stockholm University (AK) and the University of Greifswald (KD), grants from the Swedish Research Council (K2015-57-21980-04-4) and the Carl Tryggers Foundation (CTS17:111) as well as Federal funds from the National Institute of Allergy and Infectious Diseases, the National Institutes of Health, and the United States Department of Health and Human Services, under the CEIRS contract number HHSN272201400005C to RD.

REFERENCES

Arkowitz, R. A., Joly, J. C., and Wickner, W. (1993). Translocation can drive the unfolding of a preprotein domain. *EMBO J.* 12, 243–253. doi: 10.1002/j.1460-2075.1993.tb05650.x

Baneyx, F., and Mujacic, M. (2004). Recombinant protein folding and misfolding in *Escherichia coli*. *Nat. Biotechnol.* 22, 1399–1408. doi: 10.1038/nbt1029

Bardwell, J. C., Mcgovern, K., and Beckwith, J. (1991). Identification of a protein required for disulfide bond formation *in vivo*. *Cell* 67, 581–589. doi: 10.1016/0092-8674(91)90532-4

Baumgarten, T., Ytterberg, A. J., Zubarev, R. A., and De Gier, J. W. (2018). Optimizing recombinant protein production in the *Escherichia coli* periplasm alleviates stress. *Appl. Environ. Microbiol.* 84, e00270-18. doi: 10.1128/AEM.00270-18

Browning, D. F., Richards, K. L., Peswani, A. R., Roobol, J., Busby, S. J. W., and Robinson, C. (2017). *Escherichia coli* "TatExpress" strains super-secrete human growth hormone into the bacterial periplasm by the Tat pathway. *Biotechnol. Bioeng.* 114, 2828–2836. doi: 10.1002/bit.26434

Chatzi, K. E., Sardis, M. F., Tsirigotaki, A., Koukaki, M., Sostaric, N., Konijnenberg, A., et al. (2017). Preprotein mature domains contain translocase targeting signals that are essential for secretion. *J. Cell Biol.* 216, 1357–1369. doi: 10.1083/jcb.201609022

Crane, J. M., and Randall, L. L. (2017). The sec system: protein export in *Escherichia coli*. *EcoSal Plus* 7, 1–44. doi: 10.1128/ecosalplus.ESP-0002-2017

Datsenko, K. A., and Wanner, B. L. (2000). One-step inactivation of chromosomal genes in *Escherichia coli* K-12 using PCR products. *Proc. Natl. Acad Sci. U.S.A.* 97, 6640–6645. doi: 10.1073/pnas.120163297

De Geyter, J., Tsirigotaki, A., Orfanoudaki, G., Zorzini, V., Economou, A., and Karamanou, S. (2016). Protein folding in the cell envelope of *Escherichia coli*. *Nat. Microbiol.* 1:16107. doi: 10.1038/nmicrobiol.2016.107

Denks, K., Vogt, A., Sachelaru, I., Petriman, N. A., Kudva, R., and Koch, H. G. (2014). The sec translocon mediated protein transport in prokaryotes and eukaryotes. *Mol. Membr. Biol.* 31, 58–84. doi: 10.3109/09687688.2014.907455

Francis, E., Daniels, R., and Hebert, D. N. (2002). Analysis of protein folding and oxidation in the endoplasmic reticulum. *Curr. Protoc. Cell Biol.* 14, 15.6.1–15.6.29. doi: 10.1002/0471143030.cb1506s14

Freudl, R. (2018). Signal peptides for recombinant protein secretion in bacterial expression systems. *Microb. Cell Fact.* 17:52. doi: 10.1186/s12934-018-0901-3

Giacalone, M. J., Gentile, A. M., Lovitt, B. T., Berkley, N. L., Gunderson, C. W., and Surber, M. W. (2006). Toxic protein expression in *Escherichia coli* using a rhamnose-based tightly regulated and tunable promoter system. *Biotechniques* 40, 355–364. doi: 10.2144/000112112

Gouridis, G., Karamanou, S., Gelis, I., Kalodimos, C. G., and Economou, A. (2009). Signal peptides are allosteric activators of the protein translocase. *Nature* 462, 363–367. doi: 10.1038/nature08559

Hjelm, A., Karyolaimos, A., Zhang, Z., Rujas, E., Vikstrom, D., Slotboom, D. J., et al. (2017). Tailoring *Escherichia coli* for the l-rhamnose PBAD promoter-based production of membrane and secretory proteins. *ACS Synth Biol.* 6, 985–994. doi: 10.1021/acssynbio.6b00321

Inaba, K. (2009). Disulfide bond formation system in *Escherichia coli*. *J. Biochem.* 146, 591–597. doi: 10.1093/jb/mvp102

Kendall, D. A., Bock, S. C., and Kaiser, E. T. (1986). Idealization of the hydrophobic segment of the alkaline phosphatase signal peptide. *Nature* 321, 706–708. doi: 10.1038/321706a0

Kim, J., Luirink, J., and Kendall, D. A. (2000). SecB dependence of an exported protein is a continuum influenced by the characteristics of the signal peptide or early mature region. *J. Bacteriol.* 182, 4108–4112. doi: 10.1128/jb.182.14.4108-4112.2000

Kim, M. J., Park, H. S., Seo, K. H., Yang, H. J., Kim, S. K., and Choi, J. H. (2013). Complete solubilization and purification of recombinant human growth hormone produced in *Escherichia coli*. *PLoS One* 8:e56168. doi: 10.1371/journal.pone.0056168

Landeta, C., Boyd, D., and Beckwith, J. (2018). Disulfide bond formation in prokaryotes. *Nat. Microbiol.* 3, 270–280. doi: 10.1038/s41564-017-0106-2

Liu, G., Topping, T. B., and Randall, L. L. (1989). Physiological role during export for the retardation of folding by the leader peptide of maltose-binding protein. *Proc. Natl. Acad Sci. U.S.A.* 86, 9213–9217. doi: 10.1073/pnas.86.23.9213

Low, K. O., Muhammad Mahadi, N., and Md Illias, R. (2013). Optimisation of signal peptide for recombinant protein secretion in bacterial hosts. *Appl. Microbiol. Biotechnol.* 97, 3811–3826. doi: 10.1007/s00253-013-4831-z

Orfanoudaki, G., Markaki, M., Chatzi, K., Tsamardinos, I., and Economou, A. (2017). MatureP: prediction of secreted proteins with exclusive information from their mature regions. *Sci. Rep.* 7:3263. doi: 10.1038/s41598-017-03557-4

Paetzel, M. (2014). Structure and mechanism of *Escherichia coli* type I signal peptidase. *Biochim. Biophys. Acta* 1843, 1497–1508. doi: 10.1016/j.bbamcr.2013.12.003

Rosano, G. L., and Ceccarelli, E. A. (2014). Recombinant protein expression in *Escherichia coli*: advances and challenges. *Front. Microbiol.* 5:172. doi: 10.3389/fmicb.2014.00172

Rusch, S. L., Mascolo, C. L., Kebir, M. O., and Kendall, D. A. (2002). Juxtaposition of signal-peptide charge and core region hydrophobicity is critical for functional signal peptides. *Arch. Microbiol.* 178, 306–310. doi: 10.1007/s00203-002-0453-z

Schierle, C. F., Berkmen, M., Huber, D., Kumamoto, C., Boyd, D., and Beckwith, J. (2003). The DsbA signal sequence directs efficient, cotranslational export of passenger proteins to the *Escherichia coli* periplasm via the signal recognition particle pathway. *J. Bacteriol.* 185, 5706–5713. doi: 10.1128/jb.185.19.5706-5713.2003

Schlegel, S., Rujas, E., Ytterberg, A. J., Zubarev, R. A., Luirink, J., and De Gier, J. W. (2013). Optimizing heterologous protein production in the periplasm of *E. coli* by regulating gene expression levels. *Microb. Cell Fact* 12:24. doi: 10.1186/1475-2859-12-24

Schmidt, A., Kochanowski, K., Vedelaar, S., Ahrne, E., Volkmer, B., Callipo, L., et al. (2016). The quantitative and condition-dependent *Escherichia coli* proteome. *Nat. Biotechnol.* 34, 104–110. doi: 10.1038/nbt.3418

Selas Castineiras, T., Williams, S. G., Hitchcock, A., Cole, J. A., Smith, D. C., and Overton, T. W. (2018). Development of a generic beta-lactamase screening system for improved signal peptides for periplasmic targeting of recombinant proteins in *Escherichia coli*. *Sci. Rep.* 8:6986. doi: 10.1038/s41598-018-25192-3

Sijbrandi, R., Urbanus, M. L., Ten Hagen-Jongman, C. M., Bernstein, H. D., Oudega, B., Otto, B. R., et al. (2003). Signal recognition particle (SRP)-mediated targeting and Sec-dependent translocation of an extracellular *Escherichia coli* protein. *J. Biol. Chem.* 278, 4654–4659. doi: 10.1074/jbc.M211630200

Singh, P., Sharma, L., Kulothungan, S. R., Adkar, B. V., Prajapati, R. S., Ali, P. S., et al. (2013). Effect of signal peptide on stability and folding of *Escherichia coli* thioredoxin. *PLoS One* 8:e63442. doi: 10.1371/journal.pone.0063442

Sonoda, H., and Sugimura, A. (2008). Improved solubilization of recombinant human growth hormone inclusion body produced in *Escherichia coli*. *Biosci. Biotechnol. Biochem.* 72, 2675–2680. doi: 10.1271/bbb.80332

Thomason, L. C., Costantino, N., and Court, D. L. (2007). *E. coli* genome manipulation by P1 transduction. *Curr. Protoc. Mol. Biol.* 79, 1.17.1–1.17.8. doi: 10.1002/0471142727.mb0117s79

Tsirigotaki, A., Chatzi, K. E., Koukaki, M., De Geyter, J., Portaliou, A. G., Orfanoudaki, G., et al. (2018). Long-lived folding intermediates predominate the targeting-competent secretome. *Structure* 26, 695.e5–707.e5. doi: 10.1016/j.str.2018.03.006

Tsirigotaki, A., De Geyter, J., Sostaric, N., Economou, A., and Karamanou, S. (2017). Protein export through the bacterial Sec pathway. *Nat. Rev. Microbiol.* 15, 21–36. doi: 10.1038/nrmicro.2016.161

Wagner, S., Klepsch, M. M., Schlegel, S., Appel, A., Draheim, R., Tarry, M., et al. (2008). Tuning *Escherichia coli* for membrane protein overexpression.

Proc. Natl. Acad Sci. U.S.A. 105, 14371–14376. doi: 10.1073/pnas.0804090105

Zhang, W., Lu, J., Zhang, S., Liu, L., Pang, X., and Lv, J. (2018). Development an effective system to expression recombinant protein in E. coli via comparison and optimization of signal peptides: expression of *Pseudomonas fluorescens* BJ-10 thermostable lipase as case study. *Microb. Cell Fact* 17:50. doi: 10.1186/s12934-018-0894-y

Zhou, Y., Liu, P., Gan, Y., Sandoval, W., Katakam, A. K., Reichelt, M., et al. (2016). Enhancing full-length antibody production by signal peptide engineering. *Microb. Cell Fact* 15:47. doi: 10.1186/s12934-016-0445-3

7

Type VI Secretion Effectors: Methodologies and Biology

Yun-Wei Lien [1,2] *and Erh-Min Lai* [1,2]*

[1] *Institute of Plant and Microbial Biology, Academia Sinica, Taipei, Taiwan,* [2] *Department of Plant Pathology and Microbiology, National Taiwan University, Taipei, Taiwan*

***Correspondence:**
Erh-Min Lai
emlai@gate.sinica.edu.tw

The type VI secretion system (T6SS) is a nanomachine deployed by many Gram-negative bacteria as a weapon against eukaryotic hosts or prokaryotic competitors. It assembles into a bacteriophage tail-like structure that can transport effector proteins into the environment or target cells for competitive survival or pathogenesis. T6SS effectors have been identified by a variety of approaches, including knowledge/hypothesis-dependent and discovery-driven approaches. Here, we review and discuss the methods that have been used to identify T6SS effectors and the biological and biochemical functions of known effectors. On the basis of the nature and transport mechanisms of T6SS effectors, we further propose potential strategies that may be applicable to identify new T6SS effectors.

Keywords: type VI secretion system, methodology, effector, toxin-immunity, proteomics, bioinformatics, library, protein-protein interaction

INTRODUCTION

Cell-to-cell communication and interaction is a central theme for all life forms including single-cell organisms such as bacteria. Gram-negative bacteria have evolved a variety of protein secretion systems to export or import macromolecules across membranes for survival and fitness. The type VI secretion system (T6SS) is a versatile injection machine that can deliver effector molecules into the environment, eukaryotic hosts and prokaryotic competitors. The T6SS mainly functions in a contact-dependent manner to target bacterial competitors for interbacterial competition and eukaryotic hosts for pathogenesis (Durand et al., 2014; Russell et al., 2014a; Cianfanelli et al., 2016b). However, recent reports also suggested that some T6SS effectors may exert their functions extracellularly rather than inside target cells (Wang et al., 2015; Lin et al., 2017; Si et al., 2017).

Typically, the T6SS gene cluster encodes 13–14 conserved core components for machinery assembly and some less conserved accessory proteins and effectors related to T6SS regulation and biological functions (Records, 2011; Basler, 2015; Cianfanelli et al., 2016b). The system consists of a TssJLM (or TssLM) trans-membrane complex (Aschtgen et al., 2008; Ma et al., 2009, 2012; Felisberto-Rodrigues et al., 2011), which serves as a docking site for the TssEFGK baseplate complex (Brunet et al., 2015). TssA, a starfish-like dodecametic complex, connects the TssEFGK baseplate to the Hcp tube and TssBC sheath components for polymerization of this tail structure (Planamente et al., 2016; Zoued et al., 2016). TssB-TssC functions as a contractile sheath, which presumably wraps around the Hcp tube and dynamically propels the Hcp-VgrG puncturing device and associated T6SS effectors across bacterial membranes (Basler et al., 2012). After the firing action, ClpV AAA+ ATPase binds to the contracted TssBC outer sheath for disassembly into subunits, which can be recycled for the next T6SS assembly (Bonemann et al., 2009; Basler and Mekalanos, 2012). Furthermore, Hcp, VgrG, and associated effectors could be translocated into bacterial target cells and reused to assemble new T6SS machineries in target cells (Vettiger and Basler, 2016).

The T6SS has evolved multiple strategies for effector delivery. On the basis of the known effector transport mechanisms, effectors can be classified as "specialized" or "cargo" effectors (Cianfanelli et al., 2016b). Specialized effectors are fused to the C-terminus of T6SS structure proteins, such as Hcp, VgrG, or PAAR (Pro-Ala-Ala-Arg)-domain-containing protein known to sharpen the VgrG spike (Shneider et al., 2013). However, cargo effectors interact directly or require a specific chaperone or adaptor protein for loading onto the lumen of the Hcp tube or VgrG spike. T6SS adaptors/chaperones including DUF4123-, DUF1795-, and potentially DUF2169-containing proteins are required for loading a specific effector onto the cognate VgrG for delivery (Alcoforado Diniz and Coulthurst, 2015; Liang et al., 2015; Unterweger et al., 2015, 2016; Whitney et al., 2015; Bondage et al., 2016; Cianfanelli et al., 2016a). The DUF4123-containing protein functioning as a chaperone/adaptor was identified in the two studies in *Vibrio cholerae* with biochemical evidence showing its interaction with cognate VgrG (VgrG-1) and T6SS effector (TseL) by co-immunoprecipitation and bacterial two-hybrid analysis as well as genetic evidence for its requirement in mediating TseL secretion and TseL-mediated antibacterial activity (Liang et al., 2015; Unterweger et al., 2015). Together with its similar characteristics (i.e., low molecular weight protein with low isoelectric point, pI $\sim$5) with chaperones in phages and the type III secretion system (T3SS), the DUF4123-containing protein was named the T6SS effector chaperone (TEC) (Liang et al., 2015). Its chaperone feature is consistent with previous observation of a DUF4123-containing protein, Atu4349 in stabilizing the Tde1 effector in *Agrobacterium tumefaciens* and forming a complex with Tde1 co-purified from *Escherichia coli* (Ma et al., 2014). However, because of its requirement for TseL binding to the VgrG-1 spike with no detectable effect on TseL stability, the DUF4123-containing protein was also named T6SS adaptor protein 1 (Tap-1) by considering its function as an adaptor for loading TseL to the VgrG-1 spike (Unterweger et al., 2015). Although Tap-1 can interact with VgrG-1 in the absence of TseL in *V. cholerae* (Unterweger et al., 2015), the *A. tumefaciens* Tap-1/TEC ortholog and Tde1 each cannot interact with its cognate VgrG (VgrG1) in the absence of each other, so the formation of this adaptor/chaperone-effector complex occurs before loading onto the VgrG spike (Bondage et al., 2016). The DUF1795-containing protein, named effector-associated gene (Eag), can specifically interact with the PAAR domain of the cognate effectors, which are stabilized by specific Eag proteins in *Serratia marcescens* and *Pseudomonas aeruginosa* (Alcoforado Diniz and Coulthurst, 2015; Whitney et al., 2015; Cianfanelli et al., 2016a). Furthermore, EagR1 and EagR2 each interact with the cognate Rhs effector with specificity (i.e., EagR1 binds to Rhs1 but not Rhs2 and vice versa) (Cianfanelli et al., 2016a). In the absence of Rhs2, EagR2 can no longer load onto its cognate VgrG spike (VirG2) (Cianfanelli et al., 2016a). These data strongly suggest that Eag functions as a chaperone perhaps with specificity for PAAR-containing effectors. Another putative chaperone family protein is the DUF2169-containing protein, which was first identified in *A. tumefaciens* Atu3641 for its role in stabilizing the PAAR-like DUF4150-containing Tde2 effector for Tde2-dependent antibacterial activity (Bondage et al., 2016). The widespread genetic linkage between DUF2169- and DUF4150/PAAR-containing genes in several T6SS$^+$ proteobacterial genomes also implies a specific interaction between the two domains DUF2169 and DUF4150/PAAR. Taken together, Tap-1/TEC, Eag, and perhaps DUF2169-containing protein mainly function for loading specific effectors onto the cognate VgrG spike for secretion, but different adaptors/chaperones may have subtle differences in their modes of action, which awaits further structural and biochemical studies to clarify.

T6SS effectors with diverse biochemical activities have been identified. The major functions include the membrane-, cell wall-, or nucleic acid-targeting antibacterial effectors and several eukaryote-targeting effectors with a variety of enzymatic activities (Russell et al., 2014a; Alcoforado Diniz et al., 2015). In addition to the effectors functioning inside target cells, a few recent examples also suggested that T6SS effectors may function extracellularly rather than inside target cells. These identified extracellular effectors mainly function to bind or facilitate metal ions for their uptake (Wang et al., 2015; Lin et al., 2017; Si et al., 2017).

T6SS is widespread in Gram-negative bacteria, which include free-living bacteria and pathogens/symbionts of plants or animals. However, the number of identified effectors remains limited, in part because the annotation based on the primary genomic sequence information is often insufficient to properly annotate the function of effector genes. Nevertheless, almost all identified effectors are encoded with genetic linkage to the *vgrG* or *hcp* locus in the T6SS main gene cluster(s) or orphan *vgrG/hcp* island (De Maayer et al., 2011; Ma et al., 2014; Shyntum et al., 2014; Liang et al., 2015; Salomon et al., 2015; Abby et al., 2016). Various approaches have been successfully used to identify T6SS effectors. In this review, we summarize past and current methods as well as proposed potential strategies for identifying T6SS effectors. We hope such information can facilitate the discovery of novel T6SS effectors and elucidate their biochemical activities and biological functions.

PAST AND CURRENT METHODS FOR IDENTIFYING T6SS EFFECTORS

With the current knowledge of biochemical characteristics and transport mechanisms of T6SS effectors, T6SS effectors can be identified via both discovery-driven and knowledge/hypothesis-based methodologies. The methods used for identifying T6SS effectors are described as follows and summarized in **Table 1**, which also lists the pros and cons of these methods.

Bioinformatics Analysis

Some VgrG and Hcp proteins belong to specialized effectors by bearing a C-terminal effector domain (Pukatzki et al., 2009; Jamet and Nassif, 2015; Cianfanelli et al., 2016b; Ma et al., 2017a). Thus, VgrG or Hcp proteins with an additional C-terminal extension region are promising effector candidates with dual structural and biological roles. VgrG proteins can be classified into two categories: the canonical VgrG with gp27-gp5 domains as a sole structure function and the

TABLE 1 | Summary of current and potential methods for discovery of T6SS effectors.

Method	Advantages	Limitation / Disadvantages
CURRENT METHOD		
Bioinformatics analysis	Fast and robust in predicting putative effector candidates without experimental work	Prior knowledge or hypothesis is required
Genetic analysis of T6SS-associated genes	Easier to conduct without large-scale omics analysis or library construction and screening	Can only reveal effectors in the T6SS gene cluster or *hcp/vgrG* island
Proteomics-based method	Revealing effectors without known features	Limited to identifying proteins with significant secretion *in vitro* or a known chaperone for stability
Mutant library screening	Revealing effectors without known features; linking the gene to certain phenotypes directly	May identify genes other than T6SS effectors but with the same phenotypes; requires a testable phenotype
POTENTIAL METHOD		
Expression library	Revealing effectors without known features	Limited to identifying genes with cytosolic toxicity or screenable phenotype; may identify genes other than T6SS effectors
Protein–protein interactions	Revealing effectors interacting with known T6SS components	Limited to effectors with direct or tight interactions with bait protein; toxic protein may be harmful for the bacterial/yeast two-hybrid host

evolved VgrG with a C-terminal extended region conferring an additional domain(s) (Pukatzki et al., 2007). Therefore, several evolved VgrG proteins are identified, such as *V. cholerae* VgrG-1 containing an actin-crosslinking domain for a role in pathogenesis (Pukatzki et al., 2007) and the C-terminus of *V. cholerae* VgrG-3 with peptidoglycan degradation capability functioning as an antibacterial effector (Brooks et al., 2013; Dong et al., 2013). Besides bearing effector functions in the C-terminal extension of these evolved VgrG proteins, in entero-aggregative *E. coli* (EAEC), the VgrG1 harbors a C-terminal extension carrying DUF2345 and transthyretin (TTR) domains that are responsible for Tle1 effector binding (Flaugnatti et al., 2016). Co-immunoprecipitation (co-IP) and bacterial two-hybrid experiments suggested that the C-terminal extension is necessary and sufficient for interacting with the Tle1 effector, with the TTR domain involved in direct interaction with Tle1, whereas DUF2345 is required to stabilize the VgrG1–Tle1 interaction. In contrast to the widespread presence of evolved VgrG proteins harboring putative effector domains or an effector-binding domain in β-Proteobacteria and γ-Proteobacteria (Pukatzki et al., 2007; Boyer et al., 2009), Hcp proteins with a C-terminal extension have been found only in Enterobacteriaceae (Ma et al., 2017a). Although the C-terminal extension of some Hcp proteins was predicted to be a toxic domain, found in *Salmonella* in 2009 (Blondel et al., 2009), several Hcp proteins with a diverse C-terminal toxic domain (Hcp-ET) including Hcp-ET1 with DNase activity were recently characterized to exhibit antibacterial activity in Shiga toxin-producing *E. coli* (Ma et al., 2017a).

Besides searching for evolved VgrG or Hcp-ET, the conserved domains residing in known effectors can also be used as a query to search for potential effectors with available genome sequences. Russell et al. identified an amidase superfamily of T6SS effectors by searching for amidase catalytic cysteine and histidine residues combined with other features of T6SS effectors in their genome (Russell et al., 2012). With similar approaches, T6SS phospholipase effectors have been identified by the existence of a conserved motif, GXSXG, HXKXXXXD (Russell et al., 2013), or GXSXG (Flaugnatti et al., 2016). T6SS peptidoglycan glycoside hydrolase effectors were also identified by searching for the lysozyme-like fold (Whitney et al., 2013). Although they were not entirely identified by bioinformatics search, several nuclease effectors with predicted or verified conserved domains, such as the HXXD catalytic site motif of *A. tumefaciens* Tde (Zhang et al., 2012; Ma et al., 2014), HNH endonuclease domain of *S. marcescens* Rhs2 (Alcoforado Diniz and Coulthurst, 2015) and *Dickeya dadantii* RhsB (Koskiniemi et al., 2013), can be used to identify new T6SS effectors in any $T6SS^+$ bacterial genome via bioinformatics tools.

T6SS effectors also possess motifs or domains related to functions involved in translocation or other structural functions rather than biochemical or biological activity. In 2012, Zhang et al. proposed that a bacterial toxin could be divided into three parts: an N-terminal trafficking associated region, a central region, and a C-terminal toxic region. Rearrangement hotspot (Rhs)/YD repeats are common features of a central region in a toxin protein transported by diverse mechanisms. However, the PAAR domain is mostly located at the N-terminus of a T6SS effector, although some of these domains are followed by an N-terminal extension region (Zhang et al., 2012; Shneider et al., 2013). Of note, some classes of PAAR-containing proteins also harbor an Rhs region and a TTR domain located at the C-terminal extension region (Zhang et al., 2012; Shneider et al., 2013). Because the TTR domain of VgrG1 protein in EAEC is necessary and sufficient for Tle1 effector binding (Flaugnatti et al., 2016), the TTR domain of some PAAR-containing proteins may function as a carrier for effectors as suggested (Shneider et al., 2013). Indeed, several T6SS effectors were identified to have these characteristics. *P. aeruginosa* Tse5/RhsP1 and RhsP2 are typical Rhs effectors (Hachani et al., 2014; Whitney et al., 2014), whereas *S. marcescens* Rhs1 and Rhs2 (Alcoforado Diniz and Coulthurst, 2015), *D. dadantii* RhsA and RhsB (Koskiniemi et al., 2013), and Rhs-CT in Shiga toxin-producing *E. coli* (Ma et al., 2017b) harbor both N-terminal PAAR and central Rhs domains. Many T6SS effectors are featured with an N-terminal

PAAR domain followed by a C-terminal effector domain. The examples are *P. aeruginosa* Tse6/PA0093 carrying the typical PAAR (Hachani et al., 2014; Whitney et al., 2014); *A. tumefaciens* Tde2 (Ma et al., 2014) and *P. aeruginosa* PA0099 (Hachani et al., 2014) with the N-terminal PAAR-like domain DUF4150; and *Francisella novicida* IglG, an effector of non-canonical T6SS, containing another PAAR-like domain DUF4280 (Rigard et al., 2016). In addition to the Rhs/YD repeat and PAAR, some T6SS effectors possess the motif named marker for type six effectors (MIX), which was first found in *Vibrio parahaemolyticus* by a proteomics approach (Salomon et al., 2014). Further bioinformatics analysis revealed that this MIX motif is present in some proteins encoded within T6SS gene clusters or the *hcp/vgrG* island in T6SS$^+$ Proteobacteria. Moreover, these MIX effectors may be horizontally transferred among marine bacteria (Salomon et al., 2015; Salomon, 2016). However, whether the MIX motif is important for the effector function or translocation remains to be tested. With these findings, PAAR- and Rhs-containing effectors were predicted to be widespread in Proteobacterial genomes, indicating that the use of bioinformatics tools to identify the presence of the Rhs/YD repeat, PAAR, TTR, and MIX motifs as a preliminary screen for identifying T6SS effectors or their interacting proteins is a promising strategy, although these motifs are not exclusively related to T6SS.

Because of an essential role required in mediating the transport of specific cargo effectors, the conserved domains of the known T6SS adaptors/chaperones including DUF4123 of Tap-1/TEC (Liang et al., 2015; Unterweger et al., 2015; Bondage et al., 2016), DUF1795 of Eag (Whitney et al., 2015; Cianfanelli et al., 2016a), and DUF2169 (Bondage et al., 2016) can be used to identify the cognate effector gene due to their close genetic linkage. Indeed, the conservation of these chaperones/adaptors with genetic linkage to the cognate effector genes has led to identification of a new T6SS effector, such as the TseC effector mediated by TEC in *A. hydrophila*, by searching for DUF4123 in the genome (Liang et al., 2015) and several putative effectors that are genetically linked to DUF4123 and DUF2169 found in many Proteobacterial genomes (Liang et al., 2015; Bondage et al., 2016). Thus, by identifying the conserved domains of these chaperone/adaptor genes, one can identify the putative effector genes and perform functional assays to confirm their effector functions and relationship with the cognate chaperone/adaptor and VgrG spike.

In addition to the above-mentioned methods specifically for T6SS effector prediction, some online bioinformatics tools or databases provide a suggestion for potential effectors, such as an effector protein predictor, secretEPDB (An et al., 2016, 2017); a T6SS database; SecReT6 (Li et al., 2015); and T346hunter, which can annotate homologs of type III, IV and VI secretion systems automatically (Martínez-García et al., 2015). For researchers working on *Burkholderia* spp., DBSecSys is an online database for the *Burkholderia mallei* and *Burkholderia pseudomallei* secretion system (Memišević et al., 2014, 2016). The Pfam database also provides a resource to identify potential effectors with conserved effector domains (Finn et al., 2014). In conclusion, the bioinformatics-based approach to identify new T6SS effectors is a convenient and powerful tool for study of organisms with a sequenced genome. However, bioinformatics is limited to our current knowledge of biochemical features and transport mechanisms of T6SS effectors.

Genetic Analysis of T6SS-Associated Genes

Some open reading frames in the T6SS main cluster encoding a hypothetical protein with unknown function could be candidate genes for T6SS effector or immunity proteins; combining deletion of these genes and phenotypic analysis such as interbacterial competition assay or virulence assay may lead to effector discovery. Hcp secretion is considered a hallmark of a functional T6SS and therefore can be used to determine the effects of the deletion mutant in T6SS assembly. In *S. marcescens*, two genes with unknown function in the T6SS gene cluster were found to be T6SS effectors, *ssp1* and *ssp2*, via bacterial competition assay and protein–protein interaction studies (English et al., 2012). The other case is the Tae and Tde1 effectors in *A. tumefaciens* C58, in which both Tae and Tde1 are secreted proteins, and their absence does not affect Hcp secretion (Lin et al., 2013; Ma et al., 2014). Follow-up experiments further demonstrated their functions as bacterial toxins and that their cognate immunity proteins are encoded from their adjacent genes (Ma et al., 2014). Some effectors are not encoded in the T6SS main cluster but are found in the *hcp/vgrG* island located outside the T6SS main cluster (Abby et al., 2016). Such effector genes include *tse5/rhsP1* and *rhsP2* in *P. aeruginosa*, found near *vgrG4* and *vgrG14*, respectively (Hachani et al., 2014; Whitney et al., 2014); *tde2* in *A. tumefaciens* C58, located in the *vgrG2* operon (Ma et al., 2014); and some effectors found located in the *vgrG/hcp* island in *Pantoea* and *Erwinia* species (De Maayer et al., 2011). Therefore, hypothetical proteins encoded in the T6SS gene cluster or orphan *vgrG/hcp* island can be analyzed genetically by virulence assay, antibacterial competition assay and/or protein secretion. Together with sequence- and structure-based search engines, such as NCBI blast searches (Ncbi Resource Coordinators, 2017) and Phyre2 (Kelley et al., 2015), one can identify T6SS effectors without large-scale omics approaches or library construction and screening.

Proteomics-Based Method

Unlike the bioinformatics-based method depending on empirical known characteristics of T6SS effectors, proteomics is a discovery-driven approach that can identify T6SS effectors without prior knowledge. A simple experimental design is to identify T6SS effector candidates via a comparative proteomics analysis between a wild-type strain and a T6SS secretion-deficient mutant. The *P. aeruginosa* H1-T6SS effectors Tse1, Tse2, and Tse3 were found by comparing the secretome between the hyper-secreted *pppA* deletion mutant and *clpV1* secretion-deficient mutant (Hood et al., 2010). Similarly, a gel-free secretome analysis comparing the wild-type strain and *clpV* mutant of *V. cholerae* revealed a TseH effector possessing an amidase domain and two YD repeats (Altindis et al., 2015). The mutant of *vasH*, encoding a transcriptional regulator of T6SS (Kitaoka et al., 2011), was used to identify the new effectors VasX and VgrG1 in *V. cholerae* and *A. hydrophila*, respectively (Suarez et al., 2010;

Miyata et al., 2011). Both of these effectors were confirmed by a virulence assay with amoeba or HeLa cells as models. The use of the *tssC* mutant as a T6SS secretion-deficient mutant to analyze the comparative secretome also led to the discovery of Tae2 in *B. thailendensis*, Bte1 and Bte2 in *Burkholderia fragilis* and Fte2 in *Flavobacterium johnsoniae* (Russell et al., 2012, 2014b; Wexler et al., 2016). Comparison of the secretomes of the *S. marcescens clpV* mutant and the wild type revealed four new effectors: Ssp3, Ssp4, Ssp5, and Ssp6 (Fritsch et al., 2013). The effector function of Ssp4 was confirmed by interbacterial competition assay; the toxic effects of others were determined by examining the survival of effector-expressing *E. coli* cells. Comparison of the secretomes of the T6SS cluster-deletion mutant and the wild type of enterohemorragic *E. coli* (EHEC) revealed a magnesium (Mn)-containing catalase, KatN, as a T6SS effector (Wan et al., 2017). This study demonstrated that KatN is secreted in a T6SS-dependent secretion to the extracellular milieu *in vitro* and into the host cell cytosol after EHEC is phagocytized by macrophages. Interestingly, *katN* is not located in the T6SS cluster or the *hcp/vgrG* island. Thus, determining the molecular mechanisms underlying how KatN is transported via T6SS through VgrG or Hcp as a carrier or via another yet-to-be discovered mechanism would be of interest. In addition, secretome analysis of each specific *vgrG* mutant in *S. marcescens* was used to identify a VgrG-dependent effector, which led to the discovery of an Slp effector in *S. marcescens* with VgrG2-dependent secretion (Cianfanelli et al., 2016a).

The characteristics of a chaperone that stabilizes the cognate effector can also be used to identify new T6SS effectors. This idea led to the identification of Tse4 in *P. aeruginosa* (Whitney et al., 2014). Because Hcp1 in *P. aeruginosa* has the chaperone activity that stabilizes effectors (Silverman et al., 2013), comparing the intracellular proteome between the wild type and Δ*hcp1* was used as a unique method to identify candidate effectors with lower accumulation in Δ*hcp1* (Whitney et al., 2014). This gave us a hint that the same principle can be used to identify the effector requiring a cognate chaperone for stability by performing comparative cellular proteome analysis between the wild type and the mutant deficient in a specific chaperone, such as a DUF4123-, DUF2169-, or DUF1795-containing protein.

The proteomics-based method is a powerful tool to identify potential effectors in a hypothesis-driven or knowledge-independent manner. Thus, this approach may identify novel effectors with conserved motifs such as MIX motif-containing effectors found in *V. parahaemolyticus* (Salomon et al., 2014). However, since not all effectors can be secreted *in vitro* in significant amounts or require a chaperone for stability, current proteomics-based methods are limited to identify effectors with these criteria.

Mutant Library Screening

In addition to proteomics approaches, mutant library screening is another powerful tool to identify T6SS effectors without known features yet is often a phenotype-dependent method. Of note, the very first T6SS reported in *V. cholerae* was identified by this approach, although discovery of a secretion system was not intentional. The key to identifying the T6SS involved in *V. chalerae* virulence is the use of a novel bacterial virulence assay with the eukaryotic amoebae organism *Dictyostelium discoiseum* as a host model system, which uptakes bacterial cells by phagocytosis, mimicking the behavior of macrophages (Steinert and Heuner, 2005; Pukatzki et al., 2006). The transposon insertion mutants, *vasH*, *vasA* (*tssF*), and *vasK* (*tssM*), were found to lose the anti-amoebae activity and later found deficient in secretion of a previously known secreted protein, Hcp (Williams et al., 1996; Pukatzki et al., 2006). Thus, a robust phenotype screening system is required for identifying T6SS effectors by mutant library screening. In *Burkholderia cenocepacia*, the T6SS effector TecA was found by selecting the transposon insertion mutants that cannot disrupt the actin of macrophages, and TecA was further demonstrated to function as a Rho-GTPase deaminase (Aubert et al., 2016). Because the immunity protein is essential for resistance against wild-type siblings with antibacterial activity but not T6SS-deficient siblings, Dong et al. used transposon insertion-site sequencing (Tn-seq) to identify the immunity gene and associated effector gene in a transposon mutant library screening (Dong et al., 2013). By this approach, they identified a new T6SS effector with lipase domain, TseL, and two known effectors: VgrG3, with peptidoglycan degradation activity, and VasX, which may interact with phospholipids. Because this study focused on identifying mutants that are absent in T6SS$^+$ but present in T6SS$^-$ *V. cholerae* strains, the mutants that are lethal in both T6SS$^+$ and T6SS$^-$ strains could not be selected. As a result, all three effectors identified are cell wall– or membrane-targeting effectors because the effectors only exhibit the function when delivered to the cell wall or membrane of target cells; thus, the respective immunity mutants remain viable for selection. For the effectors with cytosolic targets, one may identify the cognate immunity gene by directly sequencing the saturating mutant library because of the lethality of the mutants. Tn-seq technology was also used to identify a *V. cholerae* transposon insertion *tsiV3* mutant that cannot survive in the infant rabbit gut, which suggests a role for T6SS in antagonistic interbacterial interactions during infection (Fu et al., 2013). The *tsiV3* mutant of this *V. cholera* strain C6706 is viable when grown in regular growth medium because T6SS is not expressed during *in vitro* growth but is induced *in vivo* for bacterial colonization in the animal host.

Mutant library screening has been successfully used to identify T6SS effectors or cognate immunity, but mutant strains with impaired competitive growth or virulence phenotypes could be caused by mutation in genes involved in T6SS machine assembly or even other virulence-associated factors. In *B. pseudomallei*, transposon insertion mutant library screening revealed the T3SS as the main system related to host virulence, whereas T6SS-1 is important for the virulence in the Madagascar hissing cockroach model, and T6SS-5 may be associated with fitness in mice (Fisher et al., 2012; Gutierrez et al., 2015). Library screening approaches also identified T6SS associated with other phenotypes. For example, a transposon insertion in *clpV* highly attenuated the secretion of iron chelator pyoverdine from *Pseudomonas taiwanensis* with reduced antagonism ability (Chen et al., 2016). Transposon insertion mutations in *vgrG* or *hcp* operons in *Proteus mirabilis* altered swarming behavior (Alteri et al., 2013). However, whether the impact on pyoverdine secretion or swarming ability are the direct phenotype or are secondary effects due to the absence of T6SS await further validation.

POTENTIAL METHODS FOR IDENTIFYING T6SS EFFECTORS

In addition to the above-mentioned methods that have been successfully used to identify T6SS effectors directly or indirectly, we discuss two potential approaches that may help to identify T6SS effectors.

Expression Library

Most of the T6SS effectors identified to date are usually toxic to eukaryotic hosts or competitor bacteria. Thus, we can take advantage of such characteristics to identify effector genes by screening a genome-wide expression library in yeast (for eukaryotic toxin) or *E. coli* cells (for bacterial toxins). Indeed, since the 2000s, yeast has been used as a model to find bacterial effectors of other secretion systems (Siggers and Lesser, 2008). The key is to use the tightly regulated inducible promoter to express the genes of interests and screen the construct with growth inhibition or arrest under inducible condition. This kind of approach has been used to identify T3SS effectors in *Shigella* sp., *P. aeruginosa* and *Xanthomonas campestris* pv. *vesicatoria*. In *Shigella* sp., a T3SS effector, IpaJ, encoded from *Shigella* virulence plasmids, was found to be toxic and inhibit growth when expressed in yeast cells and later confirmed to have T3SS-dependent secretion (Slagowski et al., 2008). A genome-wide expression library constructed from *P. aeruginosa* was transformed into yeast cells, and the expressed proteins leading to yeast cell death or growth inhibition after induction were selected for further virulence assay by infecting *Caenorhabditis elegans* and macrophages with wild-type *P. aeruginosa* or the respective deletion mutants. Such screening indeed identified several virulence factors of *P. aeruginosa*, but the delivery mechanism of these proteins has yet to be determined (Zrieq et al., 2015). In *X. campestris* pv. *vesicatoria*, this approach was used to determine the biological function of already identified T3SS effectors (Salomon et al., 2010). In addition to performing the screening experiments under normal growth conditions, various stress conditions can be used with transformants to identify the effectors targeting the conserved pathway activated only under stress (Slagowski et al., 2008; Salomon et al., 2010).

Toxin effector screening has been also performed in prokaryotic cells. By transforming a DNA library from 388 microbial genomes into *E. coli* and sequencing all surviving transformants, the gene (toxin) that could survive only when its adjacent gene is present on the same clone (antitoxin) was identified (Sberro et al., 2013). Hence, Sorek's group identified both known (238 pairs) and newly identified (123 pairs) toxin/antitoxin pairs. Thus, in view of the toxicity of T6SS effectors and co-existence relationship of the toxin-immunity pair, one can use the inducible expression library and toxin-immunity pair for survival screening to identify T6SS effectors. However, if the effector only exerts its function on the cell membrane/surface or extracellularly, its effector function cannot be uncovered because it may require the T6SS machine for delivery to the right compartments. Therefore, expression library screening may only identify effectors with a cytoplasmic target unless a signal peptide is fused to the expressed proteins for translocation across the inner membrane. Also, the expression library is not specifically designed for T6SS effector identification. Another may also identify genes unrelated to T6SS. Therefore, after identification of putative effector candidates, further elucidation of the delivery mechanism is required to conclude the genes encoding T6SS effectors. One limitation in the expression library is that effectors with subtle effects or other physiological functions not related to cell growth may be overlooked.

Protein–Protein Interaction

As described previously, several T6SS cargo effectors are loaded to the VgrG spike via non-covalent binding with chaperone or adaptor proteins such as Eag or Tap-1/TEC (Pukatzki et al., 2009; Liang et al., 2015; Unterweger et al., 2015, 2016; Whitney et al., 2015; Bondage et al., 2016; Cianfanelli et al., 2016b). Others such as Tse2 in *P. aeruginosa* and Tae4 in *Salmonella typhimurium* are secreted by binding to the lumen of the Hcp hexamer (Silverman et al., 2013; Sana et al., 2016). Therefore, the T6SS effectors were found to be directly or indirectly associated with Hcp, VgrG, PAAR, Tap-1/TEC, or Eag in various bacteria (Shneider et al., 2013; Silverman et al., 2013; Hachani et al., 2014; Liang et al., 2015; Unterweger et al., 2015; Whitney et al., 2015; Bondage et al., 2016; Cianfanelli et al., 2016a; Rigard et al., 2016). Because the genes encoding these conserved T6SS components functioning as carriers or chaperones/adaptors for T6SS effectors are easier to be predicted than most effector genes in a sequenced genome, these proteins can be first identified and used as a bait in well-established protein–protein interaction platforms such as bacterial two-hybrid (Battesti and Bouveret, 2012), yeast two-hybrid (Mehla et al., 2015), or co-IP/pull-down assay (Brymora et al., 2004; Kaboord and Perr, 2008) to identify potential effectors. This idea was indeed proposed by Silverman et al., that interaction with the Hcp chaperone may be a novel method to identify new T6SS effectors (Silverman et al., 2013). One disadvantage of this method is that it can identify only effectors that directly or tightly bind with adaptors/chaperones or the T6SS machine. Because expression of the effector toxin may be harmful for the host used in bacterial or yeast two-hybrid assays, some key effector toxins may be missed by such screening.

BIOCHEMICAL AND BIOLOGICAL FUNCTIONS OF KNOWN T6SS EFFECTORS

Numerous T6SS effectors have been identified with experimentally proven or predicted biochemical activities. In general, the effectors can be classified by their target cells as eukaryotic hosts or bacterial competitors, although the mode of action or biochemical function of effectors can be distinct or

TABLE 2 | Known T6SS effectors with defined biochemical activities.

Effector	Organism	Biochemical activity	References
EUKARYOTIC TARGET			
VgrG1	*A. hydrophila*	ADP-ribosyltransferase	Suarez et al., 2010
TecA	*B. cenocepacia*	Deaminating Rho GTPase	Aubert et al., 2016
VgrG-5	*B. thailendensis*	Membrane fusion activity	Schwarz et al., 2014
EvpP	*E. tarda*	Inhibition of NLRP3 inflammasome	Yang et al., 2015; Chen et al., 2017
VgrG-1	*V. cholerae*	Actin-crosslinking	Pukatzki et al., 2007; Ma and Mekalanos, 2010
VgrG2b	*P. aeruginosa*	Interacting with microtubule	Sana et al., 2015
KatN	Enterohemorragic *E. coli*	Mn-containing catalase	Wan et al., 2017
DUAL TARGET			
PldA/Tle5, PldB	*P. aeruginosa*	Phospholipase, activation of Akt signaling pathway	Russell et al., 2013; Jiang et al., 2014
VasX	*V. cholerae*	Membrane-targeting activity	Dong et al., 2013; Miyata et al., 2013
TseL/Tle2	*V. cholerae*	Phospholipase	Dong et al., 2013; Russell et al., 2013
BACTERIAL TARGET—LIPASE ACTIVITY			
Tle1	*Burkholderia thailandensis*	Phospholipase	Russell et al., 2013
Tle1	Entero-aggregative *E. coli*	Phospholipase	Flaugnatti et al., 2016
BACTERIAL TARGET—CELL WALL DEGRADATION ENZYME			
Tse1	*P. aeruginosa*	Amidase	Hood et al., 2010; Russell et al., 2011
Tse3	*P. aeruginosa*	Muramidase	Hood et al., 2010; Russell et al., 2011
TseH	*V. cholerae*	Cell-wall degradation hydrolase	Altindis et al., 2015
VgrG-3	*V. cholerae*	Peptidoglycan degradation	Pukatzki et al., 2006; Brooks et al., 2013
Tge2	*P. aeruginosa*	Glycoside hydrolase	Whitney et al., 2013
BACTERIAL TARGET—DNASE			
Tde1, Tde2	*A. tumefaciens*	DNase	Ma et al., 2014
RhsA, RhsB	*D. dadantii*	DNase	Koskiniemi et al., 2013
Hcp-ET1	Shiga toxin-producing *E. coli*	DNase	Ma et al., 2017a
Rhs2	*S. marcescens*	DNase	Alcoforado Diniz and Coulthurst, 2015
BACTERIAL TARGET—OTHER BIOCHEMICAL ACTIVITY			
Tse2	*P. aeruginosa*	NAD-dependent toxicity	Hood et al., 2010; Robb et al., 2016
Tse6	*P. aeruginosa*	NAD(P)+ glycohydrolase	Whitney et al., 2014, 2015
EXTRACELLULAR			
TseM	*B. thailendensis*	Mn^{2+}-binding protein	Si et al., 2017
YezP	*Y. pseudotuberculosis*	Zinc-binding protein	Wang et al., 2015
TseF	*P. aeruginosa*	Bind OMV for iron acquisition	Lin et al., 2017

shared between eukaryotic or prokaryotic effectors (**Table 2**). So far, the major biological functions of these effectors are to increase the competitive growth advantages of environmental or host-associated bacteria or virulence of the bacterial pathogen (Kapitein and Mogk, 2013; Durand et al., 2014; Russell et al., 2014a; Alcoforado Diniz et al., 2015; Hachani et al., 2016). In addition to functioning as a contact-dependent weapon against the host or bacterial competitor, some T6SSs may secrete effectors for scavenging cofactors to survive in oxidative stress and/or facilitate metal ion acquisition (Chen et al., 2016; Lin et al., 2017; Si et al., 2017; Wan et al., 2017). These findings suggest diversified T6SS functions for bacterial survival and fitness in its ecological niches.

The biological and biochemical functions of antihost T6SS effectors are quite diverse. Some of the virulence-associated T6SS effectors can interact with the cytoskeleton of the host cells directly. *V. cholerae* VgrG-1 can cause actin crosslinking, which is associated with the intestinal inflammation (Pukatzki et al., 2007; Ma and Mekalanos, 2010). *P. aeruginosa* VgrG2b interacts with microtubules causing the successful invasion of epithelial cells (Sana et al., 2015). VgrG1 in *A. hydrophila* has ADP-ribosyltransferase activity and can disrupt the cytoskeleton of HeLa cells (Suarez et al., 2010). Others can interact with the plasma membrane of host cells. *B. thailandensis* VgrG-5 has membrane fusion activity and is required for the multinucleated giant cell formation to spread between cells (Schwarz et al., 2014). Besides targeting the host cytoskeleton or membrane, an Mn-containing catalase, KatN, was shown to be translocated from the enterohemorrhagic *E. coli* (EHEC) into host cells in a T6SS-dependent manner and reduce the reactive oxygen species concentration in host macrophages in a T6SS-dependent manner (Wan et al., 2017). Although T6SS appears to be important for EHEC virulence, as demonstrated in mouse model, no virulence phenotype could be observed in the *katN* mutant. In *B. cenocepacia*, TecA deaminates Rho GTPase in host cells and triggers the inflammation reaction and actin disruption (Aubert et al., 2016). Although infection of the *tecA* mutant will not trigger inflammation, it can kill mice, as compared with the survival of all mice infected with wild-type *B. cenocepacia*. The authors proposed that TecA may limit the bacterial cell number in the host and cause the bacteria to evolve toward a mutual relationship with the host. In contrast, the fish pathogen *Edwardsiella tarda* produces a unique T6SS effector EvpP, which does not harbor any known functional domain, to suppress inflammasome activation and promote colonization in the host (Chen et al., 2017). Thus, bacterial pathogens seem to evolve different T6SS effectors in modulating the host immunity for survival and fitness, but much more remains to be learned about their host targets and mode of actions (Yu and Lai, 2017).

The antibacterial T6SS effectors identified so far can be divided into membrane-, cell wall-, and nucleic acid-targeting effectors as well as other biological functions (Durand et al., 2014; Russell et al., 2014a; Alcoforado Diniz et al., 2015). Membrane-targeting effectors include Tle phospholipase superfamily such as *P. aeruginosa* PldA/Tle5, *V. cholerae* Tle2/TseL, and *Burkholderia thailandensis* Tle1 (Russell et al., 2013) and entero-aggregative *E. coli* Tle1 (Flaugnatti et al., 2016). VgrG3 and TseH in *V. cholerae* (Brooks et al., 2013; Altindis et al., 2015) and Tse1 amidase and Tse3 muramidase (Russell et al., 2011) in *P. aeruginosa* belong to cell wall-targeting effectors. *S. marcescens* Rhs2 (Alcoforado Diniz and Coulthurst, 2015), *D. dadantii* RhsA and RhsB (Koskiniemi et al., 2013), *A. tumefaciens* Tde1 and Tde2 (Ma et al., 2014), and Shiga toxin-producing *E. coli* Hcp-ET1 (Ma et al., 2017a) can hydrolyze DNA. Some effectors have other toxic effects in bacterial cytoplasm; for example, *P. aeruginosa* Tse6 is an NAD(P)+ glycohydrolase toxin inducing bacteriostasis by depleting cellular NAD(P)+ levels (Whitney et al., 2015) and Tse2 is likely an NAD-dependent enzyme from the structure information (Robb et al., 2016).

Many T6SS effectors can be secreted into the extracellular milieu during *in vitro* culture, but these antihost or antibacterial toxin effectors exhibit their effector functions inside the target cells. A few reports recently provided compelling biochemical evidence for some T6SS effectors functioning extracellularly rather than in target cells. These effectors so far identified mainly function to bind or facilitate metal ions for their uptake. *Yersinia pseudotuberculosis* T6SS-4 encodes a unique zinc-binding effector protein, YezP, that is involved in environmental stress resistance in the host body (Wang et al., 2015). The double mutant lacking this effector and classical zinc transporter exhibits almost no virulence toward mice. Similar to *Y. pseudotuberculosis*, TseM secreted by T6SS-4 of *Burkholderia thailandensis* functions as an Mn^{2+}-binding protein in interacting with the Mn^{2+}-specific TonB-dependent outer-membrane protein MnoT to activate transport for resistance to oxidative stress (Si et al., 2017). Deletion of *tseM* caused reduced virulence toward *Galleria mellonella* wax moth larvae. Although not functioning as a metal-binding protein, TseF secreted by H3-T6SS of *P. aeruginosa* can bind to and recruit outer-membrane vesicles for iron (Fe) acquisition (Lin et al., 2017). By engaging the Fe(III)-pyochelin receptor FptA and the porin OprF, TseF can facilitate the delivery of outer-membrane vesicle-associated Fe for bacterial cells.

Although many effectors target only eukaryotic cells (i.e., *V. cholerae* VgrG-1 with an actin cross-linking domain) or prokaryotic cells (i.e., *P. aeruginosa* Tse1 amidase and Tse3 muramidase for targeting peptidoglycan), some effectors may have dual targets. PldA/Tle5 and PldB, phospholipase D effectors secreted from *P. aeruginosa,* can target both bacterial and eukaryotic cells (Jiang et al., 2014). Both PLD effectors can inhibit bacterial growth by increasing membrane permeability when translocated into the periplasm of bacterial recipient cells while activating phosphorylation of the protein kinase B (Akt) signaling pathway to lead to the internalization of the bacteria into the host cell. TseL/Tle2 in *V. cholerae* is phospholipase, which exhibits both T6SS-dependent antibacterial activity and a virulence role toward *D. discoideum* (Dong et al., 2013; Russell et al., 2013). In addition, *V. cholerae* VasX is a membrane targeting toxin that can interact with phosphoinositides important for virulence toward the eukaryotic amoebae *D. discoideum* and compromise the integrity of the inner membrane in bacterial target cells for antibacterial activity (Miyata et al., 2011, 2013; Dong et al., 2013). The *P. aeruginosa* Tse2 toxin naturally attacks bacterial cells for interbacterial competition but can also cause toxicity when ectopically expressed in eukaryotic cells (Robb et al., 2016). Thus, although current data show that the DNase effectors such as Tde and NAD(P)+ glycohydrolase effectors such as Tse6 are involved in antibacterial activity, whether they can also be translocated into eukaryotic host cells remains to be tested. In considering the role of T6SS in bacterial colonization inside hosts such as *A. tumefaciens* T6SS for competitive growth advantage *in planta* (Ma et al., 2014), *V. cholera* T6SS for survival in the infant rabbit gut (Fu et al., 2013), and T6SS for survival of the bacterial symbionts Bacteroidetes in human guts (Wexler et al., 2016), such possibility could be explored.

PERSPECTIVES

The T6SS is known to have important roles in bacterium–host interaction, bacterium–bacterium interaction and even other functions associated with bacterial physiology. Considering the expanded and diversifying functions of the T6SS discovered since its identification more than a decade ago, the system may have more functions, especially biological significance at polymicrobial ecological niches yet to be uncovered. The methodologies and biology of T6SS effectors we discuss in this review can be a foundation for future identification and studies of the T6SS and effectors.

AUTHOR CONTRIBUTIONS

YL and EL discussed the review content and wrote the paper together.

FUNDING

Funding is provided by the Ministry of Science and Technology (MOST 104-2311-B-001-025-MY3), Taiwan to EL.

ACKNOWLEDGMENTS

We thank the members of Lai laboratory for critical readings of this manuscript.

REFERENCES

Abby, S. S., Cury, J., Guglielmini, J., Néron, B., Touchon, M., and Rocha, E. P. (2016). Identification of protein secretion systems in bacterial genomes. *Sci. Rep.* 6:23080. doi: 10.1038/srep23080

Alcoforado Diniz, J., and Coulthurst, S. J. (2015). Intraspecies competition in *Serratia marcescens* is mediated by type VI-secreted rhs effectors and a conserved effector-associated accessory protein. *J. Bacteriol.* 197, 2350–2360. doi: 10.1128/JB.00199-15

Alcoforado Diniz, J., Liu, Y. C., and Coulthurst, S. J. (2015). Molecular weaponry: diverse effectors delivered by the Type VI secretion system. *Cell. Microbiol.* 17, 1742–1751. doi: 10.1111/cmi.12532

Alteri, C. J., Himpsl, S. D., Pickens, S. R., Lindner, J. R., Zora, J. S., Miller, J. E., et al. (2013). Multicellular bacteria deploy the type VI secretion system to preemptively strike neighboring cells. *PLoS Pathog.* 9:e1003608. doi: 10.1371/journal.ppat.1003608

Altindis, E., Dong, T., Catalano, C., and Mekalanos, J. (2015). Secretome analysis of *Vibrio cholerae* type VI secretion system reveals a new effector-immunity pair. *MBio* 6, e00075-15. doi: 10.1128/mBio.00075-15

An, Y., Wang, J., Li, C., Leier, A., Marquez-Lago, T., Wilksch, J., et al. (2016). Comprehensive assessment and performance improvement of effector protein predictors for bacterial secretion systems III, IV and VI. *Brief Bioinform.* doi: 10.1093/bib/bbw100. [Epub ahead of print].

An, Y., Wang, J., Li, C., Revote, J., Zhang, Y., Naderer, T., et al. (2017). SecretEPDB: a comprehensive web-based resource for secreted effector proteins of the bacterial types III, IV and VI secretion systems. *Sci. Rep.* 7:41031. doi: 10.1038/srep41031

Aschtgen, M. S., Bernard, C. S., De Bentzmann, S., Lloubès, R., and Cascales, E. (2008). SciN is an outer membrane lipoprotein required for type VI secretion in enteroaggregative *Escherichia coli*. *J. Bacteriol.* 190, 7523–7531. doi: 10.1128/jb.00945-08

Aubert, D. F., Xu, H., Yang, J., Shi, X., Gao, W., Li, L., et al. (2016). A burkholderia type VI effector deamidates rho GTPases to activate the pyrin inflammasome and trigger inflammation. *Cell Host Microbe* 19, 664–674. doi: 10.1016/j.chom.2016.04.004

Basler, M. (2015). Type VI secretion system: secretion by a contractile nanomachine. *Philoso. Trans. R. Soc. B Biol. Sci.* 370:20150021. doi: 10.1098/rstb.2015.0021

Basler, M., and Mekalanos, J. J. (2012). Type 6 secretion dynamics within and between bacterial cells. *Science* 337, 815–815. doi: 10.1126/science.1222901

Basler, M., Pilhofer, M., Henderson, G. P., Jensen, G. J., and Mekalanos, J. J. (2012). Type VI secretion requires a dynamic contractile phage tail-like structure. *Nature* 483, 182–186. doi: 10.1038/nature10846

Battesti, A., and Bouveret, E. (2012). The bacterial two-hybrid system based on adenylate cyclase reconstitution in *Escherichia coli*. *Methods* 58, 325–334. doi: 10.1016/j.ymeth.2012.07.018

Blondel, C. J., Jiménez, J. C., Contreras, I., and Santiviago, C. A. (2009). Comparative genomic analysis uncovers 3 novel loci encoding type six secretion systems differentially distributed in Salmonella serotypes. *BMC Genomics* 10:354. doi: 10.1186/1471-2164-10-354

Bondage, D. D., Lin, J. S., Ma, L. S., Kuo, C. H., and Lai, E. M. (2016). VgrG C terminus confers the type VI effector transport specificity and is required for binding with PAAR and adaptor–effector complex. *Proc. Natl. Acad. Sci. U.S.A.* 113, E3931–E3940. doi: 10.1073/pnas.1600428113

Bonemann, G., Pietrosiuk, A., Diemand, A., Zentgraf, H., and Mogk, A. (2009). Remodelling of VipA/VipB tubules by ClpV-mediated threading is crucial for type VI protein secretion. *EMBO J.* 28, 315–325. doi: 10.1038/emboj.2008.269

Boyer, F., Fichant, G., Berthod, J., Vandenbrouck, Y., and Attree, I. (2009). Dissecting the bacterial type VI secretion system by a genome wide *in silico* analysis: what can be learnt from available microbial genomic resources? *BMC Genomics* 10:104. doi: 10.1186/1471-2164-10-104

Brooks, T. M., Unterweger, D., Bachmann, V., Kostiuk, B., and Pukatzki, S. (2013). Lytic activity of the *Vibrio cholerae* type VI secretion toxin VgrG-3 is inhibited by the antitoxin TsaB. *J. Biol. Chem.* 288, 7618–7625. doi: 10.1074/jbc.M112.436725

Brunet, Y. R., Zoued, A., Boyer, F., Douzi, B., and Cascales, E. (2015). The type VI secretion TssEFGK-VgrG phage-like baseplate is recruited to the TssJLM membrane complex via multiple contacts and serves as assembly platform for tail tube/sheath polymerization. *PLoS Genet.* 11:e1005545. doi: 10.1371/journal.pgen.1005545

Brymora, A., Valova, V. A., and Robinson, P. J. (2004). Protein-protein interactions identified by pull-down experiments and mass spectrometry. *Curr. Protoc. Cell Biol.* Chapter 17, Unit 17.15. doi: 10.1002/0471143030.cb1705s22

Chen, H., Yang, D., Han, F., Tan, J., Zhang, L., Xiao, J., et al. (2017). The bacterial T6SS effector EvpP prevents NLRP3 inflammasome activation by inhibiting the Ca^{2+}-dependent MAPK-Jnk pathway. *Cell Host Microbe* 21, 47–58. doi: 10.1016/j.chom.2016.12.004

Chen, W. J., Kuo, T. Y., Hsieh, F. C., Chen, P. Y., Wang, C. S., Shih, Y. L., et al. (2016). Involvement of type VI secretion system in secretion of iron chelator pyoverdine in *Pseudomonas taiwanensis*. *Sci. Rep.* 6:32950. doi: 10.1038/srep32950

Cianfanelli, F. R., Alcoforado Diniz, J., Guo, M., De Cesare, V., Trost, M., and Coulthurst, S. J. (2016a). VgrG and PAAR proteins define distinct versions of a functional type VI secretion system. *PLoS Pathog.* 12:e1005735. doi: 10.1371/journal.ppat.1005735

Cianfanelli, F. R., Monlezun, L., and Coulthurst, S. J. (2016b). Aim, load, fire: the type VI secretion system, a bacterial nanoweapon. *Trends Microbiol.* 24, 51–62. doi: 10.1016/j.tim.2015.10.005

De Maayer, P., Venter, S. N., Kamber, T., Duffy, B., Coutinho, T. A., and Smits, T. H. (2011). Comparative genomics of the type VI secretion systems of Pantoea and Erwinia species reveals the presence of putative effector islands that may be translocated by the VgrG and Hcp proteins. *BMC Genomics* 12:576. doi: 10.1186/1471-2164-12-576

Dong, T. G., Ho, B. T., Yoder-Himes, D. R., and Mekalanos, J. J. (2013). Identification of T6SS-dependent effector and immunity proteins by Tn-seq in *Vibrio cholerae*. *Proc. Natl. Acad. Sci. U.S.A.* 110, 2623–2628. doi: 10.1073/pnas.1222783110

Durand, E., Cambillau, C., Cascales, E., and Journet, L. (2014). VgrG, Tae, Tle, and beyond: the versatile arsenal of Type VI secretion effectors. *Trends Microbiol.* 22, 498–507. doi: 10.1016/j.tim.2014.06.004

English, G., Trunk, K., Rao, V. A., Srikannathasan, V., Hunter, W. N., and Coulthurst, S. J. (2012). New secreted toxins and immunity proteins encoded within the Type VI secretion system gene cluster of *Serratia marcescens*. *Mol. Microbiol.* 86, 921–936. doi: 10.1111/mmi.12028

Felisberto-Rodrigues, C., Durand, E., Aschtgen, M. S., Blangy, S., Ortiz-Lombardia, M., Douzi, B., et al. (2011). Towards a structural comprehension of bacterial type VI secretion systems: characterization of the TssJ-TssM complex of an *Escherichia coli* pathovar. *PLoS Pathog.* 7:e1002386. doi: 10.1371/journal.ppat.1002386

Finn, R. D., Bateman, A., Clements, J., Coggill, P., Eberhardt, R. Y., Eddy, S. R., et al. (2014). Pfam: the protein families database. *Nucleic Acids Res.* 42, D222–D230. doi: 10.1093/nar/gkt1223

Fisher, N. A., Ribot, W. J., and DeShazer, D. (2012). The Madagascar hissing cockroach as a novel surrogate host for *Burkholderia pseudomallei, B. mallei* and *B. thailandensis*. *BMC Microbiol.* 12:117. doi: 10.1186/1471-2180-12-117

Flaugnatti, N., Le, T. T., Canaan, S., Aschtgen, M. S., Nguyen, V. S., Blangy, S., et al. (2016). A phospholipase A1 antibacterial Type VI secretion effector interacts directly with the C-terminal domain of the VgrG spike protein for delivery. *Mol. Microbiol.* 99, 1099–1118. doi: 10.1111/mmi.13292

Fritsch, M. J., Trunk, K., Diniz, J. A., Guo, M., Trost, M., and Coulthurst, S. J. (2013). Proteomic identification of novel secreted antibacterial toxins of the *Serratia marcescens* type VI secretion system. *Mol. Cell. Proteomics* 12, 2735–2749. doi: 10.1074/mcp.M113.030502

Fu, Y., Waldor, M. K., and Mekalanos, J. J. (2013). Tn-Seq analysis of *Vibrio cholerae* intestinal colonization reveals a role for T6SS-mediated antibacterial activity in the host. *Cell Host Microbe* 14, 652–663. doi: 10.1016/j.chom.2013.11.001

Gutierrez, M. G., Yoder-Himes, D. R., and Warawa, J. M. (2015). Comprehensive identification of virulence factors required for respiratory melioidosis using Tn-seq mutagenesis. *Front. Cell. Infect. Microbiol.* 5:78. doi: 10.3389/fcimb.2015.00078

Hachani, A., Allsopp, L. P., Oduko, Y., and Filloux, A. (2014). The VgrG proteins are "à la Carte" delivery systems for bacterial type VI effectors. *J. Biol. Chem.* 289, 17872–17884. doi: 10.1074/jbc.M114.563429

Hachani, A., Wood, T. E., and Filloux, A. (2016). Type VI secretion and anti-host effectors. *Curr. Opin. Microbiol.* 29, 81–93. doi: 10.1016/j.mib.2015.11.006

Hood, R. D., Singh, P., Hsu, F., Güvener, T., Carl, M. A., Trinidad, R. R., et al. (2010). A Type VI SECRETION SYSTEM of *Pseudomonas aeruginosa* targets a toxin to bacteria. *Cell Host Microbe* 7, 25–37. doi: 10.1016/j.chom.2009.12.007

Jamet, A., and Nassif, X. (2015). New players in the toxin field: polymorphic toxin systems in bacteria. *MBio* 6, e00285-15. doi: 10.1128/mBio.00285-15

Jiang, F., Waterfield, N. R., Yang, J., Yang, G., and Jin, Q. (2014). A *Pseudomonas aeruginosa* type VI secretion phospholipase D effector targets both prokaryotic and eukaryotic cells. *Cell Host Microbe* 15, 600–610. doi: 10.1016/j.chom.2014.04.010

Kaboord, B., and Perr, M. (2008). Isolation of proteins and protein complexes by immunoprecipitation. *Methods Mol. Biol.* 424, 349–364. doi: 10.1007/978-1-60327-064-9_27

Kapitein, N., and Mogk, A. (2013). Deadly syringes: type VI secretion system activities in pathogenicity and interbacterial competition. *Curr. Opin. Microbiol.* 16, 52–58. doi: 10.1016/j.mib.2012.11.009

Kelley, L. A., Mezulis, S., Yates, C. M., Wass, M. N., and Sternberg, M. J. E. (2015). The Phyre2 web portal for protein modelling, prediction and analysis. *Nat. Protoc.* 10, 845–858. doi: 10.1038/nprot.2015.053

Kitaoka, M., Miyata, S. T., Brooks, T. M., Unterweger, D., and Pukatzki, S. (2011). VasH is a transcriptional regulator of the type VI secretion system functional in endemic and pandemic *Vibrio cholerae*. *J. Bacteriol.* 193, 6471–6482. doi: 10.1128/JB.05414-11

Koskiniemi, S., Lamoureux, J. G., Nikolakakis, K. C., t'Kint de Roodenbeke, C., Kaplan, M. D., Low, D. A., et al. (2013). Rhs proteins from diverse bacteria mediate intercellular competition. *Proc. Natl. Acad. Sci. U.S.A.* 110, 7032–7037. doi: 10.1073/pnas.1300627110

Li, J., Yao, Y., Xu, H. H., Hao, L., Deng, Z., Rajakumar, K., et al. (2015). SecReT6: a web-based resource for type VI secretion systems found in bacteria. *Environ. Microbiol.* 17, 2196–2202. doi: 10.1111/1462-2920.12794

Liang, X., Moore, R., Wilton, M., Wong, M. J., Lam, L., and Dong, T. G. (2015). Identification of divergent type VI secretion effectors using a conserved chaperone domain. *Proc. Natl. Acad. Sci. U.S.A.* 112, 9106–9111. doi: 10.1073/pnas.1505317112

Lin, J. S., Ma, L. S., and Lai, E. M. (2013). Systematic dissection of the agrobacterium type VI secretion system reveals machinery and secreted components for subcomplex formation. *PLoS ONE* 8:e67647. doi: 10.1371/journal.pone.0067647

Lin, J., Zhang, W., Cheng, J., Yang, X., Zhu, K., Wang, Y., et al. (2017). A Pseudomonas T6SS effector recruits PQS-containing outer membrane vesicles for iron acquisition. *Nat. Commun.* 8:14888. doi: 10.1038/ncomms14888

Ma, A. T., and Mekalanos, J. J. (2010). *In vivo* actin cross-linking induced by *Vibrio cholerae* type VI secretion system is associated with intestinal inflammation. *Proc. Natl. Acad. Sci. U.S.A.* 107, 4365–4370. doi: 10.1073/pnas.0915156107

Ma, J., Pan, Z., Huang, J., Sun, M., Lu, C., and Yao, H. (2017a). The Hcp proteins fused with diverse extended-toxin domains represent a novel pattern of antibacterial effectors in type VI secretion systems. *Virulence*. doi: 10.1080/21505594.2017.1279374. [Epub ahead of print].

Ma, J., Sun, M., Dong, W., Pan, Z., Lu, C., and Yao, H. (2017b). PAAR-Rhs proteins harbor various C-terminal toxins to diversify the antibacterial pathways of type VI secretion systems. *Environ Microbiol.* 19, 345–360. doi: 10.1111/1462-2920.13621

Ma, L. S., Hachani, A., Lin, J. S., Filloux, A., and Lai, E. M. (2014). *Agrobacterium tumefaciens* deploys a superfamily of type VI secretion dnase effectors as weapons for interbacterial competition in planta. *Cell Host Microbe* 16, 94–104. doi: 10.1016/j.chom.2014.06.002

Ma, L. S., Lin, J. S., and Lai, E. M. (2009). An IcmF family protein, ImpLM, is an integral inner membrane protein interacting with ImpKL, and its walker a motif is required for type VI secretion system-mediated Hcp secretion in *Agrobacterium tumefaciens*. *J. Bacteriol.* 191, 4316–4329. doi: 10.1128/JB.00029-09

Ma, L. S., Narberhaus, F., and Lai, E. M. (2012). IcmF family protein TssM exhibits ATPase activity and energizes type VI secretion. *J. Biol. Chem.* 287, 15610–15621. doi: 10.1074/jbc.M111.301630

Martínez-García, P. M., Ramos, C., and Rodríguez-Palenzuela, P. (2015). T346Hunter: a novel web-based tool for the prediction of type III, type IV and type VI secretion systems in bacterial genomes. *PLoS ONE* 10:e0119317. doi: 10.1371/journal.pone.0119317

Mehla, J., Caufield, J. H., and Uetz, P. (2015). The yeast two-hybrid system: a tool for mapping protein-protein interactions. *Cold Spring Harb. Protoc.* 2015, 425–430. doi: 10.1101/pdb.top083345

Memišević, V., Kumar, K., Cheng, L., Zavaljevski, N., DeShazer, D., Wallqvist, A., et al. (2014). DBSecSys: a database of *Burkholderia mallei* secretion systems. *BMC Bioinformatics* 15:244. doi: 10.1186/1471-2105-15-244

Memišević, V., Kumar, K., Zavaljevski, N., DeShazer, D., Wallqvist, A., and Reifman, J. (2016). DBSecSys 2.0: a database of *Burkholderia mallei* and *Burkholderia pseudomallei* secretion systems. *BMC Bioinformatics* 17:387. doi: 10.1186/s12859-016-1242-z

Miyata, S. T., Kitaoka, M., Brooks, T. M., McAuley, S. B., and Pukatzki, S. (2011). *Vibrio cholerae* requires the type VI secretion system virulence factor VasX To kill dictyostelium discoideum. *Infect. Immun.* 79, 2941–2949. doi: 10.1128/IAI.01266-10

Miyata, S. T., Unterweger, D., Rudko, S. P., and Pukatzki, S. (2013). Dual expression profile of Type VI secretion system immunity genes protects pandemic *Vibrio cholerae*. *PLoS Pathog.* 9:e1003752. doi: 10.1371/journal.ppat.1003752

Ncbi Resource Coordinators (2017). Database resources of the national center for biotechnology information. *Nucleic Acids Res.* 45, D12–D17. doi: 10.1093/nar/gkw1071

Planamente, S., Salih, O., Manoli, E., Albesa-Jové, D., Freemont, P. S., and Filloux, A. (2016). TssA forms a gp6-like ring attached to the type VI secretion sheath. *EMBO J.* 35, 1613–1627. doi: 10.15252/embj.201694024

Pukatzki, S., Ma, A. T., Revel, A. T., Sturtevant, D., and Mekalanos, J. J. (2007). Type VI secretion system translocates a phage tail spike-like protein into target cells where it cross-links actin. *Proc. Natl. Acad. Sci. U.S.A.* 104, 15508–15513. doi: 10.1073/pnas.0706532104

Pukatzki, S., Ma, A. T., Sturtevant, D., Krastins, B., Sarracino, D., Nelson, W. C., et al. (2006). Identification of a conserved bacterial protein secretion system in *Vibrio cholerae* using the Dictyostelium host model system. *Proc. Natl. Acad. Sci. U.S.A.* 103, 1528–1533. doi: 10.1073/pnas.0510322103

Pukatzki, S., McAuley, S. B., and Miyata, S. T. (2009). The type VI secretion system: translocation of effectors and effector-domains. *Curr. Opin. Microbiol.* 12, 11–17. doi: 10.1016/j.mib.2008.11.010

Records, A. R. (2011). The type VI secretion system: a multipurpose delivery system with a phage-like machinery. *Mol. Plant Microbe Interact.* 24, 751–757. doi: 10.1094/MPMI-11-10-0262

Rigard, M., Bröms, J. E., Mosnier, A., Hologne, M., Martin, A., Lindgren, L., et al. (2016). Francisella tularensis IglG belongs to a novel family of PAAR-Like T6SS proteins and harbors a unique N-terminal extension required for virulence. *PLoS Pathog.* 12:e1005821. doi: 10.1371/journal.ppat.1005821

Robb, C. S., Robb, M., Nano, F. E., and Boraston, A. B. (2016). The structure of the toxin and type six secretion system substrate Tse2 in complex with its immunity protein. *Structure* 24, 277–284. doi: 10.1016/j.str.2015.11.012

Russell, A. B., Hood, R. D., Bui, N. K., LeRoux, M., Vollmer, W., and Mougous, J. D. (2011). Type VI secretion delivers bacteriolytic effectors to target cells. *Nature* 475, 343–347. doi: 10.1038/nature10244

Russell, A. B., LeRoux, M., Hathazi, K., Agnello, D. M., Ishikawa, T., Wiggins, P. A., et al. (2013). Diverse type VI secretion phospholipases are functionally plastic antibacterial effectors. *Nature* 496, 508–512. doi: 10.1038/nature12074

Russell, A. B., Peterson, S. B., and Mougous, J. D. (2014a). Type VI secretion effectors: poisons with a purpose. *Nat. Rev. Microbiol.* 12, 137–148. doi: 10.1038/nrmicro3185

Russell, A. B., Singh, P., Brittnacher, M., Bui, N. K., Hood, R. D., Carl, M. A., et al. (2012). A widespread bacterial type VI secretion effector superfamily identified using a heuristic approach. *Cell Host Microbe* 11, 538–549. doi: 10.1016/j.chom.2012.04.007

Russell, A. B., Wexler, A. G., Harding, B. N., Whitney, J. C., Bohn, A. J., Goo, Y. A., et al. (2014b). A type VI secretion-related pathway in Bacteroidetes mediates interbacterial antagonism. *Cell Host Microbe* 16, 227–236. doi: 10.1016/j.chom.2014.07.007

Salomon, D. (2016). MIX and match: mobile T6SS MIX-effectors enhance bacterial fitness. *Mob. Genet. Elements* 6:e1123796. doi: 10.1080/2159256x.2015.1123796

Salomon, D., Dar, D., Sreeramulu, S., and Sessa, G. (2010). Expression of *Xanthomonas campestris* pv. vesicatoria Type III effectors in yeast affects cell growth and viability. *Mol. Plant Microbe Interact.* 24, 305–314. doi: 10.1094/MPMI-09-10-0196

Salomon, D., Kinch, L. N., Trudgian, D. C., Guo, X., Klimko, J. A., Grishin, N. V., et al. (2014). Marker for type VI secretion system effectors. *Proc. Natl. Acad. Sci. U.S.A.* 111, 9271–9276. doi: 10.1073/pnas.1406110111

Salomon, D., Klimko, J. A., Trudgian, D. C., Kinch, L. N., Grishin, N. V., Mirzaei, H., et al. (2015). Type VI secretion system toxins horizontally shared between marine bacteria. *PLoS Pathog.* 11:e1005128. doi: 10.1371/journal.ppat.1005128

Sana, T. G., Baumann, C., Merdes, A., Soscia, C., Rattei, T., Hachani, A., et al. (2015). Internalization of *Pseudomonas aeruginosa* strain PAO1 into epithelial cells is promoted by interaction of a T6SS effector with the microtubule network. *MBio* 6, e00712-15. doi: 10.1128/mBio.00712-15

Sana, T. G., Flaugnatti, N., Lugo, K. A., Lam, L. H., Jacobson, A., Baylot, V., et al. (2016). *Salmonella Typhimurium* utilizes a T6SS-mediated antibacterial weapon to establish in the host gut. *Proc. Natl. Acad. Sci. U.S.A.* 113, E5044–E5051. doi: 10.1073/pnas.1608858113

Sberro, H., Leavitt, A., Kiro, R., Koh, E., Peleg, Y., Qimron, U., et al. (2013). Discovery of functional toxin/antitoxin systems in bacteria by shotgun cloning. *Mol. Cell* 50, 136–148. doi: 10.1016/j.molcel.2013.02.002

Schwarz, S., Singh, P., Robertson, J. D., LeRoux, M., Skerrett, S. J., Goodlett, D. R., et al. (2014). VgrG-5 is a burkholderia type VI secretion system-exported protein required for multinucleated giant cell formation and virulence. *Infect. Immun.* 82, 1445–1452. doi: 10.1128/IAI.01368-13

Shneider, M. M., Buth, S. A., Ho, B. T., Basler, M., Mekalanos, J. J., and Leiman, P. G. (2013). PAAR-repeat proteins sharpen and diversify the Type VI secretion system spike. *Nature* 500, 350–353. doi: 10.1038/nature12453

Shyntum, D. Y., Venter, S. N., Moleleki, L. N., Toth, I., and Coutinho, T. A. (2014). Comparative genomics of type VI secretion systems in strains of Pantoea ananatis from different environments. *BMC Genomics* 15:163. doi: 10.1186/1471-2164-15-163

Si, M., Zhao, C., Burkinshaw, B., Zhang, B., Wei, D., Wang, Y., et al. (2017). Manganese scavenging and oxidative stress response mediated by type VI secretion system in *Burkholderia thailandensis*. *Proc. Natl. Acad. Sci. U.S.A.* 114, E2233–E2242. doi: 10.1073/pnas.1614902114

Siggers, K. A., and Lesser, C. F. (2008). The yeast Saccharomyces cerevisiae: a versatile model system for the identification and characterization of bacterial virulence proteins. *Cell Host Microbe* 4, 8–15. doi: 10.1016/j.chom.2008.06.004

Silverman, J. M., Agnello, D. M., Zheng, H., Andrews, B. T., Li, M., Catalano, C. E., et al. (2013). Haemolysin co-regulated protein is an exported receptor and chaperone of type VI secretion substrates. *Mol. Cell* 51, 584–593. doi: 10.1016/j.molcel.2013.07.025

Slagowski, N. L., Kramer, R. W., Morrison, M. F., LaBaer, J., and Lesser, C. F. (2008). A functional genomic yeast screen to identify pathogenic bacterial proteins. *PLoS Pathog.* 4:e9. doi: 10.1371/journal.ppat.0040009

Steinert, M., and Heuner, K. (2005). Dictyostelium as host model for pathogenesis. *Cell. Microbiol.* 7, 307–314. doi: 10.1111/j.1462-5822.2005.00493.x

Suarez, G., Sierra, J. C., Erova, T. E., Sha, J., Horneman, A. J., and Chopra, A. K. (2010). A type VI secretion system effector protein, VgrG1, from aeromonas hydrophila that induces host cell toxicity by ADP RIBOSYLATION OF ACTIN. *J. Bacteriol.* 192, 155–168. doi: 10.1128/JB.01260-09

Unterweger, D., Kostiuk, B., and Pukatzki, S. (2016). Adaptor proteins of type VI secretion system effectors. *Trends Microbiol.* 25, 8–10. doi: 10.1016/j.tim.2016.10.003

Unterweger, D., Kostiuk, B., Ötjengerdes, R., Wilton, A., Diaz-Satizabal, L., and Pukatzki, S. (2015). Chimeric adaptor proteins translocate diverse type VI secretion system effectors in *Vibrio cholerae*. *EMBO J.* 34, 2198–2210. doi: 10.15252/embj.201591163

Vettiger, A., and Basler, M. (2016). Type VI secretion system substrates are transferred and reused among sister cells. *Cell* 167, 99–110.e112. doi: 10.1016/j.cell.2016.08.023

Wan, B., Zhang, Q., Ni, J., Li, S., Wen, D., Li, J., et al. (2017). Type VI secretion system contributes to Enterohemorrhagic *Escherichia coli* virulence by secreting catalase against host reactive oxygen species (ROS). *PLoS Pathog.* 13:e1006246. doi: 10.1371/journal.ppat.1006246

Wang, T., Si, M., Song, Y., Zhu, W., Gao, F., Wang, Y., et al. (2015). Type VI secretion system transports $Zn^{(2+)}$ to combat multiple stresses and host immunity. *PLoS Pathog.* 11:e1005020. doi: 10.1371/journal.ppat.1005020

Wexler, A. G., Bao, Y., Whitney, J. C., Bobay, L. M., Xavier, J. B., Schofield, W. B., et al. (2016). Human symbionts inject and neutralize antibacterial toxins to persist in the gut. *Proc. Natl. Acad. Sci. U.S.A.* 113, 3639–3644. doi: 10.1073/pnas.1525637113

Whitney, J. C., Beck, C. M., Goo, Y. A., Russell, A. B., Harding, B., De Leon, J. A., et al. (2014). Genetically distinct pathways guide effector export through the type VI secretion system. *Mol. Microbiol.* 92, 529–542. doi: 10.1111/mmi.12571

Whitney, J. C., Chou, S., Russell, A. B., Biboy, J., Gardiner, T. E., Ferrin, M. A., et al. (2013). Identification, structure, and function of a novel type VI secretion peptidoglycan glycoside hydrolase effector-immunity pair. *J. Biol. Chem.* 288, 26616–26624. doi: 10.1074/jbc.M113.488320

Whitney, J. C., Quentin, D., Sawai, S., LeRoux, M., Harding, B. N., Ledvina, H. E., et al. (2015). An interbacterial NAD(P)(+) glycohydrolase toxin requires elongation factor tu for delivery to target cells. *Cell* 163, 607–619. doi: 10.1016/j.cell.2015.09.027

Williams, S. G., Varcoe, L. T., Attridge, S. R., and Manning, P. A. (1996). *Vibrio cholerae* Hcp, a secreted protein coregulated with HlyA. *Infect. Immun.* 64, 283–289.

Yang, W., Wang, L., Zhang, L., Qu, J., Wang, Q., and Zhang, Y. (2015). An invasive and low virulent Edwardsiella tarda esrB mutant promising as live attenuated vaccine in aquaculture. *Appl. Microbiol. Biotechnol.* 99, 1765–1777. doi: 10.1007/s00253-014-6214-5

Yu, M., and Lai, E. M. (2017). Warfare between host immunity and bacterial weapons. *Cell Host Microbe* 21, 3–4. doi: 10.1016/j.chom.2016.12.012

Zhang, D., de Souza, R. F., Anantharaman, V., Iyer, L. M., and Aravind, L. (2012). Polymorphic toxin systems: comprehensive characterization of trafficking modes, processing, mechanisms of action, immunity and ecology using comparative genomics. *Biol. Direct* 7, 18–18. doi: 10.1186/1745-6150-7-18

Zoued, A., Durand, E., Brunet, Y. R., Spinelli, S., Douzi, B., Guzzo, M., et al. (2016). Priming and polymerization of a bacterial contractile tail structure. *Nature* 531, 59–63. doi: 10.1038/nature17182

Zrieq, R., Sana, T. G., Vergin, S., Garvis, S., Volfson, I., Bleves, S., et al. (2015). Genome-wide screen of *Pseudomonas aeruginosa* in saccharomyces cerevisiae identifies new virulence factors. *Front. Cell. Infect. Microbiol.* 5:81. doi: 10.3389/fcimb.2015.00081

8

A Potential Late Stage Intermediate of Twin-Arginine Dependent Protein Translocation in *Escherichia coli*

Hendrik Geise, Eyleen Sabine Heidrich, Christoph Stefan Nikolin, Denise Mehner-Breitfeld and Thomas Brüser*

Institute of Microbiology, Leibniz University Hannover, Hannover, Germany

***Correspondence:**
Thomas Brüser
brueser@ifmb.uni-hannover.de

The twin-arginine translocation (Tat) system transports folded proteins across membranes of prokaryotes, plant plastids, and some mitochondria. According to blue-native polyacrylamide gel electrophoresis after solubilization with digitonin, distinct interactions between the components TatA, TatB, and TatC result in two major TatBC-containing complexes in *Escherichia coli* that can bind protein substrates. We now report the first detection of a TatABC complex that likely represents the state at which transport occurs. This complex was initially found when the photo cross-linking amino acid *p*-benzoyl-L-phenylalanine (Bpa) was introduced at position I50 on the periplasmic side of the first trans-membrane domain of TatC. Cross-linking of $TatC_{I50Bpa}$ resulted in TatC-TatC-cross-links, indicating a close proximity to neighboring TatC in the complex. However, the new complex was not caused by cross-links but rather by non-covalent side chain interactions, as it was also detectable without UV-cross-linking or with an I50Y exchange. The new complex did not contain any detectable substrate. It was slightly upshifted relative to previously reported substrate-containing TatABC complexes. In the absence of TatA, an inactive $TatBC_{I50Bpa}$ complex was formed of the size of wild-type substrate-containing TatABC complexes, suggesting that TatB occupies TatA-binding sites at $TatC_{I50Bpa}$. When substrate binding was abolished by point mutations, this $TatBC_{I50Bpa}$ complex shifted analogously to active $TatABC_{I50Bpa}$ complexes, indicating that a defect substrate-binding site further enhances TatB association to TatA-binding sites. Only TatA could shift the complex with an intact substrate-binding site, which explains the TatA requirement for substrate transport by TatABC systems.

Keywords: twin-arginine translocation, membrane protein complexes, protein translocation, *Escherichia coli*, photo cross-linking

INTRODUCTION

The twin-arginine translocation (Tat) system transports folded proteins across the cytoplasmic membrane of prokaryotes, the thylakoid membrane of plant plastids, and the inner membrane in some mitochondria (Hou and Brüser, 2011; Hamsanathan and Musser, 2018; Petrů et al., 2018). The Tat-dependent translocation is driven by the membrane potential that is generated by ionic gradients at energy-transducing membranes, which is why this system is restricted to such membranes (Cline, 2015). In *Escherichia coli*, a fully functional Tat-translocon assembles from the three components TatA, TatB, and TatC (Sargent et al., 1999). A second paralog of TatA, TatE, can mix into these systems or substitute TatA, without being specifically required for the transport of a known Tat substrate (Sargent et al., 1999). TatA/E and TatB are evolutionary related and share the same principle structural organization in their functionally important N-terminal half (Yen et al., 2002; Hu et al., 2010; Zhang et al., 2014). They are N-terminally membrane anchored by a short hydrophobic region, followed by a hinge and an amphipathic helix on the cytoplasmic surface of the membrane. This amphipathic helix is followed by some negatively charged residues and regions that are not really conserved anymore between TatA/B family proteins (Hou and Brüser, 2011). TatC consists of six transmembrane helices (TM1–6) with cytoplasmic N- and C-termini and loops on the cytoplasmic and periplasmic sides of the membrane (Rollauer et al., 2012; Ramasamy et al., 2013). Tat substrates contain N-terminal signal peptides with the eponymous highly conserved twin-arginine motif, which is recognized by a binding site that is formed by the N-terminus and the first cytoplasmic loop of TatC (Ramasamy et al., 2013). TatC binds TatB and also TatA, and corresponding binding sites have been identified (Alcock et al., 2016). TatB and TatC interact tightly, and earlier studies with the homologous TatABC systems in plants and *E. coli* were suggestive for a substrate-induced recruitment of TatA to TatBC (Cline and Mori, 2001; Mori and Cline, 2002; Alami et al., 2003). While cross-linking and co-purification analyses later demonstrated that TatA also interacts with TatBC independently of substrate binding (Aldridge et al., 2014; Behrendt and Brüser, 2014), active transport apparently increases the affinity of TatA to TatBC, as reflected by a recruitment of TatA-XFP fusion proteins to TatBC in response to substrate overproduction (Alcock et al., 2013; Rose et al., 2013). Besides TatBC, also TatA interacts with the signal peptide of precursor proteins, but this interaction contributes to transport without being involved in the RR-motif recognition by TatBC (Taubert et al., 2015). The Tat complexes are usually analyzed by BN-PAGE in combination with Western-blotting or other labeling methods. Solubilized TatA tends to form multiple homooligomers that are visible as a dense ladder in BN-PAGE (Oates et al., 2005; Richter and Brüser, 2005). These highly abundant TatA homooligomers so far prevented the direct identification of TatA as constituent of any of the TatBC-containing complexes by BN-PAGE (Behrendt and Brüser, 2014). However, cross-linking experiments showed that TatA and TatB share the same binding sites of TatC, and positions of TatA and TatB are believed to switch in course of transport (Habersetzer et al., 2017). As TatA is required for Tat transport, it is important to reveal the complexes that contain all three components.

TatB and TatC have been detected in two substrate-free and two substrate-associated complexes in the range of 400–700 kDa (Behrendt and Brüser, 2014). As the migration behavior of solubilized membrane protein complexes in BN-PAGE is influenced by the detergent and lipid content of solubilized complexes, the BN-PAGE deduced molecular masses do not permit the estimation of individual subunit numbers. Some variation of reported Tat complex molecular masses in BN-PAGE analyses may have been caused by lot variations of the commonly used mild detergent digitonin, which is enriched from extractions of foxglove (*Digitalis purpurea*). For a clear assignment, it is therefore important to include the wild-type complexes in each study. To clarify the designation of Tat complexes, we now name the smaller substrate-free TatBC-containing complex of *E. coli* Tat-complex 1 (TC1; previously termed 370, 430, or 440 kDa complex; Oates et al., 2005; Richter and Brüser, 2005; Orriss et al., 2007; Huang et al., 2017) and the larger substrate-free TatBC-containing complex Tat-complex 2 (TC2; previously termed 580 kDa or "higher molecular weight variant" TatBC complex; Richter and Brüser, 2005; Huang et al., 2017). As mentioned above, both of these complexes can in principle also contain TatA, but the continuous dissociation of TatA from TatBC during solubilization and purification and the accompanying formation of homooligomeric TatA associations in a wide range of sizes prevented so far the assignment of TatA to these complexes. Both complexes are easily identified by BN-PAGE without overproduction of Tat substrates (Richter and Brüser, 2005; Behrendt and Brüser, 2014). Due to harsher BN-PAGE conditions, other groups detected TC2 only with transport-enhancing mutations (Huang et al., 2017). The corresponding substrate-bound shifted complexes, now termed TC1S or TC2S, can so far only be detected upon substrate overproduction (Behrendt and Brüser, 2014). To facilitate comparisons, we suggest the use of this nomenclature for future studies.

During characterizations of TatC variants with individual residues substituted by the artificial cross-linking amino acid *p*-benzoyl-L-phenylalanine (Bpa), we found with a $TatC_{150Bpa}$ mutation a new TatBC-containing complex larger than TC2S. The formation of this complex did not relate to a Bpa cross-link and depended on TatA. Mutational inactivation of the substrate-binding site permitted the formation of this complex in the absence of TatA, suggesting that TatB can be recruited to TatA-binding sites in such inactive complexes. The data indicate that subtle changes in the substrate-binding site can shift the TatBC complex to a higher associated state in the absence of TatA, but TatA is required for this shift when the substrate-binding site is functional. As the $TatABC_{150Bpa}$ system is active, and as the complex does not contain substrate anymore, it is likely that the described new complex represents a late translocation state, kinetically stabilized by the $TatC_{150Bpa}$ mutation.

MATERIALS AND METHODS

Strains and Growth Conditions

The *tatABCDE* deficient *E. coli* strain DADE D6 ara^R (Lindenstrauss et al., 2010) was used for physiological and biochemical analyses, and *E. coli* XL1-Blue MRF' Tet (Agilent) was used for cloning. Strains were grown aerobically at 37°C in LB medium [1% (w/v) tryptone, 1% (w/v) NaCl, 0.5% (w/v) yeast extract] in the presence of appropriate antibiotics (100 μg/ml ampicillin, 25 μg/ml chloramphenicol, and 12.5 μg/ml tetracycline) and harvested after 5 h. Cultures carrying pABS- or pDE-derived plasmids were normalized to an OD_{600} of 1.0 and cultures carrying pZX31-derived plasmids to an OD_{600} of 2.0. Cultures containing pBW-*efeB-strep* were harvested after 3 h growth with 0.05% (w/v) rhamnose added at an OD_{600} of 0.6. For incorporation of Bpa at amber stop codons in strains carrying the pEVOL-*p*BpF system, 100 μM Bpa was added simultaneously with 100 μM arabinose. For optional cross-linking, cultures were grown for 5 h before irradiation with UV light at 365 nm for 30 min at ambient temperature, normalization, and further processing.

Plasmids and Genetic Methods

pEVOL-*p*BpF-tet, which encodes an orthogonal Bpa-specific suppressor tRNA/aminoacyl tRNA synthetase pair used for incorporation of Bpa at introduced amber stop codons, was donated by Peter G. Schultz (Young et al., 2010). The vector pDE-*tatABC-h6*, used for constitutive *tatABC* expression in pEVOL-*p*BpF-tet-containing strains, was generated by cloning the NheI-XbaI digested *ori* ColA from pCOLAduet-1 (Novagen) into the corresponding sites of pABS-*tatABC-h6*. pABS-*tatABC-h6* was generated by substituting *pspC* in pABS-*pspC-h6* (Mehner et al., 2012) with *tatABC*, using NdeI and BamHI restriction sites and the primer pair *tatA*-NdeI-F 5′-TCT TCT CAT ATG GGT GGT ATC AGT ATT TGG C-3′ and *tatC*-BamHI-R 5′-CAA GCG GAT CCT TCT TCA GTT TTT TCG CTT TCT GCT TC-3′. pABS-*tatABC-h6* results in constitutive P_{tatA}-dependent expression of the *tatABC* genes, with TatC produced as C-terminally hexahistidine-tagged protein. The plasmid pZX31-*tatBC-h6*, used for constitutive *tatBC* expression, was generated by cloning *ori* ColA into AvrII-SpeI digested pZA31-*tatBC-h6* using the primer pair ColA-AvrII-F 5′-GAT CCC TAG GAA ACG TCC TAG AAG ATG CCA GGA GGA TA-3′ and ColA-SpeI-R 5′-GAT CAC TAG TTG GTG TCG GGA ATC CGT AAA GG-3′. For generation of pZA31-*tatBC-h6*, *tatBC* was amplified using the primer pair *tatB*-EcoRI-F 5′-GAA GAC GCG AAT TCC CAC GAT AAA GAG C-3′/*tatC-h6*-PstI-R 5′-TAG CCA CTG CAG TTA ATG GTG ATG GTG ATG GTG TTC TTC AGT TTT TTC GCT TTC-3′ and cloned into the corresponding sites of pZA31MCS (Expressys, Bammental). Single amino acid exchanges in TatC were introduced by QuikChange™ mutagenesis (Stratagene) of pDE-*tatABC-h6* or pZX31-*tatBC-h6*, using the forward primers *tatC*-A47Bpa-F 5′-CAT CTA TCA CCT GGT ATC CTA GCC ATT GAT CAA GCA GTT G-3′, *tatC*-P48Bpa-F 5′-CTA TCA CCT GGT ATC CGC GTA GTT GAT CAA GCA GTT GCC G-3′, *tatC*-L49Bpa-F 5′-GGT ATC CGC GCC ATA GAT CAA GCA GTT GC-3′, *tatC*-I50Bpa-F 5′-CAC CTG GTA TCC GCG CCA TTG TAG AAG CAG TTG CCG CAA GGT TC-3′, *tatC*-I50Y-F 5′-GGT ATC CGC GCC ATT GTA TAA GCA GTT GCC GCA AG-3′, *tatC*-I50F-F 5′-GTA TCC GCG CCA TTG TTC AAG CAG TTG CCG-3′, *tatC*-I50W-F GTA TCC GCG CCA TTG TGG AAG CAG TTG CCG CAA G-3′, *tatC*-K51Bpa-F 5′-CCG CGC CAT TGA TCT AGC AGT TGC CGC AAG-3′, *tatC*-Q52Bpa-F 5′-GCG CCA TTG ATC AAG TAG TTG CCG CAA GGT TC-3′, *tatC*-F94Q-F 5′-CTA TCA GGT GTG GGC ACA GAT CGC CCC AGC GCT G-3′, and *tatC*-E103A-F 5′-CGC TGT ATA AGC ATG CGC GTC GCC TGG TGG TG-3′ in conjunction with reverse primers that cover the identical sequence region. The vector pBW-*ycdB-strep* (Sturm et al., 2006) was used for the overexpression of *efeB* (formerly called *ycdB*). All genetic constructs were confirmed by sequencing.

Biochemical Methods

BN-PAGE was performed as described previously (Richter and Brüser, 2005), but without *β*-mercaptoethanol in the buffers. TatABC complexes were solubilized with 1% digitonin and purified by Ni-NTA affinity chromatography using 50 mM Bis-Tris pH 7.0, 20% (w/v) sucrose, 10 mM $MgCl_2$, 0.1% digitonin as buffer system, and 250 mM imidazole containing buffer for elution. When indicated, an additional size exclusion chromatography purification step was applied using the same buffer. Sodium dodecyl sulfate-polyacrylamide gel electrophoresis (SDS-PAGE) analysis was carried out by standard methods (Laemmli, 1970). For immunoblots, proteins were semi-dry blotted on nitrocellulose membranes and blots were developed using antibodies directed against synthetic C-terminal peptides of TatA, TatB, TatC, or purified EfeB, using the ECL system (GE Healthcare) for signal detection. HRP-conjugated goat anti rabbit anti-bodies (Roth) served as secondary antibodies. For stripping, blots were incubated with 100 mM glycine pH 2.8, 1% SDS, 0.2% Tween 20 for 30 min, followed by three washing steps with PBS and a second blocking (PBS with 5% skim milk). Stripped blots were again incubated with the secondary antibody and developed to ensure that they gave no detectable background and thereafter used for new detections. The chain formation phenotype was assessed by phase contrast microscopy. SDS sensitivity was determined by aerobic growth in LB medium containing 4% (w/v) sodium dodecyl sulfate (SDS), using the quotient of the OD_{600} with/without SDS after 3 h of growth (Ize et al., 2003).

RESULTS

The Region Around Position I50 of TatC Contacts Neighboring TatC Protomers

In the course of *in vivo* photo-cross-linking analyses of TatC interactions with single TatC amino acids exchanged to *p*-benzoyl-L-phenylalanine (Bpa), we identified that a $TatC_{I50Bpa}$ construct efficiently cross-linked to other TatC protomers, giving rise to a strong UV-induced shifted TatC band at ~55 kDa (**Figure 1A**, compare −UV/+UV). TatB is known to tightly interact with TatC, and TatB-TatC cross-links would have resulted in shifts of similar size, but as no TatB was detectable in these bands, it is likely that the shifted bands represented TatC-TatC cross-links. TatC is only present in oligomeric complexes in *E. coli*, and the position of the I50Bpa exchange must be therefore in close proximity to neighboring TatC protomers within Tat complexes. Within the first periplasmic loop (PPL1) of TatC, Ile50 is located in an α-helix oriented almost parallel to the

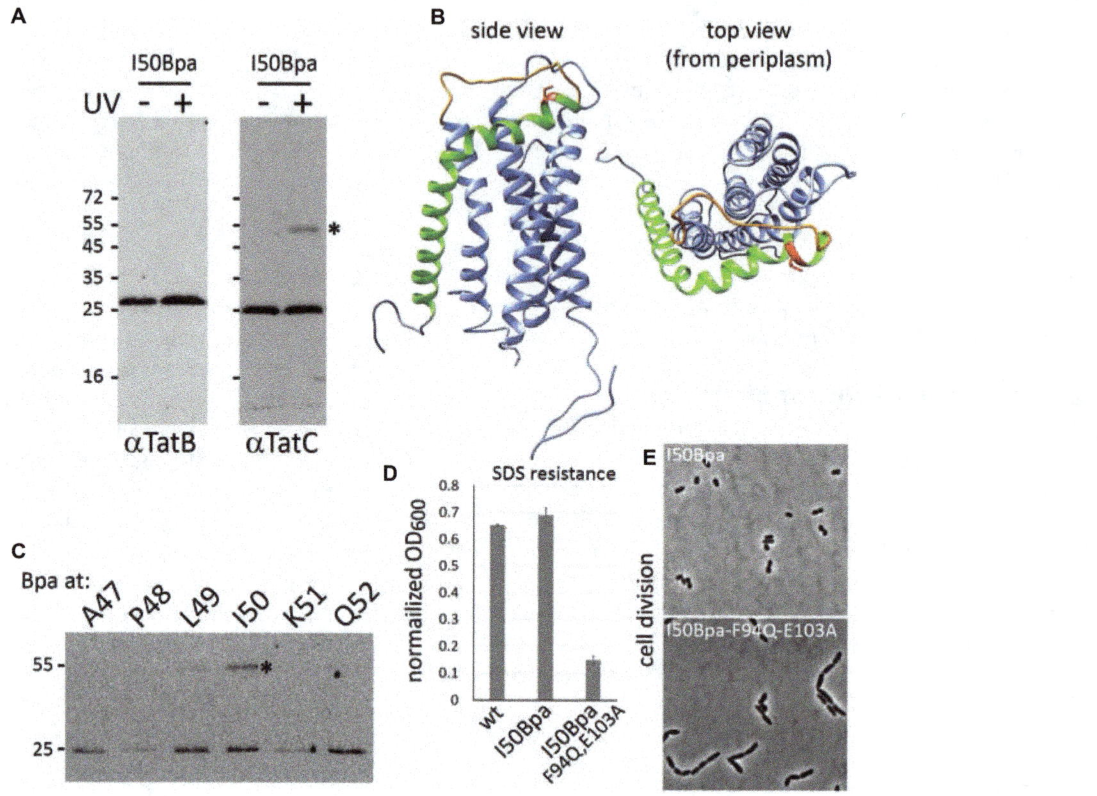

FIGURE 1 | $TatC_{I50Bpa}$ is functional and its cross-linking results in TatC dimers. **(A)** Immunoblot detection of Bpa photo cross-links in a strain producing $TatABC_{I50Bpa}$. Irradiation by UV light results in formation of a ~55 kDa TatC homodimer (*) that is detected in the TatC blot (right) but not in the TatB blot (left). **(B)** Model of *E. coli* TatC. I50 of TatC is highlighted in red and is located at the end of TM1 (green) in transition to a less structured region of the first periplasmic loop (orange). The model is based on the *Aquifex aeolicus* TatC structure (PDB 4HTS; Ramasamy et al., 2013) and was modeled using RaptorX (Peng and Xu, 2011; Källberg et al., 2014). Molecular graphics and analyses were performed using UCSF Chimera (Pettersen et al., 2004). **(C)** Analysis of photo cross-links with Bpa exchanges at indicated TatC positions in the vicinity of I50. The position of the cross-link is indicated by an asterisk. SDS-solubilized membranes were used for the detection of Tat components as described previously (Taubert et al., 2015). **(D,E)** Activity of $TatABC_{I50Bpa}$ as monitored by complementation of the SDS sensitivity **(D)** and chain formation phenotypes **(E)** of Tat deficient strains. SDS sensitivity results in lowered cell density, which is given as normalized OD_{600}, the quotient of OD_{600} after 3 h of growth in 4% SDS/OD after 3 h of growth in 0% SDS. As control for these phenotypes, the same system was analyzed with a mutated Tat substrate-binding site. The mutations F94Q/E103A completely inactivate Tat systems (Huang et al., 2017, and this study).

surface of the membrane directly following TM1 (**Figure 1B**). Previous studies showed that other residues located in this region are crucial for Tat complex assembly or functionality (Allen et al., 2002; Barrett et al., 2005; Kneuper et al., 2012). Already in early studies, an exchange of the close-by residue Pro48 to Ala has been shown to inactivate the Tat system, which has been explained by a destabilization of protomer interactions within the complex (Allen et al., 2002; Behrendt and Brüser, 2014). While Pro48 is directed inward and thus likely positions the helix at its place, Ile50 is directed outward, which likely enables the contacts to neighboring TatC subunits. To examine potential TatC cross-links with Bpa exchanges in the vicinity of Ile50, we included cross-link analyses of TatABC systems with TatC exchanges A47Bpa, P48Bpa, L49Bpa, K51Bpa, and Q52Bpa (**Figure 1C**). In case of P48Bpa and K51Bpa, the abundance of TatC was reduced, most likely due to destabilization and degradation effects. A cross-link was also observed with L49Bpa, but I50Bpa gave the strongest cross-link. In further analyses, we therefore concentrated on I50Bpa. The $TatABC_{I50Bpa}$ system was active, as shown by the complementation of SDS sensitivity and chain formation phenotypes of a *tatABCDE* mutant strain (**Figures 1D,E**).

p-Benzoylphenylalanine or Tyrosine at Position I50 of TatC Influence the Equilibrium Between Three Assembly States of Twin-Arginine Translocation Complexes

Previous studies had already indicated a strong influence of P48, which is located only half a helical turn away from I50, on the stability of TatBC complexes (Allen et al., 2002; Barrett et al., 2005; Kneuper et al., 2012; Behrendt and Brüser, 2014). To our knowledge, mutations of I50 have never been analyzed. To address complex stability of Tat systems containing I50Bpa, membranes were solubilized with 1% (w/v) digitonin and subjected to BN-PAGE/Western-blot analysis (**Figure 2**). While wild-type TatABC produced the two known TatB- and TatC-containing complexes (TC1 and TC2), $TatABC_{I50Bpa}$ predominantly formed a new complex that was clearly shifted to a higher molecular

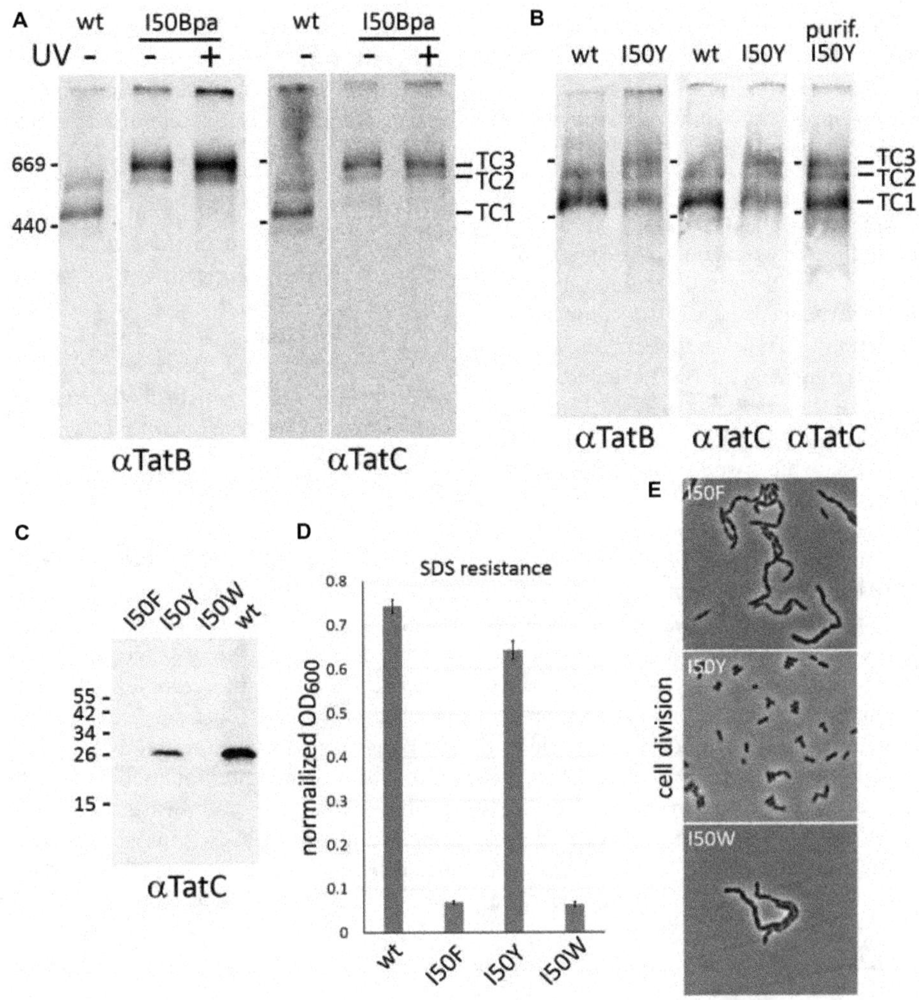

FIGURE 2 | An equilibrium between three Tat complexes is influenced by Bpa or Tyr exchanges at position I50. **(A)** BN-PAGE/Western-blot analysis of Tat complexes formed by wild-type TatABC and TatABC$_{I50Bpa}$ systems. As indicated, blots were developed using TatB or TatC specific antibodies. Positions of marker proteins are indicated on the left of the blots, positions of the three Tat complexes are indicated on the right. White lines separate lanes from one blot that were not directly neighboring. **(B)** BN-PAGE/Western-blot analysis of Tat complexes formed by wild-type TatABC and by TatABC$_{I50Y}$ systems. The Tat complexes of the latter were also enriched by affinity chromatography (lane purif. I50Y). **(C)** SDS-PAGE/Western-blot analysis of TatC with I50 exchanged by indicated aromatic natural amino acids. TatC from the non-mutated system is analyzed for comparison. **(D,E)** Activity of Tat systems with wild-type TatC or indicated TatC variants, as monitored by complementation of the SDS sensitivity **(D)** and chain formation phenotype **(E)** of Tat deficient strains. The SDS sensitivity is reflected by the normalized OD_{600} as described in **Figure 1**.

mass and also contained TatB and TatC (**Figure 2A**). In addition, a less dominant complex with the size of TC2 or TC2S of the wild-type Tat system was observed, which will be assigned later (see below). The shifted complex did not result from UV-cross-linking, as the pattern did not change without UV illumination. This suggested that most likely a non-covalent interaction of the aromatic benzophenone side chain was causing the shift. To analyze whether aromatic side chains of natural amino acids can cause the same shift, we substituted Ile50 by tyrosine, tryptophan, or phenylalanine. Before analyzing the corresponding Tat complexes, we examined the activity and formation of the mutated components (**Figures 2C–E**). The I50W and I50F substitutions resulted in inactivation and absence of TatC, and thus, most likely caused a complete degradation of TatC. Notably, an I50Y substitution lowered the abundance of TatC but did not inactivate the system. The data underlined the importance of that position for the structural integrity of the Tat system. The active TatABC$_{I50Y}$ system was then analyzed by BN-PAGE (**Figure 2B**). The usual complexes TC1 and TC2 were formed that can be also observed by non-mutated Tat systems, but there was clearly the additional band of the shifted complex (TC3) that had been previously observed with TatABC$_{I50Bpa}$. Together, these analyses indicated that an equilibrium exists between three Tat complexes in active Tat systems, and this equilibrium is strongly influenced by I50 exchanges to either Bpa or Tyr, which introduces benzophenone or hydroxyphenol moieties, respectively. Both side chains allow aromatic contacts as well as hydrogen bonding (mediated by the keto group of benzophenone or the hydroxy group in tyrosine), and this combination is likely responsible for the stabilization of TC3, which apparently otherwise is likely only a short-lived complex. As cross-links were irrelevant for TC3-formation, we did not apply cross-linking for all further analyses with Bpa-containing constructs.

TC3 Requires TatA to Be Formed

A reason for the I50Bpa-induced stabilization of a Tat complex at a higher molecular mass could have been an enhanced affinity of TatA to TatBC. A reduced TatA-dissociation would explain the observed depletion of TC1. We therefore addressed this potential role of TatA by comparing the BN-PAGE detectable complexes formed in the presence or absence of TatA (**Figure 3**). The absence of TatA resulted in the formation of a complex of the size of TC2 or TC2S, which would need a substrate detection for clear differentiation (see below), but TC3 was clearly absent. Notably, the $TatBC_{I50Bpa}$ system did not result in TC1, which is the complex usually formed in the absence of TatA (Behrendt and Brüser, 2014), indicating that the I50Bpa mutation in TatC also results in an increased affinity of TatB to TatA-binding sites. It is established that TatB can in principle bind to TatA-binding sites in active Tat systems (Habersetzer et al., 2017), and apparently, the I50Bpa mutation enhances this interaction to an extent that no TC1 is detected anymore. The data also clearly show that TatA is required for the formation of TC3 in $TatABC_{I50Bpa}$ systems, implying that TatA was likely a constituent of the new complex. To address this aspect directly, we purified TatC-h6-tagged Tat complexes in a strain producing the $TatABC_{I50Bpa}$ components, using a combination of immobilized metal affinity chromatography (IMAC) and size exclusion chromatography (SEC). Thus, the enriched complexes were analyzed by BN-PAGE/Western-blotting (**Figure 3**). Non-bound TatA, which is known to form homooligomeric assemblies that are detected as a "ladder" over a broad range of sizes in BN-PAGE (Barrett et al., 2005; Richter and Brüser, 2005; Behrendt et al., 2007), was efficiently removed by the purification. We now were able to detect TatA in a weak band of the size of the two TatBC-containing complexes, likely indicating the presence of TatA in these complexes. Based on the TatA dependence of TC3 in conjunction with the detection of TatA in a BN-PAGE band of that size, we suggest that both detected complexes contain significant amounts of TatA, which likely explains the shift of these complexes in comparison with TC1. Consequently, the I50Bpa exchange likely enhanced the stability of a TatA-assembled state of the Tat system.

Substrate Is Likely Already Transported by the Detected TC3

To clearly differentiate, whether the detected $TatBC_{I50Bpa}$ complex that is formed in the absence of TatA contains substrate or not, we examined the complex after recombinant production of the *E. coli* Tat substrate EfeB (formerly known as YcdB). EfeB has been shown to bind Tat complexes with sufficient affinity to detect the interaction by BN-PAGE/Western-blot analysis (Behrendt and Brüser, 2014). EfeB could be clearly detected in the $TatBC_{I50Bpa}$ complex, indicating that it represented the substrate-bound TC2S (**Figure 4A**). Note that, due to the lack of TatA, the $TatBC_{I50Bpa}$ system is inactive (**Figure 4B**),

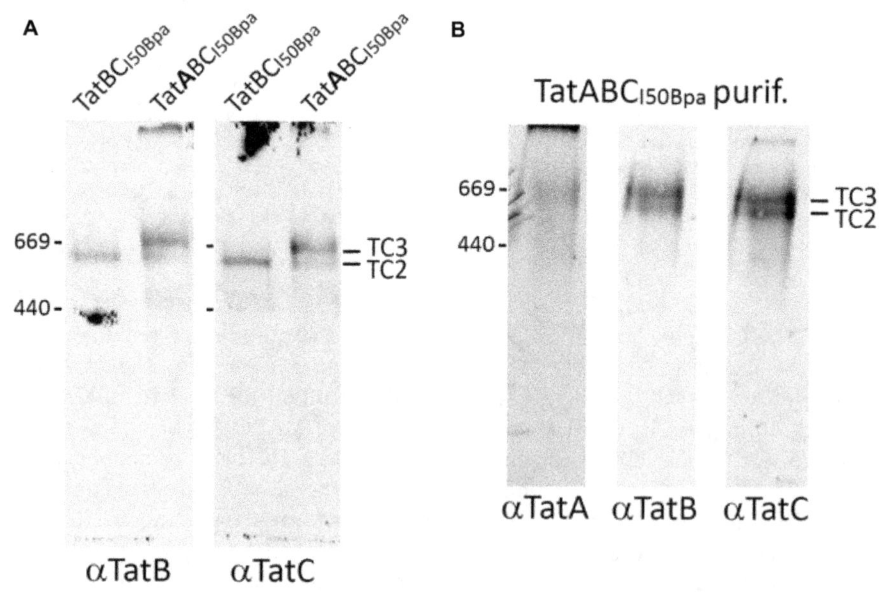

FIGURE 3 | TatA is required for the formation of TC3. **(A)** BN-PAGE/Western-blotting detection of Tat complexes in the absence or presence of TatA. Analyses of Tat complexes in strains producing $TatABC_{I50Bpa}$ or $TatBC_{I50Bpa}$. **(B)** Detection of the three Tat components within TC3 after a two-step purification (Ni-affinity purification and size exclusion chromatography using a Superose 6 column). Positions of TC2 and TC3 are indicated on the right, size markers on the left, and the used antibodies below the corresponding blots.

and EfeB accumulates strongly in the membranes, whereas the active transport by the $TatABC_{I50Bpa}$ system markedly reduces the abundance of EfeB in the membranes (**Figure 4C**, compare free EfeB signals, lanes 2 and 4). The TC2 complex was also present in $TatABC_{I50Bpa}$ systems, migrating as a second band below TC3, but the complex was less abundant, and no substrate was detectable. Also TC3 of the $TatABC_{I50Bpa}$ system contained no detectable substrate, although it was more abundant than TC2S of the $TatBC_{I50Bpa}$ system (**Figure 4C**), in line with the generally accepted idea that transport occurs upon TatA recruitment. The detected TC3 therefore likely represents a stabilized active state of the Tat system in which the complexes have already accomplished transport.

To examine whether TC3 depends also on substrate binding, we analyzed the $TatABC_{I50Bpa}$ system with an inactivated twin-arginine motif-binding site, achieved by a F94Q/E103A double exchange in TatC. The inactivity of that system has been previously established by others (Huang et al., 2017) and was confirmed in our hands (see negative control in **Figures 1D,E**). BN-PAGE analyses demonstrated that the inactivation of the substrate-binding site did not diminish TC3, clearly demonstrating that substrate binding was not a prerequisite for the shift (**Figure 4C**). When we introduced the same inactive substrate-binding site into the $TatBC_{I50Bpa}$ system, we found in analogous BN-PAGE/Western-blotting analyses that a small portion of the $TatBC_{I50Bpa/F94Q/E103A}$ complex had shifted to TC3, which is normally only found in the presence of TatA (**Figure 4C**). Apparently, TatB associated with $TatBC_{I50Bpa/F94Q/E103A}$ just like TatA to $TatBC_{I50Bpa}$. In conclusion, the F94Q/E103A mutations of the substrate-binding site caused a relaxed, not TatA or TatB differentiating enhanced association of TatA or TatB to TatC, leading to increased amounts of the TC3 with $TatABC_{I50Bpa/F94Q/E103A}$ systems and even a TC3 in $TatBC_{I50Bpa/F94Q/E103A}$ systems.

DISCUSSION

In this study, we analyzed in some detail a third Tat complex, termed TC3, that became the dominant Tat complex if only one TatC position, I50, was mutated to the artificial aromatic amino acid Bpa. We could demonstrate that side chain properties and not a photo activatable cross-link were the basis for the detection of this complex. The complex was active, suggesting

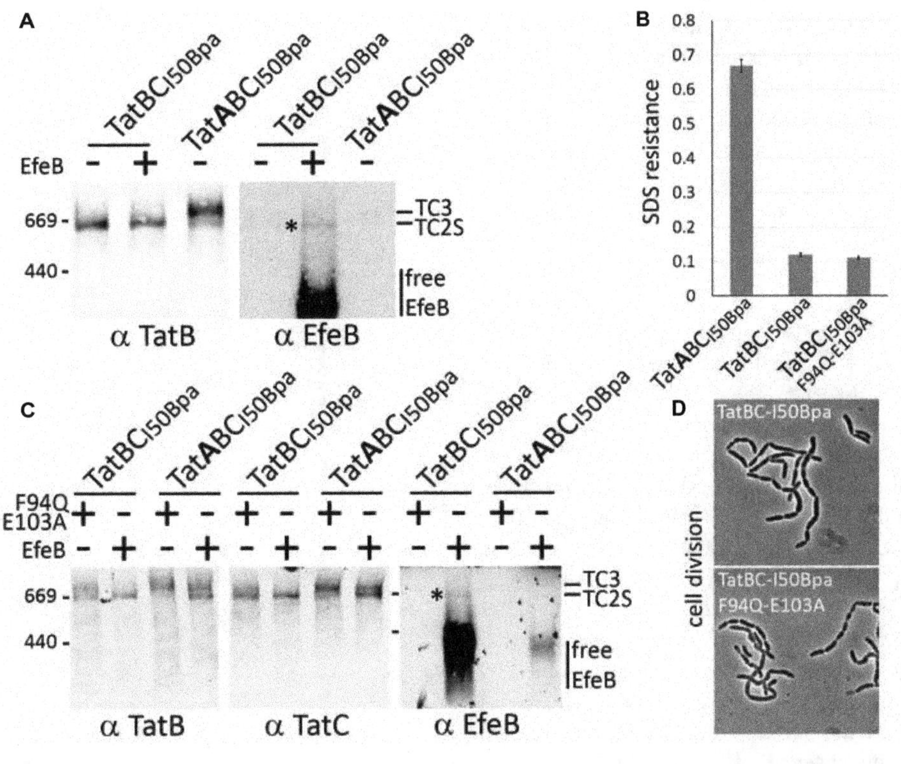

FIGURE 4 | A mutated substrate-binding site can result in TC3 formation in the absence of TatA. **(A)** Detection of recombinant Tat substrate EfeB in TC2. Overproduction of EfeB leads to the detectability of this Tat substrate in TC2 formed by $TatBC_{I50Bpa}$. After development of the blot using antibodies directed against TatB (left blot), the blot was stripped (see methods) and developed again with antibodies recognizing EfeB (right blot). **(B,D)** The absence of TatA inactivates the Tat system also in case of $TatBC_{I50Bpa}$. As monitored by SDS resistance **(B)** and the cell division **(D)**, $TatBC_{I50Bpa}$ does not complement the deficiency phenotypes of the *E. coli* Tat mutant strain DADE. The SDS sensitivity is reflected by the normalized OD_{600} as described in **Figure 1**. We used the $TatABC_{I50Bpa}$ as positive control, which behaved like the wild type (see **Figure 1D** or **2D**). **(C)** Mutational inactivation of the twin-arginine motif-binding site in TatC influences the formation of TC2 and TC3. Comparison of systems with inactivated Tat substrate-binding sites (F94Q, E103A) with systems in which the Tat substrate EfeB has been overproduced. After development of the blot using TatC antibodies, the blot was stripped and developed using antibodies recognizing EfeB. Note that EfeB is not detected in the TC2 formed by the active $TatABC_{I50Bpa}$ system, most likely due to active transport that is likely the reason for the lowered abundance of EfeB in the membranes.

that it likely represents a naturally occurring association of Tat components that is stabilized by the amino acid exchange (**Figure 1**). As additional evidence for this, we found that the exchange of I50 by the natural amino acid Tyr could stabilize TC3, too. In case of this I50Y exchange, which was active as well, all three Tat complexes were clearly detectable (**Figure 2**), suggesting that there is an equilibrium between Tat associations that can be influenced by single amino acid exchanges at position I50. The fact that the I50Bpa exchange can be photo cross-linked to neighboring TatC subunits supports the idea that certain TatC-TatC interactions stabilize the association that is detected as TC3. If there is an equilibrium between TC1, TC2, and TC3, this means that either TC3 is a lowly populated transiently formed complex in the wild-type system or it disassembles upon solubilization to TC2 or TC1, if it is not stabilized. The fact that the complex depends on TatA and that TatA can be co-purified with that complex (**Figure 3**) suggests that TC3 may be the active translocon association. In agreement with the idea that TC3 transports and therefore releases Tat substrates, we found that this complex did not contain any detectable Tat substrate (**Figure 4**). In the absence of TatA, when no transport can take place, a TC2S is formed. However, a TC2 band that is detected below the TC3 band in the active $TatABC_{I50BPA}$ system did not contain any detectable substrate. This indicates that either TC2 with or without substrate cannot be differentiated by BN-PAGE migration behavior with I50Bpa TatC variants, or substrate does not accumulate to a detectable extent at TC2 in active (TatA-containing) systems (**Figure 4**). The detection of translocon bound substrate in TC2 of $TatBC_{I50BPA}$ systems may thus be facilitated by the accumulation of substrates due to the absence of TatA, which is evidenced by the large amounts of EfeB in the membranes of that strain.

An older study questioned the physiological relevance of the Tat complexes that are detected by BN-PAGE analyses (Barrett et al., 2007). The authors used mutated TatA that significantly improved Tat functionality in systems lacking TatB (Blaudeck et al., 2005). However, the mutations were in the N-terminus of TatA and cannot be expected to influence the low detergent resistance of the TatA-TatC interaction, which is mediated mainly by the trans-membrane domain (Habersetzer et al., 2017). It is well known that *E. coli* TatA and TatC catalyze translocation with extremely low efficiency (Ize et al., 2002; Blaudeck et al., 2005). Although no easily detectable detergent-stable TatAC complexes are formed, TatA depletes TatC-only complexes in the absence of TatB, indicating that the TatA-TatC interaction engages most TatC *in vivo* and thereby does not permit the formation of TatC-only complexes (Behrendt et al., 2007). The interaction is also supported by *in vitro* TatAC cross-links in the absence of TatB (Blümmel et al., 2015). The absence of BN-PAGE-detectable detergent-solubilized TatAC complexes in systems lacking TatB therefore should not be taken as argument against the relevance of identified Tat complexes in TatABC systems. TC1 and TC2 can bind substrates, resulting in TC1S and TC2S (Behrendt and Brüser, 2014). Substrates were bound *in vivo*, supporting the physiological relevance of these complexes. Now we found a third complex (TC3) that became detectable due to stabilizing interactions of altered side chains. We propose that this is the association that actually delivered the substrate across the membrane.

The knowledge about the three *E. coli* Tat complexes detected so far can be summarized in the following assembly pathway model (**Figure 5**): TC1 can be formed without TatA and may be regarded as the core unit. TC2 is detectable when TatA is present in the system and we think that this likely represents the resting state (Behrendt and Brüser, 2014). Harsh solubilization conditions deplete TC2 in favor of TC1. When TatB is recombinantly overproduced, complexes of the size of TC2 can also be formed without TatA in the system (Behrendt et al., 2007), which agrees with the finding that TatB can occupy TatA-binding sites at TatC under certain conditions (Habersetzer et al., 2017). As both, TC1 and TC2, can bind Tat substrates (Behrendt and Brüser, 2014), it is unclear whether TatA is recruited to TatBC upon substrate binding or whether substrates

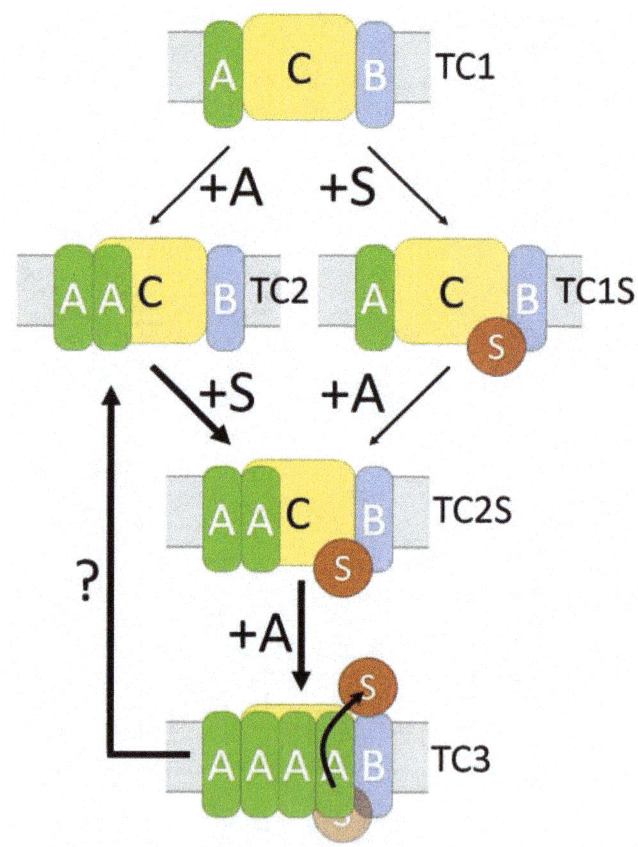

FIGURE 5 | Integrative model for the assembly of active TatABC complexes. Based on the data of this study, in combination with the published work, we propose the following assembly model for active Tat complexes: TC1 represents a Tat complex that can in principle be formed in the absence of TatA and likely is the core complex. TC2 is less stable in detergent solution but already present in wild-type TatABC systems. Both, TC1 and TC2 can in principle bind Tat substrates. Substrate-bound TC2 (TC2S) can switch to TC3, which has an increased affinity to TatA. When TC3 is formed, the substrate can be translocated through the membrane, most likely in an TatA-influenced environment. In the model, "+A" at arrows indicates that more detergent-resistant TatA is bound. "+S" indicates the binding of the twin-arginine motif to its binding site at the TatBC core.

are recruited to TC2 that already contains TatA. Both scenarios can explain the previously observed co-localization of TatA-XFP fusion proteins with TatBC upon substrate overproduction (Alcock et al., 2013; Rose et al., 2013). A stabilization of the TatA-TatBC interaction by substrate binding may compensate for a destabilization that can result from the XFP fusion to the tightly interacting smaller TatA protomers, which would also explain the observations that have been interpreted as substrate-induced assembly. Substrate binding clearly influences the interaction of TatBC with TatA, which can be monitored by cross-linking (Dabney-Smith et al., 2006), but TatA is always directly or indirectly associated with TatBC also under resting conditions (Aldridge et al., 2014; Behrendt and Brüser, 2014). The TatA-dependent rearrangement or conformational transition within Tat complexes likely paves the way for Tat transport. It could be demonstrated that a substrate interaction with TatA results in conformational transitions that destabilize the membrane, and this has been suggested to enable the translocation (Brüser and Sanders, 2003; Hou et al., 2018). All published data suggest that there exists a transiently enhanced affinity to TatA during transport, and it is a matter of kinetic stability, whether or not this active TatABC translocon can be detected. Our analyses now suggest that single amino acid exchanges at position I50, which is positioned at a TatC-TatC interface, can stabilize this active complex. This third Tat complex (TC3) likely switches back to the TC2 state when transport has taken place (question mark in **Figure 5**), and the kinetics of this transition is apparently slowed down by the amino acid exchange, thereby allowing the detection of TC3. The stabilization of TC3 does not require substrate binding and therefore is only a matter of TatABC interactions. As Bpa cross-linking indicates a close proximity of the I50 position to neighboring TatC protomers in the complex, a conformational transition that relies on TatC-TatC interactions is likely responsible for higher TatA affinity. Substrate binding may trigger the switch to this altered interaction in the native, non-mutated system, which explains why TC3 is normally very transient. The data that we obtained with TatBC systems indicate that the I50Bpa mutation enhances not only the TatA affinity but also the TatB affinity to TatC. In contrast to the $TatABC_{I50Bpa}$ system, the $TatBC_{I50Bpa}$ system is not active, indicating that TatB cannot substitute for TatA. However, an inactivated substrate-binding site can trigger the transformation of TC2 to TC3 even in the absence of TatA (**Figure 4**). The conformation of the substrate-binding site might therefore be important for a selective binding of TatA to this site, which is crucial for the transport process. In agreement with this observation, cross-linking studies with the plant TatABC system have shown in the past that TatA associates with the substrate-binding site (Aldridge et al., 2014).

AUTHOR CONTRIBUTIONS

HG obtained all presented data, except the analysis shown in **Figure 1C**, which was done by CN. DM-B and EH were involved in early stages of the project and did initial experiments. TB designed and supervised the study and wrote the manuscript together with HG. All authors contributed to the final manuscript.

FUNDING

This work was funded by the German Research Foundation (DFG grant BR2285/4-2).

ACKNOWLEDGMENTS

We thank Sybille Traupe and Inge Reupke for excellent technical assistance and Jana Behrendt for discussions and seminal contributions to the methods.

REFERENCES

Alami, M., Lüke, I., Deitermann, S., Eisner, G., Koch, H. G., Brunner, J., et al. (2003). Differential interactions between a twin-arginine signal peptide and its translocase in *Escherichia coli*. *Mol. Cell* 12, 937–946. doi: 10.1016/S1097-2765(03)00398-8

Alcock, F., Baker, M. A. B., Greene, N. P., Palmer, T., Wallace, M. I., and Berks, B. C. (2013). Live cell imaging shows reversible assembly of the TatA component of the twin-arginine protein transport system. *Proc. Natl. Acad. Sci. USA* 110, E3650–E3659. doi: 10.1073/Pnas.1306738110

Alcock, F., Stansfeld, P. J., Basit, H., Habersetzer, J., Baker, M. A., Palmer, T., et al. (2016). Assembling the tat protein translocase. *elife* 5:e20718. doi: 10.7554/eLife.20718

Aldridge, C., Ma, X., Gerard, F., and Cline, K. (2014). Substrate-gated docking of pore subunit Tha4 in the TatC cavity initiates tat translocase assembly. *J. Cell Biol.* 205, 51–65. doi: 10.1083/jcb.201311057

Allen, S. C., Barrett, C. M., Ray, N., and Robinson, C. (2002). Essential cytoplasmic domains in the *Escherichia coli* TatC protein. *J. Biol. Chem.* 277, 10362–10366. doi: 10.1074/Jbc.M109135200

Barrett, C. M. L., Freudl, R., and Robinson, C. (2007). Twin arginine translocation (tat)-dependent export in the apparent absence of TatABC or TatA complexes using modified *Escherichia coli* TatA subunits that substitute for TatB. *J. Biol. Chem.* 282, 36206–36213. doi: 10.1074/Jbc.M704127200

Barrett, C. M., Mangels, D., and Robinson, C. (2005). Mutations in subunits of the *Escherichia coli* twin-arginine translocase block function via differing effects on translocation activity or tat complex structure. *J. Mol. Biol.* 347, 453–463. doi: 10.1016/J.Jmb.2005.01.026

Behrendt, J., and Brüser, T. (2014). The TatBC complex of the tat protein translocase in *Escherichia coli* and its transition to the substrate-bound TatABC complex. *Biochemistry* 53, 2344–2354. doi: 10.1021/bi500169s

Behrendt, J., Lindenstrauss, U., and Brüser, T. (2007). The TatBC complex formation suppresses a modular TatB-multimerization in *Escherichia coli*. *FEBS Lett.* 581, 4085–4090. doi: 10.1016/j.febslet.2007.07.045

Blaudeck, N., Kreutzenbeck, P., Müller, M., Sprenger, G. A., and Freudl, R. (2005). Isolation and characterization of bifunctional *Escherichia coli* TatA mutant proteins that allow efficient tat-dependent protein translocation in the absence of TatB. *J. Biol. Chem.* 280, 3426–3432. doi: 10.1074/Jbc.M411210200

Blümmel, A.-S., Haag, L. A., Eimer, E., Müller, M., and Fröbel, J. (2015). Initial assembly steps of a translocase for folded proteins. *Nat. Commun.* 6:7234. doi: 10.1038/Ncomms8234

Brüser, T., and Sanders, C. (2003). An alternative model of the twin arginine translocation system. *Microbiol. Res.* 158, 7–17. doi: 10.1078/0944-5013-00176

Cline, K. (2015). Mechanistic aspects of folded protein transport by the twin arginine translocase (tat). *J. Biol. Chem.* 290, 16530–16538. doi: 10.1074/Jbc.R114.626820

Cline, K., and Mori, H. (2001). Thylakoid Δph-dependent precursor proteins bind to a Cptatc-Hcf106 complex before Tha4-dependent transport. *J. Cell Biol.* 154, 719–729. doi: 10.1083/Jcb.200105149

Dabney-Smith, C., Mori, H., and Cline, K. (2006). Oligomers of Tha4 organize at the thylakoid tat translocase during protein transport. *J. Biol. Chem.* 281, 5476–5483. doi: 10.1074/Jbc.M512453200

Habersetzer, J., Moore, K., Cherry, J., Buchanan, G., Stansfeld, P. J., and Palmer, T. (2017). Substrate-triggered position switching of TatA and TatB during tat transport in *Escherichia coli*. *Open Biol.* 7:170091. doi: 10.1098/rsob.170091

Hamsanathan, S., and Musser, S. M. (2018). The tat protein transport system: intriguing questions and conundrums. *FEMS Microbiol. Lett.* 365:fny123. doi: 10.1093/femsle/fny123

Hou, B., and Brüser, T. (2011). The tat-dependent protein translocation pathway. *Biomol. Concepts* 2, 507–523. doi: 10.1515/BMC.2011.040

Hou, B., Heidrich, E. S., Mehner-Breitfeld, D., and Brüser, T. (2018). The TatA component of the twin-arginine translocation system locally weakens the cytoplasmic membrane of *Escherichia coli* upon protein substrate binding. *J. Biol. Chem.* 293, 7592–7605. doi: 10.1074/jbc.RA118.002205

Hu, Y., Zhao, E., Li, H., Xia, B., and Jin, C. (2010). Solution NMR structure of the TatA component of the twin-arginine protein transport system from gram-positive bacterium *Bacillus subtilis*. *J. Am. Chem. Soc.* 132, 15942–15944. doi: 10.1021/Ja1053785

Huang, Q., Alcock, F., Kneuper, H., Deme, J. C., Rollauer, S. E., Lea, S. M., et al. (2017). A signal sequence suppressor mutant that stabilizes an assembled state of the twin arginine translocase. *Proc. Natl. Acad. Sci. USA* 114, E1958–E1967. doi: 10.1073/Pnas.1615056114

Ize, B., Gerard, F., Zhang, M., Chanal, A., Voulhoux, R., Palmer, T., et al. (2002). *In vivo* dissection of the tat translocation pathway in *Escherichia coli*. *J. Mol. Biol.* 317, 327–335. doi: 10.1006/Jmbi.2001.5431

Ize, B., Stanley, N. R., Buchanan, G., and Palmer, T. (2003). Role of the *Escherichia coli* tat pathway in outer membrane integrity. *Mol. Microbiol.* 48, 1183–1193. doi: 10.1046/j.1365-2958.2003.03504.x

Källberg, M., Margaryan, G., Wang, S., Ma, J., and Xu, J. (2014). Raptorx server: a resource for template-based protein structure modeling. *Methods Mol. Biol.* 1137, 17–27. doi: 10.1007/978-1-4939-0366-5_2

Kneuper, H., Maldonado, B., Jäger, F., Krehenbrink, M., Buchanan, G., Keller, R., et al. (2012). Molecular dissection of TatC defines critical regions essential for protein transport and a TatB-TatC contact site. *Mol. Microbiol.* 85, 945–961. doi: 10.1111/j.1365-2958.2012.08151.x

Laemmli, U. K. (1970). Cleavage of structural proteins during the assembly of the head of bacteriophage T4. *Nature* 227, 680–685. doi: 10.1038/227680a0

Lindenstrauss, U., Matos, C. F. R. O., Graubner, W., Robinson, C., and Brüser, T. (2010). Malfolded recombinant tat substrates are tat-independently degraded in *Escherichia coli*. *FEBS Lett.* 584, 3644–3648. doi: 10.1016/J.Febslet.2010.07.039

Mehner, D., Osadnik, H., Lünsdorf, H., and Brüser, T. (2012). The tat system for membrane translocation of folded proteins recruits the membrane-stabilizing Psp machinery in *Escherichia coli*. *J. Biol. Chem.* 287, 27834–27842. doi: 10.1074/jbc.M112.374983

Mori, H., and Cline, K. (2002). A twin arginine signal peptide and the Ph gradient trigger reversible assembly of the thylakoid Δph/tat translocase. *J. Cell Biol.* 157, 205–210. doi: 10.1083/jcb.200202048

Oates, J., Barrett, C. M., Barnett, J. P., Byrne, K. G., Bolhuis, A., and Robinson, C. (2005). The *Escherichia coli* twin-arginine translocation apparatus incorporates a distinct form of TatABC complex, spectrum of modular TatA complexes and minor TatAB complex. *J. Mol. Biol.* 346, 295–305. doi: 10.1016/j.jmb.2004.11.047

Orriss, G. L., Tarry, M. J., Ize, B., Sargent, F., Lea, S. M., Palmer, T., et al. (2007). TatBC, TatB, and TatC form structurally autonomous units within the twin arginine protein transport system of *Escherichia coli*. *FEBS Lett.* 581, 4091–4097. doi: 10.1016/j.febslet.2007.07.044

Peng, J., and Xu, J. (2011). Raptorx: exploiting structure information for protein alignment by statistical inference. *Proteins* 79, 161–171. doi: 10.1002/prot.23175

Petrů, M., Wideman, J., Moore, K., Alcock, F., Palmer, T., and Doležal, P. (2018). Evolution of mitochondrial tat translocases illustrates the loss of bacterial protein transport machines in mitochondria. *BMC Biol.* 16:141. doi: 10.1186/s12915-018-0607-3

Pettersen, E. F., Goddard, T. D., Huang, C. C., Couch, G. S., Greenblatt, D. M., Meng, E. C., et al. (2004). UCSF chimera–a visualization system for exploratory research and analysis. *J. Comput. Chem.* 25, 1605–1612. doi: 10.1002/jcc.20084

Ramasamy, S., Abrol, R., Suloway, C. J. M., and Clemons, W. M. (2013). The glove-like structure of the conserved membrane protein TatC provides insight into signal sequence recognition in twin-arginine translocation. *Structure* 21, 777–788. doi: 10.1016/j.str.2013.03.004

Richter, S., and Brüser, T. (2005). Targeting of unfolded Phoa to the tat translocon of *Escherichia coli*. *J. Biol. Chem.* 280, 42723–42730. doi: 10.1074/jbc.M509570200

Rollauer, S. E., Tarry, M. J., Graham, J. E., Jaaskelainen, M., Jager, F., Johnson, S., et al. (2012). Structure of the TatC core of the twin-arginine protein transport system. *Nature* 492, 210–214. doi: 10.1038/Nature11683

Rose, P., Fröbel, J., Graumann, P. L., and Müller, M. (2013). Substrate-dependent assembly of the tat translocase as observed in live *Escherichia coli* cells. *PLoS One* 8:E69488. doi: 10.1371/journal.pone.0069488

Sargent, F., Stanley, N. R., Berks, B. C., and Palmer, T. (1999). Sec-independent protein translocation in *Escherichia coli*. A distinct and pivotal role for the TatB protein. *J. Biol. Chem.* 274, 36073–36082. doi: 10.1074/Jbc.274.51.36073

Sturm, A., Schierhorn, A., Lindenstrauss, U., Lilie, H., and Brüser, T. (2006). Ycdb from *Escherichia coli* reveals a novel class of tat-dependently translocated hemoproteins. *J. Biol. Chem.* 281, 13972–13978. doi: 10.1074/Jbc.M511891200

Taubert, J., Hou, B., Risselada, H. J., Mehner, D., Lünsdorf, H., Grubmüller, H., et al. (2015). TatBC-independent TatA/tat substrate interactions contribute to transport efficiency. *PLoS One* 10:E0119761. doi: 10.1371/journal.pone.0119761

Yen, M. R., Tseng, Y. H., Nguyen, E. H., Wu, L. F., and Saier, M. H. Jr. (2002). Sequence and phylogenetic analyses of the twin-arginine targeting (tat) protein export system. *Arch. Microbiol.* 177, 441–450. doi: 10.1007/s00203-002-0408-4

Young, T. S., Ahmad, I., Yin, J. A., and Schultz, P. G. (2010). An enhanced system for unnatural amino acid mutagenesis in *E. coli*. *J. Mol. Biol.* 395, 361–374. doi: 10.1016/j.jmb.2009.10.030

Zhang, Y., Wang, L., Hu, Y., and Jin, C. (2014). Solution structure of the TatB component of the twin-arginine translocation system. *Biochim. Biophys. Acta* 1838, 1881–1888. doi: 10.1016/J.Bbamem.2014.03.015

9

Regulation of Gram-Positive Conjugation

*Verena Kohler[1], Walter Keller[1] and Elisabeth Grohmann[2]**

[1]Institute of Molecular Biosciences, BioTechMed Graz, University of Graz, Graz, Austria, [2]Life Sciences and Technology, Beuth University of Applied Sciences Berlin, Berlin, Germany

***Correspondence:**
Elisabeth Grohmann
egrohmann@beuth-hochschule.de

Type IV Secretion Systems (T4SSs) are membrane-spanning multiprotein complexes dedicated to protein secretion or conjugative DNA transport (conjugation systems) in bacteria. The prototype and best-characterized T4SS is that of the Gram-negative soil bacterium *Agrobacterium tumefaciens*. For Gram-positive bacteria, only conjugative T4SSs have been characterized in some biochemical, structural, and mechanistic details. These conjugation systems are predominantly encoded by self-transmissible plasmids but are also increasingly detected on integrative and conjugative elements (ICEs) and transposons. Here, we report regulatory details of conjugation systems from *Enterococcus* model plasmids pIP501 and pCF10, *Bacillus* plasmid pLS1, *Clostridium* plasmid pCW3, and staphylococcal plasmid pSK41. In addition, regulation of conjugative processes of ICEs (ICE*Bs*1, ICE*St*1, ICE*St*3) by master regulators belonging to diverse repressor families will be discussed. A special focus of this review lies on the comparison of regulatory mechanisms executed by proteins belonging to the RRNPP family. These regulators share a common fold and govern several essential bacterial processes, including conjugative transfer.

Keywords: Gram-positive bacteria, type IV secretion system, conjugation system, regulation, plasmid, integrative and conjugative element

INTRODUCTION

Horizontal gene transfer (HGT) is leadingly involved not only in the evolution of bacteria but also in the dissemination of antibiotic resistances and pathogenicity determinants. This process can be subdivided into three mechanisms: transformation, transduction, and conjugation (Daubin and Szöllősi, 2016), with the latter being the most common type involved in spreading of traits that are beneficial under distinct environmental conditions (Davies and Davies, 2010; Sultan et al., 2018). Conjugative transport of DNA from a donor to a recipient cell requires direct cell-to-cell contact and the formation of a pore, where the DNA molecule can be transported through (Perry and Wright, 2013). Conjugation has been described over large taxonomic distances between unrelated bacterial species (Tamminen et al., 2012). Two different conjugative mechanisms are known to date: the transport of single-stranded (ss) DNA, found in both Gram-positive (G+) and Gram-negative (G−) systems vs. the transport of double-stranded DNA. The second mechanism has been described for G+ actinomycetes, and recent work summarizes this process extensively (Thoma and Muth, 2016; Pettis, 2018). Factors needed for conjugative processes can be encoded on plasmids or integrative and conjugative elements (ICEs). The basic mechanism of conjugation is conserved in most G+ conjugative systems, which facilitate the transport of ss-DNA *via* a molecular machinery encoded by multiple genes that are mostly organized in a single operon. These systems comprise a relaxase, a coupling protein and a mating pair formation (MPF) complex. Relaxases are essential factors in the process of conjugative transfer. They initiate the process by site- and strand-specific cleavage at the *nic*-site of the origin of transfer (*oriT*), forming a covalent complex with the cleaved DNA. For G+ systems, only two relaxases have been structurally characterized so far (Edwards et al., 2013; Pluta et al., 2017). Together with potential accessory factors, the relaxase-DNA complex is called the relaxosome. The coupling protein brings the relaxosome to the MPF complex that forms the actual channel. Conjugation systems usually consist of several mating pair formation proteins, one or more ATPases and proteins facilitating the contact with recipients. In contrast to G-systems that rely on conjugative pili, the contact between donor and recipient is formed *via* surface adhesins in G+ systems. Simultaneously with transfer processes, DNA replication ensures that both donor and new host have a double-stranded version of the plasmid or ICE (Guglielmini et al., 2011; Grohmann et al., 2017).

Transfer of DNA *via* conjugative processes needs to be stringently regulated to reduce the metabolic burden on the host (Koraimann and Wagner, 2014; Singh and Meijer, 2014). Thus, gene products required for conjugation are either kept in a default "OFF" state and are induced by signaling molecules from potential recipients/the environment or conjugative genes are constitutively produced at low abundance to keep fitness costs for the host at a minimum (Frost and Koraimann, 2010; Bañuelos-Vazquez et al., 2017; Stingl and Koraimann, 2017). In this review, we will summarize the current knowledge on the regulation of conjugative processes, focusing on selected conjugation systems from G+ bacteria.

PLASMIDS VS. INTEGRATIVE AND CONJUGATIVE ELEMENTS: SIMILARITIES AND DIFFERENCES

Conjugative plasmids and ICEs harbor all necessary genetic information for conjugative transfer processes (Bañuelos-Vazquez et al., 2017). The principal difference between conjugative plasmids and ICEs lies in their respective maintenance mechanisms within a bacterial cell. While plasmids replicate autonomously, ICEs must integrate into bacterial chromosomes for stable inheritance (**Figure 1**; Perry and Wright, 2013; Burrus, 2017).

Plasmids are autonomously replicating elements that can be categorized into incompatibility (Inc) groups according to their replication and partitioning systems. The spreading of plasmids between unrelated genera is involved in the emergence of antibiotic-resistant bacteria (Sultan et al., 2018). These elements generally carry non-essential genetic features, which might become important under distinct environmental conditions, e.g. in the presence of antibiotic selection pressure (Bañuelos-Vazquez et al., 2017). Plasmids that carry all necessary factors for mobilization and transfer processes are denoted as self-transmissible or conjugative. Biofilm formation plays a substantial role in transfer and dissemination of conjugative plasmids. Conjugative transfer was shown to be considerably higher in biofilms (Kelly et al., 2009).

ICEs are omnipresent in bacterial genomes and were found to be the most abundant conjugative elements in prokaryotes (Ghinet et al., 2011; Guglielmini et al., 2011; Guédon et al., 2017). The exact processes of ICE conjugation are not completely elucidated. It is supposed that these events resemble the ss-plasmid DNA shuttling *via* conjugation systems encoded on plasmids. Since two additional steps, excision and re-integration, are required, ICEs harbor genes that resemble factors of lysogenic phages (Wozniak and Waldor, 2010). These elements show a modular structure with genes of the same/similar function clustered together and usually consist of a maintenance module (responsible for integration and excision), a dissemination module (required for conjugative transfer), and a regulation module (Burrus and Waldor, 2004). An integrated ICE shows a behavior reminiscent of prophages, with most mobility genes suppressed and passively inherited along with the chromosome. Depending on the ICE family, an intra-/intercellular/environmental signal triggers its excision and formation of a circular plasmid-like form, serving as a substrate for the conjugative transfer machinery. After successful transport, the ICE re-integrates into the recipient's chromosome. Integration into and excision from the host chromosome are catalyzed by dedicated enzymes. An integrase (frequently a tyrosine recombinase) governs the reaction between a sequence of the recombination module of the ICE (*attP*) and a sequence on the host chromosome (*attB*), yielding the attachment sites (*attL* and *attR*) that border the ICE after successful integration. In most cases, an excisionase aids in reverting this reaction, forming again an *attP* site on the circularized ICE and an *attB* site on the host's chromosome. As plasmids, ICEs also harbor genes beneficial for their host under specific conditions, e.g. mediating resistance to antimicrobial drugs, heavy metals, and infections by phages (Burrus and Waldor, 2004; Burrus, 2017).

SELECTED MOBILE GENETIC ELEMENTS AND THEIR REGULATION OF CONJUGATIVE PROCESSES

The following sections concentrate on selected plasmids or ICEs from different G+ species, ranging from broad-host range plasmids that produce their conjugative systems constitutively at low levels to inducible/repressible plasmids responding to stimuli from small peptides, called pheromones or autoinducers. These small peptides frequently regulate cellular signaling processes according to the population density, a process denoted as quorum sensing (QS). QS is described to govern essential processes, like virulence, sporulation, and gene transfer. It enables bacteria to sense information about the surrounding species composition and to adapt their expression profiles. This process encompasses the production, release, and detection of autoinducers, the latter is done by a specific sensor component

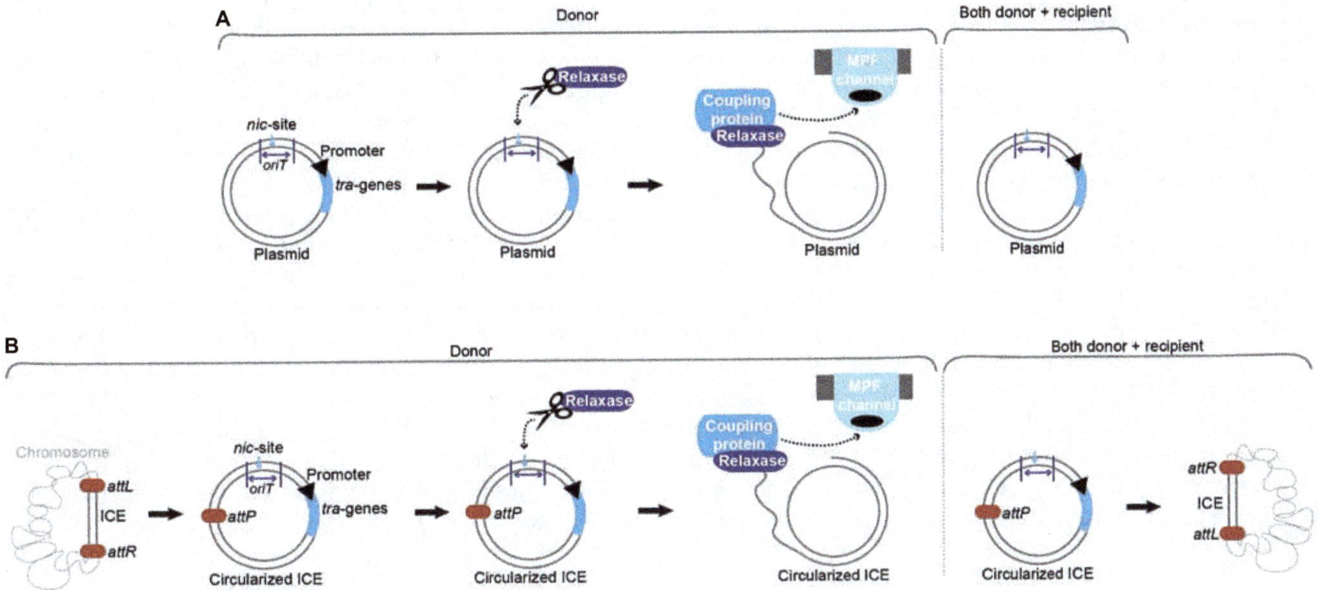

FIGURE 1 | Conjugation of plasmids and integrative and conjugative elements (ICEs). **(A)** Upon a signal (internal or external), the relaxase introduces a single-strand (ss) break at the *nic*-site of the origin of transfer (*oriT*) in the donor cell and covalently binds the ss-DNA. The coupling protein interacts with the relaxosome (relaxase, accessory factors and ss-plasmid DNA) and brings the relaxase-DNA complex to the mating pair formation (MPF) channel. The ss-DNA (most likely covalently bound by the relaxase) is shuttled to the recipient cell. In parallel, replication processes ensure that both donor and recipient harbor a double-stranded version of the conjugative plasmid. **(B)** With the aid of an excisionase and an integrase, the integrative and conjugative element (ICE) is excised from the host's chromosome at *attL* and *attR* attachment sites, forming an *attP* site on the circularized ICE and an *attB* site on the bacterial chromosome. Processing of the DNA, transport and replication follow a comparable mechanism as described for conjugative plasmids in **(A)**. After successful conjugation and replication, the ICE is again integrated into the host's chromosome.

(Papenfort and Bassler, 2016). G+ bacteria usually use oligopeptides as signaling molecules, and the receptor protein interacts directly with the signaling peptide (Monnet et al., 2016). Once a certain level of the pheromone (the quorum) in the external environment is reached, a target sensor kinase or a response regulator is either activated or repressed. Downstream processes then lead to altered expression of QS-dependent genes (Rocha-Estrada et al., 2010).

Pheromone-sensor receptors of the RRNPP-family are found in distantly related G+ bacteria among others on plasmids and ICEs harboring conjugation systems (e.g. pCF10, pLS20 and ICE*Bs*1) (Perez-Pascual et al., 2016). This protein family is named after its prototypical members, Rap, Rgg, NprR, PlcR, and PrgX that despite low sequence homology displays remarkable structural similarities. The defining feature of this group is the C-terminal domain that directly interacts with the pheromone and forms a tetratricopeptide repeat (TPR) domain-like conformation. Structural characterization of these receptors revealed a right-handed super-helical architecture, where the respective ligand is bound to an inner concave binding site of the helix-turn-helix (HTH) repeats (Zeytuni and Zarivach, 2012; Do and Kumaraswami, 2016).

An additional unifying feature of these sensor receptors is the structure of the secreted signaling peptides that are usually linear, 5–10 amino acids long, unmodified, and produced from a longer precursor protein. Pheromones are synthesized *via* the conventional path of ribosomal translation, processing/cleavage, and secretion. Since RRNPP sensor receptors are present inside a cell, the secreted pheromone must be taken up *via* oligopeptide permeases. These enzymes are sometimes aided by accessory proteins that provide high selectivity for the respective peptide (e.g. PrgZ of pCF10; Leonard et al., 1996; Neiditch et al., 2017).

Sex-Pheromone Responsive Plasmid pCF10 From *Enterococcus faecalis*

An increasing number of clinical *E. faecalis* isolates carry conjugative plasmids that are transferred upon induction of peptide pheromones and code for antibiotic resistances (Dunny and Berntsson, 2016). Sex-pheromone responsive plasmids include the tetracycline-resistance plasmid pCF10, one of the best-characterized representatives of this plasmid family, and pAD1. Both plasmids serve as model systems for pheromone-responsive conjugative systems in enterococci (Chen et al., 2017).

prgQ is the conjugative operon of pCF10. It consists of three cassettes. One cassette encodes three surface adhesins required for contacting recipients and *prgU* (Bhatty et al., 2017), encoding a regulator that will be described in further details below. The second cassette harbors the Prg/Pcf MPF and the third cassette codes for factors required for processing the pCF10 plasmid DNA. Several transcriptional and post-transcriptional processes regulate the expression of the *prgQ* operon to guarantee strict control of the Prg/Pcf conjugation system assembly and conjugative transfer (Johnson et al., 2010).

The transcriptional regulator PrgX that belongs to the RRNPP-family binds to the P_Q promoter and represses transcription (**Figure 2A**; Nakayama et al., 1994; Kozlowicz et al., 2004). PrgX further controls its own expression. Unlike other QS-systems (including the Rap-Phr cassettes from pLS20 and ICE*Bs1*), the signal that is sensed by PrgX originates from two different cell types (donor and recipient). This enables the plasmid donor to control conjugation in response to recipient population density (Kozlowicz et al., 2006). PrgX can bind two different heptapeptides, the inducer peptide cCF10 produced from *ccfA* on the bacterial chromosome and the inhibitor iCF10 that is encoded on pCF10 plasmid solely in donor cells (Antiporta and Dunny, 2002). Both inducer and inhibitor peptides are produced by cleavage of precursor proteins, secreted, and imported by the peptide-binding protein PrgZ and chromosomally encoded permeases (Leonard et al., 1996). Inside cells, these pheromones compete for PrgX binding, as both bind to the same cleft within the PrgX C-terminal dimerization domain while interacting with different residues (**Figure 3**; Shi et al., 2005). Interestingly, in contrast to canonical transcription factors that are usually low abundant and modulated by higher-abundant ligands, pCF10-harboring cells usually display an excess of PrgX (15-fold excess of PrgX to its binding site), while the pheromones are present at low concentrations (Mori et al., 1988; Nakayama et al., 1994; Caserta et al., 2012). While its apo-form is indeed able to bind DNA and repress P_Q transcriptional activity at high protein concentrations, PrgX complexed with both the inducer peptide as well as the inhibitor peptide leads to shifted/supershifted DNA complexes with much higher affinities than the unbound form (Caserta et al., 2012; Chen et al., 2017). Thus, following the import of inhibitor/inducer peptide, DNA-bound apo-PrgX is replaced by its complexed form. It is further suggested that the binding affinity of the pheromone (both inhibitor and inducer peptide) for

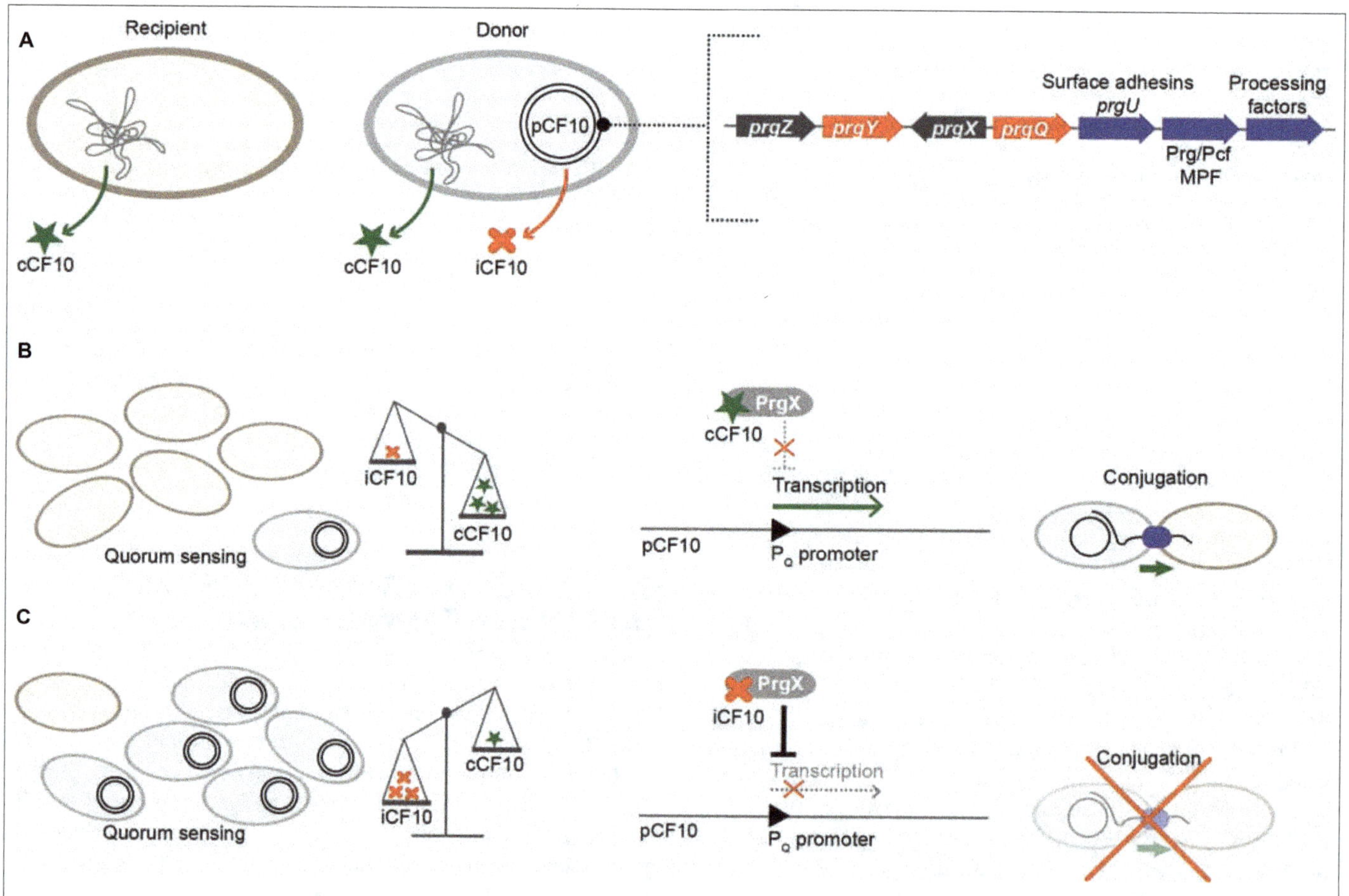

FIGURE 2 | Regulation of pCF10 conjugative transfer. **(A)** While both recipient and donor cell produce the pheromone cCF10 from *ccfA* on the chromosome, only cells harboring pCF10 secrete the inhibitor iCF10 that is produced from *prgQ*. *prgQ* lies upstream of the three cassettes responsible for conjugative transfer: The first cassette codes for the surface adhesins PrgA, PrgB, and PrgC and the regulator PrgU, the second harbors members of the Prg/Pcf conjugation system, and the third is composed of genes for processing factors (including the relaxase). The RRNPP-family protein PrgX is the master regulator and PrgY aids in reducing cCF10 pheromone activity. *prgZ* codes for a permease that imports both cCF10 and iCF10. **(B)** The relative ratio between cCF10 and iCF10 increases with a higher proportion of potential recipients present. This leads to increased import of cCF10, which binds to PrgX, thus interfering with its ability to repress the P_Q promoter. Transcription of genes required for conjugative transfer is activated, and conjugation takes place. **(C)** The cCF10/iCF10 ratio decreases with a lower proportion of potential recipients present, which leads to increased import of the inhibitor peptide iCF10. iCF10-PrgX complexes repress the P_Q promoter, thus downregulating transcription of genes required for conjugative transfer and ultimately inhibiting conjugation.

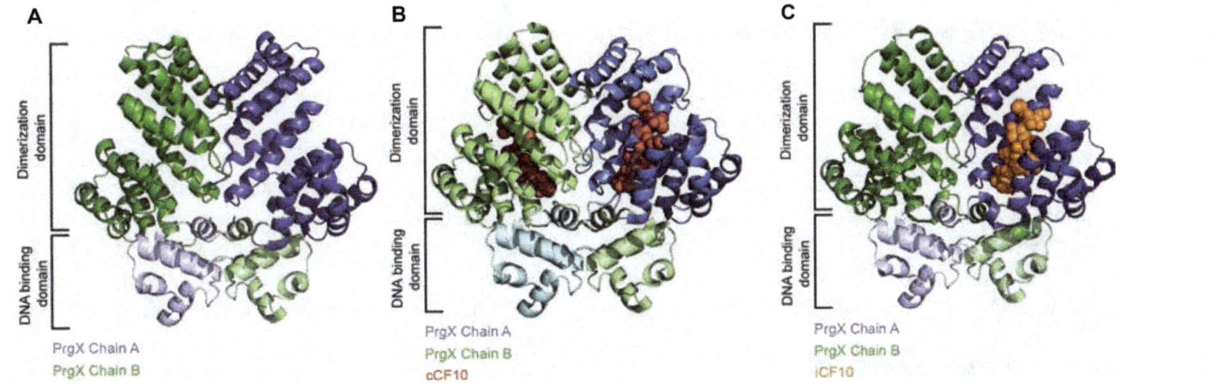

FIGURE 3 | Structural basis for the regulation of the P_Q promoter. **(A)** The PrgX protein forms tetramers, which repress transfer gene expression by binding to two adjacent operator sites, one of which overlaps with the P_Q promoter. The tetramer is built as a dimer of dimers with each dimer occupying one operator site with the N-terminal DNA-binding domain, whereas the C-terminal domain accommodates the dimerization and pheromone-binding function. The dimer of apo-PrgX (PDB: 2AXU) is shown in a cartoon presentation with the chains colored in blue and green, respectively. **(B)** Binding of the pheromone cCF10 (red spheres) leads to a conformational change in the C-terminal region of PrgX (PDB: 2AXZ), alleviating the repression of the P_Q promoter. **(C)** A surplus of iCF10 (orange spheres), which competes with cCF10 for the same binding site (PDB: 2GRL), reinstates the repressed state of the P_Q promoter.

PrgX is considerably stronger than that of complexed PrgX to its DNA binding site. Thus, changes of the donor's induction state most likely come from the exchange of the PrgX apoform for a complexed PrgX form on the DNA (Chen et al., 2017). X-ray crystallography revealed that PrgX exists as a tetramer formed by two dimers. PrgX dimers bind to two operators present on pCF10, O1 and O2. Thus, it has been proposed that the two dimers bound to O1 and O2 interact with each other and form a stable DNA loop. This DNA loop restrains the RNA polymerase from accessing the *prgQ* promoter. When cCF10 binds to the C-terminal domain of PrgX, conformational changes of PrgX are induced. These structural alterations are suggested to break up the tetramers, thus allowing the polymerase to bind to the *prgQ* promoter. By contrast, while iCF10 is thought to compete with cCF10 for the binding site, the inhibitor peptide most likely does not induce structural changes and the tetramer should not be destabilized (Shi et al., 2005). Since the iCF10 precursor is encoded within the *prgQ* locus, enhanced transcription from the *prgQ* promoter increases iCF10 levels, resulting in repression of the transcription of the *prgQ* locus (Kozlowicz et al., 2006). Interestingly, in contrast to other Rap protein-dependent pheromones, neither cCF10 nor iCF10 harbors a positively charged amino acid at the second position (Rocha-Estrada et al., 2010).

In the absence of inducer pheromone sensing, several mechanisms control transcription of the *prgQ* operon (Bae et al., 2004). These comprise PrgX-mediated repression of the P_Q promoter, elevated PrgX repression by binding of the small inhibitor peptide iCF10 that is expressed from a gene directly downstream of the P_Q promoter and production of *anti-Q*, an antisense RNA produced from the convergent P_X promoter, that binds *prgQ* transcripts and induces formation of a termination structure, which blocks transcriptional elongation of *prgQ* transcripts (Nakayama et al., 1994; Shokeen et al., 2010; Chatterjee et al., 2013). Uninduced donor cells show transcriptional activity of the *prgQ* promoter, leading to short (approximately 380 nucleotide long) transcripts. Upon cCF10 binding to PrgX and destabilization of the PrgX tetramer, the number of short *prgQ* transcripts increases, leading to transcription of the whole operon. Within the first 30 min of cCF10 pheromone-exposure, donor cells synthesize the Prg/Pcf conjugation system, form intercellular aggregates due to production of PrgB, one of the surface adhesins, and transfer pCF10 at high frequencies with up to one transconjugant per recipient (**Figure 2B**). In the following 1–2 h, *prgQ* transcription returns to pre-induction levels (Hirt et al., 2005; Chatterjee et al., 2013). Two processes ensure that pCF10-harboring cells do not undergo self-induction. PrgY, a plasmid-encoded membrane protein, reduces pheromone activity produced by donor cells. The extracellular domain of PrgY was proposed to interact with and modify/degrade cCF10 heptapeptides, thus reducing endogenous pheromone activity in donor cells (Chandler et al., 2005). Residual pheromone activity is neutralized by the inhibitor peptide iCF10 encoded by *prgQ* (Nakayama et al., 1994). This inhibitor not only plays a crucial role in returning induced donor cells to the pre-induction state but also serves as a sensor of donor cell density (**Figure 2C**; Chatterjee et al., 2013). After 30–60 min, iCF10 levels reach a certain threshold and consequently reduce transcriptional activity to pre-induction levels. Genes located between *prgQ* and *prgA* (e.g., *prgR*, *prgS*) code for factors modulating transcription and translation of genes required for conjugation (Chung and Dunny, 1992; Bensing et al., 1997).

It was recently demonstrated that cCF10 induction is highly toxic for cells without *prgU*, a small gene downstream of *prgB*, encoding an essential surface adhesin (Bhatty et al., 2017). These *prgU* mutants displayed impaired cell envelope integrity and overproduction of Prg adhesins. By contrast, PrgU overproduction rendered cells insensitive to the sex pheromone and blocked surface adhesin production. PrgU was found to belong to a novel class of RNA binding regulators, reducing toxicity by overproduced surface adhesins. Thus, PrgU exerts another layer of negative regulation of pCF10 conjugative processes. Modeling studies showed that PrgU most likely exists as a tetramer and comprises a PUA (pseudouridine synthase and archaeosine transglycosylase) fold, domains widely distributed and usually interacting with RNA substrates (Pérez-Arellano et al., 2007). Thus, it is hypothesized that PrgU controls Prg adhesin production by binding of RNA substrates, likely regulating trans-acting sRNAs or *prgQ* transcripts. Interestingly, *prgB-prgU* gene pairs were identified in many *E. faecalis* strains and several other enterococci and staphylococci, suggesting that this genetic linkage has evolved to regulate the production of PrgB-like adhesins (Bhatty et al., 2017).

pLS20 From *Bacillus subtilis*

pLS20 is a 65-kbp conjugative plasmid originally isolated from *B. subtilis natto.* It was shown to considerably influence the physiology of its host, e.g. by inhibition of natural competence by the plasmid-encoded repressor ComK (Singh et al., 2012). An operon encoding more than 40 genes responsible for conjugative processes lies downstream of a divergently oriented gene, encoding Rco_{LS20}, the master regulator that keeps conjugative processes in a default "OFF" state, reminiscent of PrgX from pCF10 (**Figure 4A**). Activation of conjugation requires the RRNPP-family protein Rap_{LS20}, the anti-repressor, which is regulated by the signaling pentapeptide Phr^*_{LS20}. Conjugative transfer of pLS20 takes place only during exponential growth (Itaya et al., 2006). pLS20-encoded conjugative proteins are regulated on three levels: first, expression of conjugative proteins and the key transcriptional regulator Rco_{LS20} is controlled by two overlapping divergent promoters of different strengths. Second, Rco_{LS20} exerts three different functions. It is not only a repressor of the main promoter but also an autoregulator of its own promoter, either negatively or positively depending on its abundance. Third, a DNA loop is formed by binding of tetrameric Rco_{LS20} to two operators, overlapping with the divergent promoters (**Figure 4B**; Ramachandran et al., 2014). This scenario is similar to that described for PrgX in the pCF10 system. In contrast to sex-pheromone responsive plasmids like pCF10, pLS20 is not activated by recipient's signaling (*via* an activator peptide like cCF10), but by factors encoded on the plasmid itself. A Rap-Phr module regulates the transfer of pLS20 (Singh et al., 2013). Most Rap-Phr cassettes known to date govern processes like sporulation, competence, or protease/antibiotic synthesis, where the RRNPP-family protein Rap inhibits these events by interacting with a factor required for activating genes involved in these processes. The *phr* gene codes for a protein that is processed into a penta/hexapeptide after Sec-dependent secretion. Following re-uptake, the peptide binds and inactivates its cognate Rap protein (Pottathil and Lazazzera, 2003). The anti-repressor Rap_{LS20} directly binds to the helix-turn-helix domain of Rco_{LS20} in an equimolar stoichiometry, thus most likely interfering with DNA binding of Rco_{LS20} (**Figure 4C**). The pheromone Phr^*_{LS20}, produced as a precursor molecule, directly interacts with Rap_{LS20}, inducing a conformational change of this regulator, which leads to dissociation from Rco_{LS20} (**Figure 4D**). This pheromone is secreted and re-imported into the cells; thus, it is a signal that underlies cell density (Rösch and Graumann, 2015). Interestingly, in contrast to pheromone-induction of pCF10 and processes where QS activates gene expression, genes required for pLS20 conjugation are repressed when the signaling molecule produced by the conjugative plasmid itself reaches a certain quorum. Consequently, when a large number of donor cells is present in the population and thus high levels of the pheromone Phr^*_{LS20}, conjugative processes are inhibited, while plasmid dissemination is activated when more recipient cells are around (and thus lower pheromone levels; Singh and Meijer, 2014). Further, Phr^*_{LS20} plays an essential role in returning conjugative processes to the default "OFF"-state (Singh et al., 2013). This results in heterogeneity of the bacterial population, where up to 30% of the cells induce expression of the operon responsible for conjugation. Several other plasmids, including pX01 of *B. anthracis*, and pBS32 as well as pTA1060 from *B. subtilis*, carry Rap-Phr modules, pointing toward a similar mode of regulation (Koetje et al., 2003; Bongiorni et al., 2006; Parashar et al., 2013).

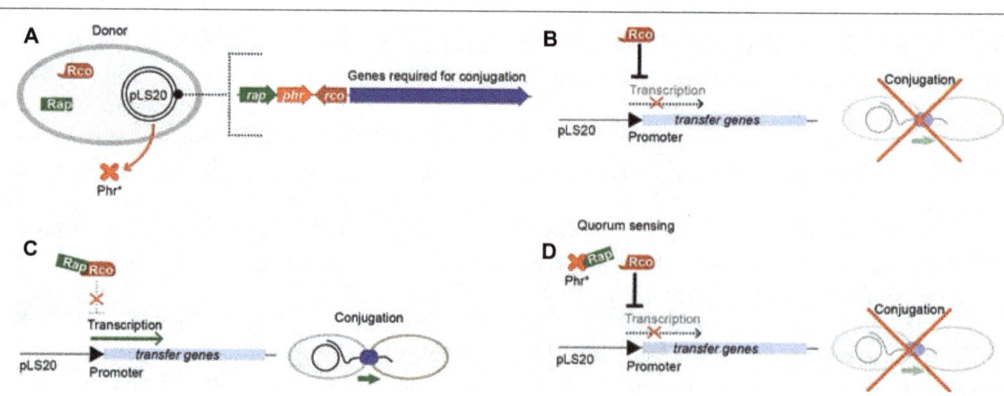

FIGURE 4 | Regulation of pLS20 conjugative transfer. **(A)** Key players involved in regulation are encoded on the pLS20 plasmid, including the master regulator Rco_{LS20} and the Rap_{LS20}-Phr_{LS20} sensor-pheromone cassette upstream of genes required for conjugation. **(B)** The master regulator Rco_{LS20} represses transcription of transfer genes, thus ultimately inhibiting conjugation. **(C)** Binding of Rap_{LS20} to the master regulator Rco_{LS20} leads to conformational changes of Rco_{LS20}, interfering with transcriptional repression of transfer genes, thus conjugation can take place. **(D)** Upon a distinct pheromone concentration (the quorum), the pheromone Phr^*_{LS20} interacts with Rap_{LS20} and interferes with its binding to the master regulator Rco_{LS20}. In consequence, Rco_{LS20} represses transcription of transfer genes and inhibits conjugation.

ICE*Bs*1 From *Bacillus subtilis*

ICE*Bs*1 is a 20.5-kbp ICE that is found on the chromosome of diverse *B. subtilis* strains. The genes required for conjugative transfer are related to those from ICE*St*1 and Tn*916*. This ICE has one of the highest transfer rates in Firmicutes. ICE*Bs*1 harbors more than 20 ORFs and integrates into a locus coding for a tRNA. ICE*Bs*1 cannot only transfer itself, but it can mobilize non-conjugative plasmids as well (Lee et al., 2012). ICE*Bs*1's transfer rate was reported to be considerably higher in biofilms, even though the presence of donor cells in a biofilm did not change the frequency of ICE excision (Lécuyer et al., 2018). Interestingly, regulatory processes of ICE*Bs*1 resemble those of plasmid pLS20. In both systems, conjugation is kept in a default "OFF" state by a master regulator that represses expression of the conjugation genes. ImmR is the master regulator of ICE*Bs*1 that can be modulated by the RapI-PhrI cassette, reminiscent of the Rap-Phr module in pLS20 (**Figures 5A,B**). ImmR inhibits the expression of the excisionase and further downstream genes that are required for ICE*Bs*1 excision and transfer. The RRNPP-family protein RapI induces the production of proteins governing conjugation by interfering with ImmR-mediated repression *via* the anti-repressor ImmA. RapI has been proposed to increase the specific activity of the metalloprotease ImmA that cleaves and thus inactivates ImmR (Bose et al., 2008). In turn, the signaling peptide produced from PhrI inhibits RapI activity. PhrI is encoded downstream of the *rapI* gene within ICE*Bs1* (Auchtung et al., 2005). *phrI* is both expressed from the *rapI* promoter and also produced from its own promoter that is regulated by the sigma factor σ^H (McQuade et al., 2001). Thus, *phrI* transcription increases with enhanced cell density. PhrI is secreted and cleaved by host-encoded factors. After pheromone import *via* the oligopeptide permease Opp, PhrI binds RapI, thus inhibiting its activity and subsequently reducing ICE*Bs1* excision and transfer (Auchtung et al., 2005). Similar to iCF10 from pCF10, extracellular concentrations of PhrI correlate with the number of cells harboring ICE*Bs1*. Thus, when only donors harboring ICE*Bs1* are present, PhrI blocks activation of conjugation *via* interaction with RapI. When potential recipient cells without ICE*Bs1* are present, they take up the pheromone PhrI. This then leads to reduced pheromone levels in donor cells, resulting in RapI-dependent activation of excision, *tra*-gene transcription, and consequently conjugative transfer (Auchtung et al., 2005). RapI not only activates conjugative transfer of ICE*Bs*1 but also inhibits sporulation (Auchtung et al., 2005; Bose et al., 2008). De-repression of ICE*Bs*1, followed by excision and conjugative processes, also takes place upon global DNA damage and is mediated by the DNA-repair protein RecA (Auchtung et al., 2005; Bose and Grossman, 2011), the key mediator of the SOS response. It remains elusive, whether RecA can directly influence the activity of the protease ImmA or acts as stabilization factor (**Figures 5C,D**; Bose and Grossman, 2011).

pIP501 From *Enterococcus faecalis*

The Inc18-family plasmid pIP501 was originally isolated from a clinical *Streptococcus agalactiae* strain and – due to its small size and simplicity – has become the paradigm to study broad host-range plasmids in G+ bacteria with a low G + C content. Inc18 plasmids, including pRE25, pAMß1 and pIP501, have been isolated from clinically relevant *E. faecalis* and *E. faecium* strains and are thought to disseminate resistance to the last-line antimicrobial drug vancomycin, to methicillin-resistant lineages of *Staphylococcus aureus* (Kohler et al., 2018b). The region responsible for conjugative transfer processes is organized as a single operon of 14 kbp, and seven transfer proteins were identified as functional/structural homologs of the G- prototype *A. tumefaciens* T4SS. The *tra*-genes of the pIP501 plasmid were found to be co-transcribed, and mRNA levels remained mostly unchanged

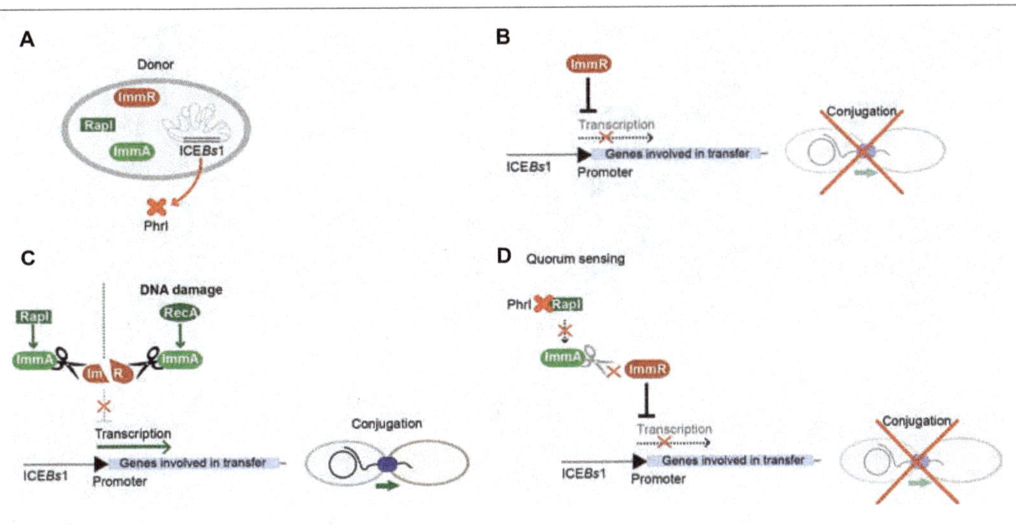

FIGURE 5 | Regulation of ICE*Bs*1 conjugative transfer. **(A)** ICE*Bs*1 harbors most key players involved in regulation of conjugative processes, including the master-regulator ImmR, the metalloprotease ImmA, and the RapI-PhrI sensor-pheromone cassette. **(B)** Similar to processes in pLS20, the master-regulator ImmR represses transcription of genes involved in mobilization and transfer, which ultimately leads to inhibition of conjugation. **(C)** RapI or RecA that is activated by general DNA damage control proteolytic cleavage of ImmR by the metalloprotease ImmA. Thus, transcription is activated and conjugation can take place. **(D)** Upon a distinct concentration (the quorum), the signal peptide PhrI binds to RapI, interfering with its ability to activate ImmA; thus, ImmR can exert repression of the promoter as described in **(A)**.

until late stationary phase. The relaxase TraA, encoded as the first *tra*-gene of the operon, was described to be leadingly involved in regulation of conjugative transfer. TraA was shown to bind to the P_{tra} promoter, thereby negatively regulating transcription of the *tra*-operon (Kurenbach et al., 2006). Recently, TraN, a small cytosolic transfer protein, was identified as additional repressor of the pIP501 conjugation system by binding to its cognate binding site upstream of the P_{tra} promoter and the *oriT nic*-site (Kohler et al., 2018a). TraN is an internal dimer containing two structurally equivalent domains, which belong to the family of winged-helix fold proteins. Its recognition helices protrude into two adjoining major grooves. The wings are required for formation of the interface between the two domains of the internal dimer and insert into the central minor groove. This composition differs from the DNA binding mode of homo-dimeric winged-helix transcription factors (e.g., LysR-type transcriptional regulators; **Figure 6A**; Brown et al., 2003). In contrast to the relaxase TraA that shows autoregulation, TraN's regulatory processes appear to be multi-layered. In addition to the P_{tra} promoter, a second promoter P_{traNO} upstream of the *traN* gene was identified. It was shown to be also negatively regulated by TraN (**Figure 7**). Thus, it was hypothesized that while the P_{tra} promoter was most likely controlled by a concerted action of the relaxase TraA and TraN, TraN binding to P_{traNO} might not only tune its own production but might also be required to regulate levels of TraO, the proposed surface adhesin needed for contacting potential recipients (Kohler et al., 2018a). Toxicity due to overproduction of surface adhesins was demonstrated for other G+ conjugation systems (Bhatty et al., 2017). Nevertheless, the nature of the signal either from potential recipients and/or the environment preceding the nicking of the plasmid DNA by TraA has not been identified so far. Since TraN-homologs and potential TraN binding sites were identified on several Inc18-like and other related multi-resistance plasmids, a similar mechanism of repression was postulated for those plasmids highlighting the potential applicability of TraN as a pharmacological target to combat the dissemination of antibiotic resistances (Kohler et al., 2018a).

pCW3 From *Clostridium perfringens*

The tetracycline resistance plasmid pCW3 from *Clostridium perfringens* belongs to a class of related antibiotic resistance

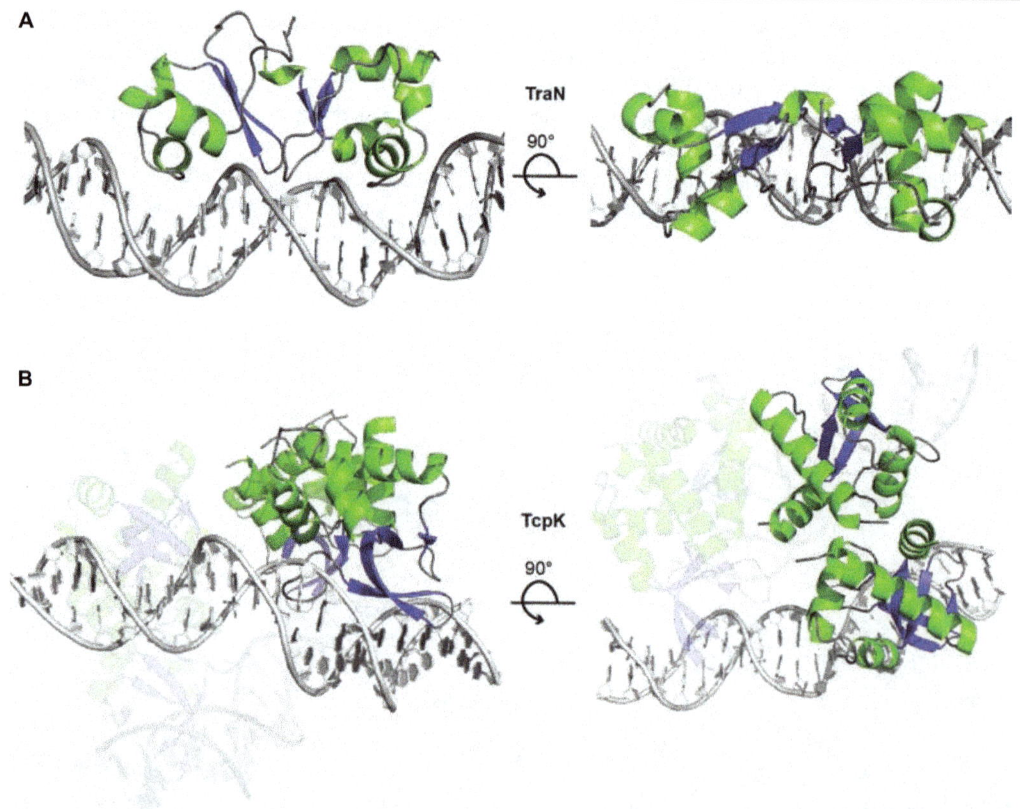

FIGURE 6 | TraN and TcpK are small regulatory proteins exhibiting a winged-helix-turn-helix (wHTH) fold motif. **(A)** TraN is a potent repressor of transcription of the pIP501 *tra*-operon. The TraN protein forms an internal dimer with one wHTH motif in each the N- and C-terminal domain. The TraN-DNA complex (PDB: 6G1T) is shown in cartoon representation, with TraN colored according to secondary structure (α-helices green, β-strands blue) and the DNA in grey. The view is along the recognition helices inserted in the major grooves of the recognition site (left panel) and along the pseudo-two-fold axis of the TraN molecule (right panel).
(B) The TcpK-DNA complex (PDB: 5VFX) contains two protein dimers bridging two double-stranded DNA molecules with two direct-repeat recognition sequences. The color scheme is the same as in panel **(A)**, and the orientation is chosen to show the DNA-binding mode with the β-ribbon of the wing inserted in the major groove of the double-stranded DNA (left panel) and the bridging of the two double-stranded DNA pieces by one of the TcpK dimers (right panel).

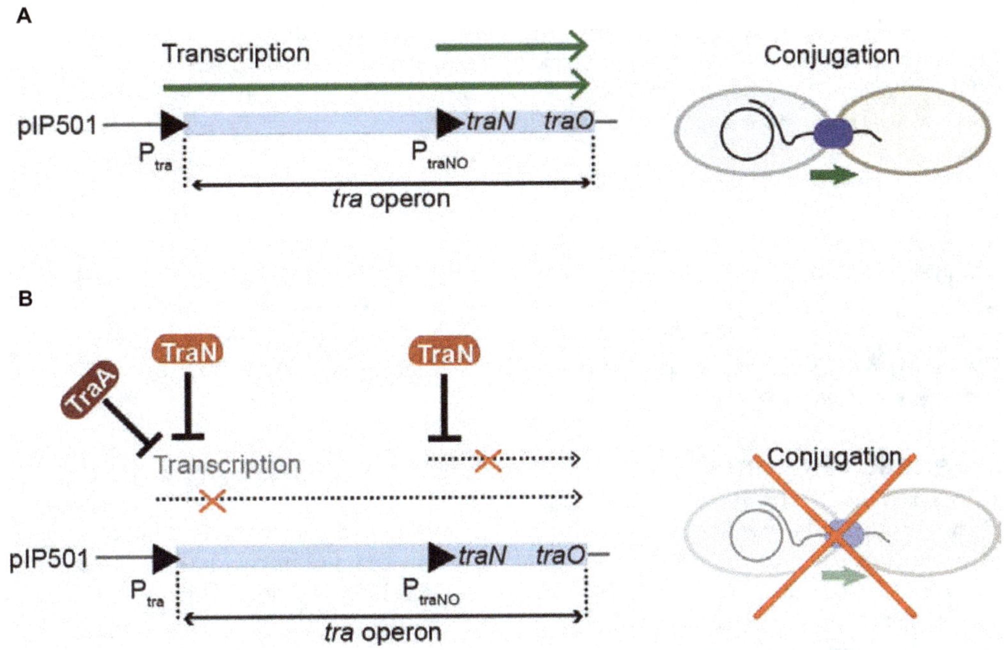

FIGURE 7 | Regulation of pIP501 conjugative transfer. **(A)** Transcription of the pIP501 transfer genes is executed by two promoters, P_{tra} and P_{traNO}, leading to the production of factors required for conjugative transfer. **(B)** By a concerted action between the relaxase TraA and TraN, transcription from the P_{tra} promoter is reduced. Further, TraN can repress the P_{traNO} promoter upstream of its own gene. These regulative processes lead to inhibition of conjugative transfer.

and toxin plasmids (Li et al., 2013). The transfer of clostridial plasmids (*tcp*) locus encoding 11 genes mediates conjugative transfer, with eight proteins essential for these processes (Wisniewski and Rood, 2017). Recently, the structure of TcpK was solved, an essential protein encoded upstream of the *oriT* and shown to be involved in efficient conjugation of pCW3 (Traore et al., 2018). Similar to TraN from pIP501, TcpK is a member of the winged helix-turn-helix protein family and binds specifically to tandem repeats within the pCW3 *oriT*. The complex structure of TcpK with its binding site DNA revealed a binding mode, completely different from other known winged helix-turn-helix proteins. Interestingly, while the wing protrudes into the major groove, the recognition helix makes only a single contact to the binding site DNA. Further, each TcpK dimer binds two binding boxes on different DNA molecules, thus suggesting that TcpK dimers bridge across two DNA molecules (**Figure 6B**). It was suggested that TcpK is an accessory factor of the pCW3 relaxosome, most likely binding to the *oriT* by directly interacting with sequences only present in this region of the plasmid, thus aiding in the proper recruitment of the relaxase TcpM (Traore et al., 2018). Even though the binding mode of TcpK has been uncovered and described in extensive detail, the exact mechanism of regulation remains to be elucidated.

pMV158 From *Streptococcus agalactiae*

pMV158 is a rolling-circle plasmid and was originally isolated from *S. agalactiae* (Burdett, 1980). This 5.5-kbp plasmid is not conjugative but can be mobilized among diverse G+ and G- species by several Inc18-family plasmids including pIP501 and pAMß1 (Priebe and Lacks, 1989; Van der Lelie et al., 1990; Grohmann et al., 1999). MobM is the relaxase of pMV158 and belongs to the MOB_V family of relaxases (Garcillán-Barcia et al., 2009). The full-length MobM as well as the relaxase domain specifically bind to $oriT_{pMV158}$ and are able to perform ss-cleavage of supercoiled pMV158 DNA at the *nic*-site (Grohmann et al., 1999; de Antonio et al., 2004; Lorenzo-Díaz et al., 2011, 2014). Recently, the structure of MobM in complex with different DNA substrates was solved (Pluta et al., 2017). The structures reveal a tight network of protein-DNA interactions involving base-specific as well as backbone interactions. Besides its role in mobilization, MobM is able to repress its own transcription by binding to the *oriT* region, which contains two promoter sequences, one directly overlapping the *oriT* and the second adjacent to the *oriT* (Lorenzo-Díaz et al., 2012), in a similar mode as observed for TraA of pIP501 (Kurenbach et al., 2006). MobM not only autoregulates its own synthesis but is also involved in regulating the pMV158 copy number by binding to the promoter region of the antisense RNAII consequently alleviating the repression of the replication initiator RepB (Lorenzo-Díaz et al., 2017).

pSK41 From *Staphylococcus aureus*

pSK41 and pGO1 are two well-characterized representatives of a large family of low-copy number multi-resistance plasmids from *S. aureus* (Liu et al., 2013). pSK41-like plasmids are self-transmissible and comprise a compact conjugation system with a 14-kbp *tra* region, which consists of two operons, *traA-K* and *traL-M* (Firth et al., 1993), and an additional, divergently transcribed gene, *artA* (*trsN* in pGO1), which will be discussed in detail below (Ni et al., 2009). However, in contrast to the architecture of Inc18-like plasmids, encoding the relaxase as the first open reading frame of the operon, the relaxase gene of pSK41-like plasmids is outside of the *tra* region. The product of the conversely oriented gene, ArtA, is a global transcriptional regulator of pSK41 with six binding sites present on the plasmid, repressing the transcription of conjugative genes as well as those of the segregation system. All three *tra* promoters (P_{artA}, P_{traA}, and P_{traL}) contain these specific ArtA recognition sites and exhibit ArtA binding affinities in the nanomolar range. The crystal structure of ArtA in complex with its cognate DNA binding site reveals that ArtA belongs to the family of ribbon-helix-helix DNA-binding proteins with a lysine-rich N-terminal stretch, which supposedly contributes to additional binding strength (**Figure 8**; Ni et al., 2009).

Other Mobile Genetic Elements

ICEs similar to ICE*Bs*1 include Tn*916*, ICE*St*1, and ICE*St*3. Tn*916* shows a wide host range and is frequently encountered in clinical isolates of *E. faecalis*, *Clostridium difficile*, and *Streptococcus pneumoniae* (Roberts and Mullany, 2009). This ICE is one of the smallest and least complex conjugative elements and carries a tetracycline resistance gene. It can replicate autonomously in *B. subtilis*, which depends on its relaxase. Tn*916* shows its maximal excision frequency during late exponential phase (Celli et al., 1997), thus activation of the element depends on the respective growth phase. Circularization of Tn*916* is required for conjugative transfer, since distinct transcripts can only be produced when the *att* sites are covalently joined (Celli and Trieu-Cuot, 1998). Interestingly, when exposed to tetracycline conjugative transfer frequencies increase 19-fold in *B. subtilis*, while excision frequencies were apparently not affected (Showsh and Andrews, 1992; Celli et al., 1997).

ICE*St*1 with a size of 35 kbp and ICE*St*3 with 28 kbp are closely related ICEs that are found in streptococci. While ICE*St*3 was shown to be transferred to other species including *S. pyogenes* and *E. faecalis*, it is still a matter of debate whether ICE*St*1 is functional or was acquired by transformation (Bellanger et al., 2009; Fontaine et al., 2010). The dissemination module of these ICEs is a 14-kbp polycistronic operon under the control of the P_{cr} promoter (Carraro et al., 2011). It is suggested that the mobility of this ICE family relies on the activity of the P_{cr} promoter (Carraro et al., 2011). The actual DNA-processing machinery is suggested to involve a putative relaxase, which, in addition to the *oriT*, seems to be conserved between these ICEs, ICE*Bs*1 and Tn*916* (Jaworski and Clewell, 1995; Burrus et al., 2002; Rocco and Churchward, 2006). The *arp2* gene encodes a protein reminiscent of the master-regulator ImmR of ICE*Bs*1, while *orfQ* might encode an ImmA-like metalloprotease. Interestingly, while ICEs are usually controlled by one central repressor belonging to unrelated families, cI or ImmR, ICE*St*1/ICE*St*3 family members harbor both repressors (Bellanger et al., 2009; Carraro et al., 2011).

CONCLUSIONS AND PERSPECTIVES

Assembly and operation of multiprotein complexes such as bacterial conjugation systems require large amounts of energy, provided by one or more ATPases encoded by the conjugation system itself. To minimize energy costs of DNA-protein complex transport, expression and/or activity of single crucial components or the whole conjugation system need to be tightly controlled. This is exerted at different levels by different modes (1) at the transcriptional level by controlling the expression of components, such as the conjugative relaxase and/or accessory relaxosome components, or by controlling the production of surface adhesins required for contacting potential recipient cells, (2) *via* cell density sensing (QS) and sex pheromone-induced surface adhesin production, or (3) *via* a master regulator encoded by the plasmid or ICE itself that keeps the conjugative process in a default "OFF" state until the respective antirepressor gets activated, which turns "ON" the conjugative process.

Some plasmids or ICEs dispose only one of these regulatory modes, and others use combinations of them to maintain conjugative activity at an optimum level. Additionally, conjugative processes of some plasmids as well as ICEs are highly growth-phase dependent, with transfer taking place exclusively during exponential growth or exhibiting maximum rates only during (late) exponential growth. Several well-known conjugative

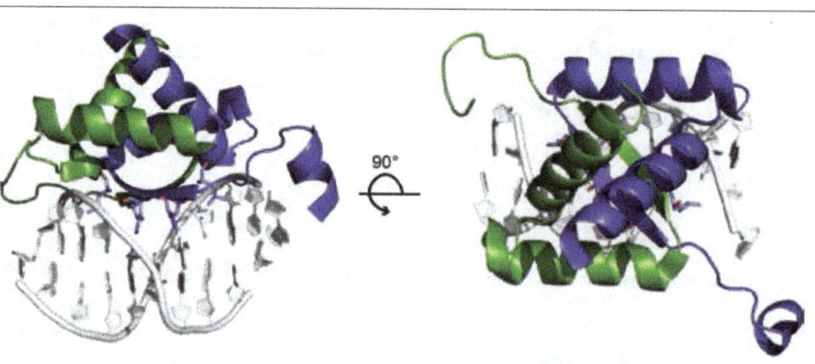

FIGURE 8 | Structure of the ArtA-DNA complex (PDB: 3GXQ). ArtA, the master regulator of the plasmid pSK41, exhibits a ribbon-helix-helix fold. The dimeric repressor is shown in cartoon representation, with chains A and B colored in green and blue, respectively. The anti-parallel β-ribbon permeates the major groove of the binding-site DNA and exerts base-specific interactions (left panel). The view along the twofold axis of the ArtA dimer and the pseudo-palindromic double-stranded DNA site is shown in the right panel.

plasmids and ICEs of G+ origin use pheromone-sensor receptors of the RRNPP family to regulate the conjugation process. Although all members of the RRNPP family display remarkable structural similarity, the downstream reactions often differ widely. Therefore, even though construction of pheromone mimicries would be doable based on the extensive information gathered for this protein family, their structural similarity combined with their differing effects on conjugation would cause problems. Precise data on the regulatory mechanism combined with extensive testing of potential pheromone analogues for diverse conjugation systems would be a first step to solve this problem.

Conjugation is one of the most important means in the dissemination of antibiotic resistance and virulence factors among pathogenic bacteria. Thus, elucidating mechanistic and regulatory details of these large nanomachines is crucial for developing novel approaches to combat multi-resistant pathogens: Important advancements in this direction have been made recently by solution of the cryo-EM structure of the first bacteria-killing T4SS core complex from the G- phytopathogen *Xanthomonas citri* (Sgro et al., 2018) and by the experimental proof that the relaxase has to unfold for efficient translocation through the conjugative T4SS complex (Trokter and Waksman, 2018). Although the abundance of ICEs seems to largely exceed that of conjugative plasmids (Guédon et al., 2017), mechanistic details of their transfer remain elusive. Thus, the aim of future research should lie on the elucidation of the spreading mechanism of ICEs to enable the development/design of specific inhibitors reducing their dissemination.

AUTHOR CONTRIBUTIONS

VK, WK, and EG drafted the manuscript. VK designed the figures. All authors approved the final version of the manuscript.

FUNDING

Work in the Grohmann lab was supported by DLR grants 50WB1166 and 50WB1466, and work in the Keller lab was supported by FWF grant P 27383.

REFERENCES

Antiporta, M. H., and Dunny, G. M. (2002). *ccfa*, the genetic determinant for the cCF10 peptide pheromone in *Enterococcus faecalis* OG1RF. *J. Bacteriol.* 184, 1155–1162. doi: 10.1128/jb.184.4.1155-1162.2002

Auchtung, J. M., Lee, C. A., Monson, R. E., Lehman, A. P., and Grossman, A. D. (2005). Regulation of a *Bacillus subtilis* mobile genetic element by intercellular signaling and the global DNA damage response. *Proc. Natl. Acad. Sci.* 102, 12554–12559. doi: 10.1073/pnas.0505835102

Bae, T., Kozlowicz, B. K., and Dunny, G. M. (2004). Characterization of cis-acting *prgQ* mutants: evidence for two distinct repression mechanisms by Qa RNA and PrgX protein in pheromone-inducible enterococcal plasmid pCF10. *Mol. Microbiol.* 51, 271–281. doi: 10.1046/j.1365-2958.2003.03832.x

Bañuelos-Vazquez, L. A., Torres Tejerizo, G., and Brom, S. (2017). Regulation of conjugative transfer of plasmids and integrative conjugative elements. *Plasmid* 91, 82–89. doi: 10.1016/j.plasmid.2017.04.002

Bellanger, X., Roberts, A. P., Morel, C., Choulet, F., Pavlovic, G., Mullany, P., et al. (2009). Conjugative transfer of the integrative conjugative elements ICE*St*1 and ICE*St*3 from *Streptococcus thermophilus*. *J. Bacteriol.* 191, 2764–2775. doi: 10.1128/JB.01412-08

Bensing, B. A., Manias, D. A., and Dunny, G. M. (1997). Pheromone cCF10 and plasmid pCF10-encoded regulatory molecules act post-transcriptionally to activate expression of downstream conjugation functions. *Mol. Microbiol.* 24, 285–294. doi: 10.1046/j.1365-2958.1997.3301710.x

Bhatty, M., Camacho, M. I., Gonzalez-Rivera, C., Frank, K. L., Dale, J. L., Manias, D. A., et al. (2017). PrgU: a suppressor of sex pheromone toxicity in *Enterococcus faecalis*. *Mol. Microbiol.* 103, 398–412. doi: 10.1111/mmi.13563

Bongiorni, C., Stoessel, R., Shoemaker, D., and Perego, M. (2006). Rap phosphatase of virulence plasmid pXO1 inhibits *Bacillus anthracis* sporulation. *J. Bacteriol.* 188, 487–498. doi: 10.1128/JB.188.2.487-498.2006

Bose, B., Auchtung, J. M., Lee, C. A., and Grossman, A. D. (2008). A conserved anti-repressor controls horizontal gene transfer by proteolysis. *Mol. Microbiol.* 70, 570–582. doi: 10.1111/j.1365-2958.2008.06414.x

Bose, B., and Grossman, A. D. (2011). Regulation of horizontal gene transfer in *Bacillus subtilis* by activation of a conserved site-specific protease. *J. Bacteriol.* 193, 22–29. doi: 10.1128/JB.01143-10

Brown, N. L., Stoyanov, J. V., Kidd, S. P., and Hobman, J. L. (2003). The MerR family of transcriptional regulators. *FEMS Microbiol. Rev.* 27, 145–163. doi: 10.1016/S0168-6445(03)00051-2

Burdett, V. (1980). Identification of tetracycline-resistant R-plasmids in *Streptococcus agalactiae* (group B). *Antimicrob. Agents Chemother.* 18, 753–760. doi: 10.1128/AAC.18.5.753

Burrus, V. (2017). Mechanisms of stabilization of integrative and conjugative elements. *Curr. Opin. Microbiol.* 38, 44–50. doi: 10.1016/j.mib.2017.03.014

Burrus, V., Pavlovic, G., Decaris, B., and Guédon, G. (2002). The ICE*St*1 element of *Streptococcus thermophilus* belongs to a large family of integrative and conjugative elements that exchange modules and change their specificity of integration. *Plasmid* 48, 77–97. doi: 10.1016/S0147-619X(02)00102-6

Burrus, V., and Waldor, M. K. (2004). Shaping bacterial genomes with integrative and conjugative elements. *Res. Microbiol.* 155, 376–386. doi: 10.1016/j.resmic.2004.01.012

Carraro, N., Libante, V., Morel, C., Decaris, B., Charron-Bourgoin, F., Leblond, P., et al. (2011). Differential regulation of two closely related integrative and conjugative elements from *Streptococcus thermophilus*. *BMC Microbiol.* 11:238. doi: 10.1186/1471-2180-11-238

Caserta, E., Haemig, H. A. H., Manias, D. A., Tomsic, J., Grundy, F. J., Henkin, T. M., et al. (2012). In vivo and in vitro analyses of regulation of the pheromone-responsive *prgQ* promoter by the PrgX pheromone receptor protein. *J. Bacteriol.* 194, 3386–3394. doi: 10.1128/JB.00364-12

Celli, J., Poyart, C., and Trieu-Cuot, P. (1997). Use of an excision reporter plasmid to study the intracellular mobility of the conjugative transposon *Tn916* in Gram-positive bacteria. *Microbiology* 143, 1253–1261. doi: 10.1099/00221287-143-4-1253

Celli, J., and Trieu-Cuot, P. (1998). Circularization of Tn*916* is required for expression of the transposon-encoded transfer functions: characterization of long tetracycline-inducible transcripts reading through the attachment site. *Mol. Microbiol.* 28, 103–117. doi: 10.1046/j.1365-2958.1998.00778.x

Chandler, J. R., Flynn, A. R., Bryan, E. M., and Dunny, G. M. (2005). Specific control of endogenous cCF10 pheromone by a conserved domain of the pCF10-encoded regulatory protein PrgY in *Enterococcus faecalis*. *J. Bacteriol.* 187, 4830–4843. doi: 10.1128/JB.187.14.4830-4843.2005

Chatterjee, A., Cook, L. C. C., Shu, C.-C., Chen, Y., Manias, D. A., Ramkrishna, D., et al. (2013). Antagonistic self-sensing and mate-sensing signaling controls antibiotic-resistance transfer. *Proc. Natl. Acad. Sci.* 110, 7086–7090. doi: 10.1073/pnas.1212256110

Chen, Y., Bandyopadhyay, A., Kozlowicz, B. K., Haemig, H. A. H., Tai, A., Hu, W. S., et al. (2017). Mechanisms of peptide sex pheromone regulation of conjugation in *Enterococcus faecalis*. *Microbiology* 6:e00492. doi: 10.1002/mbo3.492

Chung, J. W., and Dunny, G. M. (1992). Cis-acting, orientation-dependent, positive control system activates pheromone-inducible conjugation functions at distances greater than 10 kilobases upstream from its target in *Enterococcus faecalis*. *Proc. Natl. Acad. Sci.* 89, 9020–9024. doi: 10.1073/pnas.89.19.9020

Daubin, V., and Szöllősi, G. J. (2016). Horizontal gene transfer and the history of life. *Cold Spring Harb. Perspect. Biol.* 8:a018036. doi: 10.1101/cshperspect.a018036

Davies, J., and Davies, D. (2010). Origins and evolution of antibiotic resistance. *Microbiol. Mol. Biol. Rev.* 74, 417–433. doi: 10.1128/MMBR.00016-10

de Antonio, C., Farías, M. E., de Lacoba, M. G., and Espinosa, M. (2004). Features of the plasmid pMV158-encoded MobM, a protein involved in its mobilization. *J. Mol. Biol.* 335, 733–743. doi: 10.1016/J.JMB.2003.11.017

Do, H., and Kumaraswami, M. (2016). Structural mechanisms of peptide recognition and allosteric modulation of gene regulation by the RRNPP family of quorum-sensing regulators. *J. Mol. Biol.* 428, 2793–2804. doi: 10.1016/J.JMB.2016.05.026

Dunny, G. M., and Berntsson, P. (2016). Enterococcal sex pheromones: evolutionary pathways to complex, two-signal systems. *J. Bacteriol.* 198, 1556–1562. doi: 10.1128/JB.00128-16

Edwards, J. S., Betts, L., Frazier, M. L., Pollet, R. M., Kwong, S. M., Walton, W. G., et al. (2013). Molecular basis of antibiotic multiresistance transfer in *Staphylococcus aureus*. *Proc. Natl. Acad. Sci. USA* 110, 2804–2809. doi: 10.1073/pnas.1219701110

Firth, N., Ridgway, K. P., Byrne, M. E., Fink, P. D., Johnson, L., Paulsen, I. T., et al. (1993). Analysis of a transfer region from the staphylococcal conjugative plasmid pSK41. *Gene* 136, 13–25. doi: 10.1016/0378-1119(93)90442-6

Fontaine, L., Boutry, C., de Frahan, M. H., Delplace, B., Fremaux, C., Horvath, P., et al. (2010). A novel pheromone quorum-sensing system controls the development of natural competence in *Streptococcus thermophilus* and *Streptococcus salivarius*. *J. Bacteriol.* 192, 1444–1454. doi: 10.1128/JB.01251-09

Frost, L. S., and Koraimann, G. (2010). Regulation of bacterial conjugation: balancing opportunity with adversity. *Future Microbiol* 5, 1057–1071. doi: 10.2217/fmb.10.70

Garcillán-Barcia, M. P., Francia, M. V., and de La Cruz, F. (2009). The diversity of conjugative relaxases and its application in plasmid classification. *FEMS Microbiol. Rev.* 33, 657–687. doi: 10.1111/j.1574-6976.2009.00168.x

Ghinet, M. G., Bordeleau, E., Beaudin, J., Brzezinski, R., Roy, S., and Burrus, V. (2011). Uncovering the prevalence and diversity of integrating conjugative elements in actinobacteria. *PLoS One* 6:e27846. doi: 10.1371/journal.pone.0027846

Grohmann, E., Christie, P. J., Waksman, G., and Backert, S. (2017). Type IV secretion in Gram-negative and Gram-positive bacteria. *Mol. Microbiol.* 107, 455–471. doi: 10.1111/mmi.13896

Grohmann, E., Guzmán, L. M., and Espinosa, M. (1999). Mobilisation of the streptococcal plasmid pMV158: interactions of MobM protein with its cognate *oriT* DNA region. *Mol. Gen. Genet.* 261, 707–715. doi: 10.1007/s004380050014

Guédon, G., Libante, V., Coluzzi, C., Payot, S., and Leblond-Bourget, N. (2017). The obscure world of integrative and mobilizable elements, highly widespread elements that pirate bacterial conjugative systems. *Genes* 8:337. doi: 10.3390/genes8110337

Guglielmini, J., Quintais, L., Garcillán-Barcia, M. P., de la Cruz, F., and Rocha, E. P. C. (2011). The repertoire of ICE in prokaryotes underscores the unity, diversity, and ubiquity of conjugation. *PLoS Genet.* 7:e1002222. doi: 10.1371/journal.pgen.1002222

Hirt, H., Manias, D. A., Bryan, E. M., Klein, J. R., Marklund, J. K., Staddon, J. H., et al. (2005). Characterization of the pheromone response of the *Enterococcus faecalis* conjugative plasmid pCF10: complete sequence and comparative analysis of the transcriptional and phenotypic responses of pCF10-containing cells to pheromone induction. *J. Bacteriol.* 187, 1044–1054. doi: 10.1128/JB.187.3.1044-1054.2005

Itaya, M., Sakaya, N., Matsunaga, S., Fujita, K., and Kaneko, S. (2006). Conjugational transfer kinetics of pLS20 between *Bacillus subtilis* in liquid medium. *Biosci. Biotechnol. Biochem.* 70, 740–742. doi: 10.1271/bbb.70.740

Jaworski, D. D., and Clewell, D. B. (1995). A functional origin of transfer (*oriT*) on the conjugative transposon *Tn*916. *J. Bacteriol.* 177, 6644–6651. doi: 10.1128/JB.177.22.6644-6651.1995

Johnson, C. M., Manias, D. A., Haemig, H. A. H., Shokeen, S., Weaver, K. E., Henkin, T. M., et al. (2010). Direct evidence for control of the pheromone-inducible *prgQ* operon of *Enterococcus faecalis* plasmid pCF10 by a countertranscript-driven attenuation mechanism. *J. Bacteriol.* 192, 1634–1642. doi: 10.1128/JB.01525-09

Kelly, B. G., Vespermann, A., and Bolton, D. J. (2009). The role of horizontal gene transfer in the evolution of selected foodborne bacterial pathogens. *Food Chem. Toxicol.* 47, 951–968. doi: 10.1016/j.fct.2008.02.006

Koetje, E. J., Hajdo-Milasinovic, A., Kiewiet, R., Bron, S., and Tjalsma, H. (2003). A plasmid-borne Rap-Phr system of *Bacillus subtilis* can mediate cell-density controlled production of extracellular proteases. *Microbiology* 149, 19–28. doi: 10.1099/mic.0.25737-0

Kohler, V., Goessweiner-Mohr, N., Aufschnaiter, A., Fercher, C., Probst, I., Pavkov-Keller, T., et al. (2018a). TraN: a novel repressor of an enterococcus conjugative type IV secretion system. *Nucleic Acids Res.* 46, 9201–9219 doi: 10.1093/nar/gky671

Kohler, V., Vaishampayan, A., and Grohmann, E. (2018b). Broad-host-range Inc18 plasmids: occurrence, spread and transfer mechanisms. *Plasmid* 99, 11–21. doi: 10.1016/J.PLASMID.2018.06.001

Koraimann, G., and Wagner, M. A. (2014). Social behavior and decision making in bacterial conjugation. *Front. Cell. Infect. Microbiol.* 4:54. doi: 10.3389/fcimb.2014.00054

Kozlowicz, B. K., Bae, T., and Dunny, G. M. (2004). *Enterococcus faecalis* pheromone-responsive protein PrgX: genetic separation of positive autoregulatory functions from those involved in negative regulation of conjugative plasmid transfer. *Mol. Microbiol.* 54, 520–532. doi: 10.1111/j.1365-2958.2004.04286.x

Kozlowicz, B. K., Shi, K., Gu, Z. Y., Ohlendorf, D. H., Earhart, C. A., and Dunny, G. M. (2006). Molecular basis for control of conjugation by bacterial pheromone and inhibitor peptides. *Mol. Microbiol.* 62, 958–969. doi: 10.1111/j.1365-2958.2006.05434.x

Kurenbach, B., Kopeć, J., Mägdefrau, M., Andreas, K., Keller, W., Bohn, C., et al. (2006). The TraA relaxase autoregulates the putative type IV secretion-like system encoded by the broad-host-range *Streptococcus agalactiae* plasmid pIP501. *Microbiology* 152, 637–645. doi: 10.1099/mic.0.28468-0

Lécuyer, F., Bourassa, J.-S., Gélinas, M., Charron-Lamoureux, V., Burrus, V., and Beauregard, P. B. (2018). Biofilm formation drives transfer of the conjugative element ICE*Bs*1 in *Bacillus subtilis*. *mSphere* 3, 1–12. doi: 10.1128/mSphere.00473-18

Lee, C. A., Thomas, J., and Grossman, A. D. (2012). The *Bacillus subtilis* conjugative transposon ICE*Bs*1 mobilizes plasmids lacking dedicated mobilization functions. *J. Bacteriol.* 194, 3165–3172. doi: 10.1128/JB.00301-12

Leonard, B. A., Podbielski, A., Hedberg, P. J., and Dunny, G. M. (1996). *Enterococcus faecalis* pheromone binding protein, PrgZ, recruits a chromosomal oligopeptide permease system to import sex pheromone cCF10 for induction of conjugation. *Proc. Natl. Acad. Sci. USA* 93, 260–264. doi: 10.1073/PNAS.93.1.260

Li, J., Adams, V., Bannam, T. L., Miyamoto, K., Garcia, J. P., Uzal, F. A., et al. (2013). Toxin plasmids of *Clostridium perfringens*. *Microbiol. Mol. Biol. Rev.* 77, 208–233. doi: 10.1128/MMBR.00062-12

Liu, M. A., Kwong, S. M., Jensen, S. O., and Brzoska, A. J. (2013). Biology of the staphylococcal conjugative multiresistance plasmid pSK41. *Plasmid* 70, 42–51. doi: 10.1016/J.PLASMID.2013.02.001

Lorenzo-Díaz, F., Dostál, L., Coll, M., Schildbach, J. F., Menéndez, M., and Espinosa, M. (2011). The MobM relaxase domain of plasmid pMV158: thermal stability and activity upon Mn^{2+} and specific DNA binding. *Nucleic Acids Res.* 39, 4315–4329. doi: 10.1093/nar/gkr049

Lorenzo-Díaz, F., Espinosa, M., Solano-Collado, V., Garsin, D. A., Bravo, A., Ruiz-Cruz, S., et al. (2014). Mobilizable rolling-circle replicating plasmids from Gram-positive bacteria: a low-cost conjugative transfer. *Microbiol. Spectr.* 2. doi: 10.1128/microbiolspec.PLAS-0008-2013

Lorenzo-Díaz, F., Fernández-López, C., Lurz, R., Bravo, A., and Espinosa, M. (2017). Crosstalk between vertical and horizontal gene transfer: plasmid replication control by a conjugative relaxase. *Nucleic Acids Res.* 45, 7774–7785. doi: 10.1093/nar/gkx450

Lorenzo-Díaz, F., Solano-Collado, V., Lurz, R., Bravo, A., and Espinosa, M. (2012). Autoregulation of the synthesis of the MobM relaxase encoded by the promiscuous Plasmid pMV158. *J. Bacteriol.* 194, 1789–1799. doi: 10.1128/JB.06827-11

McQuade, R. S., Comella, N., and Grossman, A. D. (2001). Control of a family of phosphatase regulatory genes (*phr*) by the alternate sigma factor sigma-H of *Bacillus subtilis*. *J. Bacteriol.* 183, 4905–4909. doi: 10.1128/JB.183.16.4905-4909.2001

Monnet, V., Juillard, V., and Gardan, R. (2016). Peptide conversations in Gram-positive bacteria. *Crit. Rev. Microbiol.* 42, 339–351. doi: 10.3109/1040841X.2014.948804

Mori, M., Sakagamis, Y., Ishiis, Y., Isogais, A., Kitadaj, C., Fujinoj, M., et al. (1988). Structure of cCF10, a peptide sex pheromone which induces conjugative transfer of the *Streptococcus faecalis* tetracycline resistance plasmid, pCF10. *J. Biol. Chem.* 263, 14574–14578.

Nakayama, J., Ruhfel, R. E., Dunny, G. M., Isogai, A., and Suzuki, A. (1994). The *prgQ* gene of the *Enterococcus faecalis* tetracycline resistance plasmid pCF10 encodes a peptide inhibitor, iCF10. *J. Bacteriol.* 176, 7405–7408. doi: 10.1128/jb.176.23.7405-7408.1994

Neiditch, M. B., Capodagli, G. C., Prehna, G., and Federle, M. J. (2017). Genetic and structural analyses of RRNPP intercellular peptide signaling of Gram-positive bacteria. *Annu. Rev. Genet.* 51, 311–333. doi: 10.1146/annurev-genet-120116-023507

Ni, L., Jensen, S. O., Ky Tonthat, N., Berg, T., Kwong, S. M., Guan, F. H. X., et al. (2009). The *Staphylococcus aureus* pSK41 plasmid-encoded ArtA protein is a master regulator of plasmid transmission genes and contains a RHH motif used in alternate DNA-binding modes. *Nucleic Acids Res.* 37, 6970–6983. doi: 10.1093/nar/gkp756

Papenfort, K., and Bassler, B. L. (2016). Quorum sensing signal-response systems in Gram-negative bacteria. *Nat. Rev. Microbiol.* 14, 576–588. doi: 10.1038/nrmicro.2016.89

Parashar, V., Konkol, M. A., Kearns, D. B., and Neiditch, M. B. (2013). A plasmid-encoded phosphatase regulates *Bacillus subtilis* biofilm architecture, sporulation, and genetic competence. *J. Bacteriol.* 195, 2437–2448. doi: 10.1128/JB.02030-12

Pérez-Arellano, I., Gallego, J., and Cervera, J. (2007). The PUA domain–a structural and functional overview. *FEBS J.* 274, 4972–4984. doi: 10.1111/j.1742-4658.2007.06031.x

Perez-Pascual, D., Monnet, V., and Gardan, R. (2016). Bacterial cell-cell communication in the host via RRNPP peptide-binding regulators. *Front. Microbiol.* 7:706. doi: 10.3389/fmicb.2016.00706

Perry, J. A., and Wright, G. D. (2013). The antibiotic resistance "mobilome": searching for the link between environment and clinic. *Front. Microbiol.* 4:138. doi: 10.3389/fmicb.2013.00138

Pettis, G. S. (2018). Spreading the news about the novel conjugation mechanism in *Streptomyces* bacteria. *Environ. Microbiol. Rep.* 10, 503–510. doi: 10.1111/1758-2229.12659

Pluta, R., Boer, D. R., Lorenzo-Díaz, F., Russi, S., Gómez, H., Fernández-López, C., et al. (2017). Structural basis of a histidine-DNA nicking/joining mechanism for gene transfer and promiscuous spread of antibiotic resistance. *Proc. Natl. Acad. Sci. USA* 114, E6526–E6535. doi: 10.1073/pnas.1702971114

Pottathil, M., and Lazazzera, B. A. (2003). The extracellular phr peptide-Rap phosphatase signaling circuit of Bacillus subtilis. *Front. Biosci.* 8, 32–45. doi: 10.2741/913

Priebe, S. D., and Lacks, S. A. (1989). Region of the streptococcal plasmid pMV158 required for conjugative mobilization. *J. Bacteriol.* 171, 4778–4784. doi: 10.1128/jb.171.9.4778-4784.1989

Ramachandran, G., Singh, P. K., Luque-Ortega, J. R., Yuste, L., Alfonso, C., Rojo, F., et al. (2014). A complex genetic switch involving overlapping divergent promoters and DNA looping regulates expression of conjugation genes of a Gram-positive plasmid. *PLoS Genet.* 10:e1004733. doi: 10.1371/journal.pgen.1004733

Roberts, A. P., and Mullany, P. (2009). A modular master on the move: the Tn*916* family of mobile genetic elements. *Trends Microbiol.* 17, 251–258. doi: 10.1016/j.tim.2009.03.002

Rocco, J. M., and Churchward, G. (2006). The integrase of the conjugative transposon Tn*916* directs strand- and sequence-specific cleavage of the origin of conjugal transfer, *oriT*, by the endonuclease Orf20. *J. Bacteriol.* 188, 2207–2213. doi: 10.1128/JB.188.6.2207-2213.2006

Rocha-Estrada, J., Aceves-Diez, A. E., Guarneros, G., and De La Torre, M. (2010). The RNPP family of quorum-sensing proteins in Gram-positive bacteria. *Appl. Microbiol. Biotechnol.* 87, 913–923. doi: 10.1007/s00253-010-2651-y

Rösch, T. C., and Graumann, P. L. (2015). Induction of plasmid conjugation in *Bacillus subtilis* is bistable and driven by a direct interaction of a Rap/Phr quorum-sensing system with a master repressor. *J. Biol. Chem.* 290, 20221–20232. doi: 10.1074/jbc.M115.664110

Sgro, G. G., Costa, T. R. D., Cenens, W., Souza, D. P., Cassago, A., Coutinho de Oliveira, L., et al. (2018). Cryo-EM structure of the bacteria-killing type IV secretion system core complex from *Xanthomonas citri*. *Nat. Microbiol.* 3, 1429–1440. doi: 10.1038/s41564-018-0262-z

Shi, K., Brown, C. K., Gu, Z.-Y., Kozlowicz, B. K., Dunny, G. M., Ohlendorf, D. H., et al. (2005). Structure of peptide sex pheromone receptor PrgX and PrgX/pheromone complexes and regulation of conjugation in *Enterococcus faecalis*. *Proc. Natl. Acad. Sci.* 102, 18596–18601. doi: 10.1073/pnas.0506163102

Shokeen, S., Johnson, C. M., Greenfield, T. J., Manias, D. A., Dunny, G. M., and Weaver, K. E. (2010). Structural analysis of the Anti-Q-Qs interaction: RNA-mediated regulation of *E. faecalis* plasmid pCF10 conjugation. *Plasmid* 64, 26–35. doi: 10.1016/j.plasmid.2010.03.002

Showsh, S. A., and Andrews, R. E. (1992). Tetracycline enhances *Tn*916-mediated conjugal transfer. *Plasmid* 28, 213–224. doi: 10.1016/0147-619X(92)90053-D

Singh, P. K., and Meijer, W. J. J. (2014). Diverse regulatory circuits for transfer of conjugative elements. *FEMS Microbiol. Lett.* 358, 119–128. doi: 10.1111/1574-6968.12526

Singh, P. K., Ramachandran, G., Durán-Alcalde, L., Alonso, C., Wu, L. J., and Meijer, W. J. J. (2012). Inhibition of *Bacillus subtilis* natural competence by a native, conjugative plasmid-encoded ComK repressor protein. *Environ. Microbiol.* 14, 2812–2825. doi: 10.1111/j.1462-2920.2012.02819.x

Singh, P. K., Ramachandran, G., Ramos-Ruiz, R., Peiró-Pastor, R., Abia, D., Wu, L. J., et al. (2013). Mobility of the native *Bacillus subtilis* conjugative plasmid pLS20 is regulated by intercellular signaling. *PLoS Genet.* 9:e1003892. doi: 10.1371/journal.pgen.1003892

Stingl, K., and Koraimann, G. (2017). "Prokaryotic information games: how and when to take up and secrete DNA" in *Type IV Secretion in Gram-Negative and Gram-Positive Bacteria. Current topics in microbiology and immunology.* Vol. 413, eds. S. Backert and E. Grohmann (Cham: Springer), 61–92.

Sultan, I., Rahman, S., Jan, A. T., Siddiqui, M. T., Mondal, A. H., and Haq, Q. M. R. (2018). Antibiotics, resistome and resistance mechanisms: a bacterial perspective. *Front. Microbiol.* 9:2066. doi: 10.3389/fmicb.2018.02066

Tamminen, M., Virta, M., Fani, R., and Fondi, M. (2012). Large-scale analysis of plasmid relationships through gene-sharing networks. *Mol. Biol. Evol.* 29, 1225–1240. doi: 10.1093/molbev/msr292

Thoma, L., and Muth, G. (2016). Conjugative DNA-transfer in *Streptomyces*, a mycelial organism. *Plasmid* 87, 1–9. doi: 10.1016/j.plasmid.2016.09.004

Traore, D. A. K., Wisniewski, J. A., Flanigan, S. F., Conroy, P. J., Panjikar, S., Mok, Y. F., et al. (2018). Crystal structure of TcpK in complex with *oriT* DNA of the antibiotic resistance plasmid pCW3. *Nat. Commun.* 9:3732. doi: 10.1038/s41467-018-06096-2

Trokter, M., and Waksman, G. (2018). Translocation through the conjugative Type 4 secretion system requires unfolding of its protein substrate. *J. Bacteriol.* 200. pii: JB.00615-617. doi: 10.1128/JB.00615-17

Van der Lelie, D., Wosten, H. A. B., Bron, S., Oskam, L., and Venema, G. (1990). Conjugal mobilization of streptococcal plasmid pMV158 between strains of *Lactococcus lactis* subsp. *lactis*. *J. Bacteriol.* 172, 47–52. doi: 10.1128/jb.172.1.47-52.1990

Wisniewski, J. A., and Rood, J. I. (2017). The Tcp conjugation system of *Clostridium perfringens*. *Plasmid* 91, 28–36. doi: 10.1016/j.plasmid.2017.03.001

Wozniak, R. A. F., and Waldor, M. K. (2010). Integrative and conjugative elements: mosaic mobile genetic elements enabling dynamic lateral gene flow. *Nat. Rev. Microbiol.* 8, 552–563. doi: 10.1038/nrmicro2382

Zeytuni, N., and Zarivach, R. (2012). Structural and functional discussion of the tetra-trico-peptide repeat, a protein interaction module. *Structure* 20, 397–405. doi: 10.1016/J.STR.2012.01.006

10

Biological Functions of the Secretome of *Neisseria meningitidis*

Jan Tommassen * and Jesús Arenas

Department of Molecular Microbiology and Institute of Biomembranes, Utrecht University, Utrecht, Netherlands

***Correspondence:**
Jan Tommassen
j.p.m.tommassen@uu.nl

Neisseria meningitidis is a Gram-negative bacterial pathogen that normally resides as a commensal in the human nasopharynx but occasionally causes disease with high mortality and morbidity. To interact with its environment, it transports many proteins across the outer membrane to the bacterial cell surface and into the extracellular medium for which it deploys the common and well-characterized autotransporter, two-partner and type I secretion mechanisms, as well as a recently discovered pathway for the surface exposure of lipoproteins. The surface-exposed and secreted proteins serve roles in host-pathogen interactions, including adhesion to host cells and extracellular matrix proteins, evasion of nutritional immunity imposed by iron-binding proteins of the host, prevention of complement activation, neutralization of antimicrobial peptides, degradation of immunoglobulins, and permeabilization of epithelial layers. Furthermore, they have roles in interbacterial interactions, including the formation and dispersal of biofilms and the suppression of the growth of bacteria competing for the same niche. Here, we will review the protein secretion systems of *N. meningitidis* and focus on the functions of the secreted proteins.

Keywords: *Neisseria meningitidis*, secretome, autotransporters, two-partner secretion system, host-pathogen interactions, immune evasion, biofilms

INTRODUCTION

The gram-negative diplococcus *Neisseria meningitidis* is a commensal bacterium residing in the upper respiratory tract of humans. It is transmitted through aerosols and can asymptomatically be present in up to 30% of the population for long periods exceeding 2 years. It is a close relative of another commensal of the nasopharynx, *Neisseria lactamica*, and of the pathogen *Neisseria gonorrhoeae*, which causes infections of the urogenital system. Incidentally, also *N. meningitidis* is pathogenic when it crosses the epithelial barriers of the nasopharynx to reach the bloodstream causing septicemia. From there, it can cross the blood-brain barrier causing meningitis (Pace and Pollard, 2012; Takada et al., 2016). Meningococcal disease has an incidence rate that ranges from <1 to 1,000 cases per 100,000 per year with large geographical and temporal differences (Rouphael and Stephens, 2012). It is lethal in 10% of the cases and causes severe sequelae in 30–50% of the survivors, including neurologic disabilities, seizures, hearing or visual loss, or cognitive impairment (Pace and Pollard, 2012). The high morbidity and mortality of the disease has urged in the past decades the development of suitable vaccines.

Based on the structure of the capsular polysaccharide, *N. meningitidis* is divided into 13 serogroups, five of which (A, B, C, W, and Y) are responsible for the majority of meningococcal disease. Vaccines have been developed based on the capsular polysaccharides of serogroups A, C, W, and Y (Crum-Cianflone and Sullivan, 2016). However, the capsule of serogroup B is not immunogenic and, therefore, a capsule-based vaccine for this serogroup could not be developed. Vaccines based on outer-membrane vesicles (OMVs) have been designed against serogroup B strains, but their efficacy was restricted to determined geographic areas affected by certain clones (Bjune et al., 1991; Oster et al., 2005). In the quest for alternative vaccine candidates, new cell-surface-exposed and secreted proteins have been identified and extensively studied in recent years. These studies have resulted in two vaccine formulations, Bexsero® (4CMenB) and Trumenba® (MenB-FHbp), with an expected broader spectrum than OMV-based vaccines (Crum-Cianflone and Sullivan, 2016). In addition, they have provided extensive insights into the functions of the surface-exposed and secreted proteins and their role in the biology of *N. meningitidis*.

Although certain lineages of *N. meningitidis* are highly invasive and frequently associated with the disease, the bacteria have evolved to live in the host asymptomatically. After entering a host, the bacteria adhere to epithelial surfaces in the nasopharynx, where they form microcolonies (Sim et al., 2000). These structures resemble biofilms and help the bacteria to persist under adverse conditions. As a commensal, *N. meningitidis* has evolved different mechanisms to evade the host immune system and to compete with other members of the oral microbiome. It has to scavenge the environment for nutrients, which are restricted in this niche. In all these processes, various components of the secretome, i.e., the total of proteins that are secreted across the cell envelope to the cell surface or beyond, are implicated. It has been a decade since the last review of the secretome of *N. meningitidis* (van Ulsen and Tommassen, 2006). Since then, significant progress has been made in solving the functions of the secreted proteins. In this review, we first briefly describe the protein translocation systems that are relevant for the secretion of proteins in *N. meningitidis*. Then, we describe the current insights into the functions of the secreted proteins.

PROTEIN TRANSLOCATION SYSTEMS IN *N. MENINGITIDIS*

In Gram-negative bacteria, several conserved pathways have evolved for the translocation of proteins across the cell envelope into the extracellular milieu (Costa et al., 2015). Of these pathways, the autotransporter, two-partner secretion (TPS), and the type I secretion (T1S) systems are active in *N. meningitidis*. In addition, these bacteria use a recently discovered mechanism, which is dependent on a member of a protein family called Slam, for the cell-surface exposure of lipoproteins (Hooda et al., 2016). The T1S system (T1SS) mediates the secretion of substrates in one step from the bacterial cytoplasm across the two membranes into the milieu. In the other pathways, the substrates are first translocated across the inner membrane, after which the periplasmic intermediate is translocated across the outer membrane. To varying extent, these two-step pathways deploy general machinery for the localization of envelope proteins, i.e., the Sec and Tat systems for protein translocation across the inner membrane, the Lol system for shuttling lipoproteins across the periplasm, and the BAM system for the assembly of integral outer membrane proteins (OMPs). In this section, we briefly describe the protein transport systems that are relevant in *N. meningitidis* and emphasize what is known about these systems in this organism. It is noteworthy that fundamental research in *Neisseria* spp. has played a leading role in uncovering some of these systems, particularly the autotransporter mechanism, the BAM system, and the Slam-dependent pathway. Except for Sec and Tat, the other translocation systems have been studied at least to some extent in *N. meningitidis*. For more extensive descriptions of the pathways, we refer in each subsection to specialized recent reviews.

Transport and Assembly of Integral OMPs

Integral OMPs are not part of the secretome as they are not translocated across the entire cell envelope. Therefore, their functions are not discussed in this review. However, since their translocation and assembly machinery is deployed by protein secretion systems, this machinery will be described here.

The vast majority of integral OMPs consists of antiparallel amphipathic β-strands that form closed cylindrical structures, called β-barrels. They are synthesized in the cytoplasm as precursors with an N-terminal signal sequence, which marks them for transport across the inner membrane via the Sec system (von Heijne, 1990). The Sec system (for recent reviews, see Lycklama a Nijeholt and Driessen, 2012; Tsirigotaki et al., 2017) has, so far, not been studied in *N. meningitidis*, but homologs of all components have been identified in the available genome sequences (van Ulsen and Tommassen, 2006), and, therefore, it presumably functions similarly as in model organisms like *Escherichia coli*. In *E. coli*, the Sec system includes a chaperone, SecB, which prevents aggregation of client proteins in the cytoplasm and guides them to the inner membrane components of the system (**Figure 1A**). In the inner membrane, the heterotrimeric SecYEG complex forms the protein-translocating channel. Energy for transport is provided by the motor protein SecA, which hydrolyzes ATP, and by the proton-motive force, which is coupled to the translocation process presumably by the SecDF-YajC complex (Tsukazaki et al., 2011). After passage of the precursor through the Sec translocon, its signal sequence is removed by the signal peptidase LepB (Auclair et al., 2012).

All integral OMPs studied so far use the Sec system for translocation across the inner membrane. The channel in the SecYEG translocon is narrow and only allows for the translocation of completely unfolded proteins in a linear fashion. Some proteins have to be exported in a folded form, e.g., because they bind a co-factor in the cytoplasm. Such proteins share a consensus "twin-arginine" motif ([S/T]-R-R-x-F-L-K) in the N-terminal region of their signal sequences, and they use an alternative apparatus, the twin-arginine translocation (Tat) system, to cross the inner membrane (for recent reviews, see Palmer and Berks, 2012; Patel et al., 2014). The Tat system of *E. coli* consists of three proteins, TatA, TatB, and TatC, all of which are also present in *N. meningitidis* (van Ulsen and Tommassen, 2006).

Once OMPs are released from the Sec machinery, chaperones, such as SurA and Skp, bind the mature proteins to prevent their aggregation in the periplasm (**Figure 1A**). In *E. coli*, SurA appears

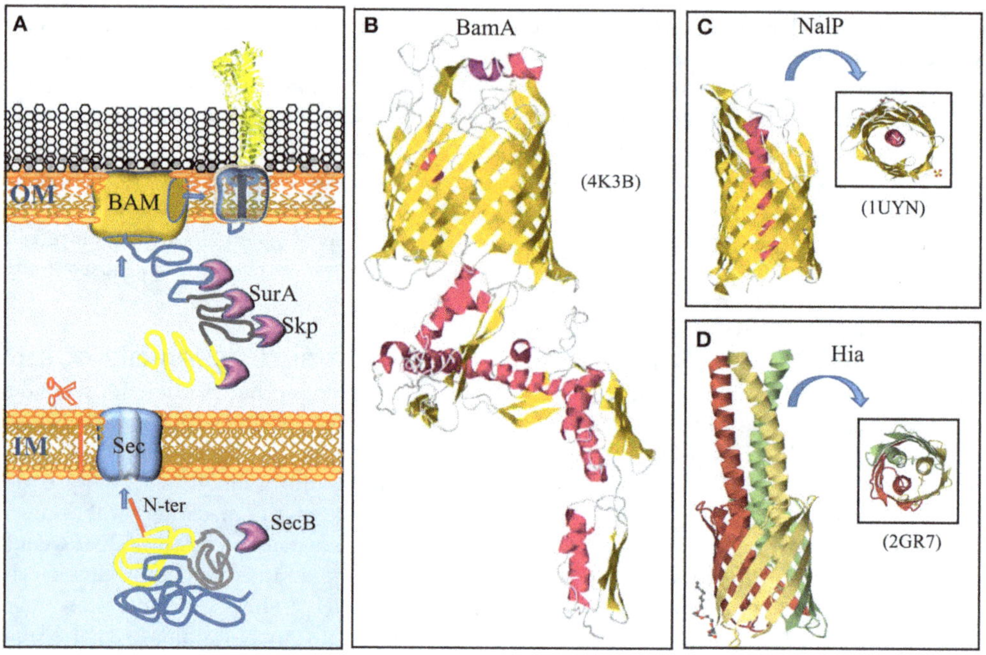

FIGURE 1 | Autotransporter secretion system. **(A)** Autotransporters use the same machinery to reach the outer membrane (OM) as integral OMPs. They are synthesized with a cleavable N-terminal signal sequence (red) for translocation across the inner membrane (IM) via the Sec translocon. After periplasmic transit, escorted by chaperones like Skp and SurA, the C-terminal β-domain (blue) is integrated as a β-barrel into the OM by the BAM complex. Presumably during this integration, the passenger domain (yellow) is transported to the cell surface. A segment that connects the passenger with the β-domain (gray) forms an α-helix that plugs the channel within the barrel. **(B)** Structure of the central component of the BAM system, BamA, from *N. gonorrhoeae* (Noinaj et al., 2013). BamA is constituted of a β-barrel in the outer membrane and five periplasmic POTRA domains. α-Helices and β-strands are shown in red and yellow, respectively. **(C)** Structure of the β-domain of the autotransporter NalP with the α-helix that plugs the β-barrel and exposes the passenger, if connected, at the cell surface, indicated in red (Oomen et al., 2004). **(D)** Structure of the β-domains of the trimeric autotransporter Hia of *H. influenzae* (Meng et al., 2006). Each subunit is indicated with a different color. In panels **(C,D)**, a top view is shown in the insets. The structural data were retrieved from the Protein Data Bank (PDB) and access codes are provided in brackets. References are provided in the text.

to be the main periplasmic chaperone for OMPs as its depletion drastically reduces the OMP content, whereas depletion of Skp had only minor effects (Sklar et al., 2007). In sharp contrast, mutational analysis indicated that SurA has no appreciable role in OMP biogenesis in *N. meningitidis*, whereas deletion of *skp* affected the major OMPs, i.e., the porins (Volokhina et al., 2011). Also the protease DegP has been suggested to have a role as a periplasmic chaperone in OMP biogenesis (Sklar et al., 2007; Krojer et al., 2008). However, its primary role in this process presumably is the degradation of toxic and membrane-damaging misfolded OMPs in the periplasm (Ge et al., 2014). Inactivation of DegQ, the DegP homolog of *N. meningitidis*, did not have any appreciable effect on OMP biogenesis (Volokhina et al., 2011).

After transit of the periplasm, OMPs are assembled into the outer membrane by the β-barrel assembly machinery (BAM) (**Figure 1A**) (for a recent review, see Noinaj et al., 2017). The central component of the BAM was originally discovered by Voulhoux et al. (2003) in *N. meningitidis*. This protein, previously known as Omp85 and now called BamA, is essential for bacterial viability and homologs are found in all Gram-negative bacteria and even in mitochondria, where it performs a similar function (Walther et al., 2009). The first complete BamA crystal structure solved was that of *N. gonorrhoeae* (Noinaj et al., 2013). The protein consists of a C-terminal 16-stranded β-barrel, which is inserted into the outer membrane, and five repeated polypeptide transport-associated (POTRA) domains extending into the periplasm (**Figure 1B**). In *E. coli*, BamA forms a complex with four lipoproteins, BamB, C, D, and E. These lipoproteins are less conserved, and only BamD is essential (Ricci and Silhavy, 2012; Noinaj et al., 2017). The BAM of *N. meningitidis* lacks the BamB component, but it contains the peptidoglycan-associated OMP RmpM, which stabilizes the complex (Volokhina et al., 2009). Also in *N. meningitidis*, BamD is essential (Volokhina et al., 2009), but a *bamD* mutant (originally described as *comL* mutant) of *N. gonorrhoeae* appeared to be viable (Fussenegger et al., 1996).

Autotransporter Secretion

The first autotransporter ever described is the IgA protease of *N. gonorrhoeae* (Pohlner et al., 1987). It is a classical autotransporter that consists of an N-terminal signal sequence for transport across the inner membrane via the Sec system, a secreted passenger domain, and a C-terminal β-domain. The β-domain was suggested to insert as a β-barrel into the outer membrane to form a pore through which the passenger domain is secreted without assistance of any other proteins, hence the name autotransporter. The first crystal structure of such a β-domain that was solved, i.e., that of the autotransporter NalP of *N. meningitidis*, seemed to confirm this idea (Oomen et al., 2004). This structure revealed a 12-stranded β-barrel with an internal hydrophilic pore filled by an N-terminal α-helix that would expose the passenger, if connected, at the cell surface (**Figure 1C**). However, it was also noticed that the internal pore was very narrow, probably too narrow to allow for the passage of a linear polypeptide that contains even small structural elements such as oligopeptide loops formed by disulfide bonds.

It is clear now that the name autotransporter is an oversimplification of the secretion pathway that is used. Autotransporters largely utilize for their secretion the machinery that is used by OMPs (**Figure 1A**) (for a recent review, see Grijpstra et al., 2013). First, they are translocated across the inner membrane by the Sec machinery. In the periplasm, they are escorted by chaperones, such as SurA, Skp, and DegP (Purdy et al., 2007; Ieva and Bernstein, 2009; Ruiz-Perez et al., 2009), although inactivation of the corresponding genes had no noticeable effect on autotransporter biogenesis in *N. meningitidis* (Volokhina et al., 2011). Also the BAM complex in the outer membrane is required for autotransporter biogenesis (Voulhoux et al., 2003), presumably for the insertion of the β-domain but possibly also for the translocation of the passenger domain, as a passenger stalled in the translocation process could be cross-linked to BamA (Ieva and Bernstein, 2009). It was proposed that the β-domain of the autotransporter and the β-barrel of BamA form a hybrid barrel with an internal diameter that is wide enough to allow for the translocation of the passenger, even if it contains small folded domains (Ieva et al., 2011). Protein folding at the cell surface probably provides the energy source to drive translocation (Drobnak et al., 2015). After outer membrane translocation, the passenger domain can remain attached to the β-domain, or it can be released into the extracellular milieu by one of several proteolytic mechanisms (Grijpstra et al., 2013). For the mechanisms relevant in *N. meningitidis*, we refer to the description of the individual autotransporters below. Besides the BAM complex, also the TAM complex may play a role in the secretion of a subset of autotransporters (Selkrig et al., 2012). The TAM complex consists of an OMP, TamA, which is a homolog of BamA, and an inner membrane protein TamB, which spans the periplasm and interacts with TamA. The precise role of the TAM complex is unclear. Reconstitution experiments showed that the BAM complex and chaperone SurA were necessary and sufficient to mediate the translocation of an autotransporter into proteoliposomes (Roman-Hernandez et al., 2014). Perhaps, TAM can substitute for BAM for some autotransporters. The role of TAM in autotransporter secretion has not been investigated in *N. meningitidis*.

Among autotransporters, four subcategories can be distinguished (Grijpstra et al., 2013). Two of them are present in *N. meningitidis*, the classical autotransporters, which include IgA protease and NalP, and the trimeric autotransporters. In trimeric autotransporters, the β-domains of each subunit contribute four β-strands to form a similar 12-stranded β-barrel as does the β-domain of the classical autotransporters (**Figure 1D**) (Meng et al., 2006). The structure of the passenger domain, however, is entirely different, as will be discussed further below.

Two-Partner Secretion (TPS) System

The TPS system facilitates the secretion of very large β-helical proteins with often virulence-related functions (for a recent review, see Jacob-Dubuisson et al., 2013). The two partners are the secreted protein, which is generically called TpsA, and a dedicated transporter in the outer membrane called TpsB (**Figure 2A**). TpsB is a homolog of BamA. From the C to the N terminus, it consists of a 16-stranded β-barrel, two POTRA domains at the periplasmic side, and an α-helix that plugs the channel in the barrel (**Figure 2B**) (Clantin et al., 2007). TpsA is transported across the inner membrane by the Sec system and escorted in the periplasm by chaperones (Jacob-Dubuisson et al., 2013). The mature TpsA contains a conserved TPS domain near the N terminus that is recognized by the POTRA domains of TpsB (**Figure 2B**). This interaction leads to removal of the plug helix and further conformational changes in TpsB (Maier et al., 2015), allowing for the transport of TpsA to the cell surface via the barrel of TpsB. *N. meningitidis* strains generally contain at least one conserved locus with *tps* genes, but genome sequence analysis indicated that various strains can contain up to five different *tpsA* genes and two *tpsB* genes (van Ulsen and Tommassen, 2006; van Ulsen et al., 2008). One of the TpsB proteins was shown to have broad substrate specificity, whereas the other one was specific for the TpsA encoded by the same operon (ur Rahman and van Ulsen, 2013). Substrate specificity was shown to be determined by the POTRA domains (ur Rahman et al., 2014).

Transport of Lipoproteins to the Outer Membrane

Besides integral OMPs, the outer membrane contains lipoproteins, which are peripherally attached to the membrane via an N-terminal lipid moiety. Lipoproteins are synthesized as precursors with an N-terminal signal sequence for export via the Sec or Tat machinery. Their signal sequences contain at the C terminus a conserved motif, [LVI][ASTVI][GAS]C, called lipobox, where the conserved cysteine represents the first residue of the mature protein (Hayashi and Wu, 1990). After translocation across the inner membrane, this cysteine is modified (for a recent review, see Buddelmeijer, 2015). First, a diacylglycerol moiety derived from phosphatidylglycerol is covalently attached to the sulfhydryl group of the cysteine by the enzyme lipoprotein diacylglyceryl transferase (Lgt) (**Figure 3**). Then, the signal sequence is cleaved by the dedicated lipoprotein signal peptidase LspA, after which the α-amino group of the cysteine is acylated by the enzyme apolipoprotein N-acyltransferase (Lnt) (**Figure 3**). In *E. coli*, these three enzymes are essential. However, the *lnt* gene could be inactivated in *N. meningitidis* and *N. gonorrhoeae* (LoVullo et al., 2015; da Silva et al., 2016). As the outer membrane contains essential lipoproteins, such as BamD (see above), this implies that these *Neisseria* spp., in contrast to *E. coli*, can transport diacylated lipoproteins to the outer membrane, which was indeed demonstrated in *N. meningitidis* (da Silva et al., 2016).

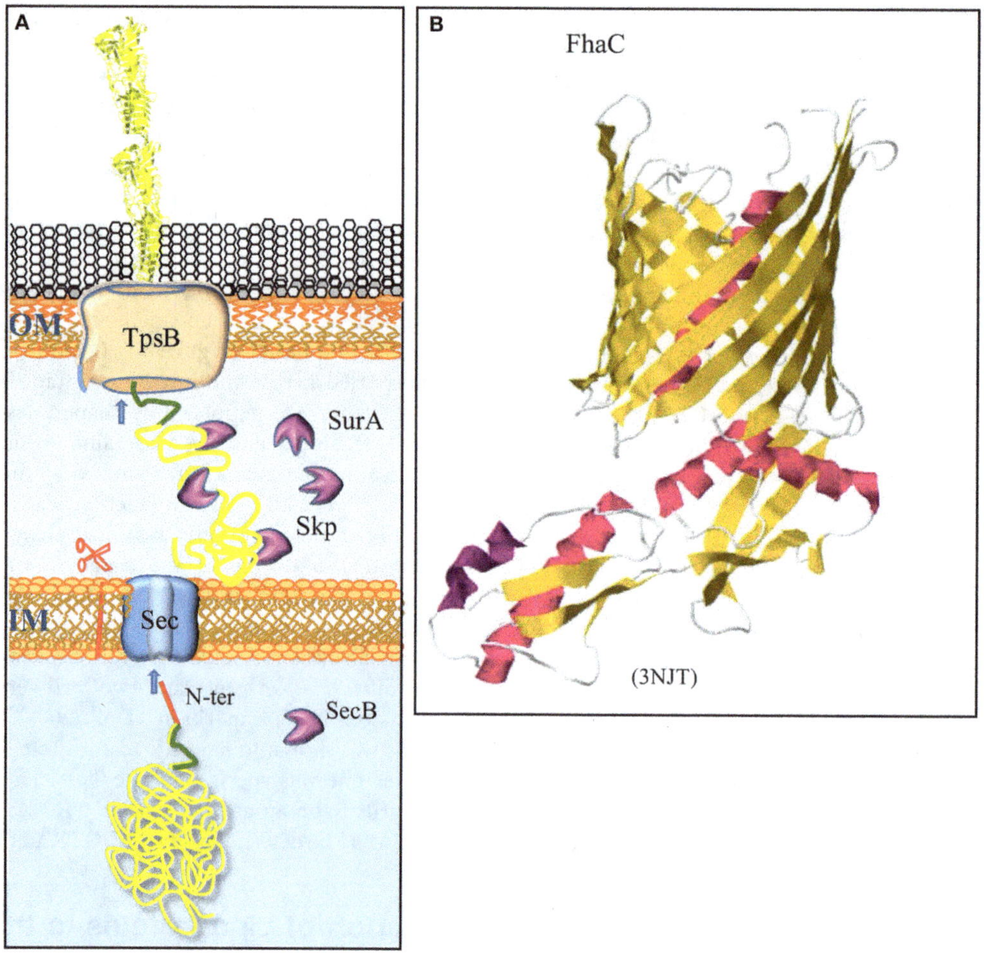

FIGURE 2 | The two-partner secretion system. **(A)** The TpsA substrates contain a cleavable signal sequence (red) for Sec-mediated translocation across the inner membrane (IM). After chaperone-escorted transit of the periplasm, the N-terminal TPS domain (green) interacts with the integral OMP TpsB, which mediates transport across the outer membrane (OM). **(B)** Structure of the TpsB FhaC of *B. pertussis* (Clantin et al., 2007). TpsB consists of a β-barrel domain embedded in the OM and two periplasmic POTRA domains that interact with the TPS domain of substrates. The first POTRA domain is connected via a linker with an N-terminal α-helix that plugs the β-barrel in the resting state. The structure of the R450A mutant of FhaC is depicted to show the plugging of the channel. β-Strands and α-helices are colored yellow and red, respectively. The PDB access code is provided in brackets.

After their maturation, lipoproteins can remain attached to the inner membrane or be transported to the outer membrane. In *E. coli*, the destination of the lipoprotein is determined by the identity of the amino-acid residue directly adjacent to the lipidated N-terminal cysteine (Yamaguchi et al., 1988). Basically, an aspartate residue at the +2 position functions as an inner-membrane-retention signal, whilst lipoproteins destined to outer membrane have a different residue at this position. However, this signal appears to be species specific as other sorting rules were reported for lipoproteins in other bacteria (Schulze and Zückert, 2006; Narita and Tokuda, 2007). The sorting rules for lipoproteins in *N. meningitidis* have not been determined, but the simple +2 rule of *E. coli* does not seem to apply as a DsbA lipoprotein with a serine at the +2 position was experimentally demonstrated to localize to the inner membrane (Tinsley et al., 2004).

Lipoproteins are targeted to the periplasmic side of the OM by the Lol system. This system consists of an ATP-binding cassette (ABC) transporter LolCDE in the inner membrane, a periplasmic chaperone LolA, and LolB, which functions as a receptor in the outer membrane (**Figure 3**) (for a recent review, see Narita and Tokuda, 2016). The LolCDE complex consists of a heterodimer of two homologous integral membrane proteins, LolC and LolE, which is associated with a dimer of the cytoplasmic ABC protein LolD. This complex selects the substrate lipoproteins that don't carry an inner-membrane retention signal and provides the energy to release them from the inner membrane by ATP hydrolysis (Yakushi et al., 2000). Interestingly, *N. meningitidis* and *N. gonorrhoeae*, as well as many other Gram-negative bacteria, each contain only a single protein that is homologous to LolC and LolE and that has specific characteristics of both of them (**Figure 3**) (LoVullo et al., 2015). This protein, which

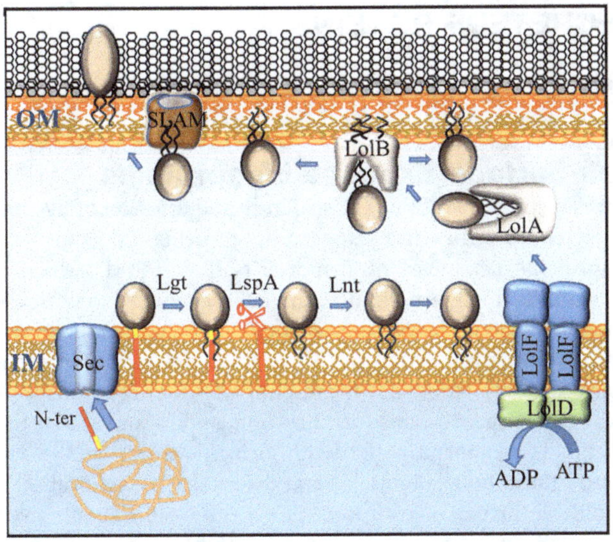

FIGURE 3 | Maturation, transport and secretion of lipoproteins. Lipoproteins contain a lipobox (yellow) at the C terminus of their signal sequence (red). After transport across the inner membrane (IM) by the Sec system, the lipoprotein precursor is modified by the addition of a diacylglyceryl moiety to the sulfhydryl group of the cysteine in the lipobox by Lgt, followed by cleavage of the signal sequence by LspA, and the addition of a third, amide-linked fatty acyl chain by Lnt. Lipoproteins that don't contain an IM-retention signal (also called Lol-avoidence motif) are expelled from the IM by an ABC transporter that, in *Neisseria*, consists of a dimer of LolF (instead of the LolCE heterodimer found in *E. coli*) and a dimer of LolD. The chaperone LolA shields the lipid moiety of the lipoprotein during periplasmic transit and shuttles it to the receptor LolB that inserts it into the inner leaflet of the OM. Some lipoproteins are translocated to the outer face of the OM via a mechanism that requires a SLAM protein. Whether SLAM forms the actual translocation channel itself is not known.

is called LolF, presumably forms a homodimer in the inner membrane, and it was suggested that the presence of a LolF instead of LolCE is related to the capacity of these bacteria to transport diacylated lipoproteins, i.e., lipoproteins not modified by Lnt, to the outer membrane (LoVullo et al., 2015). After release from the inner membrane, the lipoprotein is captured by the chaperone LolA, which shields its hydrophobic lipid moiety from the aqueous environment of the periplasm, thus forming a water-soluble complex (Matsuyama et al., 1995). From LolA, the lipoprotein is transferred to LolB, which is a structural homolog of LolA. LolB is a lipoprotein itself and it is attached with its lipid moiety in the OM (**Figure 3**). LolB then mediates the insertion of the lipoprotein into the inner leaflet of the OM (Matsuyama et al., 1997).

Transport of Lipoproteins to the Cell Surface

How lipoproteins are translocated across the outer membrane to reach the cell surface has only been studied in few exceptional cases but is an emerging field of research (for a recent review, see Wilson and Bernstein, 2017). In the best studied cases, the lipoproteins utilize conserved protein secretion systems to reach the cell surface, such as the type 2 secretion system (T2SS) (for a review, see Nivaskumar and Francetic, 2014), a system that is related to the machinery required for the biogenesis of type IV pili but is not operational as a protein secretion system in *N. meningitidis*. In this case, the lipoproteins contain an inner membrane retention signal to avoid the Lol system after their transport across the inner membrane via Sec or Tat (Pugsley and Kornacker, 1991; Putker et al., 2013). These lipoproteins are directly secreted from the periplasmic side of the inner membrane to the cell surface using the T2S machinery. Other lipoproteins, such as NalP of *N. meningitidis* (van Ulsen et al., 2003), are autotransporters. Before translocation across the outer membrane, these autotransporter lipoproteins presumably use the Lol system to reach the inner face of the outer membrane, although this has experimentally not yet been proven. Recently, a new general pathway dedicated to the transport of lipoproteins to the cell surface was uncovered in *N. meningitidis* (Hooda et al., 2016). This pathway is an extension of the Lol pathway as the substrates appear to use the Lol system to reach the outer membrane. A family of proteins named surface lipoprotein assembly modulator (Slam) was found to be involved in the subsequent translocation across the outer membrane (**Figure 3**). Slam proteins consist of an N-terminal periplasmic domain containing two tetratricopeptide repeats and a C-terminal 14-stranded β-barrel embedded in the outer membrane. There is no homolog of Slam proteins in *E. coli*, and the synthesis of cell-surface-exposed lipoproteins from *N. meningitidis* in *E. coli* does not lead to their cell-surface exposition. However, when Slam was co-expressed, these lipoproteins were transported to the *E. coli* cell surface (Hooda et al., 2016). Whether Slam forms the actual translocation channel for the lipoproteins or guides them to an alternative translocation apparatus, such as BAM, remains to be investigated. *N. meningitidis* has two Slam proteins with different substrate specificity. Slam1 has broad specificity and is involved in the translocation of at least the transferrin-binding protein B (TbpB), the lactoferrin-binding protein B (LbpB), and the factor H-binding protein (fHbp). Slam2 appears to be specific for hemoglobin-haptoglobin utilization protein A (HpuA). Interestingly, these lipoproteins, as well as another cell-surface-exposed lipoprotein, i.e., neisserial heparin-binding antigen (NHBA), show considerable structural similarity (see below), which may reflect a prerequisite for Slam-dependent transport and/or a common evolutionary origin. Slam homologs are widely distributed among proteobacteria (Hooda et al., 2016), suggesting that they represent a new general pathway dedicated to the secretion of lipoproteins. It is noteworthy, however, that TbpB expressed in *N. gonorrhoeae* without its lipid moiety was secreted into the culture supernatant (Ostberg et al., 2013), suggesting that the Slam-dependent pathway may have broader specificity than just lipoproteins.

Type 1 Secretion System (T1SS)

The T1SS is composed of three proteins: an ABC protein in the inner membrane, a TolC-type channel protein in the outer membrane, and a membrane-fusion protein that connects the integral inner and outer membrane components (**Figure 4A**) (for a recent review, see Lenders et al., 2013). The outer membrane component is trimeric protein that forms a 12-stranded β-barrel

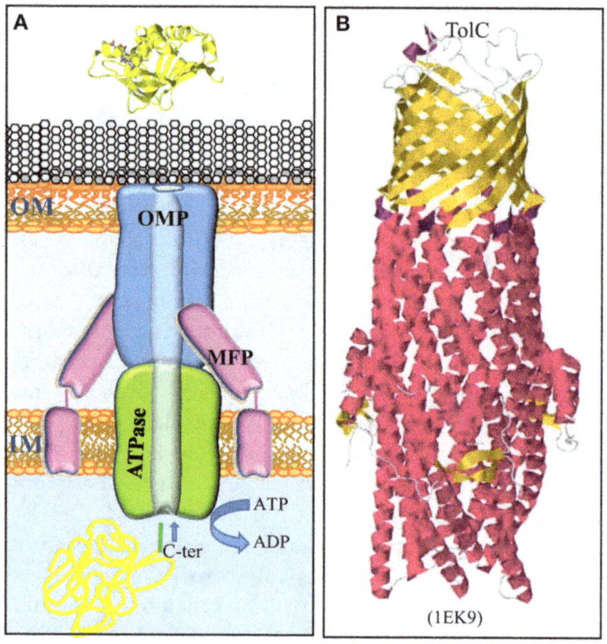

FIGURE 4 | Type 1 secretion system. **(A)** The T1SS consists of three proteins, a trimeric OMP (blue), an ABC transporter in the inner membrane (IM) (green), and an IM-anchored membrane-fusion protein (MFP) (pink), which together form a translocation channel for substrates directly from the bacterial cytoplasm to the external milieu. Substrates contain a secretion signal at their C terminus (green) for recognition by the machinery. **(B)** Structure of the homotrimeric OM T1SS component of *E. coli*, TolC (Koronakis et al., 2000). The trimer consists of a β-barrel in the OM and long α-helices, which form a tunnel that deeply penetrates into the periplasm. β-Strands and α-helices are colored yellow and red, respectively. The PDB access code is given in brackets.

in the outer membrane to which each protomer contributes four β-strands (**Figure 4B**). At the periplasmic side, the β-strands are connected by long α-helices that form together a long hollow conduit that spans the periplasm and reaches the ABC transporter in the inner membrane (Koronakis et al., 2000). This machinery mediates the secretion of substrates directly from the cytoplasm into the extracellular milieu without a periplasmic intermediate (**Figure 4A**). Thus, these substrates do not contain an N-terminal signal sequence for recognition by the Sec or Tat systems. Instead, they contain a recognition signal at the C terminus to interact with the ABC component of the T1SS. Many T1SS substrates belong to the RTX protein family, where RTX stands for repeats in toxin (Linhartová et al., 2010). These proteins contain glycine-rich repeats with the consensus sequence GGxGxDxxx (where x is any amino acid) near the secretion signal. These repeats bind calcium ions, which promotes folding in the extracellular medium where, in contrast to the cytoplasm, the concentration of calcium is high. Thus, besides ATP hydrolysis and the proton-motive force, extracellular folding might provide a driving force for translocation (Lenders et al., 2013). The components constituting the T1SS of *N. meningitidis* have been identified (Wooldridge et al., 2005).

STRUCTURE AND FUNCTION OF SECRETED PROTEINS

Table 1 summarizes the functions of the secreted proteins that are discussed in detail below.

Cell-Surface-Exposed Lipoproteins

In this section, we discuss the structure and function of five well-characterized cell-surface-exposed lipoproteins. Two other such lipoproteins, i.e., the autotransporter NalP and FrpD, which has a role in the T1SS, are discussed in subsequent sections. Several other lipoproteins have been suggested in the literature to be exposed at the cell surface in *N. meningitidis*. However, even if their function is known, it is not clear why they are transported to the cell surface to exert such function, and in some cases, their function is even incompatible with surface localization. We don't discuss these lipoproteins here and feel that their localization should be further investigated. For a discussion about caveats and pitfalls in assessing lipoprotein localization, see Wilson and Bernstein (2017).

Lipoproteins Involved in Iron Acquisition: TbpB, LbpB, and HpuA

One of the primary defense mechanisms of a host against invading bacterial pathogens is to deprive them from essential nutrients, such as iron, a defense mechanism that is known as nutritional immunity. In the human host, iron is bound by proteins, such as transferrin in serum, lactoferrin in secretions and mucosal surfaces, and hemoglobin and ferritin within cells. These proteins have very high affinities for iron, and, consequently, the concentration of free iron in the human fluids is too low to support microbial growth (Weinberg, 2009). Many bacteria produce siderophores, which are low-molecular-weight iron chelators that sequester otherwise inaccessible ferric iron from the environment. However, *Neisseria* spp. do not secrete siderophores; instead, they synthesize receptors that directly hijack iron sequestered in the iron-binding proteins of the host (Schryvers and Stojiljkovic, 1999). Several of these receptors consist of two proteins, a surface-exposed lipoprotein and an integral OMP. The integral OMP belongs to the TonB-dependent family (Tdf) of receptors, which includes also siderophore receptors. These receptors form 22-stranded β-barrels that are closed by an N-terminal plug. They interact with the TonB complex in the inner membrane, which deploys the proton-motive force to energize transport across the outer membrane (for a review, see Noinaj et al., 2010).

The transferrin receptor in *Neisseria* spp. consists of the Tdf member TbpA and the surface-exposed lipoprotein TbpB. The synthesis of these proteins is induced under iron limitation. In *N. meningitidis*, both proteins are essential for the acquisition of iron from human transferrin (Irwin et al., 1993), whilst TbpB is dispensable in *N. gonorrhoeae*, although its presence considerably enhances the efficiency of the process (Anderson et al., 1994). Both proteins can bind transferrin, but, in contrast to TbpA, TbpB preferably binds the iron-loaded form (Cornelissen and Sparling, 1996; Boulton et al., 1998). Thus, TbpB may enhance iron acquisition by selecting holo-transferrin to bind to the

TABLE 1 | Secretion mechanism and functions of secreted proteins.

Protein	Secretion system[a,b]	Biological functions[c]
TbpB	Slam	Iron acquisition from transferrin
LbpB	Slam	Iron acquisition from lactoferrin
		Protection against lactoferricin
HpuA	Slam	Heme iron acquisition from hemoglobin(-haptoglobin)
fHbp	Slam	Prevention of complement activation by binding factor H
		Protection against host defense peptide LL-37
NHBA	Slam?	Serum resistance by binding heparin
		Adhesion by binding heparan sulfate proteoglycans
		Initiation of biofilm formation
		Increase of endothelial permeability
IgA protease	Classical AT	Cleavage of IgA and other host proteins
		Immune modulation by cleavage transcription factor NF-κB
		Initiation of biofilm formation
		Binding heparin
NalP	Classical AT	Proteolytic release of bacterial cell-surface-exposed proteins
		Cleavage of complement factor C3
App, AusI	Classical AT	Binding of histones and cleavage of histone H3
		Induction of apoptosis
		Adhesion
AutA	Classical AT	Autoaggregation
		Stimulation of $CD4^{+}$ T-cells and B-cells
AutB	Classical AT	Biofilm formation
NadA	Trimeric AT	Adhesion and invasion
		Binding Hsp90
NhhA	Trimeric AT	Adhesion
		Binding laminin and heparan sulfate
		Serum resistance by binding vitronectin
		Apoptosis of macrophages
		Immune modulation by affecting differentiation monocytes
TpsA	TPS	Interbacterial competition
		Intracellular survival
		Adhesion
		Biofilm formation
FrpC	T1SS	Adhesion
FrpD	Slam?	Adhesion
MafB	Unknown	Interbacterial competition
TspB	Unknown	Binding immunoglobulins
		Biofilm formation

[a]*Question marks indicate that suggested secretion mechanism is not demonstrated.*
[b]*AT, autotransporter.*
[c]*For references, see text.*

receptor. In addition, TbpB's affinity for the holo form may help in the release of the apo form from the receptor once iron is delivered. TbpB is a bilobed protein probably resulting from gene duplication. Each lobe consists of a β-barrel domain and an adjacent handle domain, which is also rich in β-sheet structure (**Figure 5**) (Calmettes et al., 2012; Noinaj et al., 2012). Human transferrin is also a bilobed protein with a Fe^{3+}-binding site within each lobe (Hall et al., 2002). Structural analysis revealed that TbpA and TbpB both bind the C-lobe of transferrin using non-overlapping binding sites (Noinaj et al., 2012). The C-lobe of transferrin consists of two subdomains, C1 and C2, which are connected by a hinge at the base of a deep cleft that contains the Fe^{3+}-binding site. TbpB binds both the C1 and C2 subdomains of holo-transferrin via the handle and β-barrel domains in its N-lobe, respectively (Calmettes et al., 2012). This explains the selectivity of TbpB for holo-transferrin as the apo-form shows a 51° rotation of the C2 domain which would drastically reduce the TbpB-transferrin interface (Calmettes et al., 2012). In the available structures, the C-lobe of TbpB is not involved in transferrin binding, and its function is unclear. An α-helix from extracellular loop 3 of TbpA intrudes into the cleft between the C1 and C2 subdomains of transferrin, which induces a partial opening of the cleft and destabilizes the coordination site of the ferric ion (Noinaj et al., 2012), which facilitates its release. Released iron would enter a closed chamber formed by TbpA, TbpB, and transferrin, which prevents its diffusion into the medium and positions it for further transport through the barrel of TbpA (Noinaj et al., 2012).

The lactoferrin receptor is constituted by the Tdf member LbpA and the surface-exposed lipoprotein LbpB (Pettersson et al., 1994b, 1998; Prinz et al., 1999). Whilst the corresponding genes are ubiquitous in *N. meningitidis*, they are disrupted in about half of the *N. gonorrhoeae* isolates (Anderson et al., 2003). Nevertheless, expression of a functional lactoferrin receptor in *N. gonorrhoeae* provided a competitive advantage in a human infection model, and it was essential for infection by a strain lacking the transferrin receptor (Anderson et al., 2003). LbpA and LbpB both bind lactoferrin, but LbpB is not essential for its utilization as an iron source in both species (Pettersson et al., 1998; Biswas et al., 1999). Whilst lactoferrin and transferrin are homologous proteins, also LbpA and LbpB share extensive sequence similarity with TbpA and TbpB, respectively (Pettersson et al., 1994a, 1998). Crystallization of the N lobe of LbpB revealed a structure very similar to that of the corresponding part of TbpB. It also consists of two domains, a handle domain and a β-barrel domain with long loops projecting from both domains and forming an extended cap region (**Figure 5**) (Brooks et al., 2014). *In silico* docking experiments indicated that the cap region constitutes the binding site to which lactoferrin binds with its N-lobe (Brooks et al., 2014). However, recent biochemical analysis revealed that it preferentially binds the C-lobe of iron-loaded lactoferrin (Ostan et al., 2017), just like TbpB binds the C-lobe of iron-loaded transferrin (see above), suggesting a similar role for LbpB as for TbpB in selecting iron-loaded ligands. LbpB appears to have an additional role in protecting the bacteria against the bactericidal activity of lactoferricin (Morgenthau et al., 2012).

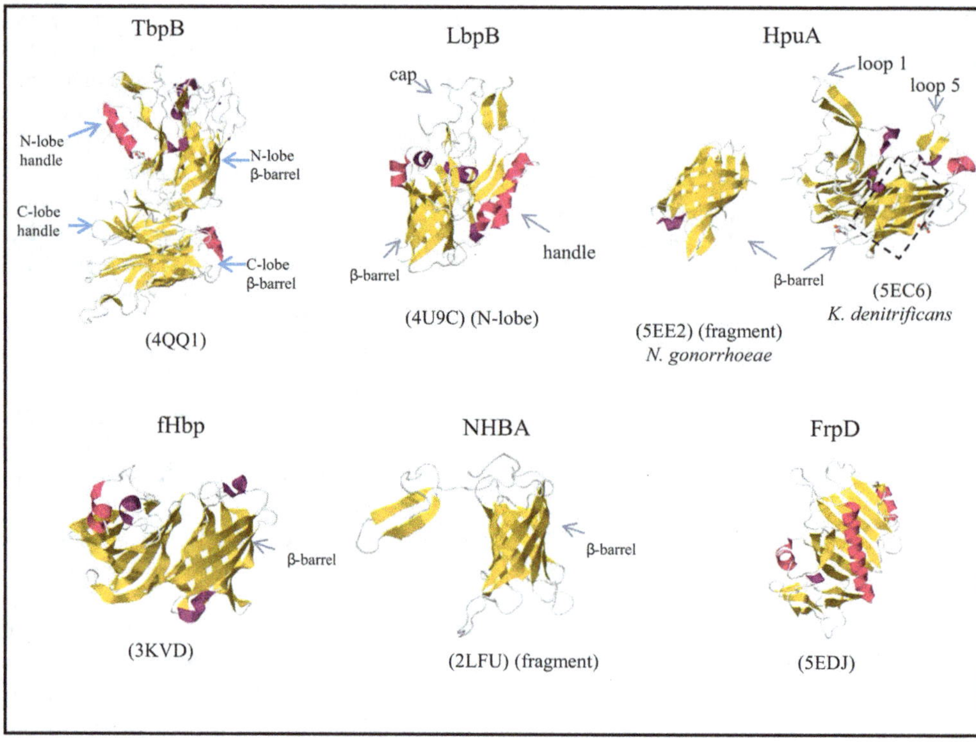

FIGURE 5 | Structures of surface-exposed lipoproteins in *N. meningitidis*. Where indicated, the structure of only a fragment of the complete protein is available. The structure of *N. meningitidis* HpuA has not been solved; instead, those of *K. denitrificans* HpuA and a C-terminal fragment of *N. gonorrhoeae* HpuA are shown. β-Strands and α-helices are colored yellow and red, respectively, and relevant domains of each protein are indicated. The PDB access codes are given in brackets and references are provided in the text.

Lactoferricin is a small cationic peptide, which is released from the N terminus of lactoferrin by proteolysis (Bellamy et al., 1992; Gifford et al., 2005). LbpB contains two highly variable stretches rich in negatively charged amino-acid residues in its C-lobe (Pettersson et al., 1999) that have been shown to mediate protection against this cationic peptide (Morgenthau et al., 2014). Thus, binding of the N-lobe of lactoferrin to the C-lobe of LbpB may prevent proteolysis events that generate lactoferricin. In addition or alternatively, the C-lobe may also be able to bind the free peptide, thus neutralizing its toxic effects (Ostan et al., 2017).

N. meningitidis can express two receptors for hemoglobin, HmbR and HpuAB, which enable it to extract heme from hemoglobin and use it as an iron source (Stojiljkovic et al., 1995; Lewis et al., 1997). Of these, the HpuAB system is also capable of extracting heme from hemoglobin-haptoglobin complexes (Lewis et al., 1997). Expression of the corresponding genes is controlled by iron availability and is prone to phase variation by slipped-strand mispairing (Lewis et al., 1999). HpuAB is a bipartite receptor, in which HpuB is a Tdf member and HpuA is a surface-exposed lipoprotein. In the absence of HpuA, HpuB still binds its ligands but with low affinity, and they cannot be used as an iron source (Rohde et al., 2002; Rohde and Dyer, 2004). Ligands did not bind to cells expressing only HpuA (Rohde and Dyer, 2004), but a direct interaction could be demonstrated in pull-down assays with purified HpuA (Wong et al., 2015). The crystal structure was solved of *Kingella denitrificans* HpuA alone and in complex with hemoglobin, and of a C-terminal fragment of *N. gonorrhoeae* HpuA (Wong et al., 2015). HpuA is about half the size of TbpB, and its structure shows a similar fold as a single lobe of TbpB (**Figure 5**). It consists of a compact C-terminal β-barrel and an N-terminal open β-sandwich domain. Two large loops, loops 1 and 5, extend from the core of the protein. These exposed loops show high sequence variability, suggesting they are under immune selection. In spite of their solvent exposure, these loops have a high content of hydrophobic amino-acid residues, and although they display high sequence variability, they are the major interaction sites of hemoglobin, with loops 1 and 5 contacting the β- and α-chains, respectively, of a hemoglobin dimer. Additional interactions involve residues on two other loops.

Factor H Binding Protein (fHbp)

The complement system plays a key role in the host's defense against microbial invaders. Host cells are protected from complement activation, amongst others by binding the soluble factor H, a major negative regulator of the alternative complement pathway. By producing fHbp (a.k.a. GNA1870 or LP2086), *N. meningitidis* hijacks factor H and thereby limits complement activation at its surface (Madico et al., 2006; Schneider et al., 2006). Indeed, fHbp expression was found to improve bacterial survival in human blood and serum (Seib et al., 2009). Additionally, fHbp synthesis was found to protect the bacteria against the cationic antimicrobial peptide LL-37, a host defense peptide that destabilizes negatively charged bacterial membranes (Seib et al., 2009). Presumably, LL-37 interacts with fHbp also by electrostatic interaction.

fHbp is highly immunogenic and induces bactericidal antibodies that activate the classical complement pathway. In addition, antibodies may prevent binding of factor H to fHbp and thereby prevent inhibition of the alternative pathway. Based on sequence variation, two different subfamilies (Fletcher et al., 2004) or three variant groups (Masignani et al., 2003) were distinguished, with very limited immunological cross-reactivity between the groups (Masignani et al., 2003). One and two variants are included as components in the recently developed vaccines Bexsero® and Trumenba®, respectively.

The level of expression of fHbp varies considerably between different strains (Biagini et al., 2016), which may affect vaccine efficacy. Synthesis of the protein is induced under oxygen limitation (Oriente et al., 2010). Furthermore, fHbp synthesis is post-transcriptionally controlled by temperature due to the presence of a secondary structure at the 5′ end of its mRNA, which could function as a thermosensor; expression is considerably lower at 30°C than at 37°C (Loh et al., 2016). Thus, the synthesis of fHbp is presumably low on the mucosal surfaces of the nasopharynx, where the temperature is relatively low and the oxygen concentration is high. Consequently, vaccines solely based on fHbp will possibly not prevent carriage and spreading of the bacteria amongst the population and offer limited herd immunity (Loh et al., 2016). The synthesis of fHbp will increase as the bacteria reach the submucosal epithelial surfaces and the bloodstream, i.e., when protection against the host's defense mechanisms becomes essential. The synthesis of fHbp is also influenced by iron availability. However, whilst expression is reduced under iron limitation in most strains, it is increased in some lineages (Sanders et al., 2012), which makes it difficult to speculate about the biological significance of this phenomenon. Whether there is a relation with the reported capacity of fHbp to bind siderophores produced by other bacteria (Veggi et al., 2012) is questionable as fHbp synthesis is reduced in most strains under iron limitation when siderophore production is induced. A physiological role for the siderophore-binding capacity of fHbp has not been demonstrated.

The structure of fHbp has been determined by NMR and crystallography (Cantini et al., 2009; Schneider et al., 2009). The overall fold resembles those of HpuA and of the individual lobes of TbpB and LbpB and consists of a C-terminal β-barrel and an N-terminal taco-shaped barrel-like structure (**Figure 5**). Factor H consists of 20 complement control protein repeats, and of these, repeats 6 and 7 were found to bind fHbp with extensive interactions in co-crystals (Schneider et al., 2009). Interestingly, this binding site overlaps with the binding site for glycosaminoglycans to which factor H binds on mammalian cells to prevent complement activation.

Neisseria Heparin-Binding Antigen (NHBA)

Like fHbp, NHBA (a.k.a. GNA2132) is a component of the Bexsero® vaccine. In spite of considerable sequence variability, particularly near the N terminus (Bambini et al., 2009), NHBA elicits broadly cross-reactive bactericidal antibodies that conferred passive protection in a rat model (Welsch et al., 2003; Giuliani et al., 2010; Serruto et al., 2010).

NHBA was shown to bind heparin and heparan sulfate (Serruto et al., 2010). It consists of two domains. The structure of the conserved C-terminal domain has been determined by NMR and shows a similar 8-stranded β-barrel as those in the lipoproteins described above (**Figure 5**) (Esposito et al., 2011). The structure of the N-terminal domain has not been solved; it was predicted to be an intrinsically unfolded polypeptide (Esposito et al., 2011). In between these domains, NHBA contains an arginine-rich segment that binds heparin and heparan sulfate by electrostatic interaction (Serruto et al., 2010). The binding of heparin and heparin-like molecules has several functions. First, heparin binding was shown to correlate with increased survival of an unencapsulated *N. meningitidis* strain in human serum. In an encapsulated strain, such effect of NHBA synthesis was not observed, presumably because of the dominant role of the capsule in mediating serum resistance. As heparin is known to interact with complement proteins, including factor H, C4b-binding protein and C1 inhibitor (Yu et al., 2005), it was suggested that heparin bound at the cell surface might impart serum resistance by its capacity to bind such complement-regulatory molecules (Serruto et al., 2010). Alternatively, it might form a negatively charged capsule-like structure resembling the classical polysaccharide capsule (Serruto et al., 2010). NHBA has also been shown to function as an adhesin by binding to heparan sulfate proteoglycans that are present on surface of epithelial cells (Vacca et al., 2016).

Besides heparin and heparin-like molecules, NHBA also binds other polyanionic structures, including DNA (Arenas et al., 2013a). *N. meningitidis* forms microcolonies on the epithelial surfaces in the nasopharynx (Sim et al., 2000). Extracellular DNA (eDNA) is an important component of the extracellular matrix of these biofilm-like structures, and it is essential in the initiation of biofilm formation in most clonal lineages (Lappann et al., 2010). NHBA binds eDNA and inactivation of the *nhbA* gene in these lineages impairs biofilm formation (Arenas et al., 2013a).

NHBA can be cleaved immediately upstream of the arginine-rich segment by the autotransporter protease NalP (see below). Consequently, a C-terminal fragment including the arginine-rich segment is released into the milieu (Serruto et al., 2010). This fragment is taken up by endothelial cells and accumulates in mitochondria, where it induces the production of reactive oxygen species. The resulting phosphorylation of the adherens junction protein VE-cadherin and its subsequent internalization leads to increased endothelial permeability (Casellato et al., 2014). Thus, the released fragment of NHBA may contribute to the extensive vascular leakage that is typical for meningococcal sepsis. The arginine-rich segment was essential for this role of the NHBA C-terminal fragment presumably by functioning as a mitochondrial targeting signal. A slightly smaller fragment that is released from NHBA by lactoferrin-mediated cleavage and that lacks the arginine-rich segment did not induce endothelial leakage (Casellato et al., 2014).

Autotransporters

IgA Protease

IgA protease is produced by both *N. meningitidis* and *N. gonorrhoeae*. The passenger domain consists of two subdomains, an N-terminal domain containing the protease activity and a polypeptide called the α-peptide; these domains are connected via a small γ-peptide (Pohlner et al., 1987). The passenger is connected to the β-domain via a linker peptide. The protease domain can be released into the extracellular milieu by autocatalytic cleavage at sites (PAPSP, PPSP, or PPAP) located in between the protease domain and the γ-peptide, between the γ-peptide and the α-peptide, and between the α-peptide and the linker peptide. The presence of the latter processing site is strain dependent (Roussel-Jazédé et al., 2014). Alternatively, the entire passenger including the linker peptide can be released after cleavage by the autotransporter protease NalP (see below) (van Ulsen et al., 2003). IgA protease cleaves human IgA1 at a site (TPPTPSPS), which resembles the autocatalytic processing sites and is located in the hinge region between the Fab and the Fc domains (Plaut et al., 1975). IgA protease does not cleave IgA2, which lacks such a cleavage site (Plaut et al., 1975). Cleavage of IgA1 may inhibit IgA-mediated agglutination and subsequent mechanical clearance of the bacteria in the nasopharynx. IgA protease has also been shown to cleave the lysosome-associated membrane protein LAMP1 (Hauck and Meyer, 1997), which was reported to promote bacterial survival within epithelial cells (Lin et al., 1997) and transcytosis across polarized epithelia (Hopper et al., 2000). In addition, IgA protease cleaves the vesicular membrane protein synaptobrevin II in chromaffin cells (Binscheck et al., 1995) and the human chorionic gonadotropin hormone (Senior et al., 2001), but the physiological implications are not clear. All these alternative substrates have a target site resembling the autocatalytic cleavage sites.

The α-peptide contains a variable number of nuclear localization signals (NLS), which are arginine-rich peptide segments (Pohlner et al., 1995; Roussel-Jazédé et al., 2014), and it was indeed demonstrated to target an attached reporter protein to the nucleus of eukaryotic cells (Pohlner et al., 1995). When NalP releases IgA protease with attached α-peptide from the bacterial cell surface, the IgA protease is targeted to the nucleus where it cleaves the p65/RelA component of transcription factor NF-κB, thus silencing the expression of several NF-κB-responsive genes, including those encoding interleukin 8 and the anti-apoptotic protein cFLIP (Besbes et al., 2015). This altered gene expression results in sustained activation of c-Jun N-terminal kinase and apoptosis. Interestingly, purified IgA protease without attached α-peptide has been shown to inhibit TNFα-induced apoptosis of monocytes, presumably by the observed cleavage of the TNF receptor II (Beck and Meyer, 2000).

When NalP is not expressed and the autocleavage site between the α-peptide and the linker peptide is absent, the α-peptide is exposed at the bacterial cell surface covalently connected to the β-domain. Since NLS are positively charged protein segments, they bind polyanions like eDNA, thus stimulating biofilm formation (Arenas et al., 2013a). The α-peptide also binds heparin (Roussel-Jazédé et al., 2014) and may have similar functions in conferring serum resistance and mediating adhesion as discussed above for NHBA. Whether the α-peptide also directly affects gene expression by binding DNA after it has been targeted to the nucleus of eukaryotic cells has not been investigated yet.

NalP

The *nalP* gene is found in the vast majority of *N. meningitidis* isolates, with the notable exception of invasive isolates of clonal complexes ST-269 and ST-461 (Oldfield et al., 2013). Its expression is prone to phase variation by slipped-strand mispairing at a poly-C tract in the coding region (Turner et al., 2002; van Ulsen et al., 2003). An analysis of its phase variation status revealed that approximately half of the strains were phased on, and no significant differences were found in this respect between carriage and invasive strains (Oldfield et al., 2013).

NalP is a subtilisin-like serine protease. Its passenger is released from the cell surface by autoproteolytic cleavage. However, after cleavage between the passenger and the β-domain, the passenger remains temporarily attached at the cell surface by an N-terminal lipid anchor (Roussel-Jazédé et al., 2013). In this position, it releases several autotransporters and lipoproteins from the bacterial cell surface. The release of these proteins has relevant consequences. As already discussed above, the NalP-mediated release of IgA protease with attached α-peptide and of the C-terminal part of NHBA from the cell surface stimulates apoptosis and increases the permeability of endothelial and epithelial cell layers. In contrast, when *nalP* is phased off, the α-peptide and complete NHBA are retained at the cell surface, which stimulates biofilm formation and adhesion and protects against host defenses. Thus, the absence of *nalP* expression appears to favor colonization and biofilm formation, whilst *nalP* expression may lead to dispersal of biofilms and invasion or spreading to a new host.

Another target of NalP is the LbpB component of the lactoferrin receptor (Roussel-Jazédé et al., 2010). LbpB is a very immunogenic protein, and its NalP-mediated release from the cell surface imparted protection against the complement-mediated killing by LbpB-specific bactericidal antibodies (Roussel-Jazédé et al., 2010). As discussed above, LbpB is not essential for the acquisition of iron from lactoferrin, although it may enhance the process by selecting holo-lactoferrin for binding to the receptor. The latter function may be lost when LbpB is cleaved from the cell surface by NalP, but this is compensated by protection against LbpB-specific bactericidal antibodies. Another function of LbpB discussed above, i.e., offering protection against the membrane-damaging effects of lactoferricin, is probably retained, even if the protein is released into the extracellular milieu. Thus, due to the phase-variable expression of *nalP*, a subpopulation of the bacterial cells will lose LbpB from the cell surface, and this subpopulation may be less efficient in iron acquisition but will have a competitive advantage when LbpB-specific antibodies are elicited in the host. It is noteworthy that in *N. gonorrhoeae*, which does not produce NalP because the gene is disrupted, the expression of *lbpB* itself is prone to phase variation (Anderson et al., 2003).

Besides bacterial proteins, NalP also cleaves at least one host protein, i.e., the complement factor C3. Thereby, it avoids complement activation and confers serum resistance (Del

Tordello et al., 2014). Consistently, *nalP* expression was earlier reported to be upregulated during growth in an *ex vivo* human whole-blood model and to be essential for survival under these conditions (Echenique-Rivera et al., 2011b).

Two Additional Proteases: App and AusI

App and AusI (a.k.a. MspA) are both chemotrypsin-like serine proteases with homology to IgA protease (Abdel Hadi et al., 2001; van Ulsen et al., 2001, 2006; Turner et al., 2006). Whereas, an intact *app* gene is ubiquitous among *N. meningitidis* strains, the *ausI* gene is disrupted by premature stop codons in several strains, and its expression is prone to phase variation by the presence of a poly-C tract in the coding region (Martin et al., 2003; van Ulsen et al., 2006). Interestingly, an extensive study in mostly serogroup B and Y isolates revealed that the *ausI* gene was phased on in 90% of both invasive and carriage strains of serogroup B. As 33% phased on would be expected in the case of random on and off switching, this suggests an important role for AusI in the biology of serogroup B (Oldfield et al., 2013). Such a preference for the on phase of *ausI* was not found in the serogroup Y isolates.

App and AusI are both released from the bacterial cell surface by autoproteolytic cleavage. However, like for IgA protease, a larger form of these proteins with a C-terminal extension, called α-peptide, can be released by NalP-mediated cleavage (van Ulsen et al., 2003, 2006; Turner et al., 2006). The α-peptide of App contains two NLS similar to those in the α-peptide of IgA protease, but such sequences are not present in the α-peptide of AusI. Nevertheless, both proteins with attached α-peptides were shown to be targeted to the nucleus after being taken up by dendritic cells and to induce apoptosis, presumably by their capacity to bind histones and to cleave histone H3 (Khairalla et al., 2015). Uptake into the dendritic cells was shown to be mediated by the mannose receptor and the transferrin receptor 1 to which the proteases could be crosslinked.

Earlier, both App and AusI were reported to function as adhesins (Serruto et al., 2003; Turner et al., 2006), a function which is difficult to reconcile with their cleavage from the cell surface. Perhaps, the proteins remain temporarily associated with the cell surface before they are released. Heterologous expression of App and AusI in *E. coli* mediated adhesion of the bacteria to Chang epithelial cells, whereas variable adhesion to other cell lines was observed (Serruto et al., 2003; Turner et al., 2006). App-mediated adhesion to Chang cells could also be demonstrated in *N. meningitidis* (Serruto et al., 2003). The presence of a capsule did not seem to interfere with the adhesin function, suggesting that this large protein (~160 kDa) extends beyond the capsule to interact with its receptor on the Chang cells. However, deletion analysis suggested that the α-peptide, which is located close to the bacterial cell surface in uncleaved App and is unlikely to extend beyond the capsule, is the main determinant for adhesion (Serruto et al., 2003). Protease treatment indicated the involvement of a proteinaceous receptor on Chang cells, but it is not known whether the transferrin receptor 1 or the mannose receptor, which mediate uptake of App and AusI in dendritic cells (see above), are involved. App was also shown to mediate interbacterial aggregation on cell layers (Serruto et al., 2003); possibly, this is mediated by the binding of the NLS in the α-peptide to eDNA, analogous to the role of the α-peptide of IgA protease in biofilm formation (see above).

Surface-Exposed Classical Autotransporters: AutA and AutB

AutA and AutB are two related, relatively small autotransporters of ~75 kDa including the β-domain. The *autA* gene is broadly distributed among *Neisseria* spp., whereas the presence of the *autB* gene is restricted to the pathogenic *Neisseria* spp., *N. meningitidis* and *N. gonorrhoeae* (Arenas et al., 2015a, 2016). Also some, but not all, strains of *Haemophilus influenzae* and closely related *Haemophilus* spp. contain an *autB* gene. Analysis of the *autB*-flanking sequences in both genera suggested that a common ancestor of the two pathogenic *Neisseria* spp. has acquired the gene from a *H. influenzae* strain by horizontal gene transfer (Davis et al., 2001; Arenas et al., 2016).

The expression of *autA* and *autB* is prone to phase variation by slipped-strand mispairing at AAGC nucleotide repeats located immediately downstream of the start codon (Peak et al., 1999). In addition, the *autA* gene was disrupted in 76% of the meningococcal genomes analyzed by premature stop codons, insertions and/or deletions (Arenas et al., 2015a) suggesting selection pressure against expression of the gene. However, the gene was phased on in ~33% of the strains with an intact *autA*, consistent with random on and off switching. In contrast to *autA*, *autB* was disrupted in only 8% of the *N. meningitidis* genomes analyzed (Arenas et al., 2016). Curiously, the gene is phased off in the vast majority of the genomes with an intact *autB*, and in only ~2% of the strains the gene could be expressed. This strongly suggests that, even if an intact *autB* is retained, its expression is under negative selection pressure probably already in the nasopharynx as no significant difference between invasive and carriage isolates was observed in this respect (Arenas et al., 2016).

The passenger domains of AutA and AutB are not proteolytically released, and they are exposed at the bacterial cell surface covalently attached to the β-domain (Arenas et al., 2015a, 2016). AutA expression was shown to cause bacterial autoaggregation (Arenas et al., 2015a), a trait associated with resistance to host immune defenses such as phagocytosis (Ochiai et al., 1993). Autoaggregation is mediated by mutual interaction of AutA proteins on neighboring cells. In addition, AutA was found to bind DNA, which acts as an adhesive between cells (Arenas et al., 2015a; Arenas and Tommassen, 2017). Autoaggregation also has consequences for the architecture of biofilms formed. Besides, AutA was reported to be a potent $CD4^+$ T-cell and B-cell stimulating antigen (Ait-Tahar et al., 2000). Expression of *autB* was shown to enhance biofilm formation, and it retarded the passage of the bacteria through epithelial cell layers (Arenas et al., 2016). This suggests that AutB could have a role during host colonization.

Trimeric Autotransporters: NadA and NhhA

NadA shows high sequence similarity to the well-known virulence factors YadA from *Yersinia enterocolitica* and UspA2 from *Moraxella catarrhalis* (van Ulsen et al., 2001). It is a component of the Bexsero® vaccine. However, the gene is present in only ~30% of *N. meningitidis* isolates with a lower prevalence in carriage strains (~15%) (Comanducci et al., 2002, 2004). Furthermore, based on sequence variability, two main groups of NadA variants can be discriminated (Bambini et al., 2014) with limited immunological cross-reactivity between the groups (Comanducci et al., 2004), and the expression levels are highly variable (Martin et al., 2003), which also may impact on vaccine efficacy (Comanducci et al., 2002). Expression levels are determined by phase variation at a TAAA tetranucleotide repeat region upstream of the *nadA* promoter (Martin et al., 2003), whilst also on/off switching can occur by phase variation at a poly-C stretch within the coding region of some *nadA* variants (Bambini et al., 2014). Expression of *nadA* is transcriptionally controlled by the repressor NadR (Schielke et al., 2009), which binds operator sequences upstream and downstream of the TAAA repeat region, the latter overlapping with the -10 region of the promoter (Metruccio et al., 2009). The metabolite 4-hydroxyphenylacetic acid was found to act as an inducer of *nadA* expression (Metruccio et al., 2009). This metabolite is secreted in human saliva, suggesting that *nadA* expression will be induced on the mucosal surfaces of the nasopharynx. Interestingly, upon derepression, the expression levels were similar independent of the number of TAAA repeats, indicating that the number of repeats determines the level of repression by NadR rather than the expression levels in the induced state.

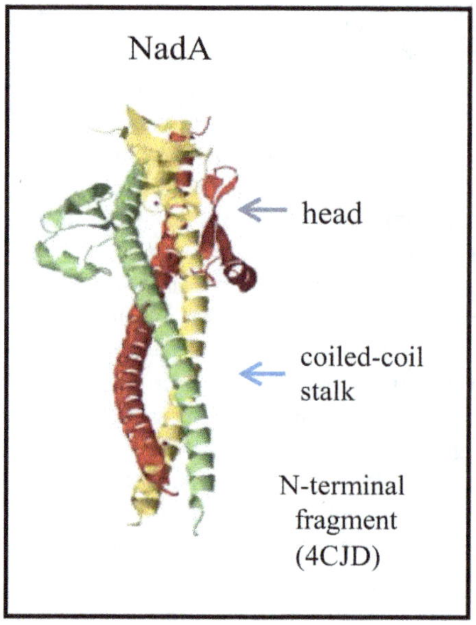

FIGURE 6 | Structure of the passenger of the trimeric autotransporter NadA. The structure of the N-terminal part of the passenger is provided (Malito et al., 2014). Each subunit is colored differently, and the major structural domains of the complex are indicated. The PDB access code is provided in brackets.

NadA functions as an adhesin and invasin. Its heterologous expression in *E. coli* promoted adherence of the bacteria to and invasion of Chang epithelial cells, and this function was confirmed in *N. meningitidis* (Capecchi et al., 2005). NadA function could also be reconstituted in a *yadA* mutant of *Y. enterocolitica*, and it was shown that β1 integrins function as its receptor on host cells (Nägele et al., 2011). Curiously, NadA was also shown to bind the heat shock protein Hsp90, a normally cytoplasmic chaperone, and purified Hsp90 was shown to inhibit NadA-mediated adhesion and invasion *in vitro* (Montanari et al., 2012). The physiological significance of these observations is not clear.

Whilst the passenger domains of classical autotransporters usually form a long β-helical structure that displays globular functional domains, trimeric autotransporters have a completely different architecture consisting of a stalk domain that displays one or several head domains (Grijpstra et al., 2013). The stalk is a coiled-coil structure that extends from the α-helices that occupy the membrane-embedded β-barrel (**Figure 1D**). NadA has a similar overall architecture but is somewhat atypical in that a separate globular head domain is lacking (Malito et al., 2014). Instead, the coiled coils extend up to the N terminus and the head is formed by insertions of 36 residues that protrude from the stalk and form wing-like structures (**Figure 6**). The adhesin activity was mapped to the head region (Capecchi et al., 2005; Tavano et al., 2011), which is apparently sufficiently exposed from the bacterial cell surface to be functional also in encapsulated strains (Capecchi et al., 2005).

NhhA (a.k.a. Msf) was identified as a homolog of the adhesin Hia of non-typeable *H. influenzae* (Peak et al., 2000; van Ulsen et al., 2001), and the *nhhA* gene was found to be ubiquitously present in all *N. meningitidis* strains examined, although disrupted in some of them (Peak et al., 2000). Expression levels varied widely and were affected in one particular lineage, i.e., ST41/44, by an amino-acid substitution in the β-domain, which prevents trimerization and surface exposition of the passenger domain (Echenique-Rivera et al., 2011a). The function of NhhA as an adhesin was demonstrated after its heterologous production in *E. coli*, which mediated adherence of the bacteria to epithelial cell lines, and was confirmed after constructing an *nhhA* mutant in an encapsulated *N. meningitidis* strain (Scarselli et al., 2006). It was demonstrated that purified NhhA and NhhA-producing cells bind the extracellular matrix components laminin and heparan sulfate (Scarselli et al., 2006). In addition, NhhA was shown to bind activated vitronectin, a complement regulatory protein, and, thus, to impart serum resistance to the bacteria by inhibiting the formation of the membrane-attack complex (Griffiths et al., 2011). Consistently, an *nhhA* mutant was less virulent and showed reduced survival in blood after intraperitoneal challenge of mice (Sjölinder et al., 2008). Expression of the protein appeared to be essential also for the colonization of the nasopharyngeal mucosa and for disease development after intranasal challenge in a murine model of meningococcal disease. The reduced colonization levels are probably related to the *in vitro* observations that wild-type bacteria are more resistant than the *nhhA* mutant to phagocytosis (Sjölinder et al., 2008) and that NhhA induces apoptosis of macrophages (Sjölinder et al., 2012). In addition, immune modulatory activities of NhhA were reported (Wang et al., 2016). NhhA blocked differentiation of monocytes into dendritic cells and induced their differentiation into macrophages that failed to produce proinflammatory mediators but produced anti-inflammatory responses. These differentiated macrophages were highly efficient in eliminating the bacteria. Hence, this response prevents dissemination and sustains asymptomatic colonization (Wang et al., 2016).

The Two-Partner Secretion (TPS) System

The substrates of the TPS systems, TpsA, are very large proteins (>2,000 amino-acid residues), which, like most classical autotransporters, display a β-helical conformation (Jacob-Dubuisson et al., 2013). *N. meningitidis* strains can produce up to five different TpsA proteins (van Ulsen and Tommassen, 2006). The genes for one TPS system are widely distributed among *N. meningitidis* isolates, whereas the genes for other TPS systems are associated with invasive clonal complexes (van Ulsen et al., 2008). Only the function of the conserved TpsA, a.k.a. HrpA, has been investigated, particularly in strains that produce this protein as the only TpsA. Substrates of TPS systems can have various, often virulence-related functions. Several functions have been attributed to the meningococcal TpsA, including a role in intracellular survival and escape from infected cells (Talà et al., 2008), adhesion to epithelial cell lines (Schmitt et al., 2007), and biofilm formation (Neil and Apicella, 2009). More recently, it was demonstrated that TpsA is involved in interbacterial competition in a process called contact-dependent growth inhibition (CDI) (Arenas et al., 2013b). CDI was originally described in *E. coli* and has evolved to inhibit the growth of closely related bacteria, usually of the same species, in niche competition (Aoki et al., 2005). In the proposed model, the surface-exposed TpsA interacts with a receptor in the outer membrane of a target cell, after which a small C-terminal part of TpsA of ~300 amino-acid residues is proteolytically released and transported into the target cell, where it exerts one of different toxic activities, e.g., by functioning as a DNase or an RNase (Aoki et al., 2010). The producing cells themselves are protected against this toxic activity by the production of a small immunity protein, here called TpsI. TpsA proteins with CDI activity are polymorphic toxins, i.e., comparison of their sequences in different strains revealed that they can display a wide variety of toxic domains at their C terminus, each of which is associated with its own cognate TpsI (Aoki et al., 2010).

The genes for the conserved TPS system of *N. meningitidis* are organized in an operon (**Figure 7**). Downstream of the *tpsI* gene, a variable number of 5′ truncated *tpsA*-related genes are located, called *tpsC* cassettes. In their 5′ end, the *tpsC* cassettes show homology with a segment immediately upstream of the 3′ end encoding the toxic domain of *tpsA*. At their 3′ end, however, the *tpsC* cassettes encode different toxic domains (**Figure 7**). Thus, the *tpsC* cassettes provide a broad repertoire of toxic domains and each of them is associated with a cognate *tpsI* gene to impart immunity to the producer. However, the products of these *tpsC* cassettes, if expressed, cannot be secreted because they lack an N-terminal signal sequence and a TPS domain, required for recognition by the Sec system and the TpsB protein, respectively. It was proposed that the 3′ end of *tpsA* along with the downstream *tpsI* can be replaced by *tpsC* cassettes and the cognate *tpsI* through genetic recombination, thus resulting in the production of a TpsA with a different toxic domain (Arenas et al., 2013b). Examination of the 3′ end of the *tpsA* genes in available genome sequences indeed suggested a high rate of recombination between *tpsA* and *tpsC* cassettes (van Ulsen and Tommassen, 2006). However, a detailed analysis of the 3′ end of the *tpsA* gene in large panels of meningococcal isolates from the same lineage collected over several decades from different parts of the world revealed a very low rate of exchange (Arenas et al., 2013b). It was suggested that the availability of a large collection of *tpsI* genes might be more important for the bacteria than the ability to replace the toxic domain of TpsA by a variety of others encoded by the *tpsC* cassettes.

The T1SS

FrpC and related proteins (some called FrpA) are the substrates of the meningococcal T1SS; they are produced under iron limitation (Thompson and Sparling, 1993; Thompson et al., 1993a,b). FrpC-like proteins are large proteins of 120–200 kDa. They are members of the RTX family of proteins and differ largely with respect to the number of glycine-rich calcium-binding repeats (see above); in addition they show considerable sequence variability in a small N-terminal domain of ~350 amino-acid residues (van Ulsen and Tommassen, 2006), possibly due to immune pressure. The RTX family includes important cytotoxins in several pathogens (Linhartová et al., 2010). However, *N. meningitidis* mutants lacking functional *frpC* genes or the T1SS machinery were not attenuated in animal infection models (Klee et al., 2000; Forman et al., 2003).

After secretion, the FrpC protein is autocatalytically cleaved at an Asp-Pro bond at physiological concentrations of calcium ions (Osička et al., 2004). The N-terminal part of ~45 kDa is retained at the bacterial cell surface by binding FrpD with high affinity (Prochazkova et al., 2005). FrpD is a cell-surface-exposed lipoprotein that is encoded by the same operon as FrpC. Its structure is different from those of the surface-exposed lipoproteins discussed above, although it also has a high β-sheet content, which may be relevant for its transport to the cell surface (**Figure 5**) (Sviridova et al., 2017). During autocatalytic processing, the released carboxyl group of the C-terminal aspartate of the N-terminal FrpC fragment forms a covalent linkage with an adjacent ε-amino group of a lysine generating a new Asp-Lys isopeptide bond (Prochazkova et al., 2005). In this way, the N-terminal FrpC fragment may be covalently linked to membrane proteins on the host epithelial cells (Sviridova et al., 2017). Thus, together with FrpD, the N-terminal fragment of FrpC may function as an adhesin, mediating the adhesion of the bacteria to epithelial cells.

Other Secretion Systems and Secreted Proteins

Besides the secretion systems discussed above, some *N. meningitidis* strains contain a type 4 secretion system (T4SS). The genes for a T4SS were first identified in gonococcal strains on a genomic island called gonococcal genomic island (Dillard and and Seifert, 2001), and the system was shown to mediate the secretion of single-stranded DNA (Hamilton et al., 2005). Subsequently, it was found that also some meningococcal strains contain genes for the T4SS (Snyder et al., 2005; Woodhams

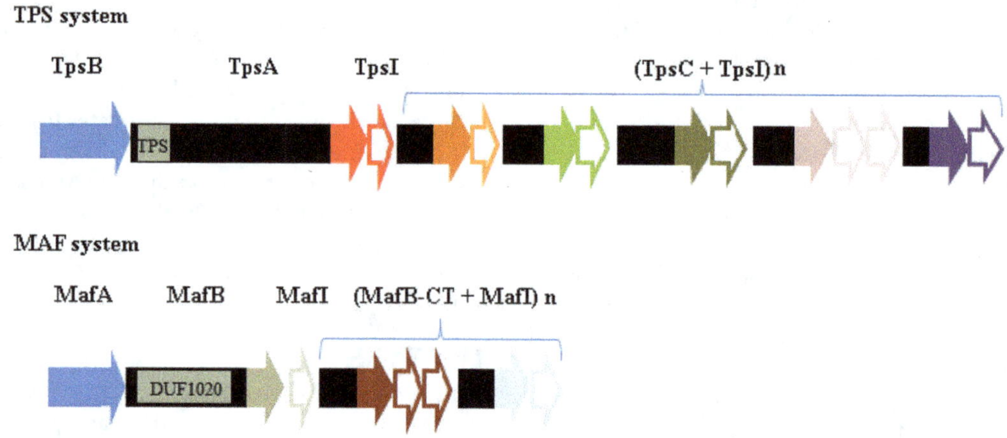

FIGURE 7 | Comparison of the genetic organization of the genes encoding a TPS system and a MAF system. In the top panel, the genetic island encoding the TPS system of strain B16B6 is depicted (Arenas et al., 2013b). TPS islands consist of genes encoding the transporter TpsB (blue), the secreted toxin TpsA, an immunity protein TpsI, and a variable repertoire of *tpsC* cassettes interspersed with genes encoding immunity proteins. The position of the gene segment of *tpsA* that encodes the TPS domain, which is recognized by TpsB, is indicated. The segments of *tpsA* and *tpsC* cassettes that encode toxic domains are colored, with different colors indicating different toxic domains. The *tpsI* genes encoding the cognate immunity proteins are indicated as open arrows with the same colors. In the bottom panel, the genetic island MGI-1 of B16B6 is depicted (Arenas et al., 2015b). The MGI islands consist of genes encoding MafA, the secreted toxin MafB, an immunity protein MafI, and a variable repertoire of *mafB-CT* cassettes interspersed with genes encoding immunity proteins. The position of the gene segment that encodes the conserved DUF1020 domain in MafB is indicated. Sequences encoding toxic domains and cognate immunity proteins are indicated as in the TPS island. Note that no evidence for MafA representing the transporter of MafB has been reported so far.

et al., 2012). However, in most of these strains, these genes are disrupted, and those strains with a complete T4SS do not appear to exhibit increased DNA release. As no function has been assigned to the T4SS in *N. meningitidis*, it won't be discussed here in detail. *N. meningitidis* also produce type IV pili, which are retractile pili involved in adhesion to host cells, interbacterial interaction, DNA uptake, and surface motility. These surface organelles are topic of several focused reviews to which we refer (e.g., see Berry and Pelicic, 2015; Mayer and Wong, 2015). In this section, we focus on a number of secreted proteins for which the secretion mechanism has not yet been solved.

MafB

Like the TpsA proteins, the MafB proteins are polymorphic toxins that display a small variable toxic domain at their C terminus (Arenas et al., 2015b; Jamet et al., 2015). The *mafB* genes are localized on the chromosome on genomic islands designated Maf Genomic Islands (MGI) (Jamet et al., 2015). Up to three MGI can be present in a single meningococcal strain. The gene organization on the MGI is reminiscent of the loci encoding the TPS systems (**Figure 7**). The *mafB* genes are usually flanked by a *mafA* and a *mafI* gene. The *mafI* gene encodes a small immunity protein, which has been shown to interact with the toxic domain of MafB and to neutralize its toxic activity (Arenas et al., 2015b; Jamet et al., 2015). Indeed, MafB producers were shown to inhibit the growth of congeners that do not produce the cognate immunity protein. Downstream of the *mafI* gene, a variable number of 5′ truncated *mafB*-related genes are located, called *mafB-CT* cassettes (**Figure 7**). Like the *tpsC* cassettes, these *mafB-CT* cassettes offer a reservoir of alternative toxic domains, which could potentially be displayed at the C terminus of MafB after a genetic recombination event.

Apart from the presence of a toxic domain, MafB proteins do not show any similarity with TpsA proteins. They are synthesized with a signal sequence for transport across the inner membrane via the Sec machinery. Immediately after the signal sequence, all MafB proteins contain a conserved domain of unknown function, DUF1020, followed by the C-terminal toxic domain (**Figure 7**). In some MafB proteins, a Hint domain, which is expected to mediate protein splicing (Amitai et al., 2003), separates the DUF1020 from the toxic domain. How MafB is translocated across the outer membrane is unclear. Considering the parallelism with the TPS loci (**Figure 7**), MafA would be an obvious candidate for a protein involved in the secretion process. Earlier studies in *N. gonorrhoeae* indicated that MafA is an adhesin that binds host glycolipids (Paruchuri et al., 1990). Accordingly, these proteins were designated members of a multiple adhesin family (maf). MafA doesn't show any similarity with TpsB. It is produced with a signal sequence containing a lipobox motif, suggesting it's a lipoprotein, and secondary structure predictions didn't suggest a β-barrel structure, which is difficult to reconcile with its possible role as a transporter (Arenas et al., 2015b). Indeed, MafB was found in the extracellular medium when expressed in a *Neisseria cinerea* mutant lacking MafA (Jamet et al., 2015), suggesting that MafB secretion is independent of MafA. Furthermore, MafB was not secreted when produced in *E. coli* suggesting that it is not a novel type of autotransporter and that the transport mechanism involved is not present in *E. coli* (Jamet et al., 2015). It was suggested that MafB may be released with OMVs, which are abundantly shed from the surface in *N. meningitidis* (Jamet et al., 2015).

T and B Cell-Stimulating Protein B (TspB)

The name TspB, for T- and B-cell-stimulating protein B, may be a misnomer as evidence for such a function has, to the best of our knowledge, never been published. TspB was identified as a protein that binds immunoglobulins (Ig) (Müller et al., 2013). The protein is encoded by a prophage that is associated with invasive-disease isolates (Bille et al., 2005; Müller et al., 2013), and its synthesis is induced in the presence of human serum (Müller et al., 2013). Up to four copies of the *tspB* gene can be found in a genome.

The TspB proteins contain variable sequences at the N- and C-terminal ends and a conserved core region followed by a variable proline-rich segment (Müller et al., 2013). The core region of TspB binds the Fcγ region of IgG2. Ig-binding proteins are produced by several human pathogens, and they prevent the activation of complement. Indeed, TspB was shown to contribute to the survival of *N. meningitidis* in normal human serum by inhibiting IgM-mediated activation of the classical complement pathway (Müller et al., 2015). Since TspB binds IgM only poorly (Müller et al., 2013), this effect is probably indirect. TspB was found to bind DNA and to the form a biofilm-like matrix consisting of TspB, IgG, and DNA surrounding bacterial aggregates (Müller et al., 2013, 2015). Possibly, the presence of IgG in the matrix stimulates the non-productive activation of complement far away from the bacterial cell surface.

The secretion mechanism of TspB remains unclear. It could be released as a minor coat protein of filamentous phage particles, which was demonstrated to be mediated by the secretin PilQ, a protein also responsible for extrusion of type IV pili from the bacteria (Bille et al., 2005). However, in the studies of Müller et al. (2013), TspB appeared not to be associated with phage particles. The protein is produced with an N-terminal signal sequence for Sec-mediated transport across the inner membrane and a hydrophobic region near the C terminus that should act as a stop-transfer signal and anchor the protein in the membrane. Thus, possibly, the protein is only released after cell lysis.

CONCLUDING REMARKS

The quest for novel vaccine antigens has enormously stimulated research into cell-surface-exposed and secreted proteins in *N. meningitidis*. These investigations have not only resulted in the development of two new registered vaccines, but they have also uncovered the functions of many of these proteins, which have roles in host-pathogen interactions, including adhesion, invasion, and immune evasion, in nutrient acquisition, and in interbacterial interactions, including biofilm formation and competition. Thus, these studies have provided important novel insights into biology of the meningococcus. In addition, they have led to the discovery of new transport mechanisms and machineries of general (micro)biological significance, such as the BAM and SLAM discussed above and the LPT machinery for the transport of lipopolysaccharides (Bos et al., 2004). Because they are essential in most Gram-negative bacteria and cell-surface exposed, particularly the integral OMPs of the BAM and LPT systems are attractive targets for the development of new antimicrobials (Srinivas et al., 2010; Urfer et al., 2016).

With respect to the transport mechanisms, there is still much to be learned. For lipoproteins, for example, it will be important to establish the sorting rules in *N. meningitidis*, which will help to predict on the basis of the sequence whether a lipoprotein is transported to the cell surface or is retained at the periplasmic side of either the inner or the outer membrane. In the MAF system, it is still unknown how MafB is transported to the cell surface, and how its toxic domain is cleaved off and imported into the target cell. Molecular insights in these mechanisms could lead to novel antimicrobial therapies.

Competition between different meningococci is a new field of study. A relatively large part of the rather small genome is occupied by genes encoding toxins and their cognate immunity proteins and transporters. It should be noted that CDI activity has, so far, only been demonstrated in competition assays between wild-type strains and their mutant derivatives lacking specific immunity proteins. What happens in a competition between two wild-type strains that express different toxins and immunity proteins is unknown. Although colonization by different strains has been demonstrated in carriage studies, it was very rare and detected in only ~1% of the carriers (Caugant et al., 2007), suggesting that competition in the nasopharynx may be very effective. In contrast, such competition was recently not observed *in vitro* in dual-strain biofilm-formation experiments (Pérez-Ortega et al., 2017).

It is evident that many components of the secretome exert similar functions. It is particularly remarkable how many of them function as adhesins. Some redundancy in function is to be expected considering the high capacity of the meningococcus to change its cell surface, e.g., by phase variation, presumably as a mechanism to escape from the immune response of the host. Thus, one protein may take over the function of another one that is switched off. However, adhesins may also work in combination or sequentially or target other receptors and, thereby, other host cells. It is also evident that many secreted proteins have multiple functions, which is conceivable considering their large size. In addition, the secreted proteases may have multiple targets and, thus, interfere with the host's metabolism and immune response in multiple ways. It is likely that new functions of the secretome will be uncovered in the coming years. In this respect, it is important to note that most functions were discovered so far in *in vitro* systems. The development of transgenic mouse models for nasopharyngeal colonization of *N. meningitidis* (Joshwich et al., 2013) certainly opens new avenues for understanding the role of secretome components in host-pathogen interactions, although limitations will persist considering the large number of host-specific components involved.

AUTHOR CONTRIBUTIONS

All authors listed, have made substantial, direct and intellectual contribution to the work, and approved it for publication.

REFERENCES

Abdel Hadi, H., Wooldridge, K. G., Robinson, K., and Ala'Aldeen, D. A. (2001). Identification and characterization of App: an immunogenic autotransporter protein of *Neisseria meningitidis*. *Mol. Microbiol.* 41, 611–623. doi: 10.1046/j.1365-2958.2001.02516.x

Ait-Tahar, K., Wooldridge, K. G., Turner, D. P., Atta, M., Todd, I., and Ala'Aldeen, D. A. (2000). Auto-transporter A protein of *Neisseria meningitidis*: a potent CD4$^+$ T-cell and B-cell stimulating antigen detected by expression cloning. *Mol. Microbiol.* 37, 1094–1105. doi: 10.1046/j.1365-2958.2000.02061.x

Amitai, G., Belenkiy, O., Dassa, B., Shainskaya, A., and Pietrokovski, S. (2003). Distribution and function of new bacterial intein-like protein domains. *Mol. Microbiol.* 47, 61–73. doi: 10.1046/j.1365-2958.2003.03283.x

Anderson, J. E., Hobbs, M. M., Biswas, G. D., and Sparling, P. F. (2003). Opposing selective forces for the expression of the gonococcal lactoferrin receptor. *Mol. Microbiol.* 48, 1325–1337. doi: 10.1046/j.1365-2958.2003.03496.x

Anderson, J. E., Sparling, P. F., and Cornelissen, C. N. (1994). Gonococcal transferrin-binding protein 2 facilitates but is not essential for transferrin utilization. *J. Bacteriol.* 176, 3162–3170.

Aoki, S. K., Diner, E. J., t'Kint de Roodenbeke, C., Burgess, B. R., Poole, S. J., Braaten, B. A., et al. (2010). A widespread family of polymorphic contact-dependent toxin delivery systems in bacteria. *Nature* 468, 439–442. doi: 10.1038/nature09490

Aoki, S. K., Pamma, R., Hernday, A. D., Bickham, J. E., Braaten, B. A., and Low, D. A. (2005). Contact-dependent inhibition of growth in *Escherichia coli*. *Science* 309, 1245–1248. doi: 10.1126/science.1115109

Arenas, J., and Tommassen, J. (2017). Meningococcal biofilm formation: let's stick together. *Trends Microbiol.* 25, 113–124. doi: 10.1016/j.tim.2016.09.005

Arenas, J., Cano, S., Nijland, R., van Dongen, V., Rutten, L., van der Ende, A., et al. (2015a). The meningococcal autotransporter AutA is implicated in autoaggregation and biofilm formation. *Environ. Microbiol.* 17, 1321–1337. doi: 10.1111/1462-2920.12581

Arenas, J., de Maat, V., Catón, L., Krekorian, M., Herrero, J. C., Ferrara, F., et al. (2015b). Fratricide activity of MafB protein of *N. meningitidis* strain B16B6. *BMC Microbiol.* 15:156. doi: 10.1186/s12866-015-0493-6

Arenas, J., Nijland, R., Rodriguez, F. J., Bosma, T. N. P., and Tommassen, J. (2013a). Involvement of three meningococcal surface-exposed proteins, the heparin-binding protein NHBA, the α-peptide of IgA protease and the autotransporter protease NalP, in initiation of biofilm formation. *Mol. Microbiol.* 87, 254–268. doi: 10.1111/mmi.12097.

Arenas, J., Paganelli, F. L., Rodríguez-Castaño, P., Cano-Crespo, S., van der Ende, A., van Putten, J. P., et al. (2016). Expression of the gene for autotransporter AutB of *Neisseria meningitidis* affects biofilm formation and epithelial transmigration. *Front. Cell. Infect. Microbiol.* 6:162. doi: 10.3389/fcimb.2016.00162

Arenas, J., Schipper, K., van Ulsen, P., van der Ende, A., and Tommassen, J. (2013b). Domain exchange at the 3′ end of the gene encoding the fratricide meningococcal two-partner secretion protein A. *BMC Genomics* 14:622. doi: 10.1186/1471-2164-14-622

Auclair, S. M., Bhanu, M. K., and Kendall, D. A. (2012). Signal peptidase I: cleaving the way to mature proteins. *Protein Sci.* 21, 13–25. doi: 10.1002/pro.757

Bambini, S., De Chiara, M., Muzzi, A., Mora, M., Lucidarme, J., Brehony, C., et al. (2014). *Neisseria* adhesin A variation and revised nomenclature scheme. *Clin. Vaccine Immunol.* 21, 966–971. doi: 10.1128/CVI.00825-13

Bambini, S., Muzzi, A., Olcen, P., Rappuoli, R., Pizza, M., and Comanducci, M. (2009). Distribution and genetic variability of three vaccine components in a panel of strains representative of the diversity of serogroup B meningococcus. *Vaccine* 27, 2794–2803. doi: 10.1016/j.vaccine.2009.02.098

Beck, S. C., and Meyer, T. F. (2000). IgA1 protease from *Neisseria gonorrhoeae* inhibits TNFα-mediated apoptosis of human monocytic cells. *FEBS Lett.* 472, 287–292. doi: 10.1016/S0014-5793(00)01478-2

Bellamy, W., Takase, M., Wakabayashi, H., Kawase, K., and Tomita, M. (1992). Antibacterial spectrum of lactoferricin B, a potent bactericidal peptide derived from the N-terminal region of bovine lactoferrin. *J. Appl. Bacteriol.* 73, 472–479.

Berry, J. L., and Pelicic, V. (2015). Exceptionally widespread nanomachines composed of type IV pilins: the prokaryotic Swiss Army knives. *FEMS Microbiol. Rev.* 39, 134–154. doi: 10.1093/femsre/fuu001

Besbes, A., Le Goff, S., Antunes, A., Terrade, A., Hong, E., Giorgini, D., et al. (2015). Hyperinvasive meningococci induce intra-nuclear cleavage of the NF-κB protein p65/RelA by meningococcal IgA protease. *PLoS Pathog.* 11:e1005078. doi: 10.1371/journal.ppat.1005078.

Biagini, M., Spinsanti, M., De Angelis, G., Tomei, S., Ferlenghi, I., Scarselli, M., et al. (2016). Expression of factor H binding protein in meningococcal strains can vary at least 15-fold and is genetically determined. *Proc. Natl. Acad. Sci. U.S.A.* 113, 2714–2719. doi: 10.1073/pnas.1521142113

Bille, E., Zahar, J. R., Perrin, A., Morelle, S., Kriz, P., Jolley, K. A., et al. (2005). A chromosomally integrated bacteriophage in invasive meningococci. *J. Exp. Med.* 201, 1905–1913. doi: 10.1084/jem.20050112

Binscheck, T., Bartels, F., Bergel, H., Bigalke, H., Yamasaki, S., Hayashi, T., et al. (1995). IgA protease from *Neisseria gonorrhoeae* inhibits exocytosis in bovine chromaffin cells like tetanus toxin. *J. Biol. Chem.* 270, 1770–1774.

Biswas, G. D., Anderson, J. E., Chen, J. E., Cornelissen, C. N., and Sparling, P. F. (1999). Identification and functional characterization of the *Neisseria gonorrhoeae lbpB* gene product. *Infect. Immun.* 67, 455–459.

Bjune, G., Høiby, E. A., Grønnesby, J. K., Fredriksen, J. H., Halstensen, A., Holten, E., et al. (1991). Effect of outer membrane vesicle vaccine against group B meningococcal disease in Norway. *Lancet* 338, 1093–1096.

Bos, M. P., Tefsen, B., Geurtsen, J., and Tommassen, J. (2004). Identification of an outer membrane protein required for the transport of lipopolysaccharide to the bacterial cell surface. *Proc. Natl. Acad. Sci. U.S.A.* 101, 9417–9422. doi: 10.1073/pnas.0402340101

Boulton, I. C., Gorringe, A. R., Allison, N., Robinson, A., Gorinsky, B., Joannou, C. L., et al. (1998). Transferrin-binding protein B isolated from *Neisseria meningitidis* discriminates between apo and diferric human transferrin. *Biochem. J.* 15, 269–273.

Brooks, C. L., Arutyunova, E., and Lemieux, M. J. (2014). The structure of lactoferrin-binding protein B from *Neisseria meningitidis* suggests roles in iron acquisition and neutralization of host defences. *Acta Crystallogr. F Struct. Biol. Commun.* 70, 1312–1317. doi: 10.1107/S2053230X14019372.

Buddelmeijer, N. (2015). The molecular mechanism of bacterial lipoprotein modification-How, when and why? *FEMS Microbiol. Rev.* 39, 246–261. doi: 10.1093/femsre/fuu006

Calmettes, C., Alcantara, J., Yu, R. H., Schryvers, A. B., and Moraes, T. F. (2012). The structural basis of transferrin sequestration by transferrin-binding protein B. *Nat. Struct. Mol. Biol.* 19, 358–360. doi: 10.1038/nsmb.2251

Cantini, F., Veggi, D., Dragonetti, S., Savino, S., Scarselli, M., Romagnoli, G., et al. (2009). Solution structure of the factor H-binding protein, a survival factor and protective antigen of *Neisseria meningitidis*. *J. Biol. Chem.* 284, 9022–9026. doi: 10.1074/jbc.C800214200

Capecchi, B., Adu-Bobie, J., Di Marcello, F., Ciucchi, L., Masignani, V., Taddei, A., et al. (2005). *Neisseria meningitidis* NadA is a new invasin which promotes bacterial adhesion to and penetration into human epithelial cells. *Mol. Microbiol.* 55, 687–698. doi: 10.1111/j.1365-2958.2004.04423.x

Casellato, A., Rossi Paccani, S., Barrile, R., Bossi, F., Ciucchi, L., Codolo, G., et al. (2014). The C2 fragment from *Neisseria meningitidis* antigen NHBA increases endothelial permeability by destabilizing adherens junctions. *Cell. Microbiol.* 16, 925–937. doi: 10.1111/cmi.12250

Caugant, D. A., Tzanakaki, G., and Kriz, P. (2007). Lessons from meningococcal carriage studies. *FEMS Microbiol. Rev.* 31, 52–63. doi: 10.1111/j.1574-6976.2006.00052.x

Clantin, B., Delattre, A. S., Rucktooa, P., Saint, N., Méli, A. C., Locht, C., et al. (2007). Structure of the membrane protein FhaC: a member of the Omp85-TpsB transporter superfamily. *Science* 317, 957–961. doi: 10.1126/science.1143860

Comanducci, M., Bambini, S., Brunelli, B., Adu-Bobie, J., Aricò, B., Capecchi, B., et al. (2002). NadA, a novel vaccine candidate of *Neisseria meningitidis*. *J. Exp. Med.* 195, 1445–1454. doi: 10.1084/jem.20020407

Comanducci, M., Bambini, S., Caugant, D. A., Mora, M., Brunelli, B., Capecchi, B., et al. (2004). NadA diversity and carriage in *Neisseria meningitidis*. *Infect. Immun.* 72, 4217–4223. doi: 10.1128/IAI.72.7.4217-4223.2004

Cornelissen, C. N., and Sparling, P. F. (1996). Binding and surface exposure characteristics of the gonococcal transferrin receptor are dependent on both transferrin-binding proteins. *J. Bacteriol.* 178, 1437–1444.

Costa, T. R., Felisberto-Rodrigues, C., Meir, A., Prevost, M. S., Redzej, A., Trokter, M., et al. (2015). Secretion systems in Gram-negative bacteria:

structural and mechanistic insights. *Nat. Rev. Microbiol.* 13, 343–359. doi: 10.1038/nrmicro3456

Crum-Cianflone, N., and Sullivan, E. (2016). Meningococcal vaccinations. *Infect. Dis. Ther.* 5, 89–112. doi: 10.1007/s40121-016-0107-0.

da Silva, R. A., Churchward, C. P., Karlyshev, A. V., Eleftheriadou, O., Snabaitis, A. K., Longman, M. R., et al. (2016). The role of apolipoprotein N-acyl transferase, Lnt, in the lipidation of factor H binding protein of *Neisseria meningitidis* strain MC58 and its potential as a drug target. *Br. J. Pharmacol.* doi: 10.1111/bph.13660. [Epub ahead of print].

Davis, J., Smith, A. L., Hughes, W. R., and Golomb, M. (2001). Evolution of an autotransporter: domain shuffling and lateral transfer from pathogenic *Haemophilus* to *Neisseria*. *J. Bacteriol.* 183, 4626–4635. doi: 10.1128/JB.183.15.000-000.2001

Del Tordello, E., Vacca, I., Ram, S., Rappuoli, R., and Serruto, D. (2014). *Neisseria meningitidis* NalP cleaves human complement C3, facilitating degradation of C3b and survival in human serum. *Proc. Natl. Acad. Sci. U.S.A.* 111, 427–432. doi: 10.1073/pnas.1321556111

Dillard, J. P., and Seifert, H. S. (2001). A variable genetic island specific for *Neisseria gonorrhoeae* is involved in providing DNA for natural transformation and is found more often in disseminated infection isolates. *Mol. Microbiol.* 41, 263–277. doi: 10.1046/j.1365-2958.2001.02520.x

Drobnak, I., Braselmann, E., and Clark, P. L. (2015). Multiple driving forces required for efficient secretion of autotransporter virulence proteins. *J. Biol. Chem.* 2090, 10104–10116. doi: 10.1074/jbc.M114.629170

Echenique-Rivera, H., Brunelli, B., Scarcelli, M., Taddei, A. R., Pizza, M., Aricò, B., et al. (2011a). A naturally occurring single-residue mutation in the translocator domain of *Neisseria meningitidis* NhhA affects trimerization, surface localization, and adhesive capacities. *Infect. Immun.* 79, 4308–4321. doi: 10.1128/IAI.00198-11

Echenique-Rivera, H., Muzzi, A., Del Tordello, E., Seib, K. L., Francois, P., Rappuoli, R., et al. (2011b). Transcriptome analysis of *Neisseria meningitidis* in human whole blood and mutagenesis studies identify virulence factors involved in blood survival. *PLoS Pathog.* 7:e1002027. doi: 10.1371/journal.ppat.1002027

Esposito, V., Musi, V., de Chiara, C., Veggi, D., Serruto, D., Scarselli, M., et al. (2011). Structure of the C-terminal domain of *Neisseria* heparin binding antigen (NHBA), one of the main antigens of a novel vaccine against *Neisseria meningitidis*. *J. Biol. Chem.* 286, 41767–41775. doi: 10.1074/jbc.M111.289314

Fletcher, L. D., Bernfield, L., Barniak, V., Farley, J. E., Howell, A., Knauf, M., et al. (2004). Vaccine potential of the *Neisseria meningitidis* 2086 lipoprotein. *Infect. Immun.* 72, 2088–2100. doi: 10.1128/IAI.72.4.2088-2100.2004

Forman, S., Linhartova, I., Osicka, R., Nassif, X., Sebo, P., and Pelicic, V. (2003). *Neisseria meningitidis* RTX proteins are not required for virulence in infant rats. *Infect. Immun.* 71, 2253–2257. doi: 10.1128/IAI.71.4.2253-2257.2003

Fussenegger, M., Facius, D., Meier, J., and Meyer, T. F. (1996). A novel peptidoglycan-linked lipoprotein (ComL) that functions in natural transformation competence of *Neisseria gonorrhoeae*. *Mol. Microbiol.* 19, 1095–1105. doi: 10.1046/j.1365-2958.1996.457984.x

Ge, X., Wang, R., Ma, J., Liu, Y., Ezemaduka, A. N., Chen, P. R., et al. (2014). DegP primarily functions as a protease for the biogenesis of β-barrel outer membrane proteins in the gram-negative bacterium *Escherichia coli*. *FEBS J.* 281, 1226–1240. doi: 10.1111/febs.12701

Gifford, J. L., Hunter, H. N., and Vogel, H. J. (2005). Lactoferricin: a lactoferrin-derived peptide with antimicrobial, antiviral, antitumor and immunological properties. *Cell. Mol. Life Sci.* 62, 2588–2598. doi: 10.1007/s00018-005-5373-z

Giuliani, M. M., Biolchi, A., Serruto, D., Ferlicca, F., Vienken, K., Oster, P., et al. (2010). Measuring bactericidal antigen-specific responses to a multicomponent vaccine against serogroup B meningococcus. *Vaccine* 28, 5023–5030. doi: 10.1016/j.vaccine.2010.05.014

Griffiths, N. J., Hill, D. J., Borodina, E., Sessions, R. B., Devos, N. I., Feron, C. M., et al. (2011). Meningococcal surface fibril (Msf) binds to activated vitronectin and inhibits the terminal complement pathway to increase serum resistance. *Mol. Microbiol.* 82, 1129–1149. doi: 10.1111/j.1365-2958.2011.07876.x

Grijpstra, J., Arenas, J., Rutten, L., and Tommassen, J. (2013). Autotransporter secretion: varying on a theme. *Res. Microbiol.* 146, 562–582. doi: 10.1016/j.resmic.2013.03.010

Hall, D. R., Hadden, J. M., Leonard, G. A., Bailey, S., Neu, M., Winn, M., et al. (2002). The crystal and molecular structures of diferric porcine and rabbit serum transferrins at resolutions of 2.15 and 2.60 Å, respectively. *Acta Crystallogr. D Biol. Crystallogr.* 58, 70–80. doi: 10.1107/S0907444901017309

Hamilton, H. L., Domínguez, N. M., Schwartz, K. J., Hackett, K. T., and Dillard, J. P. (2005). *Neisseria gonorrhoeae* secretes chromosomal DNA via a novel type IV secretion system. *Mol. Microbiol.* 55, 1704–1721. doi: 10.1111/j.1365-2958.2005.04521.x

Hauck, C. R., and Meyer, T. F. (1997). The lysosomal/phagosomal membrane protein h-lamp-1 is a target of the IgA1 protease of *Neisseria gonorrhoeae*. *FEBS Lett.* 405, 86–90.

Hayashi, S., and Wu, H. C. (1990). Lipoproteins in bacteria. *J. Bioenerg. Biomembr.* 22, 451–471.

Hooda, Y., Lai, C. C., Judd, A., Buckwalter, C. M., Shin, H. E., Gray-Owen, S. D., et al. (2016). Slam is an outer membrane protein that is required for the surface display of lipidated virulence factors in *Neisseria*. *Nat. Microbiol.* 1:16009. doi: 10.1038/nmicrobiol.2016.9

Hopper, S., Vasquez, B., Merz, A., Clary, S., Wilbur, J. S., and So, M. (2000). Effects of immunoglobulin A1 protease on *Neisseria gonorrhoeae* trafficking across polarized T84 epithelial monolayers. *Infect. Immun.* 68, 906–911. doi: 10.1128/IAI.68.2.906-911.2000

Ieva, R., and Bernstein, H. D. (2009). Interaction of an autotransporter passenger domain with BamA during its translocation across the bacterial outer membrane. *Proc. Natl. Acad. Sci. U.S.A.* 106, 19120–19125. doi: 10.1073/pnas.0907912106

Ieva, R., Tian, P., Peterson, J. H., and Bernstein, H. D. (2011). Sequential and spatially restricted interactions of assembly factors with an autotransporter β domain. *Proc. Natl. Acad. Sci. U.S.A.* 108, E383–E391. doi: 10.1073/pnas.1103827108

Irwin, S. W., Averil, N., Cheng, C. Y., and Schryvers, A. B. (1993). Preparation and analysis of isogenic mutants in the transferrin receptor protein genes, *tbpA* and *tbpB*, from *Neisseria meningitidis*. *Mol. Microbiol.* 8, 1125–1133.

Jacob-Dubuisson, F., Guérin, J., Baelen, S., and Clantin, B. (2013). Two-partner secretion: as simple as it sounds? *Res. Microbiol.* 164, 583–595. doi: 10.1016/j.resmic.2013.03.009

Jamet, A., Jousset, A. B., Euphrasie, D., Mukorako, P., Boucharlat, A., Ducousso, A., et al. (2015). A new family of secreted toxins in pathogenic *Neisseria* species. *PLoS Pathog.* 11:e1004592. doi: 10.1371/journal.ppat.1004592

Joshwich, K. O., McCaw, S. E., Islam, E., Sintsova, A., Gu, A., Shively, J. E., et al. (2013). *In vivo* adaptation and persistence of *Neisseria meningitidis* within the nasopharyngeal mucosa. *PLoS Pathog.* 9:e1003509. doi: 10.1371/journal.ppat.1003509

Khairalla, A. S., Omer, S. A., Mahdavi, J., Aslam, A., Dufailu, O. A., Self, T., et al. (2015). Nuclear trafficking, histone cleavage and induction of apoptosis by the meningococcal App and MspA autotransporters. *Cell. Microbiol.* 17, 1008–1020. doi: 10.1111/cmi.12417.

Klee, S. R., Nassif, X., Kusecek, B., Merker, P., Beretti, J. L., Achtman, M., et al. (2000). Molecular and biological analysis of eight genetic islands that distinguish *Neisseria meningitidis* from the closely related pathogen *Neisseria gonorrhoeae*. *Infect. Immun.* 68, 2082–2095. doi: 10.1128/IAI.68.4.2082-2095.2000

Koronakis, V., Sharff, A., Koronakis, E., Luisi, B., and Hughes, C. (2000). Crystal structure of the bacterial membrane protein TolC central to multidrug efflux and protein exports. *Nature* 405, 914–919. doi: 10.1038/35016007

Krojer, T., Sawa, J., Schäfer, E., Saibil, H. R., Ehrmann, M., and Clausen, T. (2008). Structural basis for the regulated protease and chaperone function of DegP. *Nature* 453, 885–890. doi: 10.1038/nature07004

Lappann, M., Claus, H., van Alen, T., Harmsen, M., Elias, J., Molin, S., et al. (2010). A dual role of extracellular DNA during biofilm formation in *Neisseria meningitidis*. *Mol. Microbiol.* 75, 1355–1371. doi: 10.1111/j.1365-2958.2010.07054.x

Lenders, M. H. H., Reimann, S., Smits, S. H., and Schmitt, L. (2013). Molecular insights into type I secretion systems. *Biol. Chem.* 394, 1371–1384. doi: 10.1515/hsz-2013-0171.

Lewis, L. A., Gipson, M., Hartman, K., Ownbey, T., Vaughn, J., and Dyer, D. W. (1999). Phase variation of HpuAB and HmbR, two distinct haemoglobin receptors of *Neisseria meningitidis* DNM2. *Mol. Microbiol.* 32, 977–989.

Lewis, L. A., Gray, E., Wang, Y. P., Roe, B. A., and Dyer, D. W. (1997). Molecular characterization of *hpuAB*, the haemoglobin-haptoglobin-utilization operon of *Neisseria meningitidis*. *Mol. Microbiol.* 23, 737–749.

Lin, L., Ayala, P., Larson, J., Mulks, M., Fukuda, M., Carlsson, S. R., et al. (1997). The *Neisseria* type 2 IgA1 protease cleaves LAMP1 and promotes survival of bacteria within epithelial cells. *Mol. Microbiol.* 24, 1083–1094.

Linhartová, I., Bumba, L., Mašín, J., Basler, M., Osička, R., Kamanová, J., et al. (2010). RTX proteins: a highly diverse family secreted by a common mechanism. *FEMS Microbiol. Rev.* 34, 1076–1112. doi: 10.1111/j.1574-6976.2010.00231.x

Loh, E., Lavender, H., Tan, F., Tracy, A., and Tang, C. M. (2016). Thermoregulation of meningococcal fHbp, an important virulence factor and vaccine antigen, is mediated by anti-ribosomal binding site sequences in the open reading frame. *PLoS Pathog.* 12:e1005794. doi: 10.1371/journal.ppat.1005794

LoVullo, E. D., Wright, L. F., Isabella, V., Huntley, J. F., and Pavelka, M. S. Jr. (2015). Revisiting the Gram-negative lipoprotein paradigm. *J. Bacteriol.* 197, 1705–1715. doi: 10.1128/JB.02414-14

Lycklama a Nijeholt, J. A., and Driessen, A. J. (2012). The bacterial Sec-translocase: structure and mechanism. *Philos. Trans. R. Soc. Lond. B* 367, 1016–1028. doi: 10.1098/rstb.2011.0201

Madico, G., Welsch, J. A., Lewis, L. A., McNaughton, A., Perlman, D. H., Costello, C. E., et al. (2006). The meningococcal vaccine candidate GNA1870 binds the complement regulatory protein factor H and enhances serum resistance. *J. Immunol.* 177, 501–510. doi: 10.4049/jimmunol.177.1.501

Maier, T., Clantin, B., Gruss, F., Dewitte, F., Delattre, A. S., Jacob-Dubuisson, F., et al. (2015). Conserved Omp85 lid-lock structure and substrate recognition in FhaC. *Nat. Commun.* 6:7452. doi: 10.1038/ncomms8452

Malito, E., Biancucci, M., Faleri, A., Ferlenghi, I., Scarselli, M., Maruggi, G., et al. (2014). Structure of the meningococcal vaccine antigen NadA and epitope mapping of a bactericidal antibody. *Proc. Natl. Acad. Sci. U.S.A.* 111, 17128–17133. doi: 10.1073/pnas.1419686111

Martin, P., van de Ven, T., Mouchel, N., Jeffries, A. C., Hood, D. W., and Moxon, E. R. (2003). Experimentally revised repertoire of putative contingency loci in *Neisseria meningitidis* strain MC58: evidence for a novel mechanism of phase variation. *Mol. Microbiol.* 50, 245–257. doi: 10.1046/j.1365-2958.2003.03678.x

Masignani, V., Comanducci, M., Giuliani, M. M., Bambini, S., Adu-Bobie, J., Arico, B., et al. (2003). Vaccination against *Neisseria meningitidis* using three variants of the lipoprotein GNA1870. *J. Exp. Med.* 197, 789–799. doi: 10.1084/jem.20021911

Matsuyama, S., Tajima, T., and Tokuda, H. (1995). A novel periplasmic carrier protein involved in the sorting and transport of *Escherichia coli* lipoproteins destined for the outer membrane *EMBO J.* 14, 3365–3372.

Matsuyama, S., Yokota, N., and Tokuda, H. (1997). A novel outer membrane lipoprotein, LolB (HemM), involved in the LolA (p20)-dependent localization of lipoproteins to the outer membrane of *Escherichia coli. EMBO J.* 16, 6947–6955.

Mayer, B., and Wong, G. C. (2015). How bacteria use type IV pili machinery on surfaces. *Trends Microbiol.* 23, 775–788. doi: 10.1016/j.tim.2015.09.002

Meng, G., Surana, N. K., St Geme, J. W. III., and Waksman, G. (2006). Structure of the outer membrane translocator domain of the *Haemophilus influenzae* Hia trimeric autotransporter. *EMBO J.* 25, 2297–2304. doi: 10.1038/sj.emboj.7601132

Metruccio, M. M., Pigozzi, E., Roncarati, D., Berlanda Scorza, F., Norais, N., Hill, S. A., et al. (2009). A novel phase variation mechanism in the meningococcus driven by a ligand-responsive repressor and differential spacing of distal promoter elements. *PLoS Pathog.* 5:e1000710. doi: 10.1371/journal.ppat.1000710

Montanari, P., Bozza, G., Capecchi, B., Caproni, E., Barrile, R., Norais, N., et al. (2012). Human heat shock protein (Hsp) 90 interferes with *Neisseria meningitidis* adhesin A (NadA)-mediated adhesion and invasion. *Cell. Microbiol.* 14, 368–385. doi: 10.1111/j.1462-5822.2011.01722.x

Morgenthau, A., Beddek, A., and Schryvers, A. B. (2014). The negatively charged regions of lactoferrin binding protein B, an adaptation against anti-microbial peptides. *PLoS ONE* 9:e86243. doi: 10.1371/journal.pone.0086243

Morgenthau, A., Livingstone, M., Adamiak, P., and Schryvers, A. B. (2012). The role of lactoferrin binding protein B in mediating protection against human lactoferricin. *Biochem. Cell. Biol.* 90, 417–423. doi: 10.1139/o11-074.

Müller, M. G., Ing, J. Y., Cheng, M. K., Flitter, B. A., and Moe, G. R. (2013). Identification of a phage-encoded Ig-binding protein from invasive *Neisseria meningitidis. J. Immunol.* 191, 3287–3296. doi: 10.4049/jimmunol.1301153

Müller, M. G., Moe, N. E., Richards, P. Q., and Moe, G. R. (2015). Resistance of *Neisseria meningitidis* to human serum depends on T and B cell stimulating protein B. *Infect. Immun.* 83, 1257–1264. doi: 10.1128/IAI.03134-14

Nägele, V., Heesemann, J., Schielke, S., Jiménez-Soto, L. F., Kurzai, O., and Ackermann, N. (2011). *Neisseria meningitidis* adhesin NadA targets β1 integrins: functional similarity to *Yersinia* invasin. *J. Biol. Chem.* 286, 20536–20546. doi: 10.1074/jbc.M110.188326

Narita, S. I., and Tokuda, H. (2007). Amino acids at positions 3 and 4 determine the membrane specificity of *Pseudomonas aeruginosa* lipoproteins. *J. Biol. Chem.* 4, 13372–13378. doi: 10.1074/jbc.M611839200

Narita, S., and Tokuda, H. (2016). Bacterial lipoproteins: biogenesis, sorting and quality control. *Biochim. Biophys. Acta.* doi: 10.1016/j.bbalip.2016.11.009. [Epub ahead of print].

Neil, R. B., and Apicella, M. A. (2009). Role of HrpA in biofilm formation of *Neisseria meningitidis* and regulation of the *hrpBAS* transcripts. *Infect. Immun.* 77, 2285–2293. doi: 10.1128/IAI.01502-08

Nivaskumar, M., and Francetic, O. (2014). Type II secretion system: a magic beanstalk or a protein escalator. *Biochim. Biophys. Acta* 1843, 1568–1577. doi: 10.1016/j.bbamcr.2013.12.020

Noinaj, N., Easley, N. C., Oke, M., Mizuno, N., Gumbart, J., Boura, E., et al. (2012). Structural basis for iron piracy by pathogenic *Neisseria. Nature* 483, 53–58. doi: 10.1038/nature10823

Noinaj, N., Guillier, M., Barnard, T. J., and Buchanan, S. K. (2010). TonB-dependent transporters: regulation, structure, and function. *Annu. Rev. Microbiol.* 64, 43–60. doi: 10.1146/annurev.micro.112408.134247

Noinaj, N., Gumbart, J. C., and Buchanan, S. K. (2017). The β-barrel assembly machinery in motion. *Nat. Rev. Microbiol.* 15, 197–204. doi: 10.1038/nrmicro.2016.191

Noinaj, N., Kuszak, A. J., Gumbart, J. C., Lukacik, P., Chang, H., Easley, N. C., et al. (2013). Structural insight into the biogenesis of β-barrel membrane proteins. *Nature* 501, 385–390. doi: 10.1038/nature12521

Ochiai, K., Kurita-Ochiai, T., Kamino, Y., and Ikeda, T. (1993). Effect of co-aggregation on the pathogenicity of oral bacteria. *J. Med. Microbiol.* 39, 183–190. doi: 10.1099/00222615-39-3-183

Oldfield, N. J., Matar, S., Bidmos, F. A., Alamro, M., Neal, K. R., Turner, D. P., et al. (2013). Prevalence and phase variable expression status of two autotransporters, NalP and MspA, in carriage and disease isolates of *Neisseria meningitidis. PLoS ONE* 25:e69746. doi: 10.1371/journal.pone.0069746

Oomen, C. J., van Ulsen, P., van Gelder, P., Feijen, M., Tommassen, J., and Gros, P. (2004). Structure of the translocator domain of a bacterial autotransporter. *EMBO J.* 23, 1257–1266. doi: 10.1038/sj.emboj.7600148

Oriente, F., Scarlato, V., and Delany, I. (2010). Expression of factor H binding protein of meningococcus responds to oxygen limitation through a dedicated FNR-regulated promoter. *J. Bacteriol.* 192, 691–701. doi: 10.1128/JB.01308-09

Osička, R., Procházková, K., Šulc, M., Linhartová, I., Havlíček, V., and Šebo, P. (2004). A novel "clip-and-link" activity of repeat in toxin (RTX) proteins from Gram-negative pathogens. Covalent protein cross-linking by and Asp-Lys isopeptide bond upon calcium-dependent processing at an Asp-Pro bond. *J. Biol. Chem.* 279, 24944–24956. doi: 10.1074/jbc.M314013200

Ostan, N. K. H., Yu, R.-H., Ng, D., Lai, C. C.-L., Pogoutse, A. K., Sarpe, V., et al. (2017). Lactoferrin binding protein B – a bifunctional bacterial receptor protein. *PLoS Pathog.* 13:e1006244. doi: 10.1371/journal.ppat.1006244

Ostberg, K. L., DeRocco, A. J., Mistry, S. D., Dickinson, M. K., and Cornelissen, C. N. (2013). Conserved regions of gonococcal TbpB are critical for surface exposure and transferrin utilization. *Infect. Immun.* 81, 3442–3450. doi: 10.1128/IAI.00280-13

Oster, P., Lennon, D., O'Hallahan, J., Mulholland, K., Reid, S., and Martin, D. (2005). MeNZB^{TM}: a safe and highly immunogenic taylor-made vaccine against the New Zealand *Neisseria meningitidis* serogroup B disease epidemic strain. *Vaccine* 23, 2191–2196. doi: 10.1016/j.vaccine.2005.01.063

Pace, D., and Pollard, A. J. (2012). Meningococcal disease: clinical presentation and sequelae. *Vaccine* 30, B3–B9. doi: 10.1016/j.vaccine.2011.12.062

Palmer, T., and Berks, B. C. (2012). The twin-arginine translocation (Tat) protein export pathway. *Nat. Rev. Microbiol.* 10, 483–496. doi: 10.1038/nrmicro2814

Paruchuri, D. K., Seifert, H. S., Ajioka, R. S., Karlsson, K. A., and So, M. (1990). Identification and characterization of a *Neisseria gonorrhoeae* gene encoding a glycolipid-binding adhesin. *Proc. Natl. Acad. Sci. U.S.A.* 87, 333–337.

Patel, R., Smith, S. M., and Robinson, C. (2014). Protein transport by the bacterial Tat pathway. *Biochim. Biophys. Acta* 1843, 1620–1628. doi: 10.1016/j.bbamcr.2014.02.013

Peak, I. R., Jennings, M. P., Hood, D. W., and Moxon, E. R. (1999). Tetranucleotide repeats identify novel virulence determinant homologues in *Neisseria meningitidis*. *Microb. Pathog.* 26, 13–23.

Peak, I. R., Srikhanta, Y., Dieckelmann, M., Moxon, E. R., and Jennings, M. P. (2000). Identification and characterisation of a novel conserved outer membrane protein from *Neisseria meningitidis*. *FEMS Immunol. Med. Microbiol.* 28, 329–334. doi: 10.1111/j.1574-695X.2000.tb01494.x

Pérez-Ortega, J., Rodríguez, A., Ribes, E., Tommassen, J., and Arenas, J. (2017). Interstrain cooperation in meningococcal biofilms: role of the autotransporters NalP and AutA. *Front. Microbiol.* 8:434. doi: 10.3389/fmicb.2017.00434

Pettersson, A., Klarenbeek, V., van Deurzen, J., Poolman, J. T., and Tommassen, J. (1994a). Molecular characterization of the structural gene for the lactoferrin receptor of the meningococcal strain H44/76. *Microb. Pathog.* 17, 395–408. doi: 10.1006/mpat.1994.1085

Pettersson, A., Maas, A., and Tommassen, J. (1994b). Identification of the *iroA* gene product of *Neisseria meningitidis* as a lactoferrin receptor. *J. Bacteriol.* 176, 1764–1766.

Pettersson, A., Prinz, T., Umar, A., van der Biezen, J., and Tommassen, J. (1998). Molecular characterization of LbpB, the second lactoferrin-binding protein of *Neisseria meningitidis*. *Mol. Microbiol.* 27, 599–610.

Pettersson, A., van der Biezen, J., Joosten, V., Hendriksen, J., and Tommassen, J. (1999). Sequence variability of the meningococcal lactoferrin-binding protein LbpB. *Gene* 231, 105–110.

Plaut, A. G., Gilbert, J. V., Artenstein, M. S., and Capra, J. D. (1975). *Neisseria gonorrhoeae* and *Neisseria meningitidis*: extracellular enzyme cleaves human immunoglobulin A. *Science* 190, 1103–1105.

Pohlner, J., Halter, R., Beyreuther, K., and Meyer, T. F. (1987). Gene structure and extracellular secretion of *Neisseria gonorrhoeae* IgA protease. *Nature* 325, 458–462. doi: 10.1038/325458a0

Pohlner, J., Langenberg, U., Wölk, U., Beck, S. C., and Meyer, T. F. (1995). Uptake and nuclear transport of *Neisseria* IgA1 protease-associated α-proteins in human cells. *Mol. Microbiol.* 17, 1073–1083. doi: 10.1111/j.1365-2958.1995.mmi_17061073.x

Prinz, T., Meyer, M., Pettersson, A., and Tommassen, J. (1999). Structural characterization of the lactoferrin receptor from *Neisseria meningitidis*. *J. Bacteriol.* 181, 4417–4419.

Prochazkova, K., Osicka, R., Linhartova, I., Halada, P., Sulc, M., and Sebo, P. (2005). The *Neisseria meningitidis* outer membrane lipoprotein FrpD binds the RTX protein FrpC. *J. Biol. Chem.* 280, 3251–3258. doi: 10.1074/jbc.M411232200

Pugsley, A. P., and Kornacker, M. G. (1991). Secretion of the cell surface lipoprotein pullulanase in *Escherichia coli*. Cooperation or competition between the specific secretion pathway and the lipoprotein sorting pathway? *J. Biol. Chem.* 266, 13640–13645.

Purdy, G. E., Fisher, C. R., and Payne, S. M. (2007). IcsA surface presentation in *Shigella flexneri* requires the periplasmic chaperones DegP, Skp, and SurA. *J. Bacteriol.* 189, 5566–5573. doi: 10.1128/JB.00483-07

Putker, F., Tommassen-van Boxtel, R., Stork, M., Rodríguez-Herva, J. J., Koster, M., and Tommassen, J. (2013). The type II secretion system (Xcp) of *Pseudomonas putida* is active and involved in the secretion of phosphatases. *Environ. Microbiol.* 15, 2658–2671. doi: 10.1111/1462-2920.12115

Ricci, D. P., and Silhavy, T. J. (2012). The Bam machine: a molecular cooper. *Biochim. Biophys. Acta* 1818, 1067–1084. doi: 10.1016/j.bbamem.2011.08.020

Rohde, K. H., and Dyer, D. W. (2004). Analysis of haptoglobin and hemoglobin-haptoglobin interactions with the *Neisseria meningitidis* TonB-dependent receptor HpuAB by flow cytometry. *Infect. Immun.* 72, 2494–2506. doi: 10.1128/IAI.72.5.2494-2506.2004

Rohde, K. H., Gillaspy, A. F., Hatfield, M. D., Lewis, L. A., and Dyer, D. W. (2002). Interactions of haemoglobin with the *Neisseria meningitidis* receptor HpuAB: the role of TonB and an intact proton motive force. *Mol. Microbiol.* 43, 335–354. doi: 10.1046/j.1365-2958.2002.02745.x

Roman-Hernandez, G., Peterson, J. H., and Bernstein, H. (2014). Reconstitution of bacterial autotransporter assembly using purified components. *Elife* 3:e04234. doi: 10.7554/eLife.04234

Rouphael, N. G., and Stephens, D. S. (2012). *Neisseria meningitidis*: biology, microbiology, and epidemiology. *Methods Mol. Biol.* 799, 1–20. doi: 10.1007/978-1-61779-346-2_1

Roussel-Jazédé, V., Arenas, J., Langereis, J. D., Tommassen, J., and van Ulsen, P. (2014). Variable processing of the IgA protease autotransporter at the cell surface of *Neisseria meningitidis*. *Microbiology* 160, 2421–2431. doi: 10.1099/mic.0.082511

Roussel-Jazédé, V., Grijpstra, J., van Dam, V., Tommassen, J., and van Ulsen, P. (2013). Lipidation of the autotransporter NalP of *Neisseria meningitidis* is required for its function in the release of cell-surface-exposed proteins. *Microbiology* 159, 286–295. doi: 10.1099/mic.0.063982-0

Roussel-Jazédé, V., Jongerius, I., Bos, M. P., Tommassen, J., and van Ulsen, P. (2010). NalP-mediated proteolytic release of lactoferrin-binding protein B from the meningococcal cell surface. *Infect. Immun.* 78, 3083–3089. doi: 10.1128/IAI.01193-09

Ruiz-Perez, F., Henderson, I. R., Leyton, D. L., Rossiter, A. E., Zhang, Y., and Nataro, J. P. (2009). Role of the periplasmic chaperone proteins in the biogenesis of serine protease autotransporters of *Enterobacteriaceae*. *J. Bacteriol.* 191, 6571–6583. doi: 10.1128/JB.00754-09

Sanders, H., Brehony, C., Maiden, M. C., Vipond, C., and Feavers, I. M. (2012). The effect of iron availability on transcription of the *Neisseria meningitidis* fHbp gene varies among clonal complexes. *Microbiology* 158, 869–876. doi: 10.1099/mic.0.054957-0.

Scarselli, M., Serruto, D., Montanari, P., Capecchi, B., Adu-Bobie, J., Veggi, D., et al. (2006). *Neisseria meningitidis* NhhA is a multifunctional trimeric autotransporter adhesin. *Mol. Microbiol.* 61, 631–644. doi: 10.1111/j.1365-2958.2006.05261.x

Schielke, S., Huebner, C., Spatz, C., Nägele, V., Ackermann, N., Frosch, M., et al. (2009). Expression of the meningococcal adhesin NadA is controlled by a transcriptional regulator of the MarR family. *Mol. Microbiol.* 72, 1054–1067. doi: 10.1111/j.1365-2958.2009.06710.x

Schmitt, C., Turner, D., Boesl, M., Abele, M., Frosch, M., and Kurzai, O. (2007). A functional two-partner secretion system contributes to adhesion of *Neisseria meningitidis* to epithelial cells. *J. Bacteriol.* 189, 7968–7976. doi: 10.1128/JB.00851-07

Schneider, M. C., Exley, R. M., Chan, H., Feavers, I., Kang, Y. H., Sim, R. B., et al. (2006). Functional significance of factor H binding to *Neisseria meningitidis*. *J. Immunol.* 176, 7566–7575. doi: 10.4049/jimmunol.176.12.7566

Schneider, M. C., Prosser, B. E., Caesar, J. J., Kugelberg, E., Li, S., Zhang, Q., et al. (2009). *Neisseria meningitidis* recruits factor H using protein mimicry of host carbohydrates. *Nature* 458, 890–893. doi: 10.1038/nature07769

Schryvers, A. B., and Stojiljkovic, I. (1999). Iron acquisition systems in the pathogenic *Neisseria*. *Mol. Microbiol.* 32, 1117–1123. doi: 10.1099/mic.0.068874-0

Schulze, R. J., and Zückert, W. R. (2006). *Borrelia burgdorferi* lipoproteins are secreted to the outer surface by default. *Mol. Microbiol.* 59, 1473–1484. doi: 10.1111/j.1365-2958.2006.05039.x

Seib, K. L., Serruto, D., Oriente, F., Delany, I., Adu-Bobie, J., Veggi, D., et al. (2009). Factor H-binding protein is important for meningococcal survival in human whole blood and serum and in the presence of the antimicrobial peptide LL-37. *Infect. Immun.* 77, 292–299. doi: 10.1128/IAI.01071-08

Selkrig, J., Mosbahi, K., Webb, C. T., Belousoff, M. J., Perry, A. J., Wells, T. J., et al. (2012). Discovery of an archetypical protein transport system in bacterial outer membranes. *Nat. Struct. Mol. Biol.* 19, 506–510. doi: 10.1038/nsmb.2261

Senior, B. W., Stewart, W. W., Galloway, C., and Kerr, M. A. (2001). Cleavage of the hormone human chorionic gonadotropin, by the Type 1 IgA1 protease of *Neisseria gonorrhoeae*, and its implications. *J. Infect. Dis.* 184, 922–925. doi: 10.1086/323397

Serruto, D., Adu-Bobie, J., Scarselli, M., Veggi, D., Pizza, M., Rappuoli, R., et al. (2003). *Neisseria meningitidis* App, a new adhesin with autocatalytic serine protease activity. *Mol. Microbiol.* 48, 323–334. doi: 10.1046/j.1365-2958.2003.03420.x

Serruto, D., Spadafina, T., Ciucchi, L., Lewis, L. A., Ram, S., Tontini, M., et al. (2010). *Neisseria meningitidis* GNA2132, a heparin-binding protein that induces protective immunity in humans. *Proc. Natl. Acad. Sci. U.S.A.* 107, 3770–3775. doi: 10.1073/pnas.0915162107

Sim, R. J., Harrison, M. M., Moxon, E. R., and Tang, C. M. (2000). Underestimation of meningococci in tonsillar tissue by nasopharyngeal swabbing. *Lancet* 356, 1653–1654. doi: 10.1016/S0140-6736(00)03162-7

Sjölinder, H., Eriksson, J., Maudsdotter, L., Aro, H., and Jonsson, A. B. (2008). Meningococcal outer membrane protein NhhA is essential for colonization and disease by preventing phagocytosis and complement attack. *Infect. Immun.* 76, 5412–5420. doi: 10.1128/IAI.00478-08

Sjölinder, M., Altenbacher, G., Hagner, M., Sun, W., Schedin-Weiss, S., and Sjölinder, H. (2012). Meningococcal outer membrane protein NhhA triggers apoptosis in macrophages. *PLoS ONE* 7:e29586. doi: 10.1371/journal.pone.0029586

Sklar, J. G., Wu, T., Kahne, D., and Silhavy, T. J. (2007). Defining the roles of the periplasmic chaperones SurA, Skp, and DegP in *Escherichia coli*. *Genes Dev.* 21, 2473–2484. doi: 10.1101/gad.1581007

Snyder, L. A., Jarvis, S. A., and Saunders, N. J. (2005). Complete and variant forms of the 'gonococcal genetic island' in *Neisseria meningitidis*. *Microbiology* 151, 4005–4013. doi: 10.1099/mic.0.27925-0

Srinivas, N., Jetter, P., Ueberbacher, B. J., Werneburg, M., Zerbe, K., Steinmann, J., et al. (2010). Peptidomimetic antibiotics target outer-membrane biogenesis in *Pseudomonas aeruginosa*. *Nature* 327, 1010–1013. doi: 10.1126/science.1182749

Stojiljkovic, I., Hwa, V., de Saint Martin, L., O'Gaora, P., Nassif, X., Heffron, F., et al. (1995). The *Neisseria meningitidis* haemoglobin receptor: its role in iron utilization and virulence. *Mol. Microbiol.* 15, 531–541.

Sviridova, E., Rezacova, P., Bondar, A., Veverka, V., Novak, P., Schenk, G., et al. (2017). Structuraal basis of the interaction between the putative adhesion-involved and iron-regulated FrpD and FrpC proteins of *Neisseria meningitidis*. *Sci. Rep.* 7:40408. doi: 10.1038/srep40408

Takada, S., Fujiwara, S., Inoue, T., Kataoka, Y., Hadano, Y., Matsumoto, K., et al. (2016). Meningococcemia in adults: a review of the literature. *Intern. Med.* 55, 567–572. doi: 10.2169/internalmedicine.55.3272

Talà, A., Progida, C., De Stefano, M., Cogli, L., Spinosa, M. R., Bucci, C., et al. (2008). The HrpB-HrpA two-partner secretion system is essential for intracellular survival of *Neisseria meningitidis*. *Cell. Microbiol.* 10, 2461–2482. doi: 10.1111/j.1462-5822.2008.01222.x

Tavano, R., Capecchi, B., Montanari, P., Franzoso, S., Marin, O., Sztukowska, M., et al. (2011). Mapping of the *Neisseria meningitidis* NadA cell-binding site: relevance of predicted α-helices in the NH_2-terminal and dimeric coiled-coil regions. *J. Bacteriol.* 193, 107–115. doi: 10.1128/JB.00430-10

Thompson, S. A., and Sparling, P. F. (1993). The RTX cytotoxin-related FrpA protein of *Neisseria meningitidis* is secreted extracellularly by meningococci and by $HlyBD^+$ *Escherichia coli*. *Infect. Immun.* 61, 2906–2911.

Thompson, S. A., Wang, L. L., and Sparling, P. F. (1993a). Cloning and nucleotide sequence of *frpC*, a second gene from *Neisseria meningitidis* encoding a protein similar to RTX cytotoxins. *Mol. Microbiol.* 9, 85–96. doi: 10.1111/j.1365-2958.1993.tb01671.x

Thompson, S. A., Wang, L. L., West, A., and Sparling, P. F. (1993b). *Neisseria meningitidis* produces iron-regulated proteins related to the RTX family of exoproteins. *J. Bacteriol.* 175, 811–818. doi: 10.1128/jb.175.3.811-818.1993

Tinsley, C. R., Voulhoux, R., Beretti, J. L., Tommassen, J., and Nassif, X. (2004). Three homologues, including two membrane-bound proteins, of the disulfide oxidoreductase DsbA in *Neisseria meningitidis*: Effects on bacterial growth and biogenesis of functional type IV pili. *J. Biol. Chem.* 279, 27078–27087. doi: 10.1074/jbc.M313404200

Tsirigotaki, A., de Geyter, J., Šoštarik, N., Economou, A., and Karamanou, S. (2017). Protein export through the bacterial Sec pathway. *Nat. Microbiol. Rev.* 15, 21–36. doi: 10.1038/nrmicro.2016.161

Tsukazaki, T., Mori, H., Echizen, Y., Ishitan, R., Fukai, S., Tanaka, T., et al. (2011). Structure and function of a membrane component SecDF that enhances protein export. *Nature* 474, 235–238. doi: 10.1038/nature09980

Turner, D. P., Marietou, A. G., Johnston, L., Ho, K. K., Rogers, A. J., Wooldridge, K. G., et al. (2006). Characterization of MspA, an immunogenic autotransporter protein that mediates adhesion to epithelial and endothelial cells in *Neisseria meningitidis*. *Infect. Immun.* 74, 2957–2964. doi: 10.1128/IAI.74.5.2957-2964.2006

Turner, D. P., Wooldridge, K. G., and Ala'Aldeen, D. A. (2002). Autotransported serine protease A of *Neisseria meningitidis*: an immunogenic, surface-exposed outer membrane, and secreted protein. *Infect. Immun.* 70, 4447–4461. doi: 10.1128/IAI.70.8.4447-4461.2002

ur Rahman, S., and van Ulsen, P. (2013). System specificity of the TpsB transporters of coexpressed two-partner secretion systems of *Neisseria meningitidis*. *J. Bacteriol.* 195, 788–797. doi: 10.1128/JB.01355-12

ur Rahman, S., Arenas, J., Öztürk, H., Dekker, N., and van Ulsen, P. (2014). The polypeptide transport-associated (POTRA) domains of TpsB transporters determine the system specificity of two-partner secretion systems. *J. Biol. Chem.* 289, 19799–19809. doi: 10.1074/jbc.M113.544627

Urfer, M., Bogdanovic, J., Lo Monte, F., Moehle, K., Zerbe, K., Omasits, U., et al. (2016). A peptidomimetic antibiotic targets outer membrane proteins and disrupts selectively the outer membrane in *Escherichia coli*. *J. Biol. Chem.* 291, 1921–1932. doi: 10.1074/jbc.M115.691725

Vacca, I., Del Tordello, E., Gasperini, G., Pezzicoli, A., Di Fede, M., Rossi Paccani, S., et al. (2016). Neisserial heparin binding antigen (NHBA) contributes to the adhesion of *Neisseria meningitidis* to human epithelial cells. *PLoS ONE* 11:e0162878. doi: 10.1371/journal.pone.0162878

van Ulsen, P., Adler, B., Fassler, P., Gilbert, M., van Schilfgaarde, M., van der Ley, P., et al. (2006). A novel phase-variable autotransporter serine protease, AusI, of *Neisseria meningitidis*. *Microbes Infect.* 8, 2088–2097. doi: 10.1016/j.micinf.2006.03.007

van Ulsen, P., and Tommassen, J. (2006). Protein secretion and secreted proteins in pathogenic *Neisseriaceae*. *FEMS Microbiol. Rev.* 30, 292–319. doi: 10.1111/j.1574-6976.2006.00013.x

van Ulsen, P., Rutten, L., Feller, M., Tommassen, J., and van der Ende, A. (2008). Two-partner secretion systems of *Neisseria meningitidis* associated with invasive clonal complexes. *Infect. Immun.* 76, 4649–4658. doi: 10.1128/IAI.00393-08.

van Ulsen, P., van Alphen, L., Hopman, C. T., van der Ende, A., and Tommassen, J. (2001). *In vivo* expression of *Neisseria meningitidis* proteins homologous to the *Haemophilus influenzae* Hap and Hia autotransporters. *FEMS Immunol. Med. Microbiol.* 32, 53–64. doi: 10.1111/j.1574-695X.2001.tb00534.x

van Ulsen, P., van Alphen, L., ten Hove, J., Fransen, F., van der Ley, P., and Tommassen, J. (2003). A Neisserial autotransporter NalP modulating the processing of other autotransporters. *Mol. Microbiol.* 50, 1017–1030. doi: 10.1046/j.1365-2958.2003.03773.x

Veggi, D., Gentile, M. A., Cantini, F., Lo Surdo, P., Nardi-Dei, V., Seib, K. L., et al. (2012). The factor H binding protein of *Neisseria meningitidis* interacts with xenosiderophores *in vitro*. *Biochemistry* 51, 9384–9393. doi: 10.1021/bi301161w

Volokhina, E. B., Beckers, F., Tommassen, J., and Bos, M. P. (2009). The β-barrel outer membrane protein assembly complex of *Neisseria meningitidis*. *J. Bacteriol.* 191, 7074–7085. doi: 10.1128/JB.00737-09

Volokhina, E. B., Grijpstra, J., Stork, M., Schilders, I., Tommassen, J., and Bos, M. P. (2011). Role of the periplasmic chaperones Skp, SurA and DegQ in outer membrane protein biogenesis in *Neisseria meningitidis*. *J. Bacteriol.* 193, 1612–1621. doi: 10.1128/JB.00532-10

von Heijne, G. (1990). The signal peptide. *J. Membr. Biol.* 115, 195–201.

Voulhoux, R., Bos, M. P., Geurtsen, J., Mols, M., and Tommassen, J. (2003). Role of a highly conserved bacterial protein in outer membrane protein assembly. *Science* 299, 262–265. doi: 10.1126/science.1078973

Walther, D. M., Rapaport, D., and Tommassen, J. (2009). Biogenesis of β-barrel membrane proteins in bacteria and eukaryotes: evolutionary conservation and divergence. *Cell. Mol. Life Sci.* 66, 2789–2804. doi: 10.1007/s00018-009-0029-z

Wang, X., Sjölinder, M., Gao, Y., Wan, Y., and Sjölinder, H. (2016). Immune homeostatic macrophages programmed by the bacterial surface protein NhhA potentiate nasopharyngeal carriage of *Neisseria meningitidis*. *mBio* 7, e01670–e01615. doi: 10.1128/mBio.01670-15

Weinberg, E. D. (2009). Iron availability and infection. *Biochim. Biophys. Acta* 1790, 600–605. doi: 10.1016/j.bbagen.2008.07.002

Welsch, J. A., Moe, G. R., Rossi, R., Adu-Bobie, J., Rappuoli, R., and Granoff, D. M. (2003). Antibody to genome-derived neisserial antigen 2132, a *Neisseria meningitidis* candidate vaccine, confers protection against bacteremia in the absence of complement-mediated bactericidal activity. *J. Infect. Dis.* 188, 1730–1740. doi: 10.1086/379375

Wilson, M. M., and Bernstein, H. D. (2017). Surface-exposed lipoproteins: an emerging secretion phenomenon in Gram-negative bacteria. *Trends Microbiol.* 24, 198–208. doi: 10.1016/j.tim.2015.11.006

Wong, C. T., Xu, Y., Gupta, A., Garnett, J. A., Matthews, S. J., and Hare, S. A. (2015). Structural analysis of haemoglobin binding by HpuA from the *Neisseriaceae* family. *Nat. Commun.* 6:10172. doi: 10.1038/ncomms 10172

Woodhams, K. L., Benet, Z. L., Blonsky, S. E., Hackett, K. T., and Dillard, J. P. (2012). Prevalence and detailed mapping of the gonococcal genetic island in *Neisseria meningitidis*. *J. Bacteriol.* 194, 2275–2285. doi: 10.1128/JB. 00094-12.

Wooldridge, K. G., Kizil, M., Wells, D. B., and Ala'Aldeen, D. A. (2005). Unusual genetic organization of a functional type I protein secretion system in *Neisseria meningitidis*. *Infect. Immun.* 73, 5554–5567. doi: 10.1128/IAI.73.9.5554-5567.2005

Yakushi, T., Masuda, K., Narita, S., Matsuyama, S., and Tokuda, H. (2000). A new ABC transporter mediating the detachment of lipid-modified proteins from membranes. *Nat. Cell Biol.* 2, 212–218. doi: 10.1038/35008635

Yamaguchi, K., Yu, F., and Inouye, M. (1988). A single amino acid determinant of the membrane localization of lipoproteins in *E. coli*. *Cell* 53, 423–432.

Yu, H., Muñoz, E. M., Edens, R. E., and Linhardt, R. J. (2005). Kinetic studies on the interactions of heparin and complement proteins using surface plasmon resonance. *Biochim. Biophys. Acta* 1726, 168–176. doi: 10.1016/j.bbagen.2005.08.003

11

A Type VI Secretion System Trans-Kingdom Effector is Required for the Delivery of a Novel Antibacterial Toxin in *Pseudomonas aeruginosa*

Benjamin Berni, Chantal Soscia, Sarah Djermoun, Bérengère Ize and Sophie Bleves*

LISM, IMM (Institut de Microbiologie de la Méditerranée), CNRS and Aix-Marseille Univ, Marseille, France

***Correspondence:** *Sophie Bleves*
bleves@imm.cnrs.fr

Pseudomonas aeruginosa has evolved multiple strategies to disarm and take advantage of its host. For this purpose, this opportunist pathogen has particularly developed protein secretion in the surrounding medium or injection into host cells. Among this, the type VI secretion system (T6SS) is utilized to deliver effectors into eukaryotic host as well as target bacteria. It assembles into a contractile bacteriophage tail-like structure that functions like a crossbow, injecting an arrow loaded with effectors into the target cell. The repertoire of T6SS antibacterial effectors of *P. aeruginosa* is remarkably broad to promote environmental adaptation and survival in various bacterial communities, and presumably in the eukaryotic host too. Here, we report the discovery of a novel pair of antibacterial effector and immunity of *P. aeruginosa*, Tle3 and Tli3. Tli3 neutralizes the toxicity of Tle3 in the periplasm to protect from fratricide intoxication. The characterization of the secretion mechanism of Tle3 indicates that it requires a cytoplasmic adaptor, Tla3, to be targeted and loaded onto the VgrG2b spike and thus delivered by the H2-T6SS machinery. Tla3 is different from the other adaptors discovered so far and defines a novel family among T6SS with a DUF2875. Interestingly, this led us to discover that VgrG2b that we previously characterized as an anti-eukaryotic effector possesses an antibacterial activity as well, as it is toxic towards *Escherichia coli*. Excitingly Tli3 can counteract VgrG2b toxicity. VgrG2b is thus a novel trans-kingdom effector targeting both bacteria and eukaryotes. VgrG2b represents an interesting target for fighting against *P. aeruginosa* in the environment and in the context of host infection.

Keywords: virulence, type VI secretion system, bacterial competition, VgrG, Tle, immunity, adaptor, chaperone

INTRODUCTION

Pseudomonas aeruginosa is one of the most virulent opportunistic pathogens, being responsible for various diseases such as acute infections of lungs and burned skin that can lead to septicemia more particularly in immunocompromized patients, or broncho-alveolar colonization in cystic fibrosis sufferers. *P. aeruginosa* has been classified in 2017 as critical by the WHO in the top three lists of antibiotic resistant bacteria (WHO, 2019). *P. aeruginosa* has developed various pathogenicity strategies among which protein secretion or protein delivery into target cells is key. Indeed, this pathogen possesses five of the six secretion systems so far identified among Gram-negative bacteria, if we exclude the T9SS (type IX secretion system) restricted to one phylum, and remarkably in several copies for most of them (Bleves et al., 2010).

The type VI secretion system (T6SS) was first discovered in the context of eukaryotic host infection (Mougous et al., 2006; Pukatzki et al., 2006) and later during bacterial competition (Hood et al., 2010), which seems to be its primary function (Cianfanelli et al., 2016b). The T6SS confers a fitness advantage (1) in environmental niches against rival bacteria (inter- and intraspecies competitiveness have been described) and (2) in the eukaryotic host towards commensal bacteria (Cianfanelli et al., 2016b). Indeed, recent studies have highlighted a novel role for T6SS-dependent antibacterial responses in interbacterial competition in the mammalian gut (Hecht et al., 2016; Sana et al., 2016b), suggesting that T6SSs may be important not only in shaping microbial community composition but also in governing interactions between the microbiota and invading pathogens. Interestingly, several T6SSs are also known to target both cell type genus such as the T6SS of *Vibrio cholera* (Pukatzki et al., 2006; MacIntyre et al., 2010) and *P. aeruginosa* (Sana et al., 2012) (Russell et al., 2013). Even more remarkably, three T6SS effectors of *P. aeruginosa*, namely, PldA (also called Tle5a), PldB (Tle5b), and TplE (Tle4) (Jiang et al., 2014, 2016) have been called "trans-kingdom effectors" since these toxins can target both prokaryotic and eukaryotic cells (Bleves, 2016). Indeed, toxins are usually directed against eukaryotic cells (like AB toxins or RTX pore-forming toxins) or against rival bacteria (like bacteriocins).

The T6SS functions as a dynamic contractile phage tail-like structure anchored in the bacterial cell envelope that delivers effector proteins directly into the target cell in a one-step manner. T6SS includes a contractile sheath that cover a nanotube of stacked Hcp topped with a membrane-puncturing spike made of VgrG and PAAR (proline-alanine-alanine-arginine repeat) proteins (Basler, 2015). The sheath can contract and inject the arrow loaded with effectors into the target cell. Characterizing the repertoire of effectors delivered by the T6SS has highlighted a great diversity in terms of effector activities, host cell targets, and mode of recruitment by the T6SS machinery. In brief, there are two broad effector categories: the "specialized" effectors fused to components of the machinery (evolved VgrG, evolved PAAR, and evolved Hcp have been described so far) and the "cargo" effectors (Cianfanelli et al., 2016b). The later are addressed to the T6SS machinery by binding directly one of the arrow components (VgrG, PAAR, and Hcp) or by being targeted through cytoplasmic adaptor proteins also called chaperones. To date, three families of adaptor have been described, the first one harboring a DUF4123 (Miyata et al., 2013; Salomon et al., 2014; Liang et al., 2015; Unterweger et al., 2015), the second a DUF1795 (Alcoforado Diniz and Coulthurst, 2015; Whitney et al., 2015), and the third one a DUF2169 (Bondage et al., 2016). In line with this, many effector-encoding genes are found in close proximity to *vgrG*, *hcp*, and *paar* or adaptor genes. Finally to protect themselves from self-intoxication or from antibacterial toxins injected by neighboring sibling cells, bacteria always synthesize immunity proteins, which are encoded by adjacent genes (Benz and Meinhart, 2014).

P. aeruginosa encodes three distinct T6SS loci, H1- to H3-T6SS. While H1-T6SS has only been involved in antibacterial activity so far (Sana et al., 2016a; LaCourse et al., 2018), H2-T6SS and H3-T6SS can target both bacterial and eukaryotic cells possessing even as said earlier trans-kingdom effectors (Sana et al., 2012, 2015; Russell et al., 2013; Jiang et al., 2014, 2016; Burkinshaw et al., 2018). We discovered the anti-eukaryotic function of the H2-T6SS machinery that promotes the uptake of *P. aeruginosa* by non-phagocytic cells (Sana et al., 2012). The two phospholipases D mentioned earlier, PldA (Tle5a) and PldB (Tle5b), delivered, respectively, by H2-T6SS and H3-T6SS machineries participate in the host kinase pathway hijacking that facilitates further entry of *P. aeruginosa* (Jiang et al., 2014). The evolved VgrG2b effector (Sana et al., 2015) is delivered by H2-T6SS into epithelial cells where it targets the γ-tubulin ring complex, a microtubule-nucleating multiprotein complex to promote a microtubule-dependent internalization of *P. aeruginosa*. Finally TplE (Tle4), which is secreted by the H2-T6SS machinery, promotes autophagy in epithelial cells once localized to the endoplasmic reticulum (Jiang et al., 2016). Interestingly, PldA (Tle5a), PldB (Tle5b), and TplE (Tle4) have also been identified as antibacterial phospholipases of the Tle (type VI lipase effectors) family (Russell et al., 2013). They work by affecting membrane integrity of the rival bacteria (Russell et al., 2013; Jiang et al., 2014, 2016). More precisely, PldA degrades the major constituent of bacterial membranes, the phosphatidylethanolamine (Russell et al., 2013).

In the present study, we have discovered a novel antibacterial toxin, Tle3, and its cognate immunity, Tli3, whose genes are encoded downstream of *vrgG2b*. By characterizing the secretion mechanism of Tle3 by H2-T6SS, we showed that it requires Tla3, a cytoplasmic adaptor of a unique family, to be targeted to the VgrG2b spike. Interestingly, we also found that the C-terminal extension of VgrG2b is toxic towards *Escherichia coli* making VgrG2b a new trans-kingdom effector of *P. aeruginosa*.

MATERIALS AND METHODS

Bacterial Strains, Growth Conditions, and Plasmid Construction

All *P. aeruginosa* and *E. coli* strains used in this study are described in **Supplementary Table S1**. Briefly, the *E. coli* K-12 DH5α and CC118λPir were used for cloning procedures. The BL21(DE3)pLysS and BTH101 were used for protein production and BACTH analyses, respectively. Strains were grown in LB or in TSB medium (for *P. aeruginosa*) at 37 or 30°C. Specific growth conditions are specified in the text when necessary. Recombinant plasmids were introduced into *P. aeruginosa* by triparental mating using the conjugative properties of the helper plasmid pRK2013 (**Supplementary Table S1**). Plasmids were maintained by the addition of ampicillin (50 μg/ml), kanamycin (50 μg/ml), chloramphenicol (30 μg/ml), streptomycin (30 μg/ml for *E. coli* and 2,000 μg/ml for *P. aeruginosa*), or gentamicin (30 μg/ml for *E. coli* and 115 μg/ml for *P. aeruginosa*). Expression of genes from pT7 in BL21(DE3)pLysS was blocked with 0.4% of glucose and induced in exponential phase (OD_{600} = 0.4–0.6) for 3 h with 1 mM of IPTG. Cloning procedures were described in Sana et al. (2015). The plasmids used and constructed are described in **Supplementary Table S1**, the list of oligonucleotides (synthesized by Eurogentec or IDT) is given in **Supplementary Table S2**.

Cloning Procedures for *P. aeruginosa* Mutants

To generate *P. aeruginosa* mutants, 500 bp upstream and 500 bp downstream of the gene to be deleted were amplified by overlapping PCR with Q5 high fidelity DNA polymerase (NEB) using primers listed in **Supplementary Table S1**. The PCR product was cloned in pKNG101 suicide vector by one-step sequence and ligation-independent cloning (SLIC) (Jeong et al., 2012), which was then sequenced. pKNG101 derivatives, maintained in the *E. coli* CC118λpir strain, were mobilized in *P. aeruginosa* strains. The mutants, in which the double recombination events occurred, were confirmed by PCR analysis.

Heterologous Toxicity Assays

E. coli BL21(DE3)pLysS containing plasmids producing cytoplasmic or periplasmic targeted proteins were grown overnight at 37°C in LB with 0.4% of glucose. About 10 μl drops of bacterial suspensions serially diluted were spotted onto LB agar plates containing 0.1 mM IPTG or 0.4% glucose and cells were grown for 16 h at 37°C.

Bacterial Two-Hybrid Assay

Protein-protein interactions were assessed with the adenylate cyclase-based two-hybrid technique using protocols published previously (Karimova et al., 1998; Battesti and Bouveret, 2012). Briefly, the proteins to be tested were fused to the isolated T18 and T25 catalytic domains of the *Bordetella* adenylate cyclase. After introduction of the two plasmids producing the fusion proteins into the reporter BTH101 strain, plates were incubated at 30°C for 24 h. Three independent colonies for each transformation were inoculated into 600 μl of LB medium supplemented with ampicillin (50 μg/ml), kanamycin (50 μg/ml), and IPTG (0.5 mM). After overnight growth at 30°C, 5 μl of each culture was spotted onto LB agar plates supplemented with ampicillin, kanamycin, IPTG, and 5-bromo-4-chloro-3-indonyl-D-galactopyrannoside (X-gal, 40 μg/ml) and incubated for 16 h at 30°C.

Protein Purification by Affinity Chromatography

E. coli BL21(DE3)plysS cells carrying the pRSFDUET-1 and pETDUET-1 derivatives were grown at 37°C in LB to an OD_{600}~0.5 and the expression of the PA0262, PA0261, PA0260, or PA0259 genes was induced with IPTG (1 mM) for 3 h at 37°C. Cells were harvested by centrifugation at 1,914 × *g* for 30 min at 4°C. The cell pellet was resuspended in Tris–HCl 50 mM pH 8.0, NaCl 150 mM, Triton X-100 0.1%, lysozyme 0.5 mg/ml, and EDTA 1 mM and stored at −80°C. Cells were supplemented with DNase (20 μg/ml), $MgCl_2$, and phenylmethylsulfonyl fluoride 1 mM and cells were lysed by three passages at the Emulsiflex-C5 (Avestin), and lysates were clarified by centrifugation at 16,000 × *g* for 30 min. The supernatant was loaded onto a 5-ml StrepTrap HP (GE Healthcare) column and then washed with 50 mM Tris-HCl pH 8.0, 150 mM NaCl at 4°C. The fusion protein was eluted in the affinity buffer supplemented with 2.5 mM desthiobiotin. Peak fractions were pooled and loaded onto a Superose 200 10/300 column (GE Healthcare) equilibrated in 50 mM Tris-HCl pH 8.0, 50 mM NaCl.

Fractionation of *P. aeruginosa*

Fractionation of cells into spheroplasts (cytoplasm and membranes) and periplasmic fractions were done as previously described (Ize et al., 2014). Proteins corresponding to the cytoplasm and periplasm fractions or to insoluble material were resuspended in loading buffer.

Protein Secretion

P. aeruginosa strains were grown at 25°C in TBS for 24 h. Cells corresponding to 10 U DO_{600} and extracellular medium were separated by centrifugation at 2,000 × *g* for 10 min at room temperature. 2/3 of the supernatants were collected and centrifuged at 13,000 × *g* for 5 min at room temperature. Proteins contained in the supernatant were precipitated with tricholoro-acetic acid (TCA, 15%) for 3 h at 4°C. Samples were centrifuged at 13,000 × *g* for 30 min at 4°C, pellets washed with 90% acetone, and resuspended in loading buffer.

SDS-PAGE and Western-Blot

Protein samples derived from equivalent amounts of culture (i.e., optical density equivalents) resuspended in loading buffer were boiled and separated by SDS-PAGE. Proteins were then stained by Coomassie-blue or immunodetected as described before (Sana et al., 2015) using primary polyclonal antibodies directed against His6 epitope-tag (Penta His, Qiagen, dilution 1:1,000), V5 epitope-tag (Bethyl Laboratories, dilution 1:1,000), Strep epitope-tag (IBA StrepMAB Classic, dilution 1:000), DsbA (kindly gifted by K. E. Jaeger – university of Heinrich-Heine, dilution 1:25,000), or monoclonal antibodies directed against EF-Tu (Hycult-biotech, dilution 1:20,000), XcpY (laboratory collection, dilution 1:5,000), and TolB (laboratory collection, dilution 1:500). Peroxidase-conjugated anti-Mouse or anti-Rabbit IgGs (Sigma, dilution 1:5,000) were used as secondary antibodies. Nitrocellulose membranes were revealed with homemade enhanced chemiluminescence and were scanned using ImageQuant LAS4000 analysis software (GE Healthcare Life sciences).

Protein samples equivalent to 0.1 OD_{600} units were loaded for whole cell and spheroplasts analysis while protein samples equivalent to 0.2 OD_{600} units were used for cytoplasm or periplasm analysis and protein samples equivalent to 1 OD_{600} units were used for extracellular medium analysis.

Bacterial Competition Assays

Intraspecific competition assays between *P. aeruginosa* strains were performed as previously described (Jiang et al., 2014) with modifications. The prey cells carry pJN105 vector (Gm^R) to allow counterselection. Overnight cultures of *P. aeruginosa* attacker and prey cells were mixed in a 10:1 (attacker:prey) ratio and harvested by centrifugation at 3,724 × *g* for 5 min. The pellet was resuspended in 200 μl of PBS 1X and spotted onto 0.45-μm nitrocellulose membranes overlaid on a 1% bactoagar plate. After 24 h of incubation at 37°C, cells were resuspended in 2 ml of PBS 1X, normalized to an OD_{600nm} of 0.5, and 10 μl of bacterial serially diluted (10^{-1} to 10^{-6}) were spotted onto selective LB agar plates containing gentamicin (125 μg/ml). Significant growth difference of the prey bacteria for each competition assay was computed by one-way ANOVA (Stat Plus) and unpaired Student's Test (Excel).

RESULTS

Tle3 Is a Novel Antibacterial Toxin of *P. aeruginosa*

The analysis of the *vgrG2b* genetic environment revealed the presence of the PA0260 gene encoding a protein with a α/β hydrolase domain (PF00561) and a putative Ser-Asp-His catalytic triad used by various esterase enzymes and has thus been classified in the Tle3 family of antibacterial Tle toxins (**Figure 1A**; Russell et al., 2013). The immunities of Tle proteins, which are lipolytic toxins active in the periplasm of the prey bacterium, are localized in or exposed to the periplasm where they neutralize the cognate toxin (Russell et al., 2013; Jiang et al., 2014, 2016; Flaugnatti et al., 2016). The two genes surrounding *tle3*, PA0259 and PA0261, are good candidates as Tle3 immunity. Indeed, sequence comparisons of PA0259 and PA0261 showed that PA0261 is homologous to Tsi6, the immunity protein of a H1-T6SS effector called Tse6, and PA0259 to TplEi (Tli4), the immunity of TlpE (Tle4), a H2-T6SS effector (Lu et al., 2014; Jiang et al., 2016). We used the SignalP 4.1 server (Nielsen, 2017) to predict the cellular localization of the two immunity candidates. While a Sec signal sequence is predicted at the N-terminal extremity of PA0261, the analysis of the PA0259 sequence did not reveal any. However, three upstream ATG can be found in frame with the annotated ATG of PA0259 (**Supplementary Figure S1**). The sequence of the proteins synthesized from the two first ATG (ATG_1 and ATG_2) presents then a N-terminal signal peptide, while the protein synthesized from the last codon (ATG_3) does not. Moreover, a ribosome-binding sequence (RBS) can be found only upstream of ATG_1 and with a significant Kolaskar score that indicates a strong probability to be used as an initiation codon (Kolaskar and Reddy, 1985). Altogether, these data tend to indicate an incorrect annotation of the start codon of PA0259 and that ATG_1 should be considered for the initiation of PA0259 translation. Such a correction has been already seen for Tli5 (PA3488) the immunity protein of PldA (Tle5a) of *P. aeruginosa* (Russell et al., 2013). In conclusion of this *in silico* analysis, the two putative immunities may harbor a Sec signal peptide that suggests a periplasmic localization in agreement with Tle3 activity in this compartment.

To demonstrate the antibacterial activity of Tle3 and to identify its immunity protein, we developed a heterologous toxicity assay in *E. coli* on the basis that Tle3 should be toxic when produced in the periplasm of *E. coli* and that it should be counteracted by the co-production of its immunity protein. In order to artificially address Tle3 to the periplasm of *E. coli* the *tle3* sequence has been cloned in frame with the sequence coding the PelB signal peptide on the pET22b vector under a P_{T7} promoter. PA0259 from ATG_1 and PA0261 were respectively cloned in pRSF-DUET, a higher copy-number vector, to allow a maximal co-expression with *ss-tle3* in the *E. coli* BL21(DE3) pLysS strain. The correct production and localization of all the recombinant proteins in *E. coli* has been verified by western blot after cell fractionation (**Supplementary Figure S2**). The results presented in **Figure 1B** indicate that whereas the cytoplasmic production of Tle3 was not toxic, Tle3 targeting to the periplasm led to *E. coli* killing. Moreover, while PA0259 had no effect, the coproduction of PA0261 in the periplasm protected the cells against the toxicity of Tle3. We verified that the sole overproduction of PA0259 and PA0261 was not toxic to *E. coli*. As PA0261 neutralized Tle3 toxicity, we called it "Tli3" for Type VI lipase immunity (**Figure 1A**). During this study, we observed that the protection conferred by Tli3 coproduction could be sometimes partial and we solved this issue by cloning in tandem *tli3* and *tle3* on the same plasmid like they are organized on *P. aeruginosa* genome (**Figure 1B**, line 8). The importance of this genetic link is the proof of the close connection between these two proteins as a pair of toxin-antitoxin.

Tli3 (PA0261) Is the Immunity Protein of Tle3

As the immunities bind specifically their effector, which is suggested by the release of Tle3 toxicity by Tli3, we tested the physical interaction between both proteins by co-purification with affinity chromatography (**Figure 2A** and **Supplementary Figure S6**). A cytoplasmic Strep-tagged version of Tli3 was engineered by fusing the tag to the mature domain of Tli3, lacking its signal peptide. The recombinant protein was coproduced in *E. coli* BL21(DE3) pLysS with the cytoplasmic His_{10}-tagged Tle3. The bacterial lysate was loaded to a StrepTactin matrix (see section "Materials and Methods"), and $Tli3^{C}_{Strep}$ was eluted with desthiobiotin. The presence of $Tle3^{C}_{His}$ was controlled in the elution fraction with anti-His antibodies. As showed in **Figure 2A** and **Supplementary Figure S6**, $Tle3^{C}_{His}$ was found in the eluted fraction only upon coproduction with $Tli3^{C}_{Strep}$ (left panel). Indeed when produced alone in *E. coli*, $Tle3^{C}_{His}$ was not purified by affinity chromatography (right panel). As expected for an immunity protein, Tli3 directly interacts with Tle3.

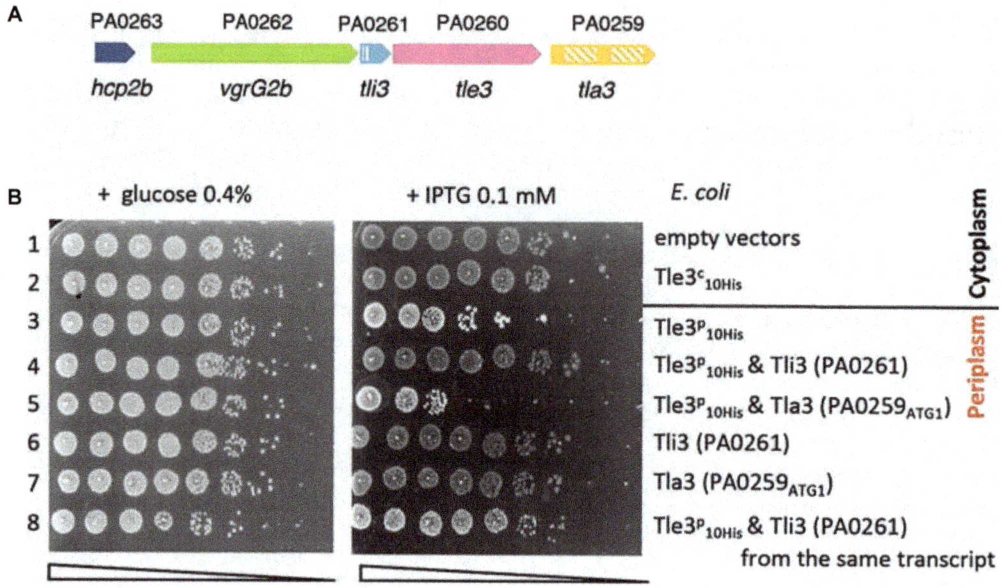

FIGURE 1 | (A) *vgrG2b* island organization. The genes are labeled with the given name (i.e., *hcp2b*) and are indicated by their annotation number (e.g., PA0263). The Sec signal peptide of Tli3 (PA0261) and the two DUF2875 of Tla3 (PA0259) are represented with stripped boxes. **(B)** The Tle3 periplasmic toxicity is counteracted by Tli3 (PA0261). Serial dilutions (from non-diluted to 10^{-7}) of normalized cultures of *E. coli* BL21(DE3)pLysS producing the wild-type Tle3 in the cytoplasm, called Tle3^{c} (from pVT1, a pETDuet-1 derivative) or in the periplasm, called Tle3^{p} [from pSBC81, a pET22b(+) derivative yielding a fusion of Tle3 with a Sec signal peptide] were spotted on LB agar plates supplemented (left panel) with 0.4% glucose or (right panel) with 0.1 mM IPTG. Glucose and IPTG allow respectively repression and induction of the gene encoding the T7 RNA polymerase. When indicated Tli3 (PA0261) or Tla3 (PA0259) were produced in the periplasm from pVT8, pSBC107 and pVT9, respectively, pRSFDuet-1 derivatives. Line 1: pET22b(+) and pRSFDuet-1, line 2: pVT1 and pRSFDuet-1, line 3: pSBC81 and pRSFDuet-1, line 4: pSBC81 and pVT8, line 5: pSBC81 and pVT9, line 6: pVT8 and pET22b(+), line 7: pVT9 and pET22b(+), and line 8: pSBC107 is a pRSFDuet-1 derivative producing Tli3 (PA0261) and Tle3^{p} from the same transcript.

To go further into Tli3 characterization, we chose to determine its cellular localization in *P. aeruginosa*. All the immunity proteins identified so far for Tle proteins are localized in the periplasm or associated to the periplasmic side of the outer membrane (Russell et al., 2013; Jiang et al., 2014, 2016; Flaugnatti et al., 2016) in order to counteract their cognate toxin. A chromosomally encoded Tli3$_{V5}$ translational fusion was engineered in order to specifically immunodetect the protein in *P. aeruginosa* (**Supplementary Table S1**). After fractionation of *P. aeruginosa* (**Figure 2B**), Tli3 was readily observed in the same fraction as DsbA that catalyzes intrachain disulfide bond formation as peptides emerge into the periplasm. This indicates a periplasmic localization for Tli3 in *P. aeruginosa* in agreement with the presence of a Sec signal peptide and our working hypothesis suggesting an immunity function.

Tle3 Protein-Protein Interaction Network

To further characterize Tle3, we performed a bacterial two-hybrid (BACTH) assay with the other gene products of the *vgrG2b* operon hypothesizing that a genetic link could reflect protein-protein interaction. The sequences coding PA0259 and Tli3 (PA0261) after their signal sequences and Tle3 were cloned downstream and upstream the T18 or T25 domains of the *Bordetella adenylate* cyclase. Because of the high molecular weight of VgrG2b and since the interaction of another Tle with a VgrG in entero-aggregative *E. coli* (EAEC) was previously delimitated to the C-terminal domain of VgrG (Flaugnatti et al., 2016), we cloned the sequences encoding the C-terminal extension (domains 1, 2, and 3) and a truncated version harboring only the DUF2345 and the transthyretin-like (TTR) domains (domains 1 and 2) of VgrG2b downstream the T18 and T25 domains (**Figure 3A**).

Unexpectedly, the sole interaction revealed by the BACTH assay for Tle3 was with PA0259 (**Figure 3B**) since only the T18/T25-Tle3 and T18/T25-PA0259 fusion proteins coproduction activated the expression of the reporter gene. This assay did not confirm the interaction between Tle3 and Tli3 observed by copurification and did not evidence an interaction with VgrG2b. Interestingly, Tle3 and PA0259 did not interact anymore when they were fused upstream the T18 and T25 domains suggesting that both proteins interact *via* their C-terminal domains (**Supplementary Figure S3A**). To go further in characterizing the interaction between Tle3 and PA0259, we constructed truncated variants taking into account their domain organization (**Figure 3A**). We further delimitated the interacting domain of Tle3 to its extreme C-terminus (the DUF3274 domain) since only the truncated T18-Tle3D2 construct still interacts with T25-PA0259 (**Figure 3C**). We did not find the domain of interaction within PA0259 as none of the DUF2875 domains alone interacts with Tle3 or this may suggest that both of them are required for the interaction (**Figure 3C**).

Next, we took advantage of all the constructs we made to test other interactions. The BACTH assay also showed that

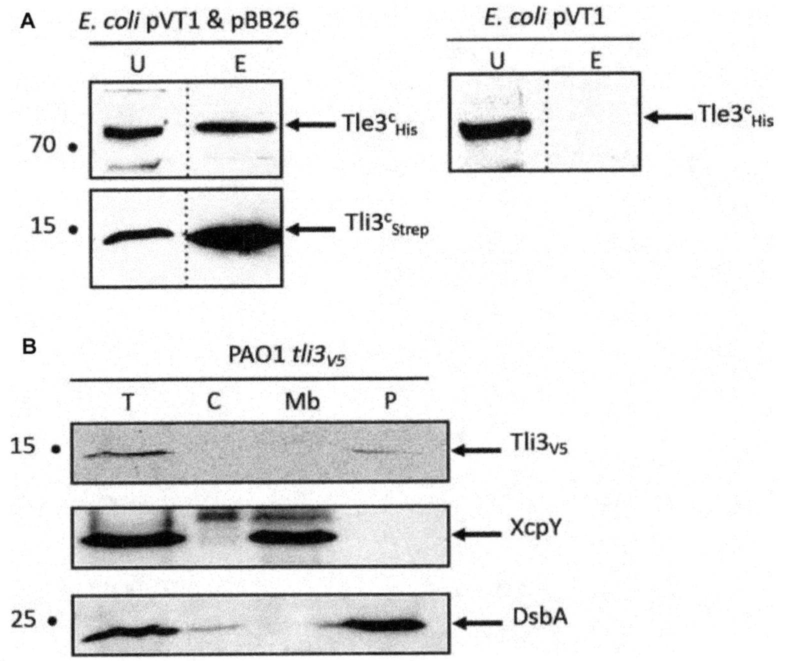

FIGURE 2 | (A) Tli3 (PA0261) interacts with Tle3. Copurification assay on StrepTactin column of Tli3c$_{Strep}$ with Tle3c10$_{His}$ produced in *E. coli* BL21(DE3)pLysS produced from pBB26 and pVT1 respectively. The unbound (U) and eluted (E) fractions were collected and subjected to SDS-PAGE (10.5%) and Western blot analyses using anti-His antibody (Upper) and anti-streptavidin antibody (Lower). The position of the proteins and the molecular mass markers (in kDa) are indicated. **(B)** Tli3 (PA0261) is a periplasmic protein in *P. aeruginosa*. Cells of *P. aeruginosa* PAO1 tli3$_{V5}$ were subjected to fractionation and immunoblotting using antibodies directed against the V5 tag, XcpY and DsbA. XcpY and DsbA were used as membrane and periplasmic controls, respectively. T, whole cell; C, cytoplasm; Mb, total membrane; P, periplasm. The position of the proteins and the molecular mass markers (in kDa) are indicated.

PA0259 interacts with both forms of the VgrG2b C-terminal extension (**Figure 3B**). The domain of interaction on VgrG2b is thus at least constituted by the DUF2345 and TTR domains. We also observed that PA0259 at least dimerizes since all the PA0259 constructs interact with each other whatever the orientation of PA0259 (**Figures 3B,D; Supplementary Figure S3B**). VgrG2b and Tli3 (PA0261) are also able to homomultimerize since all the constructs interact with each other (**Figure 3D; Supplementary Figure S3B**).

Taking into account the interactions revealed by the BACTH assay, we propose that the Tle3 toxin can be addressed to the H2-T6SS machinery VgrG2b component via PA0259. We thus named PA0259 "Tla3" for Type VI lipase adaptor protein.

Tla3 (PA0259) Characterization

To gain insight into the role of Tla3 during Tle3 secretion, we first validated the interactions of Tla3 with Tle3 and VgrG2b by a complementary approach of co-purification by affinity chromatography. Two different tagged versions of Tla3 were engineered by fusing a Strep-tag or a 10His-tag to the mature domain of Tla3 this leading to cytoplasmic tagged Tla3 proteins. The recombinant Tla3$^{C}_{Strep}$ was coproduced in *E. coli* BL21(DE3) pLysS with Tle3$^{C}_{His}$, and the Tla3$^{C}_{His}$ with 3 recombinant forms of Strep-tagged VgrG2b consisting in the full-length VgrG2b, or VgrG2b truncated for the extreme C-terminus (deletion of domain 3 in **Figure 3A**) or VgrG2b truncated for the extreme C-terminus and the TTR domain (deletion of domains 2 and 3 in **Figure 3A**). We initially tried with His-tagged VgrG2b but a problem of protein instability led us to shift for Strep-tagged VgrG2b. The bacterial lysates were loaded on a StrepTactin matrix, and Tla3$^{C}_{Strep}$ or the three recombinant VgrG2b$_{Strep}$ were eluted with desthiobiotin. The presence of Tle3$^{C}_{His}$ and of Tla3$^{C}_{His}$ was visualized in the elution fractions with anti-His antibodies. As shown in **Figures 4A, 2A** right panel and **Supplementary Figure S6**, Tle3$^{C}_{His}$ was only found in the eluted fraction upon coproduction with Tla3$^{C}_{Strep}$. We observed that Tla3$^{C}_{His}$ is copurified only upon coproduction with the full length VgrG2b (**Figure 4B** and **Supplementary Figure S6**) or the VgrG2b truncated for the extreme C-terminus (**Figure 4C** and **Supplementary Figure S6**). Indeed when produced alone (**Figure 4B** and **Supplementary Figure S6**) or with VgrG2b truncated for the extreme C-terminus and the TTR domain (**Figure 4D** and **Supplementary Figure S6**), Tla3$^{C}_{His}$ was not purified by affinity chromatography. Since VgrG2b truncated for the extreme C-terminus still copurified with Tla3 we can exclude that this domain is required for the interaction. This is in line with the BACTH assay that showed an interaction between Tla3 and a truncated VgrG2b consisting in only the DUF2345 and TTR domains (domains 1 and 2, **Figure 3A**). Moreover, the deletion of the TTR domain affecting the copurification, one can conclude that this domain is key for the interaction.

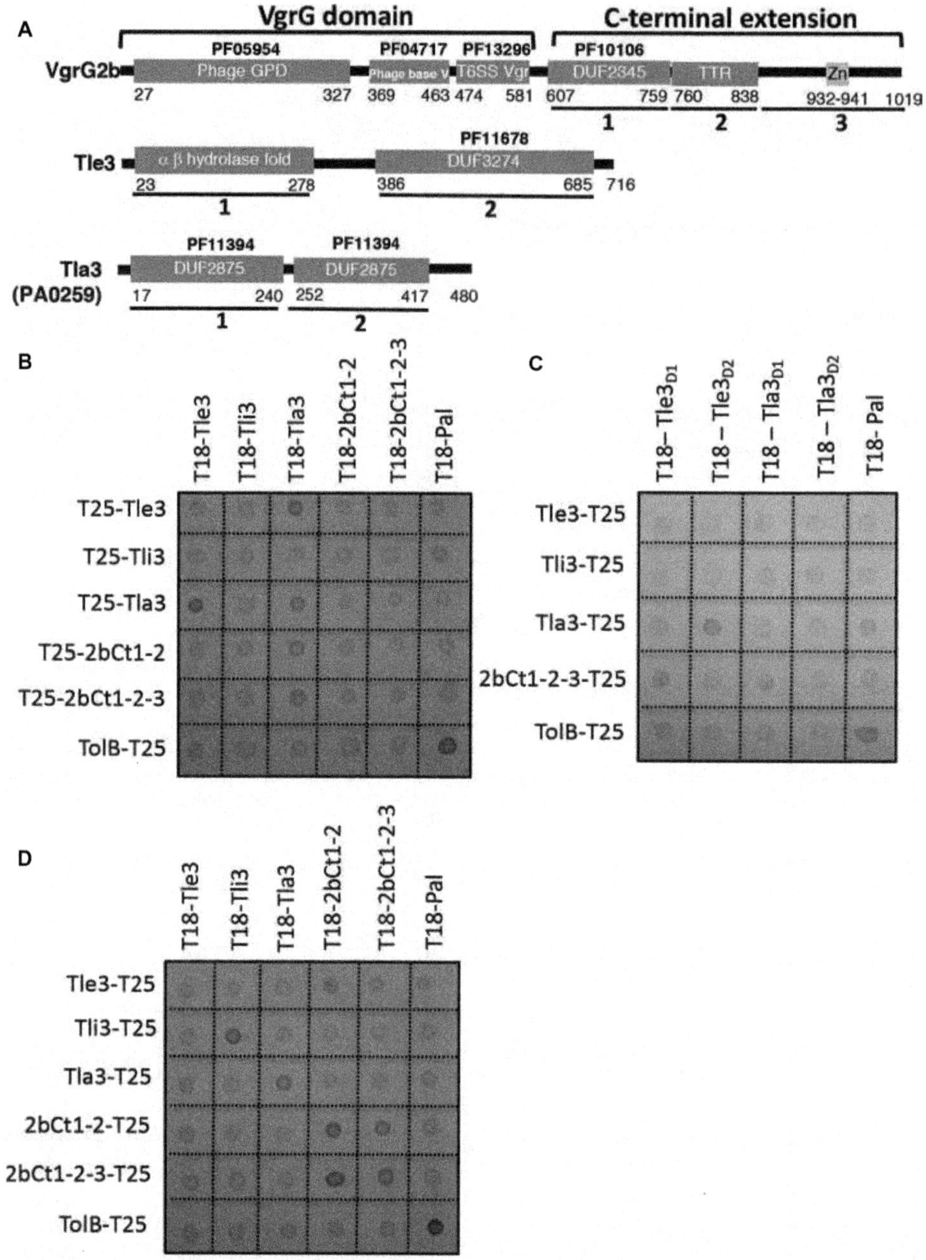

FIGURE 3 | (A) Tla3 (PA0259) interacts with the Tle3 toxin and with VgrG2b. Domain organization of VgrG2b, Tle3, and Tla3 (PA0259). The first 581 residues of VgrG2b carry the VgrG domain homologous to gp27 and gp5 phage tail proteins and consisting of three sub-domains. This is followed by the C-terminal extension, composed of a conserved domain of uncharacterized proteins (DUF2345, PF10106) (2bCt1), a TTR (transthyretin-like region) (2bCt2), and a putative zinc-dependent metallopeptidase pattern (LFIHEMTHVW signature, PS00142) (2bCt3). Tle3 architecture consists of an α,β hydrolase fold domain (Tle3$_{D1}$) followed by a DUF3274 (Tle3$_{D2}$), Tla3 of a tandem of DUF2875 (Tla3$_{D1}$ and Tla3$_{D2}$). **(B–D)** Bacterial two-hybrid assay. BTH101 reporter cells producing the indicated proteins or domains fused to the T18 or T25 domain of the *Bordetella* adenylate cyclase were spotted on X-gal indicator plates. The blue color of the colony reflects the interaction between the two proteins. TolB and Pal are two proteins known to interact but unrelated to the T6SS. The experiment was performed in triplicate and a representative result is shown.

Taken together these data confirmed a direct interaction of Tla3 with Tle3 on one side and with VgrG2b on the other side. By taking into account the BACTH data and the copurification with two truncated forms of VgrG2b, the domain of interaction of VgrG2b with Tla3 can be delimitated to the TTR domain.

We then analyzed the cellular localization of Tla3 in *P. aeruginosa* that, according to its interactions with Tle3 and VgrG2b, should be cytoplasmic. As for Tli3, we engineered a chromosomally encoded Tla3$_{V5}$ translational fusion (**Supplementary Table S1**). Tla3 was indeed immunodetected in the cytoplasmic fraction (**Figure 5A**). One could note that

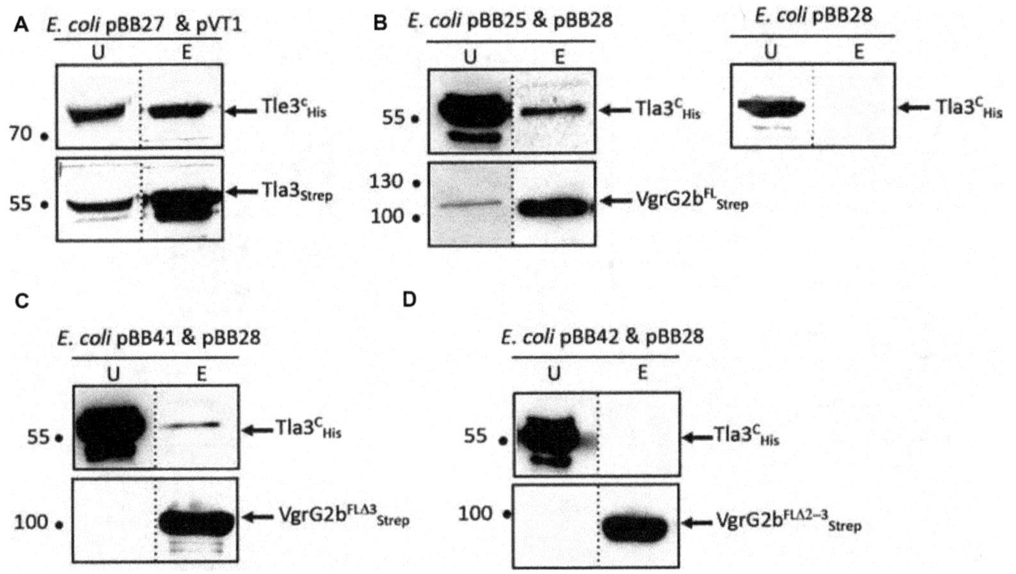

FIGURE 4 | Tla3 interaction with Tle3 **(A)** and VgrG2b **(B–D)**. Copurification assay on StrepTactin column of Tla3$^{c}_{STREP}$ with Tle3$^{c}_{His}$ produced *in E. coli* BL21(DE3) pLysS from pBB27 and pVT1, respectively, **(A)** or of Tla3$^{c}_{His}$ with VgrG2b$^{FL}_{STREP}$ **(B)** or VgrG2b$^{FLD3}_{STREP}$ **(C)** or VgrG2b$^{FLD2-3}_{STREP}$ **(D)** produced from pBB28 and pBB25 **(B)** or pBB41**(C)** or pBB42 **(D)**. The unbound (U) and eluted (E) fractions were collected and subjected to SDS-PAGE (10.5%) and Western blot analysis using anti-His antibody (upper) and anti-streptavidin antibody (lower). The dashed line separates lanes from non-adjacent part of the same gel. The position of the proteins and the molecular mass markers (in kDa) are indicated.

in contrast with our first hypothesis suggesting an incorrect start codon for *tla3* and our observation of the recombinant protein in the periplasm of *E. coli* (**Supplementary Figure S1**), Tla3 was totally absent from the periplasmic fraction of *P. aeruginosa*. To strengthen this result, each putative ATG was individually mutated on the chromosome of the PAO1 strain encoding the Tla3$_{V5}$ translational fusion (**Figure 5B**). In agreement with its cytoplasmic localization in *P. aeruginosa*, Tla3 was only produced if the fourth ATG was intact. Accordingly the Tla3 protein synthesized from this ATG is not predicted to possess a N-terminal signal peptide (**Supplementary Figure S1**). In conclusion, Tla3 is a cytoplasmic protein synthesized from the annotated translation start[1] and this localization is in agreement with a role in the targeting of the toxin to the secretion machinery.

Next, we asked whether Tla3 is specific for Tle3 or if it can be required for the secretion of other substrates of the H2-T6SS machinery. Since Hcp secretion is the hallmark of a functional secretion system, we studied the secretion of Hcp2b whose gene is upstream *vgrG2b* (**Figure 1A**). Like Allsopp et al. (2017), we deleted the *rsmA* gene to enable Hcp2b production and thus secretion by a PAO1 strain encoding a Hcp2b$_{His}$ translational fusion (**Figure 5C**, compare lines 1 and 2). RsmA is a posttranscriptional regulator known to repress all three T6SS clusters of *P. aeruginosa* (Allsopp et al., 2017). This results in a massive secretion since Hcp2b$_{His}$ can be observed in the extracellular protein samples by Coomassie-blue staining (**Figure 5C**, lower panel). While Hcp2b$_{His}$ secretion was abolished in a *rsmA clpV2* mutant, a H2-T6SS mutant, we observed that Hcp2b$_{His}$ is still secreted in the absence of Tla3 (**Figure 5C**, compare line 3 with line 4), suggesting that Tla3 is specific for the secretion of Tle3 but not for other H2-T6SS proteins. In line with this specific adaptor-toxin pair, Tla3$^{C}_{Strep}$ did not copurify TplE (Tle4), another antibacterial phospholipase delivered by the H2-T6SS machinery (**Figure 5D** and **Supplementary Figure S6**).

Finally as the interaction with a VgrG can suggest that Tla3 is itself a T6SS effector, we studied whether Tla3 is secreted by *P. aeruginosa*. To this end, the *rsmA* gene was deleted from the PAO1 strain encoding a Tla3$_{V5}$ translational fusion. Whereas Tla3$_{V5}$ was better produced upon *rsmA* deletion (**Figure 5E**), this did not lead to immunodetection of Tla3$_{V5}$ in the extracellular medium although Hcp2b was readily observed by Coomassie-blue staining of the same samples. Tla3 is thus not an effector *per se*.

In conclusion, the cytoplasmic localization of Tla3 in *P. aeruginosa* is appropriate with a recruiting role of Tle3 to the H2-T6SS machinery through an interaction with the TTR domain of VgrG2b. Tla3 is not co-secreted with Tle3. Tla3 role seems specific to Tle3 since it is not required for a functional H2-T6SS machinery and does not interact with TplE (Tle4), another H2-T6SS effector.

Tle3 Secretion Mechanism by the H2-T6SS Machinery of *P. aeruginosa*

In order to study the antibacterial role and the secretion of Tle3 by *P. aeruginosa*, we performed intra-species bacterial competition assays between *P. aeruginosa* strains. This consists in studying the survival of prey bacteria lacking the *tli3* immunity gene (by CFU counting of antibiotic resistant bacteria) co-cultivated

[1]pseudomonas.com

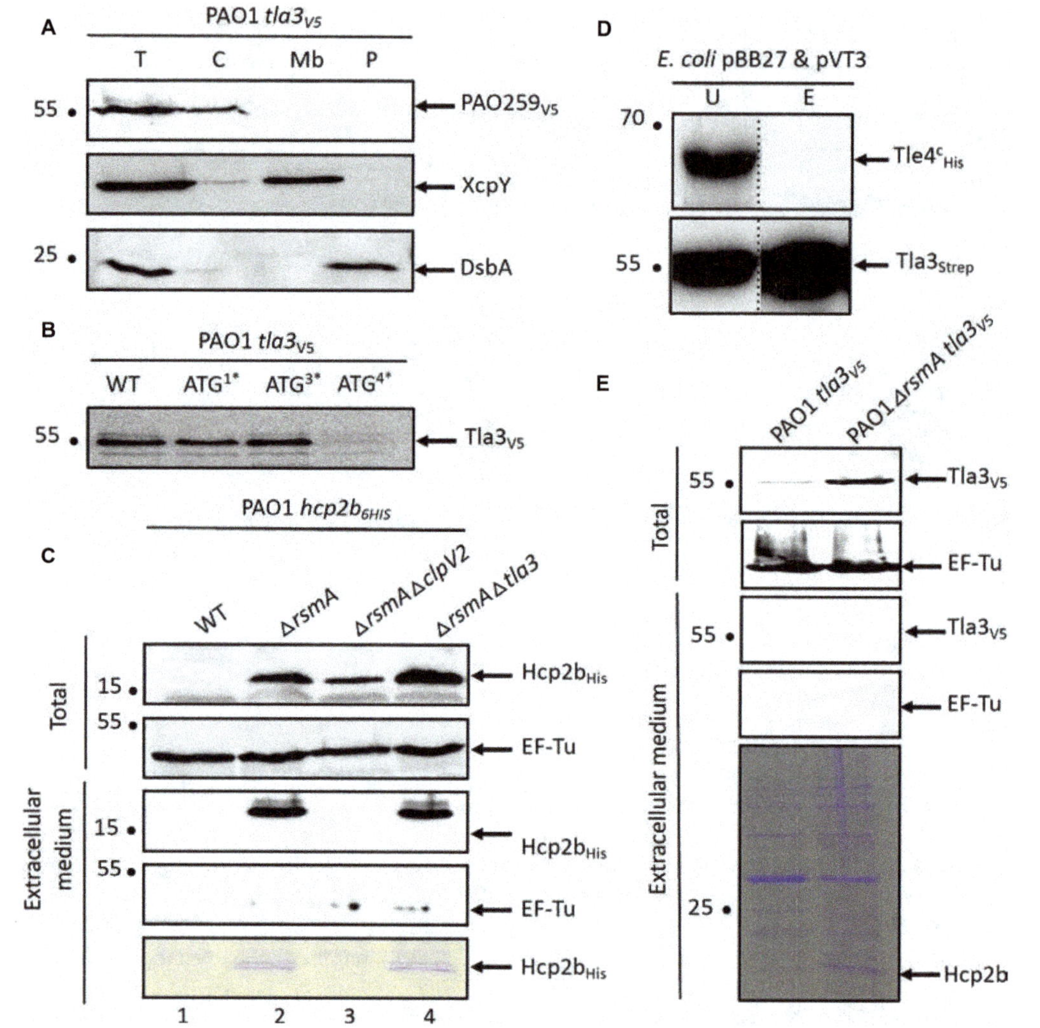

FIGURE 5 | (A) Tla3(PA0259) is a cytoplasmic protein of *P. aeruginosa*. Cells of *P. aeruginosa* PAO1 *tla3*$_{V5}$ were subjected to fractionation and immunoblotting using antibodies directed against the V5 tag, XcpY and DsbA. XcpY and DsbA were used as membrane and periplasmic controls, respectively. T, whole cell; C, cytoplasm; Mb, total membrane; P, periplasm. **(B)** The annotated ATG drives the initiation of translation of *tla3* in *P. aeruginosa*. Immunodetection of Tla3$_{V5}$ with anti-V5 antibodies produced in a WT background or in strains in which one of the four predicted ATG in *tla3* have been substituted in ATA. The number followed by a star indicates which ATG from **Supplementary Figure S1** ATG$_4$ corresponds to the annotated ATG. **(C)** Tla3 is not required for Hcp2b secretion. Immunodetection of Hcp2b$_{6His}$ with anti-His antibodies produced in a WT background (line 1) or in strains deleted for *rsmA* (lines 2–4) and *clpV2* (line 3) or *tla3* (line 4). The strains were grown at 25°C for 24 h and total bacteria were separated from extracellular medium. Anti-EF-Tu is used as a lysis control. The extracellular medium proteins were also stained with Coomassie-blue. **(D)** Tla3 does not interact with Tle4. Copurification assay on StrepTactin column of Tla3$^{c}_{STREP}$ with Tle4$_{His}$ produced *in E. coli* BL21(DE3)pLysS from pBB27 and pVT3 respectively. Legend is as in **Figure 4**. **(E)** Tla3 is not secreted. Immunodetection of Tla3$_{V5}$ with anti-HV5 antibodies produced in a WT background (line 1) or in strain deleted for *rsmA* (line 2). The strains were grown at 25°C for 24 h and total bacteria were separated from extracellular medium. Anti-EF-Tu is used as a lysis control. The extracellular medium proteins were also stained with Coomassie-blue. **(A-E)** The position of the proteins and the molecular mass markers (in kDa) are indicated.

24 hours at 37°C on plate with various attackers. As immunity genes are essential genes (protection from fratricide intoxication), both toxin and immunity genes have been deleted in order to construct a *via*ble Δ*tli3*Δ*tle3* mutant strain (**Supplementary Table S1**). **Figure 6** first confirms the antibacterial function of Tle3 and the immunity role of Tli3 since the growth of the immunity mutant was affected by the WT strain because it cannot resist Tle3 toxicity, while the Δ*tle3* mutant had no effect (**Figure 6** compare line 1 with line 2). The use of the Δ*clpV2* strain, a H2-T6SS mutant, and of Δ*vgrG2b* and Δ*tla3* mutants allowed to demonstrate that Tle3 is delivered to prey bacteria through the H2-T6SS machinery and confirms that VgrG2b and Tla3 participate in Tle3 targeting to the H2-T6SS machinery (**Figure 6** compare line 1 with lines 4, 5, and 6). Indeed Δ*clpV2*, Δ*vgrG2b* and Δ*tla3* mutants had no effect on the immunity mutant growth since they cannot deliver Tle3 into prey bacteria. The complementation *in cis* of *tle3* and *tla3* deletions (the mutants have been constructed for this study) restored a WT phenotype (**Figure 6** compare line 2 with line 3, and line 6 with line 7) this demonstrating no polar effect on downstream

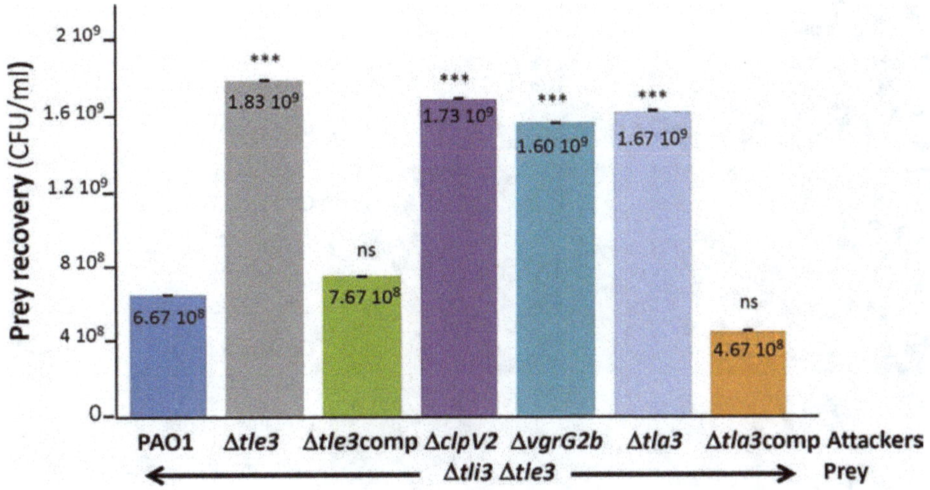

FIGURE 6 | *P. aeruginosa* growth competition. The *P. aeruginosa* prey strain (Δ*tli3*Δ*tle3*) was incubated with various *P. aeruginosa* attacker strains as indicated in the figure for 24 h at 37°C. The number of recovered prey bacteria is indicated in CFU/ml. "comp" stand for *cis* complementation of the corresponding mutation with a wild-type copy inserted at the *attB* site on *P. aeruginosa* chromosome. Error bars represent ± SEM ($n = 3$); $^{***}p < 0.001$, ns not significant.

genes. Furthermore, the introduction of a wild-type copy of *tli3* at the *attB* site on *P. aeruginosa* chromosome of the Δ*tli3*Δ*tle3* mutant restores wild-type competition capacity to this strain (**Supplementary Figure S4A**). This confirms that the absence of the immunity was responsible of the phenotypes observed for the Δ*tli3*Δ*tle3* strain (**Figure 6**). Finally, the Δ*tla3* mutant has been used as a prey and was not affected by the WT strain or any of the mutants excluding definitively a role of immunity as proposed by its annotation (see text footnote 1) and confirming its adaptor function (**Supplementary Figure S4B**).

Taken together, these results demonstrated that Tle3 is an effective H2-T6SS-dependent antibacterial toxin loaded onto the VgrG2b puncturing device *via* Tla3 and neutralized by Tli3 in resistant prey bacteria.

VgrG2b Is a Trans-Kingdom Toxin

A putative neutral zinc metallopeptidase domain has been predicted at the extreme C-terminus of VgrG2b by Pukatzki and colleagues (**Figure 3A**; Pukatzki et al., 2007). This motif (Prosite PS00142, PFAM04298) consists in a metal-binding consensus motif HExxH, the two histidine residues being ligands of the catalytic Zn^{2+} and the glutamic acid residue involved in nucleophilic attack. As an effector with a protease activity can target both eukaryotic and bacterial proteins, we searched for an antibacterial activity of the VgrG2b C-terminal extension. To do this, we performed the same heterologous toxicity assay in *E. coli* as with Tle3. As shown in **Figure 7**, whereas the production of $VgrG2b_{Cter}$ in the cytoplasm did not impact *E. coli* growth, its periplasmic production killed *E. coli* (**Figure 7**, compare lines 2 and 5). Moreover substitution of the histidine in position 935 and of the glutamic acid 936 for an alanine relieves $VgrG2b_{Cter}$ toxicity this showing that VgrG2b is a novel antibacterial protease active in the periplasm.

Finally because of their genetic proximity, we tested whether Tli3 can be the immunity of $VgrG2b_{Cter}$. Indeed, the coproduction of Tli3 counteracts $VgrG2b_{Cter}$ toxicity whether it is coproduced from a second plasmid (**Figure 7**, line 6) or from the same transcript (**Figure 7**, line 7). In conclusion Tli3 serves two toxins and protects also from the antibacterial activity of VgrG2b.

DISCUSSION

Here, we report the existence of a novel pair of antibacterial effector and immunity of the H2-T6SS of *P. aeruginosa*, Tle3 (PA0260) and Tli3 (PA0261), and we propose a chronology of Tle3 secretion process that includes a cytoplasmic adaptor protein, Tla3 (PA0259) to load the toxin onto the VgrG2b spike (a model is proposed in **Figure 8**). Through heterologous toxicity assay and bacterial competition, we show that Tle3 was toxic once delivered in the periplasm of prey bacteria and that Tli3 can neutralize the toxin in this compartment. Interestingly, this led us to discover that VgrG2b that we previously recognized as an anti-eukaryotic effector possesses an antibacterial activity as well.

The VgrG-recruitment of cargo effectors has been previously evidenced for several antibacterial effectors among them two toxins of the Tle family, TseL (Tle2) of *V. cholerae* (Dong et al., 2013; Liang et al., 2015; Unterweger et al., 2015) and Tle1 of EAEC (Flaugnatti et al., 2016). In both cases, the *tle* genes were just downstream the *vgrG* genes like the organization of *vgrG2b* and *tle3* of *P. aeruginosa*. TseL and Tle1 have been shown to directly bind a dedicated VgrG, VgrG3 and VgrG1 respectively (Dong et al., 2013; Flaugnatti et al., 2016). The domain of interaction within EAEC VgrG1 has been delimitated to the TTR domain and may also include the DUF2345, both of which are present in the *P. aeruginosa* VgrG2b. In line with this, and despite Tle3 requiring an adaptor to be targeted to VgrG2b, we have shown that Tla3 interacts with the TTR domain of VgrG2b. Taken together these data demonstrate that TTR domains of VgrGs are involved in recruitment

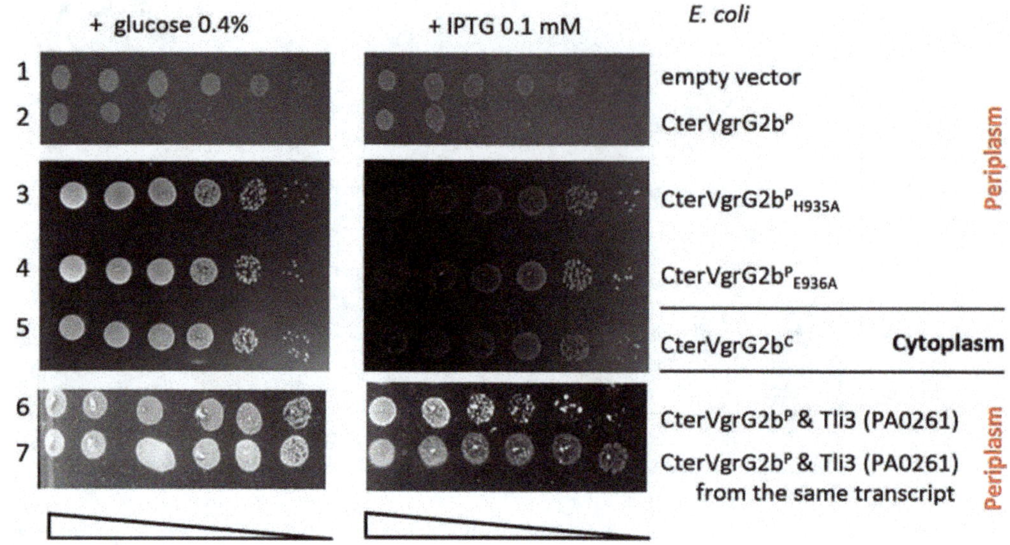

FIGURE 7 | The protease domain of VgrG2b is required for its antibacterial activity. Serial dilutions (from non-diluted to 10^{-5}) of normalized cultures of *E. coli* BL21(DE3)pLys producing the wild-type C-terminal domain of VgrG2b in the cytoplasm, called CterVgrG2b^{C} (from pBB43, a pETDuet-1 derivative, line 5) or in the periplasm, called CterVgrG2b^{P} (from pBB44, a pET22b(+) derivative yielding a fusion of the C-terminal domain with a Sec signal peptide, line 2) or two variants CterVgr2b$^{P}_{H935A}$ (from pBB45, line 3) and CterVgr2b$^{P}_{E936A}$ (from pBB46, line 4) were spotted on LB agar plates supplemented (left panel) with 0.4% glucose or (right panel) with 0.1 mM IPTG. In line 6, the CterVgrG2b^{P} (from pBB44) was coproduced with Tli3 (from pVT8) and in line 7, the CterVgrG2b^{P} and Tli3 were coproduced from the same transcript (from pSBC110). Glucose and IPTG allow respectively repression and induction of the gene encoding the T7 RNA polymerase.

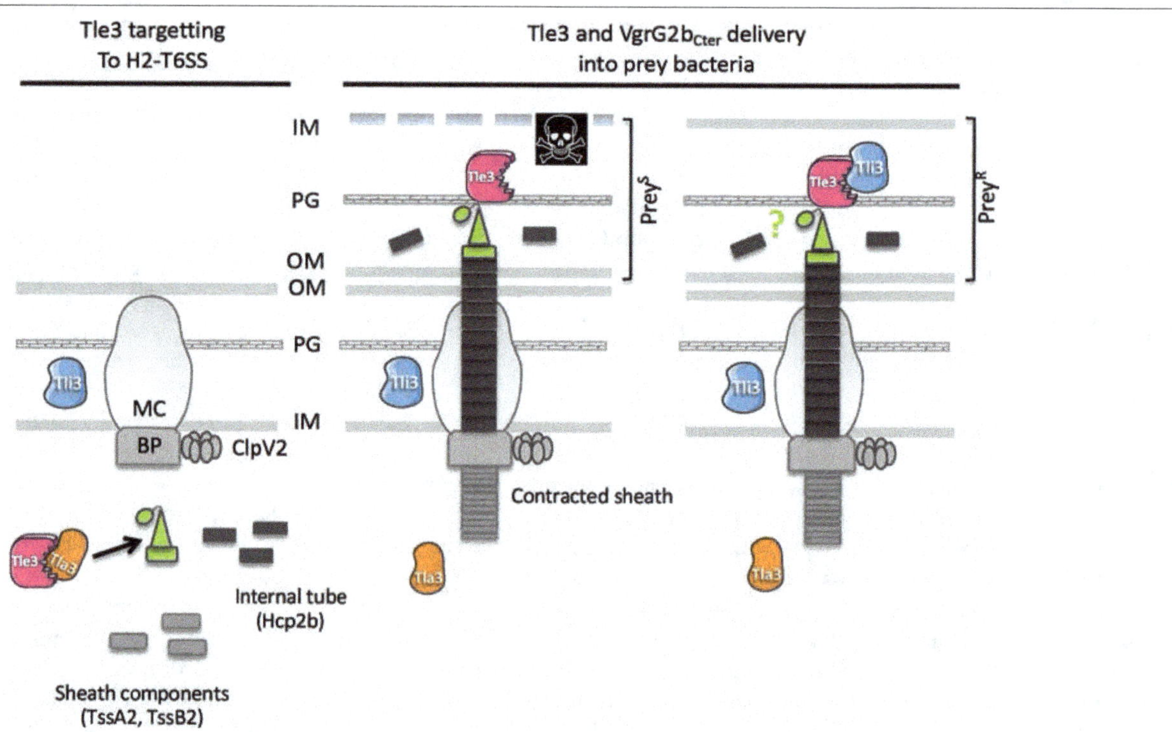

FIGURE 8 | Working model: Tle3 targeting to the H2-T6SS machinery (left panel) Tle3 (in pink) is taken in charge in the cytoplasm by Tla3 (in orange) that binds VgrG2b (green). Upon Hcp2b (in dark grey) assembly into the growing sheath (light grey), the VgrG2b loaded with the Tle3 is placed at the tip of the Hcp arrow. Tle3 and VgrG2b$_{Cter}$ delivery into bacteria (right panel). The sheath contraction in the cytoplasm propels the Hcp arrow towards the target bacterium. Tle3 associated with this expelled structure is thus translocated into target cells, as well as VgrG2b$_{Cter}$ (green circle). The prey is killed (Prey^{S}) except if it has the Tli3 immunity (in blue) in the periplasm (Prey^{R}). The mechanism of resistance to VgrG2b$_{Cter}$ is still unknown. OM, outer membrane; PG, peptidoglycan; IM, inner membrane; BP, base plate; MC, membrane complex. Prey^{S}, sensitive prey; Prey^{R}, resistant prey.

and transport of Tle effectors, directly or through adaptor. Likewise, C-terminal extensions of VgrG1 and VgrG2 of *Agrobacterium tumefaciens* were identified as specifically required for the delivery of each cognate DNAse toxins, named Tde1 and Tde2, respectively (Bondage et al., 2016). C-terminal domains of VgrGs can thus be considered more generally as specificity determinants for T6SS effector loading and transport.

Interestingly, TseL of *V. cholerae* requires also Tap-1 (Tec) as an adaptor protein to be delivered to another VgrG, called VgrG1 (Liang et al., 2015; Unterweger et al., 2015), showing that a sole toxin can be targeted directly and indirectly to two different VgrG proteins. Tap-1 (Tec) belongs to the DUF4123 family of adaptor proteins that contains also VasW of *V. cholerae* (Miyata et al., 2013) and several uncharacterized gene products linked to effector genes with a MIX (marker for type VI effectors) motif in *Proteus mirabilis* or *B. thailandensis* for instance (Salomon et al., 2014). Interestingly TecT, a DUF4123 adaptor of *P. aeruginosa*, has been shown to require a co-adaptor, called co-TecT, to deliver the TseT effector to the PAAR4 protein (Burkinshaw et al., 2018). This is the first example of an adaptor-co-adaptor module. Taken together, these data suggest a conserved role for DUF4123 adaptors in the recruitment of a number of T6SS effectors. Remarkably Tla3 of *P. aeruginosa* does not belong to the DUF4123 adaptor family, or to that of the two other unrelated families, the DUF1795 adaptor family, reported with EagR (effector-associated gene with Rhs) in *Serratia marcescens* (Alcoforado Diniz and Coulthurst, 2015) or EagT6 in *P. aeruginosa* (Whitney et al., 2015), and the DUF2169 adaptor family reported with Atu3641 in *A. tumefaciens* (Bondage et al., 2016). Nor Tla3 is a PAAR protein, the last class of effector targeting mode to a VgrG (Shneider et al., 2013; Whitney et al., 2014; Bondage et al., 2016; Cianfanelli et al., 2016a; Burkinshaw et al., 2018). Instead, we find that Tla3 harbors two DUF2875 domains (**Figure 3A**) that are both required for the interaction with the toxin. Moreover, genes coding DUF2875 containing proteins can be find at the vicinity of *tle*, *tli*, *vgrG*, or PAAR genes, but they are restricted to α and β proteobacteria (**Supplementary Figure S5**). We thus hypothesize that DUF2875 might assist in T6SS-mediated effector delivery. Like the three other adaptor families (DUF4123, DUF1795, and DUF2169), we have observed that Tla3 is not required for the H2-T6SS functionality since the Δ*tla3* mutant still secrete Hcp2b and can compete with a WT strain, and that Tla3 is specific for the Tle3 toxin since it did not interact with another H2-T6SS effector, TplE (Tle4). Finally, as we did not detect Tla3 secretion under constitutive H2-T6SS condition, we propose that Tla3 hands over Tle3 to VgrG2b in the cytoplasm prior to its loading to the baseplate and further recruitment to the central Hcp tube in preparation (**Figure 8**). Overall, the existence of various modes of effector recruitment, further refined with adaptors, likely explains how the T6SS is able to deliver numerous and structurally diverse proteins.

Five families of Tle, Tle1–5, have been identified among Gram-negative bacteria (Russell et al., 2013) and four Tle have been studied in *P. aeruginosa* so far. Our demonstration of the activity of Tle3 in the periplasm is consistent with the observations that the heterologous periplasmic production of PldA (Tle5a) (Russell et al., 2013; Jiang et al., 2014), PldB (Tle5b) (Jiang et al., 2014), Tle1 (Hu et al., 2014), and TplE (Tle4) (Jiang et al., 2016) is toxic. The reason of the periplasmic activity of Tle proteins is still unclear although several hypothesis have been proposed (Flaugnatti et al., 2016), the most likely being an activation of the toxin in this compartment. Very recently this has nicely been exemplified with the hijacking of DsbA in the target cells of *S. marcescens* for the activation of incoming effectors (Mariano et al., 2018). No member of the Tle3 family has yet been enzymatically characterized. Our attempts to efficiently purify Tle3 from *E. coli* or from *P. aeruginosa* have been unsuccessful, even if we have noticed that the presence of the Tla3 adaptor stabilized Tle3, it still formed inclusion bodies. In the future, we will decipher the enzymatic activity of Tle3, which is presumably active on membrane phospholipids as our preliminary data of thin-layer chromatography tend to show.

The periplasmic activity of Tle toxins is counteracted by the synthesis of a cognate immunity protein that is usually a periplasmic soluble protein, as we showed for Tli3 in *P. aeruginosa*, or a membrane-anchored lipoprotein (Russell et al., 2013). Interestingly the genetic organization of the *tli3* gene upstream of the *tle3* gene observed in *E. coli*, *K. pneumoniae*, *B. cenocepacia*, or *R. solanacearum* (Russell et al., 2013) is conserved in *P. aeruginosa*. The fact that the two genes are co-transcribed (the immunity being the first) is key for the protection against toxicity. Indeed, we have observed systematic protection against the periplasmic toxicity of Tle3 in *E. coli* when the two genes were expressed from the same promoter under the same plasmid, whereas it was not as efficient when the genes were on two plasmids. This genetic link reinforces the connection within the toxin/immunity pair. This has previously been noticed with a T7SS antibacterial toxin and its immunity in *Staphylococcus aureus* (Cao et al., 2016). Other immunities of Tle characterized so far have been shown to inhibit the action of the effector by direct protein-protein contacts (Russell et al., 2013; Jiang et al., 2014; Flaugnatti et al., 2016). Our copurification assay in *E. coli* demonstrates a direct interaction between Tle3 and Tli3 that was already suggested with the release of Tle3 toxicity upon coproduction of Tli3 in the periplasm. A crystal structure of the *P. aeruginosa* TplE (Tle4) effector in complex with its immunity protein TplEI (Tli4) revealed that the immunity uses a grasp mechanism to prevent the interfacial activation of the toxin (Lu et al., 2014).

Like two other H2-T6SS-related orphan *vgrG* loci, the *vgrG4b* cluster encoding PldA (Tle5a) and the *vgrG2a* encoding TplE (Tle4), we show here that the *vgrG2b* cluster has both antibacterial activities (through Tle3 and VgrG2b) and anti-eukaryotic (through VgrG2b; Sana et al., 2015). Interestingly, VgrG2b is thus (1) a structural component of the H2-T6SS puncturing device since our bacterial competition showed its requirement for Tle3 delivery, (2) an anti-eukaryotic effector through an interaction with the microtubule nucleating complex (Sana et al., 2015), and (3) an antibacterial effector as suggested by our toxicity assay in *E. coli*. We have shown that two conserved residues of the putative metallopeptidase motif (a histidine and a glutamic acid) are essential for the VgrG2b antibacterial activity. This is consistent with an antibacterial protease activity for VgrG2b that will be, to our knowledge, the first case in the T6SS effector literature.

The discovery of the VgrG2b immunity is even more exciting since we observed that Tli3 also relieves $VgrG2b_{Cter}$ toxicity. To our knowledge, Tli3 is the first example of a T6SS immunity serving two different proteins. But one can cite the chaperones of the type III secretion translocators that form a pore in the eukaryotic target cell membrane (Neyt and Cornelis, 1999). Indeed the chaperone SycD from *Yersinia* enterocolitica and its homologues from other Gram-negative bacteria bind the two translocators to prevent their premature interaction in the inner membrane and avoid further toxicity.

AUTHOR CONTRIBUTIONS

BB and SB designed and conceived the experiments. BB, CS, and SD performed the experiments. SB supervised the execution of the experiments. BB, CS, BI, and SB analyzed and discussed the data. SB wrote the paper with contribution from BB and reading from BI.

FUNDING

This work was supported by recurrent funding from the CNRS and Aix-Marseille University. The project leading to this publication has received funding from the Excellence Initiative of Aix-Marseille University-A*Midex, a French "Investissements d'Avenir" program ("Emergence & Innovation" A-M-AAP-EI-17-139-170301-10.31-BLEVES-HLS).

ACKNOWLEDGMENTS

We thank V. Tutagata for pVT1, pVT8, and pVT9 constructs, and B. Douzi for all the advices during protein purification and members of B.B. PhD committee for helpful discussion and support. We are grateful to M. Ba, I. Bringer, A. Brun, and O. Uderso for technical assistance. BB was financed with a PhD fellowship from the French Research Ministry.

REFERENCES

Alcoforado Diniz, J., and Coulthurst, S. J. (2015). Intraspecies competition in *Serratia marcescens* is mediated by type VI-secreted Rhs effectors and a conserved effector-associated accessory protein. *J. Bacteriol.* 197, 2350–2360. doi: 10.1128/JB.00199-15

Allsopp, L. P., Wood, T. E., Howard, S. A., Maggiorelli, F., Nolan, L. M., Wettstadt, S., et al. (2017). RsmA and AmrZ orchestrate the assembly of all three type VI secretion systems in *Pseudomonas aeruginosa*. *Proc. Natl. Acad. Sci. USA* 114, 7707–7712. doi: 10.1073/pnas.1700286114

Basler, M. (2015). Type VI secretion system: secretion by a contractile nanomachine. *Philos. Trans. R. Soc. Lond. Ser. B Biol. Sci.* 370. doi: 10.1098/rstb.2015.0021

Battesti, A., and Bouveret, E. (2012). The bacterial two-hybrid system based on adenylate cyclase reconstitution in *Escherichia coli*. *Methods* 58, 325–334. doi: 10.1016/j.ymeth.2012.07.018

Benz, J., and Meinhart, A. (2014). Antibacterial effector/immunity systems: it's just the tip of the iceberg. *Curr. Opin. Microbiol.* 17, 1–10. doi: 10.1016/j.mib.2013.11.002

Bleves, S. (2016). Game of trans-kingdom effectors. *Trends Microbiol.* 24, 773–774. doi: 10.1016/j.tim.2016.08.002

Bleves, S., Viarre, V., Salacha, R., Michel, G. P. F., Filloux, A., and Voulhoux, R. (2010). Protein secretion systems in *Pseudomonas aeruginosa*: a wealth of pathogenic weapons. *Int. J. Med. Microbiol.* 300, 534–543. doi: 10.1016/j.ijmm.2010.08.005

Bondage, D. D., Lin, J.-S., Ma, L.-S., Kuo, C.-H., and Lai, E.-M. (2016). VgrG C terminus confers the type VI effector transport specificity and is required for binding with PAAR and adaptor-effector complex. *Proc. Natl. Acad. Sci. USA* 113, E3931–E3940. doi: 10.1073/pnas.1600428113

Burkinshaw, B. J., Liang, X., Wong, M., Le, A. N. H., Lam, L., and Dong, T. G. (2018). A type VI secretion system effector delivery mechanism dependent on PAAR and a chaperone-co-chaperone complex. *Nat. Microbiol.* 3, 632–640. doi: 10.1038/s41564-018-0144-4

Cao, Z., Casabona, M. G., Kneuper, H., Chalmers, J. D., and Palmer, T. (2016). The type VII secretion system of *Staphylococcus aureus* secretes a nuclease toxin that targets competitor bacteria. *Nat. Microbiol.* 2:16183. doi: 10.1038/nmicrobiol.2016.183

Cianfanelli, F. R., Alcoforado Diniz, J., Guo, M., De Cesare, V., Trost, M., and Coulthurst, S. J. (2016a). VgrG and PAAR proteins define distinct versions of a functional type VI secretion system. *PLoS Pathog.* 12:e1005735. doi: 10.1371/journal.ppat.1005735

Cianfanelli, F. R., Monlezun, L., and Coulthurst, S. J. (2016b). Aim, load, fire: the type VI secretion system, a bacterial nanoweapon. *Trends Microbiol.* 24, 51–62. doi: 10.1016/j.tim.2015.10.005

Dong, T. G., Ho, B. T., Yoder-Himes, D. R., and Mekalanos, J. J. (2013). Identification of T6SS-dependent effector and immunity proteins by Tn-seq in *Vibrio cholerae*. *Proc. Natl. Acad. Sci. USA* 110, 2623–2628. doi: 10.1073/pnas.1222783110

Flaugnatti, N., Le, T. T. H., Canaan, S., Aschtgen, M.-S., Nguyen, V. S., Blangy, S., et al. (2016). A phospholipase A1 antibacterial Type VI secretion effector interacts directly with the C-terminal domain of the VgrG spike protein for delivery. *Mol. Microbiol.* 99, 1099–1118. doi: 10.1111/mmi.13292

Hecht, A. L., Casterline, B. W., Earley, Z. M., Goo, Y. A., Goodlett, D. R., and Bubeck Wardenburg, J. (2016). Strain competition restricts colonization of an enteric pathogen and prevents colitis. *EMBO Rep.* 17, 1281–1291. doi: 10.15252/embr.201642282

Hood, R. D., Singh, P., Hsu, F., Güvener, T., Carl, M. A., Trinidad, R. R. S., et al. (2010). A type VI secretion system of *Pseudomonas aeruginosa* targets a toxin to bacteria. *Cell Host Microbe* 7, 25–37. doi: 10.1016/j.chom.2009.12.007

Hu, H., Zhang, H., Gao, Z., Wang, D., Liu, G., Xu, J., et al. (2014). Structure of the type VI secretion phospholipase effector Tle1 provides insight into its hydrolysis and membrane targeting. *Acta Crystallogr. D Biol. Crystallogr.* 70, 2175–2185. doi: 10.1107/S1399004714012899

Ize, B., Viarre, V., and Voulhoux, R. (2014). Cell fractionation. *Methods Mol. Biol.* 1149, 185–191. doi: 10.1007/978-1-4939-0473-0_15

Jeong, J.-Y., Yim, H.-S., Ryu, J.-Y., Lee, H. S., Lee, J.-H., Seen, D.-S., et al. (2012). One-step sequence- and ligation-independent cloning as a rapid and versatile cloning method for functional genomics studies. *Appl. Environ. Microbiol.* 78, 5440–5443. doi: 10.1128/AEM.00844-12

Jiang, F., Wang, X., Wang, B., Chen, L., Zhao, Z., Waterfield, N. R., et al. (2016). The *Pseudomonas aeruginosa* type VI secretion PGAP1-like effector induces host autophagy by activating endoplasmic reticulum stress. *Cell Rep.* 16, 1502–1509. doi: 10.1016/j.celrep.2016.07.012

Jiang, F., Waterfield, N. R., Yang, J., Yang, G., and Jin, Q. (2014). A *Pseudomonas aeruginosa* type VI secretion phospholipase D effector targets both prokaryotic and eukaryotic cells. *Cell Host Microbe* 15, 600–610. doi: 10.1016/j.chom.2014.04.010

Karimova, G., Pidoux, J., Ullmann, A., and Ladant, D. (1998). A bacterial two-hybrid system based on a reconstituted signal transduction pathway. *Proc. Natl. Acad. Sci. USA* 95, 5752–5756.

Kolaskar, A. S., and Reddy, B. V. (1985). A method to locate protein coding sequences in DNA of prokaryotic systems. *Nucleic Acids Res.* 13, 185–194. doi: 10.1093/nar/13.1.185

LaCourse, K. D., Peterson, S. B., Kulasekara, H. D., Radey, M. C., Kim, J., and Mougous, J. D. (2018). Conditional toxicity and synergy drive diversity among antibacterial effectors. *Nat. Microbiol.* 3, 440–446. doi: 10.1038/s41564-018-0113-y

Liang, X., Moore, R., Wilton, M., Wong, M. J. Q., Lam, L., and Dong, T. G. (2015). Identification of divergent type VI secretion effectors using a conserved chaperone domain. *Proc. Natl. Acad. Sci. USA* 112, 9106–9111. doi: 10.1073/pnas.1505317112

Lu, D., Zheng, Y., Liao, N., Wei, L., Xu, B., Liu, X., et al. (2014). The structural basis of the Tle4-Tli4 complex reveals the self-protection mechanism of H2-T6SS in *Pseudomonas aeruginosa. Acta Crystallogr. D Biol. Crystallogr.* 70, 3233–3243. doi: 10.1107/S1399004714023967

MacIntyre, D. L., Miyata, S. T., Kitaoka, M., and Pukatzki, S. (2010). The *Vibrio cholerae* type VI secretion system displays antimicrobial properties. *Proc. Natl. Acad. Sci. USA* 107, 19520–19524. doi: 10.1073/pnas.1012931107

Mariano, G., Monlezun, L., and Coulthurst, S. J. (2018). Dual role for DsbA in attacking and targeted bacterial cells during type VI secretion system-mediated competition. *Cell Rep.* 22, 774–785. doi: 10.1016/j.celrep.2017.12.075

Miyata, S. T., Unterweger, D., Rudko, S. P., and Pukatzki, S. (2013). Dual expression profile of type VI secretion system immunity genes protects pandemic *Vibrio cholerae. PLoS Pathog.* 9:e1003752. doi: 10.1371/journal.ppat.1003752

Mougous, J. D., Cuff, M. E., Raunser, S., Shen, A., Zhou, M., Gifford, C. A., et al. (2006). A virulence locus of *Pseudomonas aeruginosa* encodes a protein secretion apparatus. *Science* 312, 1526–1530. doi: 10.1126/science.1128393

Neyt, C., and Cornelis, G. R. (1999). Role of SycD, the chaperone of the *Yersinia* Yop translocators YopB and YopD. *Mol. Microbiol.* 31, 143–156. doi: 10.1046/j.1365-2958.1999.01154.x

Nielsen, H. (2017). "Predicting secretory proteins with SignalP" in *Protein function prediction*. ed. D. Kihara (New York, NY: Springer), 59–73.

Pukatzki, S., Ma, A. T., Revel, A. T., Sturtevant, D., and Mekalanos, J. J. (2007). Type VI secretion system translocates a phage tail spike-like protein into target cells where it cross-links actin. *Proc. Natl. Acad. Sci. USA* 104, 15508–15513. doi: 10.1073/pnas.0706532104

Pukatzki, S., Ma, A. T., Sturtevant, D., Krastins, B., Sarracino, D., Nelson, W. C., et al. (2006). Identification of a conserved bacterial protein secretion system in *Vibrio cholerae* using the Dictyostelium host model system. *Proc. Natl. Acad. Sci. USA* 103, 1528–1533. doi: 10.1073/pnas.0510322103

Russell, A. B., LeRoux, M., Hathazi, K., Agnello, D. M., Ishikawa, T., Wiggins, P. A., et al. (2013). Diverse type VI secretion phospholipases are functionally plastic antibacterial effectors. *Nature* 496, 508–512. doi: 10.1038/nature12074

Salomon, D., Kinch, L. N., Trudgian, D. C., Guo, X., Klimko, J. A., Grishin, N. V., et al. (2014). Marker for type VI secretion system effectors. *Proc. Natl. Acad. Sci. USA* 111, 9271–9276. doi: 10.1073/pnas.1406110111

Sana, T. G., Baumann, C., Merdes, A., Soscia, C., Rattei, T., Hachani, A., et al. (2015). Internalization of *Pseudomonas aeruginosa* strain PAO1 into epithelial cells is promoted by interaction of a T6SS effector with the microtubule network. *MBio* 6:e00712. doi: 10.1128/mBio.00712-15

Sana, T. G., Berni, B., and Bleves, S. (2016a). The T6SSs of *Pseudomonas aeruginosa* strain PAO1 and their effectors: beyond bacterial-cell targeting. *Front. Cell. Infect. Microbiol.* 6:61. doi: 10.3389/fcimb.2016.00061

Sana, T. G., Flaugnatti, N., Lugo, K. A., Lam, L. H., Jacobson, A., Baylot, V., et al. (2016b). Salmonella Typhimurium utilizes a T6SS-mediated antibacterial weapon to establish in the host gut. *Proc. Natl. Acad. Sci. USA* 113, E5044–E5051. doi: 10.1073/pnas.1608858113

Sana, T. G., Hachani, A., Bucior, I., Soscia, C., Garvis, S., Termine, E., et al. (2012). The second type VI secretion system of *Pseudomonas aeruginosa* strain PAO1 is regulated by quorum sensing and Fur and modulates internalization in epithelial cells. *J. Biol. Chem.* 287, 27095–27105. doi: 10.1074/jbc.M112.376368

Shneider, M. M., Buth, S. A., Ho, B. T., Basler, M., Mekalanos, J. J., and Leiman, P. G. (2013). PAAR-repeat proteins sharpen and diversify the type VI secretion system spike. *Nature* 500, 350–353. doi: 10.1038/nature12453

Unterweger, D., Kostiuk, B., Ötjengerdes, R., Wilton, A., Diaz-Satizabal, L., and Pukatzki, S. (2015). Chimeric adaptor proteins translocate diverse type VI secretion system effectors in *Vibrio cholerae. EMBO J.* 34, 2198–2210. doi: 10.15252/embj.201591163

Whitney, J. C., Beck, C. M., Goo, Y. A., Russell, A. B., Harding, B. N., De Leon, J. A., et al. (2014). Genetically distinct pathways guide effector export through the type VI secretion system. *Mol. Microbiol.* 92, 529–542. doi: 10.1111/mmi.12571

Whitney, J. C., Quentin, D., Sawai, S., LeRoux, M., Harding, B. N., Ledvina, H. E., et al. (2015). An interbacterial NAD(P)(+) glycohydrolase toxin requires elongation factor Tu for delivery to target cells. *Cell* 163, 607–619. doi: 10.1016/j.cell.2015.09.027

WHO (2019). WHO publishes list of bacteria for which new antibiotics are urgently needed. Available at: https://www.who.int/news-room/detail/27-02-2017-who-publishes-list-of-bacteria-for-which-new-antibiotics-are-urgently-needed (Accessed January 21, 2019).

12

Baseplate Component TssK and Spatio-Temporal Assembly of T6SS in *Pseudomonas aeruginosa*

David Liebl[1*†], Mylène Robert-Genthon[1], Viviana Job[1,2], Valentina Cogoni[1] and Ina Attrée[1*]

[1] Univ. Grenoble Alpes, CNRS, Bacterial Pathogenesis and Cellular Responses, ERL 5261, INSERM, UMR-S 1036, CEA, Grenoble, France, [2] Univ. Grenoble Alpes, CEA, CNRS, Institut de Biologie Structurale (IBS), Grenoble, France

***Correspondence:**
David Liebl
David_Liebl@sris.a-star.edu.sg
Ina Attrée
ina.attree-delic@cea.fr

The Gram-negative bacteria use the contractile multi-molecular structure, called the Type VI Secretion System (T6SS) to inject toxic products into eukaryotic and prokaryotic cells. In this study, we use fluorescent protein fusions and time-lapse microscopy imaging to study the assembly dynamics of the baseplate protein TssK in *Pseudomonas aeruginosa* T6SS. TssK formed transient higher-order structures that correlated with dynamics of sheath component TssB. Assembly of peri-membrane TssK structures occurred *de novo* upon contact with competing bacteria. We show that this assembly required presence of TagQ-TagR envelope sensors, activity of PpkA kinase and anchoring to the inner membrane *via* TssM. Disassembly and repositioning of TssK component was dependent on PppA phosphatase and indispensable for repositioning and deployment of the entire contractile apparatus toward a new target cell. We also show that TssE is necessary for correct elongation and stability of TssB-sheath, but not for TssK assembly. Therefore, in *P. aeruginosa*, assembly of the TssK-containing structure relays on the post-translational regulatory envelope module and acts as spatio-temporal marker for further recruitment and subsequent assembly of the contractile apparatus.

Keywords: molecular nanomachine, T4 bacteriophage, bacterial competition, T6SS, baseplate assembly

INTRODUCTION

Gram-negative bacteria deploy a number of complex envelope-embedded multi-protein nanomachines adjusted exclusively for protein export across bacterial membrane and interaction with eukaryotic and prokaryotic species (Filloux et al., 2008; Hayes et al., 2010). One of these is Type VI secretion system (T6SS), originally identified in *Vibrio cholerae* using *Dictyostelium* host (Pukatzki et al., 2006). The T6SS is a versatile protein exporting machinery implicated in virulence as well as competition and cooperation in microbial communities and it is capable of targeting both bacteria and eukaryotic cells (Alcoforado Diniz et al., 2015; Bleves, 2016; Jani and Cotter, 2010). Thirteen essential core proteins are conserved in all T6SSs (Shalom et al., 2007; Cascales, 2008; Filloux et al., 2008; Boyer et al., 2009); the membrane associated complex TssJ-TssL-TssM, the baseplate proteins TssE, TssF, TssG, and TssK, the bacteriophage-related puncturing complex composed of the tube (Hcp), the tip/puncturing device VgrG, and the contractile sheath structure

(TssB and TssC) (Coulthurst, 2013; Ho et al., 2014; Cianfanelli et al., 2016). Finally, the starfish-shaped dodecameric protein, TssA, limits contractile sheath polymerization at its distal part when TagA captures TssA (Planamente et al., 2016; Santin et al., 2018).

T6SS function involves a highly dynamic process of TssB/TssC sheath assembly, contraction and disassembly (Basler and Mekalanos, 2012; Basler et al., 2012; Forster et al., 2014), where recycling of the sheath subunits depends on ClpV ATPase (Bonemann et al., 2009; Pietrosiuk et al., 2011). Analogous to the T4 bacteriophage tail structure, the T6SS sheath assembles around an inner Hcp tube with membrane-puncturing device (VgrG) positioned at its proximal end (Leiman et al., 2009; Shneider et al., 2013; Silverman et al., 2013). Current model assumes that the positioning and contraction of the apparatus and subsequent ejection of effectors is strictly dependent on transient yet firm anchoring of the contractile device to precise loci within the bacteria membrane. Upon activation of the system, certain components such as Hcp and VgrG, could be exchanged between neighboring bacteria to launch rapid response to external stimuli (Vettiger and Basler, 2016).

Biogenesis of T6SS baseplate start with the association of two molecules of TssF, one molecule of TssG and two TssK trimers to form a stable complex, that then oligomerize further around the VrgG hub (Cherrak et al., 2018; Park et al., 2018). The composition of the whole 1.15 megadalton baseplate is proposed to be $(TssE)_6$-$(TssF)_{12}$-$(TssG)_6$ with eighteen TssK subunits (Nazarov et al., 2018) connecting the whole assembly to the membrane-embedded 1.7 megadalton TssJLM complex (Brunet et al., 2015; Durand et al., 2015), where it stabilizes and initiates tube/sheath polymerization probably through interactions with TssE, as it is the case for its homolog gp25 in the phage (Yap et al., 2010; Nazarov et al., 2018).

Although the structural components of the T6SS machinery are conserved in several organisms, regulatory pathways controlling the expression and assembly of the apparatus differ. *Pseudomonas aeruginosa*, a major human opportunistic pathogen and causative agent of chronic infections in cystic fibrosis patients possesses three T6SS-encoding loci and a number of effector-encoding genes spread through the genome (Filloux et al., 2008). The machinery encoded by the Hcp secretion island I-encoded T6SS (H1-T6SS) (Mougous et al., 2006) is used for bacterial warfare and can translocate multiple effectors into target bacteria (Russell et al., 2011, 2012). The expression and activity of H1-T6SS is controlled at the transcriptional and translational level by a regulatory network including membrane sensors (RetS, LadS, and GacS) (Brencic and Lory, 2009; Brencic et al., 2009), small regulatory RNAs (Brencic and Lory, 2009; Brencic et al., 2009) and di-cGMP signaling (Moscoso et al., 2011). At the post-translational level, eukaryotic-like threonine kinase/phosphatase pair (PpkA-PppA) and a signal-transmission module (TagQ-TagR-TagS-TagT) embedded within the bacterial envelope are required for the T6SS activity (Mougous et al., 2007; Hsu et al., 2009; Basler et al., 2013). The Tag proteins are found in the so-called "defensive" T6SS and modulate the T6SS activity in response to an attack from a competitor, while the "offensive" system is always active, ready to fire on any surrounding bacteria as in the case of *V. cholera* (Basler et al., 2013; Ostrowski et al., 2018).

The aim of this study was to examine the role of signal-transmission module and threonine kinase/phosphatase pair on initial T6SS baseplate assembly. To that purpose, we used quantitative time-lapse microscopy imaging to investigate the spatio-temporal recruitment and assembly of two baseplate proteins, TssK, TssE and the sheath component, TssB. Although the fusion constructs compromised the very last step of T6SS action, i.e., prey killing, they reflected the dynamics of all initial steps of machinery function, including sensing, baseplate and contractile apparatus assembly and firing. As the exploring behavior of TssK/TssB fusions was possible only in the wild-type background, we propose that assembly of baseplate/sheet structure still required a pool of endogenous (non-tagged) TssK/TssB to co-assemble with exogenous (tagged) TssK/TssB. Our results show dynamic assembly of the T6SS baseplate structure that depends on *P. aeruginosa* post-translational signaling cascade, as well as the key functions the baseplate provides for assembly and relocation of the T6SS nanomachinery to a site of contact with competing bacteria.

RESULTS

TssK Assembles Into Peri-Membrane Higher-Order Structures in Response to TagQRST-Mediated Activation of T6SS

To examine whether the environmental signals, transmitted by TagQRST, influence the baseplate assembly we engineered fusions between genes encoding several baseplate proteins (*tssE*, *tssF*, *tssG* and *tssK*) with *sfGFP* and *tssB-GFP*. The plasmids carrying the fusion under arabinose-inducible promoter were introduced in the wild-type *P. aeruginosa* strain PAO1. We were unable to exploit any florescent signal with TssG-sfGFP and TssF-sfGFP. However, in a subpopulation of bacteria, TssK-sfGFP assembled into discernible foci localized to the bacterial periphery, reminiscent to TssK foci observed in *Escherichia coli* (Brunet et al., 2015; Cherrak et al., 2018), whereas TssB-GFP assembled into elongated rod-like structures with substantially higher fluorescence intensity relative to TssK foci (**Figure 1A**). In *P. aeruginosa*, T6SS assemble preferentially upon the attack by neighboring bacteria in a behavior named "T6SS dueling" (Basler and Mekalanos, 2012; Basler et al., 2013). We analyzed *P. aeruginosa* cells expressing TssK-sfGFP grown in presence of *Acetinobacter baumannii* and found that spots formed by TssK-sfGFP were mostly positioned specifically toward contacts with *A. baumannii* (**Figure 1B**), as would be expected for a "defensive" T6SS apparatus. Notably, quantification revealed that incidence of TssK-sfGFP spots in *P. aeruginosa* increased significantly in mixed culture with wild type *A baumannii* but not with a *tssM* mutant (*Ab*Δ*tssM*) with an inactive T6SS (Weber et al., 2013); showing that TssK-sfGFP spot formation correlates specifically with a response of *P. aeruginosa* to T6SS-mediated attack by competitor bacteria. Thus, TssK has the capacity to assemble from the cytosolic pool into perimembrane higher order structures

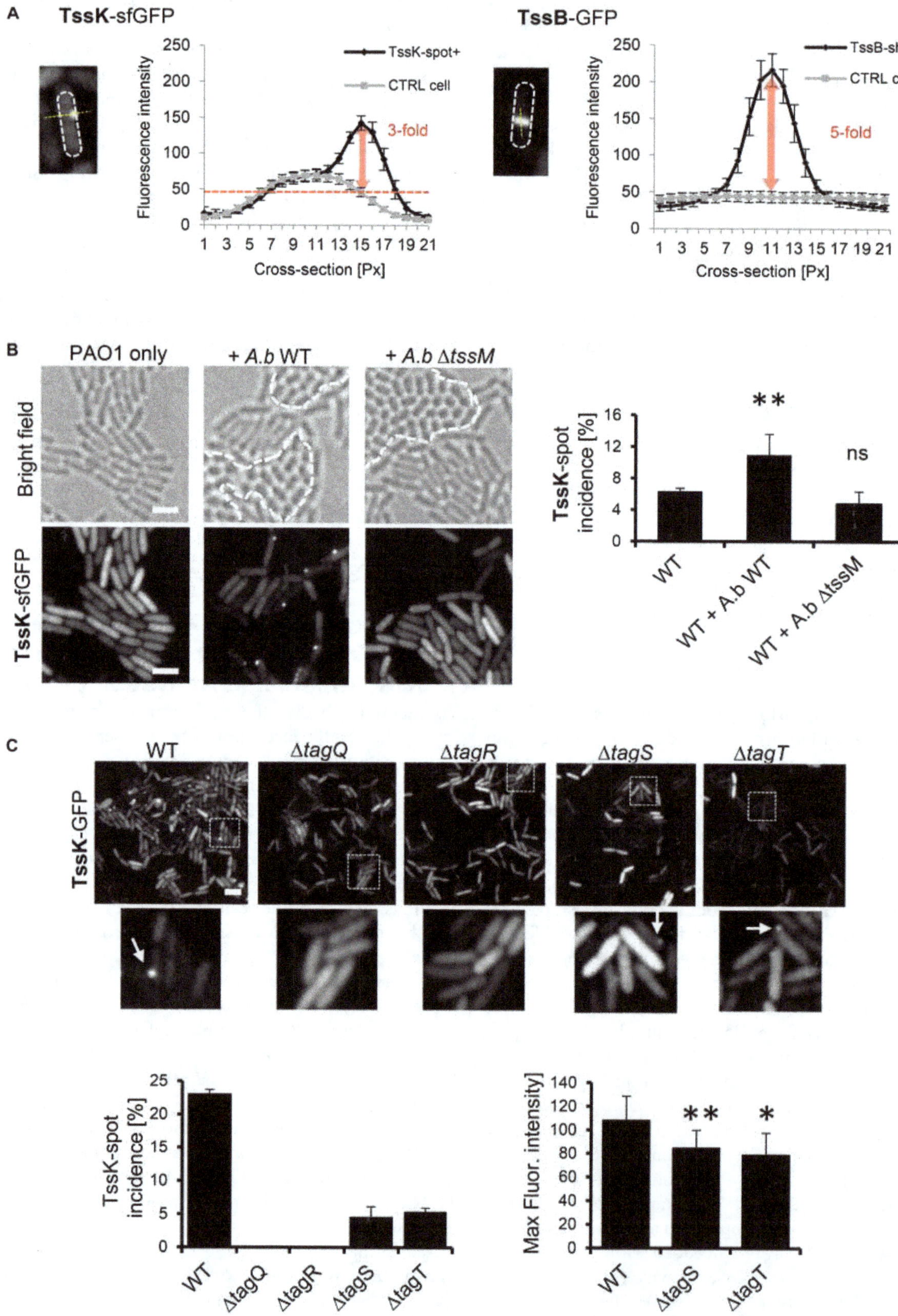

FIGURE 1 | Continued

FIGURE 1 | TssK assembles into discernible perimembrane structures in response to T6SS activation. **(A)** Fluorescent intensities of TssK-sfGFP rounded foci and TssB-GFP elongated rods observed in PAO1 strain after arabinose-induction. Note the 3-fold and 5-fold higher intensity compared to the cytoplasmic levels for TssK and TssB, respectively. **(B)** Assembly of TssK-sfGFP increases in mixed culture with wild type *A. baumannii* but not with T6SS-inactive *A. baumannii* mutant Δ*tssM*. Note that *A. baumannii* can be discriminated from *P. aeruginosa* by its rounded-cell morphology in bright field images (borders between two species are depicted by dashed lines). Representative bright field and fluorescent images are shown with quantification of TssK-spot incidence. Average and SD values were calculated for each group from 4 to 6 ROI each containing between 400 and 1000 of cells. Two-tail *t*-test, **P* = 0.038. **(C)** Role of TagQRST sensor module in TssK assembly. Fluorescence images are shown for each *tag* mutant. Histograms showing quantification of TssK-spot (white arrows in images) incidence and fluorescence intensity measurements of TssK-sfGFP structures are shown below. No spots of TssK-sfGFP were found in Δ*tagQ* and Δ*tagR* mutants and deletion of *tagS* or *tagT* lead to a significant reduction of the incidence of TssK-spot formation. Note that fluorescence intensity of TssK-sfGFP spots in Δ*tagS* and Δ*tagT* mutants was significantly decreased. Average and SD values were calculated from 20 spots for each sample. Two-tail *t*-test, ***P* = 0.03 (Δ*tagS*) and **P* = 0.09 (Δ*tagT*). ns, non significant. Bar = 2 μm.

with the incidence that increases specifically during interaction with competitor bacteria. In competition assays with *E. coli* DH5α, the killing capacity of the PAO1 strain expressing TssK-sfGFP toward the pray was diminished compared to PAO1 (wild type) (**Supplementary Figure S1**). This suggest that TssK-sfGFP incorporates within the baseplate with native TssK but limits the capacity of T6SS to inject effectors. Similar dominant-negative effect on T6SS activity was already observed for single domains of TssK in enteroaggregative *E. coli* (Cherrak et al., 2018).

However, the TssK-sfGFP protein fusion construct could be used to study T6SS assembly kinetics, as demonstrated in the following experiments.

A unique feature of post-translational regulation of T6SS activity in *P. aeruginosa* is the sensory module composed of TagQ, TagR, TagS, and TagT proteins encoded together with other T6SS core components within the HSI-1 locus (Cascales, 2008; Filloux et al., 2008; Boyer et al., 2009). TagQ is an outer-membrane lipoprotein required for recruitment of periplasmic TagR which in turn activates a Threonine Protein Phosphorylation (TPP) pathway promoting assembly of the T6SS apparatus (Mougous et al., 2007; Hsu et al., 2009). TagS and TagT form a classical ABC transporter embedded within the inner membrane of bacteria, and all four proteins are necessary for optimal T6SS activation *in vitro* (Casabona et al., 2013). We used previously characterized Δ*tagQ*, Δ*tagR*, Δ*tagS* and Δ*tagT* mutants (Casabona et al., 2013) and analyzed TssK-sfGFP localization and its incidence in these cells mixed with *A. baumannii*. Unlike the wild type, no TssK-assemblies were detected in Δ*tagQ* and Δ*tagR* mutants (**Figure 1C**). Interestingly, deletion of *tagS* or *tagT* did not lead to complete block of TssK assembly, yet significantly reduced the incidence of TssK-assembling cells so as the overall spot size (fluorescence intensity) of TssK-structures relative to the control (**Figure 1C**). Consequently, we found that assembly of TssK into a baseplate structure within a particular site at the plasma membrane is dependent on TagQ and TagR. We established that presence of TagS and TagT was dispensable for TssK assembly but may have modulatory/accessory function in the process of TssK recruitment into functional entities, in agreement with previous findings that TagT was required for "dueling" behavior and reposition of the apparatus toward neighboring sister cells (Basler et al., 2013). We conclude that observed TssK-sfGFP foci represent protein complexes within T6SS apparatus that are assembled in post-translational manner and can be used as a marker for monitoring T6SS baseplate assembly.

TssK Undergoes Dynamic Assembly-Disassembly

We then analyzed *P. aeruginosa* cells expressing TssK-sfGFP by time-lapse imaging to examine its dynamic features including assembly/disassembly and/or lateral displacement or motility. Real-time detection of transient appearance-growth-decay-disappearance of discernible perimembrane-localized fluorescent spot(s) was measured as described in Section "Materials and Methods," and reported as assembly/disassembly. Intensity measurements started prior structure appearance (except for TssK in *pppA* mutant) where values at T0 (normalized to 1) correspond to fluorescence intensity of cytosolic pool of the fusion protein. We typically observed multiple rounds of assembly disassembly occurring at the same location before assembly was detected at another location of the same bacteria. Interestingly, no more than one TssK-sfGFP spot was found per single cell at a given time. Kinetic measurements of fluorescence intensity revealed that assembly of the TssK-sfGFP occurred within 39 ± 6 s whereas disassembly of the structures lasted in average 61 ± 10 s (**Figure 2A**), implying that these transient structures undergo frequent assembly-disassembly as a dynamic component of T6SS baseplate structure.

Downstream of TagQRST sensory module, the T6SS activity is post-translationally regulated by the threonine kinase PpkA whose activity is antagonized by the phosphatase PppA (Mougous et al., 2007; Ostrowski et al., 2018). We detected TssK-sfGFP assemblies in the Δ*pppA* mutant but never in the Δ*ppkA* mutant of *P. aeruginosa* (**Figure 2A** and **Supplementary Figure S2**), despite the same expression levels of corresponding constructs as revealed by immunoblot analysis (**Supplementary Figure S3**). Similar result was reported for TssB and Fha foci in Δ*pppA* and Δ*ppkA* mutants in *S. marcescens* (Ostrowski et al., 2018). Kinetic measurements showed that in Δ*pppA*, structures formed by TssK-sfGFP were static, permanently retained at one perimembrane site without *de novo* assembly at another location (**Figure 2A**). This suggests that the dephosphorylation of Fha, a target of PpkA kinase and PppA dephosphatase, is necessary for repositioning of the TssK-containing baseplate, but not for its assembly *per se*. Otherwise, PppA could have another, yet unidentified target. To note, in the wild-type *E. coli*, TssK foci also seem to be static at a given time but several of them were detected per cell, which may be explained by the absence of post-translational regulatory module in this system (Brunet et al., 2015).

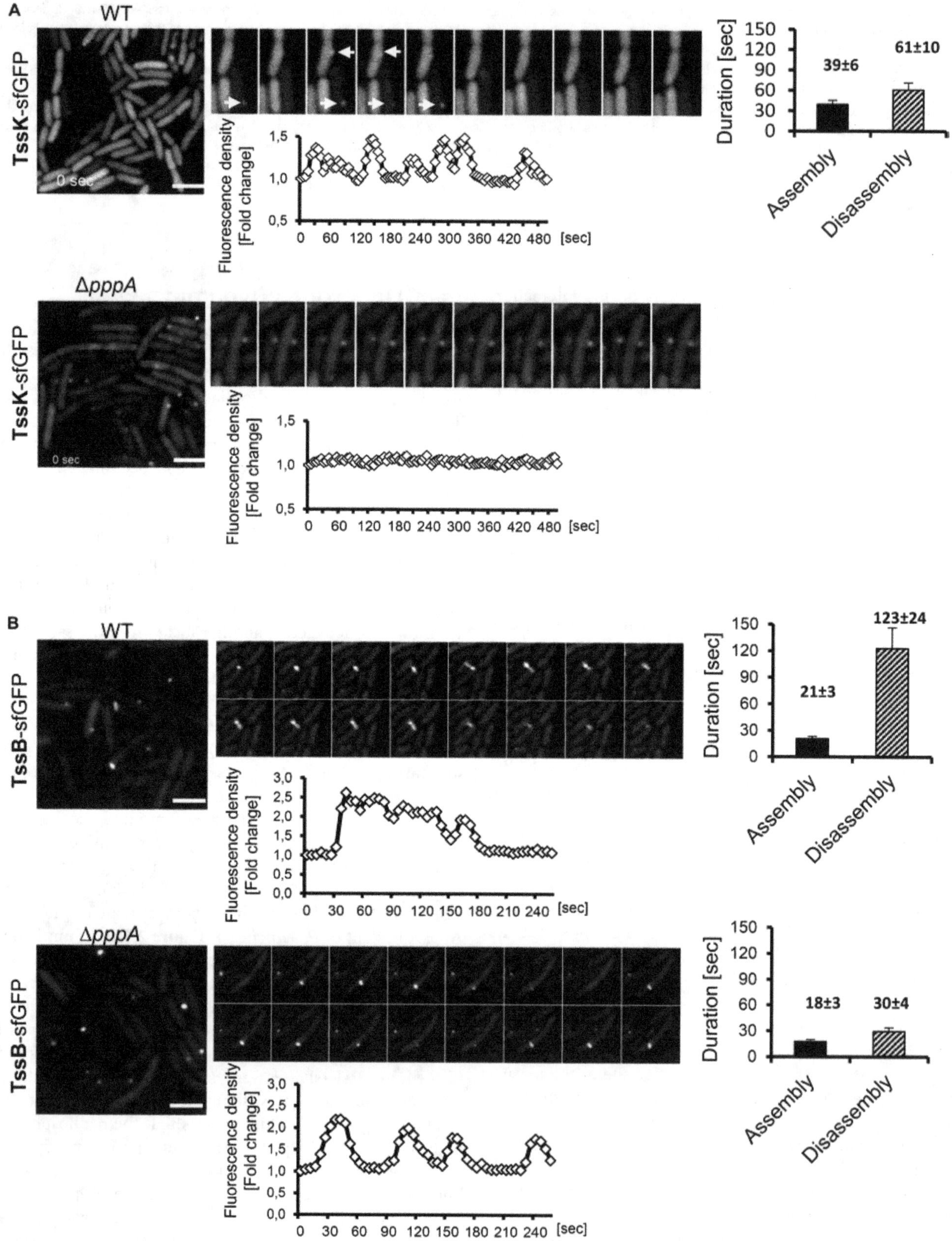

FIGURE 2 | Continued

FIGURE 2 | Assembly of TssK baseplate component requires activity of PpkA kinase while disassembly and repositioning is dependent on PppA phosphatase. Time-lapse series of *P. aeruginosa* PA01 (WT) and PppA-deficient mutant (Δ*pppA*) expressing TssK-sfGFP **(A)** or TssB-GFP **(B)**. Series of fluorescence images are shown (10 s between frames) with kinetic measurements of relative fluorescence intensity in ROI containing TssK-sfGFP/TssB-GFP structures (below) and quantification of an average duration of assembly/disassembly of these structures (on right). Bars = 2 μm. **(A)** In wild type cells, individual TssK-sfGFP structures (white arrows) assemble/disassemble within an average 100 ± 10 s. Note that lack of PppA impaired recycling and re-positioning of TssK structures but not their assembly *per se*. **(B)** A single round of assembly-contraction-disassembly of individual TssB-GFP sheaths in wild type occurs within an average 144 ± 24 s period. Note that in Δ*pppA*, the TssB-sheath assembles repeatedly at the same location and exhibits an aberrant kinetics with significantly faster rate of disassembly (30 ± 4 s) relative to the wild type control (123 ± 24 s).

To investigate how TssK dynamics relates to the assembly, contraction and disassembly of the contractile sheath in *P. aeruginosa* we first analyzed the sheath dynamics in *P. aeruginosa* expressing TssB-GFP construct. Similar to TssK, the TssB-GFP exhibited homogenous cytosolic localization in a vast majority of cells but elongated sheath-like structures were readily detected in mixed cultures with *A. baumanni*. As for TssK foci, we have never detected more than one TssB-sheath structure per individual cell at a given time. Quantitative measurements of assembly-disassembly revealed distinct kinetics features for TssB-sheaths, relative to TssK structures, where assembly occurred within 21 ± 3 s followed by disassembly at a slower rate of 123 ± 24 s resulting in an average 140 s lifetime for an individual TssB-structure (**Figure 2B**), similar to value found by others (Basler and Mekalanos, 2012; Basler et al., 2012). Kinetic profiles of TssK-baseplate in homogenous culture of *P. aeruginosa* were similar to corresponding profiles in competition conditions where the assembly of TssK structures was targeted preferentially toward the contact with competing *A. baumannii* (**Supplementary Figure S4**). Like TssK assemblies, several rounds of assembly-contraction of TssB-sheaths were detected at the same site of the bacteria prior complete disassembly of the structure. The dynamic feature of the contractile sheath (TssB-GFP) was not lost in *P. aeruginosa* mutant Δ*pppA* but led to significantly faster (likely premature) disassembly of the sheath relative to wild-type (30 ± 4 vs. 123 ± 24 s, respectively).

We conclude that similarly to contractile sheath component TssB, the baseplate component TssK also undergoes regulated periodic assembly and disassembly during deployment of the T6SS apparatus. This implies that in cells primed for T6SS activity, TssK does not assemble into multiple rigid baseplate structures per cell, but instead assembles transiently *ad hoc* at a specific site of the inner membrane toward the contact with neighbor attacking bacteria. We provide evidence that PppA phosphatase functions downstream of this TssK assembly and is essential for its disassembly and relocation to a new site for subsequent *de novo* assembly of the contractile apparatus.

TssK Functions as a Spatio-Temporal Marker for Assembly of Contractile Apparatus

Because both, TssK-sfGFP structures and TssB-GFP sheaths exhibited similar yet distinct dynamic phenotype, we examined TssK as a spatio-temporal cue for recruitment of contractile sheath components TssB. We analyzed colocalization between TssK and TssB and measured the kinetics of their assembly-disassembly in *P. aeruginosa* co-expressing TssK-RFPT and TssB-GFP. Correspondent with our previous observations with individually expressed proteins, TssK only assembled into rounded spots at the periphery of bacteria, whereas TssB assembled into rod-like structure often spanning the entire width of the cell (in an extended conformation) (**Figure 3A**). It is noteworthy that in bacteria with discernible TssK and TssB assemblies the TssK structure co-localized exclusively with the proximal tip of the TssB structure at the perimembrane site (**Figures 3A,B**), in agreement with proposed baseplate localization and function as a platform for sheath polymerization (Brunet et al., 2015; Cianfanelli et al., 2016; Cherrak et al., 2018; Nazarov et al., 2018).

Importantly, in the wild-type strain co-expressing both fusions the incidence of discernible TssK structures at random time points (as captured on still images) was more frequent than that of TssB-sheaths and consequently a vast majority (95%) of TssB sheaths was marked by discernible TssK-structure, while only a subpopulation of cells with TssK structure (40%) also exhibited assembled TssB-sheath (not shown). This suggests that both components assemble sequentially rather than simultaneously into the apparatus and that only a fraction of TssK assemblies gives rise to TssB-sheath assembly, similarly to a study in *E. coli* (Santin et al., 2018). We then examined relative timing of TssK and TssB assembly by time-lapse analysis of individual bacteria followed by quantitative measurement of fluorescence intensity within a region of interest (ROI) covering the site of TssK/TssB assembly. Although the incidence of TssK and TssB assemblies exhibited a spatial correlation within individual cells, their temporal co-localization was mutually exclusive. Quantitative fluorescence intensity measurements also revealed consistent interval of 20 ± 10 s between kinetic fluorescence maxima of TssK and TssB assemblies (**Figure 3B**), suggesting that TssK component of the baseplate is involved in the initial stage of TssB-sheath nucleation rather than in subsequent steps of sheath elongation, contraction or disassembly.

TssK Requires TssM to Trigger TssB-Assembly

To determine the impact of the membrane component TssM on the kinetics of TssK assembly, we analyzed the incidence of TssK and TssB assemblies in *P. aeruginosa* PAO1Δ*tssM*. We found that although deletion of *tssM* significantly decreased the incidence of bacteria positive for TssK foci (from 10% in the parental strain to 3.7% in Δ*tssM*) the assembly of TssK structures and their localization to the membrane was not completely abolished (**Figure 4A**). In contrast, elongated TssB rod-shaped assemblies were detected only in wild-type bacteria

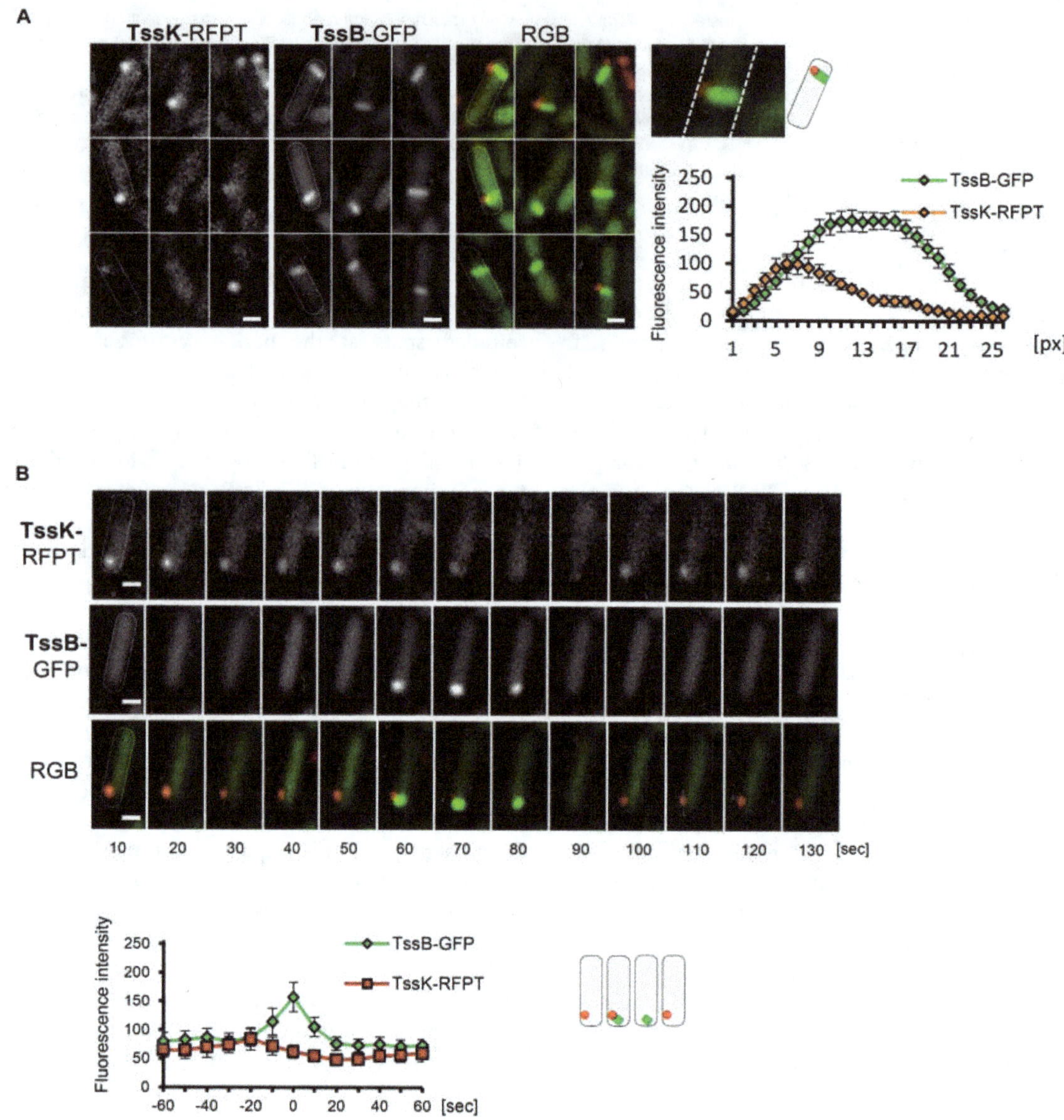

FIGURE 3 | TssK functions as essential positional marker for recruitment, nucleation and assembly of contractile apparatus. **(A)** TssK assembles into discrete structure at the proximal tip of TssB-sheath. Several representative fluorescence images of WT cells co-expressing TssK-RFPT and TssB-GFP. Note that fully elongated TssB-sheath spans the entire width of bacteria. Fluorescence intensity profile of structures positive for both TssK-RFPT and TssB-GFP measured on line cross-section from proximal to distal pole with average values and SD calculated from 20 structures (*N* = 20). Schematic representation of bacteria with TssK (in red) and TssB (in green). **(B)** TssK and TssB assemblies within individual bacteria exhibit spatial correlation, but differential temporal co-localization. Time-lapse of fluorescence images of representative bacteria co-expressing TssB-GFP and TssK-RFPT (top panel). Note that TssK structure transiently disappears after the onset of TssB assembly at that site. Kinetics of fluorescence intensity measured at ROI containing nascent TssK/TssB structure indicates an interval (20 ± 10 s) between fluorescence maxima of TssK-RFPT and TssB-GFP. SD were calculated from ROI of 20 bacteria (*N* = 20). Bars = 0.5 μm. Schematic representation of bacteria from time lapse in B, with TssK (in red) and TssB (in green).

and never in Δ*tssM*. We noted however, that in less than 0.6% of Δ*tssM* bacteria the TssB assembled into low-intensity round-shaped fluorescent spot(s) localizing to the membrane but these structures never extended into a shape of an extended sheath-like structure (**Figure 4A**). To assess possible impact of *tssM* deletion on dynamic properties of TssK we quantified an average time span of TssK structure (period between onset of assembly and complete disassembly/disappearance) by measuring the TssK-GFP fluorescence intensity in time-lapse of Δ*tssM* relative to wild-type bacteria. We found kinetics of TssK assembly-disassembly to be almost identical in both genetic backgrounds (140 ± 30 s per single round of assembly-disassembly) (**Figure 4B**). However, in a contrast to the wild-type, TssK-assemblies in Δ*tssM* often "detached" from the original perimembrane site and exhibited lateral movement along the membrane before their complete disassembly (**Figure 4C**).

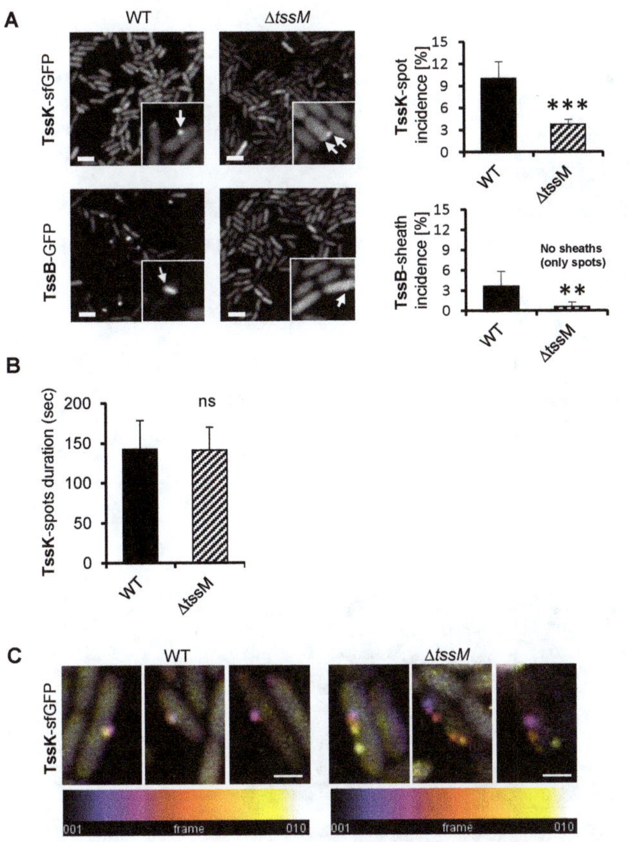

FIGURE 4 | Assembly of TssK baseplate is essential but not sufficient to trigger TssB-assembly and requires stable anchor *via* TssM. **(A)** Assembly of TssK-baseplate and TssB-sheath in *P. aeruginosa* PAO1 lacking transmembrane component of T6SS apparatus, TssM. Representative fluorescence images (left) and quantification of TssK or TssB assemblies (right). Note that incidence of TssK structures (white arrows) in Δ*tssM* cells is significantly reduced, but not completely abolished, while elongated TssB-sheaths (white arrows) are only visible in PAO1 and only minor subpopulation of Δ*tssM* bacteria (<0.6%) was found with a spot-like structure of TssB-GFP. Two-tail *t*-test, ***$P = 0.002$ (TssK) and **$P = 0.015$ (TssB). Bars = 2 μm. **(B)** Quantification of TssK-spot duration as a time between appearance and disappearance of discernible TssK-sfGFP structure over the background fluorescence intensity of the fusion construct. Note that kinetics of TssK-sfGFP baseplate assembly-disassembly is not impaired in Δ*tssM* relative to the WT cells (140 ± 30 s). Average values and SD were calculated from 30 cells ($N = 30$). Two-tail *t*-test. ns, non-significant. **(C)** Three representative time-lapse series (10 s between frames, 10 frames) of WT and Δ*tssM* cells expressing TssK-sfGFP. Each series were superimposed and color-coded (fire LUT) to visualize TssK structure stability/displacement over time. Note that TssK-structure in Δ*tssM* undergoes lateral displacement. Bars = 0.5 μm.

Together, these results show that (i) TssM is not necessary for assembly of TssK *per se*, (ii) TssM is essential for stabilization of assembled TssK structure at a fixed perimembrane site and (iii) in the absence of TssM, the TssB protein is still recruited to the structure containing TssK, but the assembly of TssB-sheath is aborted (likely at the nucleation stage). This implies that TssM provides a necessary anchor, probably through the outer membrane lipoprotein TssJ (Felisberto-Rodrigues et al., 2011) for transient assembly of TssK-oligomeric structure. This could restrict its lateral movement along cell periphery, which in turn is required for subsequent recruitment, assembly and elongation of TssB-sheath.

TssE Functions Downstream of TssK and Prevents Detachment of TssB-Sheath and Its Premature Disassembly

TssE is the structural homolog of the T4 phage baseplate components gp25 (Leiman et al., 2009; Lossi et al., 2011). TssE oligomerises within the baseplate together with TssF and TssG, interacts with the sheath element TssB (Cherrak et al., 2018; Nazarov et al., 2018) and was proposed to act as an initiator of sheath polymerization (Kudryashev et al., 2015; Yap et al., 2010). To check contribution of TssE in TssK-mediated assembly of TssB-sheath in *P. aeruginosa*, we generated *P. aeruginosa* mutant Δ*tssE* expressing TssK-sfGFP and analyzed TssK-assemblies. TssK-sfGFP retained the capacity to assemble into foci in Δ*tssE* mutant and in the strain Δ*tssE/tssE* (**Supplementary Figure S5**). Similar to TssK-GFP, the TssE-sfGFP fusion protein also exhibited homogenous cytosolic localization in a majority of bacteria with a minor subpopulation of cells where TssE-sfGFP assembled into discernible perimembrane spots (**Figure 5A**). To confirm that TssK and TssE assemble into the same spot-like structure we generated bacteria co-expressing TssE-sfGFP with TssK-RFPT. Analysis of *P. aeruginosa* co-expressing TssE-sfGFP and TssK-RFPT (**Figure 5A**) revealed that all discernible TssE assemblies colocalized with TssK (upper panel) while not all TssK-RFPT assemblies were positive for TssE-sfGFP (lower panel). We generated also a Δ*tssE* strain expressing TssB-GFP. As already reported (Basler and Mekalanos, 2012; Vettiger and Basler, 2016), very low number of TssB-structures were seen in the mutant bacteria, and this was corrected by complementation (**Supplementary Figure S5C**). The time-lapse analysis of TssB-sheaths in Δ*tssE* revealed an aberrant phenotype, where the sheaths were unusually elongated and curved, not perpendicular to the membrane, and their dissociation from the inner membrane followed by disintegration into fragments was readily observed (**Figure 5B**), suggesting a weak or inefficient connection to the baseplate. This aberrant phenotype may explain the extremely low activity of the T6SS machinery in the absence of TssE observed in *Vibrio* (Basler and Mekalanos, 2012; Vettiger and Basler, 2016).

From these results, we conclude that (i) TssK component of the baseplate assembles independently of TssE, (ii) TssE is recruited to the baseplate complex at later stages as already suggested, and (iii) TssE is dispensable for TssB-sheath assembly *per se* but is required for stabilization of the connection between the sheath and the baseplate.

DISCUSSION

The access to assembly of multi-protein, membrane-bound architectures in live cells is extremely challenging in the context of spatial and temporal resolution. Here we focused on dynamic and regulatory aspects of assembly/disassembly

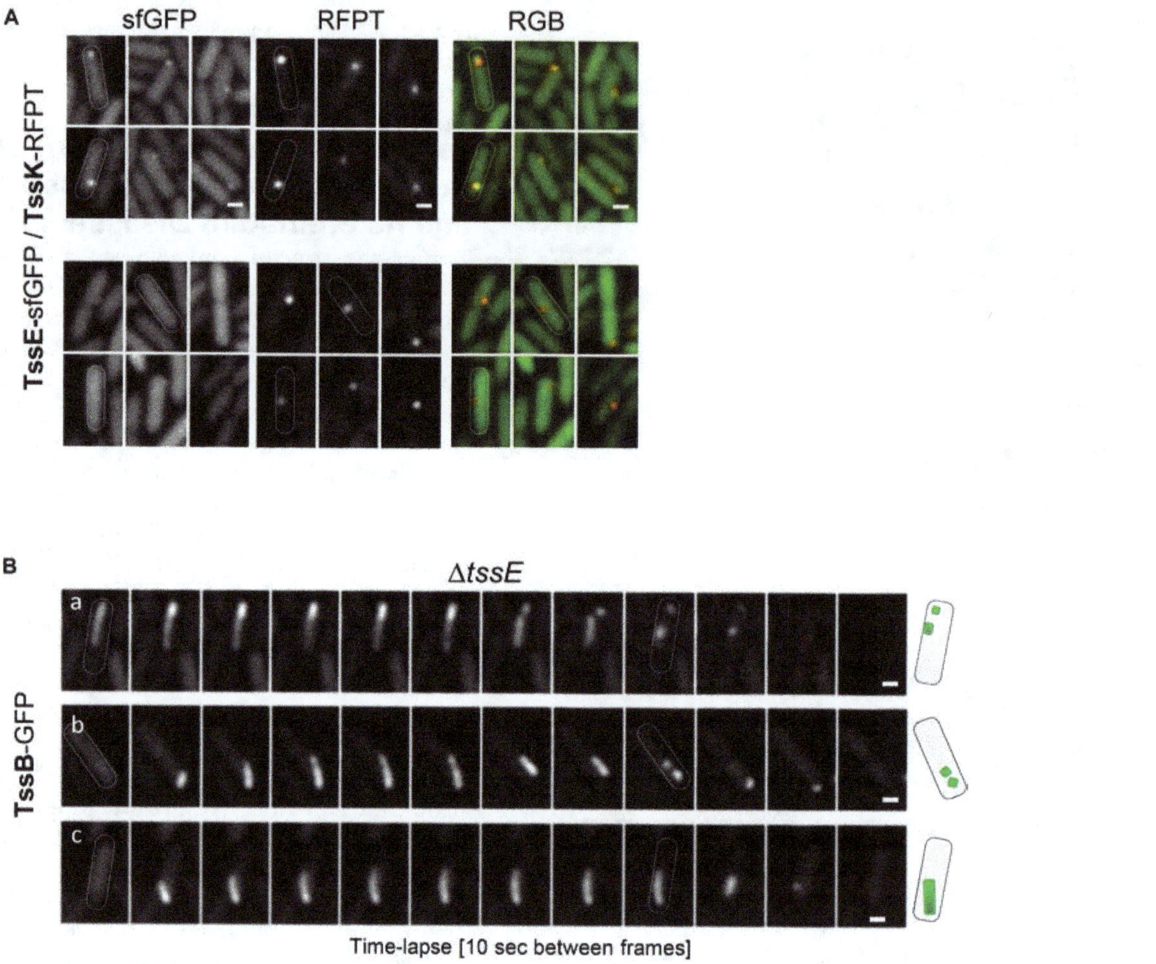

FIGURE 5 | Role of TssE in baseplate assembly and elongation of TssB-sheath. **(A)** TssE-sfGFP assembles into discernible perimembrane structures and colocalizes with TssK in *P. aeruginosa* PAO1 co-expressing *tssE-sfGFP* with *tssK-RFPT* (upper panel). In some cases, TssK-RFPT assembles into spots without TssE-sfGFP (lower panel). Six representative fluorescence images are shown. Bars = 0.5 μm. **(B)** Aberrant phenotype of contractile sheath (TssB-GFP) in PA01 Δ*tssE* strain. Fluorescence images from three representative time-lapse series of Δ*tssE* cells are shown. Note that the TssB-sheath is unusually elongated and curved and breaks off from the membrane. Bars = 0.5 μm. Schematic representation of bacteria with TssB (green) in PAO1 Δ*tssE*.

of the T6SS baseplate components, TssK and TssE, and TssB by microscopy imaging of intact bacteria. We followed their capability to assemble into discernible higher order structure(s) depending on post-transcriptional regulatory module, their subcellular localization and spatial orientation and their dynamic characteristics (assembly/disassembly).

We found that TssK assembles into higher-ordered structure localizing to discrete foci at the cell periphery in the proximity of exogenous attacker and initiates further assembly of the contractile sheath, in agreement with its higher-ordered oligomerization capacity within the T6SS baseplate. In the complex baseplate-extended sheath forms TssK the connector ring between the membrane complex TssJLM and the rest of the baseplate. The overall structure has a six fold symmetry in the center, and is composed by two TssF interconnected by one TssG. TssG then contacts two TssK trimers trough extended loops to form a final structure containing 18 subunits of TssK (Cherrak et al., 2018; Nazarov et al., 2018; Park et al., 2018).

Based on those structural studies and observations by others, the C-terminal fusion construct is not sterically incompatible and should not impede TssK binding to its partners (TssL, TssM, TssG, and TssF). TssK-sfGFP used in our work is probably not competent for *E. coli* killing because of the GFP molecule hindering structural movements in TssK oligomer required to trigger final VgrG-toxin expulsion, although we could still observe the dueling.

The main finding in this study is the transient characteristic of TssK structures suggesting that they assemble *ad hoc* as structural entities and disassemble together with the contractile sheath TssB to recycle the components for *de novo* assembly of the apparatus in another site of the cell. Results of our kinetic analysis propose that TssK assembles upstream of TssE, the major transmembrane anchoring component TssM, and contractile sheath component TssB. It also requires presence of at least two essential components of the membrane-embedded sensory module, TagQ and TagR. This is in agreement with previous findings showing that TagQ-TagR-PppK-Fha-PppA pathway recognizes an external signal to activate the T6SS (Mougous et al., 2007; Basler et al., 2012; Ho et al., 2014). Our laboratory found that the sensory module in *P. aeruginosa* has additional components TagS and TagT that form an inner membrane ABC transporter with TagT having an ATPase activity (Casabona et al., 2013). Here we show that mutants in any of these two components are still capable to assemble TssK structures even if with lower incidence within the cell population. This opens a possibility that TagQ and TagR represent the core unit of the sensory complex able alone to trigger assembly of the TssK component of the baseplate whereas TagS and TagT probably through their ATPase activity may have a role in promoting higher-order oligomerization of proteins within the baseplate to adopt and stabilize the whole sheath/tube entity. TagT was indeed required for assembly of T6SS sheaths upon activation by polymyxin B treatment (Ho et al., 2013), or upon attack by a neighboring T6SS-positive bacteria (Basler et al., 2013), as assessed by the ClpV-GFP fusion. This observation could be a consequence of TagT effect on capacity of TssK to oligomerize in higher-ordered structures. We previously found that fusion construct TagQ-mCherry expressed in *P. aeruginosa* localizes to the outer membrane, but no discernible structures could be detected (Casabona et al., 2013). We thus hypothesize that multiple TagQRST modules may be scattered throughout the surface of bacteria and function as receptors sensing physical contract with kin or prey cells to trigger rapid assembly of the T6SS apparatus by recruitment of baseplate components at close proximity to assure effective attack toward a predatory species. Intriguingly, the *E. coli* TssK assemblies are static component of the baseplate (Brunet et al., 2015). This difference in TssK behavior may reflect the difference in regulatory features between *E. coli* and *P. aeruginosa* T6SSs, notably the absence of TagQRST-PpkA-PppA signaling in the *E. coli* system (Boyer et al., 2009).

What provides the link to the transmembrane anchoring complex of the T6SS once the baseplate begins to assemble? Previous studies using bacterial two-hybrid analysis identified direct interactions between TssK and TssM and between TssK and TssL, but only the TssK/TssL interaction was confirmed by co-immunoprecipitation (Zoued et al., 2013). In our study, we showed that membrane protein TssM is essential for lateral stabilization of TssK foci but not for the TssK assembly. It is possible that TssL alone assures TssK recruitment from the cytosolic pool into the perimembrane site. Biochemistry and bacterial two-hybrid screen revealed that TssK physically interacts with N-terminal cytoplasmic domain of TssL, residues 1–184 (Zoued et al., 2013). Moreover, study of T6-secretion in *A. tumefaciens* (Lin et al., 2014) proposed a model for the TssL-TssM complex where TssL is first phosphorylated by PpkA leading to ATP-binding dependent conformational switch in TssM followed by direct binding of FHA domain of Fha to the TssL. As thus the TssM-p-TssL-Fha complex powered by TssM-mediated ATP hydrolysis (Ma et al., 2012) may initiate recruitment of tube core component Hcp for assembly. Analysis of TssK behavior in specific mutants, either lacking domains of interactions between those components, mutants deficient in TssM ATPase activity, or bearing Fha unable to be phosphorylated, should bring more insights into molecular mechanisms governing the initial recruitments of TssK.

Our data suggest that although TssE assembles into a single perimembrane structure similar to TssK, it does so independently and downstream of assembly of TssK baseplate component. TssE homolog in bacteriophage T4, gp25, is essential to initiate assembly of the tube (Wagenknecht and Bloomfield, 1977). TssE assembles in the outermost structure of the baseplate hub around the membrane-puncturing device (Cherrak et al., 2018; Nazarov et al., 2018), as does gp25 (Kostyuchenko et al., 2003; Leiman et al., 2009). In contrast to TssK, we found that, in *P. aeruginosa*, TssE was not essential for TssB assembly into a sheath-like structure, although the sheath in a Δ*tssE* mutant exhibited an aberrant phenotype and break itself in smaller pieces, in agreement with almost undetectable activity of T6SS in competition assays found by others (Vettiger and Basler, 2016). In some Δ*tssE* bacteria where the TssB-sheath elongated parallel (or in a certain angle) to the membrane, the elongation of the apparatus was not spatially limited by cell width and TssB-sheath thus often assembled to an extreme length. This implies that although TssE and TssK co-assemble into a baseplate structure they adopt distinct functions where TssE incorporation into the baseplate may fortify the physical link between tube-sheath complex and the baseplate. Access to spatio-temporal behavior of two other baseplate proteins, TssG and TssF, should provide in the future global picture of the role of the baseplate in T6SS function.

In conclusion, our work shows the role of transmembrane post-translational regulatory cascade in activation and/or localization of T6SS baseplate assembly and the role of TssK and TssE in spatio-temporal organization and stabilization of T6SS contractile sheath in *P. aeruginosa*. Despite the fact that due to the low fluorescence intensities we could not examine the assembly of TssF and TssG, we present a model (**Figure 6**) where trimers of TssK associate with TssF/TssG complex transiently at the perimembrane site in contact with prey bacteria. The baseplate assembly occurs downstream of TagQRST activation and PpkA activity and functions as an essential spatio-temporal cue for subsequent recruitment of other components of the machinery including TssE, Hcp, and TssB/TssC leading to assembly of functional contractile apparatus with stable anchor within bacterial membrane. Our model also highlights the role of TssK in PppA-mediated recycling and repositioning of the machinery through *de novo* assembly at different perimembrane location defined by local activation of TagQRST module upon contact with the target cell.

MATERIALS AND METHODS

Bacterial Strains and Genetic Constructions

All bacterial strains are listed in the **Supplementary Table S1**. Wild-type *Acinetobacter bauamnnii* and its Δ*tssM* mutant was kindly provided by Brent Weber (Alberta Glycomics Centre, University of Alberta, Canada) and described earlier

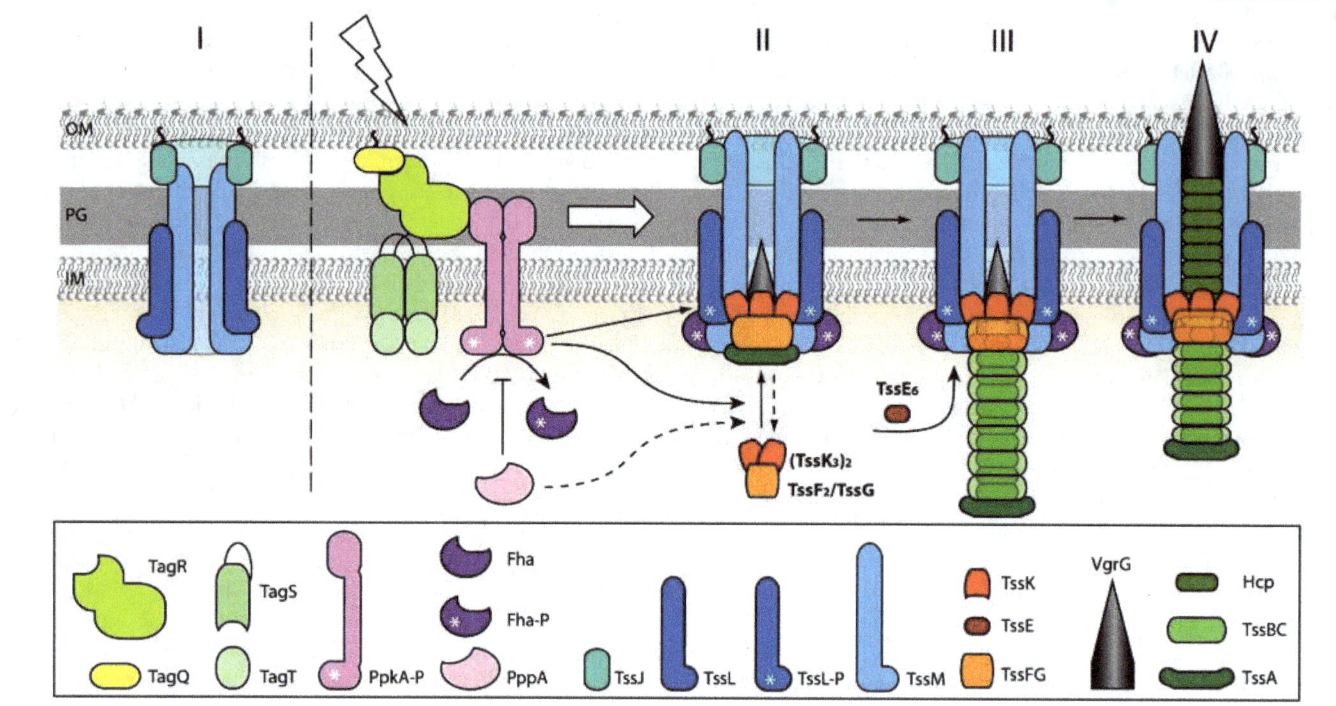

FIGURE 6 | Schematic model for TssK/TssE baseplate assembly and function. **(I)** In a steady state, the baseplate components TssK and TssE are present in the cytosol. **(II)** Upon contact with target cell and activation of TagQRST (lightning) the Fha is phosphorylated by activated PpkA and binds TssL associated with the inner membrane. Upon activation of TagQ, TagR, and TssK oligomerizes, probably concurrently with TssF and TssG, around VrgG and the whole complex (VrgG$(TssK_3)_6(TssF_2/TssG)_6$) is recruited into the baseplate and anchored to the transmembrane complex TssJLM. The TssK oligomers disassemble upon action of PppA (shown in dashed lines). Stoichiometry $(TssK_3)_2(TssF_2/TssG)$ in accordance with cryo-EM structure (PDB 6N38) is shown. **(III)** Nucleation of the tube-sheath assembly begins by recruitment of Hcp-TssB/C subunits. TssE joins the TssF-TssG complex to form a more stable tube-sheath structure. **(IV)** The fully extended state is ready to "fire". Outer membrane (OM), peptidoglycan (PG), inner membrane (IM).

(Weber et al., 2013). *P. aeruginosa* wild type and mutants Δ*tssM* (PA0077), Δ*tagQ* (PA0070), Δ*tagR* (PA0071), Δ*tagS* (PA0072), Δ*tagT* (PA0073), Δ*pppA* (PA0074), and Δ*ppkA* (PA0075) were kindly provided by J. Mougous Laboratory (University of Washington, United States) and described earlier (Silverman et al., 2011). Original plasmids and plasmid constructs used in this study are summarized in **Supplementary Table S2.** Fusion constructs *tssK*-sfGFP and *tssE*-sfGFP were generated as follows. First, the gene sequence encoding sfGFP [GenBank: HQ873313.1] was synthesized (Genscript, United States) to contain 5′ terminal *Spe*I and 3′ terminal *Sac*I restriction sites. This product was subcloned into pJN105 plasmid (Newman and Fuqua, 1999) via *SpeI*/*Sac*I restriction sites to generate pJN105-sfGFP. Second, genes encoding TssK (PA0079) and TssE (PA0087) without stop codon were amplified from *P. aeruginosa* PAO1 genomic DNA by PCR using primers listed in (**Supplementary Table S3**) and subcloned into pJN105-sfGFP via *EcoRI*/*Spe*I restriction sites. Fusion construct TssB-GFP was synthesized (ProteoGenix, France) to contain 5′ terminal *Eco*RI restriction site followed by 20nt 5′ TCGCGCGAGGGAGAAACAAG including ribosome-binding site, gene encoding TssB (PA0083), a linker sequence (5′ GCCGCCGCCGGCGGCGGC 3′), gene encoding GFP [GenBank: LN515608.1] and 3′ terminal *Sma*I restriction site. This synthetic product was subcloned into pJN105 via *Eco*RI/*Sma*I restriction sites to generate pJN105-*tssB-GFP*. Fusion construct *tssK-RFPT* was generated as follows. First, gene sequence encoding TagRFP-T (RFPT in this manuscript) [GenBank: EU582019.1] without ATG start codon was codon optimized for expression in *P. aeruginosa* and synthesized (Genescript, Unites States) to contain 5′ terminal *Spe*I restriction site followed by a linker sequence (5′ GCCGCCGCCGGCGGCGGC 3′), gene encoding TagRFP-T and 3′ terminal *Sac*I restriction site. This synthetic product was subcloned into pSW196 plasmid *via Spe*I/*Sac*I restriction sites to generate pSW196-RFPT. Second, gene encoding TssK was subcloned from pJN105-*tssK-sfGFP* into pSW196-*RFPT* via *EcoRI*/*Spe*I restriction sites to generate pSW196-*tssK-RFPT*. Plasmids encoding fusion constructs were introduced into *P. aeruginosa* strains by heat-shock transformation for pJN105-derived plasmids or by triparental mating for pSW196 integrative plasmids. To generate an in-frame deletion of *tssE* (PA0087) in *P. aeruginosa*, sequences corresponding to upstream and downstream flanking regions of the genes was synthesized (Genscript, United States) to contain *Sac*I and *Xba*I restriction sites on 5′ and 3′ terminus, respectively. This synthetic product was subcloned into pEXG2 plasmid (Rietsch et al., 2005) for subsequent *sacB*-mediated allelic exchange as described earlier (Metcalf et al., 1996). Deletion was confirmed by PCR using primers listed in **Supplementary Table S3**. The similar level of expression of fusions in different strains was confirmed by immunoblotting (**Supplementary Figure S3**). For complementation experiments, gene of interest was cloned into a pminiCTX1-derivative plasmid pSW196 containing arabinose-inducible promotor pBAD. Then, pSW196 plasmid containing *tssE* or *tssK* gene was introduced in *P. aeruginosa* by triparental mating (Hmelo et al., 2015).

Competition Assays

The competition assays were performed as previously described (Hachani et al., 2013). *P. aeruginosa* and *E. coli* (pBlueScript) were grown overnight in 3 ml LB medium supplemented with appropriate antibiotics. Diluted overnight cultures were inoculated in the same medium containing 0.025% arabinose (to induce the expression of specific fusion) and the culture was grown until OD_{600} = 1. Then 1 ml of each culture were spin down and pellet was resuspended in 100 μl LB with arabinose 0.025%. Indicated *P. aeruginosa* strains (predator) were mixed with *E. coli* (prey) in ratio 1:2 (predator:prey). Competition reactions (20 μl), realized in triplicate, were spotted onto LB agar plates containing 50 μg/ml ampicillin and incubated 5 h at 37°C. The totality of bacteria was recover in LB and dilutions were plated in triplicates onto LB plates containing Xgal (40 μg/ml) and IPTG (100 μM) to visualize *E. coli* (blue colonies).

Western Blotting

The expression levels of protein fusions in different mutants were assessed by immunoblotting. For the Western blot, total bacterial samples (OD_{600} = 1) were separated on Criterion 4–20% TGX precasted gels, BioRad and transferred onto a PVDF membrane (GE.Healthcare) by electrotransfert in 20% Laemmli buffer. After blocking step in 5% milk, polyclonal anti-GFP antibodies (diluted 1/5000 in PBS buffer with 0.1% Tween20) were incubated one hour at room temperature, followed with a second antibodies incubation (anti-rabbit HRP, dilution 1/20 000, Sigma). Detection was performed using Luminata Classico HRP-substrate (Millipore) using BioRad ChemiDoc apparatus.

Time Lapse Video Microscopy

Bacteria were grown over night in LB with appropriate selection antibiotics, then sub-cultured 1:100 in LB with arabinose (0.025–0.25%) for induction of fusion construct expression and grown for additional 3 h to reach OD_{600} = 2. To immobilize bacteria for acquisition of still images or a time-lapse series, a drop of bacterial culture was applied on top of agarose pads [1.5% Low melting agarose in Hanks' balanced salt solution (Gibco)] and transferred into the glass-bottom imaging dishes (WillCo Wells BV) so that bacteria were grown in interphase between glass bottom and semi-solid medium of agarose pad for another 1 h prior imaging. For competing conditions, culture of *P. aeruginosa* was mixed 1:1 with competing bacteria (*A. baumannii*) prior applying onto agarose pads. Still images or time-lapse image acquisitions was performed within 1–3 h post mixing competing bacteria – a time frame when we observed the peak activity of TssK/TssB assemblies. Fluorescence and bright field images were acquired at a rate of 6 frames per min (10 s between frames) for 5–15 min using OLYMPUS inverted microscope IX71 equipped with UPlanFLN 100× NA1.3 objective and 16-bit Hamamatsu ORCA-ER CCD camera, operated by Xcellence RT2.0 acquisition software (OLYMPUS). Digital image processing and analysis including measurements of fluorescent intensity and kinetic profiles was made using ImageJ 1.47v (Wayne Rusband, NIH, United States). Where necessary, the *x–y* drift in time-lapse series was corrected by StackReg plugin using Rigid body transformation algorithm. Statistic evaluation was done in MS Excel 2010.

Foci Quantification

Because rarely more than one discernible fluorescent structure of the fusion construct of TssK or TssB was detected per cell at a time, still images of live bacteria were used to quantify the incidence of assembly as a fraction (%) of cells that exhibit discernible fluorescent structure relative to the total amount of bacteria in the view field. For every sample and condition analyzed, a set of images was acquired from random fields of view using fixed exposure settings to allow quantitative comparison between samples. Because the expression levels of the fusion construct was usually heterogeneous within the same population of bacteria, automated particle counting on binary images using thresholding, filtering and segmentation algorithm was not reliable and thus manual counting from images was used instead after 16-bit to 8-bit conversion and background subtraction. Average and SD values were calculated for each group from 4 to 6 ROI each containing between 400 and 1000 of cells.

Kinetic Measurements of Fluorescence Intensity

Kinetic changes of fluorescence intensity were measured on a stack of images from a time-lapse. First, a set of ROI of fixed size (500×500 nm) covering area where discernible TssK/TssB-structures appear during the time-lapse was selected and fluorescence integrated density within these ROI was measured using ROI manager and multiple measurements plugin (ImageJ). Illustrative results for a typical profile were plotted as a fold-increase/decrease in fluorescence intensity relative to ROI intensity in frame 1 (T0) of the time-lapse (set as nominal value 1). Average duration of assembly was assessed as a time between the onset of fluorescence increase over the background level (rel. value = 1.0) and the peak fluorescence values. On contrary, average duration of disassembly was assessed as a time between fluorescence peak and its decrease to the background levels. Average values and SD were calculated from 20 to 40 ROI for each group. For an assessment of spatio-temporal colocalization between TssK-RFPT and TssB-GFP the fluorescence intensity for both channels (RFPT and GFP) was first measured at multiple ROI of several independent time lapse series. The resulting kinetic plot profiles were then aligned along *y*-axis so that maximum peak values for GFP channel (fully extended TssB-sheath) fit to $T = 0$ s and fluorescence intensity values of RFPT corresponding to each time point of GFP values were plotted accordingly. Average values and SD were calculated from 20 kinetic profiles for each channel.

Fluorescence Intensity Measurements of Protein Assemblies

For measurement of fluorescence intensity profile of TssK and TssB structures (**Figure 1A**) images were acquired with fixed exposure settings and fluorescence intensity plot profile was measured on line cross-section through the structures. Maximum fluorescence intensity (Max gray values) of TssK spots in *tagS* and *tagT* mutants (**Figure 1C**) was measured by ImageJ in fixed-size square ROI drawn around the fluorescent spots. Average and SD values were calculated from 20 plot profile measurements for each group. For an assessment of spatial co-localization between TssK-RFPT and TssB-GFP in elongated TssB-sheaths, fluorescence intensity plot profile was measured along the long axis of the structure with proximal end of the ROI crossing perimembrane TssK structure.

AUTHOR CONTRIBUTIONS

DL, MR-G, VJ, and VC performed the experiments and prepared the figures. DL and IA designed the study, analyzed the data, and wrote the manuscript.

FUNDING

This project received funding from the Laboratory of Excellence GRAL, financed within the University Grenoble Alpes graduate school (Ecoles Universitaires de Recherche) CBH-EUR-GS (ANR-17-EURE-0003). DL was funded by a grant from the Laboratory of Excellence GRAL (ANR-10-LABX-49-01). VC was funded by FINOVI (Fondation Innovation en Infectiologie).

ACKNOWLEDGMENTS

We thank to Brent Weber (Alberta Glycomics Centre, University of Alberta, Canada) and Joseph Mougous (University of Washington, United States) for providing bacterial strains and mutants used in this study. We also thank to Sylvie Elsen, Yohann Couté, and Maria Guillermina Casabona (CEA, Grenoble) for input in initial design of the project, sharing the unpublished data and interesting discussions, and to Brian T. Ho (Harvard Medical School, Boston, United States) for critical reading of the initial manuscript.

REFERENCES

Alcoforado Diniz, J., Liu, Y. C., and Coulthurst, S. J. (2015). Molecular weaponry: diverse effectors delivered by the Type VI secretion system. *Cell Microbiol.* 17, 1742–1751. doi: 10.1111/cmi.12532

Basler, M., Ho, B. T., and Mekalanos, J. J. (2013). Tit-for-tat: type VI secretion system counterattack during bacterial cell-cell interactions. *Cell* 152, 884–894. doi: 10.1016/j.cell.2013.01.042

Basler, M., and Mekalanos, J. J. (2012). Type 6 secretion dynamics within and between bacterial cells. *Science* 337:815. doi: 10.1126/science.1222901

Basler, M., Pilhofer, M., Henderson, G. P., Jensen, G. J., and Mekalanos, J. J. (2012). Type VI secretion requires a dynamic contractile phage tail-like structure. *Nature* 483, 182–186. doi: 10.1038/nature10846

Bleves, S. (2016). Game of trans-kingdom effectors. *Trends Microbiol.* 24, 773–774. doi: 10.1016/j.tim.2016.08.002

Bonemann, G., Pietrosiuk, A., Diemand, A., Zentgraf, H., and Mogk, A. (2009). Remodelling of VipA/VipB tubules by ClpV-mediated threading is crucial for type VI protein secretion. *EMBO J.* 28, 315–325. doi: 10.1038/emboj.2008.269

Boyer, F., Fichant, G., Berthod, J., Vandenbrouck, Y., and Attree, I. (2009). Dissecting the bacterial type VI secretion system by a genome wide in silico analysis: what can be learned from available microbial genomic resources? *BMC genomics* 10:104. doi: 10.1186/1471-2164-10-104

Brencic, A., and Lory, S. (2009). Determination of the regulon and identification of novel mRNA targets of *Pseudomonas aeruginosa* RsmA. *Mol. Microbiol.* 72, 612–632. doi: 10.1111/j.1365-2958.2009.06670.x

Brencic, A., McFarland, K. A., McManus, H. R., Castang, S., Mogno, I., Dove, S. L., et al. (2009). The GacS/GacA signal transduction system of *Pseudomonas aeruginosa* acts exclusively through its control over the transcription of the RsmY and RsmZ regulatory small RNAs. *Mol. Microbiol.* 73, 434–445. doi: 10.1111/j.1365-2958.2009.06782.x

Brunet, Y. R., Zoued, A., Boyer, F., Douzi, B., and Cascales, E. (2015). The type VI secretion TssEFGK-VgrG phage-like baseplate is recruited to the TssJLM membrane complex via multiple contacts and serves as assembly platform for tail tube/sheath polymerization. *PLoS Genet.* 11:e1005545. doi: 10.1371/journal.pgen.1005545

Casabona, M. G., Silverman, J. M., Sall, K. M., Boyer, F., Coute, Y., Poirel, J., et al. (2013). An ABC transporter and an outer membrane lipoprotein participate in posttranslational activation of type VI secretion in *Pseudomonas aeruginosa*. *Environ. Microbiol.* 15, 471–486. doi: 10.1111/j.1462-2920.2012.02816.x

Cascales, E. (2008). The type VI secretion toolkit. *EMBO Rep.* 9, 735–741. doi: 10.1038/embor.2008.131

Cherrak, Y., Rapisarda, C., Pellarin, R., Bouvier, G., Bardiaux, B., Allain, F., et al. (2018). Biogenesis and structure of a type VI secretion baseplate. *Nat. Microbiol.* 3, 1404–1416. doi: 10.1038/s41564-018-0260-1

Cianfanelli, F. R., Monlezun, L., and Coulthurst, S. J. (2016). Aim, load, fire: the type VI secretion system, a Bacterial Nanoweapon. *Trends Microbiol.* 24, 51–62. doi: 10.1016/j.tim.2015.10.005

Coulthurst, S. J. (2013). The type VI secretion system - a widespread and versatile cell targeting system. *Res. Microbiol.* 164, 640–654. doi: 10.1016/j.resmic.2013.03.017

Durand, E., Nguyen, V. S., Zoued, A., Logger, L., Pehau-Arnaudet, G., Aschtgen, M. S., et al. (2015). Biogenesis and structure of a type VI secretion membrane core complex. *Nature* 523, 555–560. doi: 10.1038/nature14667

Felisberto-Rodrigues, C., Durand, E., Aschtgen, M. S., Blangy, S., Ortiz-Lombardia, M., Douzi, B., et al. (2011). Towards a structural comprehension of bacterial type VI secretion systems: characterization of the TssJ-TssM complex of an *Escherichia coli* pathovar. *PLoS Pathogens.* 7:e1002386. doi: 10.1371/journal.ppat.1002386

Filloux, A., Hachani, A., and Bleves, S. (2008). The bacterial type VI secretion machine: yet another player for protein transport across membranes. *Microbiology* 154(Pt 6), 1570–1583. doi: 10.1099/mic.0.2008/016840-0

Forster, A., Planamente, S., Manoli, E., Lossi, N. S., Freemont, P. S., and Filloux, A. (2014). Coevolution of the ATPase ClpV, the sheath proteins TssB and TssC, and the accessory protein TagJ/HsiE1 distinguishes type VI secretion classes. *J. Biol. Chem.* 289, 33032–33043. doi: 10.1074/jbc.M114.600510

Hachani, A., Lossi, N. S., and Filloux, A. (2013). A visual assay to monitor T6SS-mediated bacterial competition. *J. Vis. Exp.* 73:e50103. doi: 10.3791/50103

Hayes, C. S., Aoki, S. K., and Low, D. A. (2010). Bacterial contact-dependent delivery systems. *Annu. Rev. Genet.* 44, 71–90. doi: 10.1146/annurev.genet.42.110807.091449

Hmelo, L. R., Borlee, B. R., Almblad, H., Love, M. E., Randall, T. E., Tseng, B. S., et al. (2015). Precision-engineering the *Pseudomonas aeruginosa* genome with two-step allelic exchange. *Nat. protoc.* 10, 1820–1841. doi: 10.1038/nprot.2015.115

Ho, B. T., Basler, M., and Mekalanos, J. J. (2013). Type 6 secretion system-mediated immunity to type 4 secretion system-mediated gene transfer. *Science* 342, 250–253. doi: 10.1126/science.1243745

Ho, B. T., Dong, T. G., and Mekalanos, J. J. (2014). A view to a kill: the bacterial type VI secretion system. *Cell Host Microbe* 15, 9–21. doi: 10.1016/j.chom.2013.11.008

Hsu, F., Schwarz, S., and Mougous, J. D. (2009). TagR promotes PpkA-catalysed type VI secretion activation in *Pseudomonas aeruginosa*. *Mol. Microbiol.* 72, 1111–1125. doi: 10.1111/j.1365-2958.2009.06701.x

Jani, A. J., and Cotter, P. A. (2010). Type VI secretion: not just for pathogenesis anymore. *Cell Host Microbe* 8, 2–6. doi: 10.1016/j.chom.2010.06.012

Kostyuchenko, V. A., Leiman, P. G., Chipman, P. R., Kanamaru, S., van Raaij, M. J., Arisaka, F., et al. (2003). Three-dimensional structure of bacteriophage T4 baseplate. *Nat. Struct. Biol.* 10, 688–693.

Kudryashev, M., Wang, R. Y., Brackmann, M., Scherer, S., Maier, T., Baker, D., et al. (2015). Structure of the type VI secretion system contractile sheath. *Cell* 160, 952–962. doi: 10.1016/j.cell.2015.01.037

Leiman, P. G., Basler, M., Ramagopal, U. A., Bonanno, J. B., Sauder, J. M., Pukatzki, S., et al. (2009). Type VI secretion apparatus and phage tail-associated protein complexes share a common evolutionary origin. *Proc. Natl. Acad. Sci. U.S.A.* 106, 4154–4159. doi: 10.1073/pnas.0813360106

Lin, J. S., Wu, H. H., Hsu, P. H., Ma, L. S., Pang, Y. Y., Tsai, M. D., et al. (2014). Fha interaction with phosphothreonine of TssL activates type VI secretion in *Agrobacterium tumefaciens*. *PLoS Pathog.* 10:e1003991. doi: 10.1371/journal.ppat.1003991

Lossi, N. S., Dajani, R., Freemont, P., and Filloux, A. (2011). Structure-function analysis of HsiF, a gp25-like component of the type VI secretion system, in *Pseudomonas aeruginosa*. *Microbiology* 157(Pt 12), 3292–3305. doi: 10.1099/mic.0.051987-0

Ma, L. S., Narberhaus, F., and Lai, E. M. (2012). IcmF family protein TssM exhibits ATPase activity and energizes type VI secretion. *J. Biol. Chem.* 287, 15610–15621. doi: 10.1074/jbc.M111.301630

Metcalf, W. W., Jiang, W., Daniels, L. L., Kim, S. K., Haldimann, A., and Wanner, B. L. (1996). Conditionally replicative and conjugative plasmids carrying lacZ alpha for cloning, mutagenesis, and allele replacement in bacteria. *Plasmid* 35, 1–13. doi: 10.1006/plas.1996.0001

Moscoso, J. A., Mikkelsen, H., Heeb, S., Williams, P., and Filloux, A. (2011). The *Pseudomonas aeruginosa* sensor RetS switches type III and type VI secretion via c-di-GMP signalling. *Environ. Microbiol.* 13, 3128–3138. doi: 10.1111/j.1462-2920.2011.02595.x

Mougous, J. D., Cuff, M. E., Raunser, S., Shen, A., Zhou, M., Gifford, C. A., et al. (2006). A virulence locus of *Pseudomonas aeruginosa* encodes a protein secretion apparatus. *Science* 312, 1526–1530. doi: 10.1126/science.1128393

Mougous, J. D., Gifford, C. A., Ramsdell, T. L., and Mekalanos, J. J. (2007). Threonine phosphorylation post-translationally regulates protein secretion in *Pseudomonas aeruginosa*. *Nat. Cell Biol.* 9, 797–803. doi: 10.1038/ncb1605

Nazarov, S., Schneider, J. P., Brackmann, M., Goldie, K. N., Stahlberg, H., and Basler, M. (2018). Cryo-EM reconstruction of Type VI secretion system baseplate and sheath distal end. *EMBO J.* 37:e97103. doi: 10.15252/embj.201797103

Newman, J. R., and Fuqua, C. (1999). Broad-host-range expression vectors that carry the L-arabinose-inducible *Escherichia coli* araBAD promoter and the araC regulator. *Gene* 227, 197–203. doi: 10.1016/s0378-1119(98)00601-5

Ostrowski, A., Cianfanelli, F. R., Porter, M., Mariano, G., Peltier, J., Wong, J. J., et al. (2018). Killing with proficiency: integrated post-translational regulation of an offensive Type VI secretion system. *PLoS pathogens* 14:e1007230. doi: 10.1371/journal.ppat.1007230

Park, Y. J., Lacourse, K. D., Cambillau, C., DiMaio, F., Mougous, J. D., and Veesler, D. (2018). Structure of the type VI secretion system TssK-TssF-TssG baseplate subcomplex revealed by cryo-electron microscopy. *Nat. Commun.* 9:5385. doi: 10.1038/s41467-018-07796-5

Pietrosiuk, A., Lenherr, E. D., Falk, S., Bonemann, G., Kopp, J., Zentgraf, H., et al. (2011). Molecular basis for the unique role of the AAA + chaperone ClpV in type VI protein secretion. *J. Biol. Chem.* 286, 30010–30021. doi: 10.1074/jbc.M111.253377

Planamente, S., Salih, O., Manoli, E., Albesa-Jove, D., Freemont, P. S., and Filloux, A. (2016). TssA forms a gp6-like ring attached to the type VI secretion sheath. *EMBO J.* 35, 1613–1627. doi: 10.15252/embj.201694024

Pukatzki, S., Ma, A. T., Sturtevant, D., Krastins, B., Sarracino, D., Nelson, W. C., et al. (2006). Identification of a conserved bacterial protein secretion system in *Vibrio cholerae* using the Dictyostelium host model system. *Proc. Natl. Acad. Sci. U.S.A.* 103, 1528–1533. doi: 10.1073/pnas.0510322103

Rietsch, A., Vallet-Gely, I., Dove, S. L., and Mekalanos, J. J. (2005). ExsE, a secreted regulator of type III secretion genes in *Pseudomonas aeruginosa*. *Proc. Natl. Acad. Sci. U.S.A.* 102, 8006–8011. doi: 10.1073/pnas.0503005102

Russell, A. B., Hood, R. D., Bui, N. K., LeRoux, M., Vollmer, W., and Mougous, J. D. (2011). Type VI secretion delivers bacteriolytic effectors to target cells. *Nature* 475, 343–347. doi: 10.1038/nature10244

Russell, A. B., Singh, P., Brittnacher, M., Bui, N. K., Hood, R. D., Carl, M. A., et al. (2012). A widespread bacterial type VI secretion effector superfamily identified using a heuristic approach. *Cell Host Microbe* 11, 538–549. doi: 10.1016/j.chom.2012.04.007

Santin, Y. G., Doan, T., Lebrun, R., Espinosa, L., Journet, L., and Cascales, E. (2018). In vivo TssA proximity labelling during type VI secretion biogenesis reveals TagA as a protein that stops and holds the sheath. *Nat. Microbiol.* 3, 1304–1313. doi: 10.1038/s41564-018-0234-3

Shalom, G., Shaw, J. G., and Thomas, M. S. (2007). In vivo expression technology identifies a type VI secretion system locus in Burkholderia pseudomallei that is induced upon invasion of macrophages. *Microbiology* 153(Pt 8), 2689–2699. doi: 10.1099/mic.0.2007/006585-0

Shneider, M. M., Buth, S. A., Ho, B. T., Basler, M., Mekalanos, J. J., and Leiman, P. G. (2013). PAAR-repeat proteins sharpen and diversify the type VI secretion system spike. *Nature* 500, 350–353. doi: 10.1038/nature12453

Silverman, J. M., Agnello, D. M., Zheng, H., Andrews, B. T., Li, M., Catalano, C. E., et al. (2013). Haemolysin coregulated protein is an exported receptor and chaperone of type VI secretion substrates. *Mol. Cell.* 51, 584–593. doi: 10.1016/j.molcel.2013.07.025

Silverman, J. M., Austin, L. S., Hsu, F., Hicks, K. G., Hood, R. D., and Mougous, J. D. (2011). Separate inputs modulate phosphorylation-dependent and -independent type VI secretion activation. *Mol. Microbiol.* 82, 1277–1290. doi: 10.1111/j.1365-2958.2011.07889.x

Vettiger, A., and Basler, M. (2016). Type VI secretion system substrates are transferred and reused among sister cells. *Cell* 167, 99–110.e112. doi: 10.1016/j.cell.2016.08.023

Wagenknecht, T., and Bloomfield, V. A. (1977). In vitro polymerization of bacteriophage T4D tail core subunits. *J. Mol. Biol.* 116, 347–359. doi: 10.1016/0022-2836(77)90074-2

Weber, B. S., Miyata, S. T., Iwashkiw, J. A., Mortensen, B. L., Skaar, E. P., Pukatzki, S., et al. (2013). Genomic and functional analysis of the type VI secretion system in Acinetobacter. *PLoS One* 8:e55142. doi: 10.1371/journal.pone.0055142

Yap, M. L., Mio, K., Leiman, P. G., Kanamaru, S., and Arisaka, F. (2010). The baseplate wedges of bacteriophage T4 spontaneously assemble into hubless baseplate-like structure in vitro. *J. Mol. Biol.* 395, 349–360. doi: 10.1016/j.jmb.2009.10.071

Zoued, A., Durand, E., Bebeacua, C., Brunet, Y. R., Douzi, B., Cambillau, C., et al. (2013). TssK is a trimeric cytoplasmic protein interacting with components of both phage-like and membrane anchoring complexes of the type VI secretion system. *J. Biol. Chem.* 288, 27031–27041. doi: 10.1074/jbc.M113.499772

13

Type III Secretion in the Melioidosis Pathogen *Burkholderia pseudomallei*

Charles W. Vander Broek and Joanne M. Stevens *

The Roslin Institute and Royal (Dick) School of Veterinary Studies, University of Edinburgh, Midlothian, United Kingdom

***Correspondence:**
Joanne M. Stevens
jo.stevens@roslin.ed.ac.uk

Burkholderia pseudomallei is a Gram-negative intracellular pathogen and the causative agent of melioidosis, a severe disease of both humans and animals. Melioidosis is an emerging disease which is predicted to be vastly under-reported. Type III Secretion Systems (T3SSs) are critical virulence factors in Gram negative pathogens of plants and animals. The genome of *B. pseudomallei* encodes three T3SSs. T3SS-1 and -2, of which little is known, are homologous to Hrp2 secretion systems of the plant pathogens *Ralstonia* and *Xanthomonas*. T3SS-3 is better characterized and is homologous to the Inv/Mxi-Spa secretion systems of *Salmonella* spp. and *Shigella flexneri*, respectively. Upon entry into the host cell, *B. pseudomallei* requires T3SS-3 for efficient escape from the endosome. T3SS-3 is also required for full virulence in both hamster and murine models of infection. The regulatory cascade which controls T3SS-3 expression and the secretome of T3SS-3 have been described, as well as the effect of mutations of some of the structural proteins. Yet only a few effector proteins have been functionally characterized to date and very little work has been carried out to understand the hierarchy of assembly, secretion and temporal regulation of T3SS-3. This review aims to frame current knowledge of *B. pseudomallei* T3SSs in the context of other well characterized model T3SSs, particularly those of *Salmonella* and *Shigella*.

Keywords: T3SS, effector, translocator, *Burkholderia pseudomallei*, melioidosis

INTRODUCTION

Bacteria are required to adapt to and survive in constantly changing and harsh environments. In order to respond to and alter their environment, secretion systems have evolved in bacteria to export proteins into the surrounding milieu (reviewed in Costa et al., 2015). One secretion system in Gram-negative bacteria that has been the focus of much research in the last three decades is the Type III secretion system (T3SS) (reviewed in Galán et al., 2014). Type III Secretion Systems (T3SSs) have been shown to be important for virulence in many Gram-negative bacterial pathogens of animals and plants; including *Pseudomonas syringae*, *Xanthomonas*, *Ralstonia solanacearum*, *Erwinia*, pathogenic *Escherichia coli*, *Salmonella*, *Shigella*, *Yersinia*, and *Burkholderia* (Gemski et al., 1980; Maurelli et al., 1985; Galán and Curtiss, 1989; Jarvis et al., 1995; Stevens et al., 2002; Büttner and He, 2009). T3SSs span the bacterial inner and outer membranes forming a "molecular syringe" which allows bacteria to export proteins, called effectors, from the bacterial cytoplasm into a target eukaryotic cell (reviewed in Galán et al., 2014).

The focus of this review is T3SSs in the pathogenic bacterium *Burkholderia pseudomallei*, and to some extent the closely related species *B. mallei* and *B. thailandensis*. Originally described in drug addicts in Rangoon in the early twentieth century by Alfred Whitmore (Whitmore, 1913), *Burkholderia pseudomallei* is a facultative intracellular pathogen (Pruksachartvuthi et al., 1990) that is the causative agent of melioidosis, or Whitmore's disease. Melioidosis is a severe disease of humans and animals, causing an estimated 165,000 cases of human melioidosis per year resulting in a predicted 89,000 deaths (Limmathurotsakul et al., 2016). Infection with *B. pseudomallei* is usually associated with environmental exposure and can occur through breaks in the skin, inhalation or ingestion (reviewed in Cheng and Currie, 2005). In the majority of cases, the incubation period for melioidosis is between 1 and 21 days following infection (Ngauy et al., 2005). About 50% of melioidosis cases affect people with diabetes and other important risk factors include lung disease, cystic fibrosis and excessive alcohol consumption (Currie et al., 2010). There are varied clinical presentations of *B. pseudomallei* infection ranging from skin infections to pneumonia and septic shock, which hampers accurate diagnosis in a clinical setting (Currie et al., 2010). *B. pseudomallei* is reported to be able to reactivate after remaining latent following a primary infection. The longest reported period between infection and reactivation occurred in a World War II veteran who manifested symptoms 62 years after exposure (Ngauy et al., 2005). *B. pseudomallei* has been classified as a bioterrorism agent by both the UK government and the US Centres for Disease Control and Prevention (reviewed in Rotz et al., 2002; Cheng and Currie, 2005).

B. mallei, the causative agent of glanders in horses and other solipeds, is a zoonotic pathogen with restricted host range (Yabuuchi et al., 1992; Srinivasan et al., 2001). In humans, *B. mallei* causes a disease similar to melioidosis and has been similarly classified as a potential bioterrorism agent in the UK and US (Rotz et al., 2002; Van Zandt et al., 2013). The soil saprophyte *B. thailandensis* is non-pathogenic and present in high numbers in the soils and standing waters of endemic areas. This species is commonly used as an alternative model system for *B. pseudomallei* and *B. mallei* studies as its genome encodes many homologs of virulence factors from these pathogenic species (Brett et al., 1998; Moore et al., 2004; Yu et al., 2006; Haraga et al., 2008).

TYPE THREE SECRETION SYSTEMS IN *B. PSEUDOMALLEI*

The *B. pseudomallei* genome encodes three T3SSs which are referred to as T3SS-1, T3SS-2, and T3SS-3. The genome of *B. pseudomallei* consists of two circular chromosomes, with all three T3SSs residing on chromosome 2 (Holden et al., 2004). T3SS-2 and T3SS-3 are present in the genomes of *B. mallei* and *B. thailandensis*, whereas T3SS-1 is absent from both (Rainbow et al., 2002). T3SS-1 and T3SS-2 are relatively poorly characterized and share homology with the Hrp2 family of T3SSs found in plant pathogens (Winstanley et al., 1999; Rainbow et al., 2002). The best characterized of the *B. pseudomallei* T3SSs, T3SS-3, is also known as the Burkholderia secretion apparatus (Bsa) T3SS. It is a member of the Inv-Mxi-Spa family of T3SSs from *Salmonella* spp. (SPI-1) and *Shigella flexneri* (Attree and Attree, 2001; Stevens et al., 2002; Egan et al., 2014).

Burkholderia pseudomallei T3SS-1 (BPSS1390-BPSS1410) and T3SS-2 (BPSS1610-BPSS1629) show closest homology to the Hrp2 T3SS of the plant pathogen *Ralstonia solanacearum* (Angus et al., 2014). *B. pseudomallei* T3SS-2 expression is activated by the AraC-type regulator HrpB (BPSS1610) (Lipscomb and Schell, 2011). HrpB also regulates the expression of a type IV pilus encoded directly upstream of T3SS-2, but does not appear to regulate the other T3SSs in *B. pseudomallei* (Lipscomb and Schell, 2011). In order to investigate the role that T3SS-1 and T3SS-2 play in plants, a tomato plant infection model was established for *B. pseudomallei* and *B. thailandensis* (Lee et al., 2010). *B. pseudomallei* KHW T3SS-1 and T3SS-2 mutants were reported to be attenuated in tomato plants (Lee et al., 2010). However, in a more recent study, *B. thailandensis* did not display phytopathogenic activity in tomato plants which were treated identically to those in Lee et al. (2010), Lipscomb and Schell (2011). The question of whether *B. pseudomallei* is capable of infecting plants, and what if any role T3SS-1 and -2 have in this process, remain important unanswered questions.

The function of T3SS-1 and -2 in mammalian systems has also been investigated to some extent. T3SSs-1 and -2 do not appear to be required for vacuole escape of the bacterium into the cytoplasm of infected macrophages (Burtnick et al., 2008), and are dispensable in a Syrian hamster model of infection (Warawa and Woods, 2005). However, a *B. pseudomallei* T3SS-1 mutant displayed increased co-localisation with the autophagy marker LC3 and a reduction in intracellular survival in RAW264.7 cells (D'Cruze et al., 2011). In the same study, T3SS-1 was required for full virulence in a respiratory murine model of melioidosis (D'Cruze et al., 2011). Both earlier studies inactivated T3SS-1 by mutating the structural auto-protease component (BpscU, **Table 1**) (Warawa and Woods, 2005; Burtnick et al., 2008) while the latter study generated a system knockout by mutation of the ATPase (BpscN, **Table 1**), which may account for the different phenotypes observed.

T3SS-3

The best characterized of the three T3SSs, T3SS-3 (BPSS1516-BPSS1552, **Figure 1**) is a member of the Inv/Mxi-Spa family of T3SSs (Egan et al., 2014) of which the prototypic systems are found in *Salmonella* Spp. and *Shigella flexneri*. The *Salmonella* and *Shigella* prototypic systems are required in these bacteria for host cell invasion and escape from the endocytic vacuole into the cytosol, respectively (reviewed in Galán et al., 2014). *B. pseudomallei* is a facultative intracellular pathogen capable of survival in both phagocytic and non-phagocytic cell lines (Jones et al., 1996). T3SS-3 is required for *B. pseudomallei* to efficiently escape the endocytic vesicle (Stevens et al., 2002). The T3SS-3 is also required for full virulence in both murine and Syrian hamster models of infection (Stevens et al., 2004; Warawa and Woods, 2005; Gutierrez et al., 2015a). T3SS-3 deficient mutants are also impaired in their ability to disseminate from the lungs of mice infected intra-nasally (Gutierrez et al., 2015a). A recent study used Tn-seq to identify genes required for respiratory melioidosis in mice and identified the following T3SS-3 genes as being required: *bprA* (BPSS1530), *bipC* (BPSS1531), *bipB* (BPSS1532), *bicA* (BPSS1533), *bsaZ* (BPSS1534), *bsaW* (BPSS1537), *bsaV* (BPSS1538), *bsaO* (BPSS1545), *bsaM* (BPSS1547), *bsaL* (BPSS1548), *bsaK* (BPSS1549), and *bsaJ*

TABLE 1 | *B. pseudomallei* T3SS-1 and -2 genes, corresponding proteins and predicted functions.

B. pseudomallei K96243		*B. pseudomallei* K96243		*R. solanacearum*	Universal Nomenclature	Predicted function
T3SS-1		T3SS-2		Hrp2	Protein name	
Locus tag	Protein name	Locus tag	Protein name	Protein name		
BPSS1388	Similar to Hrpk1 from *P. syringae*					Translocator
BPSS1390	BpscC	BPSS1592	BpscC2	HrcC	SctC	Outer membrane ring / secretin
BPSS1391	BpspB	BPSS1610	BpspB2	HrpB		T3SS Regulator
BPSS1392	BpscT	BPSS1629	BpscT2	HrcT	SctT	Inner membrane export apparatus
BPSS1393	BpspD	BPSS1628	BpspD2	HrpD	SctO	Stalk protein
BPSS1394	BpscN	BPSS1627	BpscN2	HrcN	SctN	ATPase
BPSS1395	BpscL	BPSS1626	BpscL2	HrcL	SctL	Stator protein
BPSS1396	BpspH	BPSS1625	BpspH2	HrpH		Inner membrane component
BPSS1397	BpscJ	BPSS1624	BpscJ2	HrcJ	SctJ	Inner membrane ring component
BPSS1398	BpspJ	BPSS1623	BpspJ2	HrpJ		Inner rod component
BPSS1399	BpspK	BPSS1622	BpspK2	HrpK		Inner rod component
BPSS1400	BpscU	BPSS1621	BpscU2	HrcU	SctU	Autoprotease, early/middle substrate switch
BPSS1401	BpscV	BPSS1620	BpscV2	HrcV	SctV	Export gate
BPSS1402	BpsaP	BPSS1619	BpsaP2	HpaP	SctP	Regulates needle length / translocator secretion
BPSS1403	BpscQ	BPSS1618	BpscQ2	HrcQ	SctQ	Cytoplasmic ring
BPSS1404	BpscR	BPSS1617	BpscR2	HrcR	SctR	Inner membrane export apparatus
BPSS1405	BpscS	BPSS1616	BpscS2	HrcS	SctS	Inner membrane export apparatus
BPSS1406	BpspV	BPSS1615	BpspV2	HrpV		Secreted regulator
BPSS1407	BpscD	BPSS1614	BpscD2	HrcD	SctD	Inner membrane ring component
BPSS1410	BpsaB	BPSS1611	BpsaB2	HpaB		Chaperone

Homologs from R. solanacearum Hrp2 T3SS are given for reference. Where applicable, the universal nomenclature for T3SS structural proteins is also listed to allow for comparison to genes from T3SS-3 listed in ***Table 2****.*

(BPSS1550) (Gutierrez et al., 2015b). This highlights the importance of T3SS-3 in melioidosis.

TRANSCRIPTIONAL REGULATION OF T3SS-3

The transcriptional regulation of T3SS-3 has been elucidated (Sun et al., 2010). At the top of the regulatory hierarchy is the gene *bspR* (BPSL1105), which, when disrupted led to a reduced level of expression of genes in the T3SS-3 locus as shown by microarray and real time PCR (Sun et al., 2010). *bspR* signals through the membrane bound regulator *bprP* (BPSS1553) which controls expression levels of both structural and secreted components of T3SS-3 (Sun et al., 2010). *bprP* further signals through *bsaN* (BPSS1546) and its co-activator, the chaperone *bicA* (BPSS1533), controlling transcription of the known effectors *bopC, bopE* and *bopA*, as well as the chaperone *bicP* (BPSS1523) and the regulators *bprB-D* (BPSS1520-22) (Sun et al., 2010; Chen et al., 2014). *bsaN* also relays the regulation signal to the virulence-associated Type 6 secretion system and virulence factors such as *bimA* and *virAG* (Sun et al., 2010; Chen et al., 2014).

The presence of genes allowing for arabinose assimilation was one of the first methods used to differentiate between virulent *B. pseudomallei* and avirulent *B. thailandensis* (Smith et al., 1997; Moore et al., 2004). Expressing the *B. thailandensis* arabinose assimilation operon in *B. pseudomallei* causes a down-regulation of expression of the T3SS-3 genes, notably the T3SS regulator *bsaN,* and also results in a reduction in virulence in a Syrian hamster model of infection (Moore et al., 2004). This suggests that loss of the arabinose assimilation operon may account for some of the differential virulence observed between these two species of *Burkholderia* (Moore et al., 2004).

STRUCTURE OF T3SS-3

The structure of the T3SS is well conserved and is similar to that of the bacterial flagella system (Kubori et al., 1998; Young et al., 1999; Gophna et al., 2003). It is thought that the flagella and the T3SS evolved from a similar ancestor, but the T3SS has been the product of a large amount of horizontal gene transfer (Gophna et al., 2003). T3SSs are separated into seven different families named after the archetype system in each family, each with slight differences in structure and host cell target, with multiple types of T3SS present in some bacterial species (e.g., plant or animal) (reviewed by Büttner, 2012). In recent years the structure of the *Salmonella* T3SS has been solved using cryo-EM (Schraidt and Marlovits, 2011) and cryo-ET (Hu et al., 2017). The structure of the *Salmonella* T3SS (reviewed by Galán et al., 2014) consists of inner and outer membrane rings connected by a rod, the extracellular needle and a secreted translocation pore which

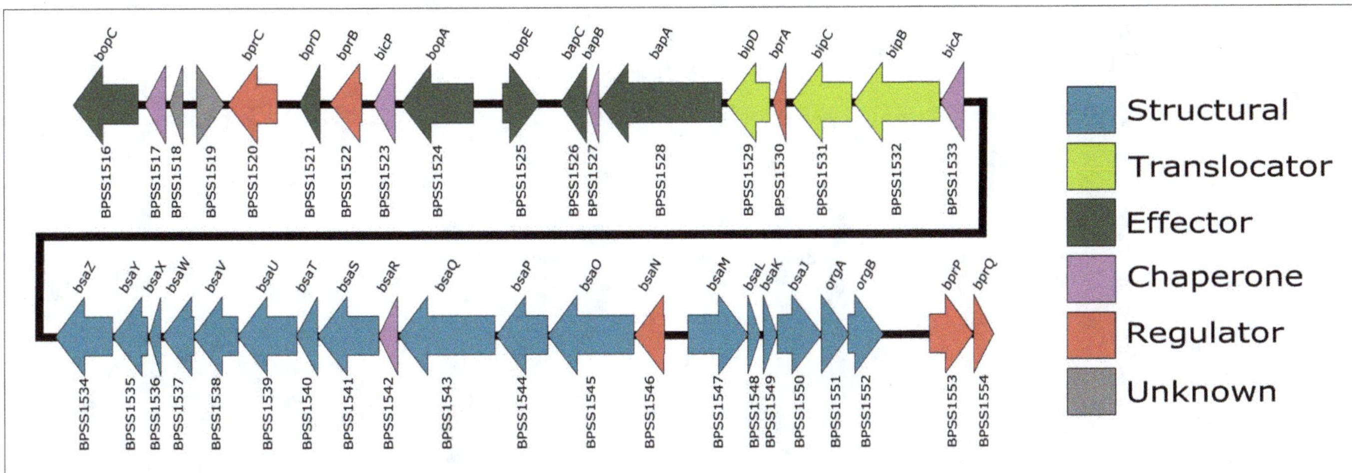

FIGURE 1 | *B. pseudomallei* K96243 T3SS-3 gene locus. Also known as the *bsa* locus, the T3SS-3 genes are encoded by chromosome 2 (BPSS1516-BPSS1554). Arrows which point right represent genes encoded on the forward strand and arrows which point left represent genes encoded on the reverse strand. Gene locus tags are listed below each arrow and gene names are listed above. The arrows are color coded according to their predicted function.

spans the target cell membrane. Assembly of the T3SS appears to be hierarchical and is the subject of multiple recent reviews (Büttner, 2012; Diepold and Wagner, 2014). Based on evidence from *Yersinia* and *Salmonella*, it appears construction of the T3SS begins with the formation of the outer-membrane/structural ring (Secretin) and independently the inner membrane export machinery, which are then linked by YscJ (*Yersinia*) or PrgK (*Salmonella*), followed by assembly of the ATPase/C-ring and the formation of the mature needle complex (Diepold et al., 2010, 2011; Wagner et al., 2010). The structural proteins in the *B. pseudomallei* T3SS-3 (**Figure 2**) that have been studied specifically are described below.

Components of the Export Apparatus

BsaZ (BPSS1534)

BsaZ (BPSS1534) is a structural component with homology to SpaS of *Salmonella* (Stevens et al., 2002). SpaS is a component of the export apparatus and has been shown to undergo auto-cleavage causing a switch between secretion of structural components (early substrates) to secretion of the needle complex and translocator proteins (intermediate substrates) (Zarivach et al., 2008). *B. pseudomallei bsaZ* mutants are unable to secrete effector and translocator proteins (Stevens et al., 2003; Muangman et al., 2011; Vander Broek et al., 2015). Mutations in *bsaZ* result in a significant delay in escape from the phagosome and reduced intracellular survival in J774.2 cells (Stevens et al., 2002; Burtnick et al., 2008). *bsaZ* mutants are also attenuated in murine (Stevens et al., 2004; Burtnick et al., 2008) and Syrian hamster (Warawa and Woods, 2005) models of melioidosis. *bsaZ* mutants have proven to be a useful tool in understanding the role of T3SS-3 in *B. pseudomallei* pathogenesis, however, the effect of these mutations other than an inability to secrete effector proteins is not well understood. For example, it is unclear if BsaZ undergoes auto-cleavage and is involved in secretion hierarchy in a similar manner to SpaS. It has also not been determined if *bsaZ* mutants still assemble the external needle appendage, or at what stage in formation of the mature T3SS complex they are impaired.

BsaQ (BPSS1543)

Another component of the export apparatus of the *B. pseudomallei* T3SS-3 that has been described is BsaQ (BPSS1543), which is homologous to InvA from *Salmonella* (Sun et al., 2005). InvA is required for the formation of the mature T3SS in *Salmonella* as well as secretion of effector proteins (Kubori et al., 2000). *bsaQ* mutants display a similar phenotype to other T3SS-3 structural mutants in that they are delayed in phagosome escape, resulting in reduced intracellular survival (Muangsombut et al., 2008). The *bsaQ* mutants are unable to secrete the effector protein BopE or translocator tip protein BipD, and show significant defects in cell invasion (Muangsombut et al., 2008). Similarly to BsaZ, there is a gap in our knowledge of whether BsaQ truly functions in a manner similar to InvA.

Inner Membrane Ring

BsaM (BPSS1547)

The inner membrane component BsaM (BPSS1547) is homologous to *Salmonella* PrgH, which forms the inner membrane ring of the *Salmonella* T3SS and is required for secretion of effector proteins (Kubori et al., 2000; Marlovits et al., 2004). A *B. pseudomallei bsaM* mutant induced lower levels of NF-κB signaling in HEK 293T cells when compared to the isogenic parent strain (Teh et al., 2014). The *bsaM* mutant activated NF-κB at time points corresponding to a delayed escape from the phagosome, suggesting it has similar effects to the other T3SS-3 structural knockouts (Teh et al., 2014).

Needle Components

BsaL (BPSS1548)

A crystal structure of the protein BsaL (BPSS1548) demonstrated significant structural similarity with MxiH (*Shigella*) and PrgI (*Salmonella*), which form the external needle structure (major needle subunit) of the T3SS (Zhang et al., 2006; Wang et al., 2007; Barrett et al., 2008). BsaL is recognized by the host cell neuronal inhibitory protein (NAIP) leading to activation of the NLRC4 inflammasome (Yang et al., 2013). NLRC4 has been shown to

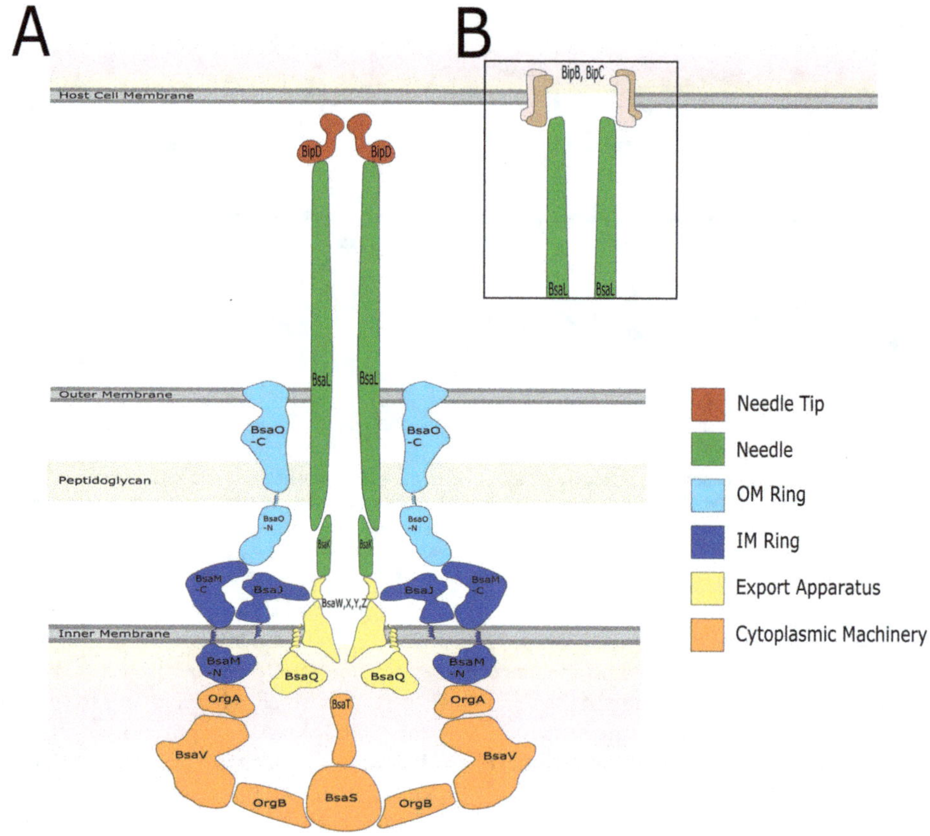

FIGURE 2 | The predicted structure of the *B. pseudomallei* K96243 T3SS-3 based on *Salmonella* SPI-1. The name of the protein which is predicted to constitute each structural component, as well as colors to represent different portions of the T3SS structure, are shown. **(A)** Before host cell contact, the T3SS is fully assembled spanning both the inner and outer bacterial membrane. The needle forms a channel extending out of the bacterium which is capped by the needle tip protein. **(B)** After contact with the host cell, a signal is relayed through the T3SS and translocator proteins form a pore in the host cell membrane allowing for injection of bacterial effector proteins.

be important in a murine model of respiratory melioidosis and a human *NLRC4* polymorphism is associated with survival in melioidosis patients (West et al., 2014).

BsaK (BPSS1549)

The cellular protein Nod-like receptor NLRC4 recognizes T3SS minor needle component proteins from a range of Gram-negative bacteria, including BsaK (BPSS1549) from *B. pseudomallei*, activating an innate immune response through casapase-1 (Miao et al., 2010; Zhao et al., 2011; Bast et al., 2014). A *B. pseudomallei bsaK* mutant was highly attenuated in an intranasal murine model of melioidosis (Bast et al., 2014).

Non-structural Proteins

BsaU (BPSS1539)

BsaU (BPSS1539) is homologous to InvJ from *Salmonella* which is the "molecular ruler" which determines the length of the T3SS needle (Kubori et al., 2000). Mutation of *bsaU* in *B. pseudomallei* resulted in delay in phagosome escape and reduced virulence in a BALB/c intranasal mouse infection model (Pilatz et al., 2006). The mutant was also deficient in its ability to secrete the effector protein BopE and the translocator tip protein BipD (Pilatz et al., 2006). It can be hypothesized that without BsaU, the needle complex of T3SS-3 forms incorrectly, accounting for the lack of effector and translocator secretion.

BsaP (BPSS1544)

A family of T3SS proteins called the "gatekeeper" proteins (InvE/MxiC/SepL/YopN-TyeA) are involved in the control of effector and translocator protein secretion and the switch from intermediate to late substrates for secretion in *Salmonella/Shigella/E. coli/Yersinia*, respectively (reviewed in Büttner, 2012). These proteins are involved in the temporal regulation of their respective T3SSs and deletion of all of these proteins causes an increase in levels of secreted effectors, but has differing effects on the secretion of translocators. For example deletion of *invE* (Kubori and Galán, 2002) and *sepL* (Kresse et al., 2000; Deng et al., 2004, 2005) causes a reduction in translocator secretion, deletion of *mxiC* (Botteaux et al., 2009) has no effect on translocator secretion, and deletion of *yopN* (Forsberg et al., 1991; Iriarte et al., 1998) increases levels of secreted translocators. The closest homolog to this family of proteins in *B. pseudomallei* is BsaP (BPSS1544), which we have demonstrated functions as a gatekeeper protein for effectors in a manner most similar to *Salmonella invE* (Vander Broek et al., 2015). Deletion of *bsaP* creates a phenotype in which effector proteins are hyper-secreted with a concomitant decrease in translocator secretion (Vander Broek et al., 2015). Further studies are warranted to determine the molecular interactions of BsaP with other components of the T3SS-3, in order to fully understand BsaP in the context of the other members of the gatekeeper family of proteins.

Structural Components for Which No Data Is Available

Proteins for which no published work is available are OrgA (BPSS1551), OrgB (BPSS1552), BsaJ (BPSS1550), BsaO (BPSS1545), BsaT (BPSS1540), BsaV (BPSS1538), BsaW (BPSS1537), BsaX (BPSS1536), and BsaY (BPSS1535). The putative function of these proteins can be found in **Table 2** and their predicted location in the structure of T3SS-3 is shown in **Figure 2**.

T3SS-3 has been well studied in terms of its role in virulence however very little has been reported on the structure and the molecular interactions between individual components of the T3SS itself. Determining the finer points of structure and hierarchy of assembly represent a major unanswered research question in terms of understanding T3SS-3 in the context of other well studied T3SSs.

ENERGIZING SECRETION AND UNFOLDING OF SUBSTRATES

Proteins are secreted in an unfolded state due to the narrow width of the needle channel which is 2–3 nm in *S. flexneri* (Blocker et al., 2001; Radics et al., 2014). This was proven by cryo-EM imaging of a substrate which was engineered to become trapped in the secretion channel (Radics et al., 2014). The ATPase at the base of the secretion system has been shown to dissociate proteins from their chaperones and is also thought to be involved in unfolding of the protein (Akeda and Galán, 2005). The source of energy for exporting proteins is the subject of some debate. In the flagellar T3SS of *Yersinia* and *Salmonella*, secretion can take place in the absence of the ATPase (Wilharm et al., 2004; Erhardt et al., 2014). This evidence, combined with a study which showed flagellar T3S was halted in the absence of a proton gradient, has led to the hypothesis that the ATPase is primarily required for protein unfolding and that proton motive force energizes secretion of the unfolded substrate (Paul et al., 2008).

The *B. pseudomallei* T3SS-3 ATPase is BsaS (BPSS1541). A *B. pseudomallei bsaS* mutant was unable to secrete the known effector protein BopE, demonstrated a defect in intracellular survival in RAW264.7 cells and was attenuated in BALB/c mice infected intra-nasally, demonstrating that the ATPase is required for T3S of effector proteins and has a similar phenotype to other T3SS-3 structural mutants (Gong et al., 2015). This agrees with a previous study in which *B. pseudomallei* and *B. thailandensis* containing in-frame deletions in *bsaS*, were impaired in their ability to escape the endosomal compartment and form plaques in host cell monolayers as a result of cell-to-cell spread (French et al., 2011). *Salmonella* lacking a functional ATPase will still assemble a mature T3SS, and will still secrete effector proteins, though at a reduced level (Erhardt et al., 2014). The effect that a *bsaS* mutation in *B. pseudomallei* T3SS-3 may have on either assembly of the T3SS apparatus or the secretion of early/middle substrates, such as the major needle component (BsaL) and the needle tip protein (BipD) remains an interesting open question.

NEEDLE TIP AND TRANSLOCATOR PROTEINS-SENSING HOST CONTACT

Upon host cell contact, the translocators of the T3SS are inserted into the host cell membrane forming a pore through which effector proteins may pass. *Shigella* is able to lyse red blood cells by inserting the translocators IpaB and IpaC into the RBC membrane forming a 25 angstrom pore (Blocker et al., 1999). Secretion of the translocator proteins is controlled by the needle tip protein. A *Shigella* needle tip protein mutant (*ipaD*), constitutively secreted the translocators IpaB and IpaC (Picking et al., 2005). Immunofluorescence microscopy demonstrated the presence of the *Shigella* IpaD protein on the surface of the bacteria in the absence of a host cell membrane, a finding that was further confirmed by electron microscopy (Espina et al., 2006). The needle tip protein probably acts to sense host cell contact, inhibiting premature secretion of translocators (Espina et al., 2006). In the same study it was shown that antibodies to the tip protein disrupted the haemolysis of sheep erythrocytes indicating the functional importance of IpaD in the insertion of the translocator complex into eukaryotic cell membranes, as well as regulating effector protein secretion (Espina et al., 2006). Next, the T3SS senses host cell contact when IpaD binds to bile salts causing IpaB to be exposed at the needle tip (Olive et al., 2007; Stensrud et al., 2008).

The final step in secretion of the translocation pore is dependent on *Shigella* interacting with host cell lipids to induce IpaC secretion (Epler et al., 2009). When cultured in the presence of liposomes, IpaC localizes to the bacterial surface and is secreted (Epler et al., 2009). This agrees with earlier evidence that cholesterol is bound by translocation components (*Salmonella* SipB and *Shigella* IpaB), is important for entry into host cells and is required for efficient translocation of effector proteins into host cells (Lafont et al., 2002; Hayward et al., 2005). There is some evidence to suggest that the translocator proteins, along with the needle tip and a functional T3SS, may be sufficient to determine the intracellular niche of *Salmonella* and *Shigella* in the absence of any of the effector proteins (Du et al., 2016).

BipD (BPSS1529)

BipD (BPSS1529) is the needle tip protein of the *B. pseudomallei* T3SS-3. It is homologous to SipD (*Salmonella*), IpaD (*Shigella*), and LcrV (*Yersinia*). BipD has been confirmed to be secreted by the *B. pseudomallei* T3SS-3 (Stevens et al., 2003; Vander Broek et al., 2015). The crystal structure of BipD was solved (Erskine et al., 2006; Knight et al., 2006; Roversi et al., 2006; Johnson et al., 2007; Pal et al., 2010) and the 3-dimensional

TABLE 2 | *B. pseudomallei* T3SS-3 genes, corresponding proteins and predicted functions.

***B. pseudomallei* K96243**		**Homologs**		**Universal nomenclature**	**Function**
T3SS-3		***Salmonella* SPI-1**	***Shigella***	**Protein name**	
Locus tag	**Protein name**	**Protein name**	**Protein name**		
BPSS1516	BopC				Effector
BPSS1517	–				Chaperone of BopC
BPSS1518	–				
BPSS1519	–				
BPSS1520	BprC				Regulator of T6SS-1
BPSS1521	BprD				Effector/Regulator
BPSS1522	BprB				Putative regulator
BPSS1523	BicP	SicP			Chaperone, shown to bind to BapA
BPSS1524	BopA	SptP	IcsB		Effector protein
BPSS1525	BopE	SopE			Effector protein
BPSS1526	BapC	IagB	IpgF		Effector protein
BPSS1527	BapB	IacP			Negative regulator, predicted chaperone
BPSS1528	BapA				Effector protein
BPSS1529	BipD	SipD	IpaD	SctA	Needle tip
BPSS1530	BprA				Putative regulator
BPSS1531	BipC	SipC	IpaC	SctB	Minor translocator
BPSS1532	BipB	SipB	IpaB	SctE	Major translocator
BPSS1533	BicA	SicA	IpgC		Chaperone/Co-activator of BsaN
BPSS1534	BsaZ	SpaS	Spa40	SctU	Autoprotease, early/middle substrate switch
BPSS1535	BsaY	SpaR	Spa29	SctT	Inner membrane export apparatus
BPSS1536	BsaX	SpaQ	Spa9	SctS	Inner membrane export apparatus
BPSS1537	BsaW	SpaP	Spa24	SctR	Inner membrane export apparatus
BPSS1538	BsaV	SpaO	Spa33	SctQ	Cytoplasmic ring
BPSS1539	BsaU	InvJ	Spa32	SctP	Needle length control protein
BPSS1540	BsaT	InvI	Spa13	SctO	Stalk protein
BPSS1541	BsaS	InvC	Spa47	SctN	ATPase
BPSS1542	BsaR	InvB	Spa15		Chaperone, predicted to bind BopE
BPSS1543	BsaQ	InvA	MxiA	SctV	Export gate
BPSS1544	BsaP	InvE	MxiC	SctW	Middle/late substrate switch
BPSS1545	BsaO	InvG	MxiD	SctC	Outer membrane ring / secretin
BPSS1546	BsaN	InvF	MxiE		Regulator of T3SS-3 effector proteins
BPSS1547	BsaM	PrgH	MxiG	SctD	Inner membrane ring component
BPSS1548	BsaL	PrgI	MxiH	SctF	Outer rod/ major needle component
BPSS1549	BsaK	PrgJ	MxiI	SctI	Inner rod/minor needle component
BPSS1550	BsaJ	PrgK	MxiJ	SctJ	Inner membrane ring component
BPSS1551	OrgA	OrgA	MxiK	SctK	ATPase cofactor
BPSS1552	OrgB	OrgB	MxiN	SctL	Stator protein
BPSS1553	BprP				Regulator of T3SS-3
BPSS1554	BprQ				Regulator of T3SS-3

Homologs from Salmonella SPI-1 and Shigella are given for reference. Where applicable, the universal nomenclature for T3SS structural proteins is also listed.

structure is highly similar to IpaD of *Shigella* (Johnson et al., 2007) and *Salmonella* SipD (Espina et al., 2007), but *bipD* cannot functionally complement *sipD* in *Salmonella* (Klein et al., 2017). It was demonstrated that the structure of BipD, as well as IpaD and SipD, is dependent on pH changes (Markham et al., 2008). Interestingly when cultured in a more acidic pH of 4.5, *B. thailandensis* secretes increased amounts of BipD as well as BopE (Jitprasutwit et al., 2010), though the study also described amino acid differences between BipD of *B. pseudomallei* and *B. thailandensis* (Jitprasutwit et al., 2010), raising the question as to what effect pH may have on T3SS-3 in *B. pseudomallei*.

In a *B. pseudomallei bipD* mutant, the levels of both translocators and effectors secreted into the culture supernatant are increased (Stevens et al., 2003; Vander Broek et al., 2015), in agreement with data published for the homologous *Shigella* protein IpaD (Parsot et al., 1995; Picking et al., 2005). IpaD blocks secretion of effector proteins until host cell contact has taken place (Roehrich et al., 2013), and because of its similarity in both sequence and structure (Erskine et al., 2006) it is perhaps unsurprising that BipD would function in a similar manner. More recent evidence suggests that *Shigella* IpaD may be involved in controlling the secretion of translocator and effector proteins through an interaction with the gatekeeper protein MxiC (Roehrich et al., 2016). This would suggest an interaction between BipD and BsaP in *B. pseudomallei* which may have a similar activity.

Needle tip proteins are of particular interest because of their potential use as subunit vaccines. Most notably the *Yersinia pestis* needle tip protein LcrV (V antigen), especially when combined with the Fraction 1 (F1) protein, is an effective vaccine that has been tested in human clinical trials (reviewed by Williamson, 2009). Vaccination of mice with the *Shigella* needle tip protein IpaD, along with the translocator IpaB, induced high levels of protection upon subsequent challenge (Martinez-Becerra et al., 2012, 2013; Heine et al., 2013). Some 15 years ago, it was described that convalescent serum from a melioidosis patient reacted specifically with a recombinant GST-tagged BipD protein (Stevens et al., 2002). CD4+ T cells taken from mice infected with an attenuated strain of *B. pseudomallei* showed specificity for BipD (Haque et al., 2006). Similarly, human monocyte-derived dendritic cells from healthy *B. pseudomallei* seropositive donors were pulsed with purified BipD after which CD4+ T cells were able to recognize the recombinant protein (Tippayawat et al., 2011). A *B. pseudomallei bipD* mutant was significantly attenuated in a BALB/c intranasal murine infection model, demonstrating the importance of this protein *in vivo* (Stevens et al., 2004). Attempts have been made to use BipD as a recombinant subunit vaccine in a BALB/c murine intraperitoneal model of infection, but in both studies the vaccine showed no protection upon subsequent challenge (Stevens et al., 2004; Druar et al., 2008).

BipB (BPSS1532)

BipB (BPSS1532) shares 46% amino acid identity with the *Salmonella* translocator protein SipB, and is secreted by T3SS-3 in a *bsaZ* dependant manner (Vander Broek et al., 2015). In a *Salmonella sipB* mutant, *bipB* is unable to complement *sipB* demonstrating evolutionary separation of these proteins (Klein et al., 2017). This is an important reminder that although T3SSs are similar in structure, T3SS proteins do not always function identically.

A *B. pseudomallei* K96243 *bipB* insertion mutant showed reduced invasion and cell-to-cell spread in HeLa cells, and a reduced ability to form multi-nucleated giant cells in J774 cells (Suparak et al., 2005). *In vivo*, in BALB/c mice infected intranasally, the *bipB* mutant was greatly attenuated and showed a phenotype similar to that of the *bipD* translocator mutant (Stevens et al., 2004; Suparak et al., 2005). This is likely due to the inability of the T3SS to function correctly without the formation of the translocation pore, and any proposed secondary function would require further investigation. The N-terminal region of BipB has been tested as a protective antigen in a murine model of melioidosis, but as with BipD, showed no protection upon subsequent challenge (Druar et al., 2008).

BipC (BPSS1521)

BipC (BPSS1531) is a homolog of the *Salmonella* SipC and *Shigella* IpaC translocator proteins. The *Salmonella* SipC protein has been shown to interact with SipB to form the translocon pore (Myeni et al., 2013). Beyond their role as a translocator protein, SipC and IpaC also function as effector proteins within the eukaryotic cell. SipC has actin nucleation activity and bundles F-actin (Hayward and Koronakis, 1999). The ability of SipC to nucleate and bundle F-actin as well as form the translocation pore, are all dependant on the C-terminal 209 amino acids of the 409 amino acid protein (Hayward and Koronakis, 1999; Chang et al., 2005; Myeni and Zhou, 2010). Actin bundling appears to be important for cell invasion, as *Salmonella* containing a *sipC* mutation abolishing its actin bundling activity, but not translocator function, was less invasive than the parental strain (Myeni and Zhou, 2010). Internalization of *B. pseudomallei* can be blocked by the actin polymerization inhibitor cytochalasin D (Jones et al., 1996), indicating the importance of actin cytoskeletal rearrangements in the uptake of *B. pseudomallei*. Similarly to SipC, BipC is also able to polymerise actin *in vitro* and stabilizes F-actin indicating possible actin bundling activity (Kang et al., 2016a and our own unpublished observations). Interestingly, *Salmonella* SipA protein enhances the ability of SipC to nucleate and bundle F-actin (McGhie et al., 2001), but a homolog of SipA is not encoded by the genome of *B. pseudomallei*. SipC has also been shown to bind to host Syntaxin 6 and thereby recruit LAMP1 to the *Salmonella* containing vacuole (SCV), helping to stabilize its membrane (Madan et al., 2012). The functional relevance of the actin-binding activity of BipC and whether it binds any other host cell proteins requires further study.

In *Salmonella* and *Shigella*, the SipC/IpaC family of translocator/effector proteins may play a crucial role in determining the intracellular niche of the bacteria. There is evidence to suggest that *sipC* and *ipaC* cannot fully complement each other (Osiecki et al., 2001; Klein et al., 2017). The authors suggest the proteins, while both translocators, may have some divergent functions which may parallel the different intracellular lifestyles of the pathogens, with *Salmonella* residing within the vacuole and *Shigella* rapidly escaping into the cytosol (Osiecki et al., 2001). This is supported by a recent study in which *Salmonella* expressing *ipaC*, was shown to be capable of vacuole escape (Du et al., 2016). Because of closer parallels between the intracellular lifestyles of *B. pseudomallei* and *Shigella*, we would predict that BipC would function in a manner more similar to IpaC than SipC, mediating the exit of the bacterium from the endocytic compartment into the host cell cytosol. In a *Salmonella sipC* mutant, *bipC* is unable to complement *sipC* (Klein et al., 2017), further supporting a possible difference in accessory function of these proteins.

The C-terminal and N-terminal regions of BipC have previously been separately tested as a protective antigen in mice, but neither antigen showed any protection upon challenge with

B. pseudomallei (Druar et al., 2008). In another study, a *B. pseudomallei bipC* mutant showed reduced cell adhesion and invasion of A549 cells (Kang et al., 2015). This *bipC* mutant also showed a delay in escape from the phagosome, leading to a delay in formation of actin tails in the host cell cytoplasm and intracellular replication (Kang et al., 2015). The mutant was also attenuated in BALB/c mice infected intraperitoneally (Kang et al., 2015). The transcriptome of livers of mice infected with the *bipC* mutant showed a lower expression of genes involved in actin cytoskeleton signaling, MAPK signaling, integrin signaling and TNF when compared to mice infected with the parental *B. pseudomallei* strain (Kang et al., 2016b). While this shows the importance of BipC in virulence, both *in vivo* and *in vitro*, it does not separate the role of BipC as a translocator necessary for a functional T3SS and rapid escape from the endosome, from its role as an effector protein.

SECRETION SIGNALS AND CHAPERONES

Type III secretion signals are located at the N-terminus of a protein, are not cleaved and do not share primary sequence identity with each other (Michiels and Cornelis, 1991; Schesser et al., 1996). In *Yersinia*, it was demonstrated that as little as 15 amino acids of the N-terminus of YopE were required for secretion (Sory et al., 1995). Interestingly, substituting the alanines at position 2 and 15 in the YopE secretion signal with glutamic acid, did not affect secretion (Anderson and Schneewind, 1997). Even shifting the reading frame did not prevent the secretion of YopE, which the authors suggested may indicate a secretion signal located in the mRNA encoding the effector protein (Anderson and Schneewind, 1997).

A second important signal in the N-terminal region of T3SS effector proteins is the chaperone binding domain (CBD) which is required for the specific secretion of effector proteins (Abe et al., 1999; Ehrbar et al., 2003, 2006; Lee and Galán, 2003, 2004) as well as for their stability in the bacterial cytoplasm (Frithz-Lindsten et al., 1995; Abe et al., 1999). By determining the crystal structure of an effector protein bound to its chaperone, it is known that T3SS chaperones in *Yersinia* and *Salmonella* bind the N-terminal amino acids of their effector protein, just after the T3S signal, but before any functional domains, and maintain the bound effector protein in a partially unfolded state that may be more competent for secretion (Birtalan and Ghosh, 2001; Stebbins and Galán, 2001).

The T3S signal also appears to be promiscuous between systems, allowing secretion of effector proteins from bacteria in an unrelated host bacterium containing a T3SS (Rossier et al., 1999; Hovis et al., 2013). This is even true of less similar T3SSs, as demonstrated by the ability of the Hrp Plant T3SS of *Xanthomonas* to secrete the mammalian effector protein *Yersinia* YopE (Rossier et al., 1999). Perhaps because of the similarity between the two, virulence-associated T3SS effectors can be secreted by the bacterial flagellar T3SS system in *Yersinia* and *Salmonella* (Young and Young, 2002; Lee and Galán, 2004; Warren and Young, 2005; Ehrbar et al., 2006). It is only through interaction with the appropriate chaperone that specific secretion through a single system is achieved. Inhibition of the ability of *Salmonella* SopE to bind its chaperone InvB, caused secretion through both the flagellar and SPI-1 virulence associated T3SS (Lee and Galán, 2004; Ehrbar et al., 2006). The genome of *B. pseudomallei* encodes five putative T3SS-3 chaperones, though little work concerning their function has yet been described.

BPSS1517

The protein encoded by the gene BPSS1517 is predicted to be a putative chaperone (Panina et al., 2005) and has been shown to interact with the downstream effector protein BopC (BPSS1516) (Muangman et al., 2011).

BicP (BPSS1523)

BicP (BPSS1523) shares homology with *Salmonella* SicP, a specific chaperone for the effector protein SptP (Fu and Galán, 1998). The closest homolog to SptP in *B. pseudomallei* is BopA (BPSS1524) which was predicted to be the binding partner for BicP (Panina et al., 2005), but the interaction between the two proteins has not formally been demonstrated.

BapB (BPSS1526)

BapB (BPSS1526) is homologous to *Salmonella* IacP (Stevens et al., 2002). IacP is important for *Salmonella* invasion into host cells by playing a role in regulating SopA, SopB and SopD secretion (Kim et al., 2011). Formerly considered a possible candidate effector protein, BapB was not found to be secreted by T3SS-3 (Treerat et al., 2015; Vander Broek et al., 2015). Deletion of *bapB* caused an increase in the transcription and secretion of BopE, indicating it may be a negative regulator of effector transcription (Treerat et al., 2015). In the same study the authors performed a phylogenetic analysis which suggested that BapB may be closely related to the *Salmonella* FliT chaperone protein (Treerat et al., 2015).

BicA (BPSS1533)

BicA (BPSS1533) is homologous to the *Salmonella* chaperone SicA. SicA binds to and prevents the association and resulting degradation of SipC and SipB in the bacterial cytoplasm (Tucker and Galán, 2000). As SipC is secreted by a mature needle complex, SicA is freed and interacts with the transcriptional regulator InvF to increase the expression of effector proteins (Tucker and Galán, 2000; Darwin and Miller, 2001). *B. pseudomallei* BicA is required for the secretion of the known effector proteins BopE and BopA (Sun et al., 2010), and BicA along with the regulator BsaN activate transcription of T3SS-3 effector proteins, translocators and chaperones (Chen et al., 2014). The *bicA* gene is also able to partially complement a *Salmonella sicA* mutant (Klein et al., 2017), further supporting a homologous function. BicA is required for respiratory melioidosis in mice (Gutierrez et al., 2015b).

BsaR (BPSS1542)

BsaR (BPSS1542) is homologous to the *Salmonella* chaperone InvB which has been shown to be required for secretion of the effector proteins SopE and SopA (Ehrbar et al., 2003, 2004; Lee and Galán, 2003). BsaR is predicted to be the chaperone for BopE using a computational screen for chaperones in *B. pseudomallei*

K96243 (Panina et al., 2005), although this has not yet been experimentally validated.

T3SS-3 EFFECTOR PROTEINS

The role of the T3SS is to deliver an array of effector proteins into the target cell to subvert host cell functions. The function of different effector proteins is extremely varied, ranging from blocking apoptosis (*E. coli* NleH) (Hemrajani et al., 2010), prevention of phagocytosis (*Yersinia* YopH) (Persson et al., 1997), cytotoxic activity (*Pseudomonas aeruginosa* ExoU) (Sato et al., 2003) and disruption of the actin cytoskeleton (*Salmonella* SopE) (Hardt et al., 1998). *B. pseudomallei* encodes seven effector proteins known to be secreted by T3SS-3 (CHBP, BopC, BopA, BapA, BprD, BapC, and BopE), as well as one hypothetical T3SS effector BopB (Pumirat et al., 2014; Vander Broek et al., 2015).

CHBP/Cif (BPSS1385)

CHBP (BPSS1385) is a homolog of *E. coli* cell cycle inhibiting factor (Cif) and is the only identified T3SS-3 effector protein that is encoded outside of the T3SS-3 locus. Cif and CHBP are able to deamidate cellular NEDD8 causing cell cycle arrest (Nougayrède et al., 2001; Cui et al., 2010) and CHBP causes cell cycle arrest when expressed in *B. thailandensis* (Cui et al., 2010). CHBP has also been shown to activate the cellular kinase ERK independent of its ability to deamidate NEDD8 (Ng et al., 2017).

The gene encoding CHBP is present in ~76% of the available *B. pseudomallei* genomes and a Western blot assay used to probe for the presence of the CHBP protein in *B. pseudomallei* clinical isolates from the endemic region detected CHBP in 47% of isolates tested (Pumirat et al., 2014). Interestingly, CHBP is not secreted under standard growth conditions, but *B. pseudomallei* secretes CHBP in U937 cells in a *bsaQ*–dependent manner (Pumirat et al., 2014). A *B. pseudomallei* K96243 *chbP* insertion mutant was impaired in its ability to form plaques in HeLa cells at 24 h and demonstrated lower cytotoxicity at 6 h as assessed by LDH release assays (Pumirat et al., 2014). Both phenotypes could be complemented by expression of *chbP in trans* indicating that these phenotypes were due to disruption of the *chbP* gene and not due to unexpected polar effects of the insertion mutation (Pumirat et al., 2014).

BopB/FolE (BPSS1514)

BopB (BPSS1514) or FolE, is annotated to be a GTP cyclohydrolase I and was thought to be a candidate effector protein (Stevens et al., 2004). Yet a *bopB* mutant did not display a significantly reduced time to death in a BALB/c intraperitoneal infection model (Stevens et al., 2004). Similarly, mutation of *bopB* did not affect invasion and intracellular replication of host cells (Chen et al., 2014). Expression of *bopB* is co-regulated with the other T3SS-3 effectors by BsaN, but the role BopB plays in infection is still unknown (Chen et al., 2014).

BopC (BPSS1516)

BopC (BPSS1516) is a 509 amino acid protein with no significant sequence homology to proteins from other species besides *B. mallei*. It is encoded just before the T3SS-3 locus along with BPSS1517, its chaperone (Muangman et al., 2011). BopC was detected in the culture supernatants of WT *B. pseudomallei* 10276, but not in supernatants of a *bsaZ* insertion mutant, indicating that BopC is secreted by T3SS-3 (Muangman et al., 2011; Vander Broek et al., 2015). The first 20 amino acids of BopC fused to the β-lactamase gene TEM1, was sufficient for translocation into HeLa cells in a T3SS-dependant manner (Muangman et al., 2011). A *B. pseudomallei* K96243 *bopC* mutant was hindered in its ability to invade A549 cells (Muangman et al., 2011) and displayed reduced levels of intracellular survival (Srinon et al., 2013). The *bopC* mutant also demonstrated delayed phagosome escape in J774A.1 cells as shown by staining for co-localisation of the bacteria with the cellular lysosomal marker protein LAMP-1, which likely explains the defect in intracellular survival (Srinon et al., 2013).

BprD (BPSS1521)

BprD (BPSS1521) has no known homology to proteins outside of *B. pseudomallei* and the closely related *Burkholderia* species. It is labeled as a putative regulator of T3SS-3, though a knockout of the *bpr* operon (*bprB-D*) showed no effect on the expression of T3SS-3 genes (Sun et al., 2010). Its expression is regulated by BsaN along with the known effector proteins BopA, BopC and BopE (Sun et al., 2010; Chen et al., 2014) and *bipD* gene expression is significantly up-regulated in tissues of infected mice (Chirakul et al., 2014). The same study demonstrated attenuation of the *bprD* mutant in BALB/c mice infected intraperitoneally, which the authors speculate may be due to the up-regulation of the T6SS-1 through effects on *bprC* (Chirakul et al., 2014).

Our own work has demonstrated that BprD is secreted into the supernatant in a T3SS-3 *bsaZ*-dependant manner (Vander Broek et al., 2015). While this may seem surprising, it is not without precedent that a regulator of the T3SS is also a substrate for secretion, for example, *Yersinia* LcrQ (Cambronne et al., 2000). It is thought that LcrQ acts as a feedback inhibitor of the expression of *Yersinia* effectors (Cambronne et al., 2000). When LcrQ is secreted into host cells and the levels of LcrQ in the bacterium are depleted, inhibition is relieved and T3S can progress (Cambronne et al., 2000). Whether BprD is secreted into host cells, whether it acts as a true effector protein and the mechanisms by which it regulates virulence and T6S present interesting research questions for the field.

BopA (BPSS1524)

BopA (BPSS1524) shares 23% amino acid identity with *Shigella* IcsB (Cullinane et al., 2008). It has been predicted to contain a Rho GTPase inactivation domain (RID) similar to that found in *Vibrio cholerae* VcRtxA and other MARTX toxins which indirectly inactivate Rho GTPases (Pei and Grishin, 2009). IcsB, along with its chaperone IpgA, are important for *Shigella*'s ability to escape LC3-positive autophagosomes once inside the host cell, and this activity is dependent on the IcsB cholesterol-binding domain (Kayath et al., 2010; Campbell-Valois et al., 2015). BopA also contains a functional cholesterol-binding domain (Kayath et al., 2010). The *B. pseudomallei* homolog of IpgA is BicP, which co-purifies with BopA and helps to prevent its degradation, indicating that it is the chaperone for BopA (Kayath et al., 2010).

BopA is secreted by T3SS-3 in a *bsaZ*-dependant manner (Vander Broek et al., 2015) and the first 58 amino acids of *B. mallei* BopA fused to the *Yersinia enterolitica* phospholipase YplA, has been shown to be secreted in a surrogate enteropathogenic *E. coli* host (Whitlock et al., 2008).

BopA is important for the intracellular survival of *B. pseudomallei* in phagocytic cells (Cullinane et al., 2008). A *B. pseudomallei* K96243 *bopA* mutant displayed reduced intracellular survival and an increased localisation with GFP-LC3, an indicator of autophagy stimulation, in RAW 264.7 cells (Cullinane et al., 2008). This reduction in intracellular survival was overcome when cells were treated with the autophagy inhibitor wortmannin (Cullinane et al., 2008). Another study demonstrated that BopA is important for escape of the bacterium from the phagosome (Gong et al., 2011). A *B. mallei* ATCC 23344 *bopA* mutant demonstrated reduced intracellular survival in J774A.1 cells (Whitlock et al., 2008). Interestingly, in the murine alveolar macrophage cell line MH-S, the same *B. mallei bopA* mutant exhibited increased intracellular survival when compared to the isogenic parental strain, indicating that different cell types may rely on different mechanisms to control intracellular *B. mallei* (Whitlock et al., 2009). BALB/c mice infected intraperitoneally with a *B. pseudomallei* 576 *bopA* insertion mutant were significantly delayed in time to death when compared to the parental strain (Stevens et al., 2004). Similarly, BALB/c mice infected intra-nasally with a *B. mallei* ATCC 23344 *bopA* insertion mutant also showed a delayed time to death. No bacteria were recovered from the lung tissue of animals infected with the *bopA* mutant while 10^8 bacteria were recovered from the lungs of animals infected with *B. mallei* (Whitlock et al., 2009). In mice immunized with recombinant BopA and challenged intra-nasally with *B. mallei* ATCC23344 or *B. pseudomallei* 1026b, the BopA vaccine protected 100 and 60%, respectively of animals 21 days post infection (Whitlock et al., 2010).

BopE (BPSS1525)

BopE (BPSS1525) is the best characterized of the *B. pseudomallei* effector proteins and is commonly used as a readout for the ability of the T3SS-3 to secrete effector proteins. BopE is 27% identical over a region of 168 amino acids to the *Salmonella* guanine nucleotide exchange factor (GEF) SopE (Stevens et al., 2003). BopE is secreted by *B. pseudomallei* in a manner dependent on the T3SS-3 (Stevens et al., 2003; Muangsombut et al., 2008; Vander Broek et al., 2015) and is required for efficient invasion of non-phagocytic cells (Stevens et al., 2003). Also, ectopic expression of BopE in HeLa cells causes significant actin cytoskeletal rearrangements with similarity to ectopic SipC expression (Stevens et al., 2003). Using fluorescence spectrometry, it was demonstrated that purified BopE, similar to SopE, is a functional GEF for both Rac1 and Cdc42, but with about 10 fold lower activity than SopE (Stevens et al., 2003). This lower activity may be explained by differences in the catalytic domains, as the SopE catalytic domain stays in an open conformation, while BopE adopts a closed conformation which requires interaction with Cdc42 to allow for GEF activity (Upadhyay et al., 2008). HEK 293T cells transfected with plasmids expressing BopE and caspase–1, resulted in increased activation of caspase–1 and –7 (Bast et al., 2014). When the active site of the transfected BopE was mutated, levels of activation of caspase–1 and –7 returned to basal levels indicating BopE's GEF activity is required for the activation of caspase–1 and –7 (Bast et al., 2014).

BopE is a potent T cell antigen in mice (Haque et al., 2006) and in sero-positive recovered melioidosis patients (Tippayawat et al., 2009). However, in an intraperitoneal BALB/c murine model of melioidosis, ablation of the *bopE* gene did not affect the median time to death of the animals when compared to the parental strain (Stevens et al., 2004).

BapC (BPSS1527)

BapC (BPSS1526) is homologous to *Salmonella* IagB (Stevens et al., 2002). IagB is thought to be a lytic transglycosylase involved in the breakdown of the bacterial peptidoglycan layer, allowing connection of the inner and outer membrane components of the secretion system, though this function has not been formally demonstrated (Zahrl et al., 2005). BapC is secreted by T3SS-3 in a manner dependant on *bsaS* (Treerat et al., 2015). A *B. pseudomallei* K96243 *bapC* mutant showed a slight attenuation in a competitive growth assay in an acute BALB/c model of infection (Treerat et al., 2015), whereas a *B. pseudomallei* 1026b *bapC* mutant showed no significant attenuation in a Syrian hamster model of melioidosis (Warawa and Woods, 2005).

BapA (BPSS1528)

BapA (BPSS1528) has no known homology to any other bacterial or host cell proteins except the BapA orthologues of *B. pseudomallei* and the closely related species *B. thailandensis* and *B. mallei*. BapA is secreted by T3SS-3 in both a BsaS and BsaZ dependant manner (Treerat et al., 2015; Vander Broek et al., 2015). A 1026b *B. pseudomallei bapA* mutant showed no attenuation in a Syrian hamster model of infection (Warawa and Woods, 2005), whereas a *B. pseudomallei* K96243 *bapA* mutant was attenuated in a competitive growth assay in an acute BALB/c model of infection (Treerat et al., 2015).

FUTURE PERSPECTIVES

While there has been a wealth of new information concerning *B. pseudomallei* T3S, there are still many significant gaps in knowledge, particularly with regard to the importance of T3SS-1 and T3SS-2 in melioidosis. It is possible that the plant pathogen-like T3SS-1 and T3SS-2 add fitness to *B. pseudomallei* in the environment, allowing infection/colonization of adjacent plant life. However, it is also possible that these systems are relevant in mammalian hosts, given their conservation amongst the *B. pseudomallei* strains sequenced to date, and the finding that T3SS-1 is required for full virulence in a murine model of melioidosis (D'Cruze et al., 2011).

It is still unclear whether T3SS-3 plays a significant role in invasion of host cells, particularly non-phagocytic cells. There have been many reports of T3SS-3 or its effector proteins playing a role in invasion (Stevens et al., 2003; Suparak et al., 2005; Muangsombut et al., 2008; Muangman et al., 2011; Kang et al., 2015). Yet in another study, T3SS-3 ATPase (*bsaS*) mutants did not show a decrease in invasion efficiency in HEK293 or HeLa cells (French et al., 2011). This observation is important because it questions the involvement of T3SS-3 in invasion as well as highlighting a general lack of understanding of the pathways and mechanisms involved in *B. pseudomallei* entry into host cells.

To date CHBP is the only effector protein secreted by the *B. pseudomallei* T3SS-3 that is encoded outside of the T3SS-3 locus, and which is not present in the genome of all strains (Pumirat et al., 2014). This raises the question of whether other effector proteins secreted by T3SS-3, but encoded outside of the T3SS-3 locus, may be present in other strains. Full genome comparisons of almost 100 strains of *B. pseudomallei* identified 86% of genes as being conserved and present in all strains (Sim et al., 2008). The other 14% of genes, considered accessory genes, were disproportionally present in genomic islands and were associated with clinical isolates (Sim et al., 2008). The GIs in *B. pseudomallei* are highly variable between strains. A study of five clinical *B. pseudomallei* strains identified a total of 71 GIs distributed between the strains, with at least half being unique to the strain in which they were identified (Tuanyok et al., 2008). The variability in these regions is largely due to horizontal gene transfer and *B. pseudomallei* has a relatively high rate of lateral gene transfer compared to the mutation rates of other bacterial species (Pearson et al., 2009). As *B. pseudomallei* has a large amount of genomic diversity in its accessory genome and a high rate of lateral gene transfer, there is a strong possibility that other novel effector proteins are encoded in other strains of *B. pseudomallei* that have yet to be identified. Study of these effector proteins could provide important insights into strain differences as well as the potential for novel effector biology, but it is a difficult task as the secretome of *B. pseudomallei* is very complex (Vander Broek et al., 2015). Also, due to the temporal and hierarchical control of the T3SS, there is always the possibility that an effector protein will not be secreted under the conditions used in a given experiment (reviewed in Büttner, 2012). Previous methods of identifying T3SS-3 effector proteins have relied on initial bioinformatics prediction (Stevens et al., 2003; Muangman et al., 2011) or high throughput screens (Vander Broek et al., 2015), both of which may prove to be useful tools for further studies of *B. pseudomallei* T3S. Indeed it is timely to apply these approaches to the identification and characterization of effector proteins secreted by the lesser studied T3SS-1 and T3SS-2.

Another important outstanding area of research is characterizing the functions of those effector proteins that have already been identified. One common characteristic of many T3SS effector proteins that complicates this task is their functional redundancy (reviewed in Galán, 2009). Commonly, deletion of one effector protein yields little or no phenotype in infection models because other effector proteins target either the same host cell protein or pathway (reviewed in Galán, 2009). In a biological system this would increase the chance of effectors successfully carrying out their intended function and decrease the likelihood of host cell interference. This functional redundancy also highlights the evolutionary importance of dysregulating specific cellular pathways from the standpoint of the bacterium. Even where redundancy does not exist, the function of effector proteins are often subtle when compared to bacterial toxins and may not be easily measurable in *in vitro* or *in vivo* models of infection (reviewed in Dean, 2011). This lack of a phenotype to inform focused studies presents a challenge for investigators, creating the need for high throughput "fishing" assays, such as protein immunoprecipitation/pull-downs and yeast two-hybrid assays. These assays have been used successfully to identify host cell binding partners and the subsequent functions of effector proteins from other bacteria (Zhou et al., 2013; Pallett et al., 2014). Understanding the functions of T3SS-3 effector proteins may provide new insights into host-pathogen interactions in *B. pseudomallei* infection.

Finally, the T3SS-3 has been shown to be one of the most important virulence factors in *B. pseudomallei* models of infection, raising the question of whether T3SS-3 could be a useful target for protective vaccines or therapeutic intervention in melioidosis patients. Although several attempts have been made to use live-attenuated vaccines based on mutation of key T3SS-3 genes in murine models of melioidosis, none have proven to provide sterilizing immunity. Despite being a potent B- and T-cell antigen, attempts to utilize BipD as a subunit vaccine in murine models of melioidosis have shown little promise (Stevens et al., 2004; Druar et al., 2008). There have also been attempts to use the translocator proteins of T3SS-3 as subunit vaccines. The N-terminal region of BipB was tested as a protective antigen in mice, but showed no protection against subsequent challenge (Druar et al., 2008). The C-terminal and N-terminal regions of BipC have also been separately tested as a subunit vaccine in mice, but neither antigen showed any protection (Druar et al., 2008). Some studies have focused on the use of effector proteins as subunit vaccines. Out of these studies the BopA protein shows most promise, since mice immunized with recombinant BopA were protected against subsequent intranasal challenge with both *B. mallei* and *B. pseudomallei* (Whitlock et al., 2010). More recently interest in the use of small molecule inhibitors of the T3SS-3 has arisen (Gong et al., 2015). Treatment of *B. pseudomallei* infected RAW264.7 cells with a small molecule inhibitor targeting BsaS of the T3SS-3, resulted in a decrease in bacterial intracellular survival (Gong et al., 2015). However, the use of such inhibitors is still very much in its infancy, with important *in vivo* studies being required to determine whether such small molecules would be effective in murine models of melioidosis.

AUTHOR CONTRIBUTIONS

CV and JS contributed equally to the writing of this review article.

ACKNOWLEDGMENTS

The authors are supported by an Institute Strategic Programme Grant from the BBSRC.

REFERENCES

Abe, A., De Grado, M., Pfuetzner, R. A., Sánchez-SanMartin, C., DeVinney, R., Puente, J. L. et al. (1999). Enteropathogenic *Escherichia coli* translocated intimin receptor. Tir, requires a specific chaperone for stable secretion. *Mol. Microbiol.* 33, 1162–1175. doi: 10.1046/j.1365-2958.1999.01558.x

Akeda, Y., and Galán, J. E. (2005). Chaperone release and unfolding of substrates in type III secretion. *Nature* 437, 911–915. doi: 10.1038/nature03992

Anderson, D. M., and Schneewind, O. (1997). A mRNA signal for the type III secretion of Yop proteins by *Yersinia enterocolitica*. *Science* 278, 1140–1143. doi: 10.1126/science.278.5340.1140

Angus, A. A., Agapakis, C. M., Fong, S., Yerrapragada, S., Estrada-De Los Santos, P., Yang, P., et al. (2014). Plant-associated symbiotic *Burkholderia* species lack hallmark strategies required in mammalian pathogenesis. *PLoS ONE* 9:e83779. doi: 10.1371/journal.pone.0083779

Attree, O., and Attree, I. (2001). A second type III secretion system in *Burkholderia pseudomallei*: who is the real culprit?. *Microbiology* 147, 3197–3199. doi: 10.1099/00221287-147-12-3197

Barrett, B. S., Picking, W. L., Picking, W. D., and Middaugh, C. R. (2008). The response of type three secretion system needle proteins MxiHΔ5, BsaLΔ5, and PrgIΔ5 to temperature and pH. *Proteins* 73, 632–643. doi: 10.1002/prot.22085

Bast, A., Krause, K., Schmidt, I. H., Pudla, M., Brakopp, S., Hopf, V., et al. (2014). Caspase-1-dependent and -independent cell death pathways in *Burkholderia pseudomallei* infection of macrophages. *PLoS Pathog.* 10:e1003986. doi: 10.1371/journal.ppat.1003986

Birtalan, S., and Ghosh, P. (2001). Structure of the Yersinia type III secretory system chaperone SycE. *Nat. Struct. Mol. Biol.* 8, 974–978. doi: 10.1038/nsb1101-974

Blocker, A., Gounon, P., Larquet, E., Niebuhr, K., Cabiaux, V., Parsot, C., et al. (1999). The tripartite type III secreton of Shigella flexneri inserts IpaB and IpaC into host membranes. *J. Cell Biol.* 147, 683–693. doi: 10.1083/jcb.147.3.683

Blocker, A., Jouihri, N., Larquet, E., Gounon, P., Ebel, F., Parsot, C., et al. (2001). Structure and composition of the Shigella flexneri 'needle complex', a part of its type III secreton. *Mol. Microbiol.* 39, 652–663. doi: 10.1046/j.1365-2958.2001.02200.x

Botteaux, A., Sory, M. P., Biskri, L., Parsot, C., and Allaoui, A. (2009). MxiC is secreted by and controls the substrate specificity of the Shigella flexneri type III secretion apparatus. *Mol. Microbiol.* 71, 449–460. doi: 10.1111/j.1365-2958.2008.06537.x

Brett, P. J., DeShazer, D., and Woods, D. E. (1998). Note *Burkholderia thailandensis* sp. nov., a *Burkholderia pseudomallei*-like species. *Int. J. Syst. Bacteriol.* 48, 317–320. doi: 10.1099/00207713-48-1-317

Burtnick, M. N., Brett, P. J., Nair, V., Warawa, J. M., Woods, D. E., and Gherardini, F. C. (2008). *Burkholderia pseudomallei* type III secretion system mutants exhibit delayed vacuolar escape phenotypes in RAW 264.7 murine macrophages. *Infect. Immun.* 76, 2991–3000. doi: 10.1128/IAI.00263-08

Büttner, D. (2012). Protein export according to schedule: architecture, assembly, and regulation of type III secretion systems from plant-and animal-pathogenic bacteria. *Microbiol. Mol. Biol. Rev.* 76, 262–310. doi: 10.1128/MMBR.05017-11

Büttner, D., and He, S. Y. (2009). Type III protein secretion in plant pathogenic bacteria. *Plant Physiol.* 150, 1656–1664. doi: 10.1104/pp.109.139089

Cambronne, E. D., Cheng, L. W., and Schneewind, O. (2000). LcrQ/YscM1, regulators of the Yersinia yop virulon, are injected into host cells by a chaperone-dependent mechanism. *Mol. Microbiol.* 37, 263–273. doi: 10.1046/j.1365-2958.2000.01974.x

Campbell-Valois, F.-X., Sachse, M., Sansonetti, P. J., and Parsot, C. (2015). Escape of actively secreting *Shigella flexneri* from ATG8/LC3-positive vacuoles formed during Cell-To-Cell spread is facilitated by IcsB and VirA. *MBio* 6:e02567-14. doi: 10.1128/mBio.02567-14

Chang, J., Chen, J., and Zhou, D. (2005). Delineation and characterization of the actin nucleation and effector translocation activities of Salmonella SipC. *Mol. Microbiol.* 55, 1379–1389. doi: 10.1111/j.1365-2958.2004.04480.x

Chen, Y., Schröder, I., French, C. T., Jaroszewicz, A., Yee, X. J., The, B. E., et al. (2014). Characterization and analysis of the *Burkholderia pseudomallei* BsaN virulence regulon. *BMC Microbiol.* 14:206. doi: 10.1186/s12866-014-0206-6

Cheng, A. C., and Currie, B. J. (2005). Melioidosis: epidemiology, pathophysiology, and management. *Clin. Microbiol. Rev.* 18, 383-416. doi: 10.1128/CMR.18.2.383-416.2005

Chirakul, S., Bartpho, T., Wongsurawat, T., Taweechaisupapong, S., Karoonutaisiri, N., Talaat, A. M., et al. (2014). Characterization of BPSS1521 (bprD), a regulator of *Burkholderia pseudomallei* virulence gene expression in the mouse model. *PLoS ONE* 9:e104313. doi: 10.1371/journal.pone.0104313

Costa, T. R. D., Felisberto-Rodrigues, C., Meir, A., Prevost, M. S., Redzej, A., Trokter, M., et al. (2015). Secretion systems in Gram-negative bacteria: structural and mechanistic insights. *Nat. Rev. Microbiol.* 13, 343–359. doi: 10.1038/nrmicro3456

Cui, J., Yao, Q., Li, S., Ding, X., Lu, Q., Mao, H., et al. (2010). Glutamine deamidation and dysfunction of ubiquitin/NEDD8 induced by a bacterial effector family. *Science* 329, 1215–1218. doi: 10.1126/science.1193844

Cullinane, M., Gong, L., Li, X., Adler, N.-L., Tra, T., Adler, B., et al. (2008). Stimulation of autophagy suppresses the intracellular survival of *Burkholderia pseudomallei* in mammalian cell lines. *Autophagy* 4, 744–753. doi: 10.4161/auto.6246

Currie, B. J., Ward, L., and Cheng, A. C. (2010). The epidemiology and clinical spectrum of melioidosis: 540 cases from the 20 year Darwin prospective study. *PLoS Negl. Trop. Dis.* 4:e900. doi: 10.1371/journal.pntd.0000900

D'Cruze, T., Gong, L., Treerat, P., Ramm, G., Boyce, J. D., Prescott, M., et al. (2011). Role for the *Burkholderia pseudomallei* type three secretion system cluster 1 bpscN gene in virulence. *Infect. Immun.* 79, 3659–3664. doi: 10.1128/IAI.01351-10

Darwin, K. H., and Miller, V. L. (2001). Type III secretion chaperone-dependent regulation: activation of virulence genes by SicA and InvF in *Salmonella typhimurium*. *EMBO J.* 20, 1850–1862. doi: 10.1093/emboj/20.8.1850

Dean, P. (2011). Functional domains and motifs of bacterial type III effector proteins and their roles in infection. *FEMS Microbiol. Rev.* 35, 1100–1125. doi: 10.1111/j.1574-6976.2011.00271.x

Deng, W., Li, Y., Hardwidge, P. R., Frey, E. A., Pfuetzner, R. A., Lee, S., et al. (2005). Regulation of type III secretion hierarchy of translocators and effectors in attaching and effacing bacterial pathogens. *Infect. Immun.* 73, 2135–2146. doi: 10.1128/IAI.73.4.2135-2146.2005

Deng, W., Puente, J. L., Gruenheid, S., Li, Y., Vallance, B. A., Vázquez, A., et al. (2004). Dissecting virulence: systematic and functional analyses of a pathogenicity island. *Proc. Natl. Acad. Sci. U.S.A.* 101, 3597–3602. doi: 10.1073/pnas.0400326101

Diepold, A., Amstutz, M., Abel, S., Sorg, I., Jenal, U., and Cornelis, G. R. (2010). Deciphering the assembly of the Yersinia type III secretion injectisome. *EMBO J.* 29, 1928–1940. doi: 10.1038/emboj.2010.84

Diepold, A., and Wagner, S. (2014). Assembly of the bacterial type III secretion machinery. *FEMS Microbiol. Rev.* 38, 802–822. doi: 10.1111/1574-6976.12061

Diepold, A., Wiesand, U., and Cornelis, G. R. (2011). The assembly of the export apparatus (YscR. S, T, U, V) of the Yersinia type III secretion apparatus occurs independently of other structural components and involves the formation of an YscV oligomer*Mol. Microbiol.* 82, 502–514. doi: 10.1111/j.1365-2958.2011.07830.x

Druar, C., Yu, F., Barnes, J. L., Okinaka, R. T., Chantratita, N., Beg, S., et al. (2008). Evaluating *Burkholderia pseudomallei* Bip proteins as vaccines and Bip antibodies as detection agents. *FEMS Immunol. Med. Microbiol.* 52, 78–87. doi: 10.1111/j.1574-695X.2007.00345.x

Du, J., Reeves, A. Z., Klein, J. A., Twedt, D. J., Knodler, L. A., and Lesser, C. F. (2016). The type III secretion system apparatus determines the intracellular niche of bacterial pathogens.. *Proc. Natl. Acad. Sci. U.S.A.* 113, 4794–4799. doi: 10.1073/pnas.1520699113

Egan, F., Barret, M., and O'Gara, F. (2014). The SPI-1-like Type III secretion system: more roles than you think.. *Front. Plant Sci.* 5:34. doi: 10.3389/fpls.2014.00034

Ehrbar, K., Friebel, A., Miller, S. I., and Hardt, W.-D. (2003). Role of the Salmonella pathogenicity island 1 (SPI-1) protein InvB in type III secretion of SopE and SopE2, two Salmonella effector proteins encoded outside of SPI-1. *J. Bacteriol.* 185, 6950–6967. doi: 10.1128/JB.185.23.6950-6967.2003

Ehrbar, K., Hapfelmeier, S., Stecher, B., andHardt, W.-D. (2004). InvB is required for type III-dependent secretion of SopA in *Salmonella enterica* serovar Typhimurium. *J. Bacteriol.* 186, 1215–1219. doi: 10.1128/JB.186.4.1215-1219.2004

Ehrbar, K., Winnen, B., and Hardt, W.-D. (2006). The chaperone binding domain of SopE inhibits transport via flagellar and SPI-1 TTSS in the absence of InvB. *Mol. Microbiol.* 59, 248–264. doi: 10.1111/j.1365-2958.2005.04931.x

Epler, C. R., Dickenson, N. E., Olive, A. J., Picking, W. L., and Picking, W. D. (2009). Liposomes recruit IpaC to the *Shigella flexneri* type III secretion apparatus needle as a final step in secretion induction. *Infect. Immun.* 77, 2754–2761. doi: 10.1128/IAI.00190-09

Erhardt, M., Mertens, M. E., Fabiani, F. D., and Hughes, K. T. (2014). ATPase-independent type-III protein secretion in *Salmonella enterica*. *PLoS Genet.* 10:e1004800. doi: 10.1371/journal.pgen.1004800

Erskine, P. T., Knight, M. J., Ruaux, A., Mikolajek, H., Sang, N. W. F., Withers, J., et al. (2006). High resolution structure of BipD: an invasion protein associated with the type III secretion system of *Burkholderia pseudomallei*. *J. Mol. Biol.* 363, 125–136. doi: 10.1016/j.jmb.2006.07.069

Espina, M., Ausar, S. F., Middaugh, C. R., Baxter, M. A., Picking, W. D., and Picking, W. L. (2007). Conformational stability and differential structural analysis of LcrV, PcrV, BipD, and SipD from type III secretion systems. *Protein Sci.* 16, 704–714. doi: 10.1110/ps.062645007

Espina, M., Olive, A. J., Kenjale, R., Moore, D. S., Ausar, S. F., Kaminski, R. W., et al. (2006). IpaD localizes to the tip of the type III secretion system needle of *Shigella flexneri*. *Infect. Immun.* 74, 4391–4400. doi: 10.1128/IAI.00440-06

Forsberg, Å., Viitanen, A.-M., Skurnik, M., and Wolf-Watz, H. (1991). The surface-located YopN protein is involved in calcium signal transduction in *Yersinia pseudotuberculosis*. *Mol. Microbiol.* 5, 977–986. doi: 10.1111/j.1365-2958.1991.tb00773.x

French, C. T., Toesca, I. J., Wu, T.-H., Teslaa, T., Beaty, S. M., Wong, W., et al. (2011). Dissection of the Burkholderia intracellular life cycle using a photothermal nanoblade. *Proc. Natl. Acad. Sci. U.S.A.* 108, 12095–12100. doi: 10.1073/pnas.1107183108

Frithz-Lindsten, E., Rosqvist, R., Johansson, L., and Forsberg, Å. (1995). The chaperone-like protein YerA of *Yersinia pseudotuberculosis* stabilizes YopE in the cytoplasm but is dispensable for targeting to the secretion loci. *Mol. Microbiol.* 16, 635–647. doi: 10.1111/j.1365-2958.1995.tb02426.x

Fu, Y., and Galán, J. E. (1998). Identification of a specific chaperone for SptP, a substrate of the centisome 63 type III secretion system of *Salmonella typhimurium*. *J. Bacteriol.* 180, 3393–3399.

Galán, J. E. (2009). Common themes in the design and function of bacterial effectors. *Cell Host Microbe* 5, 571–579. doi: 10.1016/j.chom.2009.04.008

Galán, J. E., and Curtiss, R. (1989). Cloning and molecular characterization of genes whose products allow *Salmonella Typhimurium* to penetrate tissue culture cells. *Proc. Natl. Acad. Sci. U.S.A.* 86, 6383–6387. doi: 10.1073/pnas.86.16.6383

Galán,. J. E., Lara-Tejero, M., Marlovits, T. C., and Wagner, S. (2014). Bacterial type III secretion systems: specialized nanomachines for protein delivery into target cells. *Annu. Rev. Microbiol.* 68:415. doi: 10.1146/annurev-micro-092412-155725

Gemski, P., Lazere, J. R., Casey, T., and Wohlhieter, J. A. (1980). Presence of a virulence-associated plasmid in *Yersinia pseudotuberculosis*. *Infect. Immun.* 28, 1044–1047.

Gong, L., Cullinane, M., Treerat, P., Ramm, G., Prescott, M., Adler, B., et al. (2011). The *Burkholderia pseudomallei* type III secretion system and BopA are required for evasion of LC3-associated phagocytosis. *PLoS ONE* 6:e17852. doi: 10.1371/journal.pone.0017852

Gong, L., Lai, S.-C., Treerat, P., Prescott, M., Adler, B., Boyce, J. D., et al. (2015). *Burkholderia pseudomallei* Type III secretion system cluster 3 ATPase BsaS, a chemotherapeutic Target for Small-Molecule ATPase Inhibitors. *Infect. Immun.* 83, 1276–1285. doi: 10.1128/IAI.03070-14

Gophna, U., Ron, E. Z., and Graur, D. (2003). Bacterial type III secretion systems are ancient and evolved by multiple horizontal-transfer events. *Gene* 312, 151–163. doi: 10.1016/S0378-1119(03)00612-7

Gutierrez, M. G., Pfeffer, T. L., and Warawa, J. M. (2015a). Type 3 secretion system cluster 3 is a critical virulence determinant for lung-specific melioidosis. *PLoS Negl. Trop. Dis.* 9:e3441. doi: 10.1371/journal.pntd.0003441

Gutierrez, M. G., Yoder-Himes, D. R., and Warawa, J. M. (2015b). Comprehensive identification of virulence factors required for respiratory melioidosis using Tn-seq mutagenesis. *Front. Cell. Infect. Microbiol.* 5:78. doi: 10.3389/fcimb.2015.00078

Haque, A., Chu, K., Easton, A., Stevens, M. P., Galyov, E. E., Atkins, T., et al. (2006). A live experimental vaccine against *Burkholderia pseudomallei* Elicits CD4+ T Cell-Mediated Immunity. Priming T Cells Specific for 2 Type III Secretion System Proteins. *J. Infect. Dis.* 194, 1241-1248. doi: 10.1086/508217

Haraga, A., West, T. E., Brittnacher, M. J., Skerrett, S. J., and Miller, S. I. (2008). *Burkholderia thailandensis* as a model system for the study of the virulence-associated type III secretion system of *Burkholderia pseudomallei*. *Infect. Immun.* 76, 5402–5411. doi: 10.1128/IAI.00626-08

Hardt, W.-D., Chen, L.-M., Schuebel, K. E., Bustelo, X. R., and Galán, J. E. (1998). *S. Typhimurium* encodes an activator of Rho GTPases that induces membrane ruffling and nuclear responses in host cells. *Cell* 93, 815–826. doi: 10.1016/S0092-8674(00)81442-7

Hayward, R. D., Cain, R. J., McGhie, E. J., Phillips, N., Garner, M. J., and Koronakis, V. (2005). Cholesterol binding by the bacterial type III translocon is essential for virulence effector delivery into mammalian cells. *Mol. Microbiol.* 56, 590–603. doi: 10.1111/j.1365-2958.2005.04568.x

Hayward, R. D., and Koronakis, V. (1999). Direct nucleation and bundling of actin by the SipC protein of invasive Salmonella. *EMBO J.* 18, 4926–4934. doi: 10.1093/emboj/18.18.4926

Heine, S. J., Diaz-McNair, J., Martinez-Becerra, F. J., Choudhari, S. P., Clements, J. D., Picking, W. L., et al. (2013). Evaluation of immunogenicity and protective efficacy of orally delivered Shigella type III secretion system proteins IpaB and IpaD. *Vaccine* 31, 2919–2929. doi: 10.1016/j.vaccine.2013.04.045

Hemrajani, C., Berger, C. N., Robinson, K. S., Marchès, O., Mousnier, A., and Frankel, G. (2010). NleH effectors interact with Bax inhibitor-1 to block apoptosis during enteropathogenic *Escherichia coli* infection. *Proc. Natl. Acad. Sci. U.S.A.* 107, 3129–3134. doi: 10.1073/pnas.0911609106

Holden, M. T. G., Titball, R. W., Peacock, S. J., Cerdeño-Tárraga, A. M., Atkins, T., Crossman, L. C., et al. (2004). Genomic plasticity of the causative agent of melioidosis. *Burkholderia pseudomallei. Proc. Natil. Acad. Sci. U.S.A.* 101, 14240–14245. doi: 10.1073/pnas.0403302101

Hovis, K. M., Mojica, S., McDermott, J. E., Pedersen, L., Simhi, C., Rank, R. G., et al. (2013). Genus-optimized strategy for the identification of Chlamydial type III secretion substrates. *Pathog. Dis.* 69, 213–222. doi: 10.1111/2049-632X.12070

Hu, B., Lara-Tejero, M., Kong, Q., Galán, J. E., and Liu, J. (2017). *In situ* molecular architecture of the salmonella Type III secretion machine. *Cell* 168, 1065–1074. doi: 10.1016/j.cell.2017.02.022

Iriarte, M., Sory, M.-P., Boland, A., Boyd, A. P., Mills, S. D., Cornelis, G. R. et al. (1998). TyeA, a protein involved in control of Yop release and in translocation of Yersinia Yop effectors. *EMBO J.* 17, 1907–1918. doi: 10.1093/emboj/17.7.1907

Jarvis, K. G., Giron, J. A., Jerse, A. E., McDaniel, T. K., Donnenberg, M. S., and Kaper, J. B. (1995). Enteropathogenic *Escherichia coli* contains a putative type III secretion system necessary for the export of proteins involved in attaching and effacing lesion formation. *Proc. Natl. Acad. Sci. U.S.A.* 92, 7996–8000. doi: 10.1073/pnas.92.17.7996

Jitprasutwit, S., Thaewpia, W., Muangsombut, V., Lulitanond, A., Leelayuwat, C., Lertmemongkolchai, G., et al. (2010). Effect of acidic pH on the invasion efficiency and the type III secretion system of *Burkholderia thailandensis*. *J. Microbiol.* 48, 526–532. doi: 10.1007/s12275-010-0078-x

Johnson, S., Roversi, P., Espina, M., Olive, A., Deane, J. E., Birket, S., et al. (2007). Self-chaperoning of the type III secretion system needle tip proteins IpaD and BipD. *J. Biol. Chem.* 282, 4035–4044. doi: 10.1074/jbc.M607945200

Jones, A. L., Beveridge, T. J., and Woods, D. E. (1996). Intracellular survival of *Burkholderia pseudomallei*. *Infect. Immun.* 64, 782–790.

Kang, W. T., Vellasamy, K. M., Chua, E.-G., and Vadivelu, J. (2015). Functional characterizations of effector protein BipC, a type III secretion system protein, in *Burkholderia pseudomallei* pathogenesis. *J. Infect. Dis.* 211, 827–834. doi: 10.1093/infdis/jiu492

Kang, W. T., Vellasamy, K. M., Rajamani, L., Beuerman, R. W., and Vadivelu, J. (2016a). *Burkholderia pseudomallei* type III secreted protein BipC: role in actin modulation and translocation activities required for the bacterial intracellular lifecycle. *Peer J.* 4:e2532. doi: 10.7717/peerj.2532

Kang, W. T., Vellasamy, K. M., and Vadivelu, J. (2016b). Eukaryotic pathways targeted by the type III secretion system effector protein. BipC, involved in the intracellular lifecycle of Burkholderia pseudomallei. *Sci. Rep.* 6:33528. doi: 10.1038/srep33528

Kayath, C. A., Hussey, S., Nagra, K., Philpott, D., and Allaoui, A. (2010). Escape of intracellular Shigella from autophagy requires binding to cholesterol through the type III effector, IcsB. *Microbes Infect.* 12, 956–966. doi: 10.1016/j.micinf.2010.06.006

Kim, J. S., Eom, J. S., Im Jang, J., Kim, H. G., Seo, D. W. Bang, I. S., et al (2011). Role of Salmonella pathogenicity island 1 protein IacP in *Salmonella enterica* serovar Typhimurium pathogenesis. *Infect. Immun.* 79, 1440–1450. doi: 10.1128/IAI.01231-10

Klein, J. A., Dave, B. M., Raphenya, A. R., McArthur, A. G., and Knodler, L. A. (2017). Functional relatedness in the Inv/Mxi-Spa type III secretion system family. *Mol. Microbiol.* 103, 973–991. doi: 10.1111/mmi.13602

Knight, M. J., Ruaux, A., Mikolajek, H., Erskine, P. T., Gill, R., Wood, S. P., et al. (2006). Crystallization and preliminary X-ray diffraction analysis of BipD, a virulence factor from *Burkholderia pseudomallei*. *Acta Crystallogr. Sect. F Struct. Biol. Cryst. Commun.* 62, 761–764. doi: 10.1107/S1744309106024857

Kresse, A. U., Beltrametti, F., Müller, A., Ebel, F., and Guzmán, C. A. (2000). Characterization of SepL of enterohemorrhagic *Escherichia coli*. *J. Bacteriol.* 182, 6490–6498. doi: 10.1128/JB.182.22.6490-6498.2000

Kubori, T., and Galán, J. E. (2002). Salmonella type III secretion-associated protein InvE controls translocation of effector proteins into host cells. *J. Bacteriol.* 184, 4699–4708. doi: 10.1128/JB.184.17.4699-4708.2002

Kubori, T., Matsushima, Y., Nakamura, D., Uralil, J., Lara-Tejero, M., Sukhan, A., et al. (1998). Supramolecular structure of the *Salmonella Typhimurium* type III protein secretion system. *Science* 280, 602–605. doi: 10.1126/science.280.5363.602

Kubori, T., Sukhan, A., Aizawa, S.-I., and Galán, J. E. (2000). Molecular characterization and assembly of the needle complex of the Salmonella Typhimurium type III protein secretion system. *Proc. Natl. Acad. Sci. U.S.A.* 97, 10225–10230. doi: 10.1073/pnas.170128997

Lafont, F., Tran Van Nhieu, G., Hanada, K., Sansonetti, P., and van der Goot, G. F. (2002). Initial steps of Shigella infection depend on the cholesterol/sphingolipid raft-mediated CD44-IpaB interaction. *EMBO J.* 21, 4449–4457. doi: 10.1093/emboj/cdf457

Lee, S. H., and Galán, J. E. (2003). InvB is a type III secretion-associated chaperone for the *Salmonella enterica* effector protein SopE. *J. Bacteriol.* 185, 7279–7284. doi: 10.1128/JB.185.24.7279-7284.2003

Lee, S. H., and Galán, J. E. (2004). Salmonella type III secretion-associated chaperones confer secretion-pathway specificity. *Mol. Microbiol.* 51, 483–495. doi: 10.1046/j.1365-2958.2003.03840.x

Lee, Y. H., Chen, Y., Ouyang, X., and Gan, Y.-H. (2010). Identification of tomato plant as a novel host model for *Burkholderia pseudomallei*. *BMC Microbiol.* 10:28. doi: 10.1186/1471-2180-10-28

Limmathurotsakul, D., Golding, N., Dance, D. A., Messina, J. P., Pigott, D. M., Moyes, C. L., et al. (2016). Predicted global distribution of *Burkholderia pseudomallei* and burden of melioidosis. *Nat. Microbiol.* 1:15008. doi: 10.1038/nmicrobiol.2015.8

Lipscomb, L., and Schell, M. A. (2011). Elucidation of the regulon and cis-acting regulatory element of HrpB, the AraC-type regulator of a plant pathogen-like type III secretion system in *Burkholderia pseudomallei*. *J. Bacteriol.* 193, 1991–2001. doi: 10.1128/jb.01379-10

Madan, R., Rastogi, R., Parashuraman, S., and Mukhopadhyay, A. (2012). *Salmonella acquires* lysosome-associated membrane protein 1 (LAMP1) on phagosomes from Golgi via SipC protein-mediated recruitment of host Syntaxin6. *J. Biol. Chem.* 287, 5574–5587. doi: 10.1074/jbc.M111.286120

Markham, A. P., Birket, S. E., Picking, W. D., Picking, W. L., and Middaugh, C. R. (2008). pH sensitivity of type III secretion system tip proteins. *Proteins* 71, 1830–1842. doi: 10.1002/prot.21864

Marlovits, T. C., Kubori, T., Sukhan, A., Thomas, D. R., Galán, J. E., and Unger, V. M. (2004). Structural insights into the assembly of the type III secretion needle complex. *Science* 306, 1040–1042. doi: 10.1126/science.1102610

Martinez-Becerra, F. J., Chen, X., Dickenson, N. E., Choudhari, S. P., Harrison, K., Clements, J. D., et al. (2013). Characterization of a novel fusion protein from IpaB and IpaD of Shigella sand its potential as a pan-Shigella vaccine. *Infect. Immun.* 81, 4470–4477. doi: 10.1128/IAI.00859-13

Martinez-Becerra, F. J., Kissmann, J. M., Diaz-McNair, J., Choudhari, S. P., Quick, A. M., Mellado-Sanchez, G., et al. (2012). Broadly protective Shigella vaccine based on type III secretion apparatus proteins. *Infect. Immun.* 80, 1222–1231. doi: 10.1128/IAI.06174-11

Maurelli, A. T., Baudry, B. H., d'Hauteville, H., Hale, T. L., and Sansonetti, P. J. (1985). Cloning of plasmid DNA sequences involved in invasion of HeLa cells by *Shigella flexneri*. *Infect. Immun.* 49, 164–171.

McGhie, E. J., Hayward, R. D., and Koronakis, V. (2001). Cooperation between actin-binding proteins of invasive Salmonella: SipA potentiates SipC nucleation and bundling of actin. *EMBO J.* 20, 2131–2139. doi: 10.1093/emboj/20.9.2131

Miao, E. A., Mao, D. P., Yudkovsky, N., Bonneau, R., Lorang, C. G., Warren, S. E., et al. (2010). Innate immune detection of the type III secretion apparatus through the NLRC4 inflammasome. *Proc. Natl. Acad. Sci. U.S.A.* 107, 3076–3080. doi: 10.1073/pnas.0913087107

Michiels, T., and Cornelis, G. R. (1991). Secretion of hybrid proteins by the Yersinia Yop export system. *J. Bacteriol.* 173, 1677–1685. doi: 10.1128/jb.173.5.1677-1685.1991

Moore, R. A., Reckseidler-Zenteno, S., Kim, H., Nierman, W., Yu, Y., Tuanyok, A., et al. (2004). Contribution of gene loss to the pathogenic evolution of *Burkholderia pseudomallei* and *Burkholderia mallei*. *Infect. Immun.* 72, 4172–4187. doi: 10.1128/IAI.72.7.4172-4187.2004

Muangman, S., Korbsrisate, S., Muangsombut, V., Srinon, V., Adler, N. L., Schroeder, G. N., et al. (2011). BopC is a type III secreted effector protein of *Burkholderia pseudomallei*. *FEMS Microbiol. Lett.* 323, 75–82. doi: 10.1111/j.1574-6968.2011.02359.x

Muangsombut, V., Suparak, S., Pumirat, P., Damnin, S., Vattanaviboon, P., Thongboonkerd, V., et al. (2008). Inactivation of *Burkholderia pseudomallei* bsaQ results in decreased invasion efficiency and delayed escape of bacteria from endocytic vesicles. *Arch. Microbiol.* 190, 623–631. doi: 10.1007/s00203-008-0413-3

Myeni, S. K., Wang, L., and Zhou, D. (2013). SipB-SipC complex is essential for translocon formation. *PLoS ONE* 8:e60499. doi: 10.1371/journal.pone.0060499

Myeni, S. K., and Zhou, D. (2010). The C terminus of SipC binds and bundles F-actin to promote Salmonella invasion. *J. Biol. Chem.* 285, 13357–13363. doi: 10.1074/jbc.M109.094045

Ng, M. Y., Wang, M., Casey, P. J., Gan, Y.-H., and Hagen, T. (2017). Activation of MAPK/ERK signaling by *Burkholderia pseudomallei* cycle inhibiting factor (Cif). *PLoS ONE* 12:e0171464. doi: 10.1371/journal.pone.0171464

Ngauy, V., Lemeshev, Y., Sadkowski, L., and Crawford, G. (2005). Cutaneous melioidosis in a man who was taken as a prisoner of war by the Japanese during World War II. *J. Clin. Microbiol.* 43, 970–972. doi: 10.1128/JCM.43.2.970-972.2005

Nougayrède, J.-P., Boury, M., Tasca, C., Marchès, O., Milon, A., Oswald, E., et al. (2001). Type III secretion-dependent cell cycle block caused in HeLa cells by enteropathogenic *Escherichia coli* O103. *Infect. Immun.* 69, 6785–6795. doi: 10.1128/IAI.69.11.6785-6795.2001

Olive, A. J., Kenjale, R., Espina, M., Moore, D. S., Picking, W. L., and Picking, W. D. (2007). Bile salts stimulate recruitment of IpaB to the *Shigella flexneri* surface, where it colocalizes with IpaD at the tip of the type III secretion needle. *Infect. Immun.* 75, 2626–2629. doi: 10.1128/IAI.01599-06

Osiecki, J. C., Barker, J., Picking, W. L., Serfis, A. B., Berring, E., Shah, S., et al. (2001). IpaC from Shigella and SipC from Salmonella possess similar biochemical properties but are functionally distinct. *Mol. Microbiol.* 42, 469–481. doi: 10.1046/j.1365-2958.2001.02654.x

Pal, M., Erskine, P. T., Gill, R. S., Wood, S. P., and Cooper, J. B. (2010). Near-atomic resolution analysis of BipD, a component of the type III secretion system of *Burkholderia pseudomallei*. *Acta Crystallogr. Sect. F Struct. Biol. Cryst. Commun.* 66, 990–993. doi: 10.1107/S1744309110026333

Pallett, M. A., Berger, C. N., Pearson, J. S., Hartland, E. L., and Frankel, G. (2014). The type III secretion effector NleF of enteropathogenic *Escherichia coli* activates NF-κB early during infection. *Infect. Immun.* 82, 4878–4888. doi: 10.1128/IAI.02131-14

Panina, E. M., Mattoo, S., Griffith, N., Kozak, N. A., Yuk, M. H., and Miller, J. F. (2005). A genome-wide screen identifies a Bordetella type III secretion effector and candidate effectors in other species. *Mol. Microbiol.* 58, 267–279. doi: 10.1111/j.1365-2958.2005.04823.x

Parsot, C., Ménard, R., Gounon, P., and Sansonetti, P. J. (1995). Enhanced secretion through the *Shigella flexneri* Mxi-Spa translocon leads to assembly of extracellular proteins into macromolecular structures. *Mol. Microbiol.* 16, 291–300. doi: 10.1111/j.1365-2958.1995.tb02301.x

Paul, K., Erhardt, M., Hirano, T., Blair, D. F., and Hughes, K. T. (2008). Energy source of flagellar type III secretion. *Nature* 451, 489–492. doi: 10.1038/nature06497

Pearson, T., Giffard, P., Beckstrom-Sternberg, S., Auerbach, R., Hornstra, H., Tuanyok, A., et al. (2009). Phylogeographic reconstruction of a bacterial species with high levels of lateral gene transfer. *BMC Biol.* 7:78. doi: 10.1186/1741-7007-7-78

Pei, J., and Grishin, N. V. (2009). The Rho GTPase inactivation domain in *Vibrio cholerae* MARTX toxin has a circularly permuted papain-like thiol protease fold. *Proteins* 77, 413–419. doi: 10.1002/prot.22447

Persson, C., Carballeira, N., Wolf-Watz, H., and Fällman, M. (1997). The PTPase YopH inhibits uptake of Yersinia, tyrosine phosphorylation of p130Cas and FAK, and the associated accumulation of these proteins in peripheral focal adhesions. *EMBO J.* 16, 2307–2318. doi: 10.1093/emboj/16.9.2307

Picking, W. L., Nishioka, H., Hearn, P. D., Baxter, M. A., Harrington, A. T., Blocker, A., et al. (2005). IpaD of *Shigella flexneri* is independently required for regulation of Ipa protein secretion and efficient insertion of IpaB and IpaC into host membranes. *Infect. Immun.* 73, 1432–1440. doi: 10.1128/IAI.73.3.1432-1440.2005

Pilatz, S., Breitbach, K., Hein, N., Fehlhaber, B., Schulze, J., Brenneke, B., et al. (2006). Identification of *Burkholderia pseudomallei* genes required for the intracellular life cycle and *in vivo* virulence. *Infect. Immun.* 74, 3576–3586. doi: 10.1128/IAI.01262-05

Pruksachartvuthi, S., Aswapokee, N., and Thankerngpol, K. (1990). Survival of *Pseudomonas pseudomallei* in human phagocytes. *J. Med. Microbiol.* 31, 109–114. doi: 10.1099/00222615-31-2-109

Pumirat, P., Vander Broek, C., Juntawieng, N., Muangsombut, V., Kiratisin, P., Pattanapanyasat, K., et al. (2014). Analysis of the prevalence, secretion and function of a cell cycle-inhibiting factor in the melioidosis pathogen *Burkholderia pseudomallei*. *PLoS ONE* 9:e96298. doi: 10.1371/journal.pone.0096298

Radics, J., Königsmaier, L., and Marlovits, T. C. (2014). Structure of a pathogenic type 3 secretion system in action. *Nat. Struct. Mol. Biol.* 21, 82–87. doi: 10.1038/nsmb.2722

Rainbow, L., Hart, C. A., and Winstanley, C. (2002). Distribution of type III secretion gene clusters in *Burkholderia pseudomallei*. *B. thailandensis* and *B. mallei*. *J. Med. Microbiol.* 51, 374–384. doi: 10.1099/0022-1317-51-5-374

Roehrich, A. D., Bordignon, E., Mode, S., Shen, D.-K., Liu, X., Blocker, A. J. et al. (2016). Steps for Shigella Gatekeeper MxiC Function in Hierarchical Type III Secretion Regulation. *J. Biol. Chem.* 292, 1705–1723. doi: 10.1074/jbc.M116.746826

Roehrich, A. D., Guillossou, E., Blocker, A. J., and Martinez-Argudo, I. (2013). Shigella IpaD has a dual role: signal transduction from the type III secretion system needle tip and intracellular secretion regulation. *Mol. Microbiol.* 87, 690–706. doi: 10.1111/mmi.12124

Rossier, O., Wengelnik, K., Hahn, K., and Bonas, U. (1999). The Xanthomonas Hrp type III system secretes proteins from plant and mammalian bacterial pathogens. *Proc. Natl. Acad. Sci. U.S.A.* 96, 9368–9373. doi: 10.1073/pnas.96.16.9368

Rotz, L. D., Khan, A. S., Lillibridge, S. R., Ostroff, S. M., and Hughes, J. M. (2002). Public health assessment of potential biological terrorism agents. *Emerg. Infect. Dis.* 8, 225–230. doi: 10.3201/eid0802.010164

Roversi, P., Johnson, S., Field, T., Deane, J. E., Galyov, E. E., and Lea, S. M. (2006). Expression, purification, crystallization and preliminary crystallographic analysis of BipD, a component of the *Burkholderia pseudomallei* type III secretion system. *Acta Crystallogr. Sect. F Struct. Biol. Cryst. Commun.* 62, 861–864. doi: 10.1107/S1744309106027035

Sato, H., Frank, D. W., Hillard, C. J., Feix, J. B., Pankhaniya, R. R., Moriyama, K., et al. (2003). The mechanism of action of the *Pseudomonas aeruginosa*-encoded type III cytotoxin, ExoU. *EMBO J.* 22, 2959–2969. doi: 10.1093/emboj/cdg290

Schesser, K., Frithz-Lindsten, E., and Wolf-Watz, H. (1996). Delineation and mutational analysis of the *Yersinia pseudotuberculosis* YopE domains which mediate translocation across bacterial and eukaryotic cellular membranes. *J. Bacteriol.* 178, 7227–7233. doi: 10.1128/jb.178.24.7227-7233.1996

Schraidt, O., and Marlovits, T. C. (2011). Three-dimensional model of Salmonella's needle complex at subnanometer resolution. *Science* 331, 1192–1195. doi: 10.1126/science.1199358

Sim, S. H., Yu, Y., Lin, C. H., Karuturi, R. K., Wuthiekanun, V., Tuanyok, A., et al. (2008). The core and accessory genomes of *Burkholderia pseudomallei*: implications for human melioidosis. *PLoS Pathog.* 4:e1000178. doi: 10.1371/journal.ppat.1000178

Smith, M. D., Angus, B. J., Wuthiekanun, V., and White, N. J. (1997). Arabinose assimilation defines a nonvirulent biotype of *Burkholderia pseudomallei*. *Infect. Immun.* 65, 4319–4321.

Sory, M.-P., Boland, A., Lambermont, I., and Cornelis, G. R. (1995). Identification of the YopE and YopH domains required for secretion and internalization into the cytosol of macrophages, using the cyaA gene fusion approach. *Proc. Natl. Acad. Sci. U.S.A.* 92, 11998–12002.

Srinivasan, A., Kraus, C. N., DeShazer, D., Becker, P. M., Dick, J. D., Spacek, L., et al. (2001). Glanders in a military research microbiologist. *N. Engl. J. Med.* 345, 256–258. doi: 10.1056/NEJM200107263450404

Srinon, V., Muangman, S., Imyaem, N., Muangsombut, V. L. A., Adler, N. R., Galyov, E. E., et al. (2013). Comparative assessment of the intracellular survival of the *Burkholderia pseudomallei* bopC mutant. *J. Microbiol.* 51, 522–526. doi: 10.1007/s12275-013-2557-3

Stebbins, C. E., and Galán, J. E. (2001). Maintenance of an unfolded polypeptide by a cognate chaperone in bacterial type III secretion. *Nature* 414, 77–81. doi: 10.1038/35102073

Stensrud, K. F., Adam, P. R., La Mar, C. D., Olive, A. J., Lushington, G. H., Sudharsan, R., et al. (2008). Deoxycholate interacts with IpaD of Shigella flexneri in inducing the recruitment of IpaB to the type III secretion apparatus needle tip. *J. Biol. Chem.* 283, 18646–18654. doi: 10.1074/jbc.M802799200

Stevens, M. P., Friebel, A., Taylor, L. A., Wood, M. W., Brown, P. J., Galyov, E. E. et al. (2003). A Burkholderia pseudomallei type III secreted protein. BopE, facilitates bacterial invasion of epithelial cells and exhibits guanine nucleotide exchange factor activity. *J. Bacteriol.* 185, 4992–4996. doi: 10.1128/JB.185.16.4992-4996.2003

Stevens, M. P., Haque, A., Atkins, T., Hill, J., Wood, M. W., Easton, A., et al. (2004). Attenuated virulence and protective efficacy of a *Burkholderia pseudomallei* bsa type III secretion mutant in murine models of melioidosis. *Microbiology* 150, 2669–2676. doi: 10.1099/mic.0.27146-0

Stevens, M. P., Wood, M. W., Taylor, L. A., Monaghan, P., Hawes, P., Jones, P. W., et al. (2002). An Inv/Mxi-Spa-like type III protein secretion system in *Burkholderia pseudomallei* modulates intracellular behaviour of the pathogen. *Mol. Microbiol.* 46, 649–659. doi: 10.1046/j.1365-2958.2002.03190.x

Sun, G. W., Chen, Y., Liu, Y., Tan, G.-Y., Ong, C., Gan, H., et al. (2010). Identification of a regulatory cascade controlling Type III Secretion System 3 gene expression in *Burkholderia pseudomallei*. *Mol. Microbiol.* 76, 677–689. doi: 10.1111/j.1365-2958.2010.07124.x

Sun, G. W., Lu, J., Pervaiz, S., Cao, W. P., and Gan, Y.-H. (2005). Caspase-1 dependent macrophage death induced by *Burkholderia pseudomallei*. *Cell. Microbiol.* 7, 1447–1458. doi: 10.1111/j.1462-5822.2005.00569.x

Suparak, S., Kespichayawattana, W., Haque, A., Easton, A., Damnin, S., Lertmemongkolchai, G., et al. (2005). Multinucleated giant cell formation and apoptosis in infected host cells is mediated by *Burkholderia pseudomallei* type III secretion protein BipB. *J. Bacteriol.* 187, 6556–6560. doi: 10.1128/JB.187.18.6556-6560.2005

Teh, B. E., French, C. T., Chen, Y., Chen, I. G. J., Wu, T.-H., Gan, H., et al. (2014). Type three secretion system-mediated escape of *Burkholderia pseudomallei* into the host cytosol is critical for the activation of NFκB. *BMC Microbiol.* 14:115. doi: 10.1186/1471-2180-14-115

Tippayawat, P., Pinsiri, M., Rinchai, D., Riyapa, D., Romphruk, A., Gan, Y. H., et al. (2011). *Burkholderia pseudomallei* proteins presented by monocyte-derived dendritic cells stimulate human memory T cells *in vitro*. *Infect. Immun.* 79, 305–313. doi: 10.1128/IAI.00803-10

Tippayawat, P., Saenwongsa, W., Mahawantung, J., Suwannasaen, D., Chetchotisakd, P., Limmathurotsakul, D., et al. (2009). Phenotypic and functional characterization of human memory T cell responses to *Burkholderia pseudomallei*. *PLoS Negl. Trop. Dis.* 3:e407. doi: 10.1371/journal.pntd.0000407

Treerat, P., Alwis, P., D'Cruze, T., Cullinane, M., Vadivelu, J., Devenish, R. J., et al. (2015). The *Burkholderia pseudomallei* Proteins BapA and BapC Are Secreted TTSS3 Effectors and BapB levels modulate expression of Bop. *PLoS ONE* 10:e0143916. doi: 10.1371/journal.pone.0143916

Tuanyok, A., Leadem, B. R., Auerbach, R. K., Beckstrom-Sternberg, S. M., Beckstrom-Sternberg, J. S., Mayo, M., et al. (2008). Genomic islands from five strains of *Burkholderia pseudomallei*. *BMC Genomics* 9:566. doi: 10.1186/1471-2164-9-566

Tucker, S. C., and Galán, J. E. (2000). omplex function for SicA, a *Salmonella enterica* serovar typhimurium type III secretion-associated chaperone. *J. Bacteriol.* 182, 2262–2268. doi: 10.1128/JB.182.8.2262-2268.2000

Upadhyay, A., Wu, H., Williams, C., Field, T., Galyov, E. J., et al. (2008). The guanine-nucleotide-exchange factor BopE from *Burkholderia pseudomallei* adopts a compact version of the Salmonella SopE/SopE2 fold and undergoes a closed-to-open conformational change upon interaction with Cdc42. *Biochem. J.* 411, 485–493. doi: 10.1042/BJ20071546

Vander Broek, C. W., Chalmers, K. J., Stevens, M. P., and Stevens, J. M. (2015). Quantitative Proteomic analysis of *Burkholderia pseudomallei* Bsa Type III secretion system effectors using hypersecreting mutants. *Mol. Cell. Proteomics* 14, 905–916. doi: 10.1074/mcp.M114.044875

Van Zandt, K. E., Greer, M. T., and Gelhaus, H. C. (2013). Glanders: an overview of infection in humans. *Orphanet J. Rare Dis.* 8, 131. doi: 10.1186/1750-1172-8-131

Wagner, S., Königsmaier, L., Lara-Tejero, M., Lefebre, M., Marlovits, T. C., and Galán, J. E. (2010). Organization and coordinated assembly of the type III secretion export apparatus. *Proc. Natl. Acad. Sci. U.S.A.* 107, 17745–17750. doi: 10.1073/pnas.1008053107

Wang, Y., Ouellette, A. N., Egan, C. W., Rathinavelan, T., Im, W., and Da Guzman, R. N. (2007). Differences in the electrostatic surfaces of the type III secretion needle proteins PrgI, BsaL, and MxiH. *J. Mol. Biol.* 371, 1304–1314. doi: 10.1016/j.jmb.2007.06.034

Warawa, J., and Woods, D. E. (2005). Type III secretion system cluster 3 is required for maximal virulence of *Burkholderia pseudomallei* in a hamster infection model. *FEMS Microbiol. Lett.* 242, 101–108. doi: 10.1016/j.femsle.2004.10.045

Warren, S. M., and Young, G. M. (2005). An amino-terminal secretion signal is required for YplA export by the Ysa, Ysc, and flagellar type III secretion systems of *Yersinia enterocolitica* biovar 1B. *J. Bacteriol.* 187, 6075–6083. doi: 10.1128/JB.187.17.6075-6083.2005

West, T. E., Myers, N. D., Chantratita, N., Chierakul, W., Limmathurotsakul, D., Wuthiekanun, V., et al. (2014). NLRC4 and TLR5 each contribute to host defense in respiratory melioidosis. *PLoS Negl. Trop. Dis.* 8:e3178. doi: 10.1371/journal.pntd.0003178

Whitlock, G. C., Deeraksa, A., Qazi, O., Judy, B. M., Taylor, K., Propst, K. L., et al. (2010). Protective response to subunit vaccination against intranasal *Burkholderia mallei* and *B. pseudomallei* challenge. *Procedia Vaccinol.* 2, 73–77. doi: 10.1016/j.provac.2010.03.013

Whitlock,. G. C., Estes, D. M., Young, G. M., Young, B., and Torres, A. G. (2008). Construction of a reporter system to study *Burkholderia mallei* type III secretion and identification of the BopA effector protein function in intracellular survival. *Trans. R. Soc. Trop. Med. Hyg.* 102, S127–S133. doi: 10.1016/S0035-9203(08)70029-4

Whitlock, G. C., Valbuena, G. A., Popov, V. L., Judy, B. M., Estes, D. M., and Torres, A. G. (2009). *Burkholderia mallei* cellular interactions in a respiratory cell model. *J. Med. Microbiol.* 58, 554–562. doi: 10.1099/jmm.0.007724-0

Whitmore, A. (1913). An account of a glanders-like disease occurring in Rangoon. *J. Hyg.* 13, 1–34. doi: 10.1017/S0022172400005234

Wilharm, G., Lehmann, V., Krauss, K., Lehnert, B., Richter, S., Ruckdeschel, K., et al. (2004). *Yersinia enterocolitica* type III secretion depends on the proton motive force but not on the flagellar motor components MotA and MotB. *Infect. Immun.* 72, 4004–4009. doi: 10.1128/IAI.72.7.4004-4009.2004

Williamson, E. D. (2009). Plague. *Vaccine* 27, D56–D60. doi: 10.1016/j.vaccine.2009.07.068

Winstanley, C., Hales, B. A., and Hart, C. A. (1999). Evidence for the presence in *Burkholderia pseudomallei* of a type III secretion system-associated gene cluster. *J. Med. Microbiol.* 48, 649–656. doi: 10.1099/00222615-48-7-649

Yabuuchi, E., Kosako, Y., Oyaizu, H., Yano, I., Hotta, H., Hashimoto, Y., et al. (1992). Proposal of *Burkholderia gen.* nov. and transfer of seven species of the genus *Pseudomonas homology* group II to the new genus, with the type species *Burkholderia cepacia* (Palleroni and Holmes 1981) comb. nov. *Microbiol. Immunol.* 36, 1251–1275. doi: 10.1111/j.1348-0421.1992.tb02129.x

Yang, J., Zhao, Y., Shi, J., and Shao, F. (2013). Human NAIP and mouse NAIP1 recognize bacterial type III secretion needle protein for inflammasome activation. *Proc. Natl. Acad. Sci. U.S.A.* 110, 14408–14413. doi: 10.1073/pnas.1306376110

Young, B. M., and Young, G. M. (2002). YplA is exported by the Ysc, Ysa, and flagellar type III secretion systems of *Yersinia enterocolitica*. *J. Bacteriol.* 184, 1324–1334. doi: 10.1128/JB.184.5.1324-1334.2002

Young, G. M., Schmiel, D. H., and Miller, V. L. (1999). A new pathway for the secretion of virulence factors by bacteria: the flagellar export apparatus functions as a protein-secretion system. *Proc. Natl. Acad. Sci. U.S.A.* 96, 6456–6461. doi: 10.1073/pnas.96.11.6456

Yu, Y., Kim, H. S., Chua, H. H., Lin, C. H., Sim, S. H., Lin, D., et al. (2006). Genomic patterns of pathogen evolution revealed by comparison of *Burkholderia pseudomallei*, the causative agent of melioidosis, to avirulent *Burkholderia thailandensis*. *BMC Microbiol.* 6:46. doi: 10.1186/1471-2180-6-46

Zahrl, D., Wagner, M., Bischof, K., Bayer, M., Zavecz, B., Beranek, A., et al. (2005). Peptidoglycan degradation by specialized lytic transglycosylases associated with type III and type IV secretion systems. *Microbiology* 151, 3455–3467. doi: 10.1099/mic.0.28141-0

Zarivach, R., Deng, W., Vuckovic, M., Felise, H. B., Nguyen, H. V., Miller, S. I., et al. (2008). Structural analysis of the essential self-cleaving type III secretion proteins EscU and SpaS. *Nature* 453, 124–127. doi: 10.1038/nature06832

Zhang, L., Wang, Y., Picking, W. L., Picking, W. D., and Da Guzman, R. N. (2006). Solution structure of monomeric BsaL, the type III secretion needle protein of *Burkholderia pseudomallei*. *J. Mol. Biol.* 359, 322–330. doi: 10.1016/j.jmb.2006.03.028

Zhao, Y., Yang, J., Shi, J., Gong, Y.-N., Lu, Q., Xu, H., et al. (2011). The NLRC4 inflammasome receptors for bacterial flagellin and type III secretion apparatus. *Nature* 477, 596–600. doi: 10.1038/nature10510

Zhou, Y., Dong, N., Hu, L., and Shao, F. (2013). The Shigella type three secretion system effector OspG directly and specifically binds to host ubiquitin for activation. *PLoS ONE* 8:e57558. doi: 10.1371/journal.pone.0057558

14

The Dynamic ATP-Driven Mechanism of Bacterial Protein Translocation and the Critical Role of Phosphologipids

*Ian Collinson**

School of Biochemistry, University of Bristol, Bristol, United Kingdom

***Correspondence:**
Ian Collinson
ian.collinson@bristol.ac.uk

Protein secretion from the cell cytoplasm to the outside is essential for life. Bacteria do so for a range of membrane associated and extracellular activities, including envelope biogenesis, surface adherence, pathogenicity, and degradation of noxious chemicals such as antibiotics. The major route for this process is *via* the ubiquitous Sec system, residing in the plasma membrane. Translocation across (secretion) or into (insertion) the membrane is driven through the translocon by the action of associated energy-transducing factors or translating ribosomes. This review seeks to summarize the recent advances in the dynamic mechanisms of protein transport and the critical role played by lipids in this process. The article will include an exploration of how lipids are actively involved in protein translocation and the consequences of these interactions for energy transduction from ATP hydrolysis and the trans-membrane proton-motive-force (PMF).

Keywords: SecA ATPase, bacteria, SecY translocon, lipids, protein secretion

PROTEIN SECRETION THROUGH THE BACTERIAL SEC MACHINERY

Protein secretion, from the cell cytoplasm to the outside, is essential for life. Bacteria secrete proteins for enveloping biogenesis, surface adherence, and pathogenicity and degrading noxious chemicals (including antibiotics), among a range of many other membrane and extracellular activities. The major route for protein secretion is *via* the ubiquitous Sec translocon: a conserved hetero-trimeric core-complex of the inner membrane. This machinery is also responsible for membrane protein insertion, whereby proteins containing Trans-Membrane α-Helices (TMH) are threaded laterally into the bilayer rather than across it. Therefore, the interaction of the machinery with lipids is critical in that they interface with the protein complex through which proteins cross and enter the membrane. Moreover, it turns out that phospholipids play direct and critical roles in the active energy transducing process driving protein transport.

Secretory and membrane proteins are targeted to the Sec machinery with the aid of an N-terminal signal sequence (Chang et al., 1978; Walter et al., 1981b). These proteins then translocate through the apparatus in an unfolded conformation (Arkowitz et al., 1993), either during their synthesis (co-translationally), or afterwards (post-translationally). The former is a ubiquitous process in which the signal sequence at the N-terminus of the nascent polypeptide emerging from the ribosome exit tunnel is recognized by the signal recognition particle (SRP) and targeted to the membrane-associated SRP-receptor (Walter et al., 1981a,b,c; Poritz et al., 1990). The ribosome nascent chain complex (RNC) is then shuttled to the Sec complex for translocation

(Jomaa et al., 2016), where the growing polypeptide is forced through the membrane. In the bacterial model *Escherichia coli* (**Figure 1**), the co-translational pathway is generally utilized for membrane protein insertion (Ulbrandt et al., 1997; Müller et al., 2001), while transport across the membrane (secretion) tends to be post-translational. In the latter process, cytosolic chaperones, such as SecB, deliver the pre-protein in a translocation competent (unfolded) state to the SecA motor protein and channel complex SecYEG for secretion (Hartl et al., 1990; Cranford-Smith and Huber, 2018).

STRUCTURE OF THE CORE-TRANSLOCON

The pathway for translocation of the mature region of the pre-protein – the protein-conducting channel – is formed through the center of SecY, between two pseudo-symmetrical halves, each with five TMHs (Van den Berg et al., 2004; Cannon et al., 2005). When at rest, the channel is kept sealed by a short α-helix (the plug) and a sphincter of six hydrophobic residues, usually isoleucine (Van den Berg et al., 2004). Separation of the two halves opens a channel across the membrane (for secretion) as well as a lateral gate (LG) for entry of TMHs into the bilayer (insertion). Multiple rounds of ATP hydrolysis and the trans-membrane proton-motive-force (PMF) then drive protein transport across the membrane (Brundage et al., 1990; Schiebel et al., 1991; Economou and Wickner, 1994).

MECHANISM OF SECA-DRIVEN PROTEIN TRANSLOCATION

Today, we understand a great deal about the structure and activity of Sec machinery, particularly the bacterial counterpart (almost exclusively through the study of the *E. coli* system). During SecA-driven secretion, the association of SecA with the pre-protein causes the signal sequence to bind at the LG of SecY, at the interface with the lipid bilayer (Hizlan et al., 2012; Li et al., 2016). Many studies have shown that this interaction causes a conformational change in both SecA and the protein channel (Hizlan et al., 2012; Corey et al., 2016; Li et al., 2016), and a priming of SecA for increased ATPase activity (Gouridis et al., 2009; Robson et al., 2009; Gold et al., 2013). In this "unlocked" state, the channel opens slightly and the plug, which helps keep the Sec-complex closed, retracts from its central position (Zimmer et al., 2008; Hizlan et al., 2012; Corey et al., 2016; Li et al., 2016).

Our understanding of this reaction has been aided by recent single-molecule fluorescence studies (Allen et al., 2016;

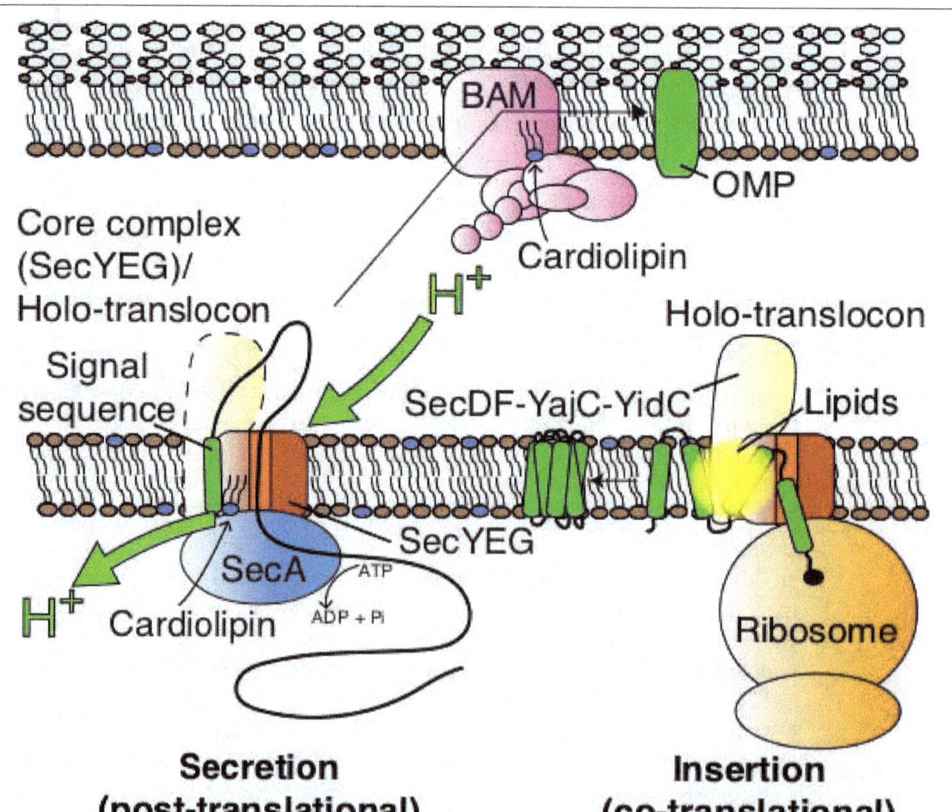

FIGURE 1 | Pathways for protein transport across and into the Gram-negative inner membrane and further across the periplasm for outer-membrane insertion and folding. The figure also highlights the action of lipids on the SecYEG-SecA complex (see **Figure 3**), the holo-translocon (HTL), and the BAM complex.

Fessl et al., 2018). These applications enable the dissection and analysis of different stages of the reaction, which would otherwise be blurred in the reaction ensemble. The analysis demonstrates that the initiation process requires the signal sequence and mature regions of the pre-protein, as well as ATP (Fessl et al., 2018). The initiation involves an ATP-driven transport step, independent of pre-protein length, likely to be the intercalation of the signal sequence and early mature regions of the polypeptide (presumably as a loop with the N-terminus pointing towards the cytosol; **Figures 1, 2**) into the "unlocked" translocon, as was previously proposed (Hizlan et al., 2012). At this point, the ATPase becomes fully activated and translocation across the membrane can begin through SecYEG; the kinetics of which depend on the length of the substrate (see Fessl et al., 2018 and Figure 6 therein; Tomkiewicz et al., 2006).

MODEL FOR PROTEIN TRANSLOCATION

For the transport process *per* se, a stochastic Brownian ratchet model for the ATP-driven reaction has been proposed wherein the free energy available from ATP binding and hydrolysis helps bias the random diffusional flow of polypeptide to favor an outward direction (see Figure 8 and associated movie in Allen et al., 2016).

A follow-up study proposed how this stochastic process could be further enhanced. ATP-dependent control of protein folding has been well documented in the protein chaperone field (Clarke, 1996). We have introduced this concept into the protein transport field whereby the translocon utilizes the hydrolytic cycle of ATP to exert an asymmetric control of pre-protein folding (Corey et al., 2019). Preventing partial folding at the cytosolic interface of the ATPase SecA with SecY, while enabling it at the outward facing exit site, would prevent back sliding of the translocating polypeptide (**Figure 2**), thus favoring outward flow of the pre-protein. This concept is consistent with independent findings that the folding propensity (or lack of it) of the mature regions of the pre-protein has profound effects on the secretion efficiency (Gonsberg et al., 2017; Jung and Tatzelt, 2018; Tsirigotaki et al., 2018).

An alternative "push-and-slide" model invokes both diffusional and ATP-driven power-stroke components, involving the 2-helix finger (2HF) motif of SecA moving up and down to physically push polypeptides across the membrane (Erlandson et al., 2008; Zimmer et al., 2008; Bauer et al., 2014). A recent follow-up study, based on single-molecule fluorescence, confirms that the 2HF is indeed conformationally mobile throughout the ATP-driven transport cycle (Catipovic et al., 2019). The differences in fluorescence were equated to a very large apparent change in energy transfer efficiency (~0.1 to ~0.9). The main contention here is if the observations are really due to Förster **R**esonance **E**nergy **T**ransfer (FRET), or to **P**rotein-**I**nduced **F**luorescent **E**nhancement (PIFE; Stennett et al., 2015). In the former case, this would require an extraordinarily large movement of >20 Å. Otherwise, the affect may be due to changes of the dye environment, for example, the formation of steric constraints at alternative conformations. This phenomenon is a known feature of some fluorescent reporters, particularly Cy3 (Stennett et al., 2015), used in this new study (Catipovic et al., 2019). Consequently, more subtle movements could also be responsible for these large fluorescent fluctuations, which would be more concordant with the limited space available for the 2HF to move (Zimmer et al., 2008; Li et al., 2016).

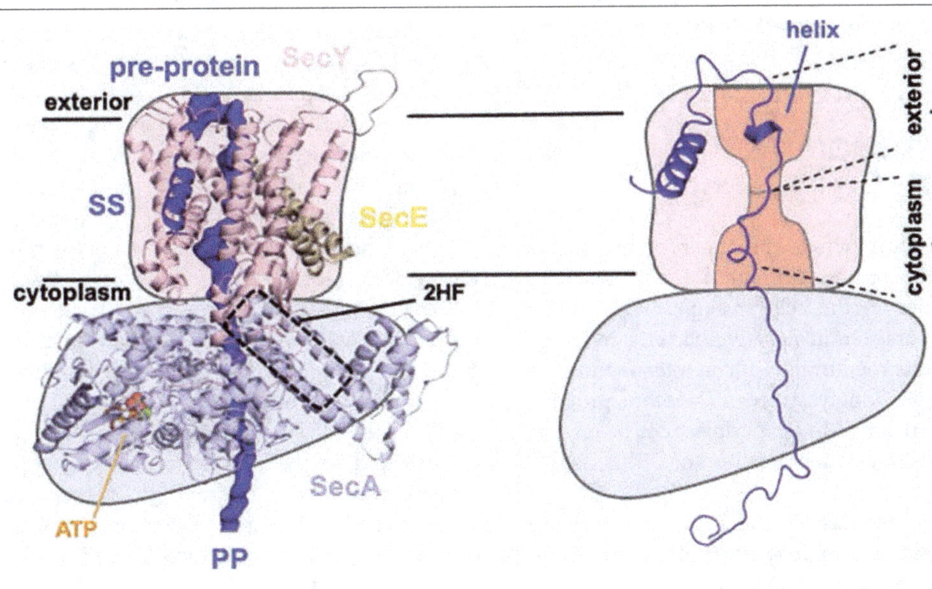

FIGURE 2 | Sec control of protein folding as a driver for transport. Model structure of SecYEG-SecA^{ATP} based on Li et al. (2016) with key features labeled accordingly; SS, signal sequence; 2HF, 2 helix finger (left). Schematic of same (right), showing only the simulated structure of the pre-protein, with a partial folding in the exterior cavity (helix) and unfolded regions in the cytoplasmic cavity and hence favoring transport to the exterior, as required. Adapted from figures shown previously, used with permission (Corey et al., 2019).

Either way, whether a diffusional ratchet (Allen et al., 2016) or power stroke/diffusional hybrid (Bauer et al., 2014; Catipovic et al., 2019) is at play, the core ATP-driven process is further stimulated by the PMF (Brundage et al., 1990) in order to achieve the high rates of secretion required for rapid growth. Presently, it is not known if or how the electrical ($\Delta\psi$) or chemical (ΔpH) components of the PMF achieve this enhancement. The PMF may indeed also operate to favor the outward flow of polypeptide in a Brownian ratchet-type mechanism.

During the final stages of transport, the signal sequence of the pre-protein is proteolytically cleaved to release the mature protein on the other side of the membrane (Josefsson and Randall, 1981). The terminal closure of the translocon is apparently independent of ATP (Fessl et al., 2018).

THE HOLO-TRANSLOCON (HTL)

To complicate matters further, the SecYEG core complex also associates with a number of accessory proteins: the membrane protein "insertase" YidC and the sub-complex SecDF (Duong and Wickner, 1997; Schulze et al., 2014). In a large number of cases, including *E. coli*, this complex also contains the YajC protein of obscure function (Duong and Wickner, 1997; Schulze et al., 2014). YidC facilitates the lateral insertion of TMHs from SecY into the bilayer (Houben et al., 2000; Samuelson et al., 2000; Scotti et al., 2000; Kumazaki et al., 2014), while SecDF makes an additional use of the PMF to help drive the transport of secretory proteins (Arkowitz and Wickner, 1994; Duong and Wickner, 1997; Tsukazaki et al., 2011; Botte et al., 2016; Furukawa et al., 2017). Thus, the resultant super-complex – the **H**olo-**T**rans**L**ocon (HTL) – associates with co-translating ribosomes for efficient membrane protein insertion and SecA for ATP/PMF-driven secretion (**Figure 1**; Schulze et al., 2014; Komar et al., 2016).

TRANSPORT THROUGH THE CELL ENVELOPE

For many proteins, transport across the inner membrane is only the first step. Following passage through the Sec-translocon proteins are either retained in the cell envelope or find their way to the external medium. Gram-negative bacteria have the added complexity of an outer membrane with an inter-membrane periplasm containing the peptidoglycan layer. Therefore, proteins must either be folded and retained in the periplasm or be further trafficked into or across the outer membrane. This is no mean feat.

There are a number of periplasmic shock proteins that are transported to the periplasm and folded in enormous quantities, e.g., Spy and HdeI (Tapley et al., 2009; Quan et al., 2011). Moreover, the demand for the insertion and folding of β-barrel **O**uter **M**embrane **P**roteins (OMPs) in rapidly dividing cells is vast. The process is facilitated by the periplasmic chaperones SurA and Skp (Sklar et al., 2007; McMorran et al., 2013), which presumably collect proteins as they emerge from the translocon for folding or for delivery to the outer membrane. How the chaperoned OMPs negotiate the peptidoglycan layer is unclear. We know that when they get there, they are welcomed by the β-**B**arrel **A**ssembly **M**achinery (BAM) – a complex of five proteins BamABCDE, responsible for the insertion and folding of OMPs (**Figure 1**; Voulhoux et al., 2003; Wu et al., 2005). But how this is achieved for very large fluxes of proteins, without aggregation, and in the absence of an energy source is not easily reconciled. Many structures of the BAM complex have been determined (Bakelar et al., 2016; Gu et al., 2016; Iadanza et al., 2016); despite this, the mechanism for energy-independent OMP insertion and folding has yet to emerge.

THE ROLE OF SPECIFIC PHOSPHOLIPIDS IN PROTEIN TRANSPORT

While the dynamic mechanism for protein secretion through the protein machinery of the bacterial energy conserving, inner membrane has been the focus of our attention, it is becoming increasingly clear that the resident lipids also play a critical role in the transport proteins across, as well as into the membrane (**Figure 1**).

This post-translational reaction has been known for many years to require acidic phospholipids. Mutants defective in acidic phospholipid – cardiolipin (CL) and phosphatidyl-glycerol (PG) – synthesis have protein export deficiencies (Tommassen et al., 1989). Moreover, these lipids are required for functional association of SecA to the inner membrane (Hendrick and Wickner, 1991). Later work showed that the CL and, to a lesser extent, PG are important for stability of the SecYEG complex and to stimulate SecA ATPase activity (Bessonneau et al., 2002; Gold et al., 2010) – but the mechanism of action was unclear. Recent progress on this subject is beginning to unravel the mysterious action of this unique lipid.

Course-Grain Molecular Dynamics (CGMD) simulations have identified SecYEG sites, which transiently interacted with CL, which were validated empirically (**Figure 3**; Corey et al., 2018): native mass spectrometry demonstrated that variants in which the positive surface charges of the putative binding sites were diminished, bound CL less effectively. Remarkably, it turns out that these specific CL interactions confer the stimulation of SecA ATPase activity and PMF enhancement of secretion (**Figures 1, 3**; Corey et al., 2018); the latter may be achieved by proton carriage by the lipid itself. If true, this would be the first description of a direct involvement of a phospholipid in the process of energy coupling.

Furthermore, the interaction of the translocon with CL has profound consequences for the structure of the protein complex. Interestingly, specific sites on SecYEG, used to monitor the opening and closure of the protein channel (Fessl et al., 2018), are strongly dependent on CL for SecA promoted channel opening (Figure 5 and Supplementary Figure 2 in Corey et al., 2019).

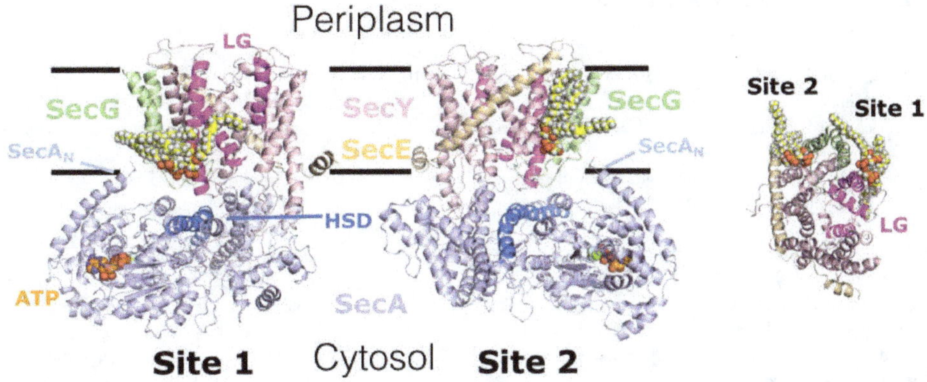

FIGURE 3 | Interaction of cardiolipin (CL) with the bacterial translocon. 2 CL (yellow, white, and red spheres) – binding sites 1 & 2 – on SecYEG; identified by CGMD, validated and shown to be critical for ATP and PMF-driven secretion (Corey et al., 2018). Left and middle, side view of SecYEG bound to the motor ATPase SecA; right, cytosolic view of SecYEG alone. In SecA: ATP, SecA$_N$ – N-terminus; HSD (including the 2HF) – helical scaffold domain, at the entrance to the protein channel in SecY. In SecYEG: LG – lateral gate, which opens during transport of proteins across and into the membrane. Adapted from Corey et al. (2018), used with permission.

A LIPID POOL IN THE HTL

Co-translational transport of proteins into the membrane occurs through the HTL (see above; **Figure 1**). Lipids, again CL in particular, are required to stabilize the holo-complex (Schulze et al., 2014), and critically, lipids also form an encapsulated pool at its center (**Figure 1**; Botte et al., 2016; Martin et al., 2019). This remarkable feature could provide an enclosed lipidic environment to promote efficient membrane protein insertion and assembly, protecting the translocating membrane protein from aggregation and proteolysis. This is a familiar concept for promoting folding of globular proteins within a chamber of the chaperonin GroEL (Ranson et al., 1997; Xu et al., 1997).

OTHER PROTEIN TRANSLOCATION SYSTEMS

It is very interesting that other protein translocation systems have also been implicated in the association with CL, including the BAM complex of the outer membrane (**Figure 1**; Chorev et al., 2018), and the mitochondrial Tim23 import machinery (Malhotra et al., 2017), but apparently not the TAT machinery responsible for the export of fully folded proteins in bacteria (Rathmann et al., 2017). Given the well-known dependence of CL for many proton translocating energy transducing systems, such as the ATP synthase (Duncan et al., 2016), and the electron transfer chain complexes I, III, and IV (Fry and Green, 1981; Pfeiffer et al., 2003; Fiedorczuk et al., 2016; Malkamäki and Sharma, 2019), one could surmise a critical role for proton-driven protein transport too.

CONCLUSION

Finally, our understanding of the dynamic mechanism underlying ATP-driven secretion through the Sec machinery is approaching clarity, while its augmentation by the PMF is a mystery shortly to be resolved, after nearly 30 years since its discovery (Brundage et al., 1990; Schiebel et al., 1991). In this context, CL seems to play essential and multifarious roles, for the structure and for both ATP and PMF-driven protein translocation activity. This warrants further investigation and exploitation. The essential lipid-protein interface could be a prime target for infiltration by small molecules for prospective antibiotic development.

AUTHOR CONTRIBUTIONS

The author confirms being the sole contributor of this work and has approved it for publication.

FUNDING

Our work on bacterial secretion is funded by the BBSRC (BB/S008349/1, BB/N015126/1, and BB/M003604/1). The author is indebted to Drs. Robin Corey, Phillip Stansfeld, and Daniel Watkins for preparing the figures.

REFERENCES

Allen, W. J., Corey, R. A., Oatley, P., Sessions, R. B., Baldwin, S. A., Radford, S. E., et al. (2016). Two-way communication between SecY and SecA suggests a Brownian ratchet mechanism for protein translocation. *eLife* 5. doi: 10.7554/eLife.15598

Arkowitz, R. A., Joly, J. C., and Wickner, W. (1993). Translocation can drive the unfolding of a preprotein domain. *EMBO J.* 12, 243–253. doi: 10.1002/j.1460-2075.1993.tb05650.x

Arkowitz, R. A., and Wickner, W. (1994). SecD and SecF are required for the proton electrochemical gradient stimulation of preprotein translocation. *EMBO J.* 13, 954–963. doi: 10.1002/j.1460-2075.1994.tb06340.x

Bakelar, J., Buchanan, S. K., and Noinaj, N. (2016). The structure of the β-barrel assembly machinery complex. *Science* 351, 180–186. doi: 10.1126/science.aad3460

Bauer, B. W., Shemesh, T., Chen, Y., and Rapoport, T. A. (2014). A "push and slide" mechanism allows sequence-insensitive translocation of secretory proteins by the SecA ATPase. *Cell* 157, 1416–1429. doi: 10.1016/j.cell.2014.03.063

Bessonneau, P., Besson, V., Collinson, I., and Duong, F. (2002). The SecYEG preprotein translocation channel is a conformationally dynamic and dimeric structure. *EMBO J.* 21, 995–1003. doi: 10.1093/emboj/21.5.995

Botte, M., Zaccai, N. R., Nijeholt, J. L. À., Martin, R., Knoops, K., Papai, G., et al. (2016). A central cavity within the holo-translocon suggests a mechanism for membrane protein insertion. *Sci. Rep.* 6:38399. doi: 10.1038/srep38399

Brundage, L., Hendrick, J. P., Schiebel, E., Driessen, A. J., and Wickner, W. (1990). The purified *E. coli* integral membrane protein SecY/E is sufficient for reconstitution of SecA-dependent precursor protein translocation. *Cell* 62, 649–657. doi: 10.1016/0092-8674(90)90111-Q

Cannon, K. S., Or, E., Clemons, W. M., Shibata, Y., and Rapoport, T. A. (2005). Disulfide bridge formation between SecY and a translocating polypeptide localizes the translocation pore to the center of SecY. *J. Cell Biol.* 169, 219–225. doi: 10.1083/jcb.200412019

Catipovic, M. A., Bauer, B. W., Loparo, J. J., and Rapoport, T. A. (2019). Protein translocation by the SecA ATPase occurs by a power-stroke mechanism. *EMBO J.* 38. doi: 10.15252/embj.2018101140

Chang, C. N., Blobel, G., and Model, P. (1978). Detection of prokaryotic signal peptidase in an *Escherichia coli* membrane fraction: endoproteolytic cleavage of nascent f1 pre-coat protein. *Proc. Natl. Acad. Sci. USA* 75, 361–365.

Chorev, D. S., Baker, L. A., Wu, D., Beilsten-Edmands, V., Rouse, S. L., Zeev-Ben-Mordehai, T., et al. (2018). Protein assemblies ejected directly from native membranes yield complexes for mass spectrometry. *Science* 362, 829–834. doi: 10.1126/science.aau0976

Clarke, A. R. (1996). Molecular chaperones in protein folding and translocation. *Curr. Opin. Struct. Biol.* 6, 43–50. doi: 10.1016/S0959-440X(96)80093-5

Corey, R. A., Ahdash, Z., Shah, A., Pyle, E., Allen, W. J., Fessl, T., et al. (2019). ATP-induced asymmetric pre-protein folding as a driver of protein translocation through the sec machinery. *eLife* 8. doi: 10.7554/eLife.41803

Corey, R. A., Allen, W. J., Komar, J., Masiulis, S., Menzies, S., Robson, A., et al. (2016). Unlocking the bacterial SecY translocon. *Structure* 24, 518–527. doi: 10.1016/j.str.2016.02.001

Corey, R. A., Pyle, E., Allen, W. J., Watkins, D. W., Casiraghi, M., Miroux, B., et al. (2018). Specific cardiolipin-SecY interactions are required for proton-motive force stimulation of protein secretion. *Proc. Natl. Acad. Sci. USA* 115, 7967–7972. doi: 10.1073/pnas.1721536115

Cranford-Smith, T., and Huber, D. (2018). The way is the goal: how SecA transports proteins across the cytoplasmic membrane in bacteria. *FEMS Microbiol. Lett.* 365. doi: 10.1093/femsle/fny093

Duncan, A. L., Robinson, A. J., and Walker, J. E. (2016). Cardiolipin binds selectively but transiently to conserved lysine residues in the rotor of metazoan ATP synthases. *Proc. Natl. Acad. Sci. USA* 113, 8687–8692. doi: 10.1073/pnas.1608396113

Duong, F., and Wickner, W. (1997). Distinct catalytic roles of the SecYE, SecG and SecDFyajC subunits of preprotein translocase holoenzyme. *EMBO J.* 16, 2756–2768. doi: 10.1093/emboj/16.10.2756

Economou, A., and Wickner, W. (1994). SecA promotes preprotein translocation by undergoing ATP-driven cycles of membrane insertion and deinsertion. *Cell* 78, 835–843. doi: 10.1016/S0092-8674(94)90582-7

Erlandson, K. J., Miller, S. B. M., Nam, Y., Osborne, A. R., Zimmer, J., and Rapoport, T. A. (2008). A role for the two-helix finger of the SecA ATPase in protein translocation. *Nature* 455, 984–987. doi: 10.1038/nature07439

Fessl, T., Watkins, D., Oatley, P., Allen, W. J., Corey, R. A., Horne, J., et al. (2018). Dynamic action of the Sec machinery during initiation, protein translocation and termination. *eLife* 7. doi: 10.7554/eLife.35112

Fiedorczuk, K., Letts, J. A., Degliesposti, G., Kaszuba, K., Skehel, M., and Sazanov, L. A. (2016). Atomic structure of the entire mammalian mitochondrial complex I. *Nature* 538, 406–410. doi: 10.1038/nature19794

Fry, M., and Green, D. E. (1981). Cardiolipin requirement for electron transfer in complex I and III of the mitochondrial respiratory chain. *J. Biol. Chem.* 256, 1874–1880.

Furukawa, A., Yoshikaie, K., Mori, T., Mori, H., Morimoto, Y. V., Sugano, Y., et al. (2017). Tunnel formation inferred from the I-form structures of the proton-driven protein secretion motor SecDF. *Cell Rep.* 19, 895–901. doi: 10.1016/j.celrep.2017.04.030

Gold, V. A. M., Robson, A., Bao, H., Romantsov, T., Duong, F., and Collinson, I. (2010). The action of cardiolipin on the bacterial translocon. *Proc. Natl. Acad. Sci. USA* 107, 10044–10049. doi: 10.1073/pnas.0914680107

Gold, V. A. M., Whitehouse, S., Robson, A., and Collinson, I. (2013). The dynamic action of SecA during the initiation of protein translocation. *Biochem. J.* 449, 695–705. doi: 10.1042/BJ20121314

Gonsberg, A., Jung, S., Ulbrich, S., Origi, A., Ziska, A., Baier, M., et al. (2017). The Sec61/SecY complex is inherently deficient in translocating intrinsically disordered proteins. *J. Biol. Chem.* 292, 21383–21396. doi: 10.1074/jbc.M117.788067

Gouridis, G., Karamanou, S., Gelis, I., Kalodimos, C. G., and Economou, A. (2009). Signal peptides are allosteric activators of the protein translocase. *Nature* 462, 363–367. doi: 10.1038/nature08559

Gu, Y., Li, H., Dong, H., Zeng, Y., Zhang, Z., Paterson, N. G., et al. (2016). Structural basis of outer membrane protein insertion by the BAM complex. *Nature* 531, 64–69. doi: 10.1038/nature17199

Hartl, F. U., Lecker, S., Schiebel, E., Hendrick, J. P., and Wickner, W. (1990). The binding cascade of SecB to SecA to SecY/E mediates preprotein targeting to the *E. coli* plasma membrane. *Cell* 63, 269–279. doi: 10.1016/0092-8674(90)90160-G

Hendrick, J. P., and Wickner, W. (1991). SecA protein needs both acidic phospholipids and SecY/E protein for functional high-affinity binding to the *Escherichia coli* plasma membrane. *J. Biol. Chem.* 266, 24596–24600.

Hizlan, D., Robson, A., Whitehouse, S., Gold, V. A., Vonck, J., Mills, D., et al. (2012). Structure of the SecY complex unlocked by a preprotein mimic. *Cell Rep.* 1, 21–28. doi: 10.1016/j.celrep.2011.11.003

Houben, E. N., Scotti, P. A., Valent, Q. A., Brunner, J., de Gier, J. L., Oudega, B., et al. (2000). Nascent Lep inserts into the *Escherichia coli* inner membrane in the vicinity of YidC, SecY and SecA. *FEBS Lett.* 476, 229–233. doi: 10.1016/S0014-5793(00)01735-X

Iadanza, M. G., Higgins, A. J., Schiffrin, B., Calabrese, A. N., Brockwell, D. J., Ashcroft, A. E., et al. (2016). Lateral opening in the intact β-barrel assembly machinery captured by cryo-EM. *Nat. Commun.* 7:12865. doi: 10.1038/ncomms12865

Jomaa, A., Boehringer, D., Leibundgut, M., and Ban, N. (2016). Structures of the *E. coli* translating ribosome with SRP and its receptor and with the translocon. *Nat. Commun.* 7:10471. doi: 10.1038/ncomms10471

Josefsson, L. G., and Randall, L. L. (1981). Different exported proteins in *E. coli* show differences in the temporal mode of processing in vivo. *Cell* 25, 151–157. doi: 10.1016/0092-8674(81)90239-7

Jung, S., and Tatzelt, J. (2018). Impaired transport of intrinsically disordered proteins through the Sec61 and SecY translocon; implications for prion diseases. *Prion* 12, 88–92. doi: 10.1080/19336896.2018.1435936

Komar, J., Alvira, S., Schulze, R. J., Martin, R., Lycklama A Nijeholt, J. A., Lee, S. C., et al. (2016). Membrane protein insertion and assembly by the bacterial holo-translocon SecYEG-SecDF-YajC-YidC. *Biochem. J.* 473, 3341–3354. doi: 10.1042/BCJ20160545

Kumazaki, K., Chiba, S., Takemoto, M., Furukawa, A., Nishiyama, K.-I., Sugano, Y., et al. (2014). Structural basis of sec-independent membrane protein insertion by YidC. *Nature* 509, 516–520. doi: 10.1038/nature13167

Li, L., Park, E., Ling, J., Ingram, J., Ploegh, H., and Rapoport, T. A. (2016). Crystal structure of a substrate-engaged SecY protein-translocation channel. *Nature* 531, 395–399. doi: 10.1038/nature17163

Malhotra, K., Modak, A., Nangia, S., Daman, T. H., Gunsel, U., Robinson, V. L., et al. (2017). Cardiolipin mediates membrane and channel interactions of the mitochondrial TIM23 protein import complex receptor Tim50. *Sci. Adv.* 3:e1700532. doi: 10.1126/sciadv.1700532

Malkamäki, A., and Sharma, V. (2019). Atomistic insights into cardiolipin binding sites of cytochrome c oxidase. *Biochim. Biophys. Acta Bioenerg.* 1860, 224–232. doi: 10.1016/j.bbabio.2018.11.004

Martin, R., Larsen, A. H., Corey, R. A., Midtgaard, S. R., Frielinghaus, H., Schaffitzel, C., et al. (2019). Structure and dynamics of the central lipid pool and proteins of the bacterial holo-translocon. *Biophys. J.* 116, 1931–1940. doi: 10.1016/j.bpj.2019.04.002

McMorran, L. M., Bartlett, A. I., Huysmans, G. H. M., Radford, S. E., and Brockwell, D. J. (2013). Dissecting the effects of periplasmic chaperones on the in vitro folding of the outer membrane protein PagP. *J. Mol. Biol.* 425, 3178–3191. doi: 10.1016/j.jmb.2013.06.017

Müller, M., Koch, H. G., Beck, K., and Schäfer, U. (2001). Protein traffic in bacteria: multiple routes from the ribosome to and across the membrane. *Prog. Nucleic Acid Res. Mol. Biol.* 66, 107–157.

Pfeiffer, K., Gohil, V., Stuart, R. A., Hunte, C., Brandt, U., Greenberg, M. L., et al. (2003). Cardiolipin stabilizes respiratory chain supercomplexes. *J. Biol. Chem.* 278, 52873–52880. doi: 10.1074/jbc.M308366200

Poritz, M. A., Bernstein, H. D., Strub, K., Zopf, D., Wilhelm, H., and Walter, P. (1990). An *E. coli* ribonucleoprotein containing 4.5S RNA resembles mammalian signal recognition particle. *Science* 250, 1111–1117. doi: 10.1126/science.1701272

Quan, S., Koldewey, P., Tapley, T., Kirsch, N., Ruane, K. M., Pfizenmaier, J., et al. (2011). Genetic selection designed to stabilize proteins uncovers a chaperone called Spy. *Nat. Struct. Mol. Biol.* 18, 262–269. doi: 10.1038/nsmb.2016

Ranson, N. A., Burston, S. G., and Clarke, A. R. (1997). Binding, encapsulation and ejection: substrate dynamics during a chaperonin-assisted folding reaction. *J. Mol. Biol.* 266, 656–664. doi: 10.1006/jmbi.1996.0815

Rathmann, C., Schlösser, A. S., Schiller, J., Bogdanov, M., and Brüser, T. (2017). Tat transport in *Escherichia coli* requires zwitterionic phosphatidylethanolamine but no specific negatively charged phospholipid. *FEBS Lett.* 591, 2848–2858. doi: 10.1002/1873-3468.12794

Robson, A., Gold, V. A. M., Hodson, S., Clarke, A. R., and Collinson, I. (2009). Energy transduction in protein transport and the ATP hydrolytic cycle of SecA. *Proc. Natl. Acad. Sci. USA* 106, 5111–5116. doi: 10.1073/pnas.0809592106

Samuelson, J. C., Chen, M., Jiang, F., Möller, I., Wiedmann, M., Kuhn, A., et al. (2000). YidC mediates membrane protein insertion in bacteria. *Nature* 406, 637–641. doi: 10.1038/35020586

Schiebel, E., Driessen, A. J., Hartl, F. U., and Wickner, W. (1991). Delta mu H+ and ATP function at different steps of the catalytic cycle of preprotein translocase. *Cell* 64, 927–939. doi: 10.1016/0092-8674(91)90317-R

Schulze, R. J., Komar, J., Botte, M., Allen, W. J., Whitehouse, S., Gold, V. A. M., et al. (2014). Membrane protein insertion and proton-motive-force-dependent secretion through the bacterial holo-translocon SecYEG-SecDF-YajC-YidC. *Proc. Natl. Acad. Sci. USA* 111, 4844–4849. doi: 10.1073/pnas.1315901111

Scotti, P. A., Urbanus, M. L., Brunner, J., de Gier, J. W., Heijne, von, G., van der Does, C., et al. (2000). YidC, the *Escherichia coli* homologue of mitochondrial Oxa1p, is a component of the Sec translocase. *EMBO J.* 19, 542–549. doi: 10.1093/emboj/19.4.542

Sklar, J. G., Wu, T., Kahne, D., and Silhavy, T. J. (2007). Defining the roles of the periplasmic chaperones SurA, Skp, and DegP in *Escherichia coli*. *Genes Dev.* 21, 2473–2484. doi: 10.1101/gad.1581007

Stennett, E. M. S., Ciuba, M. A., Lin, S., and Levitus, M. (2015). Demystifying PIFE: the photophysics behind the protein-induced fluorescence enhancement phenomenon in Cy3. *J. Phys. Chem. Lett.* 6, 1819–1823. doi: 10.1021/acs.jpclett.5b00613

Tapley, T. L., Körner, J. L., Barge, M. T., Hupfeld, J., Schauerte, J. A., Gafni, A., et al. (2009). Structural plasticity of an acid-activated chaperone allows promiscuous substrate binding. *Proc. Natl. Acad. Sci. USA* 106, 5557–5562. doi: 10.1073/pnas.0811811106

Tomkiewicz, D., Nouwen, N., van Leeuwen, R., Tans, S., and Driessen, A. J. M. (2006). SecA supports a constant rate of preprotein translocation. *J. Biol. Chem.* 281, 15709–15713. doi: 10.1074/jbc.M600205200

Tommassen, J., de Vrije, T., de Cock, H., Bosch, D., and de Kruijff, B. (1989). Involvement of membrane lipids in protein export in Escherichia coli. *J. Cell Sci.* 11, 73–83.

Tsirigotaki, A., Chatzi, K. E., Koukaki, M., De Geyter, J., Portaliou, A. G., Orfanoudaki, G., et al. (2018). Long-lived folding intermediates predominate the targeting-competent Secretome. *Structure* 26, 695.e5–707.e5. doi: 10.1016/j.str.2018.03.006

Tsukazaki, T., Mori, H., Echizen, Y., Ishitani, R., Fukai, S., Tanaka, T., et al. (2011). Structure and function of a membrane component SecDF that enhances protein export. *Nature* 474, 235–238. doi: 10.1038/nature09980

Ulbrandt, N. D., Newitt, J. A., and Bernstein, H. D. (1997). The *E. coli* signal recognition particle is required for the insertion of a subset of inner membrane proteins. *Cell* 88, 187–196. doi: 10.1016/S0092-8674(00)81839-5

Van den Berg, B., Clemons, W. M., Collinson, I., Modis, Y., Hartmann, E., Harrison, S. C., et al. (2004). X-ray structure of a protein-conducting channel. *Nature* 427, 36–44. doi: 10.1038/nature02218

Voulhoux, R., Bos, M. P., Geurtsen, J., Mols, M., and Tommassen, J. (2003). Role of a highly conserved bacterial protein in outer membrane protein assembly. *Science* 299, 262–265. doi: 10.1126/science.1078973

Walter, P., Ibrahimi, I., and Blobel, G. (1981a). Translocation of proteins across the endoplasmic reticulum. I. Signal recognition protein (SRP) binds to in-vitro-assembled polysomes synthesizing secretory protein. *J. Cell Biol.* 91, 545–550.

Walter, P., Walter, P., Blobel, G., and Blobel, G. (1981b). Translocation of proteins across the endoplasmic reticulum III. Signal recognition protein (SRP) causes signal sequence-dependent and site-specific arrest of chain elongation that is released by microsomal membranes. *J. Cell Biol.* 91, 557–561.

Walter, P., Walter, P., Blobel, G., and Blobel, G. (1981c). Translocation of proteins across the endoplasmic reticulum. II. Signal recognition protein (SRP) mediates the selective binding to microsomal membranes of in-vitro-assembled polysomes synthesizing secretory protein. *J. Cell Biol.* 91, 551–556.

Wu, T., Malinverni, J., Ruiz, N., Kim, S., Silhavy, T. J., and Kahne, D. (2005). Identification of a multicomponent complex required for outer membrane biogenesis in *Escherichia coli*. *Cell* 121, 235–245. doi: 10.1016/j.cell.2005.02.015

Xu, Z., Horwich, A. L., and Sigler, P. B. (1997). The crystal structure of the asymmetric GroEL-GroES-(ADP)7 chaperonin complex. *Nature* 388, 741–750. doi: 10.1038/41944

Zimmer, J., Nam, Y., and Rapoport, T. A. (2008). Structure of a complex of the ATPase SecA and the protein-translocation channel. *Nature* 455, 936–943. doi: 10.1038/nature07335

15

Distribution, Function and Regulation of Type 6 Secretion Systems of Xanthomonadales

*Ethel Bayer-Santos[1], Lucas de Moraes Ceseti[2], Chuck Shaker Farah[3] and Cristina Elisa Alvarez-Martinez[2]**

[1] *Departamento de Microbiologia, Instituto de Ciências Biomédicas, Universidade de São Paulo, São Paulo, Brazil,* [2] *Departamento de Genética, Evolução, Microbiologia e Imunologia, Instituto de Biologia, Universidade Estadual de Campinas, Campinas, Brazil,* [3] *Departamento de Bioquímica, Instituto de Química, Universidade de São Paulo, São Paulo, Brazil*

Correspondence:
Cristina Elisa Alvarez-Martinez
ceamarti@unicamp.br

Members of the Xanthomonadales order include several plant pathogens of significant economic and agricultural impact, such as *Xanthomonas* spp. Type 6 secretion systems (T6SSs) are contractile nanomachines used by many bacterial species to inject protein effectors into target prokaryotic and eukaryotic cells and provide a competitive advantage for bacteria in different environments. Effectors with antibacterial properties include peptidoglycan hydrolases, lipases and phospholipases that break down structural components of the cell envelope, promoting target-cell lysis; and RNases, DNAses, and NADases that affect target-cell metabolism, arresting growth. Effectors with anti-eukaryotic properties are functionally more diverse. The T6SS of *Xanthomonas citri* is the only example experimentally characterized so far within the Xanthomonadales order and displays anti-eukaryotic function by providing resistance to predation by amoeba. This T6SS is regulated at the transcriptional level by a signaling cascade involving a Ser/Thr kinase and an extracytoplasmic function (ECF) sigma factor. In this review, we performed *in silico* analyses of 35 genomes of Xanthomonadales and showed that T6SSs are widely distributed and phylogenetically classified into three major groups. *In silico* predictions identified a series of proteins with known toxic domains as putative T6SS effectors, suggesting that the T6SSs of Xanthomonadales display both anti-prokaryotic and anti-eukaryotic properties depending on the phylogenetic group and bacterial species.

Keywords: Xanthomonadales, *Xanthomonas*, T6SS, bacterial killing, amoeba predation, effectors, toxins

INTRODUCTION

The order Xanthomonadales includes many Gram-negative rod-shaped bacteria with very diverse physiological characteristics and habitats. Members of this group range from plant and human pathogens to non-pathogenic environmental bacteria that are able to survive in adverse conditions such as contaminated soil and hot springs (Saddler and Bradbury, 2005). Xanthomonadales is an early diverging branch of the Gammaproteobacteria (Williams et al., 2010). The taxonomy of the order is controversial, but a recent phylogenetic analysis has divided Xanthomonadales into two main branches comprising the families *Xanthomonadaceae* and *Rhodanobacteraceae* (Naushad et al., 2015). *Xanthomonadaceae* includes genera *Xanthomonas, Xylella, Stenotrophomonas, Pseudoxanthomonas, Luteimonas, Lysobacter, Thermomonas, Arenimonas,* and *Silanimonas*; while *Rhodanobacteraceae* includes genera *Rhodanobacter, Dyella, Frateuria, Luteibacter, Fulvimonas, Pseudofulvimonas, Aquimonas, Dokdonella,* and *Rudaea* (Naushad et al., 2015).

Species of the *Xanthomonadaceae* family have been by far the most studied due to their importance as plant pathogens. The genera *Xanthomonas* and *Xylella* contain species that promote disease in more than 400 economically important crops, including citrus, tomato, rice, cabbage, pepper, coffee, grapes, and olives (Leyns et al., 1984; Rapicavoli et al., 2018). Species within both genera vary in their ability to colonize different plant tissues and show a high degree of host specificity (Ryan et al., 2011). *Stenotrophomonas* is another important genus in *Xanthomonadaceae. Stenotrophomonas maltophilia* include several strains that are nosocomial pathogens, causing bacteremia, endocarditis and pneumonia in immunocompromised and cystic fibrosis patients (Adegoke et al., 2017). Conversely, species like *Stenotrophomonas rhizophila* are environmental bacteria found in association with plants and have a well-documented ability to promote plant growth, suppress colonization by plant pathogens and degrade a wide variety of xenobiotics, making them potential agents for biocontrol and bioremediation (Berg and Martinez, 2015). The genus *Lysobacter* comprises gliding predatory bacteria that display broad-spectrum lytic activity against nematodes, fungi, Gram-negative and Gram-positive bacteria (Christensen and Cook, 2009), including species of significant biotechnological and biocontrol interest (Ko et al., 2009; Hayward et al., 2010).

Availability of genomic data from an increasing number of Xanthomonadales species has provided important insights into environmental adaptations and physiological diversity. Gene clusters encoding bacterial secretion systems are recognized as key virulence factors of pathogenic species within the order (Büttner and Bonas, 2010). The type 6 secretion system (T6SS) is a molecular nanomachine that provides increased fitness to bacteria by firing a series of toxic effector proteins into neighbor competitor species, thus shaping bacterial communities. The T6SS of the biocontrol agent *Pseudomonas putida* kills phytopathogens upon co-infection *in planta*, and the *Agrobacterium tumefaciens* T6SS promotes plant colonization by providing a competitive advantage (Ma et al., 2014; Bernal et al., 2017). Anti-eukaryotic T6SSs are important for virulence in mammalian hosts, as well as for bacterial survival in the environment by providing resistance to predation by amoebas and exhibiting killing activity against fungi (Hachani et al., 2016; Bayer-Santos et al., 2018; Trunk et al., 2018). A role of T6SS in nutrient acquisition by the secretion of metal-scavenging proteins in the extracellular milieu has also been reported (Si et al., 2017a,b).

The sole T6SS representative of the Xanthomonadales order experimentally characterized to date from *Xanthomonas citri* pv. *citri* is required for resistance against predation by the soil amoeba *Dictyostelium discoideum* (Bayer-Santos et al., 2018), a yet unexplored aspect of xanthomonad biology that can be expected to be an important factor for environmental survival and dissemination. The secreted effectors and dynamics of *X. citri*-amoeba interactions are still elusive. The *X. citri* T6SS does not confer a competitive advantage in encounters with other Gram-negative bacteria and *X. citri* antibacterial activity is dependent on a type 4 secretion system (T4SS) (Souza et al., 2015; Bayer-Santos et al., 2018).

T6SSs are encoded in the genome of several species within Xanthomonadales. In this review, we performed *in silico* analyses of Xanthomonadales T6SSs to describe the distribution and genomic organization of T6SSs clusters in these species. Furthermore, we identified putative T6SS effectors that provided clues about the function of uncharacterized T6SS clusters in several Xanthomonadales species.

TYPE 6 SECRETION SYSTEM

The T6SS is a contractile machinery composed of 13 core structural proteins. This system is evolutionarily related to the tail of bacteriophages (Basler and Mekalanos, 2012) and assembles into three major complexes: the *trans*-membrane complex, the baseplate and the tail. The *trans*-membrane complex is composed of three proteins TssM, TssL, and TssJ. The baseplate is formed by TssE, TssF, TssG, and TssK, and represents an adaptor between the *trans*-membrane complex and the tail. The tail has an internal tube formed by Hcp topped with VgrG and is enveloped by a contractile sheath composed of TssB and TssC (Nguyen et al., 2018). The assembly of the tail requires TssA, which interacts with baseplate, inner tube and sheath components and stabilizes the distal extremity of the tube (Planamente et al., 2016; Zoued et al., 2016; Dix et al., 2018). After contraction, the T6SS tail is recycled via the ATPase ClpV that disassembles the sheath into monomeric components (Kapitein et al., 2013). In addition to the core structural proteins described above, T6SS gene clusters also encode accessory proteins, which comprise components required for the assembly of the secretion apparatus, regulatory subunits acting transcriptionally or post-translationally to control the expression or the activity of the T6SS, and effectors and immunity proteins required for its function (Silverman et al., 2012).

T6SSs deliver protein effectors into diverse cell types including prokaryotic and eukaryotic cells in a contact-dependent manner (Cianfanelli et al., 2016; Hachani et al., 2016). T6SSs were also reported to display contact-independent functions in which secreted effectors facilitate the acquisition of nutrients (Wang et al., 2015; Si et al., 2017a,b). T6SS gene clusters have been classified into four subtypes (T6SS^{i-iv}) (Boyer et al., 2009; Bröms et al., 2010; Russell et al., 2014; Bock et al., 2017): (i) the majority of T6SSs belong to subtype T6SS^{i} and are present in Proteobacteria; (ii) the *Francisella* pathogenicity island-like systems were classified as T6SS[ii] (Bröms et al., 2010); (iii) Bacteroidetes T6SSs are distinct from the first two and were classified as T6SS[iii] (Russell et al., 2014); and (iv) a contractile system from *Amoebophilus asiaticus* was classified T6SS[iv] (Bock et al., 2017). Proteobacteria T6SS[i] are the most diverse and have been further subdivided into five phylogenetic clades (Boyer et al., 2009). Xanthomonadales harbor three subtypes of T6SS[i] belonging to clades 1, 3, and 4, which will be discussed below.

GENOMIC ARCHITECTURE OF XANTHOMONADALES T6SS CLUSTERS

From 71 Xanthomonadales species genomes retrieved from the KEGG database (Kanehisa and Goto, 2000), we identified 35 genomes harboring one or two T6SS clusters (**Supplementary Table S1**). Distribution of T6SS does not show a clear correlation with species lifestyles and they are found in several environmental bacteria and phytopathogenic species that colonize distinct plant tissues (vascular and non-vascular pathogens) (**Supplementary Table S1**). T6SS clusters are absent in xylem-limited phytopathogens with reduced genomes, including *X. albilineans* (3.78 Mb) and all members of the *Xylella* genus (2.5 Mb).

According to phylogenetic analyses using the sheath component TssC, T6SSs clusters separate into three groups matching clades/groups 1, 3, and 4 proposed by Boyer et al. (2009) (**Figure 1A**). Similarly, analysis of T6SS distribution in plant-associated bacteria that included several members of the *Xanthomonas* genus has shown an overrepresentation of these three clades (Bernal et al., 2018). Group 3 presents the most heterogeneous distribution among plant-associated bacteria and the Xanthomonadales representatives are clustered in a clade that includes *Burkholderia* species (Bernal et al., 2018). Xanthomonadales group 4 T6SSs belong to subgroup 4B2 described by Bernal et al. (2018), which also includes *Ralstonia* and *Burkholderia* species. The sole member of group 1 is found in a *Stenotrophomonas* sp. isolated from the phyllosphere and a phylogenetic analysis showed that it belongs to the subclade 1.2A described by Bernal et al. (2018), which includes *P. putida* species. Each group displays a unique genetic architecture, contains different T6SS-associated genes (Tag proteins) and present variable regions within or in the vicinity of the structural gene clusters, which contain putative effectors and/or regulatory proteins (**Figure 1B**). The characteristics of each group are further described in detail below.

Group 3 presents two main clusters of structural genes separated by an insertion in which the content varies depending on the bacterial species, ranging from ~2.2 to 15.6 kb (**Figure 1B**). Group 3 was further divided into three subclades (hereafter referred to as subgroup 3*, 3**, and 3***) (**Figures 1A,B**). Division of *Xanthomonas* group 3 T6SSs in two subclades is also observed in the phylogenetic analysis by Bernal et al. (2018), which did not include *Dyella* species. Group 3* contains the only Xanthomonadales T6SS functionally characterized to date from *X. citri* (**Figure 1A**) (Bayer-Santos et al., 2018) and is restricted to *Xanthomonas* species, a few of them also containing a second T6SS from subgroup 3*** (**Figure 1**). The presence of one representative from each subgroup in some species may indicate distinct functions in bacterial physiology, despite similarities in cluster organization.

Group 3 shows unique features such as the presence of components from the PpkA-PppA-FHA post-translational phosphorylation pathway (Mougous et al., 2007) and its repressor TagF (Silverman et al., 2011; Lin et al., 2018). It also contains TagJ, which interacts with the ATPase ClpV and the sheath component TssB to control the disassembly of the contracted sheath (Forster et al., 2014). TagJ co-evolved with a specific subset of the TssB/TssC/ClpV gene cluster (Forster et al., 2014). The TssA component from group 3 is closely related to homologs from group 4 T6SSs, belonging to the same phylogenetic clade 1 (TssA1) (Dix et al., 2018), while TssA from group 1 T6SS shows a divergent C-terminal region and belongs to clade 2 (Dix et al., 2018). In *X. citri*, the variable region is ~15.6 kb and contains the extracytoplasmic function (ECF) sigma factor EcfK and its cognate kinase PknS, which were shown to be required for T6SS activation and function (Bayer-Santos et al., 2018). Curiously, EcfK/PknS are not present in the most similar T6SS cluster from *Dyella japonica* and *Dyella thiooxydans* (**Figure 1A**), which has an intermediate-size variable region of ~5.2 kb. Despite the differences in the variable region, T6SS from *D. japonica* and *D. thiooxydans* are very similar to *X. citri* T6SS and present conserved genes of unknown function downstream of the *PAAR* and *vgrG* genes (**Figure 1B**, white arrows). A third genetic architecture within group 3 can be observed in a branch comprising *Xanthomonas oryzae* pv. *oryzae*, *Xanthomonas oryzae* pv. *oryzicola*, and *Lysobacter enzymogenes* (subgroup 3***). These species contain even smaller insertions in the variable region (~2.2 kb) and *tssA* appears in a different orientation (**Figure 1B**). Bacterial species carrying this subtype of T6SS usually have a second T6SS belonging to group 4 or subgroup 3* (**Figure 1A**, blue).

T6SSs classified as group 4 display two genetic architectures. In the majority of species, the structural core genes are organized in conserved clusters and variability is found in the vicinity of *vgrG* genes. These variable regions are very large – ranging from 40 kb to 75 kb – and contain several duplications of *vgrG* (up to 6), proteins with domains of unknown functions (DUFs) and proteins with putative toxic domains (**Table 1**). In some species, such as *Xanthomonas fragariae*, the cluster of structural genes composed of *tssM*, *tagF*, *tagN*, and *tssA* are located ~300 kb from the other structural cluster composed of *tssJ*, *tssK*, *tssL*, *vgrG*. Another interesting and unique feature of T6SSs from group 4 is the presence of a gene encoding TagX, which is a membrane-associated peptidoglycan hydrolase proposed to help degrade the bacterial wall for the insertion of T6SS machinery (Weber et al., 2016). Group 4 T6SSs also contain the associated genes *tagN* and *tagM*, but their role in T6SS biogenesis or regulation was not yet clarified. *Lysobacter gummosus* is the only example belonging to group 4 that contains

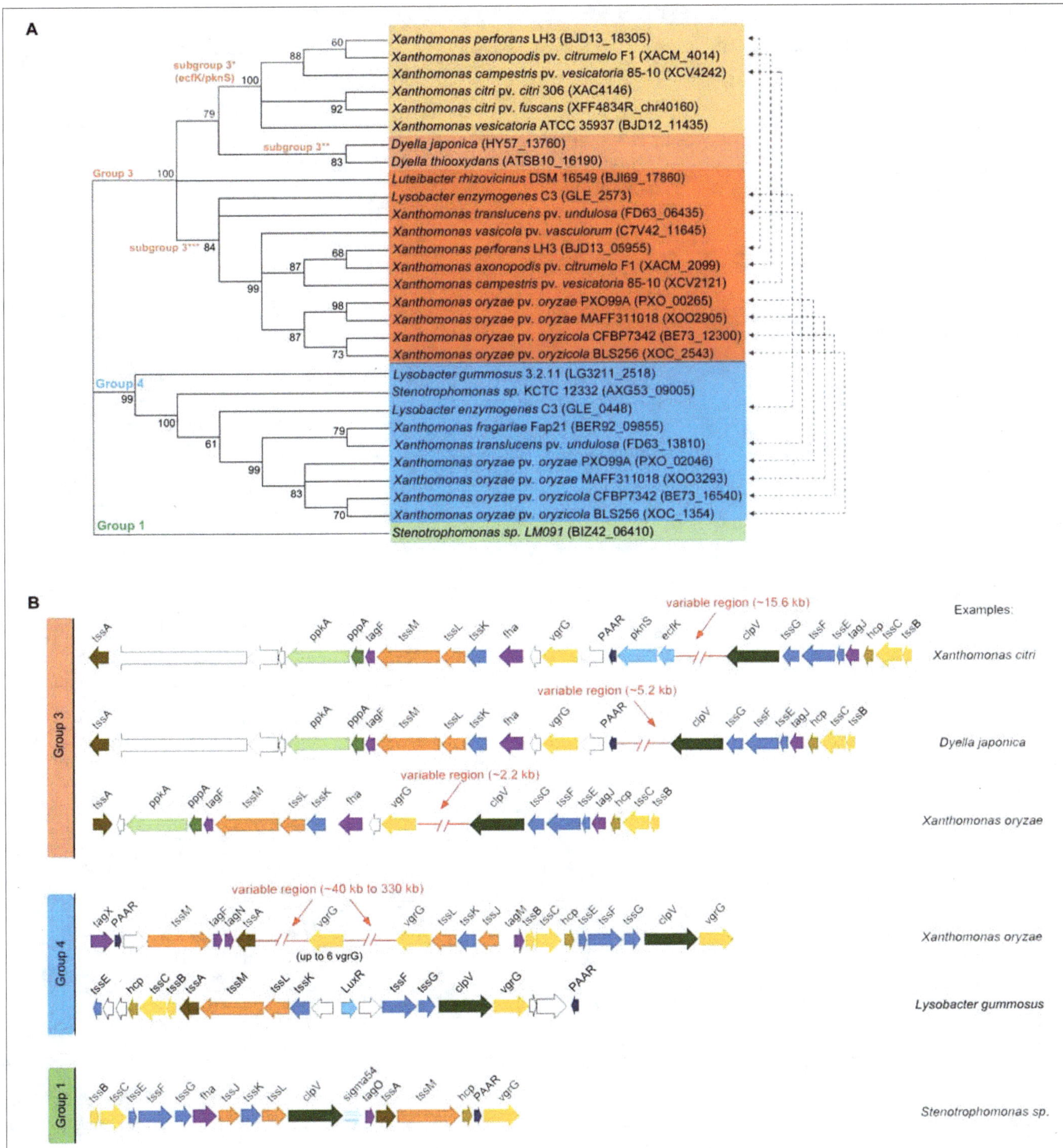

FIGURE 1 | Phylogenetic classification of T6SSs within Xanthomonadales and genomic organization of T6SS gene clusters. **(A)** Phylogenetic distribution of Xanthomonadales T6SS clusters. Maximum-likelihood consensus tree with 1000 bootstrap replicates built with amino acid sequence of TssC (XAC4146) homologs aligned and grouped using MEGA 7.0 (Kumar et al., 2016). **(B)** Schematic representation of genomic regions encoding the different subtypes of T6SSs.

a divergent version of T6SS, displaying a different genetic architecture (**Figure 1B**).

The only example of a Group 1 T6SS is found in *Stenotrophomonas* sp. LM91. This system is very similar to the T6SS from *Vibrio cholerae* and contains the sigma factor σ^{54} transcriptional regulator, which was reported to control the expression of T6SS in other bacteria (Bernard et al., 2011). Group 1 T6SS also contains a gene

TABLE 1 | List of PAAR and VgrG proteins from Xanthomonadales and putative toxic effectors identified by Bastion6 software.

Species	T6SS Group	PAAR	Orphan PAAR	VgrG (class)	Orphan VgrG (class)	Bastion6 predicted T6SS effectors	
						N	Description[1,2]
X. vesicatoria ATCC 35937	3*	1	0	1 (I)	0	4	Hypothetical
X. vasicola pv. *vasculorum* SAM119	3***	0	0	1 (I)	0	1	Hypothetical
X. translucens pv. *undulosa* Xtu 4699	3***	0	2	1 (I)	0	1	Hypothetical
	4	1		2 (II)		12#	Tle1-like, DUF3304 (2), neuraminidase
						2 orphans	Tox-REase-5 (2)
X. perforans LH3	3*	1	0	1 (I)	0	3	Acid phosphatase, EEP endo/exonuclease/phosphatase
	3***	0		1 (I)		2	TIR_2 superfamily
X. oryzae pv. *oryzicola* CFBP7342	3***	0	3	1 (I) and 1 (II)	1 (I) and 10 (II)	3	Hypothetical
	4	1		4 (II)		19	Tle1-like (4), DUF3304 (5), hydrolase
						55 orphans	Tle1-like (2), Tle3-like (2), Tle4-like (2), DUF2875 (5), DUF3304 (4), DUF1800, DUF1501, catalase, peptidase (2), PAAR/Rhs MafB19-deaminase, haemolysin, Tox-REase-5
X. oryzae pv. *oryzicola* BLS256	3***	0	1	2 (I) and 1 (II)	3 (I) and 5 (II)	1	Hypothetical
	4	1		2 (I) and 3 (II)		14	Tle1-like (4), DUF3304 (3), hydrolase
						39 orphans	Tle1-like (2), Tle3-like (4), Tle4-like (3), DUF2875 (5), DUF3304 (5), DUF493, peptidase, transglycosylase, carboxypeptidase, Rhs repeat proteins (2), oxidoreductase
X. oryzae pv. *oryzae* PXO99A	3***	0	5	1 (I) and 1 (II)	1 (I) and 5 (II)	2	Hypothetical
	4	1		3 (II)		12	Tle1-like (2), DUF3304 (2), transglycosylase
						13 orphans	Tle3-like (5), DUF2875 (4)
X. oryzae pv. *oryzae* MAFF 311018	3***	0	3	1 (I)	2 (I) and 7 (II)	0	
	4	1		5 (II)		14	Tle1-like (2), DUF3304 (5), DUF2345
						51 orphans	Tle1-like (4), DUF3304 (6), Tle3-like (5), DUF2875 (5), murein hydrolase D, muraminidase (2), carboxypeptidase, peptidase, PAAR/Rhs XOO_2897-like deaminase, Tox-REase-5 (2)
X. fragariae Fap21	4	1	0	4 (II)	2 (I) and 4 (II)	12	Hypothetical
						27 orphans	Tle1-like (2), DUF3304 (6), glycoside hydrolase, Rhs repeat protein, PAAR/DUF4150 Colicin-DNAse
X. citri pv. *fuscans*	3*	1	0	1 (I)	0	4	Acid phosphatase, EEP endo/exonuclease/phosphatase
X. citri pv. *citri* 306	3*	1	0	1 (I)	0	4	Acid phosphatase, EEP endo/exonuclease/phosphatase, adhesin
X. campestris pv. *vesicatoria* 85-10	3*	1	0	1 (I)	0	4	Acid phosphatase, EEP endo/exonuclease/phosphatase
	3***	0		1 (I)		4	TIR_2 superfamily
X. axonopodis pv. *citrumelo* F1	3*	1	0	1 (I)	0	4	Acid phosphatase, EEP endo/exonuclease/phosphatase
	3***	0		1 (I)		3	TIR_2 superfamily
Stenotrophomonas sp. LM091	1	1	0	1 (II)	0	5	Tle3-like (2), DUF4274, DUF2875
Stenotrophomonas sp. KCTC 12332	4	1	2	1 (I) and 1 (II)	1 (II)	4	Hypothetical
						6 orphans	Tle1-like, Tle4-like, DUF4287
L. rhizovicinus DSM 16549	3***	0	0	1 (I)	0	1	Hypothetical
L. gummosus 3.2.11	4	1	2	1 (I)	1 (I)	4	Transglycosylase, peptidoglycan-binding protein
						7 orphans	DUF4157, peptidase, DUF3244/DUF3218

(Continued)

TABLE 1 | Continued

Species	T6SS Group	PAAR	Orphan PAAR	VgrG (class)	Orphan VgrG (class)	Bastion6 predicted T6SS effectors	
						N	**Description[1,2]**
L. enzymogenes C3	3***	0	3	1 (I)	1 (I)	0	
	4	0		3 (II)		7	Tle4-like protein, hydrolase, peptidase, phospholipase
						3 orphans	GTP-binding protein, TPR-repeat protein
D. thyooxidans ATSB10	3**	1	0	1 (I)	0	1	Hypothetical
D. japonica A8	3**	1	0	1 (I)	2 (II)	2	DUF5636, amidase
						7 orphans	Phospholipase, muraminidase, peptidase, TPR repeat proteins (3)

#All genes flanking the two VgrGs were analyzed. [1]Only proteins with putative toxic domains or DUFs previously associated with T6SS are described. [2]Number of proteins with identical domains are indicated in brackets when >1.

encoding the associated protein TagO, but its function is still unknown.

FUNCTION OF PUTATIVE XANTHOMONADALES T6SS EFFECTORS

T6SSs translocate effectors by decorating the Hcp-VgrG-PAAR puncturing device that is propelled against target cells, thus delivering a cocktail of effectors after each contraction event. The current model suggests that effectors can either interact with one of these three proteins, named "cargo" effectors, or be presented as an extra domain within the same proteins, named "specialized" effectors (Durand et al., 2014; Cianfanelli et al., 2016).

In order to assess the repertoire of Xanthomonadales T6SSs effectors, we manually analyzed the genomic regions encoding the structural components Hcp, VgrG, and PAAR to search for specialized effectors. All Xanthomonadales species analyzed display only one copy of Hcp that is associated with a T6SS structural cluster, and these Hcps do not present a C-terminal extension (**Supplementary Table S2A**). The VgrG repertoire of Xanthomonadales seems to be more diverse. VgrGs are categorized into three classes (De Maayer et al., 2011): the first is composed of proteins with a N-terminal VgrG domain; the second class is formed by VgrG proteins carrying a C-terminal domain of unknown function DUF2345, which is required for interaction with cargo effectors (Flaugnatti et al., 2016); and the third class comprises the evolved- or specialized-VgrG that carry a C-terminal toxic domain. In Xanthomonadales, class I VgrGs seem to be predominantly associated with the T6SSs belonging to group 3 (**Table 1** and **Supplementary Table S2B**). Class II VgrGs are more abundant in species that carry a T6SS belonging to group 4 (**Table 1** and **Supplementary Table S2B**). In addition, Xanthomonadales species that contain T6SS clusters from group 4 usually have orphan VgrGs scattered in the genome (**Supplementary Table S2B**), while species with T6SSs from subgroup 3* do not harbor orphan VgrGs. Interestingly, a high number of orphan VgrGs are found in the genomes of *X. oryzae* from different strains. Moreover, no bona fide specialized VgrGs (class III) were detected in the analyzed genomes (**Table 1** and **Supplementary Table S2B**). PAAR proteins were associated with T6SS structural gene clusters in all groups, except species from subgroup 3*** (**Figure 1**, **Table 1**, and **Supplementary Table S2C**). Most species harboring a subgroup 3*** T6SS that lack a PAAR protein also encode an additional T6SS cluster in their genomes, either from subgroup 3* or 4 (**Figure 1A**). At this point, it is unclear whether PAAR-like proteins could be shared between two systems or whether subgroup 3*** systems are non-functional due to the loss of an associated PAAR protein. Nevertheless, none of the Xanthomonadales PAAR proteins associated with T6SS clusters have extended domains coding for putative toxic proteins (**Supplementary Table S2C**). Orphan PAAR proteins are present in genomes that contain a group 4 T6SS cluster, some of them with extended sizes that might encode effector functions (**Supplementary Table S2C**). Among them, three proteins that belong to the modular PAAR-Rhs-toxin group of antibacterial toxins (Ma et al., 2017) are present in the genomes of *X. fragarie*, *X. oryzae* pv. *oryzae*, and *X. oryzae* pv. *oryzicola* (**Supplementary Tables S2C, S3**).

To search for cargo effectors using *in silico* analyses, we arbitrarily chose to analyze a fixed number of 10 genes immediately flanking all T6SS VgrGs using Bastion6 software (Wang et al., 2018). We also searched for genes encoding DUF4123, DUF2169 or DUF1795-containing proteins, which act as adaptors for effector recruitment by T6SSs (Liang et al., 2015; Unterweger et al., 2015; Bondage et al., 2016). Genes encoding DUF1795 are not present in Xanthomonadales genomes, while DUF2169 genes are located in association with VgrG (**Supplementary Table S2D**), as previously described (Liang et al., 2015; Unterweger et al., 2015). Fourteen DUF4123-containing proteins were found encoded in Xanthomonadales genomes, not associated with structural clusters or orphan VgrGs (**Supplementary Table S2E**). We retrieved the two genes located immediately downstream from each DUF4123-encoding gene for effector prediction using Bastion6. Antibacterial cargo effectors are usually encoded in bicistronic units with genes encoding cognate immunity proteins, targeting components of the bacterial cell-envelope and/or nucleic acids, such as peptidoglycan hydrolases, amidases, lipases, and nucleases (Lien and Lai, 2017).

Effectors with anti-eukaryotic properties are less studied and functionally more diverse and include proteins involved in actin-crosslinking, lipases, deaminases and catalases (Jiang et al., 2014; Lien and Lai, 2017).

Analysis of the set of predicted T6SS effectors from Xanthomonadales showed a high number of putative antibacterial toxins associated with group 4 T6SS, which display hydrolase, lipase, carboxypeptidase and muraminidase domains (**Table 1** and **Supplementary Table S3**). Interestingly, most genomes harbouring a group 4 T6SS present multiple copies of members of the superfamily of antibacterial T6SS lipase effectors (Tle), more specifically from families Tle1, Tle3, and Tle 4 (Russell et al., 2013) (**Table 1** and **Supplementary Table S3**). Furthermore, Tle1 copies are found in association with T6SS clusters from group 4 in most genomes (**Table 1** and **Supplementary Table S3**). Tle1 homologs from *Burkholdeia thailandensis* and *Escherichia coli* EAEC Sci-1 T6SS were shown to have antibacterial activity (Russell et al., 2013; Flaugnatti et al., 2016). In *X. fragariae*, putative antibacterial toxins containing hydrolase domains and a colicin-DNAse domain are associated with orphan VgrGs (**Table 1** and **Supplementary Table S3**). Similarly, a variety of putative effectors with hydrolase and phospholipase domains were found associated to the variable regions of *L. enzymogenes* group 4 cluster, which has 3 class II *vgrG* genes (**Table 1** and **Supplementary Table S3**). Interestingly, five putative predicted effectors belonging to the Tox-REase-5 family of restriction endonucleases that includes TseT, a previously described antibacterial toxin from *Pseudomonas aeruginosa* H2-T6SS (Burkinshaw et al., 2018) are present in association with orphan DUF4123-containing proteins in *X. oryzae* strains and *Xanthomonas translucens* pv. *undulosa* (**Table 1**). As no example belonging to Xanthomonadales group 4 T6SS has been functionally characterized to date, based on the domains of putative effectors secreted by these systems, we hypothesize that members of group 4 might display antibacterial activity rather than anti-eukaryotic activity as observed for the *X. citri* T6SS from group 3* (Bayer-Santos et al., 2018). Interestingly, group 4 T6SSs are mostly found in Xanthomonadales genomes that lack the antibacterial T4SS (Souza et al., 2015; Sgro et al., 2019), as exemplified by all *X. oryzae* strains.

Prediction of T6SS effectors associated with VgrGs from group 3 identified a limited number of proteins of unknown function (**Table 1** and **Supplementary Table S3**). For the T6SSs from subgroup 3*, which are homologous to the anti-amoeba T6SS from *X. citri*, two proteins were frequently identified in the vicinity of VgrGs: an acid phosphatase and a protein with lectin and phosphodiesterase domains from the EEP family (endo/exonuclease/phosphatase) (**Table 1**). A distinct set of conserved hypothetical proteins were identified in subgroup 3***, including some with a TIR_2 domain typical of Toll-like receptors. Bacterial proteins containing these domains have been originally implicated in virulence by subversion of host immune responses, but recent work showed their roles as NADases that cleave NAD^+ and interfere with cellular metabolism (Cirl et al., 2008; Essuman et al., 2018). On the other hand, *D. japonica* group 3 T6SS is the sole representative in the group that presents several predicted antibacterial T6SS effectors (proteins containing amidase, muraminidase, and phospholipase domains) in the vicinity of VgrG, suggesting a role as an anti-prokaryotic weapon (**Table 1**). These observations lead us to speculate that distinct subgroups within group 3 might display diverse functions in Xanthomonadales.

REGULATION OF T6SSS

Regulation of T6SSs assembly and firing events occur at different levels: transcriptional, posttranscriptional, and post-translational. Transcriptional regulation of T6SS genes is highly variable among species and several components have been implicated in this activity, including the nucleoid-structuring protein H-NS, σ^{54} and regulators of nutrient acquisition pathways such as Fur and PhoB-PhoR (Brunet et al., 2011, 2015; Silverman et al., 2012). The *X. citri* T6SS is regulated at the transcriptional level by a mechanism involving an alternative sigma factor of the ECF family named EcfK and a transmembrane eukaryotic-like serine-threonine kinase (PknS), which is required for activation of EcfK (Bayer-Santos et al., 2018).

Interestingly, T6SS clusters from subgroup 3*** that do not contain EcfK/PknS carry a gene for a LysR-type transcriptional regulator that is predicted to be co-transcribed with the *tssA* gene (**Figure 1** and **Supplementary Table S4**). Genes encoding a VirA/VirG-like two-component system are also found in a putative operon associated with *lysR-tssA* in all subgroup 3*** T6SS, except for *Luteibacter rhizovicinus* that form a distinct branch within the group (**Figure 1** and **Supplementary Table S4**). This conservation in genome organization and the absence of *ecfK/pknS* suggest that these regulators might be involved in the control of subgroup 3*** T6SS gene expression. T6SS from *Dyella* spp. form a distinct branch in group 3 and do not present *ecfK/pknS* homologs or *lysR*-type genes near their T6SS clusters. No conserved gene encoding a transcriptional regulator was identified in T6SS clusters from group 4, except for a LuxR homolog that is present in the divergent *L. gummosus* cluster.

Distribution of T6SS putative post-translational regulators also differ among distinct clades. T6SS clusters from group 3 contain *tagF-pppA-ppkA-fha* genes, suggesting that activation of T6SS assembly and firing is dependent on the activation of the kinase PpkA, as originally described in *P. aeruginosa* (Mougous et al., 2007; Casabona et al., 2013). Recent work has demonstrated that PpkA acts by counteracting the inhibitory effect of TagF on T6SS activity, and the distinction between defensive and offensive T6SSs is mainly determined by the upstream input from protein(s) responsible for activation of PpkA (Lin et al., 2018; Ostrowski et al., 2018). In the defensive T6SS of *P. aeruginosa*, an incoming attack is sensed in the periplasm by TagQRST, while in the offensive *Serratia marcensens* T6SS, the signal is sensed by RtkS (Ostrowski et al., 2018). The offensive/defensive model has been described in only a few antibacterial T6SS and the mechanism of post-translational activation of anti-eukaryotic machines is possibly different.

Whether the post-translational regulatory cascade depending on tagF-pppA-ppkA-fha genes are functional in Xanthomonadales remains to be determined experimentally. Interestingly, T6SS clusters from group 4 do not encode *ppkA* or *pppA*, but instead contain a *tagF* gene (**Figure 1**), suggesting that a yet undescribed mechanism may be involved in relieving repression imposed by TagF in this group. The group 1 T6SS clusters from *Stenotrophomonas* sp. do not encode any component of the post-translational regulatory pathway, suggesting that activation occurs at the transcriptional level through σ^{54}.

PERSPECTIVES

Studies on the function, regulation and characterization of effector repertoire of T6SSs from Xanthomonadales is only just beginning. Much work is still required to understand how these systems contribute to the biology and pathogenesis of members of these important groups of bacteria. Further experimental evidence is needed to clarify whether the different subtypes of T6SSs pointed out in this review function as antibacterial and/or anti-eukaryotic weapons and whether the predicted putative effectors identified here are bona fide cargo proteins.

AUTHOR CONTRIBUTIONS

EB-S and CA-M performed *in silico* analysis, wrote and edited the review. LC performed *in silico* analysis and wrote the review. CF wrote and edited the review.

FUNDING

This work was supported by grants from the São Paulo Research Foundation (FAPESP) to EB-S (2017/02178-2), CF (2017/17303-7), and CA-M (2018/01852-4). EB-S (2018/04553-8) and LC (2017/02318-9) was supported by FAPESP. CF received a research fellowship award from the National Council for Scientific and Technological Development (CNPq). This study was also supported partly by the Coordenação de Aperfeiçoamento de Pessoal de Nível Superior – Brasil (CAPES) – Finance Code 001.

REFERENCES

Adegoke, A. A., Stenström, T. A., and Okoh, A. I. (2017). *Stenotrophomonas maltophilia* as an emerging ubiquitous pathogen: looking beyond contemporary antibiotic therapy. *Front. Microbiol.* 8:2276. doi: 10.3389/fmicb.2017.02276

Basler, M., and Mekalanos, J. (2012). Type 6 secretion dynamics within and between bacterial cells. *Science* 337:815. doi: 10.1126/science.1222901

Bayer-Santos, E., Lidia dos Passos, L., de Moraes Ceseti, L., Ratagami, C. Y., de Santana, E. S., da Silva, A. M., et al. (2018). Xanthomonas citri T6SS mediates resistance to *Dictyostelium* predation and is regulated by an ECF σ factor and cognate Ser/Thr kinase. *Environ. Microbiol.* 20, 1562–1575. doi: 10.1111/1462-2920.14085

Berg, G., and Martinez, J. L. (2015). Friends or foes: can we make a distinction between beneficial and harmful strains of the *Stenotrophomonas maltophilia* complex? *Front. Microbiol.* 6:241. doi: 10.3389/fmicb.2015.00241

Bernal, P., Allsopp, L. P., Filloux, A., and Llamas, M. A. (2017). The *Pseudomonas putida* T6SS is a plant warden against phytopathogens. *ISME J.* 11, 972–987. doi: 10.1038/ismej.2016.169

Bernal, P., Llamas, M. A., and Filloux, A. (2018). Type VI secretion systems in plant-associated bacteria. *Environ. Microbiol.* 20, 1–15. doi: 10.1111/1462-2920.13956

Bernard, C. S., Brunet, Y. R., Gavioli, M., Lloubès, R., and Cascales, E. (2011). Regulation of type VI secretion gene clusters by σ^{54} and cognate enhancer binding proteins. *J Bacteriol.* 193, 2158–2167. doi: 10.1128/JB.00029-11.

Bock, D., Medeiros, J., Tsao, H., Penz, T., Weiss, G., Aistleitner, K., et al. (2017). *In situ* architecture, function, and evolution of a contractile injection system. *Science* 357, 713–717. doi: 10.1126/science.aan7904

Bondage, D. D., Lin, J., Ma, L., Kuo, C., and Lai, E. (2016). VgrG C terminus confers the type VI effector transport specificity and is required for binding with PAAR and adaptor – effector complex. *Proc. Natl. Acad. Sci. U.S.A.* 113, E3931–E3940. doi: 10.1073/pnas.1600428113

Boyer, F., Fichant, G., Berthod, J., Vandenbrouck, Y., and Attree, I. (2009). Dissecting the bacterial type VI secretion system by a genome wide *in silico* analysis: what can be learned from available microbial genomic resources? *BMC Genomics* 10:104. doi: 10.1186/1471-2164-10-104

Bröms, J. E., Sjöstedt, A., and Lavander, M. (2010). The role of the *Francisella tularensis* pathogenicity island in type VI secretion, intracellular survival, and modulation of host cell signaling. *Front. Microbiol.* 1:136. doi: 10.3389/fmicb.2010.00136

Brunet, Y. R., Bernard, C. S., Gavioli, M., Lloubès, R., and Cascales, E. (2011). An epigenetic switch involving overlapping fur and DNA methylation optimizes expression of a type VI secretion gene cluster. *PLoS Genet.* 7:e1002205. doi: 10.1371/journal.pgen.1002205

Brunet, Y. R., Khodr, A., Logger, L., Aussel, L., Mignot, T., Rimsky, S., et al. (2015). H-NS silencing of the *salmonella* pathogenicity island 6-encoded type VI secretion system limits *Salmonella enterica* serovar typhimurium interbacterial killing. *Infect. Immun.* 83, 2738–2750. doi: 10.1128/IAI.00198-15

Burkinshaw, B. J., Liang, X., Wong, M., Le, A. N. H., Lam, L., and Dong, T. G. (2018). A type VI secretion system effector delivery mechanism dependent on PAAR and a chaperone–co-chaperone complex. *Nat. Microbiol.* 3, 632–640. doi: 10.1038/s41564-018-0144-4.

Büttner, D., and Bonas, U. (2010). Regulation and secretion of *Xanthomonas* virulence factors. *FEMS Microbiol. Rev.* 34, 107–133. doi: 10.1111/j.1574-6976.2009.00192.x

Casabona, M. G., Silverman, J. M., Sall, K. M., Boyer, F., Couté, Y., Poirel, J., et al. (2013). An ABC transporter and an outer membrane lipoprotein participate in posttranslational activation of type VI secretion in *Pseudomonas aeruginosa*. *Environ. Microbiol.* 15, 471–486. doi: 10.1111/j.1462-2920.2012.02816.x

Christensen, P., and Cook, F. D. (2009). *Lysobacter*, a new genus of nonfruiting. gliding bacteria with a high base ratio. *Int. J. Syst. Bacteriol.* 28, 367–393. doi: 10.1099/00207713-28-3-367

Cianfanelli, F. R., Monlezun, L., and Coulthurst, S. J. (2016). Aim. load, fire: the type VI secretion system, a bacterial nanoweapon. *Trends Microbiol.* 24, 51–62. doi: 10.1016/j.tim.2015.10.005

Cirl, C., Wieser, A., Yadav, M., Duerr, S., Schubert, S., Fischer, H., et al. (2008). Subversion of toll-like receptor signaling by a unique family of bacterial Toll/interleukin-1 receptor domain-containing proteins. *Nat. Med.* 14:399. doi: 10.1038/nm1734

De Maayer, P., Venter, S., Kamber, T., Duffy, B., Coutinho, T., and Smits, T. (2011). Comparative genomics of the Type VI secretion systems of *Pantoea* and *Erwinia* species reveals the presence of putative effector islands that may be translocated by the VgrG and Hcp proteins. *BMC Genomics* 12:576. doi: 10.1186/1471-2164-12-576

Dix, S. R., Owen, H. J., Sun, R., Ahmad, A., Shastri, S., Spiewak, H. L., et al. (2018). Structural insights into the function of type VI secretion system TssA subunits. *Nat. Commun.* 9:4765. doi: 10.1038/s41467-018-07247-1

Durand, E., Cambillau, C., Cascales, E., and Journet, L. (2014). VgrG. Tae, Tle, and beyond: the versatile arsenal of Type VI secretion effectors. *Trends Microbiol.* 22, 498–507. doi: 10.1016/j.tim.2014.06.004

Essuman, K., Summers, D., Sasaki, Y., Mao, X., Yim, A., DiAntonio, A., et al. (2018). TIR domain proteins are an ancient family of NAD(+)-consuming enzymes. *Curr. Biol.* 28, 421–430. doi: 10.1016/j.cub.2017.12.024

Flaugnatti, N., Le, T., Canaan, S., Aschtgen, M., Nguyen, V., Blangy, S., et al.

(2016). A phospholipase A1 antibacterial type VI secretion effector interacts directly with the C-terminal domain of the VgrG spike protein for delivery. *Mol. Microbiol.* 99, 1099–1118. doi: 10.1111/mmi.13292

Forster, A., Planamente, S., Manoli, E., Lossi, N. S., Freemont, P. S., and Filloux, A. (2014). Coevolution of the ATPase ClpV, the sheath proteins TssB and TssC and the accessory protein TagJ/HsiE1 distinguishes type VI secretion classes. *J. Biol. Chem.* 289, 33032–33043. doi: 10.1074/jbc.M114.600510

Hachani, A., Wood, T. E., and Filloux, A. (2016). Type VI secretion and anti-host effectors. *Curr. Opin. Microbiol.* 29, 81–93. doi: 10.1016/j.mib.2015.11.006

Hayward, A. C., Fegan, N., Fegan, M., and Stirling, G. R. (2010). *Stenotrophomonas* and *Lysobacter*: ubiquitous plant-associated gamma-*proteobacteria* of developing significance in applied microbiology. *J. Appl. Microbiol.* 108, 756–770. doi: 10.1111/j.1365-2672.2009.04471.x

Jiang, F., Waterfield, N. R., Yang, J., Yang, G., and Jin, Q. (2014). A *Pseudomonas aeruginosa* type VI secretion phospholipase D effector targets both prokaryotic and eukaryotic cells. *Cell Host Microbe* 15, 600–610. doi: 10.1016/j.chom.2014.04.010

Kanehisa, M., and Goto, S. (2000). KEGG: kyoto encyclopedia of genes and genomes. *Nucleic Acids Res.* 28, 27–30. doi: 10.1093/nar/28.1.27

Kapitein, N., Bönemann, G., Pietrosiuk, A., Seyffer, F., Hausser, I., Locker, J. K., et al. (2013). ClpV recycles VipA/VipB tubules and prevents non-productive tubule formation to ensure efficient type VI protein secretion. *Mol. Microbiol.* 87, 1013–1028. doi: 10.1111/mmi.12147

Ko, H. S., Jin, R., De Krishnan, H. B., Lee, S. B., and Kim, K. Y. (2009). Biocontrol ability of *Lysobacter antibioticus* HS124 against phytophthora blight is mediated by the production of 4-hydroxyphenylacetic acid and several lytic enzymes. *Curr. Microbiol.* 59, 608–615. doi: 10.1007/s00284-009-9481-0

Kumar, S., Stecher, G., and Tamura, K. (2016). MEGA7: molecular evolutionary genetics analysis version 7.0 for bigger datasets. *Mol. Biol. Evol.* 33, 1870–1874. doi: 10.1093/molbev/msw054

Leyns, F., De Cleene, M., Swings, J. G., and De Ley, J. (1984). The host range of the genus *Xanthomonas*. *Bot. Rev.* 5, 308–356. doi: 10.1007/BF02862635

Liang, X., Moore, R., Wilton, M., Wong, M. J. Q., Lam, L., and Dong, T. G. (2015). Identification of divergent type VI secretion effectors using a conserved chaperone domain. *Proc. Natl. Acad. Sci. U.S.A.* 112, 9106–9111. doi: 10.1073/pnas.1505317112

Lien, Y.-W., and Lai, E.-M. (2017). Type VI secretion effectors: methodologies and biology. *Front. Cell. Infect. Microbiol.* 7:254. doi: 10.3389/fcimb.2017.00254

Lin, J. S., Pissaridou, P., Wu, H. H., Tsai, M. D., Filloux, A., and Lai, E. M. (2018). TagF-mediated repression of bacterial type VI secretion systems involves a direct interaction with the cytoplasmic protein Fha. *J. Biol. Chem.* 293, 8829–8842. doi: 10.1074/jbc.RA117.001618

Ma, J., Sun, M., Dong, W., Pan, Z., Lu, C., and Yao, H. (2017). PAAR-Rhs proteins harbor various C-terminal toxins to diversify the antibacterial pathways of type VI secretion systems. *Environ. Microbiol.* 19, 345–360. doi: 10.1111/1462-2920.13621

Ma, L.-S., Hachani, A., Lin, J.-S., Filloux, A., and Lai, E.-M. (2014). *Agrobacterium tumefaciens* deploys a superfamily of type VI secretion DNase effectors as weapons for interbacterial competition in planta. *Cell Host Microbe* 16, 94–104. doi: 10.1016/j.chom.2014.06.002

Mougous, J. D., Gifford, C. A., Ramsdell, T. L., and Mekalanos, J. J. (2007). Threonine phosphorylation post-translationally regulates protein secretion in *Pseudomonas aeruginosa*. *Nat. Cell Biol.* 9, 797–803. doi: 10.1038/ncb1605

Naushad, S., Adeolu, M., Wong, S., Sohail, M., Schellhorn, H. E., and Gupta, R. S. (2015). A phylogenomic and molecular marker based taxonomic framework for the order Xanthomonadales: proposal to transfer the families *Algiphilaceae* and *Solimonadaceae* to the order Nevskiales ord. nov. and to create a new family within the order Xanthomonadales. Antonie van Leeuwenhoek. *Int. J. Gen. Mol. Microbiol.* 107, 467–485. doi: 10.1007/s10482-014-0344-8

Nguyen, V., Douzi, B., Durand, E., Roussel, A., Cascales, E., and Cambillau, C. (2018). Towards a complete structural deciphering of type VI secretion system. *Curr. Opin. Struct. Biol.* 49, 77–84. doi: 10.1016/j.sbi.2018.01.007

Ostrowski, A., Cianfanelli, F. R., Porter, M., Mariano, G., Peltier, J., Wong, J. J., et al. (2018). Killing with proficiency: integrated post-translational regulation of an offensive Type VI secretion system. *PLoS Pathog.* 14:e1007230. doi: 10.1371/journal.ppat.1007230

Planamente, S., Salih, O., Manoli, E., Albesa-Jové, D., Freemont, P. S., and Filloux, A. (2016). TssA forms a gp6-like ring attached to the type VI secretion sheath. *EMBO J.* 35, 1613–1627. doi: 10.15252/embj.201694024

Rapicavoli, J., Ingel, B., Blanco-Ulate, B., Cantu, D., and Roper, C. (2018). *Xylella fastidiosa*: an examination of a re-emerging plant pathogen. *Mol. Plant Pathol.* 19, 786–800. doi: 10.1111/mpp.12585

Russell, A. B., Leroux, M., Hathazi, K., Agnello, D. M., Ishikawa, T., Wiggins, P. A., et al. (2013). Diverse type VI secretion phospholipases are functionally plastic antibacterial effectors. *Nature* 496, 508–512. doi: 10.1038/nature12074

Russell, A. B., Wexler, A. G., Harding, B. N., Whitney, J. C., Bohn, A. J., Goo, Y. A., et al. (2014). A Type VI secretion-related pathway in bacteroidetes mediates interbacterial antagonism. *Cell Host Microbe* 16, 227–236. doi: 10.1016/j.chom.2014.07.007

Ryan, R. P., Vorhölter, F. J., Potnis, N., Jones, J. B., Van Sluys, M. A., Bogdanove, A. J., et al. (2011). Pathogenomics of *Xanthomonas*: understanding bacterium-plant interactions. *Nat. Rev. Microbiol.* 9, 344–355. doi: 10.1038/nrmicro2558

Saddler, G., and Bradbury, J. (2005). "Order III. Xanthomonadales ord. nov," in *Bergey's Manual of Systematic Bacteriology*. ed D. J. Brenner et al. Austin: Springer, 63–122. doi: 10.1007/0-387-28022-7_3

Sgro, G. G., Oka, G. U., Souza, D. P., Cenens, W., Bayer-Santos, E., Matsuyama, B. Y., et al. (2019). Bacteria-killing type IV secretion systems. *Front. Microbiol.* 10:1078. doi: 10.3389/fmicb.2019.01078

Si, M., Wang, Y., Zhang, B., Zhao, C., Kang, Y., Bai, H., et al. (2017a). The type VI secretion system engages a redox-regulated dual-functional heme transporter for zinc acquisition. *Cell Rep.* 20, 949–959. doi: 10.1016/j.celrep.2017.06.081

Si, M., Zhao, C., Burkinshaw, B., Zhang, B., Wei, D., Wang, Y., et al. (2017b). Manganese scavenging and oxidative stress response mediated by type VI secretion system in *Burkholderia thailandensis*. *Proc. Natl. Acad. Sci. U.S.A.* 114, E2233–E2242. doi: 10.1073/pnas.1614902114

Silverman, J. M., Austin, L. S., Hsu, F., Hicks, K. G., Hood, R. D., and Mougous, J. D. (2011). Separate inputs modulate phosphorylation-dependent and -independent type VI secretion activation. *Mol. Microbiol.* 82, 1277–1290. doi: 10.1111/j.1365-2958.2011.07889.x

Silverman, J. M., Brunet, Y. R., Cascales, E., and Mougous, J. D. (2012). Structure and regulation of the type VI secretion system. *Annu. Rev. Microbiol.* 66, 453–472. doi: 10.1146/annurev-micro-121809-151619

Souza, D. P., Oka, G. U., Alvarez-Martinez, C. E., Bisson-Filho, A. W., Dunger, G., Hobeika, L., et al. (2015). Bacterial killing via a type IV secretion system. *Nat. Commun.* 6, 1–9. doi: 10.1038/ncomms7453

Trunk, K., Peltier, J., Liu, Y., Dill, B., Walker, L., Gow, N., et al. (2018). The type VI secretion system deploys antifungal effectors against microbial competitors. *Nat. Microbiol.* 3, 920–931. doi: 10.1038/s41564-018-0191-x

Unterweger, D., Kostiuk, B., Otjengerdes, R., Wilton, A., Diaz-Satizabal, L., and Pukatzki, S. (2015). Chimeric adaptor proteins translocate diverse type VI secretion system effectors in *Vibrio cholerae*. *EMBO J.* 34, 2198–2210. doi: 10.15252/embj.201591163

Wang, J., Yang, B., Leier, A., Marquez-Lago, T., Hayashida, M., Rocker, A., et al. (2018). Bastion6: a bioinformatics approach for accurate prediction of type VI secreted effectors. *Bioinformatics* 34, 2546–2555. doi: 10.1093/bioinformatics/bty155

Wang, T., Si, M., Song, Y., Zhu, W., Gao, F., Wang, Y., et al. (2015). Type VI secretion system transports Zn2+ to combat multiple stresses and host immunity. *PLoS Pathog.* 11:e1005020. doi: 10.1371/journal.ppat.1005020

Weber, B., Hennon, S., Wright, M., Scott, N., de Berardinis, V., Foster, L., et al. (2016). Genetic dissection of the type vi secretion system in *Acinetobacter* and identification of a novel peptidoglycan hydrolase. tagx, required for its biogenesis. *MBio* 7, e1253–e1216. doi: 10.1128/mBio.01253-16

Williams, K. P., Gillespie, J. J., Sobral, B. W. S., Nordberg, E. K., Snyder, E. E., Shallom, J. M., et al. (2010). Phylogeny of gamma proteobacteria. *J. Bacteriol.* 192, 2305–2314. doi: 10.1128/JB.01480-09

Zoued, A., Durand, E., Brunet, Y. R., Spinelli, S., Douzi, B., Guzzo, M., et al. (2016). Priming and polymerization of a bacterial contractile tail structure. *Nature* 531, 59–63. doi: 10.1038/nature17182

16

Targeting the Type II Secretion System: Development, Optimization and Validation of a High-Throughput Screen for the Identification of Small Molecule Inhibitors

Ursula Waack[1], Tanya L. Johnson[1,2], Khalil Chedid[1], Chuanwu Xi[3], Lyle A. Simmons[4], Harry L. T. Mobley[1] and Maria Sandkvist[1]*

[1] Department of Microbiology and Immunology, University of Michigan Medical School, Ann Arbor, MI, United States, [2] Department of Chemistry, Eastern Michigan University, Ypsilanti, MI, United States, [3] Department of Environmental Health Sciences, University of Michigan School of Public Health, Ann Arbor, MI, United States, [4] Department of Molecular, Cellular, and Developmental Biology, University of Michigan, Ann Arbor, MI, United States

***Correspondence:**
Maria Sandkvist
mariasan@umich.edu

Nosocomial pathogens that develop multidrug resistance present an increasing problem for healthcare facilities. Due to its rapid rise in antibiotic resistance, *Acinetobacter baumannii* is one of the most concerning gram-negative species. *A. baumannii* typically infects immune compromised individuals resulting in a variety of outcomes, including pneumonia and bacteremia. Using a murine model for bacteremia, we have previously shown that the type II secretion system (T2SS) contributes to *in vivo* fitness of *A. baumannii.* Here, we provide support for a role of the T2SS in protecting *A. baumannii* from human complement as deletion of the T2SS gene *gspD* resulted in a 100-fold reduction in surviving cells when incubated with human serum. This effect was abrogated in the absence of Factor B, a component of the alternative pathway of complement activation, indicating that the T2SS protects *A. baumannii* against the alternative complement pathway. Because inactivation of the T2SS results in loss of secretion of multiple enzymes, reduced *in vivo* fitness, and increased sensitivity to human complement, the T2SS may be a suitable target for therapeutic intervention. Accordingly, we developed and optimized a whole-cell high-throughput screening (HTS) assay based on secreted lipase activity to identify small molecule inhibitors of the T2SS. We tested the reproducibility of our assay using a 6,400-compound library. With small variation within controls and a dynamic range between positive and negative controls, the assay had a z-factor of 0.65, establishing its suitability for HTS. Our screen identified the lipase inhibitors Orlistat and Ebelactone B demonstrating the specificity of the assay. To eliminate inhibitors of lipase activity and lipase expression, two counter assays were developed and optimized. By implementing these assays, all seven tricyclic antidepressants present in the library were found to be inhibitors of the lipase, highlighting the potential of identifying alternative targets for approved pharmaceuticals. Although no T2SS inhibitor was identified among the compounds that reduced lipase activity by ≥30%, our small proof-of-concept pilot study indicates that the HTS regimen is simple, reproducible, and specific and that it can be used to screen larger libraries for the identification of T2SS inhibitors that may be developed into novel *A. baumannii* therapeutics.

Keywords: ***Acinetobacter baumannii*, type II secretion, high-throughput screening, small molecule inhibitors, LipA, lipase activity, antibiotic resistance, antidepressants**

INTRODUCTION

A growing concern in hospitals, nursing homes, and other healthcare facilities is the increasing frequency of antibiotic resistant infections that result in longer hospital stays, higher costs, and increased mortality. The ESKAPE pathogens *Enterococcus faecium*, *Staphylococcus aureus*, *Klebsiella pneumoniae*, *Acinetobacter baumannii*, *Pseudomonas aeruginosa*, and *Enterobacter* species have attracted considerable attention as they cause the majority of nosocomial infections (Rice, 2008). Infections caused by *A. baumannii* are prevalent with ~45,000 cases per year in the United States alone. Globally, there are about 1 million cases annually (Spellberg and Rex, 2013) and reports suggest that *A. baumannii* may be the leading cause of nosocomial infections in some countries (Wong et al., 2016). It is estimated that 50% of these infections are caused by antibiotic-resistant strains (Spellberg and Rex, 2013). Exposure to *A. baumannii* can result in a variety of infections including pneumonia, urinary tract infection, bacteremia, meningitis, skin, and wound infections that may lead to sepsis (Bergogne-Berezin and Towner, 1996; Maragakis and Perl, 2008). Considered an opportunist, *A. baumannii* typically infects immune-compromised individuals but more recently isolated strains may not be restricted to this patient population, possibly as a consequence of increased virulence (Jones et al., 2015; Paterson and Harris, 2015). The remarkable ability of *A. baumannii* to form biofilm and resist dry environments (Jawad et al., 1998; Espinal et al., 2012) may explain its prevalence in healthcare environments (Weernink et al., 1995; Catalano et al., 1999). Additional contributing factors include multi- or pan-antibiotic resistance (Maragakis and Perl, 2008; Leite et al., 2016), which is due, in part, to intrinsic properties of the outer membrane of *A. baumannii* and its notable ability to acquire foreign DNA through horizontal gene transfer (de Vries and Wackernagel, 2002).

The rise in antibiotic resistance rapidly reduces the options of effective treatment and calls for the identification of new therapeutic approaches. A recommended strategy combines antibiotics with drugs that target resistance mechanisms such as Augmentin, which consists of Amoxicillin and the β-lactamase inhibitor Clavulanate. Other feasible options include the combination of antibiotics with inhibitors of drug efflux pumps or outer membrane permeabilizers (Gill et al., 2015). Identification of new therapeutic targets is also necessary. These may include essential processes such as lipopolysaccharide synthesis and transport as well as factors that contribute to *in vivo* fitness and virulence.

One of the first studies to target virulence factors using HTS of small molecule libraries identified a compound that inhibits dimerization of ToxT, a virulence regulator in *Vibrio cholerae* (Hung et al., 2005; Shakhnovich et al., 2007). This inhibitor abolishes the production of cholera toxin and decreases TCP-mediated colonization in an infant mouse model (Hung et al., 2005). Other studies have screened for biologicals or chemical compounds that target colonization factors, such as curli and type 1 pili, toxins, protein secretion pathways or quorum sensing systems (Steadman et al., 2014; Gill et al., 2015; Ruer et al., 2015; Hauser et al., 2016). With a few exceptions, it is too soon to evaluate the outcome of these studies and their success; however, some of these potential anti-virulence drugs are in various stages of development and are being analyzed in animal models or clinical trials (Pan et al., 2009; Rasko and Sperandio, 2010; Hauser et al., 2016). An IgG antibody that targets the binding of anthrax toxin to its receptor is currently used as an antitoxin in combination therapy for the treatment of *Bacillus anthracis* infections (Hendricks et al., 2014) and demonstrates the feasibility of targeting disease-causing components of pathogens.

Secretion systems are particularly attractive targets for alternative therapeutics as their inactivation interferes with the delivery of entire batteries of secreted virulence factors. Therefore, several HTSs have been designed to identify small molecule inhibitors of the type III secretion system, which is present in many gram-negative human pathogens and secretes a wide variety of virulence effectors (Nordfelth et al., 2005; Aiello et al., 2010). Another secretion system, the type II secretion system (T2SS), is responsible for the secretion of numerous degradative enzymes and toxins that contribute to survival in the environment and the mammalian host and may also be a suitable target for alternative therapeutics (Sandkvist, 2001; Cianciotto, 2005; Cianciotto and White, 2017). As with many gram-negative pathogens, *A. baumannii* possesses a functional T2SS (Elhosseiny et al., 2016; Johnson et al., 2016). The T2SS forms an apparatus that spans both the inner and outer membrane and is encoded by 12 essential genes, *gspC-M* and *pilD* (Korotkov et al., 2012; Thomassin et al., 2017). T2S substrates are synthesized with an N-terminal signal peptide that allows for translocation from the cytoplasm to the periplasm via the general export (Sec) or twin arginine translocation (Tat) pathways. Once in the periplasm, the signal sequence is cleaved, the protein folds, and interacts with the T2SS to finally exit the cell via a gated outer membrane pore formed by GspD (Reichow et al., 2010; Douzi et al., 2011; Yan et al., 2017). GspD connects to GspC, one of the components of the inner membrane platform that also consists of the transmembrane proteins GpsF, L, and M (Sandkvist et al.,

1999; Py et al., 2001; Michel et al., 2007; Abendroth et al., 2009; Douzi et al., 2011; Korotkov et al., 2011). The pseudopilins GspG, H, I, J, and K make a pseudopilus, a structure homologous to the Type IV pilus, while PilD cleaves and methylates the pseudopilin subunits prior to their assembly (Nunn and Lory, 1993; Durand et al., 2005; Cisneros et al., 2012). The entire system is powered by a cytoplasmic ATPase, GspE (Camberg and Sandkvist, 2005; Camberg et al., 2007). All of these proteins, including their expression and interactions, are potential targets for a therapeutic compound (**Figure 1**).

Recent work by our laboratory and others has demonstrated the benefit of having a functional T2SS for colonization by *A. baumannii* and *A. nosocomialis* (Elhosseiny et al., 2016; Harding et al., 2016; Johnson et al., 2016). Inactivation of the T2SS or one of its substrates results in diminished survival in murine models for bacteremia and pneumonia, indicating that screening for compounds that target the T2SS may lead to the identification of *A. baumannii* virulence inhibitors. In this study, we describe the development and optimization of a HTS to identify small molecule inhibitors of the T2SS in *A. baumannii*. In addition, we highlight the need for inclusion of high-throughput counter-screens to remove compounds with alternative targets.

METHODS

Bacterial Strains and Plasmids

All bacterial strains and plasmids are described in **Table 1**. All strains were cultured overnight in Luria-Bertani (LB) broth. When necessary, LB broth was supplemented with carbenicillin (100 μg/ml) for plasmid maintenance. The studies were conducted under biosafety level 2 conditions.

Compound Library

All compounds tested in this study are from commercially available libraries acquired and maintained in 384-well plates in DMSO at −20°C at a concentration of 2 mM by the Center for Chemical Genomics at the University of Michigan. The five libraries include MS2400, NCC, Pathways, Prestwick, and LOPAC. MS2400 is a collection of FDA approved drugs plus compounds with known biological activity obtained from Microscource Discovery (Spectrum Collection). NCC is a library with compounds that have been used in human clinical trials. The Pathways collection is comprised of known active compounds with a variety of targets. The Prestwick library is composed of approved drugs which are safe for use in humans. Finally, the LOPAC collection is the Library of Pharmacologically Active Compounds from Sigma.

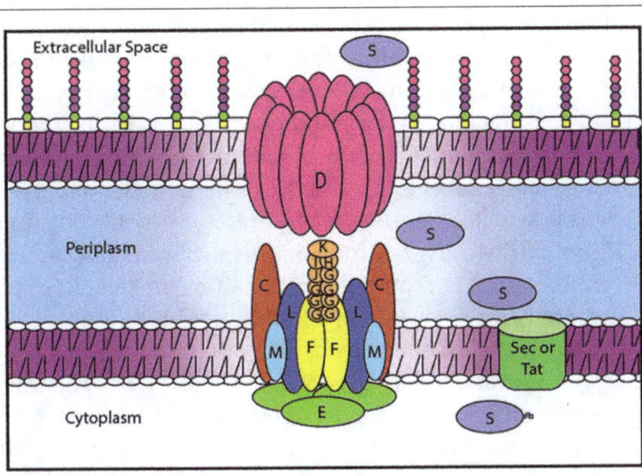

FIGURE 1 | Potential Targets of a T2SS inhibitor. T2SS inhibitors may block transcription of the T2SS genes or translation, post-translational modification, protein-protein interactions or function of the T2SS proteins. Blocking the recognition by GspC/GspD of the T2SS substrates (S) to be secreted by the T2SS system could also halt secretion.

Serum Bactericidal Assay

After cultures were grown overnight in LB, the cells were separated from supernatant by centrifugation for 10 min at 3,500 rpm. The cells were washed in PBS and diluted 1:100. Equal volumes of cells and either 100% normal human serum, heat-inactivated human serum or factor depleted sera were incubated together for 30 min at 37°C. Samples were diluted and plated on LB agar to obtain CFUs.

TABLE 1 | List of bacterial strains and plasmids used in study.

Strain or plasmid	Characteristics	Source or study
STRAIN		
17978	WT for this study	ATCC
17978Δ*gspD*	T2SS mutant	Johnson et al., 2016
17978Δ*gspN*		Johnson et al., 2016
17978Δ*lipA*		Johnson et al., 2016
AB031	Clinical strain	Tilley et al., 2014
AB031 Δ*gspD*	T2SS mutant	Waack et al., in preparation.
AB 0057	Clinical strain, tet^R, chl^R, trim^R, carb^R	Adams et al., 2008
AB 5075	Clinical strain, tet^R, rif^R, carb^R, trim^R	Jacobs et al., 2014
P020	Clinical strain, cef^R	Greene et al., 2016
C038	Clinical strain, cef^R	Greene et al., 2016
C058	Clinical strain, mer^R, azt^R, cef^R, cip^R, lev^R, imi^R	Greene et al., 2016
C076	Clinical strain, azt^R, cef^R, cip^R, lev^R, trim^R	Greene et al., 2016
P084	Clinical strain, mer^R, azt^R, cef^R, cip^R, lev^R, imi^R	Greene et al., 2016
C097	Clinical strain	Greene et al., 2016
P102	Clinical strain, mer^R, azt^R, cef^R, cip^R, lev^R, imi^R	Greene et al., 2016
P143	Clinical strain, azt^R, cef^R	Greene et al., 2016
PLASMID		
pMMB67EH	Low copy vector (Ap^r)	Fürste et al., 1986
p*lipBA*	pMMB67EH-*lipBA* (Ap^r)	Johnson et al., 2016
p*gfp*	pMMB67EH-*gfp* (Ap^r)	Scott et al., 2001

azt, aztreonam; carb, carbapenems; cef, cefazolin; chl, chloramphenicol; cip, ciprofloxacin; gent, gentamycin; imi, imipenem; lev, levofloxacin; mer, meropenem; rif, rifampin; tet, tetracycline; trim, trimethoprim-sulfate; tob, tobramycin.

High-Throughput Lipase Assay

Overnight cultures of wild-type (WT) *A. baumannii* 17978/p*lipBA* and 17978Δ*gspD*/p*lipBA* were grown in LB broth. After growth, the cultures were centrifuged for 10 min at 3500 rpm to separate cells and supernatant. The supernatant was removed and the pellet was washed in Mueller-Hinton 2 (MH) and resuspended in the original volume. 10 μL of MH was added to each well of a 384-well Greiner 784080 plate. Compounds in DMSO or the DMSO control were added to the wells (0.05 μL) using Perkin Elmer Sciclone liquid handler with a 50 nl pintool attachment (in primary assay). For concentration-response assays, the TTP LabTech Mosquito X1 was used to place variable volumes (0.02–1.2 μL) of compounds to the wells. 17978/p*lipBA* was diluted and added to all the negative control wells as well as to the sample wells while 17978Δ*gspD*/p*lipBA* was diluted and added to the positive control wells. All cultures had a starting OD_{600} of 0.005 and were supplemented with 50 μM IPTG (isopropyl-β-D-thiogalactopyranoside) to induce expression of the *lipBA* genes. Plates were centrifuged 1 min at 1,000 rpm to ensure all liquid was at the bottom of the well. Plates were incubated overnight at 24°C in a humidified incubator. After 16 h incubation, the OD_{600} value of each culture was recorded. An optimized lipase assay was used to measure lipase activity. Briefly, the cultures were incubated with 0.45 mM 4-nitrophenyl myristate, 80 mM Tris-HCl (pH 8.0), and 0.15% Triton X-100 and the absorbance at 415 nm was measured over time using the PE EnVision Multimode Plate Reader. All data were uploaded to MScreen for analysis. MScreen is a data analysis and storage system created by the Center for Chemical Genomics intended for the processing of high-throughput data generated by users of the center (Jacob et al., 2012).

Lipase Inhibitor Assay

Cultures of *A. baumannii* 17978/p*lipBA* and 17978Δ*gspD*/p*lipBA* were grown overnight in LB broth with 50 μM IPTG and the supernatant and cells were separated by centrifugation. For concentration-response assays, the TTP LabTech Mosquito X1 was used to add variable volumes (0.02–1.2 μl) of compounds. Supernatant was added to the wells, buffer with lipase substrate was added and the change in OD_{415} was recorded as above. All data were uploaded to MScreen for analysis.

GFP Expression Assay

A. baumannii 17978/p and 17978/p*gfp* cultures were grown O/N as above. Cells were washed as described above for the lipase assay. Compounds in DMSO or the DMSO control were added to the wells of a black low-volume Greiner 784073 plate using Perkin Elmer Sciclone liquid handler with a 50 nl pintool attachment calibrated to deliver 200 nM. For concentration-response assays, the TTP LabTech Mosquito X1 was used to place variable volumes (0.02–1.2 μl). Cultures were diluted to OD_{600} = 0.005 in LB supplemented with 75 μM IPTG. Plates were centrifuged for 1 min at 1,000 rpm to ensure all liquid was at the bottom of the well and incubated overnight at room temperature in a humidified incubator. Fluorescence was measured after growth using a BMGLabtech PHERAstar (485 nm excitation, 520 nm emission). All data were uploaded to MScreen.

Lipase Assay

A modified version of the lipase assay reported by Johnson et al. (2016) was used. Briefly, overnight cultures of *A. baumannii* strains were cultured in LB broth. The lipase activity of each culture was measured by a spectrophotometer after addition of 0.9 mM of the substrate, 4-nitrophenyl myristate in 80 mM Tris/0.15% Triton X-100 buffer. The absorbance at 415 nm was measured over time at 37°C. All assays were performed in triplicate with means and standard deviations presented.

RESULTS

Recent studies have demonstrated the significance of the T2SS in colonization by *A. baumannii* and *A. nosocomialis* in murine models of bacteremia and pneumonia (Elhosseiny et al., 2016; Harding et al., 2016; Johnson et al., 2016). In our study, we also revealed that one of the secreted proteins, LipA, contributes to colonization (Johnson et al., 2016), possibly by aiding in nutrient acquisition through lipid hydrolysis. It is quite likely that other T2S substrates including the lipase LipH, the phospholipase LipAN, and/or proteases and other putative enzymes identified by proteomics similarly contribute to *in vivo* survival of *A. baumannii* (Elhosseiny et al., 2016; Harding et al., 2016).

In addition, a "serum resistance factor" may be secreted by the T2SS, as a previous study aimed at identifying factors that contribute to *A. baumannii* proliferation in human serum identified a *gspN* transposon insertion mutant with diminished serum survival (Jacobs et al., 2010). Here, we followed up on this finding and tested the possibility that an intact T2SS is required for *A. baumannii* ATCC 17978 to resist serum complement. Many isolates of *A. baumannii* survive in the presence of 100% serum; however, ATCC 17978 is sensitive to this concentration and, therefore, we conducted our experiments using 50% serum. The WT and Δ*gspD* mutant strains were incubated in 50% pooled human sera, and the CFUs were determined after 30 min incubation at 37°C. As a control, we used the Δ*lipA* mutant that has an intact T2SS but lacks one of the T2S substrates, LipA. We also treated the WT and mutant strains with heat-inactivated (HI; 56°C, 30 min) serum, in which the complement system is inactivated. While no loss of viability was observed for the WT and Δ*lipA* strains, only 1% of the Δ*gspD* mutant cells survived 30 min in normal serum (**Figure 2A**). Next, we subjected the Δ*gspD* mutant cells to factor C1q-depleted and factor B-depleted human sera. The majority of Δ*gspD* mutant cells survived in the absence of factor B, which is required for activation of the alternative complement pathway; while in the C1q-depleted serum, which is deficient in the classical pathway yet contains factor B, $<0.05\%$ of the Δ*gspD* mutant cells were viable (**Figure 2B**). This result suggests that the T2SS directs the outer membrane translocation of a factor that provides protection from the alternative pathway. In contrast, the Δ*gspN* mutant was not affected by human serum (**Figure 2A**), a result that differs from the study published by Jacobs et al. (2010). The lack of a serum sensitive phenotype for the Δ*gspN* mutant is consistent with our earlier finding that GspN is not required for T2S in *A. baumannii* and with the discovery by Wang et al. that showed GspN is not needed for

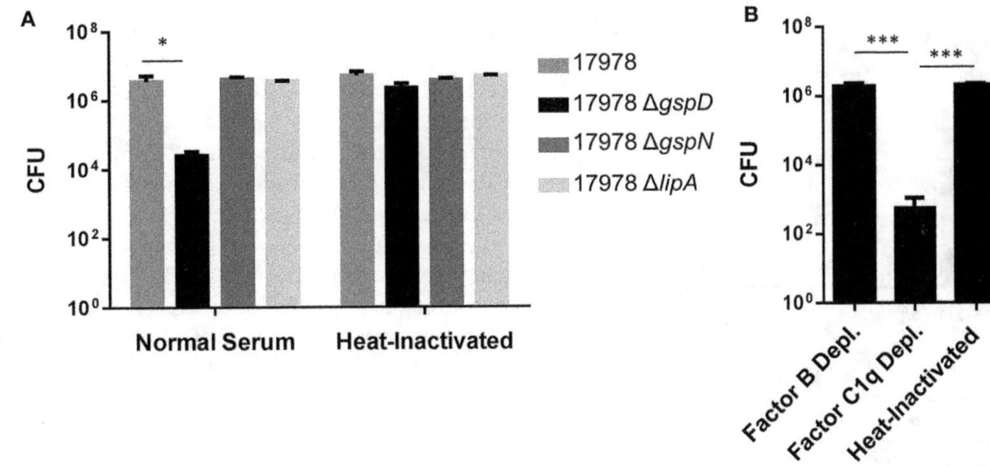

FIGURE 2 | Survival in serum depends on the T2SS. **(A)** Cells from overnight cultures were washed and incubated for 30 min at 37°C with either 50% normal human serum or heat-inactivated serum. Cells were plated for CFUs following incubation. $^{*}p < 0.05$ by Student *t*-test; $n = 3$. **(B)** Cells from overnight cultures were washed and incubated as above with Factor B deficient, C1q deficient, or heat-inactivated serum. $n = 3$, $^{***}p < 0.001$ by Student *t*-test.

survival of *A. baumannii* in a mouse model of pneumonia (Wang et al., 2014). We suggest, therefore, that the diminished growth observed for the *gspN* transposon mutant in human serum is due to a polar effect of the transposon on the downstream gene, *gspD*, which we show here is required for full protection from serum complement. Taken together with earlier findings, these results support the model that extracellular proteins secreted by the T2SS play important roles in the pathogenesis of *A. baumannii* and suggest that the T2SS may be an attractive target for therapeutic intervention.

Inhibiting the function of the T2SS would simultaneously prevent the secretion of many substrates causing a greater impact on *A. baumannii* survival than targeting individual T2S substrates. The T2SS components and their interactions (**Figure 1**) provide ideal targets for therapeutics because they are unique to the T2SS and are absent from most members of the human microbiota. Compounds could block the function of the outer membrane pore, inhibit interactions between the different components of the secretion apparatus, prevent recognition of the substrates by the apparatus, or interfere with the expression of the T2S proteins (**Figure 1**). To identify inhibitors of the T2SS, we developed a novel HTS approach.

Development of HTS Assay

In our previous study, we used a colorimetric lipase assay to measure the activity of the T2S lipase, LipA, in culture supernatant of strains overexpressing plasmid encoded *lipBA* genes. We used an overexpression strain because endogenous lipase production is very low during growth of *A. baumannii* ATCC 17978 in LB presenting a challenge for detection (Johnson et al., 2016). Further, because lipase activity is undetectable in the culture supernatant of T2SS mutants, this provides a robust assay that can be used as a readout for T2SS activity (Johnson et al., 2016) for the purpose of identifying T2SS inhibitors. However, testing the effect of a large number of compounds on LipA activity in cell-free culture supernatants would be cumbersome as it would involve an extra processing step. Thus, we compared cell-free culture supernatants and unfractionated cultures for T2SS activity (**Figure 3**). The culture supernatant and unfractionated culture of the 17978/p*lipBA* strain both showed significant lipase activity toward 4-nitrophenyl myristate, while there was little to no activity either in the supernatant or culture of the T2SS mutant, 17978Δ*gspD*/p*lipBA*. More importantly, the vast majority of the lipase activity in the unfractionated culture was generated by extracellular LipA thus allowing us to omit the step in which the culture supernatant is separated from cells. Consequently, we were able to develop an assay for HTS of small molecule inhibitors of T2SS that involved the addition of the lipase substrate directly to the bacterial cultures following growth in the presence and absence of compounds.

In our first attempt to miniaturize the assay we grew cultures of the WT 17978 strain and Δ*gspD* mutant in the wells of clear 96-well microtiter plates to which we then added 4-nitrophenyl myristate and determined the lipase activity by measuring the increase in absorbance at 415 nm over time. Prior to adding the lipase substrate, we measured the density of the cultures at 600 nm. This is an important step because many compounds, including known antibiotics, will affect the growth of the bacteria resulting in false positives. While the move to the microtiter format required titration of IPTG to induce *lipBA* expression, a reproducible difference in lipase activity could be measured in the WT and mutant cultures. We further developed the lipase secretion assay in flat-bottom 384-well plates using a MicromultiDrop liquid dispenser. As the conditions developed for the 96-well plates did not generate reproducible data in the 384-well format, we set up a systematic analysis in which we evaluated a variety of conditions to obtain the most consistent data. We varied growth media, culture volume, IPTG concentration, starting OD_{600} of the culture, growth temperature, length of growth, aeration, and

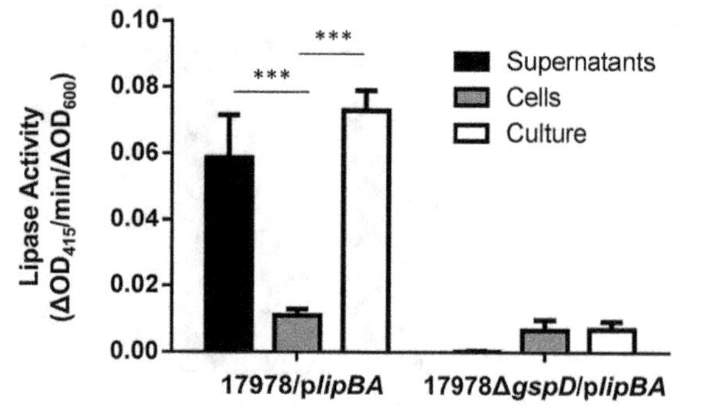

FIGURE 3 | Detection of T2S lipase activity in either culture supernatants, cells, or unfractionated bacterial cultures. Enzymatic activity of either the supernatants, cells, or unfractionated cultures of 17978/p*lipBA* or 17978Δ*gspD*/p*lipBA* strains cultured overnight in LB broth. The substrate 4-nitrophenyl-myristate was added to the samples and the change in OD_{415} was recorded over time. Activities were normalized to the OD_{600} of the original cultures. $n = 3$, ***$p < 0.001$ by One-Way Anova.

TABLE 2 | Optimization of assay conditions for development of a HTS.

Variables	Conditions tested	Final assay condition
Plate format	96 well vs. 384 well	384 well
Start OD	0.005, 0.0025, 0.001, 0.0005	0.005
Media	LB vs. Mueller-Hinton (MH)	MH
Volume	10, 15, 20, or 30 μL	10 μL
Temperature	20°, 24° (RT), 30°, 37°C	RT
Aeration	Shaking vs. non-shaking	Non-shaking
Substrate Concentration	0.225, 0.45, 0.9, 1.8 mM	0.45 mM

finally, 4-nitrophenyl myristate concentration (**Table 2**). The experimental setup that yielded the most consistent growth and reproducible lipase activity from well-to-well and plate-to-plate were obtained when cultures were grown in 10 μL MH broth (from a starting OD_{600} of 0.005) with 50 μM IPTG, in a humidified chamber at 24°C without shaking for 16 h (**Table 2**).

Pilot Screen

Following optimization, we screened 6,400 pharmacologically active compounds as well as FDA-approved drugs from the following libraries available at the University of Michigan Center for Chemical Genomics: MS2400, Prestwick, LOPAC, BioFocus NCC, and Focused Collections. This pilot screen was performed to evaluate the strength of the assay before moving on to larger compound libraries. Each 384-well plate contained 320 sample wells, 32 positive control wells, and 32 negative control wells. As the ultimate goal of the HTS is to identify T2SS inhibitors, the T2SS mutant, 17978Δ*gspD*/p*lipBA* served as our positive control while 17978/p*lipBA* served as the negative control. Both negative and positive controls were cultured in the presence of 0.5% DMSO while the sample wells containing 17978/p*lipBA* received the compounds resuspended in DMSO yielding a 0.5% final DMSO concentration. Following growth, the absorbance at 600 nm was measured for the cultures in each well. The average OD_{600} for the negative and positive controls were 0.21 ± 0.02 and 0.19 ± 0.02, respectively. The lipase substrate was then added, and the absorbance at 415 nm was measured over a period of 20 min at ambient temperature. The pilot screen yielded a z-factor of 0.65 [$z' = 1-(3^*(\sigma_p + \sigma_n)/(|\mu_p - \mu_n|))$](Zhang et al., 1999; **Figure 4A**) and coefficient of variation (CV, $CV = \sigma/\mu$) of 0.03 and 0.07 for the negative and positive controls, respectively. Initial active compounds were identified using statistical comparisons to positive and negative controls present on every plate. In the triage process, we selected compounds that resulted in all of the following: a reduction in lipase activity that was equal or >3 *SD* of the negative control, a minimum cut-off at 30% inhibition of lipase activity and an OD_{600} value >0.17 (**Figure 4**). Implementing these criteria yielded 191 compounds (3% hit rate). From these, we removed 22 compounds that are known antibiotics, such as gatifloxacin, clarithromycin, and levofloxacin and retested the remaining 169 compounds.

Each compound was re-tested twice in a concentration-dependent manner using the original DMSO stock and covering two orders of magnitude. IC_{50} values were calculated. Forty-eight compounds gave IC_{50} values of <30 μM. Following removal of compounds that affected growth, 34 active compounds remained (0.5% hit rate). Fresh powder of these compounds were ordered and retested. Of these compounds, 21 were confirmed as active. The compounds with the lowest IC_{50} values are known lipase inhibitors, Orlistat and Ebelactone B, and therefore likely had a direct effect on the lipase activity itself. Orlistat, a pancreatic lipase inhibitor, was the most potent of the compounds tested with an IC_{50} of 40.6 nM (**Figure 5A**). The other compounds exhibited IC_{50}s between 4.3 and 27 μM (**Table 3**). The titration curves of Orlistat (**Figure 5A**) and Oxyclozanide (**Figure 5B**) are shown as examples. Of note, compounds with low IC_{50} values included tricyclic antidepressants that are known to act as serotonin-norepinephrine re-uptake inhibitors. While these latter compounds may act on the secretion system, it is possible they also bind directly to the lipase via their hydrophobic rings. Our proof-of-concept pilot screen with $z' = 0.65$ and CV of 0.03 and 0.07 for the negative and positive controls, respectively, showed that our assay was reproducible and was capable of identifying compounds that result in a statistically significant reduction in lipase activity. However, the identification of compounds that are known lipase inhibitors emphasized the importance of developing counter screens and other follow-up assays to remove false positives and to assure specificity in order to identify T2S inhibitors.

Counter Screening

As the most potent compounds identified in our pilot screen described above represented lipase inhibitors, we developed a screen to eliminate these types of compounds. In this counter screen the bacterial cultures were not grown in the presence of compounds. Instead, a large batch of 17978p/*lipBA* culture was grown without compounds and following removal of cells, the

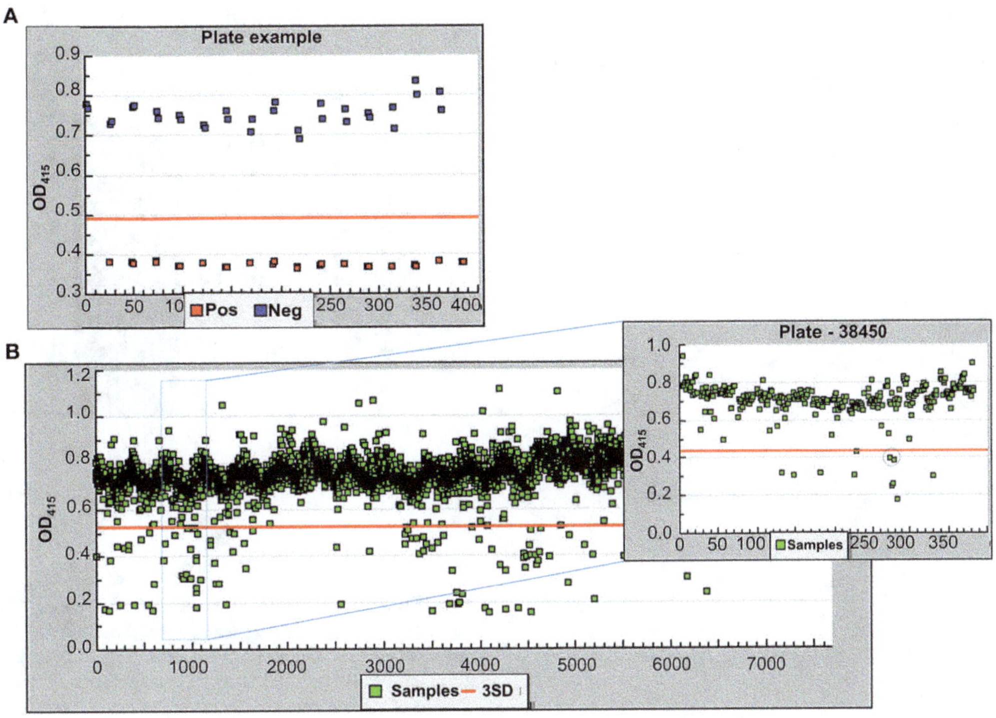

FIGURE 4 | Demonstrating the feasibility of the lipase assay for HTS. **(A)** One sample plate from the pilot screen showing the OD_{415} values for the positive (red) and negative (blue) controls. The z-factor for the entire pilot screen was calculated using $1-(3^*(\sigma_p + \sigma_n)/(|\mu_p - \mu_n|))$ where σ is the standard deviation and μ is the mean. **(B)** The results for 6,400 compounds tested in the pilot screen. All samples below the red line (3 *SD* from negative controls) were taken into consideration. A single sample plate is highlighted in the inset. The compound, Orlistat, indicated by the circle represents a hit in the primary screen and was tested further.

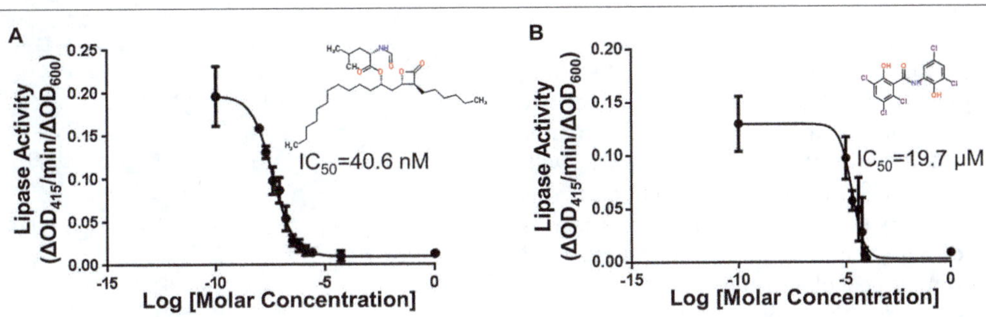

FIGURE 5 | Titrations of compounds and the corresponding IC_{50}. **(A)** Titration of Orlistat, a pancreatic lipase inhibitor that was identified in the primary screen. Increasing amounts of Orlistat was included during growth of 17978/p*lipBA*. After overnight incubation in 96-well plates, the substrate 4-nitrophenyl myristate was added and the lipase activity was measured. Activities have been normalized to OD_{600} of the cultures. $n = 3$, bars represent standard deviation from the mean. In follow-up counter screening, Orlistat was classified as an inhibitor of LipA activity **(B)** Titration of Oxyclozanide, an anthelmintic that was identified in the primary screen. Lipase activity in the presence of Oxyclozanide was measured as above. Activities have been normalized to OD_{600} of culture. $n = 3$, bars represent standard deviation from the mean. Following counter screening, Oxyclozanide was classified as an inhibitor of *lipA* expression from the plasmid.

cell-free culture supernatant containing the lipase was distributed in 384-well plates and incubated with the compounds, thus allowing us to identify compounds that inhibit the lipase itself. Following optimization, we identified the following conditions for the counter screen: (1) grow 17978p/*lipBA* with 50 μM IPTG for 16 h and remove cells by centrifugation; (2) dilute the supernatant 1:10 and add 10 μL to 384-well plates containing compounds; (3) add 10 μL 4-nitrophenyl myristate at 0.45 mM, incubate 10 min, and measure the change in absorbance at 415 nm.

An additional counter screen was developed for the elimination of compounds that interfere with plasmid-encoded *lipBA* expression. For this, we used the same plasmid backbone as p*lipBA* but substituted the *lipBA* gene with the *gfp* gene, which codes for Green Fluorescent Protein (p*gfp*). This plasmid was introduced into the WT 17978 strain (17978/p*gfp*), and without

TABLE 3 | IC_{50} values of compounds identified in primary screen and tested for concentration dependent inhibition.

Compound	IC_{50} (μM)
LIPASE INHIBITORS	
Ebelactone B ($n = 2$)	4.3
ANTI-DEPRESSANTS	
Vivactil	10
Fluoxetine	10.5
Aventyl	14.5
Duloxetine	16
Norcyclobenzaprine	16
Maprotiline	16
Desiprimine	16.5
Lofepramine	21
Sertraline	23.5
OTHER	
Perhexiline Maleate	4.7
Febuxostat	8
Alfuzosin	11
Stattic	12.5
Fendiline	13
Tomoxetine	14.5
Indatraline	27

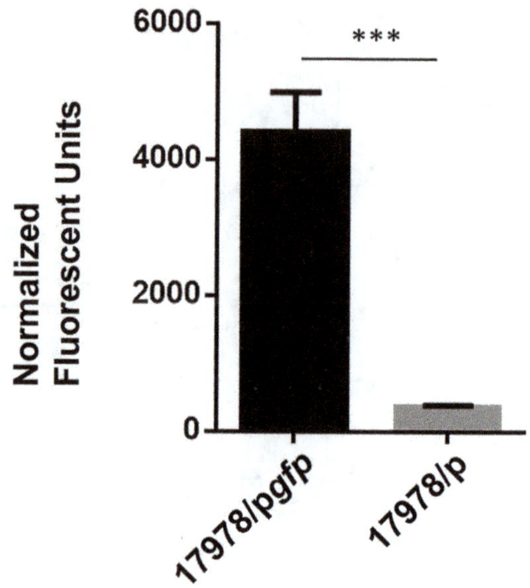

FIGURE 6 | Detection of GFP fluorescence. Strains of 17978/p and 17978/*pgfp* were grown overnight in the presence of IPTG to induce the expression of GFP. After growth, fluorescence of intact cells of each strain was measured at 485 nm excitation, 520 nm emission and normalized to the OD_{600} of the cultures. $n = 3$, $^{***}p < 0.001$ by Student *t*-test.

lysing the cells, reproducible GFP fluorescence was significantly higher than the fluorescence detected for 17978 with the vector alone negative control (17978/p; **Figure 6**). We optimized the conditions and applied them in the following counter screen: cultures were grown in 10 μL MH from a starting OD_{600} of 0.005, with 75 μM IPTG in a humid chamber at 24°C without shaking for 16 h using 17978/*pgfp* and 17978/p as negative and positive controls, respectively. For this counter screen, the compounds would be added to the cultures at the start of growth, with the fluorescence measured after growth. Any compound that is positive in both the primary screen and this counter screen is likely targeting expression of the plasmid-encoded lipase and should be removed from the pool of potential hits.

Twenty-one compounds that had responded in a concentration-dependent manner in the primary HTS and then confirmed when using fresh powders, were subjected to the lipase inhibitor and GFP counter screens. All of these compounds were identified as either lipase inhibitors ($n = 18$; **Table 3** and **Figure 5A**) or inhibitors of *lipBA* expression from the plasmid ($n = 3$; representative compound shown in **Figure 5B**) highlighting the necessity of utilizing counter screens before pursuing detailed characterization of false positives.

As we move forward to screen larger libraries to identify T2SS inhibitors, our protocol will involve the following order of analysis (**Figure 7**): the active compounds from the primary HTS (step 1) will be delivered in triplicates to three different sets of 384-well plates using the original DMSO stock solutions (step 2). The first set of plates will represent a repetition of the primary screen. The second and third sets of plates will be used to counter screen for compounds that inhibit

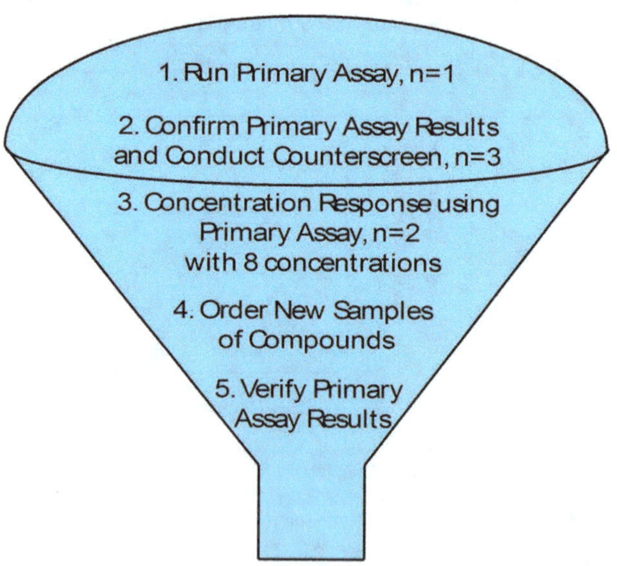

FIGURE 7 | HTS Schematic.

lipase activity or plasmid-borne gene expression, respectively (**Figure 7**). Compounds that are positive in the counter screens will be eliminated from further consideration, and the remaining compounds will be tested for their ability to prevent secretion over a range of concentrations (step 3). Fresh compounds will be reordered and tested (steps 4 and 5). Active compounds will then be analyzed for inhibition of secretion in additional *A. baumannii* strains, as T2SS inhibitors should ideally be functional against

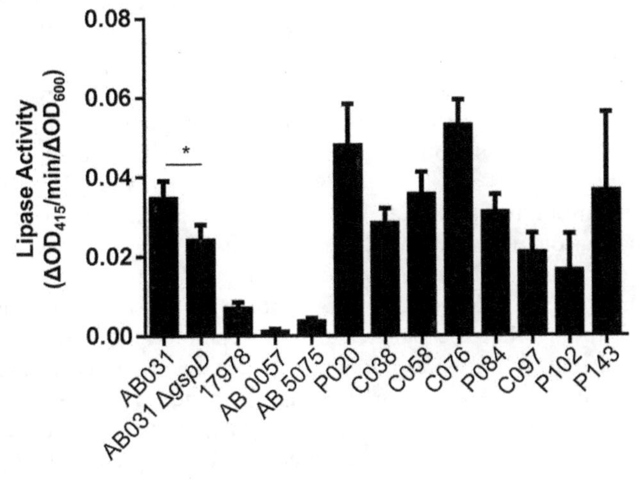

FIGURE 8 | T2S lipase activity of *A. baumannii* strains. The lipase substrate, 4-nitrophenyl myristate was added to overnight cultures of the indicated *A. baumannii* strains and the lipase activity was measured. Activities were normalized to the OD_{600} readings of each culture. Bars show standard deviations from the means, $n = 3$. $^{*}p < 0.05$ by Student *t*-test.

the T2SS of all the *A. baumannii* isolates regardless of antibiotic resistance phenotype. To this effect, we have begun to test lipase activity of other strains of *A. baumannii* that were isolated from different body sites, are resistant to different antibiotics and produce different amounts of biofilm (**Figure 8**). While we have previously shown that detection of lipase activity in the culture supernatant of ATCC 17978 grown in LB in the absence of lipids requires overexpression of LipA from a plasmid, other strains display lipase activity without the need for LipA overexpression (**Figure 8**). The difference in lipase activity among the strains may be due to differences in expression of *lipA* as well as the presence of other lipases, such as LipH, which may or may not be dependent on the T2SS for extracellular secretion. We constructed a T2SS mutant of one of the strains, AB031 (to be described elsewhere). This Δ*gspD* mutant had a statistically significant reduction in lipase activity compared to the WT AB031 strain indicating that a detectable portion of the lipase activity stems from a T2SS dependent lipase(s) (**Figure 8**). This strain, as well as any others we may find as we continue to screen *A. baumannii* isolates for extracellular lipase activity, may be used to further analyze active hits to help determine which compounds to pursue. Analysis of additional strains such as AB0057 and AB5075 that show low lipase activity, however, will likely involve plasmid-expression of LipA.

DISCUSSION

Here, we provide additional information on the T2SS in *A. baumannii*. In addition to supporting colonization, in part through the secretion of LipA, we show that the T2SS also contributes to serum resistance as there is a ≥100-fold reduction in recoverable CFUs of ATCC 17978 Δ*gspD* mutant following exposure to human complement. The mechanism by which the T2SS protects *A. baumannii* from human complement is not known, but published reports have shown that *A. baumannii* expresses secreted and surface-associated proteins that contribute to serum resistance and it is possible that they use the T2SS for outer membrane translocation. The serine protease PKF is produced with an N-terminal signal peptide, a prerequisite for T2S, and CipA, another serum resistance factor is a predicted lipoprotein (King et al., 2013; Koenigs et al., 2016). Lipoproteins, such as pullulanase and SslE, are examples of surface-associated T2S substrates, and CipA may similarly localize to the cell surface via the T2SS (Pugsley et al., 1986; Baldi et al., 2012). While PKF is a member of the HtrA family of chaperone-proteases that refold or degrade misfolded proteins in the periplasmic compartment (Hansen and Hilgenfeld, 2013), it is detected in *A. baumannii* culture supernatants possibly due to its secretion via the T2SS.

The reduced *in vivo* fitness in mouse models of bacteremia (Johnson et al., 2016) and pneumonia (Elhosseiny et al., 2016) and the increased sensitivity to complement killing of T2SS deficient *A. baumannii* (**Figure 2**) suggest that the T2SS plays an important part during infection through the action of specific T2S substrates and, thus, shows great potential for therapeutic targeting. GspD may be an especially promising target, as it forms a gated channel in the outer membrane through which transport occurs. Potential drugs, therefore, would not need to penetrate membranes to reach their target and would not be subject to the effect of drug efflux pumps. This latter point may be of particular importance in the treatment of bacteremic *A. baumannii*, as the expression of several efflux pumps are upregulated when *A. baumannii* is cultured in human serum (Jacobs et al., 2012). Here, we describe the development and optimization of a HTS to identify small molecule inhibitors of the T2SS in *A. baumannii*. Using a previously published assay, we developed, optimized, and tested a high-throughput assay on a small library of pharmacologically active compounds. Our proof-of-concept study demonstrated little fluctuations within controls and showed an acceptable dynamic range between positive and negative controls yielding a z-factor of 0.65 (Zhang et al., 1999). It also indicated that the assay is simple, straightforward, reproducible, robust, and specific as it identified the lipase inhibitors Orlistat and Ebelactone B twice and all seven tricyclic antidepressants present among the 6,400 compounds tested. In addition, the pilot study underscored the importance of including counter screens to reduce the number of false-positives. Though we are confident that our primary lipase screen and counter screens have been sufficiently optimized to be used to screen larger libraries of compounds, we may add an additional assay to reduce the number of potential hits early in the course of triage. If active compounds identified in the HTS are true inhibitors of the T2SS they should also block the secretion of other T2S substrates. Therefore, we may consider employing a double screen that simultaneously detects a reduction of the extracellular amount of LipA and another T2S enzyme. This assay has yet to be developed; however, mass spectrometry analysis of culture supernatants of *A. baumannii* and the related *A. nosocomialis* indicate that several Acinetobacter enzymes besides LipA are secreted by the T2SS (Elhosseiny et al., 2016; Harding et al., 2016), suggesting that a double screen is feasible.

Once larger libraries are subjected to this HTS and the primary assay results have been verified with fresh powders (**Figure 7**), the high-purity compounds will be further tested for specificity. While inhibitors that target the essential Sec system and signal peptidase (Tsirigotaki et al., 2017) would be detected and removed in the primary screen due to their negative impact on growth, a partial growth defect may not be observed in our end point measurement of cell density. This will be addressed in two ways. First, cultures will be grown in microtiter plates in the absence or presence of verified compounds and the OD_{600} will be measured over time. Second, the activity of a periplasmic enzyme, such as alkaline phosphatase, will be determined. This activity will be reduced when the activity of the Sec system is diminished. Other compounds may target periplasmic chaperones or disulfide isomerases such as DsbA (Goemans et al., 2014). The alkaline phosphatase assay will also be instrumental in the detection and subsequent removal of these compounds.

Other studies have implemented similar HTS approaches for the identification of secretion inhibitors with notable differences. The first screen developed and validated, used a gain-of-signal screen to identify inhibitors of SecA, an essential component of the Sec export pathway in *P. aeruginosa* (Moir et al., 2011). No inhibitors were identified to directly interfere with the Sec pathway as the transport of the periplasmic enzyme β-lactamase was not affected. However, following application of secondary assays, a set of compounds was found to reduce the extracellular activity of T2SS substrates although they had no effect on the secreted substrates themselves, suggesting that the compounds inhibit their secretion (Moir et al., 2011). Our screen differs from the Sec screen in that we use a T2SS mutant as our positive control, thus increasing the specificity of our assay. In addition, our screen is designed to include high-throughput counter screens for the removal of false positives early in the triage process. The screen developed and validated by Tran et al., utilized the plant pathogen *Dickeya dadantii* (Tran et al., 2013). Similar to our screen, the authors measured OD_{600} after growth to detect antibiotics and used the activity of a T2S substrate to measure the functionality of the T2SS. As with the Moir et al., screen, the Tran et al., screen did not have counter screens developed to be utilized in the HTS process (Moir et al., 2011; Tran et al., 2013). What is apparent in all three studies is that identified compounds have to be subjected to many additional tests, including those mentioned above, before they can be classified as true T2SS inhibitors. In addition, subcellular fractionation of cells grown in the presence of active T2S inhibitors should confirm an accumulation of T2S substrates in the periplasmic compartment, although some T2S substrates may be degraded by periplasmic proteases when their outer membrane translocation is perturbed. Once we implement our screen for the identification of T2SS inhibitors in *A. baumannii*, it will be advantageous to compare any active compounds to the compounds discovered in these two screens and search for similarities amongst the compounds.

While the primary goal of this pilot screen was to develop a HTS regimen for the identification of compounds to target the T2SS, our data indicate that our dual screen combined with counter screens also have the potential to identify compounds with antibiotic properties and reveal new targets for known pharmacological compounds already in use. An example of this is our discovery that the tricyclic antidepressants are efficient inhibitors of *A. baumannii* LipA (**Table 3**). Less surprising was the finding that the pancreatic lipase inhibitors Orlistat and Ebelactone B efficiently inhibit LipA activity. However, as our previous study has shown that LipA enhances *A. baumannii* colonization (Johnson et al., 2016) and that *A. baumannii* secretes several lipolytic enzymes including LipH and LipAN that may also support *in vivo* survival of *A. baumannii* (Elhosseiny et al., 2016; Harding et al., 2016), a lipase inhibitor could have potential for therapeutic use. Along with T2SS inhibitors, our HTS may also isolate inhibitors of LipB, the chaperone for both LipA and another T2S substrate, LipH (Harding et al., 2016), and they may also be developed for therapeutic intervention.

Identification of T2SS inhibitors for therapeutic purposes is the ultimate goal; however, we are also interested in pursuing small molecules, which can be used as tools to study T2S in multidrug resistant strains as they are often genetically intractable and very difficult to systematically study. Therefore, development of chemical probes to advance virulence studies of these antibiotic resistant isolates is also critically important.

AUTHOR CONTRIBUTIONS

MS, UW, HM, and LS conceived the experiments. CX provided clinical strains. UW, TJ, and KC performed the experiments. UW and MS analyzed the data and wrote the manuscript. All authors have read and approved the manuscript.

FUNDING

We would like to thank MCubed and the Center for Discovery of New Medicine at University of Michigan for funding. This work was supported in part by AI49294 (to MS) from the National Institute of Allergy and Infectious Diseases. HM and LS were supported by AI107184 and GM107312, respectively.

ACKNOWLEDGMENTS

The authors would like to thank the staff of the Center for Chemical Genomics, especially Thomas McQuade and Martha Larsen, for assistance with experiments and data analysis.

REFERENCES

Abendroth, J., Mitchell, D. D., Korotkov, K. V., Johnson, T. L., Kreger, A., Sandkvist, M., et al. (2009). The three-dimensional structure of the cytoplasmic domains of EpsF from the type 2 secretion system of *Vibrio cholerae*. *J. Struct. Biol.* 166, 303–315. doi: 10.1016/j.jsb.2009.03.009

Adams, M. D., Goglin, K., Molyneaux, N., Hujer, K. M., Lavender, H., Jamison, J. J., et al. (2008). Comparative genome sequence analysis of multidrug-resistant

Acinetobacter baumannii. J. Bacteriol. 190:8053. doi: 10.1128/JB.00834-08

Aiello, D., Williams, J. D., Majgier-Baranowska, H., Patel, I., Peet, N. P., Huang, J., et al. (2010). Discovery and characterization of inhibitors of *Pseudomonas aeruginosa* type III secretion. *Antimicrob. Agents Chemother.* 54, 1988–1999. doi: 10.1128/AAC.01598-09

Baldi, D. L., Higginson, E. E., Hocking, D. M., Praszkier, J., Cavaliere, R., James, C. E., et al. (2012). The Type II secretion system and its ubiquitous lipoprotein substrate SslE, are required for biofilm formation and virulence of enteropathogenic *Escherichia coli. Infect. Immun.* 80, 2042–2052. doi: 10.1128/IAI.06160-11

Bergogne-Berezin, E., and Towner, K. J. (1996). Acinetobacter spp as nosocomial pathogens: microbiological, clinical and epidemiological features. *Clin. Microbiol. Rev.* 9, 148–165.

Camberg, J. L., and Sandkvist, M. (2005). Molecular analysis of the *Vibrio cholerae* type II secretion ATPase EpsE. *J. Bacteriol.* 18, 249–256. doi: 10.1128/JB.187.1.249-256.2005

Camberg, J. L., Johnson, T., L., Patrick, M., Abendroth, J., Hol, W. G., and Sandkvist, M. (2007). Synergistic stimulation of EpsE ATP hydrolysis by EpsL and acidic phospholipids. *EMBO J.* 26, 19–27. doi: 10.1038/sj.emboj.7601481

Catalano, M., Quelle, L. S., Jeric, P. E., Di Martino, A., and Maimonet, S. M. (1999). Survival of *Acinetobacter baumannii* on bed rails during an outbreak and during sporadic cases. *J. Hosp. Infect.* 42, 27–35. doi: 10.1053/jhin.1998.0535

Cianciotto, N. P. (2005). Type II secretion: a protein secretion system for all seasons. *Trends Microbiol.* 13, 581–588. doi: 10.1016/j.tim.2005.09.005

Cianciotto, N. P., and White, R. C. (2017). Expanding role of type II secretion in bacterial pathogenesis and beyond. *Infect. Immun.* 85, e00014–00017. doi: 10.1128/IAI.00014-17

Cisneros, D. A., Bond, P. J., Pugsley, A. P., Campos, M., and Francetic, O. (2012). Minor pseudopilin self-assembly primes type II secretion pseudopilus elongation. *EMBO J.* 31, 1041–1053. doi: 10.1038/emboj.2011.454

de Vries, J., and Wackernagel, W. (2002). Integration of foreign DNA during natural transformation of *Acinetobacter* sp. by homology-facilitated illegitimate recombination. *Proc. Natl. Acad. Sci. U.S.A.* 99, 2094–2099. doi: 10.1073/pnas.042263399

Douzi, B., Ball, G., Cambillau, C., Tegoni, M., and Voulhoux, R. (2011). Deciphering the Xcp *Pseudomonas aeruginosa* Type II secretion machinery through multiple interactions with substrates. *J. Biol. Chem.* 286, 40792–40801. doi: 10.1074/jbc.M111.294843

Durand, E., Michel, G., Voulhoux, R., Kurner, J., Bernadac, A., and Filloux, A. (2005). XcpX controls biogenesis of the *Pseudomonas aeruginosa* XcpT-containing Pseudopilus. *J. Biol. Chem.* 280, 31378–31389. doi: 10.1074/jbc.M505812200

Elhosseiny, N. M., El-Tayeb, O. M., Yassin, A. S., Lory, S., and Attia, A. S. (2016). The secretome of *Acinetobacter baumannii* ATCC 17978 type II secretion system reveals a novel plasmid encoded phospholipase that could be implicated in lung colonization. *Int. J. Med. Microbiol.* 306, 633–641. doi: 10.1016/j.ijmm.2016.09.006

Espinal, P., Marti, S., and Vila, J. (2012). Effect of biofilm formation on the survival of *Acinetobacter baumannii* on dry surfaces. *J. Hosp. Infect.* 80, 56–60. doi: 10.1016/j.jhin.2011.08.013

Fürste, J. P., Pansegrau, W., Frank, R., Blöcker, H., Scholz, P., Bagdasarian, M., et al. (1986). Molecular cloning of the plasmid RP4 primase region in a multi-host-range tacP expression vector. *Gene* 48, 119–131. doi: 10.1016/0378-1119(86)90358-6

Gill, E. E., Franco, O. L., and Hancock, R. E. W. (2015). Antibiotic adjuvants: diverse strategies for controlling drug-resistant pathogens. *Chem. Biol. Drug Des.* 85, 56–78. doi: 10.1111/cbdd.12478

Goemans, C., Denoncin, K., and Collet, J.-F. (2014). Folding mechanisms of periplasmis proteins. *Biochim. Biophys. Acta* 1843, 1517–1528. doi: 10.1016/j.bbamcr.2013.10.014

Greene, C., Vadlamudi, G., Newton, D., Foxman, B., and Xi, C. (2016). The influence of biofilm formation and multidrug resistance on environmental survival of clinical and environmental isolates of *Acinetobacter baumannii. Am. J. Infect. Control* 44, e65–e71. doi: 10.1016/j.ajic.2015.12.012

Hansen, G., and Hilgenfeld, R. (2013). Architecture and regulation of HtrA-family proteins involved in protein quality control and stress response. *Cell. Mol. Life Sci.* 70, 761–775. doi: 10.1007/s00018-012-1076-4

Harding, C. M., Kinsella, R. L., Palmer, L. D., Skaar, E. P., and Feldman, M. F. (2016). Medically relevant Acinetobacter species require a type II secretion system and specific membrane-associated chaperones for the export of multiple substrates and full virulence. *PLoS Pathog.* 12:e1005391. doi: 10.1371/journal.ppat.1005391

Hauser, A. R., Mecsas, J., and Moir, D. T. (2016). Beyond antibiotics: new therapeutic approaches for bacterial infections. *Clin. Infect. Dis.* 63, 89–95. doi: 10.1093/cid/ciw200

Hendricks, K. A., Wright, M. E., Shadomy, S. V., Bradley, J. S., Morrow, M. G., Pavia, A. T., et al. (2014). Centers for disease control and prevention expert panel meetings on prevention and treatment of anthrax in adults. *Emerging Infect. Dis.* 20:e130687. doi: 10.3201/eid2002.130687

Hung, D. T., Shakhnovich, E. A., Pierson, E., and Mekalanos, J. J. (2005). Small-molecule inhibitor of *Vibrio cholerae* virulence and intestinal colonization. *Science* 310, 670–674. doi: 10.1126/science.1116739

Jacob, R. T., Larsen, M. J., Larsen, S. D., Kirchhoff, P. D., Sherman, D. H., and Neubig, R. R. (2012). MScreen. *J. Biomol. Screen.* 17, 1080–1087. doi: 10.1177/1087057112450186

Jacobs, A. C., Hood, I., Boyd, K. L., Olson, P. D., Morrison, J. M., Carson, S., et al. (2010). Inactivation of phospholipase D diminishes *Acinetobacter baumannii* pathogenesis. *Infect. Immun.* 78, 1952–1962. doi: 10.1128/IAI.00889-09

Jacobs, A. C., Sayood, K., Olmsted, S. B., Blanchard, C. E., Hinrichs, S., Russell, D., et al. (2012). Characterization of the *Acinetobacter baumannii* growth phase-dependent and serum responsive transcriptomes. *FEMS Immunol. Med. Microbiol.* 64, 403–412. doi: 10.1111/j.1574-695X.2011.00926.x

Jacobs, A. C., Thompson, M. G., Black, C. C., Kessler, J. L., Clark, L. P., McQueary, C. N., et al. (2014). AB5075, a highly virulent isolate of *Acinetobacter baumannii*, as a model strain for the evaluation of pathogenesis and antimicrobial treatments. *MBio* 5:e01076-14. doi: 10.1128/mBio.01076-14

Jawad, A., Seifert, H., Snelling, A. M., Heritage, J., and Hawkey, P. M. (1998). Survival of *Acinetobacter baumannii* on dry surfaces: comparison of outbreak and sporadic isolates. *J. Clin. Microbiol.* 36, 1938–1941.

Johnson, T. L., Waack, U., Smith, S., Mobley, H., and Sandkvist, M. (2016). *Acinetobacter baumannii* is dependent on the type II secretion system and its substrate LipA for lipid utilization and *in vivo* fitness. *J. Bacteriol.* 198, 711–719. doi: 10.1128/JB.00622-15

Jones, C. L., Clancy, M., Honnold, C., Singh, S., Snesrud, E., Onmus-Leone, F., et al. (2015). Fatal outbreak of an emerging clone of extensively drug-resistant *Acinetobacter baumannii* with enhanced virulence. *Clin. Infect. Dis.* 61, 145–154. doi: 10.1093/cid/civ225

King, L. B., Pangburn, M. K., and McDaniel, L. S. (2013). Serine protease PKF of *Acinetobacter baumannii* results in serum resistance and suprression of biofilm formation. *J. Infect. Dis.* 207, 1128–1134. doi: 10.1093/infdis/jis939

Koenigs, A., Stahl, J., Averhoff, B., Göttig, S., Wichelhaus, T. A., Wallich, R., et al. (2016). CipA of *Acinetobacter baumannii* is a novel plasminogen binding and complement inhibitory protein. *J. Infect. Dis.* 213, 1388–1399. doi: 10.1093/infdis/jiv601

Korotkov, K. V., Johnson, T. L., Jobling, M. G., Pruneda, J., Pardon, E., Héroux, A., et al. (2011). Structural and functional studies on the interaction of GspC and GspD in the type II secretion system. *PLoS Pathog.* 7:e1002228. doi: 10.1371/journal.ppat.1002228

Korotkov, K. V., Sandkvist, M., and Hol, W. G. J. (2012). The type II secretion system: biogenesis, molecular architecture and mechanism. *Nat. Rev. Microbiol.* 10, 336–351. doi: 10.1038/nrmicro2762

Leite, G. C., Oliveira, M. S., Perdigão-Neto, L. V., Rocha, C. K. D., Guimarães, T., Rizek, C., et al. (2016). Antimicrobial combinations against pan-resistant *Acinetobacter baumannii* isolates with different resistance mechanisms. *PLoS ONE* 11:e0151270. doi: 10.1371/journal.pone.0151270

Maragakis, L. L., and Perl, T. M. (2008). *Acinetobacter baumannii*: epidemiology, antimicrobial resistance, and treatment options. *Clin. Infect. Dis.* 46, 1254–1263. doi: 10.1086/529198

Michel, G. P. F., Durand, E., and Filloux, A. (2007). XphA/XqhA, a novel GspCD subunit for type II secretion in *Pseudomonas aeruginosa. J. Bacteriol.* 189, 3776–3783. doi: 10.1128/JB.00205-07

Moir, D. T., Di, M., Wong, E., Moore, R., A., Schweizer, H., P., Woods, D. E., et al. (2011). Development and application of a cellular, gain-of-signal, bioluminescent reporter screen for inhibitors of Type II secretion in

Pseudomonas aeruginosa and *Burkholderia pseudomallei*. *J. Biomol. Screen.* 16, 694–705. doi: 10.1177/1087057111408605

Nordfelth, R., Kauppi, A. M., Norberg, H. A., Wolf-Watz, H., and Elofsson, M. (2005). Small-molecule inhibitors specifically targeting type III secretion. *Infect. Immun.* 73, 3104–3114. doi: 10.1128/IAI.73.5.3104-3114.2005

Nunn, D. N., and Lory, S. (1993). Cleavage, methylation, and localization of the *Pseudomonas aeruginosa* export proteins XcpT, -U, -V, and -W. *J. Bacteriol.* 175, 4375–4382. doi: 10.1128/jb.175.14.4375-4382.1993

Pan, N. J., Brady, M. J., Leong, J. M., and Goguen, J. D. (2009). Targeting type III secretion in *Yersinia pestis*. *Antimicrob. Agents Chemother.* 53, 385–392. doi: 10.1128/AAC.00670-08

Paterson, D. L., and Harris, P. N. A. (2015). Editorial commentary: the new acinetobacter equation: hypervirulence plus antibiotic resistance equals big trouble. *Clin. Infect. Dis.* 61, 155–156. doi: 10.1093/cid/civ227

Pugsley, A. P., Chapon, C., and Schwartz, M. (1986). Extracellular pullulanase of *Klebsiella pneumonia* is a lipoprotein. *J. Bacteriol.* 166, 1083–1088. doi: 10.1128/jb.166.3.1083-1088.1986

Py, B., Loiseau, L., and Barras, F. (2001). An inner membrane platform in the type II secretion machinery of Gram-negative bacteria. *EMBO Rep.* 2, 244–248. doi: 10.1093/embo-reports/kve042

Rasko, D. A., and Sperandio, V. (2010). Anti-virulence strategies to combat bacteria-mediated disease. *Nat. Rev. Drug Discov.* 9, 117–128. doi: 10.1038/nrd3013

Reichow, S. L., Korotkov, K. V., Hol, W. G. J., and Gonen, T. (2010). Structure of the cholera toxin secretion channel in its closed state. *Nat. Struct. Mol. Biol.* 17, 1226–1232. doi: 10.1038/nsmb.1910

Rice, L. B. (2008). Federal funding for the study of antimicrobial resistance in nosocomial pathogens: no ESKAPE. *J. Infect. Dis.* 197, 1079–1081. doi: 10.1086/533452

Ruer, S., Pinotsis, N., Steadman, D., Waksman, G., and Remaut, H. (2015). Virulence-targeted antibacterials: concept, promise, and susceptibility to resistance mechanisms. *Chem. Biol. Drug Des.* 86, 379–399. doi: 10.1111/cbdd.12517

Sandkvist, M. (2001). Type II secretion and pathogenesis. *Infect. Immun.* 69, 3523–3535. doi: 10.1128/IAI.69.6.3523-3535.2001

Sandkvist, M., Hough, L. P., Bagdasarian, M. M., and Bagdasarian, M. (1999). Direct interaction of the EpsL and EpsM proteins of the general secretion apparatus in *Vibrio cholerae*. *J. Bacteriol.* 181, 3129–3135.

Scott, M. E., Dossani, Z. Y., and Sandkvist, M. (2001). Directed polar secretion of protease from single cells of *Vibrio cholerae* via the type II secretion pathway. *Proc. Natl. Acad. Sci. U.S.A.* 98, 13978–13983. doi: 10.1073/pnas.241411198

Shakhnovich, E. A., Hung, D. T., Pierson, E., Lee, K., and Mekalanos, J. J. (2007). Virstatin inhibits dimerization of the transcriptional activator ToxT. *Proc. Natl. Acad. Sci. U.S.A.* 104, 2372–2377. doi: 10.1073/pnas.0611643104

Spellberg, B., and Rex, J. H. (2013). The value of single-pathogen antibacterial agents. *Nat. Rev. Drug Discov.* 12:963. doi: 10.1038/nrd3957-c1

Steadman, D., Lo, A., Waksman, G., and Remaut, H. (2014). Bacterial surface appendages as targets for novel antibacterial therapeutics. *Future Microbiol.* 9, 887–900. doi: 10.2217/fmb.14.46

Thomassin, J.-L., Moreno, J. S., Guilvout, I., Van Nheiu, G. T., and Francetic, O. (2017). The trans-envelope architecture and function of the type 2 secretion system: new insights raising new questions. *Mol. Microbiol.* 105, 211–226. doi: 10.1111/mmi.13704

Tilley, D., Law, R., Warren, S., Samis, J. A., and Kumar, A. (2014). CpaA a novel protease from *Acinetobacter baumannii* clinical isolates deregulates blood coagulation. *FEMS Microbiol. Lett.* 356, 53–61. doi: 10.1111/1574-6968.12496

Tran, N., Zielke, R. A., Vining, O. B., Azevedo, M. D., Armstrong, D. J., Banowetz, G. M., et al. (2013). Development of a quantitative assay amenable for high-throughput screening to target the type ii secretion system for new treatments against plant-pathogenic bacteria. *J. Biomol. Screen.* 18, 921–929. doi: 10.1177/1087057113485426

Tsirigotaki, A., De Geyter, J., Sostaric, N., Economou, A., and Karamanou, S. (2017). Protein export through the bacterial Sec pathway. *Nat. Rev. Microbiol.* 15, 21–36. doi: 10.1038/nrmicro.2016.161

Wang, N., Ozer, E. A., Mandel, M. J., and Hauser, A. R. (2014). Genome-wide identification of *Acinetobacter baumannii* genes necessary for persistence in the lung. *MBio* 5:e01163-14. doi: 10.1128/mBio.01163-14

Weernink, A., Severin, W. P. J., Tjernberg, I., and Dijkshoorn, L. (1995). Pillows, an unexpected source of acinetobacter. *J. Hosp. Infect.* 29, 189–199. doi: 10.1016/0195-6701(95)90328-3

Wong, D., Nielsen, T. B., Bonomo, R. A., Pantapalangkoor, P., Luna, B., and Spellberg, B. (2016). Clinical and pathophysiological overview of acinetobacter infections: a century of challenges. *Clin. Microbiol. Rev.* 30, 409–447. doi: 10.1128/CMR.00058-16

Yan, Z., Yin, M., Xu, D., Zhu, Y., and Li, X. (2017). Structural insights into the secretin translocation channel in the type II secretion system. *Nat. Struct. Mol. Biol.* 24, 177–183. doi: 10.1038/nsmb.3350

Zhang, J.-H., Chung, T. D. Y., and Oldenburg, K. R. (1999). A simple statistical parameter for use in evaluation and validation of high throughput screening assays. *J. Biomol. Screen.* 4, 67–73. doi: 10.1177/108705719900400206

17

Prediction of T4SS Effector Proteins for *Anaplasma phagocytophilum* Using OPT4e, a New Software Tool

Zhila Esna Ashari[1*], *Kelly A. Brayton*[1,2,3†] *and Shira L. Broschat*[1,2,3†]

[1] School of Electrical Engineering and Computer Science, Washington State University, Pullman, WA, United States,
[2] Department of Veterinary Microbiology and Pathology, Washington State University, Pullman, WA, United States, [3] Paul G. Allen School for Global Animal Health, Washington State University, Pullman, WA, United States

***Correspondence:**
Zhila Esna Ashari
z.esnaashariesfahan@wsu.edu

Type IV secretion systems (T4SS) are used by a number of bacterial pathogens to attack the host cell. The complex protein structure of the T4SS is used to directly translocate effector proteins into host cells, often causing fatal diseases in humans and animals. Identification of effector proteins is the first step in understanding how they function to cause virulence and pathogenicity. Accurate prediction of effector proteins via a machine learning approach can assist in the process of their identification. The main goal of this study is to predict a set of candidate effectors for the tick-borne pathogen *Anaplasma phagocytophilum*, the causative agent of anaplasmosis in humans. To our knowledge, we present the first computational study for effector prediction with a focus on *A. phagocytophilum*. In a previous study, we systematically selected a set of optimal features from more than 1,000 possible protein characteristics for predicting T4SS effector candidates. This was followed by a study of the features using the proteome of *Legionella pneumophila* strain Philadelphia deduced from its complete genome. In this manuscript we introduce the OPT4e software package for Optimal-features Predictor for T4SS Effector proteins. An earlier version of OPT4e was verified using cross-validation tests, accuracy tests, and comparison with previous results for *L. pneumophila*. We use OPT4e to predict candidate effectors from the proteomes of *A. phagocytophilum* strains HZ and HGE-1 and predict 48 and 46 candidates, respectively, with 16 and 18 deemed most probable as effectors. These latter include the three known validated effectors for *A. phagocytophilum*.

Keywords: T4SS effector proteins, machine learning, *Anaplasma phagocytophilum*, protein prediction, OPT4e software

1. INTRODUCTION

Anaplasma phagocytophilum is a tick-borne zoonotic Gram-negative pathogen that causes Human Granulocytic Anaplasmosis (HGA). Incidence of this potentially fatal disease is rising in the United States, with the number of cases increasing from 348 in 2000 to 5,762 in 2017 and incidence rates increasing from 1.4 cases per million people in 2000 to 17.9 cases per million in 2017. The number of cases in the United States increased 39% from 2016 to 2017 alone (CDC, 2019). Moreover, the geographic range of *A. phagocytophilum* seems to be increasing along with the range expansion of the tick vector *Ixodes scapularis* (blacklegged tick). HGA is now the third most common vector-borne infection in the United States (Dumler et al., 2005; Dumler, 2012; Bakken and Dumler, 2015; Sinclair et al., 2015; CDC, 2019).

The geographic distribution of HGA is mainly focused in the upper midwest and northeastern United States, which coincides with Lyme disease and other *I. scapularis*-transmitted diseases (CDC, 2019). The agent of Lyme disease, *Borrelia burgdorferi*, and other human pathogens such as *Babesia microti*, *Borrelia mayonii*, *Borrelia miyamotoi*, and *Ehrlichia muris eauclairensis* are also transmitted by *I. scapularis*, with co-infections with *A. phagocytophilum* reported in <10% of cases (CDC, 2019).

Some Gram-negative bacteria such as *A. phagocytophilum* have evolved specialized secretion systems, secreting proteins that interact with host cells. The type IV secretion system (T4SS) is a macromolecular complex composed of proteins that are responsible for secreting effector proteins directly into the cytosol of eukaryotic host cells. The transported proteins, called effector proteins, are instrumental agents of virulence and pathogenesis and play a key role in altering environmental niches to allow pathogen replication (Voth et al., 2010, 2012; Abby et al., 2016; Han et al., 2016), yet relatively little is known about them. A critical goal is to understand how effectors cause infection in humans and animals which requires knowledge of the function of each effector. The first step toward this goal is identifying the effectors from among the entire set of proteins in the complete genome of a bacterial pathogen with a T4SS.

In addition to experimentally validating effector proteins by means of fusion protein reporter assays in translocation studies (Voth et al., 2012; Maturana et al., 2013), a time-consuming and expensive process, several computational methods have been proposed for the prediction of effectors (Burstein et al., 2009; Yu et al., 2010; Lockwood et al., 2011; Meyer et al., 2013; Zou et al., 2013; Wang et al., 2014, 2018a,b). Accurate prediction of effector proteins greatly limits the number of proteins requiring experimental verification which reduces costs. Current computational methods use either a scoring method (Meyer et al., 2013) or a machine learning approach (Burstein et al., 2009; Zou et al., 2013; Wang et al., 2014, 2018a,b) to predict a set of candidate effectors. For example, Meyer et al. (2013) used a scoring method to predict effectors for *Legionella pneumophila* and other pathogens. Burstein et al. (2009) used machine learning to focus on the *L. pneumophila* genome while (Wang et al., 2014) studied *Helicobacter pylori* effectors. In addition, there are several reviews on T4SS effector prediction and the progress made in this area (McDermott et al., 2011; Wang et al., 2017a; An et al., 2018; Zeng and Zou, 2019) as well as several databases for curating experimentally validated effector proteins for some species (Bi et al., 2013).

The computational methods previously reported for T4SS effector prediction used different sets of protein characteristics as features for their methods. We suspect that the use of these differing feature sets explains the differences in effector predictions by the different algorithms. As a result of the disparities between the results of earlier methods, we assembled all the features used in prior studies and used a multi-level, statistical approach to determine which were the most effective in predicting effector proteins (Esna Ashari et al., 2017, 2018). Because of the number of validated effectors available for *L. pneumophila*, we then ran a number of experiments on the whole genome of *L. pneumophila* using our optimal set of features (Esna Ashari et al., 2019). A comparison of our results with the list of validated effectors and those of previous studies was highly encouraging.

Although *A. phagocytophilum* employs the T4SS to invade human cells and cause anaplasmosis, a disease sometimes fatal to humans, it has just three experimentally verified effector proteins. As such, in order to conduct further research on this increasingly important human pathogen, there is a need to identify more effector proteins. Accurate prediction of effectors will assist in this identification. In this paper we turn our attention to the prediction of effector proteins in *A. phagocytophilum*. This pathogen has not been the focus of previous computational studies for effector prediction, in part because of its lack of validated effector proteins. Because of the high accuracy of the prediction results we obtained for *L. pneumophila* using a combination of validated effectors for four different pathogens, we decided to apply our method to *A. phagocytophilum*.

In addition to applying our model for T4SS effector prediction to *A. phagocytophilum*, we also improved it based on what we learned from our previous study (Esna Ashari et al., 2019) and expanded the code to make it easy for microbiologists to use for other bacteria with T4 secretion systems. We created a software package called OPT4e, for Optimal-features Predictor for T4SS Effector proteins, that performs all the steps described in our previous studies as well as incorporating new steps, including automation of feature evaluation which is very time consuming for whole proteomes. OPT4e is specifically designed for T4SS effector protein prediction and for use on Windows, Mac OS X, and Linux operating systems. One of the main characteristics of OPT4e is that it integrates all the tools, scripts, and software needed for calculation of our optimal set of features (Esna Ashari et al., 2018) and automatically creates the set of optimal features for training or test sets. OPT4e predicts candidate effectors and groups them based on their degree of likelihood of being an effector. In addition, OPT4e can be updated to become a stronger predictor over time. Finally, OPT4e has a very simple and intuitive graphical-user interface (GUI) making it easy to use.

The remainder of the manuscript is organized as follows: First, we focus on introducing OPT4e and the steps taken to create its framework and the related algorithms. Next we explain our set of optimal features and the machine learning algorithm used for OPT4e. We then introduce the datasets used in this study for the training and test sets followed by presentation of our results. In the final section, we discuss the results we obtained for OPT4e for two input proteomes.

2. MATERIALS AND METHODS

2.1. OPT4e Software

We designed and created OPT4e as a software package for the purpose of predicting effector proteins in different T4SS bacterial pathogens. OPT4e is an easy-to-use and user-friendly software package written in Python 3. Its specific features are as follows: It is based on usage of a machine learning approach for effector prediction. Each protein characteristic in a sequence is identified as a feature and is assigned the appropriate coefficient by the machine learning algorithm based on its significance as determined by the training data, and it is not necessary to determine the importance of each feature manually. Moreover, it gathers and connects multiple bioinformatic tools in order to automatically calculate and assign all the needed features and to select the best ones. Therefore, installation is simple, and there is no need to use lots of online tools or to know a specific programming language to be able to use OPT4e as is necessary for some previously developed tools (Burstein et al., 2009; Zou et al., 2013; Wang et al., 2014). In addition, OPT4e predictions are based on protein sequences and are not dependent on an entire bacterial proteome. In fact, the input to OPT4e can be a single protein sequence selected by the user. Also, OPT4e is based on predictions using a specific machine learning algorithm while taking advantage of two additional algorithms in order to present the results in three different groups of more-likely, possible, and less-likely candidate effectors. One of the most important features of OPT4e is that it can be updated over time. Thus, if a user has some new experimentally verified effectors or discovers some critical non-effectors, they can add them to the software using a few mouse clicks. The software will then include them in the training set and update the model automatically. Enriching the set of validated effectors in the software dataset will help with the accuracy of the machine learning predictions, and OPT4e will become increasingly more accurate with time.

2.1.1. Framework

The framework and Graphical User Interface (GUI) for OPT4e are presented in **Figure 1**. First we provide an explanation of the framework, shown in **Figure 1A**, as follows: In the initial step a training set of known effectors and non-effectors is provided, and values for the optimal features are calculated for them automatically (Esna Ashari et al., 2018). OPT4e uses this set as its input. Note that in the first step a user has to select the appropriate button related to the purpose for using the software. If it is being used for effector prediction, the user will need to provide the test file for a protein sequence or a set of sequences in fasta-file format for classification as effectors or non-effectors by the OPT4e software. Then the software will calculate the feature values for each of the sequences provided such that they are available for machine learning prediction. The features used in this package are explained in the next section.

In the next step, OPT4e uses a support vector machine (SVM) algorithm with a radial basis function (RBF) kernel to predict effector protein sequences. This algorithm was found to give the best results as explained in Esna Ashari et al. (2019). In addition, OPT4e uses two additional classifiers (SVM with linear kernel and logistic regression) with the test sequences and uses their results to group the initially predicted effectors into three groups of more-likely (predicted by all three classifiers), possible (predicted by an additional classifier), and less-likely (predicted by just the initial SVM RBF classifier). The predicted groups of effector sequences are given as the output of the program. It should be noted that this methodology was used in the previous version of our algorithm as well (Esna Ashari et al., 2019). However, in our earlier work we used two ensemble classifiers and divided the features into three different groups for each ensemble set in addition to using the SVM with radial basis function with all the features. We found that a single classifier used with all the optimal features gave better results (Esna Ashari et al., 2019). Hence, we have replaced the ensemble classifiers with an SVM with a linear kernel and logistic regression in order to improve the model.

If a user wants to add experimentally verified effectors or some new known non-effectors to the training set to enrich it, they should select the appropriate option when using OPT4e. Then the software will automatically calculate the corresponding feature values for the new sequences and will add them to the feature set of the older training set.

We have added an option in OPT4e in case a user has made changes to the training set incorrectly or decides they do not want to change it. When the user clicks on the last button on the GUI (**Figure 1B**), OPT4e will reset the training data back to the original version. Finally, OPT4e is an open-source package, and users can update it as they wish.

2.2. Features and Feature Selection

As described in the introduction, in our earlier study we analyzed a comprehensive set of features gathered from previous computational studies performed in the field of T4SS effector protein prediction. The total number of features, including elements of vector features, was 1,027. The complete list of these features and the tools and software needed for their computation are presented in Esna Ashari et al. (2017, 2018).

We used a multi-step feature selection algorithm, described briefly in the next paragraph, to generate a set of optimal features for prediction of effector proteins consisting of 370 features. The detailed list of selected features, including the selected vector feature elements, can be found in Esna Ashari et al. (2018). The features can be grouped into chemical properties determining the way proteins interact with their environment and how effectors enter host cells (Yu et al., 2010; Zou et al., 2013), structural properties affecting protein-protein interactions between bacterial pathogens and host cells (Yu et al., 2010; Zou et al., 2013; Wang et al., 2014), compositional properties including the amino acid and dipeptide composition of protein sequences, and position-specific scoring matrix (PSSM)-related properties including PSSM composition and PSSM auto-covariance correlation composition (Zou et al., 2013; Wang et al., 2017b). The compositional properties determine the shapes and motifs of the protein sequences and, therefore, can affect the way they interact with host cells.

The first step in determining our optimal set of features was to use a filtering selection approach. For this purpose we used the t-test as a hypothesis testing method to filter features based on their associated p-values. Next we used Principal Component Analysis and Factor Analysis for dimensional reduction and to eliminate any redundancy and correlation in our feature set. The final step in our statistical approach was designing a fast backward feature selection method based on a Hosmer-Lemeshow goodness-of-fit test and using binary logistic regression. In this fashion we were able to retrieve a set of optimal features that work well together for effector prediction, and the concordance percentage from the Hosmer-Lemeshow goodness-of-fit test was still high after removal of the less related features.

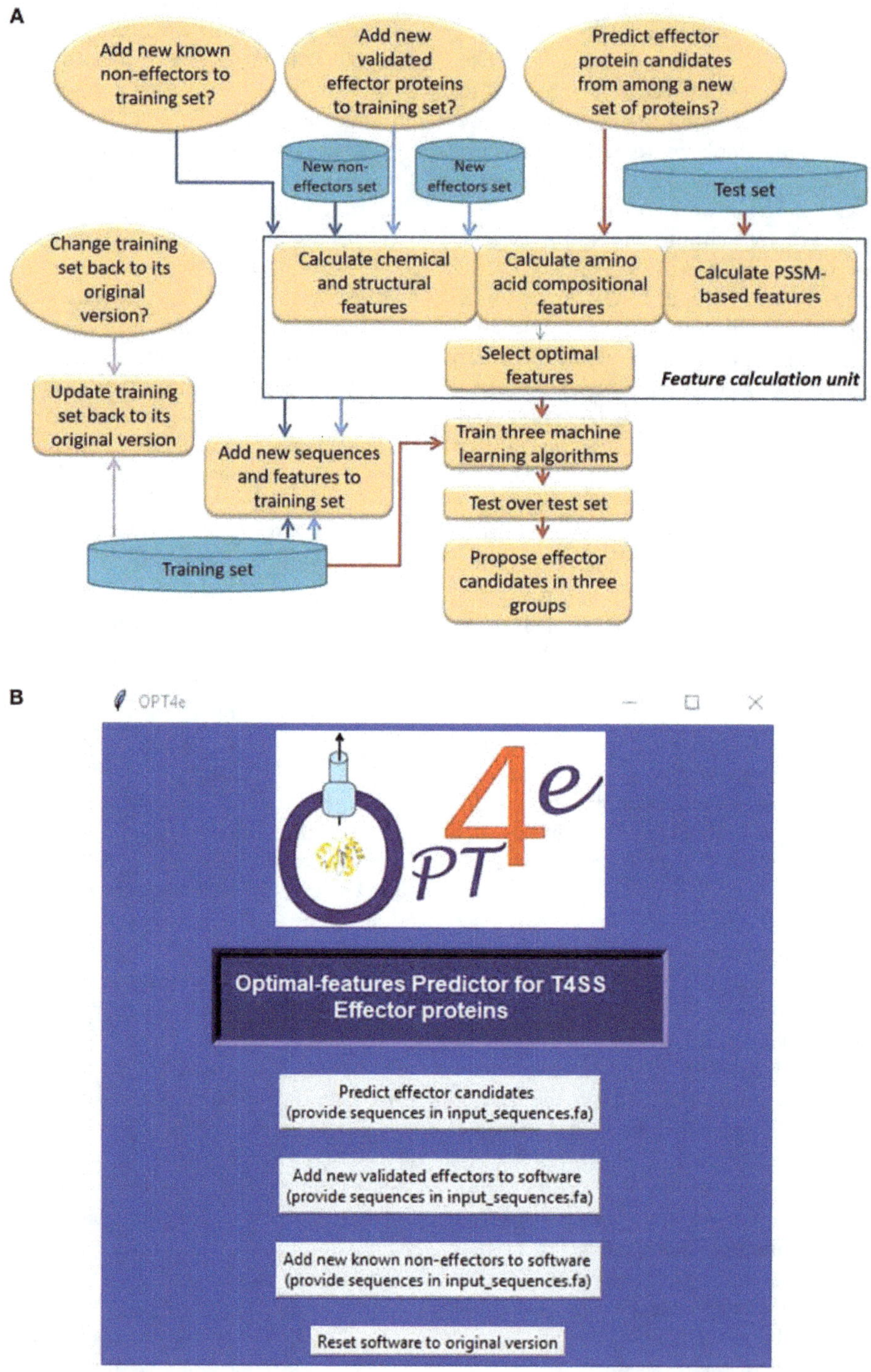

FIGURE 1 | OPT4e Software: **(A)** The framework for OPT4e. **(B)** The Graphical User Interface (GUI) for OPT4e.

2.3. Machine Learning Model

After selecting a set of optimal features, we designed multiple machine learning-based classifiers and tested them in order to select the most accurate predictor with our feature set (Esna Ashari et al., 2019). In due course, we focused on three classifiers. They included the SVM with the RBF kernel which is a well-known classifier and two ensemble classifiers (Esna Ashari et al., 2019). Based on 10-fold cross-validation results for our training set, results for our test set, and comparison with the results of previously developed methods, the SVM with the RBF kernel classifier was selected for further predictions, and it is the main classifier used in the OPT4e software package. As mentioned earlier, the ensemble classifiers were replaced by an SVM with a linear kernel and logistic regression.

2.4. Dataset

In order to create our training set, we gathered known effectors and non-effectors for four Gram-negative bacterial pathogens from the Alphaproteobacteria and Gammaproteobacteria classes. This set is composed of effectors and non-effectors from: *L. pneumophila*, *Coxiella burnettii*, *Brucella* spp., and *Bartonella* spp. Furthermore, we added the three validated effectors and multiple non-effectors from *A. phagocytophilum* to our training set. The numbers of non-effector sequences added from *A. phagocytophilum* strains HZ and HGE-1 were 115 and 120 sequences, respectively. The final training set included 1,365 sequences consisting of 432 effectors and 933 non-effectors. Moreover, we added four experimentally validated effector proteins for *Anaplasma marginale* to the training set and repeated all the experiments (Lockwood et al., 2011). Therefore, the final training set consisted of 436 effectors. The complete file of protein sequences in fasta format used in the training set is presented in **Supplementary Data Sheet S1**.

For this study we selected two strains of *A. phagocytophilum*, strain HZ (accession number CP000235) and strain HGE-1 (accession number APHH01000001), for use with OPT4e. These strains are composed of 1,352 and 1,148 protein sequences, respectively. We used these two sets of protein sequences as input files for OPT4e. In addition, the proteome for *L. pneumophila* strain Philadelphia with 2,942 sequences was examined. Results for the latter proteome are briefly described later to explain the performance of OPT4e. More details concerning these datasets are given in the next section.

3. RESULTS

In this section we present the results obtained by OPT4e for the proteomes of *A. phagocytophilum* strain HZ and strain HGE-1. First, however, we present a brief discussion on validation of our classifier for the results obtained for the proteome of *L. pneumophila* strain Philadelphia.

3.1. Validation of OPT4e

We performed a thorough validation of the earlier version of our machine learning model as described in Esna Ashari et al. (2019). Briefly, in our previous study we performed 10-fold cross-validation for our training set and achieved an average accuracy of 94.05% over all folds for the SVM with radial basis function. Also, the model was verified using other performance metrics and achieved an average precision of 92.49%, an average recall of 92.00%, an average MCC (Matthews Correlation Coefficient, a measure of correlation between real and predicted values) of 0.87, and an average AUC (area under the curve) of 0.98. For further validation of our method, we tested the algorithm using the proteome for *L. pneumophila* strain Philadelphia and compared our predictions with ones from previous computational methods. Our results for effector candidates considered to be the most likely agreed with 80.5 and 72.2% of candidate effectors predicted using previous methods developed by Burstein et al. (2009) and Meyer et al. (2013), respectively. Also, the results predicted 93.7 and 99.8% of known effectors and non-effectors, respectively, from our training set (Esna Ashari et al., 2019).

TABLE 1 | Number of effector candidate proteins for *A. phagocytophilum* strains HZ and HGE-1 before and after adding *A. marginale* validated effectors to the OPT4e training set.

	Before adding *A. marginale* effectors		After adding *A. Marginale* effectors	
	Ap strain HZ	**Ap strain HGE-1**	**Ap strain HZ**	**Ap strain HGE-1**
More likely	14	17	16	18
Possible	10	6	9	5
Less likely	22	23	23	23
Total	46	46	48	46

A. phagocytophilum is indicated by Ap.

As mentioned earlier, in our previous study we learned that using all the features with a single classifier gave more accurate results than separating the features and using them in an ensemble classifier (Esna Ashari et al., 2019). Thus, for OPT4e we replaced the ensemble classifiers in our model for determining more-likely, possible, and less-likely effectors. To ensure that changing to the SVM with linear kernel and logistic regression classifiers actually does give more accurate results, we used 10-fold cross validation with our *L. pneumophila* strain Philadelphia effector and non-effector proteins. We obtained accuracies of 93.73% for the SVM with linear kernel and 93.79% using logistic regression. This is in contrast to our previous ensemble results for which we obtained average accuracies of 93.64 and 92.44%.

3.2. Predicted Effectors for *A. phagocytophilum* Strains HZ and HGE-1

Anaplasma phagocytophilum strain HZ contains 1,352 protein sequences consisting of 115 known non-effectors including the protein sequences associated with the genes *rpoB* (DNA-directed RNA polymerase subunit beta), *rpoC* (DNA-directed RNA polymerase subunit beta'), and Msp2/P44. For this strain, 14 protein sequences were predicted to be more likely to be an effector protein.

Anaplasma phagocytophilum strain HGE-1 contains 1,148 protein sequences consisting of 120 known non-effectors including DNA pol III, delta subunit (HGE1_05467), DNA-binding protein HGE1_04712 (a helix-turn-helix DNA binding protein somewhat specific to bacteria), MerR transcriptional regulator-HGE1_05592 (a helix-turn-helix DNA binding protein somewhat specific to bacteria), type IV secretion system VirB6-HGE1_01722 (a part of the T4SS structure), putative ABC transporter, permease protein-HGE1_00015 (an outer membrane protein also found in *Escherichia coli*), thiamine biosynthesis protein ThiS-HGE1_00315 (a sulfur carrier protein common in bacterial metablolism), and Msp2/P44 sequences. For this strain, 17 protein sequences were predicted to be more likely to be an effector protein.

Table 1 lists the number of candidate effectors for both strains of *A. phagocytophilum* according to their likelihood as predicted by OPT4e.

TABLE 2 | Effector candidates predicted by OPT4e.

Suggest	HGE1	HGE1 (other)	HZ	HZ (other)	Notes
	HGE1_00145		APH_0033		
*	HGE1_00220	S, T4	APH_0049	S, T4	
*	HGE1_00312	T4	APH_0068	T4	
*	HGE1_00527	S	APH_0117	S, T4	
	HGE1_00815		APH_0189		HZ: 7-aa insert at start
*	HGE1_01135	S, T4	APH_0259	S, T4	
*	HGE1_01175	T4	APH_0267	T4	HGE1: 17-aa insert at start
	HGE1_01752		APH_0382		
*	HGE1_01772		APH_0385	T4	
	HGE1_02100		APH_0385		
	HGE1_01777		APH_0386		
	HGE1_01782		APH_0387		
	HGE1_02092		APH_0452		
	HGE1_02095		APH_0453		
	HGE1_02117		APH_0453		
	HGE1_02107		APH_0455	S	Known effector
*	HGE1_02112		APH_0457	T4	
*	HGE1_02242	T4	APH_0485	T4	
*	HGE1_02492	S, T4	APH_0546	S, T4	
*	HGE1_02802	T4	APH_0633	T4	
*	HGE1_02817	T4	APH_0636	T4	HGE1: 3-aa insert at start
	HGE1_02827		APH_0641		
	HGE1_02947		APH_0670		First aa different
*	HGE1_03022		APH_0688	T4	
*	HGE1_03117		APH_0708	T4	
*	HGE1_03122	S	APH_0709	S, T4	
*	HGE1_03182	S, T4	APH_0726	S, T4	
*	HGE1_03232	S, T4	APH_0740	S, T4	Known effector
	HGE1_03297		APH_0755		
*	HGE1_03432		APH_0792	T4	
*	HGE1_03492	S	APH_0805	S, T4	
*	HGE1_03497		APH_0807	T4	
*	HGE1_03502	T4	APH_0808		
	HGE1_03532		APH_0815		
	HGE1_03557		APH_0820		
	HGE1_03697	S, T4	APH_0859	S, T4	Known effector
*	HGE1_03707	T4	APH_0861	T4	
*	HGE1_02737	T4	APH_0863	S	
*	HGE1_03892	S, T4	APH_0914	S, T4	
*	HGE1_05072	T4	APH_1167	T4	
*	HGE1_03907		APH_0916	T4	
	HGE1_03962		APH_0928		
*	HGE1_04167	T4	APH_0976	T4	
	HGE1_03507				HZ homolog not predicted
	HGE1_05977		APH_1365		First aa different
*	HGE1_05997	T4	APH_1369	T4	HZ: 6-aa insert at start
*	HGE1_06052	S	APH_1379	S	HZ: 14-aa insert in middle
*	HGE1_06067	T4	APH_1383	T4	
			APH_0239		No equivalent sequence in HGE1
			APH_0904		No equivalent sequence in HGE1
			APH_0028		No equivalent sequence in HGE1
			APH_0640		No equivalent sequence in HGE1
			APH_0816		No equivalent sequence in HGE1

HGE1 and HZ homologs are row aligned. Blue, orange, and red text colors indicate More Likely, Possible, and Less Likely effector candidates, respectively. Black text indicates that a sequence was not predicted as an effector. The columns HGE1 (other) and HZ (other) indicate when a sequence was predicted as an effector candidate by S4TE (S) or T4EffPred (T4). The Notes column lists differences between HGE-1 and HZ homologs, and finally the Suggest column indicates effector candidates proposed for initial experimental validation based on the strength of their predictions.

Because *Anaplasma marginale* is more closely related to *A. phagocytophilum* than the bacteria used in our model, we added four experimentally verified effector proteins for *A. marginale* (Lockwood et al., 2011) to our training set and repeated our experiments. Two new candidate effectors were predicted for *A. phagocytophilum* strain HZ. Also, the more likely category of candidate effectors was increased by 2 and 1 for *A. phagocytophilum* strains HZ and HGE-1, respectively. Specific numbers are reported in **Table 1**, and all predicted candidate effectors are presented in **Tables 2–4** by locus number. In addition, **Tables 2–4** present suggestions for the order of experimental verification of candidate effectors as explained in detail in the next section.

4. DISCUSSION

The main goal of this study was predicting a set of candidate effectors for *A. phagocytophilum* using a new package called OPT4e which we developed for this purpose. In fact, OPT4e can be used to give reasonable candidate effector predictions for most T4SS bacteria from the Alphaproteobacteria and Gammaproteobacteria classes. For *A. phagocytophilum* strains HGE-1 and HZ, we predicted 48 and 46 candidate effectors, respectively, with 16 and 18 more likely to be effectors. All three experimentally-verified effector proteins were included in the 16 and 18 more-likely category.

We compared the differences between the predictions for the two strains and found that whenever there was a difference between the category in which an effector was predicted or an effector was not predicted for one of the strains, there was a difference between the homologous protein sequences of the two strains. These differences are noted in **Table 2**. In addition, five effector proteins were predicted in strain HZ for which there is no equivalent protein sequence in strain HGE-1. Strain HZ was the first *A. phagocytophilum* genome to be sequenced, and many small open reading frames (ORFs) were annotated that have not been retained in subsequent annotations (including the RefSeq for HZ). Some of these small ORFs account for the differences between the effector predictions for the two strains. Interestingly, there was one effector predicted in HGE-1 for which there was not an equivalent protein annotated in HZ. However, closer inspection of the HZ genome revealed that the sequence is present.

It should be noted that in machine learning-based prediction, an algorithm tries to fit as many training samples as it can based on the given features, and as the numbers of features and samples increase, the task increases in complexity. Also, it should be noted that the greater part of our positive training set consists of known effectors for *L. pneumophila* because it has the largest number of verified effectors. Moreover, there are only three verified effectors for *A. phagocytophilum* in our dataset. Therefore, it is possible that our set of candidate effectors for *A. phagocytophilum* include the ones that are mostly similar to *L. pneumophila* effectors. In addition, OPT4e may be detecting genes with a different signature from the rest of the genome such as those acquired by horizontal gene transfer in species where this

TABLE 3 | Groups recommended for experimental verification of effector candidates for strain HZ.

Effector candidates	Other models	Notes
Group 1		
APH_0259	S, T4	
APH_0740	S, T4	Known effector
APH_0859	S, T4	Known effector
Group 2		
APH_0239	T4	
APH_0385	T4	
APH_0457	T4	
APH_0636	T4	
APH_0904	T4	
APH_0455	S	Known effector
Group 3		
APH_0033		
APH_0382		
APH_0385		
APH_0386		
APH_0387		
APH_0452		
APH_0453		
APH_0928		
Group 4		
APH_0267	T4	
APH_0633	T4	
APH_0861	T4	
APH_1167	T4	
APH_1369	T4	
APH_0863	S	
APH_1379	S	
Group 5		
APH_0028		
APH_0640		
APH_1365		
Group 6		
APH_0049	S, T4	
APH_0546	S, T4	
APH_0709	S, T4	
APH_0726	S, T4	
APH_0805	S, T4	
APH_0914	S, T4	
Group 7		
APH_0068	T4	
APH_0485	T4	
APH_0688	T4	
APH_0708	T4	
APH_0792	T4	
APH_0807	T4	
APH_0916	T4	
APH_0976	T4	
APH_1383	T4	
APH_0117	S	

(Continued)

TABLE 3 | Continued

Effector candidates	Other models	Notes
Group 8		
APH_0641		
APH_0670		
APH_0755		
APH_0808		
APH_0815		
APH_0816		
APH_0820		
APH_1383		

Groups are based on whether effector candidates are More Likely shown in blue, Probable shown in orange, and Less Likely shown in red, followed by prediction by both S4TE (S) and T4EffPred (T4), prediction by one of them, or prediction by neither. We recommend starting with Group 1 and proceeding successively through Group 8.

occurs. Thus, caution is necessary when evaluating the output. It should be noted, however, that a strength of OPT4e is that it can be updated over time, and a user has the ability to add newly verified effectors to the training dataset. As a result, as new effectors for *A. phagocytophilum* are verified, they can be used in OPT4e to increase its accuracy for predicting effector proteins.

As a final note, we compared our effector candidates for *A. phagocytophilum* with those predicted by S4TE (Noroy et al., 2019) and T4EffPred (Zou et al., 2013) after we used these two programs to predict effectors for both *A. phagocytophilum* strains in our study. For HZ, OPT4e shared 13 of 48 predictions with S4TE and 27 of 92 predictions with T4EffPred. S4TE and T4EffPred shared ten predictions. Two of these were for known effectors. The third known effector was predicted by S4TE but not by T4EffPred. Thus, both OPT4e and S4TE predicted all three known effectors (see **Table 2**). For HGE-1, OPT4e shared 11 of 49 predictions with S4TE and 19 of 45 predictions with T4EffPred. S4TE and T4EffPred shared seven predictions. Two of these were for the homologs of known effectors. The third effector homolog was not predicted by either method; only OPT4e predicted all three.

One strategy for deciding which effector candidates to choose for experimental verification is to select from among the ones predicted by OPT4e for both strains of *A. phagocytophilum* and also predicted by one of the other two methods, S4TE or T4EffPred. There are 28 of these indicated by asterisks in **Table 2**, where HGE1 and HZ homologs are row aligned.

An alternative strategy and more systematic approach is to first group the predicted effectors on the basis of more-likely, probable, and less-likely and then based on predictions by the two methods, S4TE or T4effPred. Experimental verification would begin with Group 1 and proceed in order through successive groups as shown in **Tables 3, 4**. **Table 3** is for HZ and **Table 4** is for HGE-1, and for both strains Group 1 candidate effectors have literally been predicted by five different algorithms, the three from OPT4e plus S4TE and T4EffPred. For HZ, two of the three sequences in Group 1 are for known effectors, and the third known effector is in Group 2. For HGE-1, two of the sequences in Group 1 are for homologs of known

TABLE 4 | Groups recommended for experimental verification of effector candidates for strain HGE-1.

Effector Candidates	Other Models	Notes
Group 1		
HGE1_01135	S, T4	
HGE1_03232	S, T4	Homolog of known effector
HGE1_03697	S, T4	Homolog of known effector
Group 2		
HGE1_01175	T4	
HGE1_05997	T4	
Group 3		
HGE1_00145		
HGE1_01752		
HGE1_01772		
HGE1_01777		
HGE1_01782		
HGE1_02092		
HGE1_02095		
HGE1_02100		
HGE1_02107		Homolog of known effector
HGE1_02112		
HGE1_02117		
HGE1_03507		
HGE1_03962		
Group 4		
HGE1_02737	T4	
HGE1_02802	T4	
HGE1_02817	T4	
HGE1_03707	T4	
HGE1_05072	T4	
Group 5		
HGE1_00220	S, T4	
HGE1_02492	S, T4	
HGE1_03182	S, T4	
HGE1_03892	S, T4	
Group 6		
HGE1_00312	T4	
HGE1_02242	T4	
HGE1_03502	T4	
HGE1_04167	T4	
HGE1_06067	T4	
HGE1_00527	S	
HGE1_03122	S	
HGE1_03492	S	
HGE1_06052	S	
Group 7		
HGE1_00815		
HGE1_02827		
HGE1_03022		
HGE1_03117		
HGE1_03297		
HGE1_03432		
HGE1_03497		
HGE1_03532		
HGE1_03557		
HGE1_03907		

Groups are based on whether effector candidates are More Likely shown in blue, Probable shown in orange, and Less Likely shown in red, followed by prediction by both S4TE (S) and T4EffPred (T4), prediction by one of them, or prediction by neither. We recommend starting with Group 1 and proceeding successively through Group 7.

effectors, and the third effector homolog is in Group 3. Thus the first three groups for each strain present excellent choices for experimental verification.

AUTHOR CONTRIBUTIONS

ZE developed the OPT4e software and performed dataset preparation, machine learning predictions, computational experiments, data analysis, and drafted the manuscript. SB and KB supervised, conceived, and coordinated the study and contributed to the manuscript. All authors gave final approval for publication.

FUNDING

This study was supported by grant R01AI042792 from the National Institutes of Health and by the Carl M. Hansen Foundation. The funders had no role in the study design, data collection and analysis, decision to publish, or preparation of the manuscript.

REFERENCES

Abby, S. S., Cury, J., Guglielmini, J., Néron, B., Touchon, M., and Rocha, E. P. C. (2016). Identification of protein secretion systems in bacterial genomes. *Sci. Rep.* 6:23080. doi: 10.1038/srep23080

An, Y., Wang, J., Li, C., Leier, A., Marquez-Lago, T., Wilksch, J., et al. (2018). Comprehensive assessment and performance improvement of effector protein predictors for bacterial secretion systems III, IV and VI. *Brief. Bioinform.* 19, 148–161. doi: 10.1093/bib/bbw100

Bakken, J. S., and Dumler, S. J. (2015). Human granulocytic anaplasmosis. *Infect. Dis. Clin. North Am.* 29, 341–355. doi: 10.1016/j.idc.2015.02.007

Bi, D., Liu, L., Tai, C., Deng, Z., Rajakumar, K., and Ou, H. Y. (2013). Secret4: a web-based bacterial type iv secretion system resource. *Nucleic Acids Res.* 41, D660–D665. doi: 10.1093/nar/gks1248

Burstein, D., Zusman, T., Degtyar, E., Viner, R., Segal, G., and Pupko, T. (2009). Genome-scale identification of *Legionella pneumophila* effectors using a machine learning approach. *PLoS Pathogens* 5:e1000508. doi: 10.1371/journal.ppat.1000508

CDC (2019). *Centers for Disease Control and Prevention.* Available online at: https://www.cdc.gov/anaplasmosis/index.html

Dumler, S. J. (2012). The biological basis of severe outcomes in *Anaplasma phagocytophilum* infection. *FEMS Immunol. Med. Microbiol.* 64, 13–20. doi: 10.1111/j.1574-695X.2011.00909.x

Dumler, S. J., Choi, K.-S., Garcia-Garcia, J. C., Barat, N. S., Scorpio, D. G., Garyu, J. W., et al. (2005). Human granulocytic anaplasmosis and *Anaplasma phagocytophilum*. *Emerg. Infect. Dis.* 11, 1828–1834. doi: 10.3201/eid1112.050898

Esna Ashari, Z., Brayton, K. A., and Broschat, S. L. (2017). "Determining optimal features for predicting type IV secretion system effector proteins for *Coxiella burnetii*," in *Proceedings of The 8th ACM Conference on Bioinformatics, Computational Biology, and Health Informatics (ACM BCB)* (Boston, MA: ACM), 346–351.

Esna Ashari, Z., Brayton, K. A., and Broschat, S. L. (2018). An optimal set of features for predicting type IV secretion system effector proteins for a subset of species based on a multi-level feature selection approach. *PLoS ONE* 13:e0197041. doi: 10.1371/journal.pone.0197041

Esna Ashari, Z., Brayton, K. A., and Broschat, S. L. (2019). Using an optimal set of features with a machine learning-based approach to predict effector proteins for *Legionella pneumophila*. *PLoS ONE* 9:e0202312. doi: 10.1371/journal.pone.0202312

Han, N., Yu, W., Qiang, Y., and Zhang, W. (2016). T4sp database 2.0: an improved database for type IV secretion systems in bacterial genomes with new online analysis tools. *Comput. Math. Methods Med.* 2016:9415459. doi: 10.1155/2016/9415459

Lockwood, S., Voth, D. E., Brayton, K. A., Beare, P. A., Brown, W. C., Heinzen, R. A., et al. (2011). Identification of *Anaplasma marginale* type IV secretion system effector proteins. *PLoS ONE* 6:e27724. doi: 10.1371/journal.pone.0027724

Maturana, P., Graham, J. G., Sharma, U. M., and Voth, D. E. (2013). Refining the plasmid-encoded type IV secretion system substrate repertoire of *Coxiella burnetii*. *J. Bacteriol.* 195, 3269–3276. doi: 10.1128/JB.00180-13

McDermott, J. E., Corrigan, A., Peterson, E., Oehmen, C., Niemann, G., Cambronne, E. D., et al. (2011). Computational prediction of type III and IV secreted effectors in gram-negative bacteria. Infection and immunity. *Brief. Bioinform.* 79, 23–32. doi: 10.1128/IAI.00537-10

Meyer, D. F., Noroy, C., Moumene, A., Raffaele, S., Albina, E., and Vachiery, N. (2013). Searching algorithm for type IV secretion system effectors 1.0: a tool for predicting type IV effectors and exploring their genomic context. *Nucleic Acids Res.* 41, 9218–9229. doi: 10.1093/nar/gkt718

Noroy, C., Lefrancois, T., and Meyer, D. F. (2019). Searching algorithm for type IV effector proteins (s4te) 2.0: improved tools for type IV effector prediction, analysis and comparison. *PLoS Comput. Biol.* 15:e1006847. doi: 10.1371/journal.pcbi.1006847

Sinclair, S. H. G., Garcia-Garcia, J. C., and Dumler, J. S. (2015). Bioinformatic and mass spectrometry identification of *Anaplasma phagocytophilum* proteins translocated into host cell nuclei. *Front. Microbiol.* 6:55. doi: 10.3389/fmicb.2015.00055

Voth, D. E., Beare, P. A., Howe, D., Sharma, U. M., Samoilis, G., Cockrell, D. C., et al. (2010). The *Coxiella burnetii* cryptic plasmid is enriched in genes encoding type IV secretion system substrate. *J. Bacteriol.* 193, 1493–1503. doi: 10.1128/JB.01359-10

Voth, D. E., Broederdorf, L. J., and Graham, J. G. (2012). Bacterial type IV secretion systems: versatile virulence machines. *Fut. Microbiol.* 7, 241–257. doi: 10.2217/fmb.11.150

Wang, J., Li, J., Yang, B., Xie, R., Marquez-Lago, T. T., Leier, A., et al. (2018a). Bastion3: a two-layer ensemble predictor of type III secreted effectors. *Bioinformatics.* 35, 2017–2028. doi: 10.1093/bioinformatics/bty914

Wang, J., Yang, B., An, Y., Marquez-Lago, T., Leier, A., Wilksch, J., et al. (2017a). Systematic analysis and prediction of type IV secreted effector proteins by machine learning approaches. *Brief. Bioinform.* bbx164, 1–21. doi: 10.1093/bib/bbx164

Wang, J., Yang, B., Leier, A., Marquez-Lago, T. T., Hayashida, M., Rocker, A., et al. (2018b). Bastion6: a bioinformatics approach for accurate prediction of type VI secreted effectors. *Nat. Genet.* 34, 2546–2555. doi: 10.1093/bioinformatics/bty155

Wang, J., Yang, B., Revote, J., Leier, A., Marquez-Lago, T. T., Webb, G., et al. (2017b). POSSUM: a bioinformatics toolkit for generating numerical sequence feature descriptors based on pssm profiles. *Bioinformatics* 33, 2756–2758. doi: 10.1093/bioinformatics/btx302

Wang, Y., Wei, X., Bao, H., and Liu, S. (2014). Prediction of bacterial type IV secreted effectors by c-terminal features. *BMC Genomics* 15:50. doi: 10.1186/1471-2164-15-50

Yu, L., Guo, Y., Li, Y., Li, G., Li, M., Luo, J., et al. (2010). Secretp: identifying bacterial secreted proteins by fusing new features into chou's pseudo-amino acid composition. *J. Theor. Biol.* 267, 1–6. doi: 10.1016/j.jtbi.2010.08.001

Zeng, C., and Zou, L. (2019). An account of *in silico* identification tools of secreted effector proteins in bacteria and future challenges. *Brief. Bioinform.* 20, 110–129. doi: 10.1093/bib/bbx078

Zou, L., Nan, C., and Hu, F. (2013). Accurate prediction of bacterial type IV secreted effectors using amino acid composition and PSSM profiles. *Bioinformatics* 29, 3135–3142. doi: 10.1093/bioinformatics/btt554

18

Enterococcal PcfF is a Ribbon-Helix-Helix Protein that Recruits the Relaxase PcfG Through Binding and Bending of the *oriT* Sequence

Saima Rehman[1], Yang Grace Li[2], Andreas Schmitt[1], Lena Lassinantti[1], Peter J. Christie[2] and Ronnie P.-A. Berntsson[1,3]*

[1] Department of Medical Biochemistry and Biophysics, Umeå University, Umeå, Sweden, [2] Department of Microbiology and Molecular Genetics, McGovern Medical School, Houston, TX, United States, [3] Wallenberg Centre for Molecular Medicine, Umeå University, Umeå, Sweden

Correspondence:
Ronnie P.-A. Berntsson
ronnie.berntsson@umu.se

The conjugative plasmid pCF10 from *Enterococcus faecalis* encodes a Type 4 Secretion System required for plasmid transfer. The accessory factor PcfF and relaxase PcfG initiate pCF10 transfer by forming the catalytically active relaxosome at the plasmid's origin-of-transfer (*oriT*) sequence. Here, we report the crystal structure of the homodimeric PcfF, composed of an N-terminal DNA binding Ribbon-Helix-Helix (RHH) domain and a C-terminal stalk domain. We identified key residues in the RHH domain that are responsible for binding pCF10's *oriT* sequence *in vitro*, and further showed that PcfF bends the DNA upon *oriT* binding. By mutational analysis and pull-down experiments, we identified residues in the stalk domain that contribute to interaction with PcfG. PcfF variant proteins defective in *oriT* or PcfG binding attenuated plasmid transfer *in vivo*, but also suggested that intrinsic or extrinsic factors might modulate relaxosome assembly. We propose that PcfF initiates relaxosome assembly by binding *oriT* and inducing DNA bending, which serves to recruit PcfG as well as extrinsic factors necessary for optimal plasmid processing and engagement with the pCF10 transfer machine.

Keywords: T4SS, accessory factor, conjugation, relaxosome, X-ray crystallography, protein structural and functional analysis

INTRODUCTION

Enterococcus faecalis can transfer pheromone-inducible plasmids in a highly efficient manner upon sensing the peptide pheromones produced by recipient cells. These plasmids encode three functional modules of importance for plasmid transfer: (i) the Dtr (DNA transfer and replication) proteins responsible for processing of the plasmid for transfer, (ii) the Mpf (mating-pair formation) proteins that assemble as the translocation channel or type IV secretion system (T4SS), and (iii) cell-wall anchored adhesins that facilitate formation of donor-recipient cell mating pairs (Alvarez-Martinez and Christie, 2009). Over the past decade, studies have advanced our understanding of the mechanisms of action and structures of T4SSs and Dtr factors functioning in Gram-negative (G^-) species (Grohmann et al., 2017). Systems functioning in Gram-positive (G^+) species, however, remain less well-understood. While some mechanistic and architectural features are likely conserved among all conjugative machines, key steps of substrate processing and recruitment, mating pair formation, and substrate transfer can be expected to differ substantially between systems functioning in diderm vs. monoderm species (Bhatty et al., 2013; Grohmann et al., 2017).

The tetracycline-resistance plasmid pCF10 from *E. faecalis* is a member of the highly transmissible pheromone-responsive family of mobile genetic elements (MGEs) found in enterococci. The encoded T4SSs of these pheromone regulated MGEs are tightly regulated at the transcriptional level by sensing of peptide pheromones originating from recipient cells (Dunny, 2013; Dunny and Berntsson, 2016). The broad medical importance of this large family of pheromone-inducible plasmids is underscored by the fact that they serve as reservoirs for genes encoding many different virulence factors, adhesins and antibiotic resistance. Additionally, they can mobilize other MGEs to both enterococcal and non-enterococcal recipients (Antiporta and Dunny, 2002; Staddon et al., 2006).

In this study we focused on two of the Dtr proteins, the PcfF accessory factor and the PcfG relaxase. PcfF binds to double stranded DNA (dsDNA) and is specific for inverted repeat sequences located within pCF10's origin of transfer (*oriT*) sequence (Chen et al., 2007). PcfG exhibits no intrinsic affinity for the *oriT* sequence, or any dsDNA, in the absence of PcfF. PcfF-*oriT* complexes, however, recruit PcfG to form the relaxosome, as evidenced by supershifting of PcfF-*oriT* complexes in the presence of PcfG to higher molecular mass complexes in electrophoretic mobility shift assay (EMSA) experiments (Chen et al., 2007). PcfG then catalyzes strand-specific nicking at *oriT* and generation of the single-stranded transfer intermediate (T-strand) (Chen et al., 2007, 2008; Li et al., 2012). After cleaving the substrate, PcfG remains covalently bound to the 5′ end of the T strand and likely pilots it through the conjugation channel and into the recipient cell, as has been shown for relaxases functioning in G^- systems (Alvarez-Martinez and Christie, 2009). PcfG also catalyzes the re-joining of cleaved *nic* sites *in vitro*, a reaction thought to direct T strand re-circularization, second-strand synthesis and plasmid stabilization in the recipient cell (Chen et al., 2007; Alvarez-Martinez and Christie, 2009).

Some G^- accessory factors have been well studied, including the TraM and TraY proteins of the F plasmid, MbeC from the ColE1 plasmid, TrwA of the R388 plasmid and NikA from the R64 plasmid (Luo et al., 1994; Moncalian and de la Cruz, 2004; Yoshida et al., 2008; Varsaki et al., 2009; Wong et al., 2011). These proteins have all been shown to contain a Ribbon-Helix-Helix (RHH) domain responsible for binding to DNA. RHH domains are a well-characterized family of transcriptional repressors in bacteria, first characterized with the bacterial MetJ and Arc repressors (Somers and Phillips, 1992; Raumann et al., 1994; Schreiter and Drennan, 2007). In contrast to the common helix-turn-helix motif for DNA binding, RHH domains bind DNA via a small N-proximal β-sheet composed of one β-strand from each monomer (Schreiter and Drennan, 2007). So far, no accessory factors of G^+ origin have been structurally characterized, albeit one *in silico* analysis has indicated that many of them also contain RHH-binding domains (Miguel-Arribas et al., 2017).

Here, we determined the structure of PcfF and show that it contains an N-terminal RHH domain and a C-terminal stalk domain. PcfF is a dimer in solution and structure-guided mutational analyses identified residues involved in DNA binding and residues required for interaction with PcfG. Together, our findings expand our knowledge of how accessory factors coordinate assembly of the relaxosome in G^+ bacteria. They also suggest the importance of other intrinsic, e.g., DNA bending, and extrinsic factors for relaxosome assembly *in vivo*.

EXPERIMENTAL PROCEDURES

Bacterial Strains and Plasmids

The *Escherichia coli* pET plasmid pCY33 carrying wild-type *pcfF* listed in **Supplementary Table S1** were used to generate plasmids harboring the following *pcfF* mutations: R13L (pYGL194), R13L/I14A (pYGL196), I70S (pYGL197), 1-54 (pYGL199). Mutated genes were confirmed by sequencing using the T7F primer. The *pcfF* alleles on pYGL194, 196, 197, and 199 were then amplified with primers XhoI_pcfF_F and SphI_pcfF_R, the PCR products were digested with XhoI and SphI, and the digested products were introduced into similarly digested pDL278p23 to generate plasmids carrying the *pcfF* variants: R13L (pYGL202), R13L/I14A (pYGL203), I70S (pYGL205), 1-54 (pYGL205). Constructs were confirmed by sequencing with the M13F primer. These plasmids introduced by electroporation into *E. faecalis* strain CK104 (pCF10Δ*pcfF*).

For protein production, *pcfF* (GeneBank accession AAW51324) was PCR amplified using pCF10 as a template and cloned into pGEX-6P-2 using BamHI/XhoI. The truncated version $PcfF_{1-54}$ (lacking residues 55–118) was made by mutation of Tyr 55 to a stop codon. QuikChange mutagenesis was used to generate single, double, and triple mutations of *pcfF* (R13L, I14A, R16L, R13L/I14A, R13L/R16L, R13L/I14A/R16L, I70S, N73A/Q74A, R77S, I70S/R77S, Q105A/W) with *pcfF* expression vectors as templates. *pcfG* (GeneBank accession AAW51325) was PCR amplified from pCF10 and inserted into pBAD expression vectors via the FX cloning system (Geertsma and Dutzler, 2011).

Protein Expression and Purification

PcfG (with a C-terminal deca-histidine tag), PcfF and variants thereof (all with a N-terminal GST tag) were produced in *E. coli* BL21(DE3). For PcfF and the variants of PcfF the cells were grown at 37°C in 2 × YT medium until they reached an OD_{600} of ca 1.0. At that time, the temperature was lowered to 18°C and expression was induced by the addition of 0.4 mM IPTG. Cells were grown for 16 h before harvesting. Production of selenomethionine derivatized PcfF was carried out in *E. coli* BL21(DE3) grown in M9 minimal media supplemented with 50 mg/mL L-Selenomethionine as described previously (Vanduyne et al., 1993). Derivatized PcfF was purified as described below for wild type PcfF, with the addition of

0.5 mM TCEP [tris(2-carboxyethyl)-phosphine] to all buffers after affinity purification. PcfG was produced in the same way as PcfF, with the exception that TB medium was used instead of 2 × YT and that the cells were induced by the addition of 10^{-2}% (w/v) L-arabinose at an OD_{600} of 0.8. The cells were resuspended in lysis buffer (20 mM Tris/HCl pH 7.5, 300 mM NaCl, 0.02 mg/ml DNase I, and 0.5 mM proteinase inhibitor AEBSF) before disrupting them using a Constant Cell Disruptor (Constant Systems) at 25 kPsi and 4°C. Cell debris was removed by centrifugation at 16000 × *g* for 15 min.

Wild type and variant forms of GST-PcfF were incubated for 1 h at 4°C with Glutathione-Agarose bead (Protino). The beads were packed in a gravity flow column and washed with 20 mM Tris/HCl pH 7.5, 200 mM NaCl, before eluting the protein with elution buffer (20 mM Tris/HCl pH 8.0, 200 mM NaCl, 30 mM reduced Glutathione, and 10% Glycerol). Subsequently, the proteins were run on a Superdex 200 10/300 GL Increase column (GE Healthcare) in SEC buffer (20 mM Tris/HCl pH 7.5, 200 mM NaCl). For GST-tag cleavage, 3C protease was added in a ratio of 1:100 and incubated for 16 h at 4°C. To isolate the cleaved sample a second gel-filtration step was performed in SEC buffer. The free-GST tag coeluted with cleaved PcfF, and therefore reverse-purification was performed by passing the purified sample through pre-equilibrated Glutathione-Agarose beads. The single, double and triple variants of PcfF (R13L, I14A, R16L, R13L/I14A, R13L/R16L, R13L/I14A/R16L, I70S, N73A/Q74A, R77S, I70S/R77S, Q105A, and Q105W) all behaved virtually the same as wild type PcfF, with identical elution volume on the SEC. $PcfF_{1-54}$ has a molecular mass of 7 kDa for a monomer (PcfF residues 1-54 plus 4 extra residues left on after cleaving off the GST). $PcfF_{1-54}$ eluted with an apparent molecular mass of 17 kDa on a Superdex 75 10/300 GL column. Due to the absence of Trp in PcfF, the concentration of the protein was determined using a BCA assay (Pierce).

PcfG-His was purified via refolding as previously described (Chen et al., 2007). Briefly, the protein was dissolved in binding buffer (8 M Urea, 20 mM Tris/HCl pH 7.5, 200 mM NaCl, 15 mM Imidazole pH 7.8) and bound to Ni-NTA sepharose beads (Macherey-Nagel) at 4°C. The column was washed with 10 column volumes (CV) wash buffer (8 M Urea, 20 mM Tris/HCl pH 7.5, 200 mM NaCl, and 50 mM Imidazole pH 7.8) before being eluted. PcfG-His was then refolded via a 4-step dialysis to decrease the Urea and Imidazole concentrations. The refolded PcfG-His was subsequently run on a Superdex 200 10/300 GL Increase column, where the protein eluted at the expected volume for a monomer.

Gas-Phase Electrophoretic Mobility Macromolecule Analysis

Gas-phase electrophoretic mobility macromolecule analysis (GEMMA) on PcfF was performed as previously described (Rofougaran et al., 2008). Briefly, the peak of wild type PcfF from size exclusion chromatography (SEC) was dialyzed against 100 mM ammonium acetate, pH 7.8 and then diluted to a concentration of 0.025 mg/mL in a buffer containing 100 mM ammonium acetate, pH 7.8 and 0.005% Tween-20. This protein sample was then analyzed by GEMMA.

Crystallization and Structure Determination

Crystals of selenomethionine incorporated PcfF were grown at 20°C by sitting drop vapor diffusion in a condition containing 0.2 M Lithium sulfate, 0.1 M Bis-Tris pH 6.5, 25% (w/v) PEG 3350 with a protein concentration of 16 mg/mL and a protein:reservoir ratio of 1:1. Crystals were flash-frozen in liquid nitrogen without additional cryo-protectant. X-ray diffraction data of SeMet-PcfF was collected on ID30A-3 ESRF, France at the selenium edge. The data were processed using XDS (Kabsch, 2010). The crystallographic phase-problem was solved using the single anomalous diffraction data and the selenium sites were found and refined by the Auto-Rickshaw software (Panjikar et al., 2005) with an initial model being built by ARP/wARP (Cohen et al., 2008). The PcfF crystals belonged to space group $P2_1$ and contained 4 molecules in the asymmetric unit. The structure was further built in Coot and refined at 1.9 Å using PHENIX refine (Adams et al., 2002; Emsley and Cowtan, 2004), to R_{work}/R_{free} values of 19.0/23.0%. For complete data collection and refinement statistics see **Supplementary Table S2**. The structure has been deposited in the Protein Data Band (PDB code: 6QEQ).

Electrophoretic Mobility Shift Assay

Electrophoretic mobility shift assays were performed as described elsewhere (Hellman and Fried, 2007). Different versions of single stranded *oriT* DNA were purchased from Eurofins Genomics, with the sense strand labeled at the 5′ end with fluorescein isothiocyanate (FITC). The sequences of different DNA segments used in this study are presented in **Supplementary Table S3**. Double-stranded DNA was obtained by mixing equimolar concentration of the sense and antisense strand in annealing buffer (10 mM Tris, 1 mM EDTA, 50 mM NaCl, pH 8.0). The annealing reaction was carried out by incubation at 94°C for 2 min followed by gradual cooling. For further purification, the annealed DNA was run on a 20% polyacrylamide gel, the band containing the fluorescent double strand was excised and the DNA eluted from the gel. Duplex *oriT* DNA (30 nM) was mixed with increasing concentrations of PcfF between 0 to 960 nM in buffer containing 10 mM Tris/HCL pH 7.5, 200 mM NaCl. For the binding studies with PcfF and PcfG, 30 nM of DNA, 100 nM of the PcfF variants and 300 nM of PcfG were mixed in 10 mM Tris/HCL pH 7.5, 150 mM NaCl. The DNA-protein mixtures were incubated at room temperature for 15 min and subsequently loaded on to a 20% native-PAGE made in TAE buffer (40 mM Tris, 20 mM acetic acid, 1 mM EDTA, pH 8.0). Electrophoresis was carried out for 70 min at 110 V and 4°C. The gel was imaged on a Typhoon scanner. The bands were visualized using a 488 nm excitation filter. ImageQuant software was used to quantify the fluorescent signal of the bands, and the curves from the resulting data were fitted to a non-linear fit (Specific binding with Hill slope) using GraphPad Prism.

DNA Bending Assay

Purified PcfF/PcfF$_{1-54}$ and 120 bp long DNA fragments, each containing the *oriT* sequence at varying positions (**Supplementary Table S3**) were mixed in 20 μl of 10 mM Tris/HCl (pH 7.5), 150 mM NaCl to a final protein and DNA concentration of 300 and 30 nM, respectively. After incubation at room temperature for 20 min, the reaction mixtures were loaded onto a 5% (w/v) polyacrylamide gel (acrylamide/bisacrylamide in a ratio of 37.5:1. w/w) in buffer (40 mM Tris–acetate (pH 7.8), 1 mM EDTA) and were electrophoresed at 10 V/cm and 4°C. Following electrophoresis, the DNA fragments in the gel were stained with GelRed and visualized under UV light.

In vitro Pull-Down Assay

For GST-pull-down experiments, 2 nmol of GST-PcfF fusion protein (or variants thereof) were immobilized on GSH-Sepharose beads, while 4 nmol PcfG-His were used as pray protein. Purified proteins were incubated with pre-equilibrated 50 μl Protino® Glutathione-Agarose 4B beads. BSA and purified GST were used as controls. All proteins were dialyzed against the same buffer (20 mM Tris/HCl pH 7.5, 200 mM NaCl). After binding, the beads were washed extensively (5 × 10 CV) and subsequently eluted by elution buffer (20 mM Tris/HCl pH 8.0, 200 mM NaCl, 30 mM reduced Glutathione). Samples from wash and elution steps were analyzed on 15% SDS-PAGE and stained with Coomassie Brilliant Blue G-250.

Detection of PcfF Mutant Proteins in *E. faecalis*

Exponential-phase cultures (10 ml) of *E. faecalis* OG1RF strains carrying pCF10Δ*pcfF* without and with plasmids producing wild type or mutant PcfF proteins were normalized to an OD_{600} of 0.3. The cells were pelleted by centrifugation at 13,200 × *g* for 15 min at 4°C and washed once with cold 1X physiological buffer saline (PBS). The pellet was resuspended in 125 μl of SMM buffer (0.5 M sucrose, 0.02 M $MgCl_2$, 0.02 M maleate, pH 6.5) containing 60 μl ml^{-1} of mutanolysin (Sigma-Aldrich) and 10 mg ml^{-1} of lysozyme (Sigma-Aldrich), and the resulting mix was incubated for 1 h at 37°C with shaking. Material released from the digested cell wall was separated from cell-bound material by centrifugation at 13.200 × *g* for 15 min at 4°C. PcfF variants were detected by Western transfer and immunostaining with the anti-PcfF antibodies (Chen et al., 2008). Blots were probed with antibodies against the β-subunit of RNA polymerase as a protein loading control (Christie et al., 1988; Chen et al., 2007, 2008).

Conjugation Assays

Enterococcus faecalis donor and recipient cultures grown overnight were diluted 1:10 in BHI (Brain Heart Infusion broth; Sigma) and incubated for 1 h at 37°C without shaking. Donor and recipient cells were mixed in a ratio of 1:1 and allowed to mate in liquid without shaking for 1 h at 37°C. Mating mixtures were serially diluted in BHI, and the numbers of donors and transconjugants were obtained by plating on selective BHI agar plates. The plasmid transfer frequencies were calculated as the number of transconjugants per donor cell (Chen et al., 2008). The results are reported as an average of three replicates of each experiment.

RESULTS

PcfF Is a Dimer in Solution

Full-length PcfF (14 kDa) was produced in *E. coli* and purified to homogeneity. It eluted as a single peak on SEC, with an apparent molecular mass of ~48 kDa. The molecular mass of wild type PcfF from the SEC peak was determined by gas-phase electrophoretic mobility macromolecule analysis (GEMMA, also termed Macroion mobility spectrometer) to be ~32 kDa, very close an apparent dimer (**Figure 1A**; Kaufman et al., 1996; Bacher et al., 2001). PcfF$_{Q105A/W}$ variants that were made to probe potential differences in oligomerization state, showed no difference in elution volume on SEC as compared to wt PcfF. PcfF$_{1-54}$ eluted as a single peak on SEC, with an apparent molecular weight of 17 kDa (**Figure 1B**), close to the expected 14 kDa weight of a dimer.

We next crystallized full-length selenomethionine-incorporated PcfF for structural analysis. Crystals belonged to space group $P2_1$ and contained 4 molecules in the asymmetric unit. X-ray diffraction data was collected at the selenium edge and the phase problem solved by SAD phasing (**Supplementary Table S2**). The structure was refined at a resolution of 1.9 Å. The electron density accounted for the entire PcfF protein, with the exception of 1–7 residues that were missing at the N terminus and one residue at the C terminus, the exact number varying between each of the 4 protein chains in the structure. PcfF crystallized as a tetramer (dimer of dimers in a head to toe organization) (**Supplementary Figure S1**). However, the tetrameric interface is weak and deemed unstable by PISA calculations (Krissinel, 2015). The biologically relevant oligomer was suggested to be dimeric, in agreement with results from the SEC and GEMMA experiments. The dimeric structure is extended in one dimension, which explains why in size exclusion chromatography PcfF elutes at a higher apparent molecular mass than expected. Functional analysis described in the next section also points toward PcfF functioning as a dimer. From here onward, we base our structural analysis on one of the dimers, made up by chains A and C.

PcfF Contains a DNA-Binding RHH Domain

PcfF contains a RHH domain at the N terminus and a 2-helix bundle, here termed the stalk domain, at the C terminus. These two domains are connected by a hinge region (**Figure 2A**). In the dimer, the RHH and stalk domains are built up by secondary structure elements from both monomers in the dimer. By superimposition of PcfF's RHH domain on other RHH motifs associated with DNA, we determined that 3 residues, R13, I14, and R16, likely are involved in DNA binding (**Figures 2B,C** and **Supplementary Figure S2**). These residues were mutated and effects on DNA binding were assessed using an EMSA. Wild

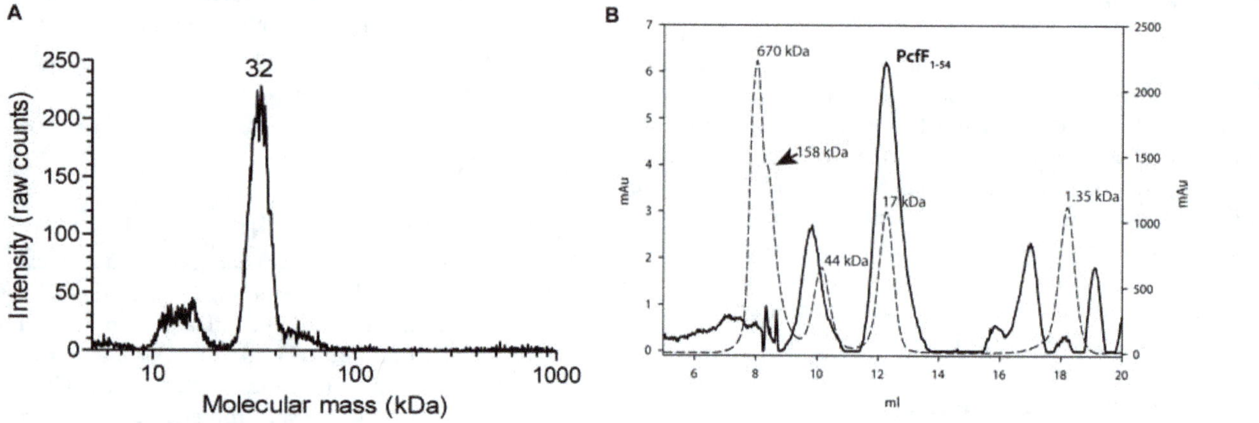

FIGURE 1 | Oligomeric state of PcfF. **(A)**: GEMMA analysis of PcfF. The sample was taken from the elution peak of the size exclusion chromatography fraction. The GEMMA analysis was performed with a protein concentration of 0.025 mg/mL. The determined molecular masses (in kDa) are written above the peak. **(B)**: SEC analysis of $PcfF_{1-54}$ on a Superdex 75 10/300 GL column, with molecular mass standards as dashed lines and $PcfF_{1-54}$ as a solid line. The molecular mass in kDa is written above each standard in the graph. Since $PcfF_{1-54}$ does not have any Tryptophan residues, the left Y axis denotes the absorbance at 254 nm for $PcfF_{1-54}$, while the right y-axis shows the absorbance at 280 nm for the standards. The main peak for $PcfF_{1-54}$ elutes at 17 kDa, with the earlier peak corresponding to free GST after the cleavage of the protein as confirmed by SDS-PAGE analysis.

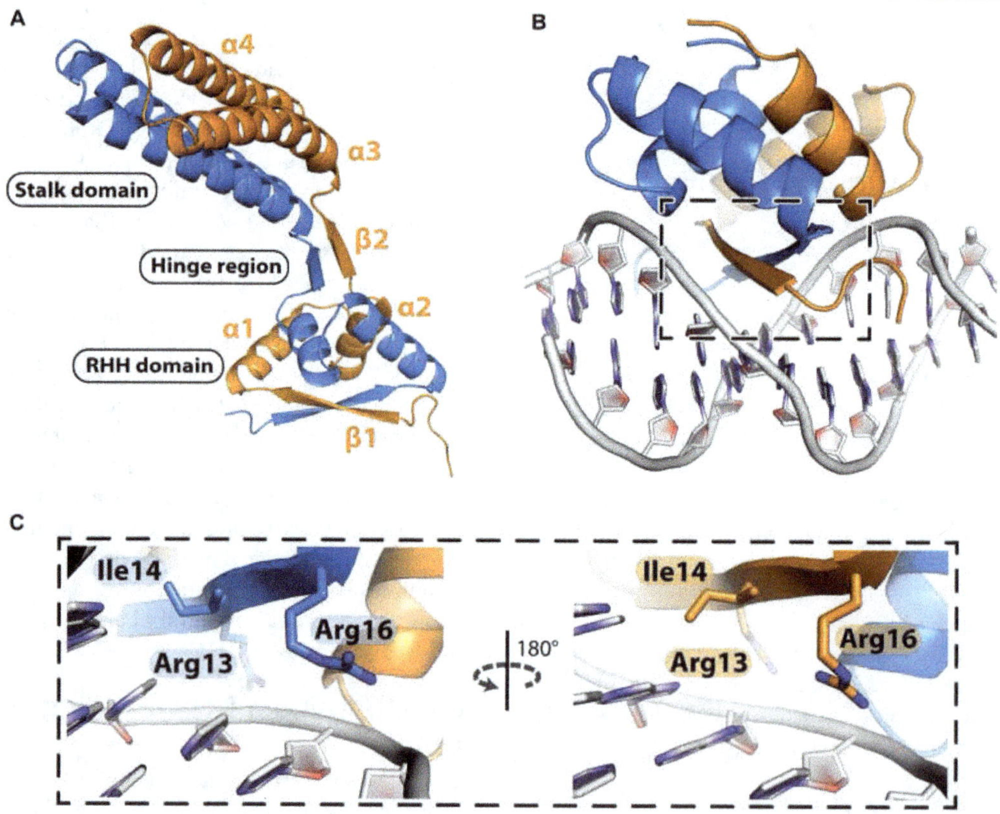

FIGURE 2 | Structure of PcfF. **(A)**: Overall representation of the dimeric PcfF structure, with the RHH and stalk domains connected by a hinge region. Both monomers contribute to the formation of the RHH domain. **(B)**: Model of PcfF with bound DNA, based on superposition with ArcA structure (**Supplementary Figure S2**). **(C)**: Enlarged view of the box from panel **B**, with the conserved DNA binding residues R13, I14, and R16 highlighted as sticks.

type PcfF bound a 40 bp *oriT* sequence, composed of double-stranded inverted repeats and the *nic*-site, with an estimated K_D of ~100 nM, in agreement with previous findings (Chen et al., 2007; **Figures 3A,B**). PcfF did not bind a random DNA sequence, verifying a specificity for *oriT* binding (**Supplementary Figure S3**). PcfF variants with single (R13L, I14A, R16L),

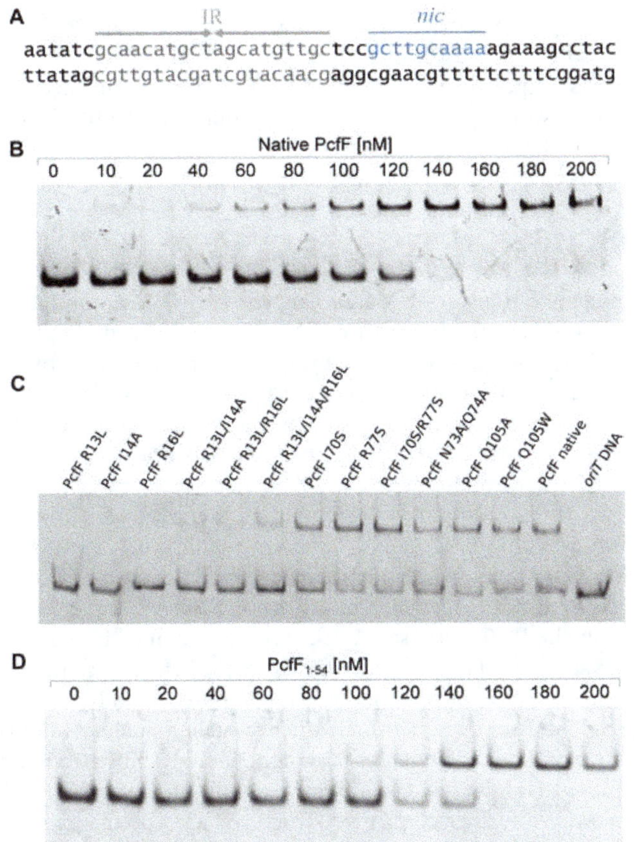

FIGURE 3 | EMSA of PcfF variants binding to *oriT*. In all cases the 30 nM of the 40 bp long double stranded *oriT* substrate was used. **(A)**: Sequence of the *oriT* (40 bp) from pCF10. The inverted repeats (IR), which PcfF binds to, and the nic-site are highlighted. **(B)**: Wild type PcfF binds to a 40 bp *oriT* substrate. The increasing concentrations of PcfF added are indicated above the lanes. **(C)**: PcfF variants bind with varying degrees to the 40 bp *oriT* sequence. Reaction mixtures contained 100 nM PcfF. **(D)**: $PcfF_{1-54}$ binds to *oriT*. The increasing concentrations of PcfF added are indicated above the gel. The reactions were analyzed on 20% native polyacrylamide gels. Protein components in each lane are shown on the top of the lane.

double (R13L/I14A, R13L/R16L), or triple (R13L/I14A/R16L) substitution mutations showed marked decreases in *oriT* binding (**Figure 3C**). We also confirmed that the RHH domain ($PcfF_{1-54}$) without the associated stalk domain bound *oriT* with little reduction in affinity (**Figure 3D**).

The Stalk Domain of PcfF Binds PcfG

PcfF binds the relaxase PcfG, as determined by EMSAs and affinity pull-down assays (Chen et al., 2007, 2008). PcfF possesses a sequence-motif NINQ in a surface-exposed region of the C-terminal stalk; this motif is semi-conserved among other T4SSs accessory proteins associated with conjugation systems in Gram-negative and -positive bacteria (**Supplementary Figure S4A**) (Varsaki et al., 2009). In the PcfF X-ray structure, the NINQ motif forms a patch within a small groove (**Supplementary Figure S4B**). We hypothesized that this motif might comprise the binding surface for PcfG. To test this

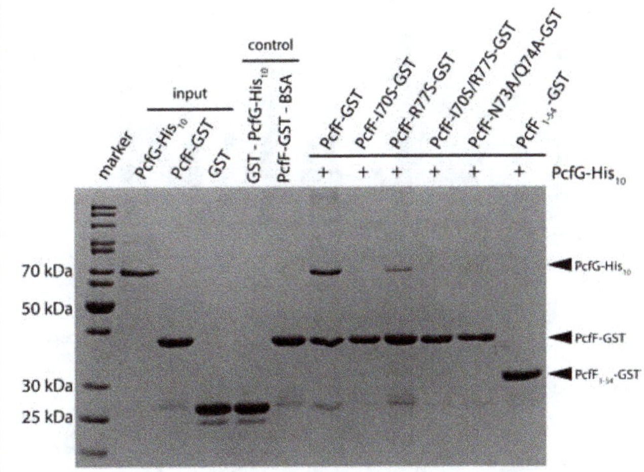

FIGURE 4 | SDS-PAGE analysis of GST pull-down of PcfG. GST-tagged PcfF and variants thereof were compared for their ability to bind $PcfG\text{-}His_{10}$. As controls, GST alone was assayed for binding of $PcfG\text{-}His_{10}$, and PcfF-GST was assayed for binding of BSA. Protein bands were visualized by staining with Coomassie Blue.

model, we introduced several mutations (I70S, R77S, I70S/R77S, N73A/Q74A) in and around this conserved surface patch. These PcfF variants, as well as $PcfF_{1-54}$ lacking the entire stalk domain, behaved as wild type PcfF with respect to purification as dimers and binding of *oriT* DNA (**Figures 3C,D**). We next tested for effects of the mutations on PcfF binding to PcfG using affinity pull-down assays. GST-PcfF and variants thereof were incubated with PcfG-His and Glutathione-Agarose beads. Following extensive washing, proteins were eluted and analyzed for the presence of GST-PcfF and PcfG-His. As shown previously, wild type GST-PcfF pulled down PcfG-His (**Figure 4**; Chen et al., 2008). None of the GST-PcfF variants detectably bound PcfG-His, except for the R77S mutant which showed a low level of binding. We further assayed for the ability of PcfG-His to bind to PcfF-*oriT* or $PcfF_{1-54}$-*oriT* complexes via EMSAs, but did not observe any additional supershifted bands upon the addition of PcfG-His (**Supplementary Figure S5**).

PcfF Binding Induces DNA Bending

Other T4SS accessory factors with RHH domains have been shown to induce bending of DNA (Yoshida et al., 2008; Wong et al., 2011). To determine if PcfF induces a bend in the pCF10 *oriT* sequence, we capitalized on findings that bent DNA fragments exhibit an anomalous electrophoretic mobility behavior, which is most pronounced when the bending locus is located close to the center of the fragment (Thompson and Landy, 1988; Levene and Zimm, 1989). We performed these DNA bending experiments using 120 bp fragments of random DNA with the *oriT* positioned at the 5′ end, middle or 3′ end of the DNA (**Supplementary Table S3**) (Crothers et al., 1991). Binding of PcfF induced a more pronounced shift in the DNA fragment containing the central *oriT* sequence compared with fragments in which *oriT* was positioned at either end. These

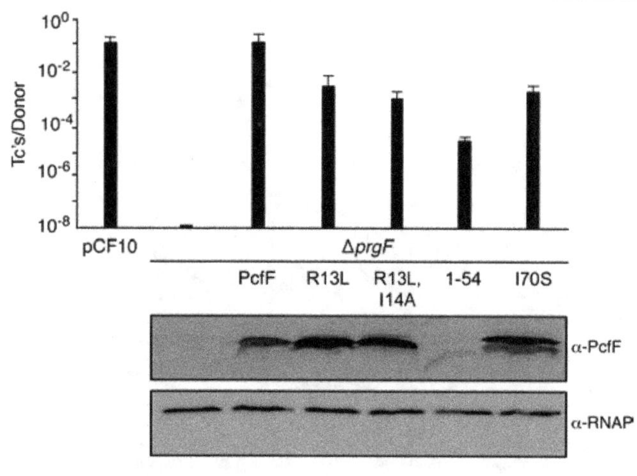

FIGURE 5 | Effects of PcfF mutations on pCF10 transfer. Upper graph: CK104 donors harboring pCF10Δ*pcfF* and producing wild type and mutant variants of PcfF from the constitutive P_{23} promoter were mated with OG1ES recipients for 1 h in BHI media at 37°C without shaking. Transfer frequencies are presented as the number of transconjugants (Tc's) per donor. Experiments were repeated at least three times in triplicate, and results from a representative experiment with standard deviations are shown. Lower panels: Steady-state levels of PcfF proteins in exponentially growing strains. Immunoblots were developed with α-PcfF antibodies for detection of PcfF variants or α-RNAP antibodies for detection of the β-subunit of RNA polymerase as a loading control. Protein extracts were loaded on a per-cell equivalent basis. Strains for both panels: CK104 (pCF10ΔpcfF) alone (ΔpcfF) or with plasmids producing PcfF (pCY33), $PcfF_{R13L}$ (pYGL202), $PcfF_{R13L/I14A}$ (pYGL203), $PcfF_{1-54}$ (pYGL204), $pcfF_{I70S}$ (pYGL205).

findings support a conclusion that PcfF does indeed induce bending of DNA at the *oriT* site (**Supplementary Figure S6**). A similar trend could be seen for $PcfF_{1-54}$ as for PcfF. However, for unknown reasons, $PcfF_{1-54}$ bound to the 120 bp long DNA yielded smeary band shifts and therefore prevent firm conclusions regarding the capacity of PcfF's RHH domain to induce DNA bending.

PcfF DNA Binding and Relaxase Recruitment Is Not Essential for Conjugation *in vivo*

Finally, we determined the effects of the PcfF mutations on pCF10 transfer *in vivo* (**Figure 5**). *E. faecalis* strain CK104(pCF10D*pcfF*) does not transfer the mutant plasmid unless it additionally carries pCY16, which produces wild type PcfF from the constitutive P_{23} promoter. Interestingly, CK104(pCF10D*pcfF*) harboring plasmids producing the RHH mutant proteins $PcfF_{R13L}$ and $PcfF_{R13L/I14A}$ which fail to bind *oriT* DNA *in vitro*, also transferred pCF10D*pcfF* albeit at reduced frequencies of 1 to 2 orders of magnitude compared with CK104(pCF10D*pcfF*, pCY16). Similarly, the CK104(pCF10D*pcfF*) donor with plasmids producing variants defective in binding PcfG *in vitro* ($PcfF_{1-54}$, $PcfF_{I70S}$) mutant also were transfer-proficient, although at reduced levels. In CK104 donor strains, the full-length mutant proteins accumulated at levels comparable to or even higher than wild type PcfF. $PcfF_{1-54}$ was detected at low levels, possibly reflecting instability or poor recognition by the anti-PcfF polyclonal antibodies. Formation of the PcfF/PcfG/*oriT* relaxosome on pCF10 thus appears to depend not only on PcfF residues responsible for binding *oriT* and PcfG *in vitro*, but on DNA structures formed *in vivo* or other unidentified host-encoded factors.

DISCUSSION

Conjugative transfer of MGEs happens by: (i) assembly of the relaxosome at *oriT* sequences, (ii) relaxase-catalyzed nicking of the DNA strand destined for transfer (T-strand), (iii) relaxosome recruitment to the type IV coupling protein (T4CP), and (iv) translocation of the relaxase/T-strand intermediate through the transfer channel (Alvarez-Martinez and Christie, 2009; Wong et al., 2012; Grohmann et al., 2017). Dtr accessory factors are known to be required for assembly of the relaxosome, but in most cases the molecular details surrounding this early stage reaction are unknown. Here, we have solved the structure of the accessory factor PcfF, which binds the pCF10 *oriT* sequence and recruits the PcfG relaxase for relaxosome assembly. Like several other accessory factors, albeit far from all, PcfF is essential for conjugation. We showed that PcfF is composed of an N-terminal RHH domain and a C-terminal a-helical stalk domain. Although residues in both of these domains contribute to dimerization of PcfF, the RHH domain also dimerizes in the absence of the stalk domain. We further confirmed structure-based predictions that β-strands within PcfF's RHH domain contribute to *oriT* binding and gained evidence that a specific patch on the C-terminal stalk of PcfF mediates binding of PcfG.

Prior to this study, only three T4SS encoded accessory factors (TraM, NikA, and VirC2) have been structurally determined to our knowledge (Yoshida et al., 2008; Lu et al., 2009; Wong et al., 2011). Each belong to the RHH superfamily whose members are best known as bacterial transcription factors. Superimposition of these proteins with PcfF reveal that their RHH domains have overall similar structures, with RMSD values of 1.5 – 4 Å (**Figure 6A**). These proteins bind DNA through intercalation of their small two-stranded β-sheet within the RHH domain into the major groove of double stranded DNA; this interaction contributes both to affinity and specificity of DNA substrate binding (Schreiter and Drennan, 2007). Structurally, this β-sheet is made up by the first β-strand of each monomer in the RHH domain. Within this β-strand, a few structurally equivalent residues are (semi)conserved among the various accessory factors (**Figure 6B**). These residues have been implicated to be important for DNA binding, as established by solved structures of TraM or ArcA bound to DNA substrates (Raumann et al., 1994; Wong et al., 2011). In agreement with findings from other RHH accessory factors, we determined that PcfF binds pCF10's *oriT* sequence via its RHH-domain, and that mutation of the surface-exposed charged residues R13 and R16, as well as I14, in the β-sheet strongly abrogate *oriT* binding *in vitro*. Deletion of the stalk domain does not impair *oriT* binding, confirming the RHH domain is both necessary and sufficient for binding the DNA substrate.

A

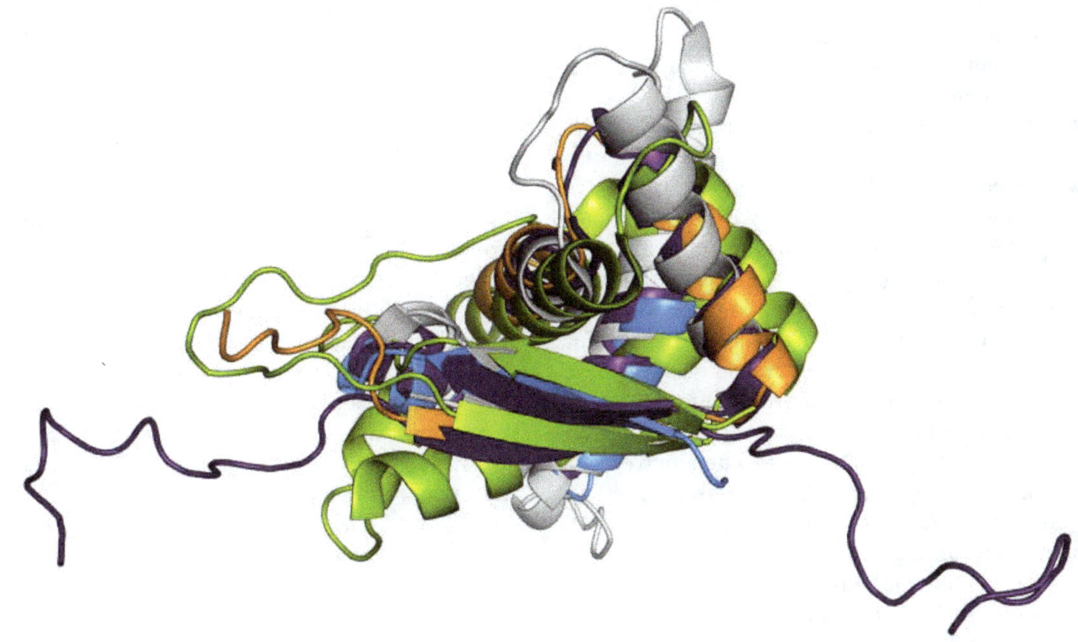

B

```
PcfF                       β1         α1                    α2
                    1        10        20        30        40        50
PcfF           MENKQKLKR...P...IQRIVRLSEEENNL...IKRKIEESFFPNFQNFALHLLIQGEIR
NikA           MSDSAVRKKSEVRQKTVVRTLRFSPVEDET...IRKKAEDSG......LTVSAYIRNAAL
TraM           MAKVNLYIS...N.DAYEKINAIIEKRRQE...GAREKDVSFSATAS.MLLELGLRVHEA
VirC2          VQTSRMFPV...S..........LIEAARNHFDPLGLETARAFGHKLATAALACFFAREKA
consensus>60   m....................e...#.........e.sf........l........
```

FIGURE 6 | Structural and sequence comparison of four T4SS accessory proteins. **(A)** Superimposed structures of the RHH domain of PcfF (blue and orange), TraM (PDB code 3ON0, gray), NikA (PDB code 2BA3, purple), and VirC2 (PDB code 2RH3, green). PcfF and NikA have similar overall RHH domain structures, with RMSD of ca 1.5 Å. TraM and VirC2 vary more with RMSDs of ca 4.5 Å, but clearly share the same overall fold. **(B)** Sequence alignment of the RHH domains of the four proteins, with the secondary structure elements indicated above the alignment. The alignment file was generated through T-COFFEE server, followed by ESPript3 for rendering alignments (Armougom et al., 2006; Robert and Gouet, 2014). Text in red denotes conserved residues within a group. Blue frame with yellow background highlights similarity across groups and lowercase character denotes the consensus residue for consensus level >0.6.

F plasmid-encoded TraM and another RHH domain-containing protein, TraY, bend and induce localized denaturation upon binding of DNA substrates (Luo et al., 1994; Karl et al., 2001; Fekete and Frost, 2002). Like PcfF, TraM comprises an RHH- and a C-terminal stalk domain. However, in contrast to what seems to be the case for PcfF, TraM assembles as a tetramer with two RHH domains and one larger 8-helical bundle stalk domain forming the tetramerization interface (**Supplementary Figure S7**). TraM binds to its cognate DNA cooperatively, which induces DNA bending (Wong et al., 2011). Here, we showed that PcfF also bends its *oriT* substrate, although seemingly by a mechanism different than TraM. Specifically, in contrast to TraM, which assembles as tetramers in solution, PcfF is dimeric as shown by GEMMA analyses and further supported by the SEC and functional assays. Besides the other evidence, the weak tetramerization interface in the crystal structure only supports a head to tail tetramer, where the two RHH domains sit rotated 180 from each other with the stalk domain in between them (**Supplementary Figure S1**). This organization of the RHH domains is unlikely to be biologically relevant. The stalk domain, which is responsible for tetramerization of TraM, likely supports only dimerization of PcfF in solution. In our EMSAs, binding induced a DNA shift to a single species with higher molecular mass, as has been shown previously (Chen et al., 2007, 2008), indicating that there is only one DNA binding site per functional PcfF molecule. If PcfF would have been a tetramer, it would have contained two RHH domains and thus be able to bind two independent *oriT* probes in the EMSA. If that was the case one would expect to see a state of two independent retarded species in the EMSA, especially around the concentration corresponding to the apparent dissociation constant (K_D). Independent of the PcfF concentration used we only observe a single retarded species, indicating that it is the dimeric PcfF that binds to *oriT*. Although we cannot exclude that PcfF can under some circumstances function as a tetramer, e.g., upon relaxosome assembly, our data points toward that the dimeric form of PcfF is the functional unit.

In view of these findings, we propose that a single PcfF dimer suffices to bind and bend pCF10's *oriT* sequence.

Structural and sequence analysis revealed that the stalk domain possesses a conserved sequence motif, NINQ. These residues, together with a few flanking residues (I70, N73, Q74, and R77), form a patch on the surface of PcfF (**Supplementary Figure S4**). We gained evidence that this patch mediates binding of PcfF to PcfG by showing that a single point mutation (I70S) abolishes PcfF-PcfG binding (**Figure 4**). Other mutations around this site also completely (N73A/Q74A) or partially (R77S) disrupt this interaction. Finally, $PcfF_{1-54}$, (lacking the entire stalk domain) does not bind PcfG, firmly establishing the importance of the stalk domain for the PcfF-PcfG interaction (**Figure 4**). In EMSAs, we do not observe any significant additional retarded species upon the addition of PcfG-His to the reaction mix of PcfF-*oriT* or $PcfF_{1-54}$-*oriT*, in contrast to what was previously shown (Chen et al., 2007). We attribute the differences observed between the previous study and our results to different experimental conditions and protein constructs.

Our finding that mutations in, or a complete deletion of, the stalk domain of PcfF attenuates but does not totally abolish pCF10 transfer *in vivo* (**Figure 5**) indicates that PcfF might be able to indirectly recruit PcfG. We speculate that this could be facilitated via bending and unwinding of the DNA upon PcfF binding. Another accessory factor, TrwA, also binds its cognate *oriT* sequence via an N-terminal RHH domain, whereas its C-terminal domain (which has not been structurally characterized) was shown to not bind to the TrwC relaxase but rather to TrwB (Tato et al., 2007). TrwB is a member of the superfamily of ATPases known as coupling proteins, which are associated with the T4SSs and function in recruitment of cognate substrates for delivery into the transfer channel (Grohmann et al., 2017). Our finding that the $PcfF_{1-54}$ variant supports pCF10 transfer *in vivo* establishes that the stalk domain is not only dispensable for relaxosome assembly but is also not required for docking of the relaxosome with the PcfC coupling protein in *E. faecalis*.

It is more difficult to reconcile the lack of strong effects of the RHH-domain mutations (R13L, R13L/I14A), which abolish PcfF-*oriT* binding *in vitro*, on pCF10 transfer *in vivo* (**Figure 5**). Equivalent mutations in conserved polar, charged residues in the β-sheets of other RHH accessory factors diminish plasmid transfer by more than 3 orders of magnitude (Moncalian and de la Cruz, 2004; Yoshida et al., 2008; Varsaki et al., 2009), although mutant accessory factors can still support a modest level (10^{-5} – 10^{-6} Tc's/D) of plasmid transfer. Even though PcfF homodimers stably bind *oriT* sequences *in vitro*, it is possible that additional binding surfaces or residues are exposed when PcfF binds PcfG, which can contribute to *oriT* binding *in vivo*. Alternatively, PcfF might be capable of binding *oriT* secondary structures that form only *in vivo* through surface-exposed residues other than those mutated in our study. Finally, we cannot exclude the possibility that other unidentified plasmid- or host-encoded factors, e.g., IHF-like proteins, contribute to relaxosome assembly *in vivo* (Nelson et al., 1995; Karl et al., 2001). Discriminating between these possibilities will require further investigations.

CONCLUSION

In summary, RHH domains of several accessory factors associated with conjugation have now been solved and functionally characterized. However, besides PcfF reported here, only one other structure exists for a full-length accessory factor. The structural basis for PcfF binding to pCF10's *oriT* sequence resembles that identified for other RHH domain containing proteins. However, in contrast to most other members, PcfF forms dimers instead of tetramers in solution and does not show cooperative binding. Furthermore, mutations in conserved charged, polar residues in the DNA binding b-sheet motif do not block *oriT* substrate binding *in vivo*. We also showed that the α-helical stalk domain contributes to binding of the relaxase PcfG, despite the fact that the dimeric RHH domain alone retains the capacity to orchestrate assembly of the functional relaxosome and support plasmid transfer *in vivo*. Together, our findings underscore both the structural conservation and functional plasticity of accessory factors as nucleators of relaxosome assembly among the conjugation systems.

AUTHOR CONTRIBUTIONS

SR, AS, and RB performed the cloning, protein purification and structure determination. SR and AS collected the diffraction data. SR, AS, and LL performed the EMSA experiments and protein size determination via GEMMA and SEC. AS performed the DNA bending experiments. YL performed the cloning in *E. faecalis* and the *in vivo* conjugation experiments. SR, PC, and RB planned the experiments, performed the data analysis and wrote the manuscript with input from all authors.

FUNDING

This work was supported by grants from the Swedish Research Council (2016–03599), Åke Wibergs Foundation, Carl Tryggers Foundation, Knut and Alice Wallenberg Foundation and Kempestiftelserna (JCK-1524) to RB and by NIH grant GM48746 to PC.

ACKNOWLEDGMENTS

We thank the beamline scientists at MaxIV, Lund for their support in data collection. We acknowledge the ESRF for provision of synchrotron radiation facilities and we also thank Guillaume Gotthard for assistance in using beamline ID30A-3. We are very grateful to Dr. Anders Hofer for his help with the GEMMA measurements.

REFERENCES

Adams, P. D., Grosse-Kunstleve, R. W., Hung, L. W., Ioerger, T. R., McCoy, A. J., Moriarty, N. W., et al. (2002). PHENIX: building new software for automated crystallographic structure determination. *Acta Crystallogr. D Biol. Crystallogr.* 58, 1948–1954.

Alvarez-Martinez, C. E., and Christie, P. J. (2009). Biological diversity of prokaryotic type IV secretion systems. *Microbiol. Mol. Biol. R.* 73, 775–808. doi: 10.1128/MMBR.00023-09

Antiporta, M., and Dunny, G. (2002). ccfA, the genetic determinant for the cCF10 peptide pheromone in *Enterococcus faecalis* OG1RF. *J. Bacteriol.* 184, 1155–1162.

Armougom, F., Moretti, S., Poirot, O., Audic, S., Dumas, P., Schaeli, B., et al. (2006). Expresso: automatic incorporation of structural information in multiple sequence alignments using 3D-Coffee. *Nucleic Acids Res.* 34, W604–W608. doi: 10.1093/nar/gkl092

Bacher, G., Szymanski, W. W., Kaufman, S. L., Zollner, P., Blaas, D., and Allmaier, G. (2001). Charge-reduced nano electrospray ionization combined with differential mobility analysis of peptides, proteins, glycoproteins, noncovalent protein complexes and viruses. *J. Mass Spectrom.* 36, 1038–1052. doi: 10.1002/jms.208

Bhatty, M., Laverde Gomez, J. A., and Christie, P. J. (2013). The Expanding bacterial type IV secretion lexicon. *Res. Microbiol.* 164, 620–639. doi: 10.1016/j.resmic.2013.03.012

Chen, Y., Staddon, J. H., and Dunny, G. M. (2007). Specificity determinants of conjugative DNA processing in the *Enterococcus faecalis* plasmid pCF10 and the *Lactococcus lactis* plasmid pRS01. *Mol. Microbiol.* 63, 1549–1564. doi: 10.1111/j.1365-2958.2007.05610.x

Chen, Y., Zhang, X., Manias, D., Yeo, H.-J., Dunny, G. M., and Christie, P. J. (2008). *Enterococcus faecalis* PcfC, a spatially localized substrate receptor for type IV secretion of the pCF10 transfer intermediate. *J. Bacteriol.* 190, 3632–3645. doi: 10.1128/JB.01999-07

Christie, P. J., Kao, S. M., Adsit, J. C., and Dunny, G. M. (1988). Cloning and expression of genes encoding pheromone-inducible antigens of *Enterococcus (Streptococcus) faecalis*. *J. Bacteriol.* 170, 5161–5168.

Cohen, S. X., Ben Jelloul, M., Long, F., Vagin, A. A., Knipscheer, P., Lebbink, J., et al. (2008). ARP/wARP and molecular replacement: the next generation. *Acta Crystallogr. D Biol. Crystallogr.* 64, 49–60.

Crothers, D. M., Gartenberg, M. R., and Shrader, T. E. (1991). Dna Bending in protein-Dna complexes. *Methods Enzymol.* 208, 118–146.

Dunny, G. M. (2013). Enterococcal sex pheromones: signaling, social behavior, and evolution. *Annu. Rev. Genet.* 47, 457–482. doi: 10.1146/annurev-genet-111212-133449

Dunny, G. M., and Berntsson, R. P.-A. (2016). Enterococcal sex pheromones: evolutionary pathways to complex, two-signal systems. *J. Bacteriol.* 198, 1556–1562. doi: 10.1128/JB.00128-16

Emsley, P., and Cowtan, K. (2004). Coot: model-building tools for molecular graphics. *Acta Crystallogr. D Biol. Crystallogr.* 60, 2126–2132.

Fekete, R. A., and Frost, L. S. (2002). Characterizing the DNA contacts and cooperative binding of F plasmid TraM to its cognate sites at oriT. *J. Biol. Chem.* 277, 16705–16711. doi: 10.1074/jbc.M111682200

Geertsma, E. R., and Dutzler, R. (2011). A versatile and efficient high-throughput cloning tool for structural biology. *Biochemistry* 50, 3272–3278. doi: 10.1021/bi200178z

Grohmann, E., Christie, P. J., Waksman, G., and Backert, S. (2017). Type IV secretion in gram-negative and gram-positive bacteria. *Mol. Microbiol.* 107, 455–471. doi: 10.1111/mmi.13896

Hellman, L. M., and Fried, M. G. (2007). Electrophoretic mobility shift assay (EMSA) for detecting protein-nucleic acid interactions. *Nat. Protoc.* 2, 1849–1861. doi: 10.1038/nprot.2007.249

Kabsch, W. (2010). XDS. *Acta Crystallogr. D Biol Crystallogr.* 66, 125–132. doi: 10.1107/S0907444909047337

Karl, W., Bamberger, M., and Zechner, E. L. (2001). Transfer protein TraY of plasmid R1 stimulates TraI-catalyzed oriT cleavage in vivo. *J. Bacteriol.* 183, 909–914. doi: 10.1128/JB.183.3.909-914.2001

Kaufman, S. L., Skogen, J. W., Dorman, F. D., Zarrin, F., and Lewis, K. C. (1996). Macromolecule analysis based on electrophoretic mobility in air: globular proteins. *Anal. Chem.* 68, 1895–1904. doi: 10.1021/ac951128f

Krissinel, E. (2015). Stock-based detection of protein oligomeric states in jsPISA. *Nucleic Acids Res.* 43, W314–W319. doi: 10.1093/nar/gkv314

Levene, S. D., and Zimm, B. H. (1989). Understanding the anomalous electrophoresis of bent DNA molecules: a reptation model. *Science* 245, 396–399.

Li, F., Alvarez-Martinez, C., Chen, Y., Choi, K.-J., Yeo, H.-J., and Christie, P. J. (2012). *Enterococcus faecalis* PrgJ, a VirB4-like ATPase, Mediates pCF10 conjugative transfer through substrate binding. *J. Bacteriol.* 194, 4041–4051. doi: 10.1128/JB.00648-12

Lu, J., den Dulk-Ras, A., Hooykaas, P. J. J., and Glover, J. N. M. (2009). *Agrobacterium tumefaciens* VirC2 enhances T-DNA transfer and virulence through its C-terminal ribbon-helix-helix DNA-binding fold. *Proc. Natl. Acad. Sci. U.S.A.* 106, 9643–9648. doi: 10.1073/pnas.0812199106

Luo, Y., Gao, Q., and Deonier, R. C. (1994). Mutational and physical analysis of F plasmid tray protein-binding to orit. *Mol. Microbiol.* 11, 459–469.

Miguel-Arribas, A., Hao, J. A., Luque-Ortega, J. R., Ramachandran, G., Val-Calvo, J., Gago-Córdoba, C., et al. (2017). The *Bacillus subtilis* conjugative plasmid pLS20 encodes two ribbon-helix-helix type auxiliary relaxosome proteins that are essential for conjugation. *Front. Microbiol.* 8:2138. doi: 10.3389/fmicb.2017.02138

Moncalian, G., and de la Cruz, F. (2004). DNA binding properties of protein TrwA, a possible structural variant of the Arc repressor superfamily. *Biochim. Biophys. Acta* 1701, 15–23. doi: 10.1016/j.bbapap.2004.05.009

Nelson, W. C., Howard, M. T., Sherman, J. A., and Matson, S. W. (1995). The traY gene product and integration host factor stimulate *Escherichia coli* DNA helicase I-catalyzed nicking at the F plasmid oriT. *J. Biol. Chem.* 270, 28374–28380.

Panjikar, S., Parthasarathy, V., Lamzin, V. S., Weiss, M. S., and Tucker, P. A. (2005). Auto-Rickshaw: an automated crystal structure determination platform as an efficient tool for the validation of an X-ray diffraction experiment. *Acta Crystallogr. D Biol. Crystallogr.* 61, 449–457. doi: 10.1107/S0907444905001307

Raumann, B. E., Rould, M. A., Pabo, C. O., and Sauer, R. T. (1994). DNA recognition by beta-sheets in the Arc repressor-operator crystal structure. *Nature* 367, 754–757. doi: 10.1038/367754a0

Robert, X., and Gouet, P. (2014). Deciphering key features in protein structures with the new ENDscript server. *Nucleic Acids Res.* 42, W320–W324. doi: 10.1093/nar/gku316

Rofougaran, R., Crona, M., Vodnala, M., Sjöberg, B.-M., and Hofer, A. (2008). Oligomerization status directs overall activity regulation of the *Escherichia coli* class Ia ribonucleotide reductase. *J. Biol. Chem.* 283, 35310–35318. doi: 10.1074/jbc.M806738200

Schreiter, E. R., and Drennan, C. L. (2007). Ribbon-helix-helix transcription factors: variations on a theme. *Nat. Rev. Microbiol.* 5, 710–720. doi: 10.1038/nrmicro1717

Somers, W. S., and Phillips, S. E. (1992). Crystal structure of the met repressor-operator complex at 2.8 A resolution reveals DNA recognition by beta-strands. *Nature* 359, 387–393. doi: 10.1038/359387a0

Staddon, J. H., Bryan, E. M., Manias, D. A., Chen, Y., and Dunny, G. M. (2006). Genetic characterization of the conjugative DNA processing system of enterococcal plasmid pCF10. *Plasmid* 56, 102–111. doi: 10.1016/j.plasmid.2006.05.001

Tato, I., Matilla, I., Arechaga, I., Zunzunegui, S., de la Cruz, F., and Cabezón, E. (2007). The ATPase activity of the DNA transporter TrwB is modulated by protein TrwA - implications for a common assembly mechanism of DNA translocating motors. *J. Biol. Chem.* 282, 25569–25576. doi: 10.1074/jbc.M703464200

Thompson, J. F., and Landy, A. (1988). Empirical estimation of protein-induced DNA bending angles: applications to lambda site-specific recombination complexes. *Nucleic Acids Res.* 16, 9687–9705.

Vanduyne, G. D., Standaert, R. F., Karplus, P. A., Schreiber, S. L., and Clardy, J. (1993). Atomic structures of the human immunophilin Fkbp-12 complexes with Fk506 and rapamycin. *J. Mol. Biol.* 229, 105–124. doi: 10.1006/jmbi.1993.1012

Varsaki, A., Moncalian, G., Garcillan-Barcia, M. D. P., Drainas, C., and de la Cruz, F. (2009). Analysis of colE1 MbeC unveils an extended ribbon-helix-helix family of nicking accessory proteins. *J. Bacteriol.* 191, 1446–1455. doi: 10.1128/JB.01342-08

Wong, J. J. W., Lu, J., Edwards, R. A., Frost, L. S., and Glover, J. N. M. (2011). Structural basis of cooperative DNA recognition by the plasmid conjugation factor, TraM. *Nucleic Acids Res.* 39, 6775–6788. doi: 10.1093/nar/gkr296

Wong, J. J. W., Lu, J., and Glover, J. N. M. (2012). Relaxosome function and conjugation regulation in F-like plasmids - a structural biology perspective. *Mol. Microbiol.* 85, 602–617. doi: 10.1111/j.1365-2958.2012.08131.x

Yoshida, H., Furuyal, N., Lin, Y.-J., Guentert, P., Komano, T., and Kainosho, M. (2008). Structural basis of the role of the NikA ribbon-helix-helix domain in initiating bacterial conjugation. *J. Mol. Biol.* 384, 690–701. doi: 10.1016/j.jmb.2008.09.067

Permissions

All chapters in this book were first published by Frontiers; hereby published with permission under the Creative Commons Attribution License or equivalent. Every chapter published in this book has been scrutinized by our experts. Their significance has been extensively debated. The topics covered herein carry significant findings which will fuel the growth of the discipline. They may even be implemented as practical applications or may be referred to as a beginning point for another development.

The contributors of this book come from diverse backgrounds, making this book a truly international effort. This book will bring forth new frontiers with its revolutionizing research information and detailed analysis of the nascent developments around the world.

We would like to thank all the contributing authors for lending their expertise to make the book truly unique. They have played a crucial role in the development of this book. Without their invaluable contributions this book wouldn't have been possible. They have made vital efforts to compile up to date information on the varied aspects of this subject to make this book a valuable addition to the collection of many professionals and students.

This book was conceptualized with the vision of imparting up-to-date information and advanced data in this field. To ensure the same, a matchless editorial board was set up. Every individual on the board went through rigorous rounds of assessment to prove their worth. After which they invested a large part of their time researching and compiling the most relevant data for our readers.

The editorial board has been involved in producing this book since its inception. They have spent rigorous hours researching and exploring the diverse topics which have resulted in the successful publishing of this book. They have passed on their knowledge of decades through this book. To expedite this challenging task, the publisher supported the team at every step. A small team of assistant editors was also appointed to further simplify the editing procedure and attain best results for the readers.

Apart from the editorial board, the designing team has also invested a significant amount of their time in understanding the subject and creating the most relevant covers. They scrutinized every image to scout for the most suitable representation of the subject and create an appropriate cover for the book.

The publishing team has been an ardent support to the editorial, designing and production team. Their endless efforts to recruit the best for this project, has resulted in the accomplishment of this book. They are a veteran in the field of academics and their pool of knowledge is as vast as their experience in printing. Their expertise and guidance has proved useful at every step. Their uncompromising quality standards have made this book an exceptional effort. Their encouragement from time to time has been an inspiration for everyone.

The publisher and the editorial board hope that this book will prove to be a valuable piece of knowledge for researchers, students, practitioners and scholars across the globe.

List of Contributors

Yogesh Hooda, Christine C. L. Lai and Trevor F. Moraes
Department of Biochemistry, University of Toronto, Toronto, ON, Canada

Can Chen
Institute of Food and Drug Inspection, College of Life Science and Agronomy, Zhoukou Normal University, Zhoukou, China

Zhuo Wang, Tietao Wang, Rui Cui, Zhenxing Zhang, Keqi Chen, Mengyun Li, Yueyue Hua, Huawei Gu, Lei Xu, Yao Wang, Yantao Yang, Xihui Shen and Xiaobing Yang
State Key Laboratory of Crop Stress Biology for Arid Areas, Shaanxi Key Laboratory of Agricultural and Environmental Microbiology, College of Life Sciences, Northwest A&F University, Yangling, China

Benoit Maffei and Agathe Subtil
Unité de Biologie Cellulaire de l'Infection Microbienne, Institut Pasteur, Paris, France
Centre National de la Recherche Scientifique UMR3691, Paris, France

Olivera Francetic
Unité de Biochimie des Interactions Macromoléculaires, Institut Pasteur, Paris, France
Centre National de la Recherche Scientifique ERL6002, Paris, France

Ina Meuskens, Athanasios Saragliadis, Jack C. Leo and Dirk Linke
Department of Biosciences, Section for Genetics and Evolutionary Biology, University of Oslo, Oslo, Norway

Alexandros Karyolaimos, Henry Ampah-Korsah, Tamara Hillenaar, Anna Mestre Borras, Robert Daniels and Jan-Willem de Gier
Department of Biochemistry and Biophysics, Center for Biomembrane Research, Stockholm University, Stockholm, Sweden

Katarzyna Magdalena Dolata, Susanne Sievers and Katharina Riedel
Institute of Microbiology, University of Greifswald, Greifswald, Germany

Yun-Wei Lien and Erh-Min Lai
Institute of Plant and Microbial Biology, Academia Sinica, Taipei, Taiwan
Department of Plant Pathology and Microbiology, National Taiwan University, Taipei, Taiwan

Hendrik Geise, Eyleen Sabine Heidrich, Christoph Stefan Nikolin, Denise Mehner-Breitfeld and Thomas Brüser
Institute of Microbiology, Leibniz University Hannover, Hannover, Germany

Verena Kohler and Walter Keller
Institute of Molecular Biosciences, BioTechMed Graz, University of Graz, Graz, Austria

Elisabeth Grohmann
Life Sciences and Technology, Beuth University of Applied Sciences Berlin, Berlin, Germany

Jan Tommassen and Jesús Arenas
Department of Molecular Microbiology and Institute of Biomembranes, Utrecht University, Utrecht, Netherlands

Benjamin Berni, Chantal Soscia, Sarah Djermoun, Bérengère Ize and Sophie Bleves
LISM, IMM (Institut de Microbiologie de la Méditerranée), CNRS and Aix-Marseille Univ, Marseille, France

David Liebl, Mylène Robert-Genthon, Valentina Cogoni and Ina Attrée
Univ. Grenoble Alpes, CNRS, Bacterial Pathogenesis and Cellular Responses, ERL 5261, INSERM, UMR-S 1036, CEA, Grenoble, France

Viviana Job
Univ. Grenoble Alpes, CNRS, Bacterial Pathogenesis and Cellular Responses, ERL 5261, INSERM, UMR-S 1036, CEA, Grenoble, France
Univ. Grenoble Alpes, CEA, CNRS, Institut de Biologie Structurale (IBS), Grenoble, France

Charles W. Vander Broek and Joanne M. Stevens
The Roslin Institute and Royal (Dick) School of Veterinary Studies, University of Edinburgh, Midlothian, United Kingdom

Ian Collinson
School of Biochemistry, University of Bristol, Bristol, United Kingdom

Ethel Bayer-Santos
Departamento de Microbiologia, Instituto de Ciências Biomédicas, Universidade de São Paulo, São Paulo, Brazil

Lucas de Moraes Ceseti and Cristina Elisa Alvarez-Martinez
Departamento de Genética, Evolução, Microbiologia e Imunologia, Instituto de Biologia, Universidade Estadual de Campinas, Campinas, Brazil

Chuck Shaker Farah
Departamento de Bioquímica, Instituto de Química, Universidade de São Paulo, São Paulo, Brazil

Ursula Waack, Khalil Chedid, Harry L. T. Mobley and Maria Sandkvist
Department of Microbiology and Immunology, University of Michigan Medical School, Ann Arbor, MI, United States

Tanya L. Johnson
Department of Microbiology and Immunology, University of Michigan Medical School, Ann Arbor, MI, United States
Department of Chemistry, Eastern Michigan University, Ypsilanti, MI, United States

Chuanwu Xi
Department of Environmental Health Sciences, University of Michigan School of Public Health, Ann Arbor, MI, United States

Lyle A. Simmons
Department of Molecular, Cellular, and Developmental Biology, University of Michigan, Ann Arbor, MI, United States

Zhila Esna Ashari
School of Electrical Engineering and Computer Science, Washington State University, Pullman, WA, United States

Kelly A. Brayton and Shira L. Broschat
School of Electrical Engineering and Computer Science, Washington State University, Pullman, WA, United States
Department of Veterinary Microbiology and Pathology, Washington State University, Pullman, WA, United States
Paul G. Allen School for Global Animal Health, Washington State University, Pullman, WA, United States

Saima Rehman, Andreas Schmitt and Lena Lassinantti
Department of Medical Biochemistry and Biophysics, Umeå University, Umeå, Sweden

Yang Grace Li and Peter J. Christie
Department of Microbiology and Molecular Genetics, McGovern Medical School, Houston, TX, United States

Ronnie P.-A. Berntsson
Department of Medical Biochemistry and Biophysics, Umeå University, Umeå, Sweden
Wallenberg Centre for Molecular Medicine, Umeå University, Umeå, Sweden

Index

A

Accessory Factor, 115, 224-225, 230, 232

Acinetobacter Baumannii, 3, 69, 204, 213

Amoeba Predation, 14, 194

Anaplasma Phagocytophilum, 42, 54, 215, 219, 223

Antibacterial Activity, 14, 21, 87-88, 90, 93-94, 142-143, 145, 151-153, 195, 200

Antibiotic Resistance, 11, 24, 30, 33-37, 114, 117-119, 203-204, 214, 225

Antidepressants, 203-204, 208, 211-212

Autotransporters, 12, 56-58, 62-64, 69, 72-74, 120, 122-123, 125, 130-133, 137-140

B

Bacterial Competition, 14, 25, 89, 142-143, 145, 151, 153, 156, 169

Bacterial Gene, 1

Bacterial Killing, 20, 52, 194, 202

Baseplate Assembly, 156-157, 159, 161, 163-166

Biofilms, 32, 51, 61, 64, 66-67, 71-72, 108, 113, 120-121, 130-131, 139

Bioinformatics, 1-2, 6, 8, 10-11, 14, 40-42, 44, 46, 74, 86-89, 95, 181, 202, 223

Bioinformatics Tools, 40-41, 88-89

Burkholderia Pseudomallei, 54, 59, 64, 73-74, 89, 94-95, 169-171, 182-186, 214

C

Chaperone, 2, 9, 38, 49, 52, 54, 61, 74, 76, 87-91, 95-96, 121-125, 132, 137, 139, 142, 154-155, 169, 172, 176, 178-179, 182-186, 189, 193, 201-202, 211-212

Colonization Resistance, 13-14, 18-19

Conjugation System, 107, 110-111, 114-116, 119

D

Diderm Bacteria, 40, 42

E

Enteric Pathogen, 13, 15, 21, 154

Enterococcus Faecalis, 109, 113, 117-119, 224, 227, 233

Escherichia Coli, 1, 11, 18, 20-25, 29, 36-38, 43, 52-55, 58-59, 63, 68-75, 84, 94, 96-98, 105-106, 121, 136-140, 142-143, 154, 157, 168-170, 182-184, 188, 192-193, 200, 213, 219, 225, 233

Eukaryotes, 15, 45, 84, 140, 142

Exoproteome, 39-40, 42-43, 52

F

Flow Cytometry, 1, 3-8, 10, 46, 50, 139

Fratricide Intoxication, 142, 150

G

Gram-negative Bacteria, 1-2, 4-5, 7, 9-12, 14, 16, 25, 30, 32, 52-53, 55-56, 61, 68-69, 71, 74-75, 86-87, 119, 121, 135-136, 140, 143, 153-154, 156, 170, 174, 182, 190, 214, 216, 223

Gram-positive Bacteria, 54, 107, 117-119, 195, 233

H

High-throughput Screening, 203-204, 214

Horizontal Gene Transfer, 16-17, 19-20, 107, 117-118, 131, 172, 204, 221

Host Immunity, 19, 25, 93, 96, 202

Host-pathogen Interactions, 120, 135, 181

I

Immune Evasion, 1, 59, 61-62, 64-67, 120, 135

Immunity Genes, 14-15, 150, 155

Inner Membrane, 2, 8, 40, 52, 58, 68, 91, 93, 95, 97, 121-126, 135, 152, 154, 156, 159, 161, 163, 165-166, 172-173, 176, 187-188, 190, 192-193, 204, 214

Integrative And Conjugative Element, 107, 109

Interbacterial Competition, 14-15, 18-19, 22, 86, 89-90, 93, 95, 127, 133, 202

Intestinal Ecosystem, 13-14, 16-17, 19-20

Intestinal Microbiota, 13-15, 19, 21, 23

L

Lipase Activity, 58, 61-63, 92, 203-204, 206-211

Lipids, 25, 42, 62, 175, 187-188, 190-191, 193, 211

Live Imaging, 39, 44, 50

M

Macromolecules, 25, 86

Mammalian Gut, 14, 16, 143

Melioidosis, 94, 170-174, 177-178, 180-186

Membrane Protein Complexes, 97-98

Microbiome, 20, 22-23, 121

Molecular Nanomachine, 156, 195

N

Neisseria Meningitidis, 1, 3, 10-12, 57, 59, 72-74, 120, 136-141

O

Outer Membrane, 1-2, 4, 7-12, 21, 37, 48, 53-56, 69-74, 94-95, 106, 120, 122-126, 133-140, 146, 152, 165-166, 168, 172, 176, 180, 190-193, 201, 204, 206-207, 211

Oxidative Stress, 15-16, 18, 22, 24-25, 30-33, 35-36, 38, 92-93, 96, 202

P

Pathogenicity, 17-18, 20, 22, 54, 66, 95, 107, 138, 143, 182, 184, 187, 196, 201, 215

Periplasm, 2, 12, 40, 42, 44-46, 48, 55-56, 58, 60, 75-76, 78-84, 93, 121-126, 142, 144-147, 149-153, 188, 190, 200, 204

Photo Cross-linking, 97

Plasmid, 6, 8, 25-26, 29-32, 50, 65, 69, 76-77, 99, 107-119, 143-145, 151, 153, 166-167, 169, 183-184, 205, 207, 209-211, 213, 223-225, 227, 230-234

Protein Injection, 41, 44, 49, 53

Protein Prediction, 215-217

Protein Production Rate, 75

Protein Production Screen, 75

Protein Secretion, 22, 27, 39-45, 48, 51-55, 69-70, 84, 86, 89, 94-95, 107, 120, 125, 127, 140-144, 154-155, 168-169, 174, 182-185, 187, 190, 192, 202, 204, 213, 223

Protein Translocation, 1, 52, 97, 105-106, 121, 187-189, 191-192

Protein-protein Interaction, 86, 89, 91, 146

Proteobacterial Genomes, 87, 89

Proteomics, 41-44, 51, 54, 69, 86, 88-90, 94, 186, 206

Proteomics-based Method, 89-90

Proton-motive-force, 187-188, 193

Pseudomonas Syringae, 15, 21, 170

R

Ralstonia Solanacearum, 49, 54, 170-171

Recombinant Protein, 75-76, 79, 84-85, 145, 177

Relaxosome, 108-109, 115-116, 224-225, 230-234

S

Salmonella, 3-4, 14, 16-18, 20, 22, 49-50, 53-55, 59, 61, 63, 69-71, 73, 88, 91, 94, 96, 155, 170-180, 182-186, 201

Secretion Machinery, 39, 41-43, 48-49, 149, 213-214

Secretion Signal, 39, 42, 44, 48, 55, 60, 126, 178, 186

Secretome, 20, 23, 54-55, 84, 89-90, 94, 120-121, 135, 170, 181, 193, 213

Shigella, 14, 18, 20, 49, 52-54, 59, 65, 69-70, 72, 91, 139, 170-171, 173-177, 179, 182-186

Signal Peptide, 40-42, 45-46, 48, 51, 53-55, 60, 69, 74-84, 91, 105-106, 113, 140, 145-146, 149, 152, 204, 211

Small Molecule Inhibitors, 181, 203-204, 211

Surface Lipoproteins, 1-2, 4, 9, 11

T

Toxin-immunity, 86, 91

Translocator, 42, 56-57, 137-138, 170, 172-178, 181

Twin-arginine Translocation, 41, 54, 97, 106, 121, 138

Two-partner Secretion System, 73, 120, 124, 139-140

V

Vibrio Cholerae, 14, 17, 19-23, 53-54, 87, 94-96, 154-156, 169, 179, 197, 202, 204, 212-214

Vibrio Parahaemolyticus, 17, 19, 23

Virulence, 1, 11, 15-20, 22-25, 33-39, 46-47, 56-57, 61, 63-74, 89-96, 108, 117, 123, 131, 133, 137-140, 142, 155-156, 169-172, 174-175, 178-186, 195, 200-201, 204-205, 212-216, 223, 225, 233

X

Xanthomonadales, 194-202

Xanthomonas, 38, 91, 95, 117, 119, 170, 178, 185, 194-196, 200-202

Y

Yersinia, 14, 17, 19, 21-25, 27, 29, 36-38, 43, 50-55, 57, 59, 63-66, 69-74, 93, 131, 138, 154-155, 170, 173-175, 177-180, 182-186, 214

Yersinia Pseudotuberculosis, 14, 17, 19, 21-25, 27, 29, 36-38, 57, 59, 69, 73, 93, 183, 185

www.ingramcontent.com/pod-product-compliance
Lightning Source LLC
LaVergne TN
LVHW081033210926
848485LV00004B/1014